LANDSLIDES AND CLIMATE CHANGE – CHALLENGES AND SOLUTIONS

BALKEMA – Proceedings and Monographs
in Engineering, Water and Earth Sciences

PROCEEDINGS OF THE INTERNATIONAL CONFERENCE ON LANDSLIDES AND CLIMATE CHANGE, VENTOR, ISLE OF WIGHT, UK, 21–24 MAY 2007

Landslides and Climate Change

Challenges and Solutions

Editors
R. McInnes, J. Jakeways, H. Fairbank & E. Mathie
Centre for the Coastal Environmental, Isle of Wight Council, UK

Taylor & Francis
Taylor & Francis Group
LONDON / LEIDEN / NEW YORK / PHILADELPHIA / SINGAPORE

The software mentioned in this book is now available for download on our Web site at: http://www.crcpress.com/e_products/downloads/default.asp

Taylor & Francis is an imprint of the Taylor & Francis Group, an informa business

Typeset by Charon Tec Ltd (A Macmillan Company), Chennai, India
Printed and bound in Great Britain by Bath Press Ltd (A CPI-group company), Bath

Published by: Taylor & Francis/Balkema
P.O. Box 447, 2300 AK Leiden, The Netherlands
e-mail: Pub.NL@tandf.co.uk
www.balkema.nl, www.taylorandfrancis.co.uk, www.crcpress.com

ISBN: 978-0-415-44318-0

Landslides and Climate Change – McInnes, Jakeways, Fairbank & Mathie (eds)
© 2007 Taylor & Francis Group, London, ISBN 978-0-415-44318-0

Table of Contents

Session 3 Advances in hazard modelling and prediction (short- and long-term)

Session 4 Experience of landslide hazard and risk management and better practices for the future

Landslides and Climate Change – McInnes, Jakeways, Fairbank & Mathie (eds)
© 2007 Taylor & Francis Group, London, ISBN 978-0-415-44318-0

Preface

The International Conference on 'Landslides and Climate Change - Challenges and Solutions' held in May 2007 in the town of Ventnor, Isle of Wight, UK, follows the very successful conference on 'Instability – Planning and Management', held in Ventnor in May 2002.

The decision to hold this Conference on 'Landslides and Climate Change' follows on from the recent completion of a three-year study, entitled 'RESPONSE: Responding to the risks from climate change', which received financial support from the European Union LIFE Environment Programme. The RESPONSE Project highlighted the importance of developing and implementing cost-effective solutions and examined the legal and administrative frameworks for hazard and risk management across Europe. Nine partner organisations in the United Kingdom, Italy, France and Poland participated, led by the Isle of Wight Council's Centre for the Coastal Environment, UK. The Conference will offer an opportunity to disseminate as widely as possible the findings from the European RESPONSE Project.

The increasing frequency of extreme weather events has highlighted our vulnerability to the impacts of climate change, and has resulted in enormous human and economic losses. The purpose of this Conference is therefore to consider the practical experiences of landslide hazard management and risk governance in a changing climate and to bring together a range of interest groups, including engineers, planners, practitioners, regional and local authorities, academics and politicians.

The Conference organisers very much hope that this Conference and the field visits will provide an opportunity to allow exchange of experiences and ideas within the unique environment of the Isle of Wight. The organising committee would like to thank all the authors whose contributions form these proceedings.

Dr. Robin McInnes OBE, FICE, FGS, FRSA
Chairman of the Organising Committee
Centre for the Coastal Environment
Isle of Wight Council
Ventnor, Isle of Wight, UK
March 2007

Landslides and Climate Change – McInnes, Jakeways, Fairbank & Mathie (eds)
© 2007 Taylor & Francis Group, London, ISBN 978-0-415-44318-0

Acknowledgements

The organisers of the International Conference on 'Landslides and Climate Change – Challenges and Solutions' (held in Ventnor, Isle of Wight, UK 21st-24th May 2007) would like to thank all those whose hard work and support have made the production of this proceedings volume and the hosting of this Conference possible, including the keynote speaker, the authors and presenters, the session chairs, the sponsors, the publishers Balkema and, in particular, the support of the Isle of Wight Council for this Conference initiative.

I would like to thank the United Nations International Strategy for Disaster Reduction (ISDR) in Geneva for its ongoing support for the work of the Isle of Wight Centre for the Coastal Environment and for the preparation of the keynote address.

These proceedings and the Conference itself could not have been arranged without the invaluable assistance of my colleagues Jenny Jakeways (Senior Coastal Geomorphologist), Helen Fairbank (Senior Coastal Scientist) and Emma Mathie (Senior Coastal Engineer) together with all the other members of staff at the Isle of Wight Centre for the Coastal Environment. The support of the Chairman and Leader of the Isle of Wight Council is gratefully acknowledged, together with the Mayor of Ventnor.

In particular I would like to thank the members of the International Organising Committee for their invaluable advice and expertise:

- Dr. Maceo-Giovanni Angeli, National Research Council, IRPI Perugia, Research Institute for Hydrogeological Protection in Central Italy.
- Dr. Christophe Bonnard, Swiss Federal Institute of Technology, Lausanne, Laboratoire de Mécanique des Sols, Switzerland.
- Dr. Alan Clark, High-Point Rendel, UK.
- Mr. Ashley Curzon, Planning Policy, Isle of Wight Council, UK.
- Mrs. Helen Fairbank, Centre for the Coastal Environment, Isle of Wight Council, UK.
- Mr. Stuart Fraser, Senior Accounting Manager, Isle of Wight Council, UK.
- Prof. Thomas Glade, University of Vienna, Austria.
- Ms. Jenny Jakeways, Centre for the Coastal Environment, Isle of Wight Council, UK.
- Dr. Mark Lee, Geohazard Risk Specialist, UK.
- Dr. Eric Leroi, Urbater, Land Use Planning and Natural Risks Division, France.
- Dr. Brian Marker OBE, Former Office of the Deputy Prime Minister, UK.
- Dr. Roger Moore, Halcrow Group Ltd, UK.
- Dr. Fabrizio Pontoni, Geoequipe, Italy.

Dr. Robin McInnes OBE, FICE, FGS, FRSA
Chairman of the Organising Committee
Centre for the Coastal Environment
Isle of Wight Council
Ventnor, Isle of Wight, UK
March 2007

Landslides and Climate Change – McInnes, Jakeways, Fairbank & Mathie (eds)
© 2007 Taylor & Francis Group, London, ISBN 978-0-415-44318-0

Organising Committee

Conference organised by the Centre for the Coastal Environment, Isle of Wight Council, UK.

Dr. Robin G. McInnes OBE
Centre for the Coastal Environment, Isle of Wight Council, UK

Dr. Maceo-Giovanni Angeli
National Research Council, IRPI Perugia, Research Institute for Hydrogeological Protection in Central Italy

Dr. Christophe Bonnard
Swiss Federal Institute of Technology, Lausanne, Laboratoire de Mécanique des Sols, Switzerland

Dr. Alan Clark
High-Point Rendel, UK

Mr. Ashley Curzon
Planning Policy, Isle of Wight Council, UK

Mrs. Helen Fairbank
Centre for the Coastal Environment, Isle of Wight Council, UK

Mr. Stuart Fraser
Senior Accounting Manager, Isle of Wight Council, UK

Prof. Thomas Glade
University of Vienna, Austria

Ms. Jenny Jakeways
Centre for the Coastal Environment, Isle of Wight Council, UK

Dr. Mark Lee
Geohazard Risk Specialist, UK

Dr. Eric Leroi
Urbater, Land Use Planning and Natural Risks Division, France

Dr. Brian Marker OBE
Former Office of the Deputy Prime Minister, UK

Dr. Roger Moore
Halcrow Group Ltd, UK

Dr. Fabrizio Pontoni
Geoequipe, Italy

Opening address

Landslides and Climate Change – McInnes, Jakeways, Fairbank & Mathie (eds)
© 2007 Taylor & Francis Group, London, ISBN 978-0-415-44318-0

Landslides and climate change: A world perspective, but a complex question

Sálvano Briceño & Pedro Basabe
United Nations, Secretariat of the International Strategy for Disaster Reduction, UN/ISDR, Geneva, Switzerland

Christophe Bonnard
Soil Mechanics Laboratory, Swiss Federal Institute of Technology, EPF Lausanne, Switzerland

ABSTRACT: The incidence of disasters around the world and the role of the United Nations International Strategy for Disaster Reduction (UNISDR) in relation to reducing vulnerability are reviewed, with reference to the relationship between climate change and landslides. This complex problem is then briefly discussed, to demonstrate trends over time and to establish the possible impacts of climate change on landslides. The ISDR secretariat supports this important event and hopes that this paper will provide an introduction to the International Conference on "Landslides and Climate Change".

1 INTRODUCTION

The Intergovernmental Panel on Climate Change (IPCC) Working Group 1 confirmed in Paris on February 2007 that our atmosphere is warming, a trend that will have an enormous impact on the frequency and severity of some natural hazards. The increase of temperature by 1.8 to 4 degrees Celsius this century will make hot extremes, heat waves and heavy precipitations events, more frequent. Similarly there will be more precipitation at high latitudes and less precipitation in most subtropical land regions and it is likely that tropical cyclones (typhoons and hurricanes) will become more intense. The IPCC's Working Group 2 will deliver more information on climate change impacts, adaptation, vulnerability and ways for mitigation in its report to be presented in Brussels by 6 April 2007.

This global trend is also confirmed by international data [1]. In the decade 1967–1985, close to one billion people were affected by disasters. But by the most recent decade, 1996–2005, the decade total had more than doubled, to nearly two and a half billion people. In the last decade alone, disasters affected 3 billion people, killed over 750,000 people and cost around US$600 billion. Such a trend cannot continue.

Disaster risk concerns every person, every community and every nation; indeed, disaster impacts are slowing down development, and their impact and actions in one region can have an impact on risks in another, and vice versa. Without taking into consideration the urgent need to reduce risk and vulnerability, the world simply cannot hope to move forward in its quest for reducing poverty and ensuring a sustainable development (Figure 1). It is necessary not to forget the wise words expressed by Prof. Laborit to describe a correct attitude in this respect: "To understand the world does not mean to possess it, but to belong to it".

The United Nations International Strategy for Disaster Reduction (UN/ISDR) aims at building disaster resilient communities by promoting increased awareness of the importance of disaster reduction as an integral component of sustainable development, with the goal of reducing human, social, economic and environmental losses due to natural hazards and related technological and environmental disasters.

Recognising that disaster reduction needs interdisciplinary and multi-sectoral action, the ISDR builds on partnerships and takes a global approach to disaster reduction. Therefore, a better cooperation between Government authorities, the international community, and the scientific community, is welcome, as such influence plays a critical role in helping people make life changing decisions about where and how they live before the disaster strikes, in particular high-risk urban areas.

Climate change will aggravate existing vulnerabilities, increase drought, flood and storm risk for millions of people and bring these risks to parts of the

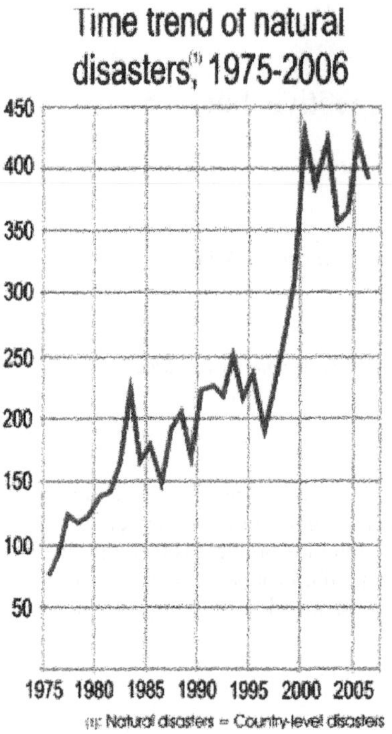

Figure 1. Time-trend of natural disasters, 1975–2006 (it includes all natural disasters at country level) (drawn from ISDR Highlights, January 2007).

world that have not felt them before. Action is therefore urgently needed to reduce people's vulnerability to climate-related hazard.

The Hyogo Framework Action adopted at the second World Conference on Disaster Reduction, Kobe, Japan, 18–22 January 2005, represents a comprehensive action-oriented policy guidance in universal understanding of disasters induced by vulnerability to natural hazards and reflects a solid commitment to implement an effective disaster reduction agenda. This framework underlines five priorities for actions:

1. Ensure that disaster risk reduction is a national and local priority with a strong institutional basis for implementation;
2. Identify, assess and monitor disaster risks and enhance early warning;
3. Use knowledge, innovation, research and education to build a culture of safety and resilience at all levels;
4. Reduce the underlying risk factors and
5. Strengthen disaster preparedness for effective response at all levels.

In order to facilitate the implementation of the Hyogo Framework and increase political commitment and financing for disaster risk reduction, a strengthened ISDR system is being proposed, as a platform to coordinate, prioritise and support activities at global, regional and national level. One key mechanism of the renewed ISDR system is the Global Platform for Disaster Risk Reduction, to be held in Geneva, 5–7 June 2007, with the participation of Governments, regional and international organisations, UN programmes, funds and specialised agencies, NGOs and civil society. The Global Platform provides a forum for devising strategies and policies to reduce disaster risk, sharing knowledge and information, monitoring progress and identifying gaps in policies and programmes and recommending remedial action.

2 OBJECTIVES OF THE CONFERENCE

The International Conference on Landslides and Climate Change to be held in the Isle of Wight on May 2007 will certainly contribute to understand the relationship between climate change and increased landslide hazards, identify measures and practices for proactive planning, management and risk mitigation. The ISDR secretariat supports this important event and looks forward for the Conference's results geared to reduce the impact of landslides, contribute to the implementation of the Hyogo Framework for Action and emphasise the use of science and technology for reducing risk and vulnerability to natural hazards.

In order to assess the relationship between landslides and climate change in a rational and appropriate way, it is important on one hand to focus not only on a world perspective, including several hazardous contexts that may be specific to various mountain ranges, hilly regions or coastal zones, but also on a long enough period of time. Indeed, since prehistoric times the world climate has experienced constant changes with differentiated intensities. The present perspective actually tends to observe the evolution of climatic factors mostly over the last two centuries, as data are available over this period, and to predict the evolution of the main parameters for the XXIst century. However the information provided by the "recent" geological or geomorphological phenomena, extending over at least ten thousand years, can also contribute to understand the relations between the inferred climatic conditions and the observed movement of slopes [2].

On the other hand, it is well known that most landslides, implying various mechanisms, are directly or indirectly triggered by adverse climatic factors (for example an analysis of more than one hundred past landslide cases in the Alps can prove that nearly 80% of them are caused by rainfall events [3]). It is thus essential to try to determine the exact relations between slope failure and climatic conditions, and not only empirical statistical correlations.

Finally the understanding of the exact relations between landslides and climate change may significantly contribute to propose appropriate remedial measures and management policies, so that the corresponding risks are reduced, as emphasised by the objectives of the UN/ISDR. Indeed, as society is evolving much faster than climate change in many regions of the world, it needs an increased protection against all natural hazards and in particular against landslides, if the authorities aim at reducing or ever maintaining constant the risk level related to such phenomena [4].

Why does the concept of landslide make its relationship with the climate complex? The English term of landslide is actually very general and implies a large number of different phenomena (fall, topple, slide, spread, flow, as defined by the International Union of Geophysical Sciences). Ever if the notion of slide alone is considered, the size of such a mechanism will imply very different behaviours, if either a local slope failure or the general long-term movement of a whole mountain slope covering dozens of km^2 are considered. This variety of phenomena must therefore prevent from drawing excessive conclusions of the type: "landslides will increase in the future due to climate change effects", as it is read too often in non-specialised texts.

As concerns large slides, an example of this "unexpected" trend has been given within a research carried out some years ago in Switzerland (National Research Project PNR 31, Disasters en Climate Change), in particular on the long-term behaviour of active mountain slopes in Switzerland [5]. Based on the monitoring data obtained through the Federal and Cantonal offices of Topography, thanks to the past and present coordinates of survey landmarks, it is possible to prove that several large landslides have hardly modified their velocity during the whole XXth century, although some variations of annual rainfall have been recorded. This is in particular the case for Lumnez landslide, in the Canton of Graubünden, where the spires of the churches of seven villages have been monitored since 1887, i.e. for more than a century. The velocity of most points is nearly constant (Figure 2); it is thus inappropriate to pretend in this case that landslide hazards are increasing with time.

On the contrary, some torrents in the Cantons of Valais and Bern, in which debris flow occasionally occur, have experienced an increased number of events during the XXth century. However, on the basis of the available data, it is difficult to establish if only the frequency of events has increased or if also the volume of each event has been growing with time [6]. The researchers from the same PNR31 project state in their final report that "In the Alpine border regions the change of climatic parameters is present, however, the changes are hardly perceptible due to the high variability of the frequency and magnitude of the debris flows". What is sure is that the impact of such phenomena is increasing, due to the important tourist development of this region, as well as to the more intense communication on disasters by the media.

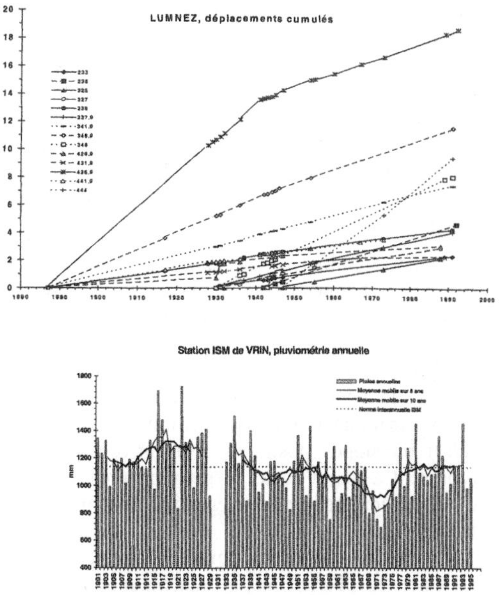

Figure 2. Evolution of displacements (in meters) of several points on Lumnez Landslide during the XXth century and annual rainfall data at a nearby station (Vrin).

Another aspect of climate change in mountain areas has also to be investigated, in relation to the expected temperature rise. As an increase in precipitation over the next decade is expected in winter months for the Alpine region, the number of landslides might tend to increase during the snowmelt period. However, due to the temperature increase during this season, in particular in the region of the Prealps, a large part of this precipitation will fall as rain and therefore not accumulate on the slopes. As the potential evapotranspiration will increase due to the temperature rise, it is also possible that the large slides will be less active in the future. This point however raises some controversy, as some researches of the PNR 31 state a relationship between climate change and landslide activity is underscored [7].

3 CONCLUSIONS

Indeed such an introductory lecture cannot propose any conclusions with respect to the trend of evolution of landslides with regards to climate change, except that this phenomenon has to be investigated in detail before any general lessons are drawn. What is sure is that several episodes of climate change have been observed in the past, in particular after the last glacier retreat. These episodes have probably influenced the behaviour of large landslides as it can be established by radiocarbon dating for instance, but the available information do not allow the exact determination of the triggering mechanisms.

It is thus fundamental to improve our knowledge in three directions:

- The understanding of the causes of the presence of critical behaviour, through a detailed hydrogeological and geomechanical modelling of the slide mass.
- The potential impact of climate change on the landscape and vegetation cover, including anthropic actions that may modify the infiltration rate of rainfall in the slopes, positively or negatively, and thus represent a major factor explaining the evolution of the preparatory and triggering mechanisms of landslides.
- The gathering of more and more monitoring data of all kinds (displacements, groundwater levels, local climatic conditions) in order to determine the real behaviour of landslide movements, in particular during pre-failure phases, that will help to prevent disaster through the development of appropriate early-warning systems.

I do hope that this conference will contribute to a better understanding of the long-term behaviour of landslides and of the possible effect of climate change on their mechanisms. We have however to keep in mind in this respect the philosophical perspective described by the Mexican writer Octavio Paz:

"Time stops being a succession and becomes what it has been and what it was originally, namely a present movement in which past and future may reconciliate".

REFERENCES

[1] Data derived from the EM-DAT: The OFDA/CRED International Disaster Database, www.en-dat.net, Université Catholique de Louvain – Brussels – Belgium, as well as by Al Gore in his book ≪An Inconvenient Truth≫ and by Sir Nicholas Stern in his study on the economic impact of climate change.

[2] Lateltin, O., Beer, Ch., Raetzo, H., Caron, C. 1997. Instabilités de pente en terrain de flysch et changements climatiques. Rapport final PNR31. v/d/f Zurich, 168 pp.

[3] Bonnard, Ch. 1994. Los deslizamientos de tierra : Fenómeno natural o fenómeno inducido por el hombre? Proc. Ist Panam. Symp. on Landslides, Guayaquil, August 1994, Vol. 2, pp. 1–15.

[4] Bonnard, Ch., Forlati, F., Scavia, C. (eds) 2004. Identification and mitigation of large landslide risks in Europe: advances in risk assessment. IMIRILAND Project. 317 p. Ed. Balkema.

[5] Noverraz, F., Bonnard, Ch., Dupraz, H., Huguenin, L. 1998. Grands glissements de versant et climat. Rapport final PNR31. v/d/f Zurich, 314 pp.

[6] Zimmermann, M., Mani, P., Gamma, P. 1997. Murganggefahr und Klimaänderung – ein GIS-basierter Ansatz. Schlussbericht PNR31. v/d/f Zurich, 161 pp.

[7] Lateltin, O., Beer, Ch., Raetzo, H., Caron, C. 1997. Instabilités de pente en terrain de flysch et changements climatiques. Rapport final PNR31. v/d/f Zurich, 168 pp.

Session 1

Experience of the historical impacts of climate change on natural hazards (including global case studies and meteorologically-induced disasters)

Recent years have seen an increased frequency of extreme weather events and there is concern that this trend is likely to continue. In order to determine how the impacts of climate change are likely to affect the timing and frequency of natural hazard events, it is important to consider evidence of how climate change and natural hazards have been linked in the past.

Gore Cliff, Isle of Wight, UK
Courtesy Wight Light Gallery, Ventnor, Isle of Wight, UK.

Landslides and Climate Change – McInnes, Jakeways, Fairbank & Mathie (eds)
© 2007 Taylor & Francis Group, London, ISBN 978-0-415-44318-0

Landslides and historic climate in northern British Columbia

M. Geertsema & V.N. Egginton
British Columbia Ministry of Forests and Range, Prince George, Canada

J.W. Schwab
British Columbia Ministry of Forests and Range, Smithers, Canada

J.J. Clague
Centre for Natural Hazard Research, Simon Fraser University, Burnaby, Canada

ABSTRACT: We have assembled inventories of landslides and climate trends in northern British Columbia (BC), Canada for the past century. Some landslides in this region occur during or soon after intense rainstorms and other severe weather events. However on the northern BC coast, only the largest storms of the last century are responsible for most landslides. Others landslides have delayed responses to such events or occur after shifts in climate. We consider the relationships between storms and shallow debris slides, debris flows, and between 20th century climate change and large soil and rock slope failures. Debuttressing of valley walls due to thinning and retreat of glaciers, thaw of permafrost under a warming climate, and increased precipitation have contributed to an increase in landslides in northern BC during the past century. We consider the implications of future climate change scenarios and the interplay of disturbance agents such as forest insect epidemics and wildfire on future landslide frequency.

1 INTRODUCTION

The climate of northern British Columbia (BC) has become wetter and warmer over the past century (Egginton 2005). In this paper we report on the relationships between climate and landslides in this large, remote, sparsely populated and under-instrumented region. We distinguish between landslides that respond rapidly to meteorological conditions and landslides that have a delayed response and are the result of longer-term climate trends.

2 SETTING

Northern BC is a vast, diverse area where plateaux are separated by generally northwest-trending mountain ranges (Figure 1). The mountainous topography is responsible for a variety of climates and ecological zones. The mountains in the interior have a continental climate with little ice cover, whereas those near the Pacific coast have a maritime climate and an extensive cover of snow and ice (Holland 1976). Some of the drier mountains and northern plateaux support permafrost. Vegetation communities range from rain forest to grassland and from boreal forest to alpine tundra (Meidinger and Pojar 1991).

All of northern BC was covered by the Cordilleran ice sheet at the peak of the last glaciation (Late Wisconsinan), and alpine glaciers in mountainous regions have fluctuated in a complex manner during the Holocene (Clague 1989). Glaciers carved deep U-shaped valleys, leaving steep, unsupported mountain walls, but they deposited thick sediment fills in other areas, burying some preglacial valleys. Today mountain glaciers that achieved their maximum Holocene extent during the Little Ice Age are rapidly receding in response to global warming. This varied landscape is subject to a wide variety of landslide types (Figure 2).

3 20TH CENTURY CLIMATE VARIABILITY

Egginton (2005) completed a detailed study of northern BC's climate variability based on the instrumental record. What follows is a brief summary of her findings.

Most of northern British Columbia has experienced an increase in annual mean temperature in the 20th century. Statistically, significant trends range from +0.6 to +1.3°C. In general, significant increases in annual extreme minimum temperatures are larger than changes in mean temperatures, ranging from +2.1

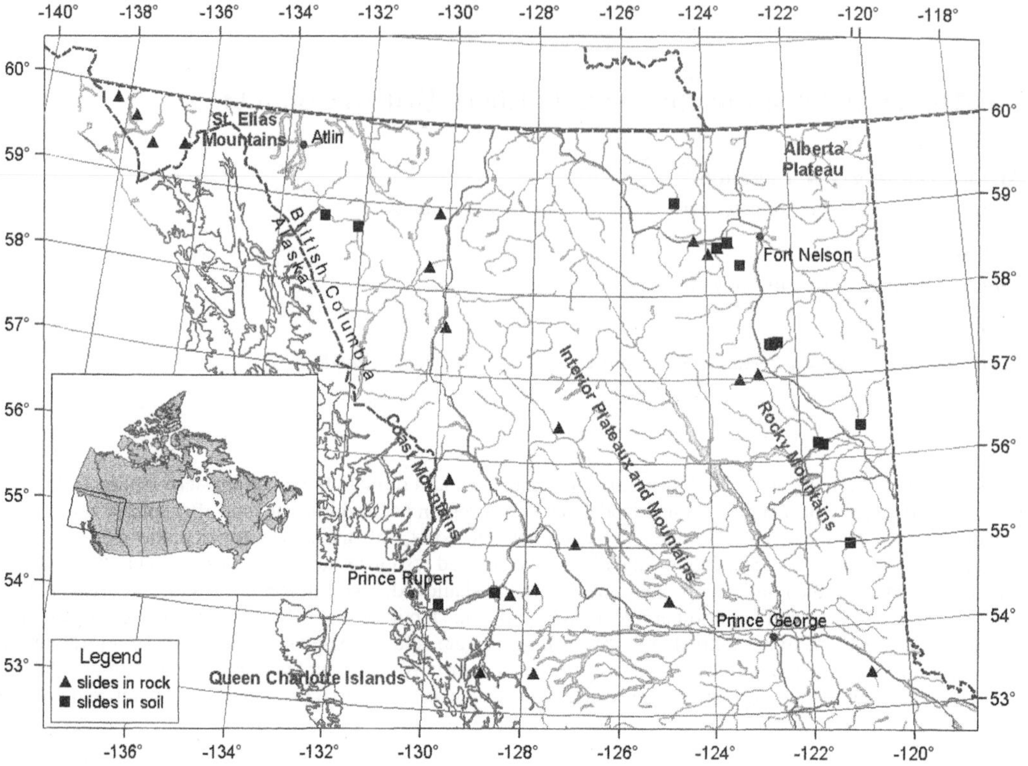

Figure 1. Large rock slides and soil slides (greater than 0.5 Mm^3) in northern BC, 1973–2006.

to +4.4°C. Significant increases in annual extreme maximum temperatures are less than changes in both minimum and mean temperatures, +0.9 to 1.1°C. Overall, mean temperatures increased in all seasons, with the largest increases in winter.

Maximum temperatures increased in the majority of the study area in winter and summer, decreased in spring, and have not changed in fall. Minimum temperatures increased in all seasons. As in the case of the annual trends, changes in seasonal minimum temperatures are larger than changes in seasonal mean temperatures, and changes in seasonal maximum temperatures are the smallest (Figure 3).

Annual precipitation has increased in most of northern BC. Trend values that are significant range from +10.2 to +18.6%. Overall, precipitation increased in spring, summer, and fall, and decreased in winter. Significant trends range from −24.5 to −41.6% in winter, 15.8 to 27.0% in spring, −18.4 to 23.4% in summer, and 12.5 to 15.5% in fall.

Ocean-atmosphere phenomena are contributing to the increases in precipitation and mean temperature in northern BC. The shift in the Pacific Decadal Oscillation (PDO) from its cool to warm phase in the mid-1970s is seen at most climate stations, but there are some regional differences. Stations with the longest records also show regime shifts in the mid-1940s (warm to cool) and the mid-1920s (cool to warm). Some stations, especially those near the Pacific coast, show dry (wet) periods corresponding to the cool (warm) PDO phases, whereas inland stations show the opposite. Effects of the El-Niño Southern Oscillation, such as strong El Niño or La Niña years, are seen in the climate data at many weather stations in northern British Columbia; however, their influence on the overall trends is much smaller than the PDO.

4 LANDSLIDES AND CLIMATE

Hillslopes have variable responses to changes in climate. Depending on the nature of the material, depth of weak zones, slope geometry, geological setting, and other factors, landslides can have rapid or delayed responses to changes in precipitation and temperature. In general soils must become saturated, allowing the build up of pore pressures, before surficial sediments will fail. Threshold pore pressures develop more rapidly in shallow soils than in deeper materials. Both rapid and delayed responses to climate occur in northern BC.

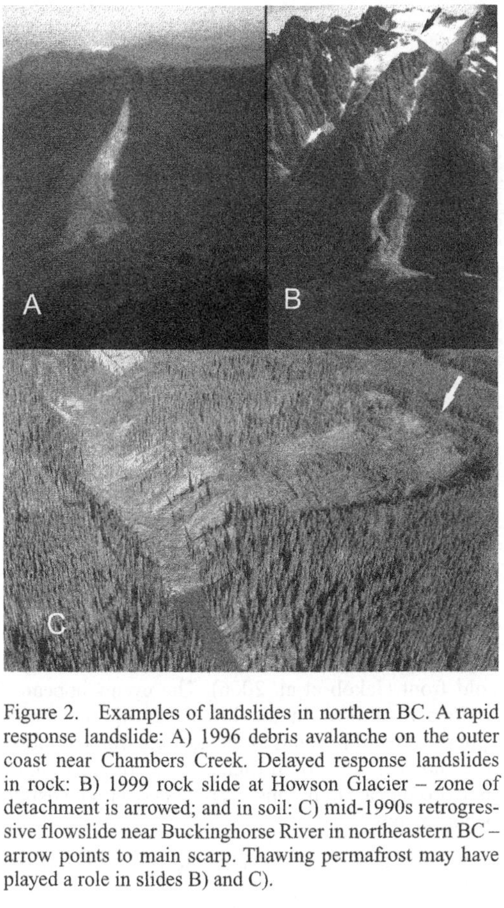

Figure 2. Examples of landslides in northern BC. A rapid response landslide: A) 1996 debris avalanche on the outer coast near Chambers Creek. Delayed response landslides in rock: B) 1999 rock slide at Howson Glacier – zone of detachment is arrowed; and in soil: C) mid-1990s retrogressive flowslide near Buckinghorse River in northeastern BC – arrow points to main scarp. Thawing permafrost may have played a role in slides B) and C).

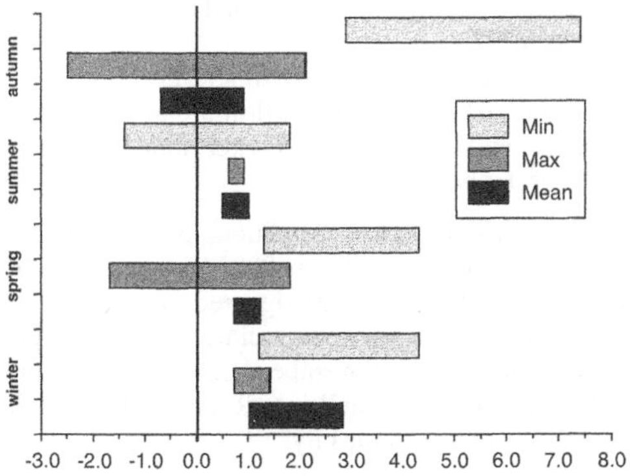

Figure 3. Range of significant changes in seasonal temperature in northern BC (°C) based on Egginton (2005).

Landslides that respond rapidly to changes in precipitation include shallow debris slides, debris flows, and rock fall. These types of landslide are especially common on the outer coast. Landslides that have delayed responses to changes in precipitation and temperature tend to be larger, deeper-seated rock slides, earth slides, and earth flows.

4.1 *Rapid response landslides*

Shallow debris slides, debris avalanches, and debris flows are typically associated with heavy rainfall, provided threshold conditions are met. Hogan and Schwab (1991) studied rainfall characteristics before and during verified debris flow events. They compared the temporal frequency of slope failures to precipitation trends for antecedent time scales ranging from years to days. Over the longer time scale, a positive correlation was found between annual moisture conditions and reported hillslope failure frequency. However, annual moisture was less significant than precipitation over shorter periods (months or days) preceding a slope failure. On the shorter time scale, only the months immediately preceding slope failure were important in conditioning the hillslope to failure. The most important situation leading to a high frequency of slope failures was continuous wet weather. More recent work by Jakob and Weatherly (2003) on the BC south coast shows that 1–4 week antecedent rainfall amounts are critical in bringing soil moisture to saturation levels conducive to landslides. Hogan and Schwab's data show that heavy rainfall is the dominant environmental factor contributing to debris slides, debris avalanches, and debris flows on the northern BC coast. Precipitation values required to trigger slope failures are regularly exceeded during fall and winter months.

Schwab (1997, 1998) used dendrochronological techniques to determine the ages of large debris slides, debris avalanches, and debris flows in selected areas along the British Columbia north coast and the Queen Charlotte Islands back to the early 1800s. A catalogue of storm information compiled by Septer and Schwab (1995) as part of the study was used to verify dates determined through tree ring analysis of increment cores and impact scars. One of the interesting conclusions of this work is that most shallow landslides occurred during a few major storms. Five storms, in 1875, 1891, 1917, 1935, and 1978, transported respectively, 1.6%, 2.9%, 13.3%, 2.1%, and 9.6% of the total volume of sediment moved by landslides in the Riley and Gregory Creek watersheds on the Queen Charlotte Islands (Figure 4). Comparable values in the Beresford Creek watershed on the Queen Charlotte Islands for major events, in 1875, 1891, 1917, and 1935, are, respectively, 16.5%, 10.8%, 36.2%, and 9.5% (Figure 5). Four major storms since 1875 moved 73% of the volume transported by landslides in the Beresford Creek watershed. Interestingly, this watershed did not experience landslides during the 1978 storm. Data from Graham Island and from the Prince Rupert area on the BC north coast (Figure 6) indicate that six storms over the past 150 years transported 76% of the

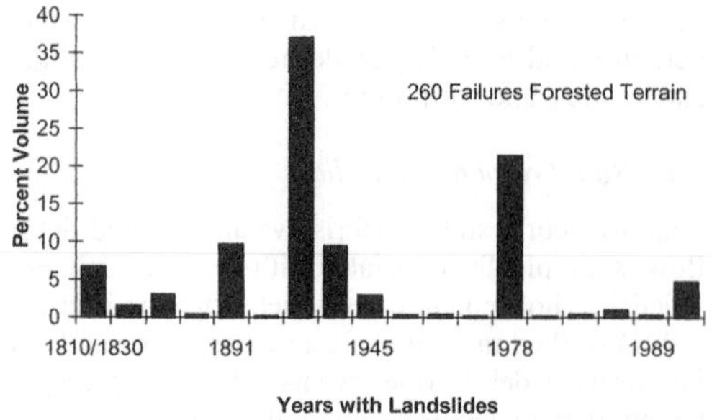

Figure 4. Percent landslide volume, Riley and Gregory creek watersheds, Queen Charlotte Islands. Most of the sediment was mobilized during a major storm in 1917.

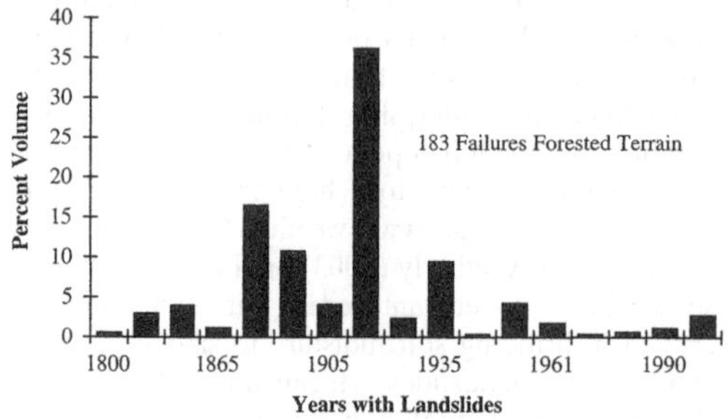

Figure 5. Percent landslide volume, Beresford Creek watershed, Queen Charlotte Islands. Most of the sediment was mobilized during a major storm in 1917.

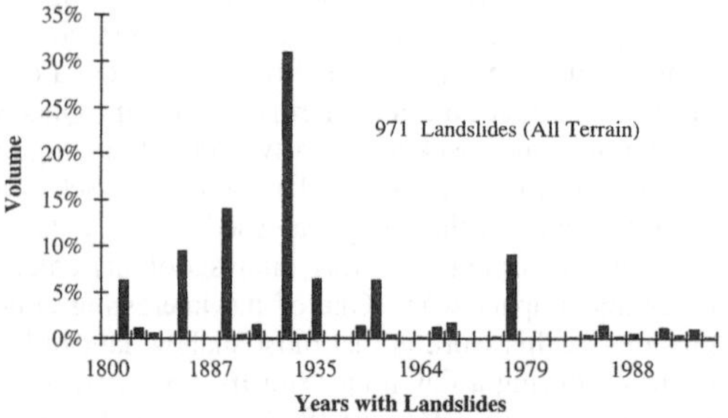

Figure 6. Percent landslide volume on the northern BC coast and islands. Most of the sediment was mobilized during a major storm in 1917.

volume of landslide debris: 9.5%, 14%, 30.9%, 6.5%, 6.4%, and 9.1%, respectively, for the years 1875, 1891, 1917, 1935, 1957, and 1978. The data suggest that the north coast has yet to experience a "Big Storm" similar to the 1917 event.

Most precipitation on the BC north coast is associated with frontal systems that pass across the region from the Pacific. Loci of warm wet conditions shift north and south depending on the prevailing storm

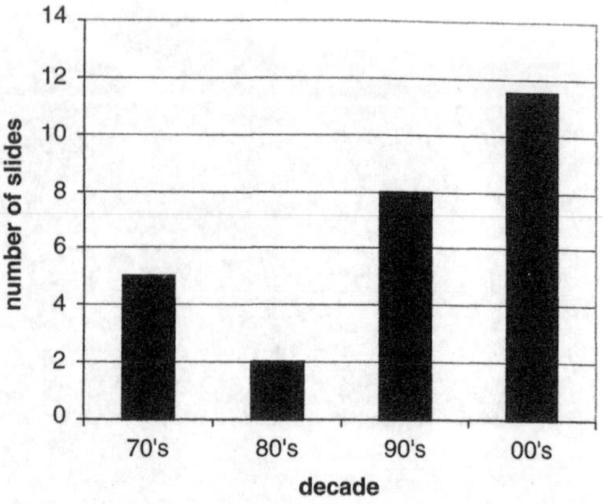

Figure 7. Frequency of large rock slides and soil slides (greater than 0.5 Mm^3) in northern BC. The numbers at the right (90s–00s) have been extrapolated (+3.5 for rockslides and +2 for soils slides) to better represent the full 20-year period. Large rock slides have nearly tripled in the last two decades.

track in the Pacific (Karanka 1986). Most landslide-triggering storms involve a warm front followed by a cold front (Jakob et al. 2006). The events appeared to involve a flow in the lower part of the atmosphere, extending south to incorporate tropical moisture. Jakob et al. (2006) found that a combination of meteorological conditions is associated with landslide occurrence:

- warm fronts with strong SE to SW winds at the 850 mbar level
- warm fronts that extend south over the Pacific Ocean, bringing much additional moisture to north coastal BC
- warm fronts with high freezing levels
- a strong jet stream with air flow arcing north of the region and W to SW winds exceeding 90 knots at the 250 mbar level.

These meteorological conditions prevailed during landslide-triggering storms in 2003 and again in 2004. The October 25, 2003, event triggered many large landslides, with 12 reported in the vicinity of Prince Rupert alone. Landslides on November 4, 2004, severed a natural gas pipeline serving Prince Rupert and closed the only highway link to the city.

4.2 *Delayed response landslides*

Having updated the data set of large landslides (greater than 0.5 Mm^3) in northern BC produced by Geertsema et al. (2006a), we can make some general comments. Numbers of both rock slides and soil slides have increased in the last two decades compared to the previous two decades (Figure 7). Egginton (2005)

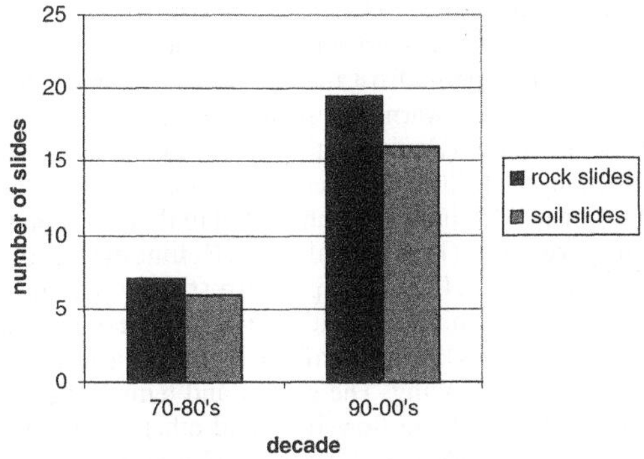

Figure 8. Frequency of large rock slides (greater than 0.5 Mm^3) in northern BC. The number for the current decade has been adjusted (+3.5 events) to represent a full decade.

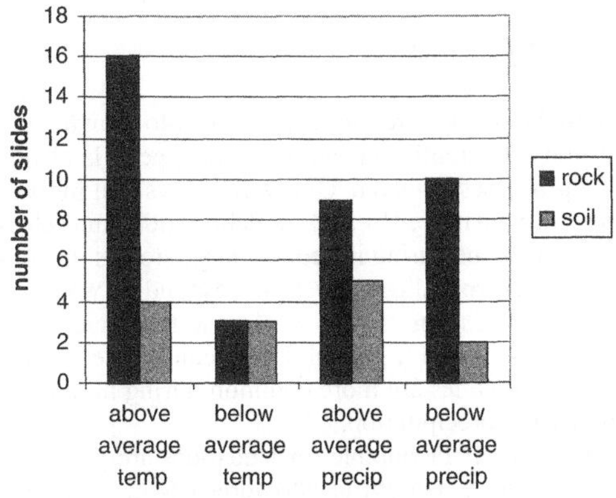

Figure 9. Number of large landslides in years of above- and below-average annual temperature and precipitation for northern BC (1971–2006) based on data from Egginton (2005) and unpublished observations.

noted that convective thunderstorms and large cyclonic storms have triggered some of the landslides in this region, but large slides typically are preceded by long periods of wetter or warmer climate. Some distinctions can be made between landslides in rock and landslides in soil.

4.2.1 *Rock slides*

Large rock slides appear to be increasing in frequency in northern BC. We have documented nearly three times the number of events in the last two decades compared to the 1970s and 1980s (Figures 7 and 8). Egginton (2005) noted that the rock slides tend to occur in years or decades of above-average temperature (Figure 9).

Four large rock slides occurred in 2002, and two others happened outside the study area in southern BC and Yukon in the same year. The spring of 2002 was

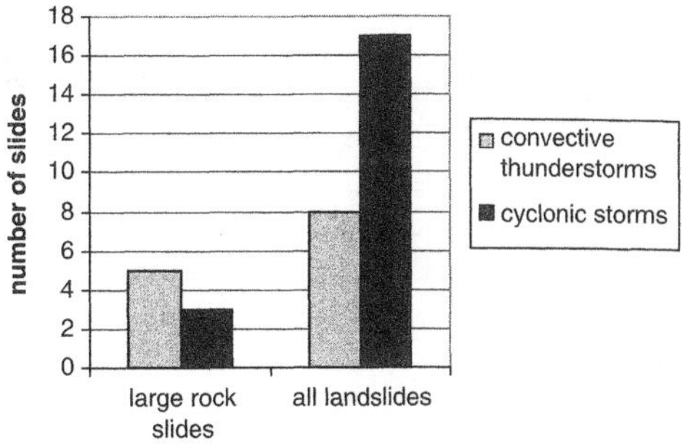

Figure 10. Landslides triggered by storms in northern BC (data from Egginton et al. 2005 and unpublished observations).

colder than normal, delaying melt of an above-average snowpack (Egginton 2005; Geertsema et al. 2006a).

One of the 2002 rock slides may have been associated with thaw of mountain permafrost. Schwab et al. (2003) reported interstitial ice in rubble at the main scarp of the Harold Price landslide.

Large rock slides have happened in recently deglaciated areas as well as areas that have been ice-free for most of the Holocene. Half of the rock slides in northern BC reported by Geertsema et al. (2006a) occurred on steep rock walls above glaciers. They suggest that debuttressing of steep rock slopes following the Little Ice Age, in response to glacier thinning under a warming climate, has destabilized many slopes.

Large rockslides may be responding to longer-term climate trends than shallow coastal debris slides, although some rockslides are triggered by torrential rainfall during convective thunderstorms. Egginton et al. (2005) used weather satellite imagery to correlate thunderstorms with large rock slides (Figure 10). Other rock slides occurred during cyclonic storms.

4.2.2 *Soil slides*

Landslides in soil have also been increasing in northern BC (Geertsema et al. 2006a) (Figure 7). Egginton (2005) noted that large landslides in soil in the 20th century occurred during years or decades of above-average precipitation, but the correlation is weaker than that between rockslides and above average temperature (Figure 9). Two glaciomarine flowslides at Mink Creek (Geertsema et al. 2006b) and Khyex River (Schwab et al. 2004; Egginton 2005) were preceded by decades and several years of increasing precipitation, respectively (Figure 11).

Recent large flowslides have occurred in cohesive till in northeastern BC, especially near Buckinghorse River (Geertsema et al. 2006a). The landslides are retrogressive and extremely mobile, with travel distances up to 1.8 km along gradients as low as 3°. The abundance of flowslides in this area in the mid-1990s suggests a link to climate. A warming trend in

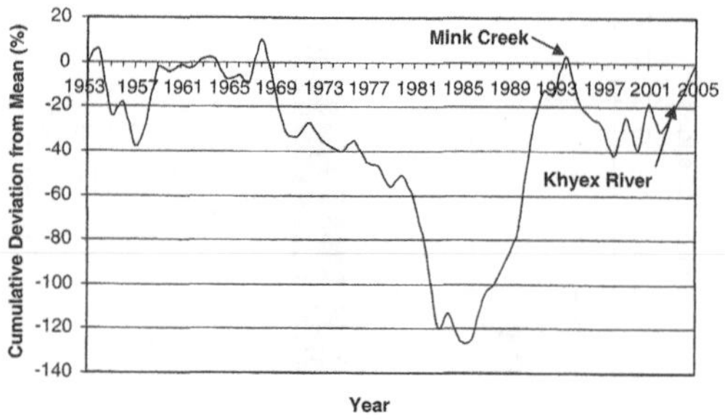

Figure 11. Percent cumulative deviation from mean precipitation at Terrace airport for the period 1953 to 2005.

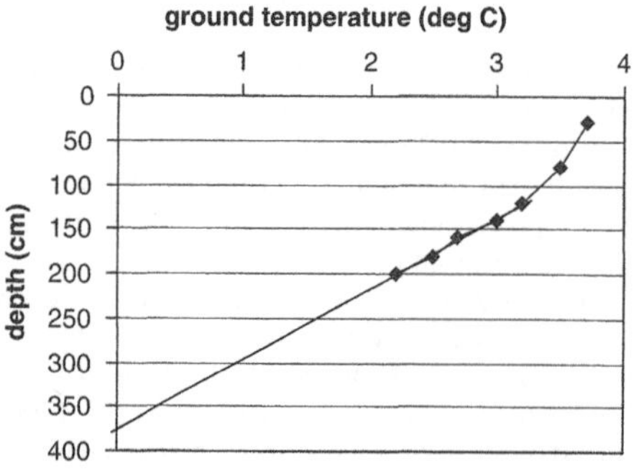

Figure 12. Ground temperature profile measured in September 2004 adjacent to a landslide in the Buckinghorse River area (modified from Geertsema, 2006). Extrapolation of the data suggests possible relict permafrost at depth.

late 20th century may have contributed to degradation of permafrost or to seasonal changes in precipitation that may have triggered landslides. Figure 12 shows a ground temperature profile from the Buckinghorse River area in September 2004. The profile indicates possible frozen ground at 3–4 m depth. An active layer this thick is unexpected in northeastern BC, as permafrost is typically present within 1 m of the ground surface (Crampton 1977, 1978). Fossil permafrost has been found in north-eastern BC at depths of 4–8 m (Geertsema, 2006), and it is possible that thaw of permafrost is contributing to the increase in flowslides.

5 PERMAFROST

Landslides in mountainous terrain are strongly influenced by climatic factors, including precipitation and temperature (Evans and Clague, 1994, 1997). Landslides at high elevations may be especially responsive to increases in temperature. Researchers have suggested that recent melting of glaciers in BC has debuttressed rock slopes adjacent to glaciers, causing deep-seated slope deformation, joint expansion, and catastrophic failure (Clague and Evans 1994, 1997; Holm et al. 2003; Geertsema et al. 2006a).

Mountain permafrost may be degrading under the influence of the warming climate, decreasing the stability of slopes (Davies et al. 2001; Harris et al. 2001; Noetzle et al. 2003). Recent large rockslides in the European Alps have been attributed to thaw of mountain permafrost (Dramis et al. 1995; Bottino et al. 2002; Noetzle et al. 2003), and permafrost thaw may also play a role in initiating landslides in northern British Columbia. Lowland permafrost in north-eastern BC may also be thawing. The spatial and temporal clustering of eight large flowslides and other small ones in the Buckinghorse River area (Geertsema et al. 2006a), combined with the ground temperature measurements (Figure 12), raises the possibility that they were triggered by thaw of permafrost.

6 CONCLUSION – THE FUTURE

Many landslides are intimately linked to climate, but respond differently depending on the type of landslide and processes involved. Our work shows that most of the sediment moved by shallow debris slides and debris flows happens during infrequent large storms. In contrast, large rockslides appear to respond to warming and debuttressing of rock walls due to glacier thinning, and may be triggered during convective storms. Large soil slides are more common during periods of increasing precipitation.

Conditions favourable for landslides are predicted for the future. Chiotti (1998) forecasts more storms for British Columbia during this century. Nearly all of the 58 global circulation models available from the Canadian Institute of Climate Studies (http://www.cics.uvic.ca) predict a warmer and wetter climate for north-western and north-eastern BC (Figure 13). Our historic analysis shows that a warmer, wetter climate is likely to be accompanied by increased landslide activity. There is some geological support for this inference. In a study of glaciomarine flowslides in the Terrace area, Geertsema et al. (2006b) noted that more than one-third of the prehistoric flowslides happened between 2000 and 3000 years ago, a period of above-average precipitation in north-western BC (Clague and Mathewes 1996).

We predict an increase in landslide activity during the 21st century due to more violent storms, continued glacial debuttressing due to glacier retreat, permafrost degradation, and generally wetter conditions.

Climate change has indirect, as well as direct, effects on landslide frequency. Recent climate warming has contributed to insect and disease epidemics that have caused unprecedented forest dieback in northern BC (Geertsema, 2006). Reduction of evapotranspiration, hydrophobic soil conditions following

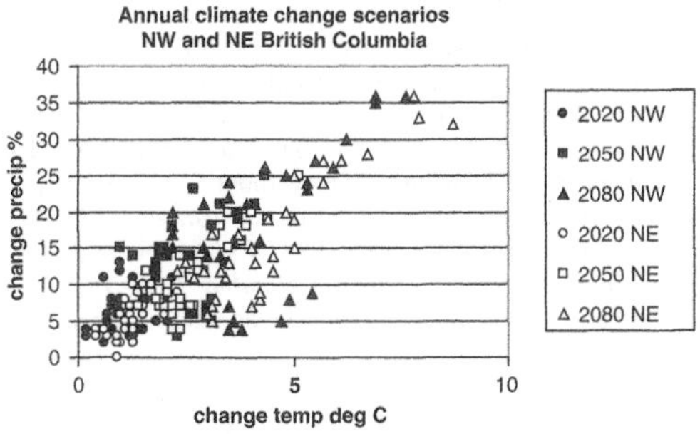

Figure 13. Annual climate change scenarios for northwestern and north-eastern BC. The 58 global circulation models predict a warmer and wetter climate later in the 21st century. (Canadian Institute for Climate Studies, http://www.cics.uvic.ca.

wildfires, and salvage harvesting predispose landscapes to increased slope failure.

More work needs to be done to define possible links between landslide events and ocean-atmosphere linkages. In northern British Columbia, the Pacific Decadal Oscillation strongly influences temperature and, to a lesser extent, precipitation patterns in 20–30 year cycles. The last decade of the current warm phase of the PDO, which started in 1976, was a period of frequent rockslides. We need to extend our comparison of landslide frequency and the PDO beyond the current phase into the past and also to investigate the link, if any, between landslides events and El Niño-Southern Oscillation.

REFERENCES

Bottino, G., Chiarle, M., Joly, A. & Mortara, G. 2002., Modelling rock avalanches and their relation to permafrost degradation in glacial environments. *Permafrost and Periglacial Processes* 13: 283–288.

Chiotti, Q. 1998. An assessment of the regional impacts and opportunities for climate change in Canada. *Canadian Geographer* 42: 380–393.

Clague, J.J. 1989. Quaternary geology of the Canadian Cordillera; Chapter 1 In R.J. Fulton (ed.), *Quaternary geology of Canada and Greenland:* 15–96. Ottawa: Geological Survey of Canada.

Clague, J.J. & Mathewes, R.W. 1996. Neoglaciation, glacier-dammed lakes, and vegetation change in northwestern British Columbia, Canada. *Arctic and Alpine Research* 28: 10–24.

Crampton, C.B. 1977. Changes in permafrost distribution in northeastern British Columbia. *Arctic* 30: 61–62.

Crampton, C.B. 1978. The distribution and thickness of icy permafrost in northeastern British Columbia. *Canadian Journal of Earth Sciences* 15: 655–659.

Davies, M.C.R., Hamza, O. & Harris, C. 2001. The effect of rise in mean annual temperature on the stability of rock slopes containing ice-filled discontinuities. *Permafrost and Periglacial Processes* 12:137–144.

Dramis, F., Govi, M., Guglielmin, M. & Mortara, G. 1995. Mountain permafrost and slope stability in the Italian Alps: the Val Pola landslide. *Permafrost and Periglacial Processes* 6: 73–82.

Egginton, V.N. 2005. *Historical climate variability from the instrumental record in northern British Columbia and its influence on slope stability.* MSc thesis. Simon Fraser University, Burnaby, BC.

Egginton, V.N., Clague, J.J. & Jackson, P.L 2005. Investigating landslide triggers in northern British Columbia using weather satellite imagery. 58th *Canadian Geotechnical Conference*, September 18–21, Saskatoon, SK.

Evans, S.G. & Clague, J.J. 1994. Recent climatic change and catastrophic geomorphic processes in mountain environments. *Geomorphology* 10: 107–128.

Evans, S.G. & Clague, J.J. 1997. The impacts of climate change on catastrophic geomorphic processes in the mountains of British Columbia, Yukon and Alberta. In E. Taylor and B. Taylor (eds.), *Responding to global climate change in British Columbia and Yukon, Vol. 1, Canada country study: climate impacts and adaptation*: 7-1 – 7-13. Vancouver: BC Ministry of Environment, Lands and Parks and Environment Canada.

Geertsema, M. 2006. *Hydrogeomorphic hazards in northern British Columbia.* Utrecht: Netherlands Geographical Studies.

Geertsema, M., Clague, J.J., Schwab, J.W. & Evans, S.G. 2006a. An overview of recent large landslides in northern British Columbia, Canada. *Engineering Geology* 83: 120–143.

Geertsema, M., Cruden, D.M. & Schwab, J.W. 2006b. A large rapid landslide in sensitive glaciomarine sediments at Mink Creek, northwestern, British Columbia, Canada. *Engineering Geology*. 8: 36–63.

Harris, C., Davies, M.C.R. & Etzelmüller, B. 2001 The assessment of potential geotechnical hazards associated with mountain permafrost in a warming global climate. *Permafrost and Periglacial Processes* 12: 145–156.

Hogan, D.L. & Schwab, J.W. 1991. *Meteorological conditions associated with hillslope failures on the Queen Charlotte Islands*. Victoria: BC Ministry of Forests.

Holland, S.S. 1976, *Landforms of British Columbia. A physiographic outline*. BC Department of Mines and Petroleum Resources Bulletin 48.

Holm, K., Bovis, M.J., & Jakob, M. 2004. The landslide response of alpine basins to post-Little Ice Age glacial thinning and retreat in southwestern British Columbia. *Geomorphology* 57: 201–216.

Jakob, M. & Wheatherly, H. 2003. A hydrometerological threshold for landslide initiation on the north shore mountains of Vancouver British Columbia. *Geomorphology* 54: 137–156.

Jakob, M., Holm, K., Lange, O. & Schwab, J.W. 2006. Hydrometeorological thresholds for landslide initiation and forest operation shutdowns on the north coast of British Columbia. *Landslides* 3: 228–238.

Karanka, E.J. 1986. *Trends and fluctuations in precipitation and stream runoff in the Queen Charlotte Islands*. Victoria: BC Ministry of Forests.

Meidinger, D.V. & Pojar, J.J. 1991. *Ecosystems of British Columbia*. Victoria: BC Ministry of Forests.

Noetzli, J., Huggel, C., Hoelzle, M. & Haeberli, W. 2003. GIS-based modelling of rock/ice avalanches from Alpine permafrost areas. *Computational Geosciences* 10: 161–178.

Schwab, J.W. 1997. Historical debris flows, British Columbia north coast. In *Proceedings, forestry geotechnique and resource engineering, Richmond, B.C.* Vancouver: BiTech Publishers.

Schwab, J.W. 1998. Landslides on the Queen Charlotte Islands: processes, rates, and climatic events. In *Carnation Creek and Queen Charlotte Islands fish/forestry workshop: applying 20 years of coast research to management solutions*, D.L. Hogan, P.J. Tschaplinski & S. Chatwin (eds.): 41–48. BC Ministry of Forests, Land Managements. Handbook 41.

Schwab, J.W., Geertsema, M. & Evans, S.G. 2003. Catastrophic rock avalanches, west-central B.C., Canada. *Proceedings, 3rd Canadian Conference on Geotechnique and Natural Hazards,* Edmonton, AB: 252–259.

Schwab, J.W., Geertsema, M. & Blais-Stevens, B. 2004. The Khyex River landslide of November 28, 2003, Prince Rupert British Columbia, Canada. *Landslides* 1: 243–246.

Septer, D. & Schwab, J.W. 1995. *Rainstorm and flood damage: Northwest British Columbia 1891–1991.* BC Ministry of Forests, BC Land Management Report 31.

Landslides and Climate Change – McInnes, Jakeways, Fairbank & Mathie (eds)

Folkestone Warren landslides and the impact of the past rainfall record

E.N. Bromhead & M.-L. Ibsen
Faculty of Engineering, Kingston University, Kingston

ABSTRACT: The Folkestone Warren landslide system occupies approximately 2 km of the coastline in south Kent, where the Weald-Boulonnais dome is cut through by the English Channel. A railway connecting Folkestone to Dover, built in 1846, crosses this infamous landslide. Periodic movements of the whole or part of the landslide affect the railway, and have done so almost from its opening. A series of major slide events in the latter part of the 19th century and early 20th century culminated in the dramatic slide of December 1915. Following this, extensive remediation works were carried out, with the construction of drainage adits, toe loading and seawall construction, with further coast defence activities continuing to the present day. This paper very briefly records the various published accounts of the landslide system, and introduces and discusses the rainfall records of a nearby weather station. It shows that the movements of the landslide are strongly influenced by periods of high rainfall, although the behaviour, not unexpectedly, changed after the construction of the extensive system of deep drainage adits. The behaviour of the High Cliff, which forms the landward extent of the landslide system, is also subject to its own local modes of instability. Mention is made of the developing slide of the High Cliff adjacent to a section, which failed in 1915. The paper concludes with observations of landslide response to rainfall, the impacts of likely climatic change and variability in the general area, and the use of deep drainage as a principal method of stabilizing deep-seated landslides.

1 INTRODUCTION

The Folkestone to Dover railway line crosses the Folkestone Warren landslide system and runs through a variety of tunnels in different geological strata. It emerges from the Martello Tunnel (at the west side of the Warren), and disappears into the Abbotscliff Tunnel at the Warren's eastern end. It then continues across the site of the former Shakespeare Colliery, more recently the scene of workings for the Channel Tunnel, and passes into a third tunnel (the Shakespeare Tunnel) before emerging from the cliffs on the Dover side. Despite ground movement problems, the railway has continued in service for over a century and a half with few interruptions (Figures 1 and 2). In the first half of this period, significant ground movements occurred, sometimes sufficient to close the railway temporarily. Although stabilization works appear to have brought the major movements under control, lesser movements have continued to affect the Warren landslide complex and the tunnels. In principle, there is no known reason why with the appropriate maintenance, new works and monitoring, the railway should not continue in safe use for the indefinite future.

Of the major movements, those of December 1915 were the most severe and spectacular. During one single evening, two major collapses of the high Chalk rear scarp occurred. One of these failures had high

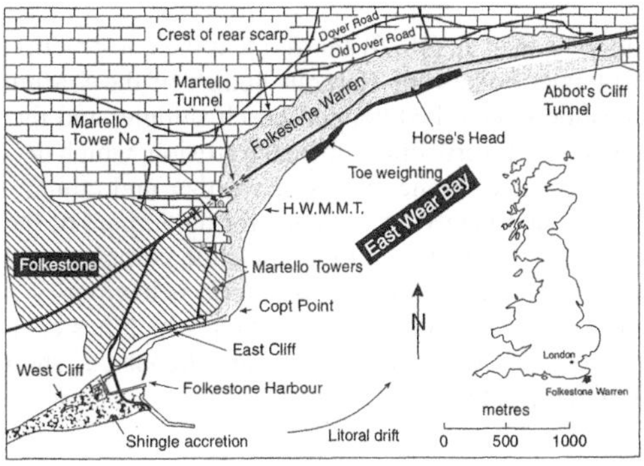

Figure 1. Location map of the Folkestone Warren landslide complex.

mobility, and flowed across the railway line, creating a major obstruction to rail traffic, and continued out to sea producing a promontory of debris; the other was contained within the landslide system, but produced head loading, and was in a large part responsible for the 70m displacement of the main deep seated landslide (Hutchinson et al. 1980). A further failure of the rear scarp carried a dwelling part way down the rear scarp cliff.

It was well known that the December 1915 landslide occurred at a time of high rainfall, although

Figure 2. Photograph of Folkestone Warren landslide complex and railway line.

Hutchinson et al. (1980) showed that the construction of piers at Folkestone Harbour updrift of the Warren landslide system had trapped beach shingle, and this had had an adverse effect on beach erosion, which perhaps downplayed the effects of weather and climate. This paper reviews the evidence for the impact of the rainfall on the activity of the landslide complex, and makes observations on the long-term behaviour of a landslide system such as the Warren, which has been stabilized by the construction of an extensive system of drainage adits.

2 FOLKESTONE WARREN LANDSLIDE COMPLEX

The geological sequence for Folkestone Warren is broadly similar to those of other major landslide complexes on the south coast of England, for example, Lyme Bay and Southern Isle of Wight, with a caprock (in this case the Chalk) overlying a clay (Gault), leading to the occurrence of bedding-controlled, compound, landslides. The Upper Greensand, which occurs between the Chalk and the Gault, is here of little significance. The underlying Lower Greensand (represented at Folkestone by the Folkestone Beds) is not involved in the landslides.

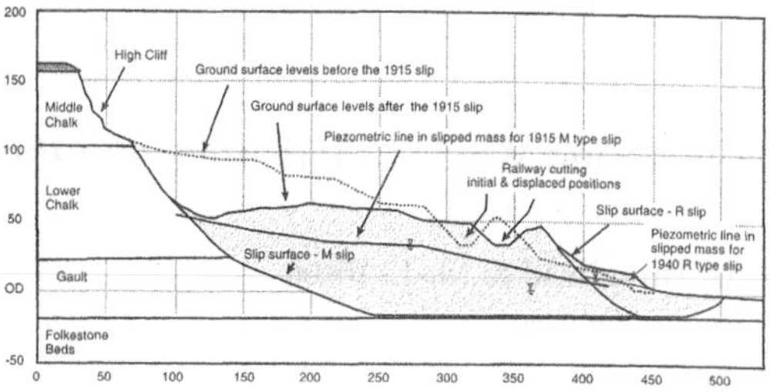

Figure 3. Folkestone Warren landslide section.

Folkestone Warren is a large, ancient but, active landslide complex, details of which are given, inter alia, by Osman (1917), Toms (1953), Wood (1957), Viner-Brady (1955), Hutchinson (1969), Hutchinson et al. (1980) and Trenter & Warren (1996). The landslide masses are primarily large blocks of landslipped Chalk and accumulations of Chalk debris, with some zones of slipped Gault Clay, moving mainly on a bedding controlled slide surface in a thin horizon of plastic clay, located close to the contact of the Gault with the underlying Folkestone Beds, part of the Lower Greensand (Figure 3). The shear strength of this slip surface is moderately well understood, as is the stability of the whole or part of the landslide complex in general terms.

The three-dimensional shape of the Warren is believed to have developed in response to coastal erosion, and the interplay of different mechanisms of landsliding with the overall geological structure. It comprises a shore platform; a sea cliff; a landslide body and a high rear scarp cliff. Landslide accumulation slides follow a bedding-controlled basal shear on a weak bed in the Gault clay and are largely made up of chalk masses and debris, which are often highly permeable. The dip of the beds to the east eventually puts the Gault Clay below sea level, and landslides occur from where the base of the Gault descends to beach level through to where it disappears below the beach.

Within the landslide system, there are various modes of ground movement. These were termed M and R type slides, and falls, type F, in Hutchinson's review of the Warren landslides Hutchinson (1969).

The M-type landslides (where "M" might equally well signify massive, major, or multiple) are compound landslides affecting the whole or most of the undercliff with a major element of their basal shear surface following the plastic layer close to the base of the Gault, i.e. they are bedding controlled Bromhead and Ibsen, 2004). R-type landslides are movements affecting the sea cliff and seaward margin of the landslide complex. They are rotational in character but, move in part along the common basal shear surface (Hutchinson et al. 1980). The M-type landslide movements affect the whole railway between the portals of the Martello and Abbotscliff tunnels, but the R-type landslides only affect the railway line locally.

A cross section (Figure 3) shows the principal details of M-type and R-type landslides. The plan shape of the landslide complex means that slide movements may not occur in a strictly landward-seaward direction, but gives rise to differential movements of varying magnitude, or even some implied rotation in plan. The landslide system is formed mainly from slipped masses and fall debris from the Lower Chalk and lowermost beds of the Middle Chalk, together with irregular masses of the Gault, although most of the clay stratum has been 'sheared out' by the rotational failures.

Falls may occur from the sea cliff, which is mainly formed from debris of previous landslides and is in places naturally recemented, as well as from the rear scarp or High Cliff. These falls can have the following behaviours:

1. Small and do not reach the railway;
2. Blocks, as debris lumps, which reach the railway;
3. Debris burying the railway, and possibly reaching the foreshore;
4. Debris which loads the head of the main slide masses and causes movement;
5. Collapse of the rear scarp causing head loading, which actually enlarges the Warren slide complex, due to foundation failures in the High Cliff that go down into the Gault Clay.

No works have been undertaken to control the incidence of such failures, although it is evident that stopping major slip activity of the Warren slide mass does have some influence on the incidence of High Cliff collapses. The principal hazard from Chalk falls is blockage of the whole or part of the permanent way, or loading which stimulates the main landslide complex and leads to distortion or dislocation of the railway track.

The above effects are understood in an approximate, qualitative and in some instances, semi-quantitative manner. However, understanding of the mechanisms behind the Chalk falls does not allow the prediction of size, location or time and date of occurrence of individual failures, only their general pattern.

2.1 *Landslide activity in the Warren*

Landslide activity in the Warren is understood to be influenced adversely and directly by marine cliff retreat and foreshore lowering. Since, major movements all followed extension of the Folkestone Harbour breakwater arm, it is believed (Hutchinson et al. 1980) that its construction in the 19th and early 20th centuries has intercepted littoral drift and intensified erosion, with consequent beach denudation. The case was made with the support of map data that indicated a massive accumulation of beach shingle on the western side of the harbour mole. However, it is also understood that movements have occurred following periods of higher than average rainfall. For instance, the year 1915 in which the last large-scale movements took place was a year of particularly heavy rainfall and hence, a time of rising ground water levels.

Table 1. List of landslide events recorded at Folkestone Warren (After Ibsen 1994)

Date	Landslide type/magnitude
1869	Minor failure
1877-1-19	Chalk fall – Major failure
1881-3	Minor failure
1885-9	Minor failure
1886-1	Minor failure
1892-11	Minor failure
1896-11-15	Major failure
1905/6?	Minor failure
1915-12-19	Minor failure
1935-8-31	Minor failure
1936	Minor failure
1937-3-16	R-type moderate failure
1940-2	R-type moderate failure
1947	Minor failure
1970	Movement
Major events occurred on similar terrain nearby	
Abbots cliff	1912-1-2
Shakespeare Cliff	1928-2-16
Hythe, Roughs	1988-2
Hythe, Roughs	1994
Hythe, Roughs	2002

The records of landslide movement are incomplete (Table 1). It is believed that M-type landslides occurred in the Warren in 1877, 1896 and 1915. Little is known about the earliest of these, although a surviving photograph shows the Chairman of the railway company in a train passing through a fresh cutting made in Chalk debris. During the 1896 landslide, the Martello tunnel was cracked near the Warren portal, and works were subsequently carried out to excavate a cutting above the tunnel so that rescuers could dig down if necessary (while leaving the line covered against falls of chalky debris). The tunnel portal was also moved westwards at the same time. This was fortunate, since the side shear of the 1915 landslide followed the alignment of the earlier cracks. Much more is known about the December 1915 landslide than about the earlier M-type landslides because of a paper by Osman in 1917 that gave a contemporary account, and a series of photographs were taken by an amateur photographer, a railway employee sent to rescue a train, which had been stopped straddling the side break of the landslide (Hutchinson et al. 1980).

At least three failures of the rear scarp, two major F-type movements and a smaller one, accompanied the December 1915 M-type landslide. Of the major movements, one created the head loading that was the main impetus for the large displacements of the slide complex (Hutchinson 1987); the other fluidized, crossed the railway, which was left blocked by debris, and ran out to sea. The minor F-type movement displaced a house, from which the three occupants escaped safely. Hutchinson et al. (1980) identified these F-type failures as highly hazardous, noting that there had been numerous but, poorly-recorded incidents where debris had failed to reach the railway line. This paper also noted that the rear scarp of the landslide had been left effectively higher and more vulnerable after the 15 collapse.

Since then, failures of the whole landslide system have not taken place. This is thought to be the result of remedial works carried out on several occasions. However, falls from the rear scarp have continued (although on a small scale) and the coastal margin R-type slips have recurred with some regularity. Documented instances of these are in 1937 and 1940, with further instances in the late 1960's, in the 1980's and in recent years.

2.2 *Preventative works*

Both the effects of rainfall and toe erosion have been addressed by the construction of seawalls and drainage adits, the results of which have been so far to control the major ground movements of the Warren. However, both systems need continual maintenance. Furthermore, the systems installed are not ideal, for example, the sea walls do not prevent foreshore lowering, only coastal cliff retreat, and the drainage adits have been excavated from above the seawall level, so that there is a limit to the drawdown they can produce.

It seems that rather than providing a total solution to stability problems so that major incidents can never occur, the stabilization works have simply lengthened the return period between major incidents so that they are less frequent (i.e. not since 1915). Although this is a considerable achievement, in a quantitative risk analysis framework, major events are not prevented, merely deferred.

The 1915 landslide was by far the most severe of the movements and after it the railway was closed until 1923 for remedial works, which included the construction of a number of drainage headings. The remedial measures also included some coast protection works and a realignment of the railway. Prior to the 1915 landslide, some drainage works (tunnelling into the slide mass) appear to have been undertaken, along with some attempts at beach erosion control taking the form of groynes that attempted to prevent beach movement (Osman 1917).

New movements in the 1930's and 1940's, taking the form of displacements of the R-type landslides, ruptured drainage tunnel linings and were evident at the ground surface causing movements in the track. Subsurface investigations of the landslide complex with boreholes provided the basis for an understanding of the landslide mechanics (Toms 1946), leading to extensive toe-weighting, sea wall construction and new drainage works (Wood 1957). Repair and extension of the sea walls and drainage tunnels have continued to the present day (Wood 1971). Hutchinson (1969) and Hutchinson et al. (1980) describe the landslide in a modern, effective stress and residual shear strength, context and the latter of these two accounts stimulated a further programme of investigation and analysis by Trenter & Warren (1996) and Warren & Palmer (2000).

2.3 *Recent developments*

Two R-type landslides have been active in recent years. The westerly one of these affects the camping site and its access roads and straddles the western end of the toe-weighting platform (Figures 3 and 5). It also affects the railway line. It has a comparatively small displacement at present. The easterly R-type landslide affects the area around Horsehead Point, where the toe-weighting is less massive. It has damaged the concrete paving of the toe weighting. Movements of this R-type slide were evident in about 1978–9 as a step of less than one metre height. When inspected in July 2005, the scarp had grown and was traceable over a large area, including where it had damaged the access road down to the shoreline defences.

Undoubtedly, movements of the R-type landslides cause damage to the drainage headings where they pass through the shear zone. The drainage headings discharge copious amounts of water, estimated to be in the range 2 to 3 litres per second when inspected in July 2005 following a very dry winter and spring. Most of this is collected in the body of the M-type landslide. Losses into the R-type landslides are unknown. The principal hazards concerning the Folkestone Warren landslide complex are dislocation or distortion of the permanent way, i.e. the railway track. Following the review published in 1980 by Hutchinson et al., the risk from Chalk falls in the Warren has been highlighted. These take three levels of severity:

1. Falls of blocks which may reach the track;
2. Larger collapses where the debris may partially fluidize, giving it the mobility to block the track, as in the 1915 Great Fall;
3. Collapses which load the head of the Warren (possibly involving shear into the Gault) and cause track dislocation.

No works have been undertaken to control the incidence of such failures, although it is evident that stopping major slip activity of the Warren slide mass does have some influence on the incidence of High Cliff collapses. The principal hazard from Chalk falls is blockage of the whole or part of the permanent way, or loading which stimulates the main landslide complex and leads to distortion or dislocation of the railway track.

In terms of the long-term stability of the Warren landslide system, the development of a zone of cracking and settlement at the cliff crest is becoming increasingly important. Hutchinson (1969) discussed a mechanism for the enlargement of the Warren by new failures of the Chalk cliffs forming its rear scarp. This area of cracking and subsidence, which has already caused major damage to a number of houses, is a possible example.

3 LANDSLIDE ACTIVITY AND RAINFALL

'At Horsehead Point a correlation appears to exist between the observed movements and rainfall a further indication that the porewater pressures acting on the slip equate to water levels in the Chalk' (Warren and Palmer, 2000)

The simplest and most immediate improvement in understanding the activity of the Warren is to quantify the water balance. Precipitation contributes to the size of the water body within the landslide system and increases the probability of failure, while drainage reduces the size of the water body, and decreases the probability of failure. The key to understanding future behaviour is to understand the interplay between the two. Movements are prevented or at least minimized if the drainage always "beats" the precipitation.

The rainfall record in Folkestone, on the south coast of Kent, dates back to about 1870, and has been measured at a site a couple of kilometres distant from the huge Folkestone Warren landslide system. It shows annual rainfall varying from a low of 400 mm to a high of 1200 mm, and a rising trend within this highly variable record with the mean rainfall now about 800 mm per annum rather than the 700 mm of the late 19th century. Rainfall data was initially obtained from Folkestone, Cherry Garden TR2100 3790, for the period 1868–1992. For the period 1993–2003 the site changed to East Bourne further west along the coast. The correlation coefficient between the two sites, for the period 1959–1992, was calculated as 0.746, which was considered a strong enough correlation to amalgamate the two data sets. However, the mean difference between Eastbourne and Folkestone was minus 24.72 indicating that Eastbourne has slightly wetter weather to that of Folkestone. In order to compensate for this variation the rainfall values for 1993–2003 were decreased by the average difference of 24.72.

Yearly rainfall is extremely erratic but the pattern of precipitation over longer periods can be shown to have both wetter and drier periods. Major movements of the Warren seem to be related to periods of accumulated rainfall. For example, 1915 was not only wet, but was the culmination of a number of wet antecedent years (Figure 4). Since, then there have been wetter years and accumulations but, these have not caused serious results – probably due to the effectiveness of the drainage system. It is clear that the Warren is too large to respond to a single day, week or month of rainfall. Several wet years are required to initiate extensive failure.

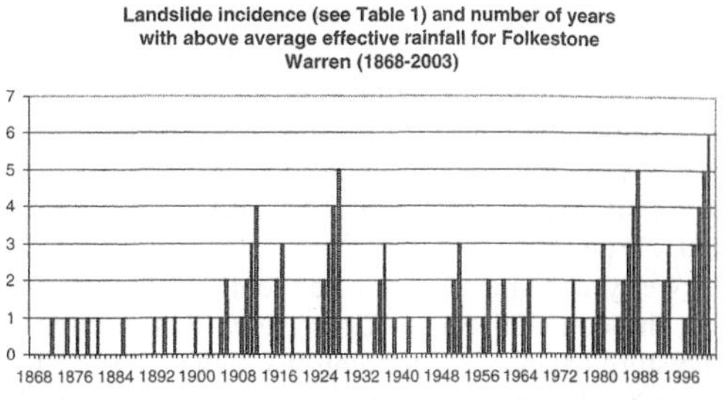

Figure 4. Wet year sequence for Folkestone Warren from 1868–2003.

Figure 5. Contemporary movements affecting the toe-weighting near Horsehead Point.

The data used in this paper is an extension of that use by Bromhead et al. (1998), which in turn is an addition to the original methodology applied by Ibsen (1994). For this paper, the moisture balance index was extended to include the period from 1991 to 2003. The Thornthwaite index was used to calculate potential evapotranspiration using the monthly mean central England temperatures from the UK Meteorological Office. Although these are average values for southern Britain the difference with the local mean is minimal varying between $\pm 1°C$. The Thornthwaite index is a simple method of estimating the maximum water loss from the surface, which is assumed to have an unrestricted supply of water and does not account for differing vegetation cover. It is less accurate than other calculations, such as the Penman formula, however, data can be determined for a significant temporal period since it requires only temperature values. These data are subsequently subtracted from the annual rainfall series to produce a moisture balance index or effective rainfall. The effective rainfall series shows a positive trend towards the present in which the average, 217 mm, increases approximately 30% over the 136 year period (1868–2003). This is the equivalent of +0.2% per year.

Wet year sequences may be identified by calculating the cumulative number of years with an annual moisture balance greater than the mean (Ibsen 1994, Bromhead et al. 1998, Ibsen & Casagli 2004). The number of years are added consecutively and reduced to zero every time the moisture balance falls below the mean. In this analysis the value of 200 mm was used to set the graph to zero, since the mean (217 mm) is a fluctuating value as the period is extended over the years. However, it should be noted that the value of 200 mm assumes that the moisture balance only becomes critical to the slope above this level. This may not be the case as the point at which moisture balance affects the slope is difficult to calculate, since it entails a comprehensive knowledge of the permeability of the geological structure and the influence of artificial drainage.

The prominent wet year periods, those with three or more consecutive years, are: 1909–1912; 1915–1917; 1922–1928; 1935–1937; 1950–1952; 1963–1966; 1979–1981; 1984–1988; 1992–1994; 1997–2002 (Figure 4). Note that due to the underlying positive trend the frequency of the wet year periods increases towards the present. To account for the magnitude of change in these periods, a series was calculated to sum the departure from the critical value, 200 mm. This cumulative departure series shows the long term trend in effective precipitation, as well as the magnitude of the changes in the moisture balance conditions. Fundamentally, a negative departure indicates that conditions are becoming drier and a positive change indicates that conditions are becoming wetter. Overall drying conditions are indicated from 1868 to 1902, then from the turn of the 20th century conditions have become increasingly wetter obviously with some fluctuations, for example, drying periods in the 1940's and early 1970's.

Landslide activity seems to coincide closely with the peaks and troughs of the annual effective precipitation data, as well as the significant wet year periods. At the turn of the last century the moisture balance index shows a definitive change from drier to wetter conditions, within which there are periodic wetter periods. Examining the graphs of effective precipitation in conjunction with the knowledge of major landslide incidence the following can be noted:

1. 1877 the slide with the famous photo of Edward Watkins. This was probably a Chalk Fall and could correspond to the peak rainfall of that year. Associated minor ground movements were noted in succeeding years in 1881, 1885, 1886 and 1892 with equally high rainfall amounts;
2. 1896, a significant landslide occurred, coinciding with a peak effective precipitation of that year, plus gradual wetting conditions over the preceding five years. As a result of this and cracking in the Martello Tunnel, a portal was moved westwards and a cutting opened out over the tunnel by the railway. Spoil was placed seaward of this location;
3. 1915, the celebrated Chalk falls and landslip occurred, following a high peak in effective precipitation (475 mm) and a general rise over the previous seven years, both factors being far more severe than previously recorded. It took over eight years to re-open the railway line within which most of the time was spent constructing adits. The creation of adits had begun sometime before 1915 but, the landslip had overwhelmed the workings;
4. 1937, post construction of some drainage adits, an R-type slip was noted, along with minor movements in 1935 and 1936, these coincided with a 3 year accumulation and peak in effective precipitation. The 1937 slip involved chalk spoil from the Martello cutting excavation. After which detailed sub-surface investigations commenced and Terzaghi was consulted. More adits and toe weighting were constructed;
5. 1940 another R-type slip occurred in the February following the highest annual effective precipitation recorded for the entire period (588 mm, 1939);
6. 1947, minor movement probably as a consequence of the high effective precipitation in 1946 (362 mm);
7. 1950s and 1960's, again, 3 and 4 year accumulations and various annual peaks of effective rainfall. Halcrow reported movements at Horsehead Point in 1970 following an extensive period of wetting over approximately the previous 20 years.
8. 1988, 1992–4, 2002 movements were noted in similar terrain some km distant from Folkestone Warren (Bromhead et al., 1998) following periods of wet weather.

Without the remedial works, 1915-scale movements could easily have occurred in 1937, 1940 or 1970 rather than the minor slip movements that did occur. Equally in 1988, 1994 and 2002 major movements may have taken place, as they did in similar terrain at the Roughs, Hythe (Bromhead et al. 1998). Remedial measures have had a major beneficial effect on stability at Folkestone Warren, although they have not prevented all ground movement.

In landslide analysis the frequency and magnitude of specific events is of major concern. The landslide data for Folkestone Warren are not sufficient enough to conduct this type of analysis, however, a simple method of extreme values was used regarding the effective rainfall data. The technique employed to statistically model the extreme event distribution of the effective rainfall data was the Gumbel method (Gumbel 1942). Gumbel uses a natural distribution that adjusts for the positive skewness common in extreme event distributions. By definition, the calculated return periods reveal standard pictures, an increase in magnitude indicates an increase in the return period. A regression line was fitted to the observed values with a high degree of correlation. From this a moisture balance value of 200 mm was shown to occur every two years. The limit of 380 mm, had a return period of every ten years, and the figure of 470 mm, associated with the 1915 event had a probability of recurrence of every 20 years. The highest value of the entire period, 1968–2003, which was 588 mm in 1939 has a return period of 100 years. With the trend for an increasingly wetter environment these return periods could be further reduced. Every decade some form of movement can be expected and the likelihood of a 1915 event has a possible recurrence of every twenty years. Although, this has not come about due to the extensive remedial works, which have been carried out on the Warren.

4 DISCUSSION

There is a good correlation between effective rainfall and landslide activity, which indicates the need for further research. The data used in this paper is an extension of that of Ibsen & Brunsden (1996, 1997), and Bromhead et al. (1998), which in turn is an addition to the original methodology used by Ibsen, 1994. The moisture balance index was calculated to include the period from 1991 to 2003. Landslide activity seems to coincide closely with the peaks and troughs of the annual effective precipitation data. At the turn of the last century the moisture balance index shows a definitive change from drier to wetter conditions, within which there are periodic wetter periods. There are no convincing cycles within the analysis of effective precipitation and landslide movement, since there are too many variables which have not been taken into account. However, conditions have obviously been worsening since the turn of the 20th century and the question is if this is due to continue or will conditions begin to pertain towards a drier environment as in the latter half of the 19th century. The overall forecast for the next 50 years is for higher temperatures, especially in the summer season, with rainfall events becoming more extreme and generally increasing in the winter months. The affect this will have on the moisture balance is as yet debatable and as a function of this the effect on landslide activity is equally unknown. One hypothesis is that the medium to small sized landslide events are most at risk, since they are affected by shorter duration, intense precipitation events (Corominas, 2001). Whereas for deeper-seated landslides, perhaps a drier summer but, wetter winter would balance out the annual effective precipitation, therefore, not having much effect.

Although to some extent, the drainage system of the Warren responds to increased water inputs by greater discharges, the innate storage capacity of the slide debris must experience raised piezometric levels, with their qualitative impact on stability. This occurs at a time when the foreshore is lowered, and lowering. It is inevitable, therefore, that some future works to enhance both the drainage system and the coastal defence will be required. Whichever of these is the best value-for-money remains to be seen or proved. Without a proper quantitative understanding of the problem, however, value for money is least likely to be attained, and risk levels could rise to an unacceptable degree.

Future infrastructure safety could be assured by the creation of an intelligent knowledge based system, based around a geographic information system, and running a real-time quantitative risk analysis, which would inform the operations staff of current and predicted short-term future levels of risk. Real-time inputs into such a system would come from the monitoring within the Folkestone Warren area, and those inputs would be interpreted within a knowledge framework based on the geotechnical and related investigations (some of which, if not all) have largely been undertaken in the past. However, such a system could not be created today with the necessary sensitivity to inputs and freedom from false alarms, because of certain deficiencies in the knowledge base, and in the systems for handling and developing information.

Potential steps to consider in the monitoring and modelling of the Folkestone Warren landslide complex could be:

(1) to assimulate the geotechnical and geomorphological information into a ground model;
(2) to input the data into stability analyses;
(3) to carry out a quantitative risk analysis;
(4) to enhance the existing rainfall record and other climatic analyses.

4.1 *Conclusions*

Although it was recognised that the event in 1915 had been an exceptionally wet year, Hutchinson et al. (1980) found a persuasive mechanism to account for the M-type failure in the progressive extension at various times in the 19th century of the Folkestone Harbour breakwater arm or mole. This intercepted littoral drift and trapped huge volumes of shingle up drift to the west of the harbour. This was thought to have left the shoreline of East Wear Bay denuded of its energy-absorbing beach and thus opening up the Warren to a greater risk of foreshore erosion. In contemporary photographs the foreshore is bare.

Work on rainfall and landslide activity (Ibsen 1994) has shown that movements in the south of England correlate well with rainfall. The major movements in the Warren are all associated with a run of wetter than average years, culminating in the 1915 event, which was a worse accumulation of wet weather than previously, bearing in mind that the railway was constructed in 1846, prior to which movements of the Warren were unrecorded. After 1915, the drainage works changed the nature of the response of the M-type landslide and rainfall has never been sufficient enough to overcome the effects of the drainage tunnels.

However, the drainage tunnels that cross the boundary shear surfaces of the R-type landslides are fractured and distorted. As a result, it is probable that they leak water into these landslides and may increase the frequency and magnitude of their movement, notwithstanding the beneficial effects of the toe loading. The precise response is complicated by foreshore denudation.

After the 1915 landslide, the High Cliff, or rear scarp, formed in Chalk was effectively higher than before. There are currently clear signs of a developing collapse within the High Cliff, which have appeared following periods of wet weather. The response of the High Cliff to toe erosion is assumed to be indirect, hence a run of wet years has the potential to cause a major collapse as significant as the 1915 event.

From this study it appears, therefore, that beach denudation, while an important factor in the evolution of the Folkestone Warren landslide complex, is a *preparatory* factor. The *trigger* factor is a run of years of higher than average rainfall. This paper has also identified the need for continued monitoring since, more movement is inevitable and current systems are merely delaying the expected.

REFERENCES

Bromhead, E.N., Hopper, A.C. & Ibsen, M-L. 1998. Landslides in the Lower Greensand escarpment in South Kent. *Bulletin of Engineering Geology and the Environment* 57(2): 131–144.

Bromhead, E.N. & Ibsen, M-L. 2004. Bedding-controlled coastal landslides in Southeast Britain between Axmouth and the Thames Estuary. *Landslides* 1(2): 131–141.

Corominas, J. 2001. Landslides and Climate. In: E.N. Bromhead (ed) *Keynote Lectures delivered at the 8th Int Symp on Landslides*, Cardiff, June 2000.

Gumbel, E.J. 1942. On the frequency distribution of extreme values in meteorological data. *Bull Am. Met. Soc.* 23: 95–105.

Hutchinson, J.N. 1969. A reconsideration of the coastal landslides at Folkestone Warren, Kent. *Geotechnique* 19: 6–38.

Hutchinson, J.N., Bromhead, E.N. & Lupini, J.F. 1980. Additional observations on the Folkestone Warren landslides. *Quarterly Journal of Engineering Geology* 13: 1–31.

Hutchinson, J.N. 1969. A reconsideration of the coastal landslides at Folkestone Warren, Kent. *Géotechnique* 19: 6–38.

Hutchinson, J.N. 1987. Mechanisms producing large displacements in landslides on pre-existing shears. *Mem. Geol. Soc. China* 9: 175–200.

Ibsen, M-L. 1994. Evaluation of the temporal distribution of landslide events along the South Coast of Britain, between Straight Point an St Margaret's Bay. *MPhil Thesis*, King's College London.

Ibsen, M-L & Brunsden, D. 1996. The nature, use and problems of historical archives for the temporal occurrence of landslides, with specific reference to the south coast of Britain, Ventnor, Isle of Wight. In: M. Soldati (ed) *Landslides in the European Union. Geomorphology* Elsevier special issue, 15 (3–4): 241–258.

Ibsen, M-L & Brunsden, D. 1997. Mass movement and climatic variation on the south coast of Great Britain. In: J.A. Matthews et al. (eds) *Rapid mass movement as a source of climatic evidence for Holocene. Paläoklimaforschung, Palaeoclimate Research, Special Issue,* ESF Project "European Palaeoclimate and Man 12", 19: 171–182.

Ibsen, M-L & Casagli, N. 2004. Rainfall patterns and related landslide incidence in the Porretta-Vegato region, Italy. *Landslides* 1(2): 143–150.

Osman, C.W. 1917. The landslips of Folkestone Warren and thickness of the Lower Chalk and Gault near Dover. *Proc. Geol. Assoc., London,* 28: 59–84.

Toms, A.H. 1946. Folkestone Warren landslips: research carried out in 1939 by the Southern Railway. *Proceedings of the Institution of Civil Engineers Railway Paper*, 19: 3–25.

Toms, A.H. 1953. Recent research into the coastal landslides of Folkestone Warren, Kent, England. *Proc. 3rd Int. Conf. on Soil Mechanics & Foundation Eng., Zurich* 2: 288–293.

Trenter, N.A. & Warren, C.D. 1996. Further investigations at the Folkestone Warren Landslide. *Geotechnique* 46(4): 589–620.

Viner-Brady, N.E.V. 1955. Folkestone Warren landslips: remedial measures, 1948–1954. *Proc. Instn civ. Engrs, Railway paper* 57: 429–41.

Warren, C.D. & Palmer, M.J. 2000. Observations on the Nature of Landslipped Strata, Folkestone Warren, United Kingdom. In: E.N. Bromhead, N. Dixon & M-L. Ibsen (eds), *Landslides, in research, theory and practice, Proc. of Eighth International Symposium on Landslides, 26–30 June 2000*, Cardiff.

Wood, A..M.M. 1955 Folkestone Warren landslips: investigations, 1948–1950. *Proc. Inst. of Civ. Engrs, Railway paper 56: 410–428.*

Wood, A.M.M. 1971. Engineering aspects of coastal landslides. *Proc. Instn Civil Engrs* 50: 256–276.

Landslides and Climate Change – McInnes, Jakeways, Fairbank & Mathie (eds)
© 2007 Taylor & Francis Group, London, ISBN 978-0-415-44318-0

Increased rockslide activity in the middle Holocene? New evidence from the Tyrolean Alps (Austria)

C. Prager & C. Zangerl
AlpS Centre for Natural Hazard Management, Innsbruck, Austria

R. Brandner
Institute of Geology and Paleontology, University of Innsbruck, Austria

G. Patzelt
Former Institute of High Mountain Research, University of Innsbruck, Austria

ABSTRACT: Some of the largest rockslides in the Alps cluster spatially in the Eastern Alps (Tyrol, Austria). A geodatabase was set up to evaluate their timing of failure. Compiled dating data of mass movements show a continuous temporal distribution with accentuations during the early Holocene and, in Tyrol, a significant emphasis of deep-seated rockslides at about 4200-3000 cal BP. Several slopes have been reactivated and show polyphase failure events. However, the majority of dated landslides did not fail immediately after late-Pleistocene glacier-retreat, but clearly a few thousand years later. The middle Holocene rockslide-activity in Tyrol coincides temporally with the progradation of some larger debris flows in the nearby main valleys and, partially, with glacier advances in the Austrian Central Alps. Based on this, deep-seated slope deformations may be induced by complex interactions of lithological, structural and morphological predisposition, fracture propagation, variable seismic activity and climatically controlled water-supply.

1 INTRODUCTION

Several well exposed scarp areas in the Tyrolean Eastern Alps (Austria) provide insights in structures and kinematics of deep-seated mass movements. Based on morphological and lithostratigraphical criteria, the ages of failure were formerly debated controversially. Generally, late-Pleistocene glacier withdrawal, causing an unbalanced relief and thus increasing the stresses within the over-steepened slopes, was assumed to be the most dominant landslides trigger (e.g. Abele 1969, 1974).

But in the majority of cases, radiometric dating of mass movements in the Alps yielded clearly Holocene ages of failure and indicates that slope instabilities are not directly controlled by deglaciation processes. However, in the Western and Southern Alps, a dependency of landslide-activity on climatic fluctuations during the Holocene was assumed already formerly (e.g. Raetzo-Brülhart 1997, Matthews et al. 1997, Dapples et al. 2003, Soldati et al. 2004).

This paper deals with the temporal distribution of dated mass movements in Tyrol (Eastern Alps, Austria) and surroundings, focusing on the Fernpass region. There, several deep-seated rockslides rank among the largest events in the Alps and show a close spatial distribution. One of them, the prominent Fernpass rockslide, was recently dated (Prager et al. 2006a) and forms a temporal cluster with its adjacent rockslides. In view of that, the first comprehensive compilation of dated mass movements in the Eastern Alps was set up and is presented herein. Based on this, several processes that may promote rock strength-degradation and slope failures during the Holocene are discussed.

2 GEOLOGICAL SETTING

The Eastern Alps are made up of complex fold- and thrust-belts of different nappe units, which have been deformed polyphase and heteroaxially. Main geological structures were formed during Cretaceous to Tertiary thrust- and extension-tectonics (Schmid et al. 2004). In Tyrol, the majority of dated mass movements are situated within the polymetamorphic Ötztal basement nappe and within detached Mesozoic cover units of the Northern Calcareous Alps.

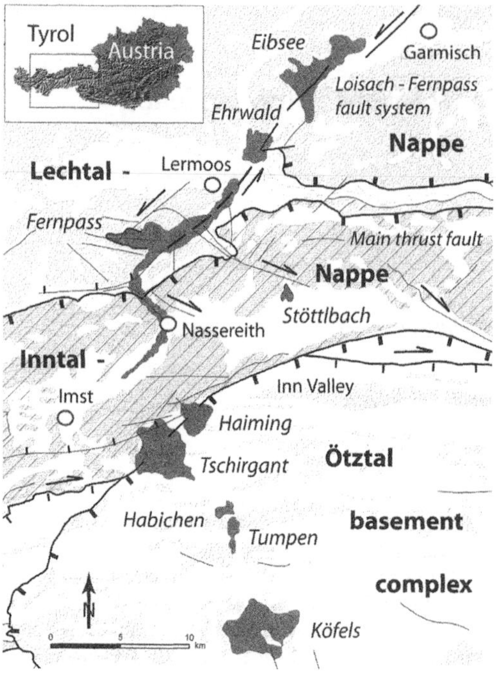

Figure 1. Sketch map of the Fernpass region showing rockslide deposits (shaded dark grey) and main geological structures.

Here detailed field studies at several instable slopes yielded evidence that fault-related valley deepening and coalescence of brittle discontinuities control progressive failure and landslide-kinematics (e.g. Brückl et al. 2004, Prager et al. 2006b, Zangerl et al. 2006). Intensive cataclasis along brittle fracture zones, e. g. the prominent Inntal- and Loisach fault systems (Fig. 1), enabled substantial fluvio-glacial erosion. This morphological change caused stress redistribution of the valley slopes and uncovered favourably oriented sliding planes, permitting subsequent slope instabilities.

3 SEISMICITY

Some major shear systems in Tyrol, e.g. the NE-orientated Inntal- and Engadiner Line, are characterised by recent seismic activity. Compiled earthquake data indicate, that the effective horizontal ground acceleration shows significant maxima of about 1 m/s^2 in the middle Inn valley and the Fernpass region (ÖNORM B 4015 2002). There, several strong earthquakes up to magnitude 5.3 and epicentral intensities I_0 7.5° MSK rank among the most intense ones ever measured in Austria (Drimmel 1980). One of these major events occurred in 1930 in the Fernpass region near the village Namlos. At least 16 main shocks and numerous aftershocks were recorded, whereby this event was subjectively registered even at distances of about 200-400 km. Locally, this earthquake changed the hydraulic flow field by dislocating springs, opened ground clefts and triggered several rockfall events nearby (Klebelsberg 1930).

4 DATA COMPILATION

The considerations presented herein base on detailed field studies of selected mass movements in Tyrol and comparative site visits in the adjacency. In order to evaluate the spatial and temporal distribution of mass movements of different types and sizes, a GIS-linked geodatabase has been set up. At present this includes various data of more than 450 different mass movements in Tyrol and surroundings, ranging from late-glacial to modern failure ages. Thereof approx. 230 events feature unknown ages of failures and/or unknown activity. About 130 post-medieval to recent active landslides have been compiled for Tyrol only and were not considered for this study. Dated fossil mass movements were implemented also from adjacent areas such as southern Germany, northern Italy and eastern Switzerland and comprise at present about 110 events, which are mainly rapid events such as rockfalls and rockslides.

Available laboratory dates of ^{14}C-dated mass movements were calibrated to calendar years (cal. BP, quoted 0 BP = 1950 AD) using the software OxCal Version 3.10 (Bronk Ramsey 2005) and its implemented calibration curve IntCal04. The ranges of the arithmetic mean ages are based on the statistical 2-sigma standard deviation (corresponding to 95.4% probability).

5 SELECTED LANDSLIDES

Some of the largest mass movement deposits in the Alps cluster spatially in the Fernpass region, in the western part of the Northern Calcareous Alps. Within an area of less than 40 × 20 km, at least 9 deep-seated failure events occurred and include the prominent rockslides at Eibsee, Fernpass, Tschirgant and Köfels (Fig. 1).

5.1 *Fernpass rockslide*

The Fernpass rockslide is characterised by two channelled Sturzstrom branches, which contain a rock mass volume of about 1 km^3 and cover excess run-out distances up to 12 and 16 km respectively. This large event was followed by a smaller rockslide of unknown age and the development of a deeply fractured slope that has not failed yet (Fig. 2).

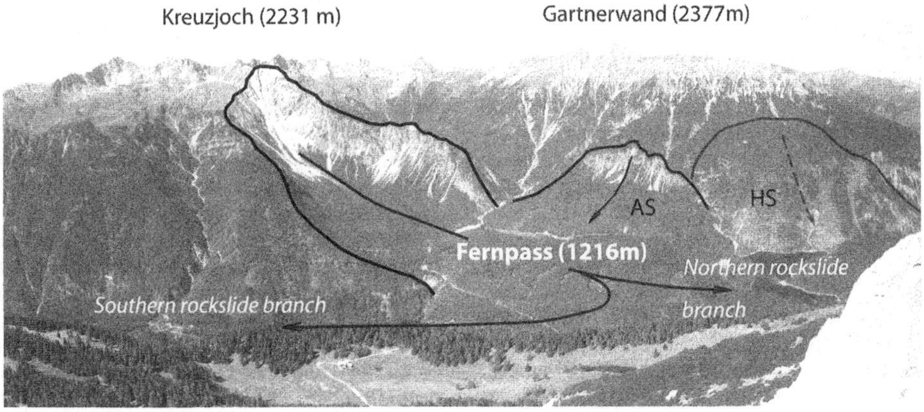

Figure 2. Oblique view to the wedge-shaped scarp of the Fernpass rockslide and its associated mass movements (secondary rockslide "Am Saum" AS, instable slope "Hohler Stein" HS).

The rockslide debris originated from a well exposed and exceptional deeply incised niche, which is made up of platy dolomites, limestones and marls of the several hundred metres thick Seefeld Formation (Norian, Upper Triassic). Polyphase and heteroaxial deformation generated fold- and fracture systems with varying orientation (Eisbacher & Brandner 1995). Thus, the failure zones of the Fernpass rockslide and its juxtaposed slopes evolved by coalescence of brittle discontinuities due to multiple step-path failure mechanisms.

Deep-seated cataclasis along the NE-orientated Loisach-Fernpass fault system (Fig. 1) is here indicated by field data and results of hybrid seismic measurements near the apex of the present Fern-Pass. This clearly revealed a steep pre-failure topography of the valley flanks with a fluvio-glacially undercut slope toe (Prager et al., unpublished data).

Due to an oblique impact of the sliding rock masses against their opposite mountain slope, they were proximally piled up as a remarkably thick debris ridge and split into two Sturzstrom branches. Their run-out was favoured by the large rockslide volume, channelling effects in the narrow valley, dynamic disintegration and, crucially, by undrained dynamic loading of the water-saturated substrate (Prager et al. 2006b).

Formerly, morphological and lithostratigraphical field criteria, e.g. moraine-like debris-ridges, funnel-shaped "dead-ice" sink-holes and the spatial distribution of Pleistocene cover rocks, were used to differentiate between a late-glacial main event and a succeeding postglacial collapse (Abele 1964, 1974). But now detailed field investigations showed that neither the rough scarp nor the intensively structured accumulation area feature any signs of a smooth morphology and argue against glacial overprints.

This was confirmed by the application of three different radiometric dating methods on individual sampling sites (Prager et al., 2006a). Close to the scarp area, rockslide-dammed torrent deposits yielded a ^{14}C minimum-age of 3380–3080 cal. BP. The chronostratigraphic base of this sequence has not been dated yet, but is assumed to date somewhat older into the middle Holocene. This coincides well with two cosmogenic radionuclide ^{36}Cl exposure ages of large-scale sliding planes at the scarp. There the sampled platy dolomites indicate a mean age of 4100 ± 1300 yrs for the failure event. Further data were gained from the curiously and strongly deflected southern rockslide branch. Post-depositional carbonate cements therein have been dated by the $^{230}Th/^{234}U$-disequilibrium method and yielded a minimum age of 4150 ± 100 yrs for the accumulation of the rockslide debris (Ostermann et al., in press).

Based on this, a temporal differentiation between two failure events, one making up the northern rockslide branch, and another, making up the southern branch, is not indicated yet. All dating coincide well and indicate the Fernpass rockslide most likely occurred about 4200–4100 yrs ago. Thus, this event was clearly not in contact with late-glacial ice and not triggered by deglaciation processes.

5.2 *Eibsee rockslide*

The Eibsee rockslide (Fig. 1) is situated 15 km northeast of the Fernpass, on the north-face of the Zugspitze massif (2961 m) the highest mountain in Germany and mobilized about 400–600 mill. m^3 of accumulated debris (Abele 1974, Golas 1996). It originated from a several hundreds metres high and subvertical cliff, built up by mainly well bedded carbonates of the Muschelkalk Group (Anisian) and the thick Wetterstein Formation (Ladinian). Due to Paleogene compression, these Triassic carbonates were thrust over incompetent Jurassic-Cretaceous limestones and marls (Eisbacher & Brandner 1995).

Stability relevant discontinuities for the Eibsee rockslide were not the bedding planes, dipping moderately inclined against the slope, but subvertical fault- and fracture systems. Tunnel constructions for the German rack railway up to the Zugspitze ran across such separation planes and cavities, whereat some of them spaced several metres wide open and occasionally showed a contact to the outside world (Knauer 1933). Field evidence of intense brittle faulting can be observed at the NW-face of the Zugspitze, where NE-orientated, subvertical faults and fractures are part of the sinistral Loisach major shear system (Fig. 1). This caused deep seated intensive fragmentation of the folded carbonates and can, within the precipitous rock walls, isolate blocks along wedge-shaped scarps.

Based on morphological field criteria, the Eibsee rockslide deposits were formerly interpreted as a "late-glacial rockslide-moraine" (Vidal 1953), but several wood samples gained in drillings yielded a mean age of around 3700 ^{14}C yrs (Jerz & Poschinger 1995). The six best fitting, presumably not redeposited, samples were calibrated to calendar years and show an arithmetic mean-age at about 4181 ± 627 cal. BP.

5.3 *Ehrwald rockslide*

At the western base of the Zugspitze massif, the Ehrwald rockslide deposits cover an area of about 2 km^2 (Abele 1974). Its lithological and structural predisposition corresponds with those of the adjacent Eibsee rockslide; both failures were clearly controlled by brittle faulting along the NE-orientated Loisach fault system (Fig. 1).

The carbonate Ehrwald deposits make up hilly scenery with several pronounced ridges and were morphologically classified as "late-glacial rockslide moraine" (Abele 1964, 1974). Thus far, radiometric dating has not been carried out here. However, the internal structure is characterised by an unstratified, coarsening-upward facies, wherein several shattered clasts feature a jig-saw-fit of grain-boundaries. These sedimentary features have not been observed in glacially derived deposits and attribute uniquely to dynamically disintegrated rockslide masses. This and the lack of Quaternary cover rocks and missing glacial smoothing of the topography clearly suggest a Holocene age for the Ehrwald rockslide.

5.4 *Tschirgant – Haiming rockslides*

About 10 km south of the Fernpass, the Tschirgant massif (2370 m) forms a steep rugged NE-SW-trending slope and released two well known, deep seated rockslides down to the river Inn: the smaller Haiming event (25–34 mill. m^3) in the northeast and the prominent Tschirgant rockslide (180–240 mill. m^3, Abele 1974) in the southwest (Fig. 1).

The scarp areas are situated at the southern margin of the Northern Calcareous Alps, which were here obliquely cut off by the NE-SW-striking Inntal fault system and separated from the metamorphic Ötztal basement complex (Eisbacher & Brandner 1995). Slope deformation was clearly structurally controlled by the complex cross-linking of medium inclined bedding planes and subvertical brittle fracture systems. Due to these densely spaced discontinuities, the deeply incised source area of the Haiming rockslide exhibits an unusually rough, stepped scarp. Lithologically this comprises dolomites of the Wetterstein Formation (Ladinian), the carbonate-siliciclastic Raibl Group (Carnian) and the Hauptdolomit Formation (Norian). At the base of the Haiming scarp, a drilling penetrated ca. 670 m subhorizontally into the slope and proved the existence of an effective water-table. Dammed to the South by low permeable siliciclastics, here high pore pressures up to 43 bar came across within the intensively fractured and thus highly permeable dolomites of the Raibl Group (Intergeo Consultants, pers. comm. 2005).

Adjacent southwest, poorly bedded dolomites of the Wetterstein Formation make up the huge Tschirgant scarp. At downslope sections, these competent and dolomites border tectonically to an incompetent succession of the Raibl Group, containing dolomites, limestones, marls and evaporates. Especially the significant carbonate-evaporitic breccias (Rauhwacken), here some decametres in thickness, make up a zone of structural weakness at the slope toe. However, the intensively fractured Tschirgant scarp exhibits several large-scale bedding- and fault planes, dipping desk-like out of the slope and enabling the translational slide of a larger rock mass volume.

The failing debris entrapped fluvio-glacial sediments from the slope toe and valley floors and penetrated into the mouth of the Ötz valley (Fig. 3). Outcrops at the riverside show here polymict fluviatile gravels overridden by carbonate rockslide-debris. Along subvertical pull-apart structures the mobilized valley fill was injected into the rockslide debris, indicating water-saturation of the substrate (Abele 1997).

Based on morphological criteria, Heuberger (1975) assumed an interaction of the Tschirgant rockslide with Late-glacial Ötztal ice. But radiometric dating indicate a Holocene age at about 2900 ^{14}C yrs (ca. 3000 cal. BP, Patzelt & Poscher 1993) for the main event. Further investigations showed that both slopes, Tschirgant and Haiming, did not release only one single event, but are characterised by multiple failures. Patzelt (2004a) differentiated here at least four significant rockslides, all occurring between 3753 ± 191 cal. BP and 3065 ± 145 cal. BP.

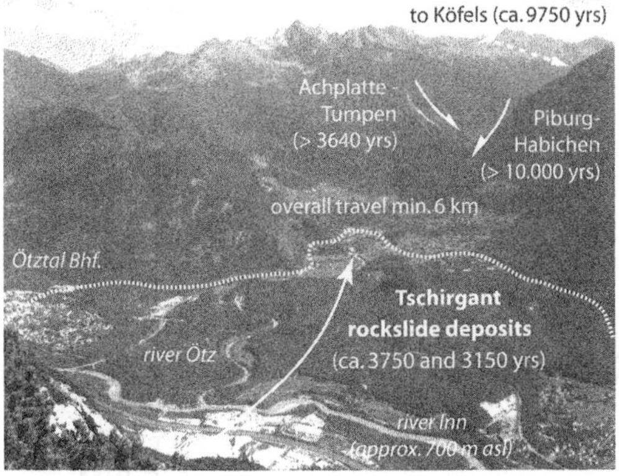

Figure 3. View from the Tschirgant massif towards Southeast to the rockslide deposits and adjacent scarp areas in the northern Ötz valley.

About 10 km to the NE of the Haiming scarp, the Stöttlbach mass movement deposits (Fig. 1) have been dated recently. There, preliminary ^{36}Cl exposure ages of accumulated limestone-boulders indicate a failure event at about 4000–3600 yrs (Kerschner & Ivy-Ochs, pers. comm. 2006).

5.5 *Northern Ötz valley*

The Ötz valley, a N-trending main tributary to the river Inn, is deeply incised in the metamorphic Ötztal basement complex. Its Quaternary valley filling is characterised by the polyphase interplay of different rockfalls, rockslides and their backwater deposits.

The Habichen rockslide deposits (Fig. 1), situated close to the distal deposits of the Tschirgant rockslide, dammed the south bay of Lake Piburg towards the valley floor beneath. Pollen analyses of aggradated lakefronts yielded a minimum age of about 10,000 yrs (Oeggl, pers. comm. 2005) and indicate a similar age for the rockslide barrier.

Adjacent to the southeast, the Tumpen plain exhibits several rockslides and rockfalls, originating from both valley slopes. Based on drillings, the several decametres thick rockslide-dammed backwater deposits show an at least two-phase fluvio-lacustrine sequence. The younger succession provided a minimum age of about 3380 ± 80 ^{14}C yrs (Poscher & Patzelt 2000), i.e. 3640 ± 200 cal. BP, for the damming rockslide. Depth-extrapolations of the existing dating data suggest the older sequence and its damming rockslide barrier date at about 6000 cal. BP (Patzelt 2001).

To the south, the Tumpen backwater deposits border to the largest crystalline mass movements in the Alps, the famous Köfels rockslide. This event features a well established early Holocene age at about 9800 cal. BP (Ivy-Ochs et al. 1998) and dammed the several decametres thick fluvio-lacustrine deposits of the Längenfeld basin.

6 TEMPORAL AND SPATIAL DISTRIBUTION OF DATED EVENTS IN THE EASTERN ALPS

Compiled dating data show a rather continuous temporal distribution of landslides and debris flows during the Holocene, without longer time-gaps (Fig. 4). However, there is no evidence for increased activity due to deglaciation processes during the late-Glacial and early Holocene. In Austria, late-glacial ages have been established for a few landslides only, e.g. an event at Pletzachkogel (Tyrol, Patzelt 2004b) and the prominent Almtal rockslide (Upper Austria, Ivy-Ochs et al. 2005a).

At about 10,000–9000 cal. BP some of the largest rockslides in the Alps failed, e.g. Flims, Kandertal (both Switzerland), Köfels and parts of the slope Gepatsch-Hochmais (both Tyrol). Between approx. 9000 to 5000 cal. BP, only a few and smaller events, with the exception of the large Wildalpen rockslide (Styria, Austria), have occurred.

In contrast, numerous deep-seated events cumulate in the middle to early Holocene, with significant emphasis in the Subboreal at about 4200–3000 cal. BP (Fig. 4). This temporal cluster comprises some of the largest rockslides in Tyrol, which, remarkably, also cluster spatially (Figs 1, 5).

Some radiometric data prove polyphase reactivations of predisposed vulnerabilities and repeated slope failures, e.g. at Tschirgant-Haiming and Pletzachkogel (Inn valley, Patzelt 2004a, b), Köfels and Tumpen (Ötz valley, Ivy-Ochs et al. 1998, Poscher & Patzelt 2000). Several modern landslides show fossil and/or historically documented precursory events, e.g. the catastrophic events at Vajont 1963 (Kilburn & Petley 2003), Val Pola 1987 (Azzoni et al 1992) and Randa 1991 (Santori et al. 2003).

Also dated debris flows show periods of fluctuating activity. Concerning the Tyrolean Inn valley and its tributaries, Patzelt (1987) established phases of raised accumulation at about 9400 ^{14}C yrs (ca. 10,630 cal. BP), between 7500–6000 (ca. 8350–6840 cal. BP) and a third at about 3500 ^{14}C yrs (ca. 3780 cal. BP). According to this, some of the largest alluvial fans in Tyrol and Northern Italy, e.g. the rivers Gadria and Weissenbach, show significant activity at about 7900–7100 cal. BP (Fig. 4). Others, e.g. the rivers Sill and Melach, show raised debris accumulation in the middle Holocene at about 3700–3600 cal. BP. In between these periods, at about 6000–4500 ^{14}C yrs (ca. 6840–5170 cal. BP), the Inn valley was affected by a distinctive phase of fluvial erosion (Patzelt 1987).

Increased fluvial dynamics and debris flow progradation in the Subboreal were established at several sites

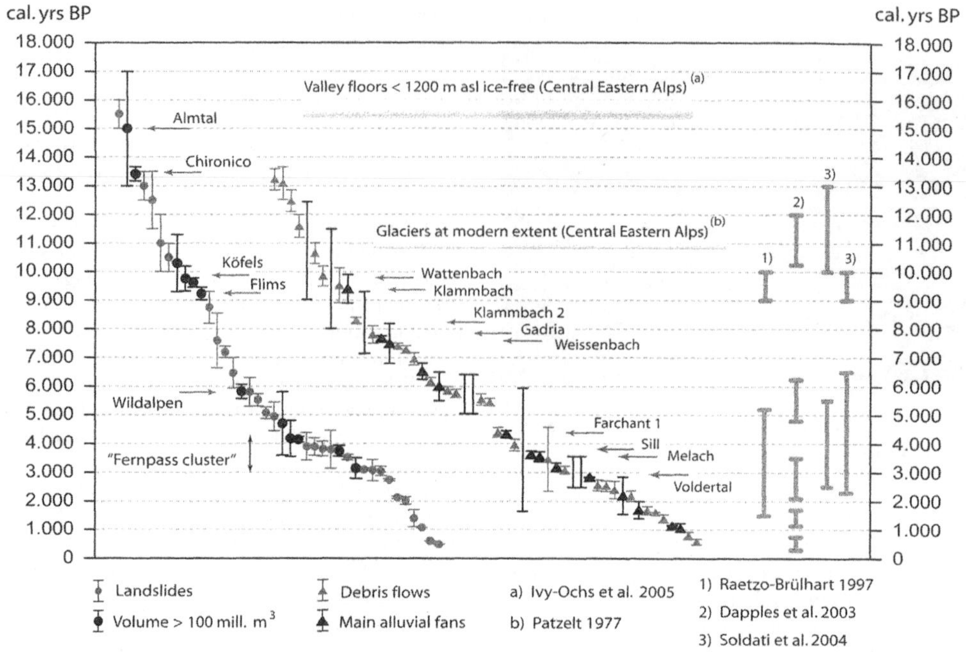

Figure 4. Temporal distribution of Late-glacial to Holocene mass movements in Tyrol and surroundings (vertical axes: calibrated years BP, horizontal axes: dimensionless sequence of dated events; vertical range bars to the right: periods of increased landslide acitivty, according to the references).

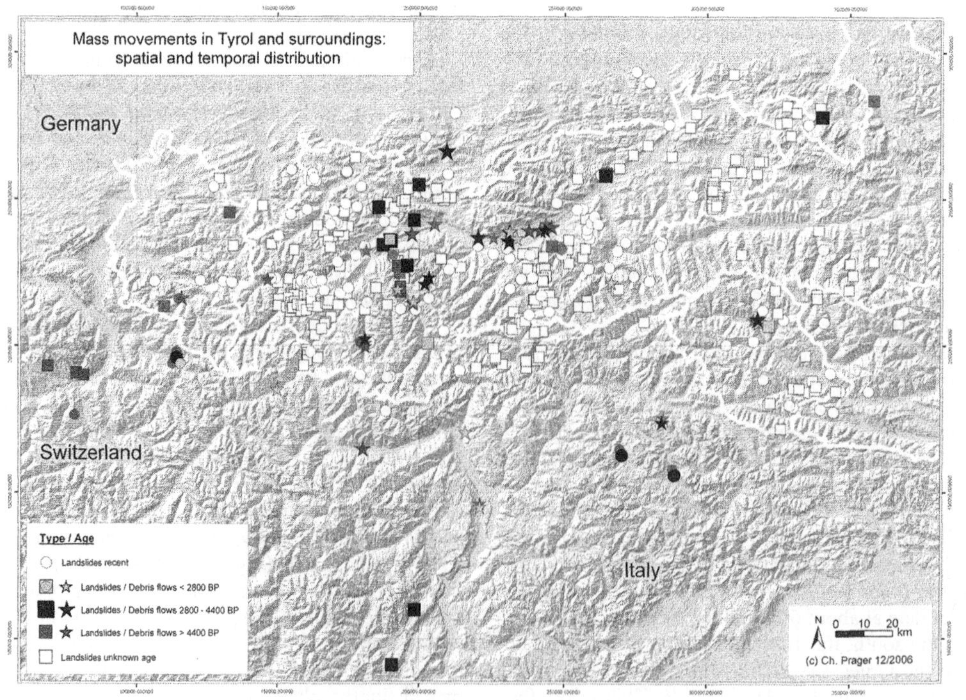

Figure 5. Spatial and temporal distribution of Lateglacial to Holocene mass movements in Tyrol and surroundings.

in Tyrol and have now been compiled in a geodatabase for the first time.

7 DISCUSSION

Detailed field surveys and compiled data indicate that rock strength degradation and slope deformation are controlled by a complex and polyphase interaction of variable processes, which may superpose each other.

7.1 *Deglaciation and decompression*

In the Eastern Alps, stability conditions within the polyphase and heteroaxially fractured rock units were fundamentally influenced by morphological changes during the Quaternary. Late-glacial Gschnitz valley-glaciers advanced, at the type locality in the central Eastern Alps, down to lowermost altitudes of about 1200 m asl not later than 15,400 ± 1400 yrs ago (Ivy-Ochs et al. 2005b). This indicates that the toes of several instable slopes, especially when East- and Southward exposed, such as e.g. at Köfels, Fernpass and Tschirgant, bordered on thin dead-ice or even ice-free valley-floors and however, were not glacial buttressed at least since the Younger Dryas. Subsequently, late-Pleistocene glaciers rapidly melted down till they reached at about 9500 ^{14}C yrs (ca. 10,850 cal. BP) for the first time modern extents (Patzelt 1972, 1977).

Fluvio-glacial erosion, valley-deepening and post-glacial debuttressing uncovered favourable oriented sliding planes and caused substantial stress redistribution within the undercut and oversteepened slopes. Therefore, the high and unbalanced relief since the early Postglacial is certainly a dominant factor for any Alpine mass movement. Subsequently some slopes, characterised by critical fracture density and thus close to their stability limit equilibrium, failed.

7.2 *Progressive failure*

Glacier retreat left oversteepened valley flanks with characteristic unloading fractures, where slope stability was continuously lowered by long-term processes due to stress redistribution. Further rock strength development has been intensively affected by stepwise interactions of pre-existing brittle discontinuities and subcritical fracture propagation (e.g. Eberhardt et al. 2004).

Thereby, complex processes of subcritical crack growth depend on the interaction of several parameters, e.g. in-situ stresses, bedrock mineralogy, fracture geometries and pore-water-characteristics. Being significantly favoured by high pore pressures, a lower bound of fracture propagation velocities ranks at about several centimetres per 1000 years (Atkinson & Meredith 1987).

7.3 *Dynamic loading*

Regional seismic data show that earthquakes close to the Fernpass feature epicentral intensities up to 7.5° and rank among the strongest ones ever measured in Austria. That some of these triggered rockfalls and changed locally the hydraulic flow field (Klebelsberg 1930) suggests, here also the release of fossil rockslides with similar ages could have been essentially favoured by seismic shaking.

But with the exception of the prominent 1348-release of the Dobratsch rockslide in Carinthia, Austria (Eisbacher & Clague 1984) and some events triggered by the 1998-earthquake in NW-Slovenia (Vidrih et al. 2001), documented case studies of seismically induced slope failures in central Europe are commonly of small dimensions. However, active fault systems can not only trigger mass movements, but do produce intensely fractured and uncemented rock masses to substantial depths, inclusive potential sliding planes.

Moreover, even less energetic earthquakes may accelerate progressive fracture propagation within the shook rock units. Comparable load tests show that component parts with discontinuities show no further fracture propagation under static loading conditions below its critical collapse load. In contrast, dynamic loading initiates fracture propagation far below the critical load (Gross 1996). Such fatigue crack growth can step-wise weaken intact rock bridges and raise the effective joint porosity. Thus, repeated seismic loading can effectively favour and prepare landslides.

7.4 *Climatic aspects*

After the Younger Dryas cold period, glaciers in the Central Eastern Alps rapidly melted down to modern extents. Subsequent glacier- and forest-line fluctuations indicate considerable changes of Holocene climate, whereat glaciers varied about modern sizes and had limited extents over longer periods in the middle and early Holocene (Patzelt 1977, 2001). During the glacial unfavourable period between ca. 10,450–3650 cal. BP both Austrian largest glaciers, Pasterze and Gepatschferner, were repeatedly and even for longer phases smaller than at present, but show from 3650 cal. BP till waning Roman age several smaller and fluctuating advances up to modern dimensions (Nicolussi & Patzelt 2001). Based on this, long periods in the Holocene showed favourable climatic conditions with average summer temperatures predominantly slightly higher than at present. Repeatedly these were interrupted by pronounced but relative short-termed deteriorations with multiple glacier advances (Fig. 6), e.g. the Löbben advance at about 3750–3250 cal. BP (Patzelt & Bortenschlager 1973).

Unstable Holocene climatic conditions are also indicated by glacier fluctuations in the Central Swiss

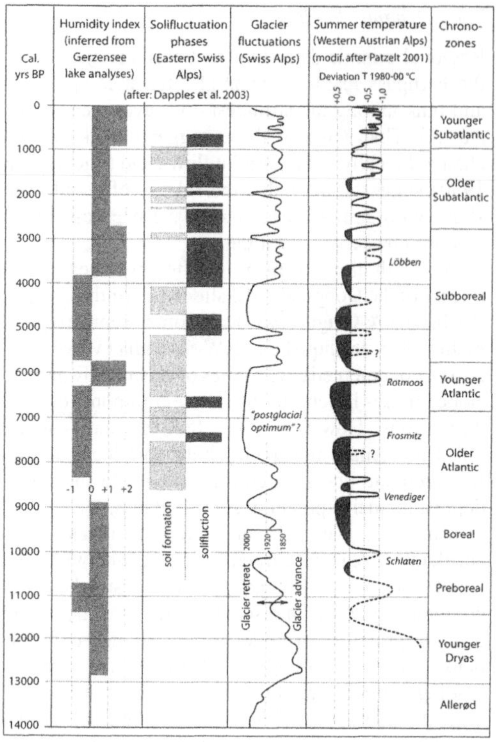

Figure 6. Late-glacial to Holocene paleoclimatic indicators of Lake Gerzensee (−1: dry, 0: normal, +1: wet, +2: very wet), solifluctuation activity and glacier fluctuations (Dapples et al. 2003) with combined glacier- and forest-line data from the Austrian Central Alps (modified after Patzelt 2001).

Alps, featuring at least eight phases of significant glacier recession with several cold-wet periods in between (Hormes et al. 2001).

Compiled dating data of mass movements in Tyrol and surroundings show a continuous temporal distribution of events during the Holocene. At least two age-clusters of enhanced slope instability are characterised by the occurrence of several deep-seated failure events: firstly in the early Postglacial and secondly in the Middle Holocene, with lower landslide activity in between (Fig. 4).

Early Holocene rockslide activity, at about 10,000-9000 cal. BP, comprises some of the largest alpine rockslides, e.g. Köfels, Kandertal and Flims, and is generally assumed to attribute to postglacial warming. This coincides well with an early phase of precipitation-controlled, raised debris accumulation of tributaries to the Tyrolean Inn valley, occurring at about 9400 ^{14}C yrs (Patzelt 1987, i.e. approx. 10,630 cal. BP). Remarkably, compiled data also indicate a significant emphasised landslide activity during the Subboreal about approx. 4200–3000 cal. BP (Fig. 4). This includes some of the largest rockslides in Tyrol, which even cluster spatially in the region Fernpass – Northern Ötz valley (Figs 1, 5). Since theirs releases were clearly not directly linked with deglaciation processes, a striking environmental change was likely to occur during this period.

The Fernpass landslide cluster correlates temporally with the activity of several debris flows in the nearby main valleys (Fig 4). At about 3500 ^{14}C yrs (approx. 3780 cal. BP), a phase of raised alluvial accumulation in the Tyrolean Inn valley was established (Patzelt 1987). This and the activity of several local torrents and debris flows nearby indicate periods of raised water supply in the catchments areas.

Similar groupings of early post-glacial and middle to young Holocene landslides, but with a phase of relative inactivity in between, were established also in the surroundings. In Switzerland, Raetzo-Brülhart (1997) attributes two distinct cluster of raised landslide activity, at about 10,000–9000 and 5200–1500 cal. BP, to warmer and/or more humid paleoclimatic conditions. Dapples et al. (2003) correlate five Lateglacial to Holocene pulses of raised landslide dynamics with climatic deteriorations, indicated by glacier advances, solifluctuation and lacustrine stratigraphical records (Fig. 6).

In the Italian dolomites, Soldati et al. (2004) differentiate also two striking age-clusters of landslides: one early postglacial at about 13,000–9000 cal. BP, which is due to deglaciation processes and was probably favoured by increased precipitation and/or permafrost meltdown, and a younger one, at about 6500–2300 cal. BP in the Subboreal, which is assumed to correlate with an increase of precipitation.

Thus, high groundwater levels, due to increased precipitation, might have climatically controlled Holocene rockslide activity. Raised pore pressures increase the velocity of subcritical crack growth (Atkinson & Meredith 1987) and lower the friction angle of weathered and water-saturated rock surfaces, which is generally lower than those of dry and unweathered ones. Coupled hydro-mechanical destabilising processes are also indicated by drilling-results from the basal Tschirgant massif (Tyrol, Austria), which is characterised by polyphase rockslide events. Here remarkably high pore pressures suggest, deep seated slope deformations could have been favoured by water-saturation of the fractured rock masses. In coincidence, also historical case studies point out that rainstorm is a dominant trigger (Eisbacher & Clague 1984) and pore pressure changes drive slope movements respectively (e.g. Weidner 2000).

8 CONCLUSIONS

Dated mass movements in Tyrol and surroundings have been compiled for the first time and show a rather continuous temporal distribution of events in the Holocene. However, there is no evidence for increased activity due to deglaciation processes during the late-Glacial and early Holocene, but at least for two phases of increased landslide-activity. One at about 10,000–9000 cal. BP and another, spatially clustered, in the middle Holocene at about 4200–3000 cal. BP. Latter comprises the prominent Fernpass rockslide and several large events nearby and coincides temporally with periods of increased debris flows activity in the nearby main valleys. All data indicate, the majority of slope collapses were not directly triggered by late Pleistocene deglaciation processes, but occurred clearly later after a preparing lag-time of several 1000 years.

Well-exposed scarp areas show that slope failures were clearly structurally controlled by fracture propagation and coalescence of brittle discontinuities. Regional earthquake data suggest a considerable neotectonical influence on slope instabilities. Active faulting can directly trigger mass movements, but above all, effectively prepare these by increasing fracture density to substantial depths.

Debris flow activity, glacier fluctuations and case studies from adjacent landslide areas are proxy of paleoclimatic conditions and indicate periods of raised precipitation and groundwater flows. These control pore pressure within the fractured rock masses and favour progressive failure.

Thus, structurally and morphologically predisposed mass movements were prepared and triggered by the complex and polyphase interaction of several rock strength degrading processes. Deep-seated slope deformations may be attributed to initiation, propagation and coalescence of brittle discontinuities, favoured by seismic activity and climatically controlled pore pressure changes. Any of these destabilising processes, even if only at subcritical thresholds, can trigger a failure event if slope stability is already close to its limit equilibrium.

ACKNOWLEDGEMENT

Discussions with Ch. Spötl and H. Kerschner (both University of Innsbruck) and financial support from ILF Consulting Engineers Ltd., TIWAG Tyrolean Hydroelectric Power Company Ltd., p+w Baugrund + Wasser Geo-ZT Ltd. and AlpECON Oeg. (all Tyrol, Austria) are gratefully acknowledged.

REFERENCES

Abele, G. 1964. Die Fernpaßtalung und ihre morphologischen Probleme. *Tübinger Geograph. Studien* 12: 1–123.

Abele, G. 1969. Vom Eis geformte Bergsturzlandschaften. *Zs. f. Geomorph. N. F.* Suppl. 8: 119–147.

Abele, G. 1974. Bergstürze in den Alpen. Ihre Verbreitung, Morphologie und Folgeerscheinungen. *Wiss. Alpenvereinshefte* 25: 1–230. München.

Abele, G. 1997. Rockslide movement supported by the mobilization of groundwater-saturated valley floor sediments. *Zs. f. Geomorph. N. F.* 41(1): 1–20.

Atkinson, B.K. & Meredith, P.G. 1987. The theory of subcritical crack growth with applications to minerals and rocks. In B.K. Atkinson (ed.), *Fracture mechanics of rock*: 111–166. London: Academic Press.

Azzoni, A., Chiesa, S., Frassoni, A. & Govi, M. 1992. The Val Pola landslide. *Eng. Geology* 33 (1): 59–70.

Bronk Ramsey, C. 2005. OxCal Version 3.10. *Computer software*. Online at: www.rlaha.ox.ac.uk/orau/oxcal.html.

Brückl, E., Zangerl, C. & Tentschert, E. 2004. Geometry and deformation mechanisms of a deep seated gravitational creep in cyrstalline rocks. In W. Schubert (ed.), *ISRM Regional Symposium Eurock 2004*: 227–230. Essen: Glückauf.

Dapples, F., Oswald, D., Raetzo, H., Lardelli, T. & Zwahlen, P. 2003. New records of Holocene landslide activity in the Western and Eastern Swiss Alps: Implication of climate and vegetation changes. *Ecl. Geol. Helv.* 96: 1–9.

Drimmel, J. 1980. Rezente Seismizität und Seismotektonik des Ostalpenraumes. In R. Oberhauser (ed.), *Der geologische Aufbau Österreichs*: 507–527. Wien: Springer.

Eberhardt, E., Stead, D. & Coggan, J.S. 2004. Numerical analysis of initiation and progressive failure in natural rock slopes – the 1991 Randa rockslide. *Int. J. Rock Mechanics Mining Sc.* 41: 69–87.

Eisbacher, G. & Clague, J. J. 1984. Destructive mass movements in high mountain: hazard and management. *Geol. Surv. Canada Paper* 84 (16): 1–230.

Eisbacher, G.H. & Brandner, R. 1995. Role of high-angle faults during heteroaxial contraction, Inntal Thrust Sheet, Northern Calcareous Alps, Western Austria. *Geol. Paläont. Mitt. Innsbruck* 20: 389–406.

Golas, B. 1996. Der Eibseebergsturz, Eine geomorphologische Studie. *Dipl. Thesis*: 1–96. Univ. Innsbruck.

Gross, D. 1996. Bruchmechanik: 1-218. Berlin: Springer.

Heuberger, H. 1975. Das Ötztal. Bergstürze und alte Gletscherstände, kulturgeographische Gliederung. *Innsbrucker Geograph. Stud.* 2: 213–249.

Hormes, A., Müller, B.U. & Schlüchter, C. 2001. The Alps with little ice: evidence for eight Holocene phases of reduced glacier extent in the Central Swiss Alps. *The Holocene* 11 (3): 255–265.

Ivy-Ochs, S., Heuberger, H., Kubik, P.W., Kerschner, H., Bonani, G., Frank, M. & Schlüchter, C. 1998. The age of the Köfels event. Relative, 14C and cosmogenic isotope dating of an early Holocene landslide in the Central Alps (Tyrol, Austria). *Zs. Gletscherkd. Glazialgeol.* 34 (1): 57–68.

Ivy-Ochs, S., Van Husen, D. & Synal, H.-A., 2005a. Exposure dating large landslides in the Alps: Almtal. 10th *Int. Conf. Accel. Mass Spectrometry*: Poster session II. Berkeley, CA.

Ivy-Ochs, S., Kerschner, H., Kubik, P.W. & Schlüchter, C. 2005b. Glacier response in the European Alps to Heinrich event 1 cooling: the Gschnitz stadial. *J. Quatern. Sc.* 21: 115–130.

Jerz, H. & Poschinger, A. 1995. Neueste Ergebnisse zum Bergsturz Eibsee-Grainau. *Geol. Bavarica* 99: 383–398.

Kilburn, R. J. & Petley, D. N. 2003. Forecasting giant, catastrophic slope collapse: lessons from Vajont, Northern Italy. *Geomorphology* 54: 21–32.
Klebelsberg, R. 1930. Das Nordalpenbeben vom 8. Oktober 1930. *Mitt. Dt. u. Österr. Alpenverein* 12: 251–254.
Knauer, J. 1933. Die geologischen Ergebnisse beim Bau der Bayerischen Zugspitz-Bahn. *Abh. Geolog. Landesunters. Bayer. Oberbergamt* 10: 23–50.
Matthews, J.A., Brunsden, B., Frenzel, B., Gläser, B. & Weiß, M.M. (eds) 1997. Rapid mass movement as a source of climatic evidence for the Holocene. *Paläoklimaforschung Spec. Iss.* 19 (1-6):1–444. Stuttgart: Fischer.
Nicolussi, K. & Patzelt, G. 2001. Untersuchungen zur Holozänen Gletscherentwicklung von Pasterze und Gepatschferner (Ostalpen). *Zs. Gletscherkde. Glazialgeol.* 36 (2000): 1–87.
ÖNORM B 4015 2002. Belastungsannahmen im Bauwesen – Außergewöhnliche Einwirkungen – Erdbebeneinwirkungen. *ÖNORM B 4015, Ausgabe 2002-06-01*: 1–59. Wien: Österreichisches Normungsinstitut.
Ostermann, M., Sanders D., Kramers, J. & Prager, C. in press. Aragonite and calcite cement "boulder-controlled" meteoric environments on the Fern Pass rockslide (Austria): implications for radiometric age-dating of catastrophic mass movements. *Facies*, in press.
Patzelt, G. 1972. Die spätglazialen Stadien und postglazialen Schwanklungen von Ostalpengletschern. *Ber. Dt. Bot. Ges.* 85: 47–57.
Patzelt, G. 1977. Der zeitliche Ablauf und das Ausmass postglazialer Klimaschwankungen in den Alpen. In B. Frenzel (ed.), *Dendrochronologie und postglaziale Klimaschwankungen in Europa*: 248–259. Wiesbaden: Steiner.
Patzelt, G. 1987. Untersuchungen zur nacheiszeitlichen Schwemmkegel- und Talentwicklung in Tirol. *Veröff. Mus. Ferdinandeum* 1987 (67): 93–123.
Patzelt, G. 2001. Natur und Mensch im Ötztaler Gebirgsraum der Nacheiszeit. *Manuscript, Inst. f. Hochgebirgsforschung*. Univ. Innsbruck.
Patzelt, G. 2004a. Tschirgant-Haiming-Pletzachkogel. Datierte Bergsturzereignisse im Inntal und ihre talgeschichtlichen Folgen. *Presentation.* alpS Symposium 13.10.2004. Galtür.
Patzelt, G. 2004b. Die Bergstürze vom Pletzachkogel bei Kramsach und ihre talgeschichtlichen Folgen. *Presentation.* Geokolloquium 11.03.2004. Univ. Innsbruck.
Patzelt, G. & Bortenschlager, S. 1973. Die postglazialen Gletscher- und Klimaschwankungen in der Venedigergruppe (Hohe Tauern, Ostalpen). *Zs. Geomorph. N. F.* Suppl. 16: 25–72.
Patzelt, G. & Poscher, G. 1993. Der Tschirgant-Bergsturz. *Arbeitstagung 1993 Geol. B.-A., Geologie des Oberinntaler Raumes:* 206–213.
Poscher, G. & Patzelt, G. 2000. Sink-hole Collapses in Soft Rocks. *Felsbau, Rock and Soil Engineering* 18 (1): 36–40.
Prager, C., Patzelt, G., Ostermann, M., Ivy-Ochs, S., Duma, G., Brandner, R. & Zangerl, C. 2006a. The age of the Fernpass rockslide (Tyrol, Austria) and its relation to dated mass movements in the surroundings. *Pangeo Austria 2006*: 258–259. Innsbruck: University Press.
Prager, C, Krainer K., Seidl V. & Chwatal, W. 2006b. Spatial features of Holocene Sturzstrom-deposits inferred from subsurface investigations (Fernpass rockslide, Tyrol, Austria). Geo.Alp 3: 147–166.
Raetzo-Brülhart, H. 1997. Massenbewegungen im Gurnigelflysch und Einfluss der Klimaänderung. *Arb.-Ber. NFP 31*: 1–256. Zürich: Hochsch.-Verl. ETH Zürich.
Sartori, M., Baillifard, F., Jaboyedoff, M. & Rouille, J.-D. 2003. Kinematics of the 1991 Randa rockslides (Valais, Switzerland). *Natural Hazards Earth System Sc.* 2003 (3): 423–433.
Schmid, S., Fügenschuh, B., Kissling, E. & Schuster, R. 2004. Tectonic map and overall architecture of the Alpine orogen. *Ecl. Geol. Helv.* 97 (1): 93–117.
Soldati, M., Corsini, A. & Pasuto, A. 2004. Landslides and climate change in the Italian Dolomites since the Late glacial. *Catena* 55: 141–161.
Vidal, H. 1953. Neue Ergebnisse zur Stratigraphie und Tektonik des nordwestlichen Wettersteingebirges und seines nördlichen Vorlandes. *Geol. Bavarica* 17: 56–88
Vidrih, R., Ribicic, M. & Suhadolc, P. 2001. Seismogeological effects on rocks during the 12 April 1998 upper Soca Territory earthquake (NW Slovenia). *Tectonophysics* 330 (3-4): 153–175.
Weidner, S. 2000. Kinematik und Mechanismus tiefgreifender alpiner Hangdeformationen unter besonderer Berücksichtigung der hydrogeologischen Verhältnisse. Ph.D. Thesis: 1–257. Univ. Erlangen-Nürnberg.
Zangerl, C., Prager, C., Volani, M. & Brandner, R. 2006. Structurally controlled failure initiation of deep seated mass movements. *Geophys. Res. Abstr.* 8: 03516.

Landslides and Climate Change – McInnes, Jakeways, Fairbank & Mathie (eds)
© 2007 Taylor & Francis Group, London, ISBN 978-0-415-44318-0

Landslide characteristics and rainfall distribution in Taiwan

C.Y. Chen
National Chiayi University, Department of Civil & Water Resources Engineering

ABSTRACT: The paper herein studies the correlation of rainfall distribution and initiated landslides in Taiwan using statistics of rain gauge data and historical landslides in recent years. Climate changes initiating landslides and debris flows are discussed. In review of the landslide characteristics and rainfall distribution in Taiwan, it is found that these were affected by the track of typhoon, the following active air currents, topographic characteristics and seismic effects, in addition to the fragile geologic conditions.

1 INTRODUCTION

An average of 3.5 typhoons invades Taiwan every year. These typhoons have initiated numerous landslides and subsequently caused hazardous debris flows in recent years. These hazardous events include the impact of Typhoon Toraji in 2001 (Cheng et al., 2004), Typhoon Mindulle and the following storms in 2004 (Chen & Petley, 2005), and Typhoon Aere and Haitang in 2005 (Chen et al., 2007a). Among these, Typhoon Toraji caused 596 landslides, and Typhoon Mindulle triggered 907 landslides. These landslides were mainly distributed in the middle of Taiwan in mountainous areas. This shows a trend for landslide hazards and climate changes.

This study collected historical rainfall and landslide data in recent years in Taiwan to clarify the trends of the rainfall distribution and the initiation of landslides. The historical of distribution landslides and the tracks of Typhoons are discussed in terms of climate changes and topographic effects. The correlation between of typhoon rainfall distribution and the initiation of landslides is emphasized.

2 LANDSLIDE HAZARDS AND ECONOMIC LOSSES

According to the Statistics of the National Disaster Prevention and Reduction Center in Taiwan, some severe typhoon rainfall events induced human life losses including:

- Typhoon Herb in 1996: 73 dead/missing and 463 injured.
- Typhoon Xangsane in 2000: 89 dead/missing and 5 injured.
- Typhoon Trami and Toraji in 2001: 219 dead/missing and 192 injured.
- Typhoon Nari in 2001: 104 dead/missing and 265 injured.
- Typhoon Mindulle and following rainstorm in 2004: 41 dead/missing and 16 injured.
- Typhoon Haitang in 2005: 15 dead/missing and 31 injured.

The average economic losses per year for typhoon induced heavy rainfall from 1980 to 1996 were 17.6 billion New Taiwan dollars. The recent six years from 1996 to 2001 were even more serious, as flood hazards caused 291 deaths, 14 injuries and 28.3 billion New Taiwan dollars in agricultural losses on average per year (source the Central Weather Bureau in Taiwan).

A secondary hazard to the economic losses was the torrential rains brought by the convective active front in addition to those induced by the typhoon. The average economic loss per year due to torrential rains is about 3 billion New Taiwan dollars according to the statistics of the Ministry of the Interior.

Figure 1 shows the location of 1,420 debris flow prone creeks published in 2003 by the Soil & Water Conservation Bureau in Taiwan after Typhoon Xangsane in 2000 and Typhoon Toraji in 2001 induced serious debris flow hazards. The number of debris flow prone creeks increased abruptly from 485 in 1996 after Typhoon Herb to 722 in 1999 after the M7.6 Chi-Chi earthquake and up to 1,420 in recent.

One of the landslide hazards is shown in Figure 2. The Typhoon Aere induced landslide event began on the August 26th, 2004, and there were 25 houses buried by the slide debris. The landslide magnitude was 430 m in length, 200 m in width and there was a slide depth of about 25 m. The debris volume was estimated at up to 0.75 million m^3. This hazard caused four deaths,

Figure 1. Site locations of the 1,420 debris flows and the seismic-induced landslides.

Figure 2. Landslide hazard in the Tuchang tribe, Hsinchu County.

two missing and one injury. The maximum rainfall intensity was 90 mm and accumulated up to 1,300 mm at the nearby rain gauge station.

The debris flow hazards at the Songher tribe, Taichung County was initiated after Typhoon Mindulle brought air currents as shown in Figure 3. The debris flow hazard began in the morning of the 3rd of July, 2004, between 8:00 to 9:00. The time of debris flow triggering coincided to the peak time of rainfall, with nearly 840 mm of accumulated rain (total rainfall 1,630 mm) and 110 mm/hr of rainfall intensity. In the hazard there were more than 40 buildings overwhelmed by the debris causing one death, and the whole tribe was isolated by the debris at that time. According to the investigation file, the debris-affected area was 0.15 km^2 for the Songher No. 1 creek and 0.065 km^2 for the No. 2 creek (Chen et al., 2007b).

3 CLIMATE CHANGE

The long term average annual rainfall from 1949 to 2004 by historical rain gauge data in Taiwan is

Figure 3. Debris flow hazard after Typhoon Mindulle at Songher tribe, Taichung County (photo by the Soil & Water Conservation Bureau).

year	long term average annual rainfall (mm)	the average annual rainfall (mm)
2000	2472	2332
2001	2483	3077
2002	2466	1572
2003	2452	1689
2004	2467	2572

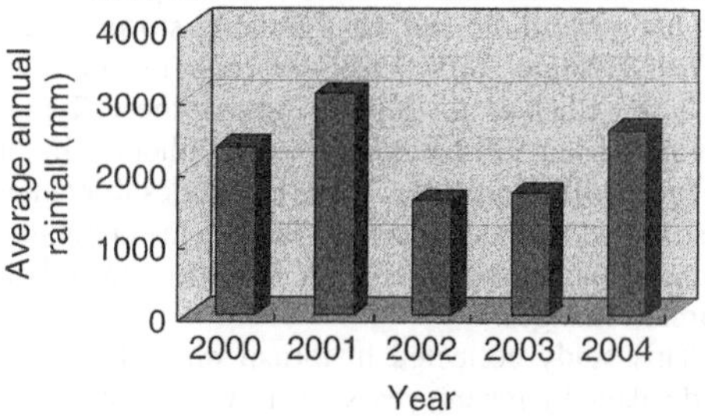

Figure 4. The average annual rainfall in five recent years from 2000 to 2004 (source the Water Resources Agency).

2,467 mm, which is lower than the average annual rainfall of 2,572 mm in 2004, mainly concentrated between April and October.

Figure 4 presents the average annual rainfall in five recent years. The average annual rainfall was extremely high in 2001, up to 3,000 mm. It was lower in 2002 and 2003. The average annual rainfall in 2000 was 2,332 mm which was 7% less than the long term average annual rainfall of 2,515 mm (1949–1990) (source the Water Resources Agency).

A total of 187 typhoons have affected Taiwan from 1897 to 2003. Among these, 84% (157 typhoons) landed on east Taiwan and only 16% (30 typhoons) landed on west Taiwan. In general, there were nine categories of typhoon tracks to Taiwan (source the Central Weather Bureau). They might move from the

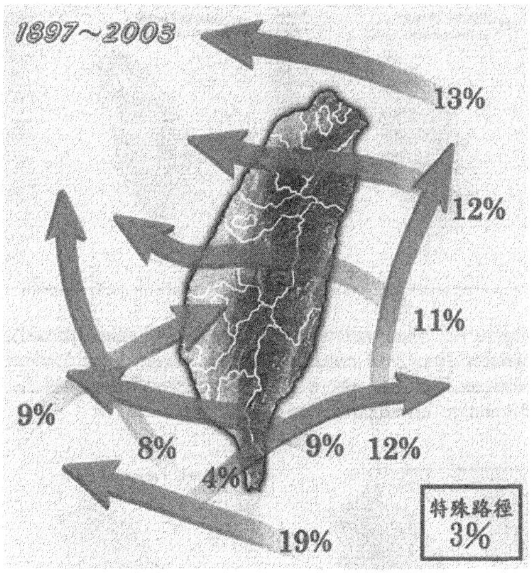

Figure 5. Nine categories of typhoon tracks to invade Taiwan (photo by Chang L. M.).

north sea (Type I), the north inland (Type II), the middle inland (Type III), the south inland (Type IV), the south sea (Type V), along the east inland or sea to the north (Type VI), along the west inland or sea to the north (Type VII), landing on the west inland to the north-east (Type VIII), of on the south inland to the north-east (Type IX) as shown in Figure 5. There could be some other special routes, for example Typhoon Nari in Figure 6.

Figure 6 plots the tracks of the hazardous typhoons in recent years for Typhoons Herb (1996), Xangsane (2000), Toraji & Nari (2001), Mindulle (2004), and Haitang (2005). These typhoons all landed on east Taiwan, but only Typhoon Herb was not blocked by the Central Range high mountain area (over 3,000 m). Typhoon Haitang was held back by the Central Range and self-rotated at sea before landing on the east Taiwan.

The climate change induced irregular rainfall events in recent years includes the extreme heavy rainfall brought by Typhoon Herb in 1996, Typhoon Toraji in 2001, and Typhoon Aere in 2004, which were 1,651 mm, 1,634 mm and 1,614 mm respectively. The accumulated rainfall during these three days was up to two-thirds of the average annual rainfall in Taiwan, resulting in hazardous landslides and debris flows.

In 2005, three intense typhoons made landfall on east Taiwan: Typhoon Haitang (16th~20th of July), Talim (30th of August~1st of September) and Longwang (1st of October). This was the first time since 1965 that three intense typhoons made landfall on Taiwan in one year. Another abnormal climate event was Typhoon Talim separating into upper and lower parts. The lower part was stopped by the Central Range area and the upper part passed onto west Taiwan and its intensity was reduced on the 1st of September, 2005.

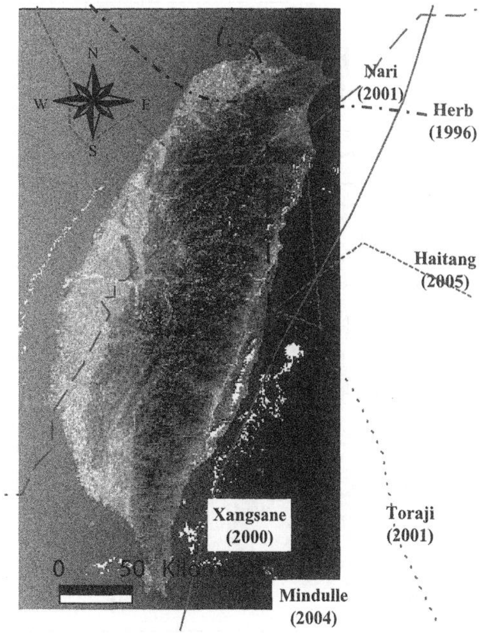

Figure 6. Track of typhoons that caused serious landslide hazards in Taiwan in recent years.

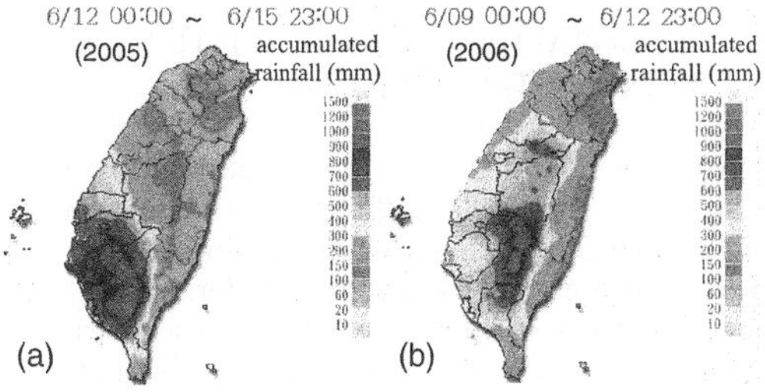

Figure 7. The distribution of torrential rains after (a) June 12th~15th, 2005 rainfall event (b) June 6th~12th, 2006 rainfall event (after the Central Weather Bureau).

In addition to the typhoon induced rainfall, the active front caused extremely heavy rainfall in Taiwan. During the 10th~17th of June, 2005, an accumulated rainfall of up to 1,681 mm was recorded in southern Taiwan (Figure 7a) which caused 17 deaths and one injury on the island. The other extremely heavy rainfall event was recorded during the 9th~12th of June, 2006 and distributed 1,708 mm in southern Taiwan (Figure 7b) which led to one death, 2 missing, 6 injuries and 138 landslides (source the SWCB).

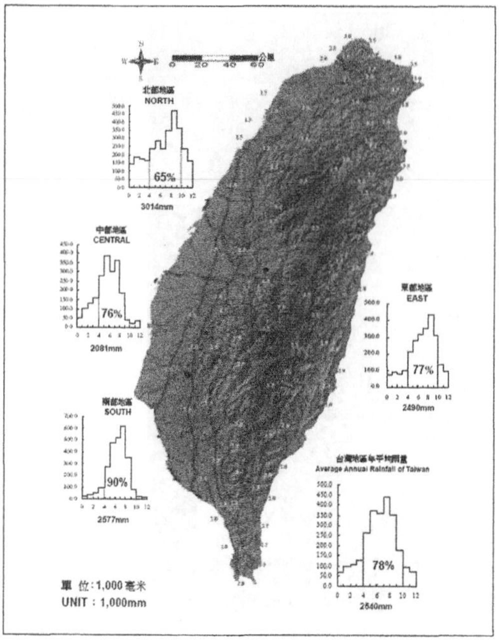

Figure 8. The average annual isohyetal map of Taiwan (1949–2004) (after the water resources agency, 2005).

4 TOPOGRAPHIC CHARACTERISTICS AND RAINFALL DISTRIBUTION

The plains area of Taiwan is separated by the Central Range area, where the mountain elevation is higher than 3,000 m. In general, it is divided into four parts by topography: the northern, eastern, middle and southern parts of Taiwan. 190 typhoons hit eastern Taiwan within the last one hundred years, and they reduced their intensity after being blocked by the Central Range (communication from the Central Weather Bureau). These topography characteristics strongly affected the intensity of the invading typhoons and their distribution of rainfall.

The tracks of historical typhoons leading to hazards were those passing through the north inland (Type II), as their intensity was not reduced by the Central Range, such as Typhoon Herb in 1996. It was also found that the typhoons that landed on the west inland and moved to the north-east (Type VIII) caused higher average rainfall (Chen & Wang, 2003).

Figure 8 shows the long term average annual isohyetal map of Taiwan from historical rain gauge stations from 1949 to 2004. It shows that the orographic uplift causing rainfall is mainly distributed around the Central Range area. The other higher rainfall distribution area is in the north-easterly areas of Taiwan, an area which was often affected by the active

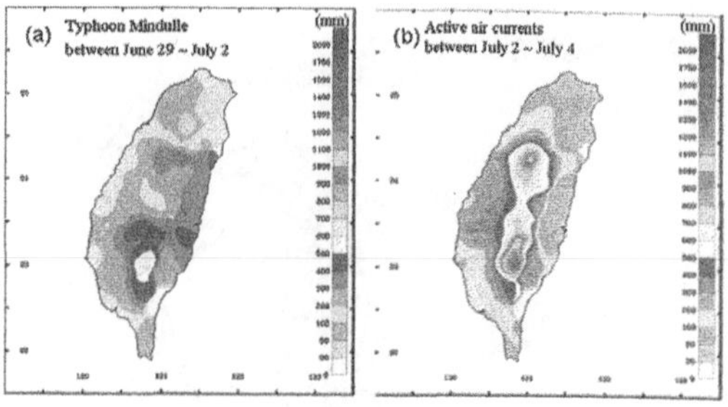

Figure 9. Accumulative rainfall of (a) Typhoon Mindulle between June 29th and July 2nd and (b) Active air currents between July 2nd~4th in 2004 (source the Central Weather Bureau in Taiwan, after Hsu et al., 2004).

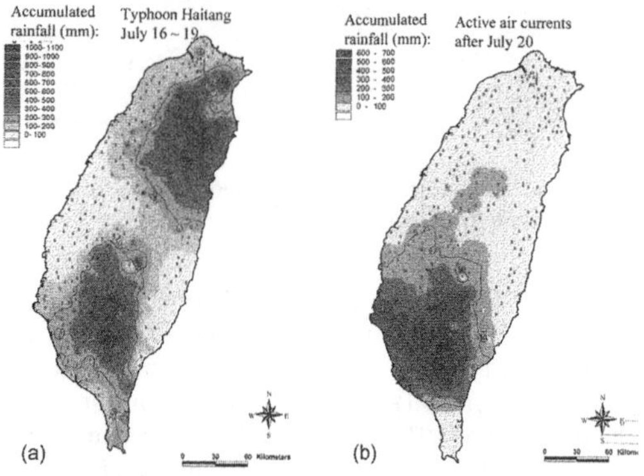

Figure 10. Typhoon Haitang July 16th~19th, 2005 total accumulated rainfall.

north-east trades induced rainfall in addition to being invaded by Type II typhoons.

Typhoons invading Taiwan might bring active air currents and cause higher levels of rainfall. Figure 9 presents the Typhoon Mindulle induced rainfall of 700 mm in southern Taiwan from June 29th to July 2nd in 2004, and brought strong active air currents causing 1,000 mm of rain in the middle mountain areas. Figure 10 shows Typhoon Haitang brought 700 mm of rain between July 16th~19th, 2005, and the strong active south-west flows caused heavy rainfall in the south-west areas of Taiwan after July 20th.

5 LANDSLIDES AND RAINFALL DISTRIBUTION

There are two main types of slopeland related hazards in Taiwan for landslides and debris flows. There were 685 landslide-induced debris flows among the published 1,420 debris flow prone creeks by the Soil & Water Conservation Bureau as shown in Figure 1. These landslide induced debris flows were mainly

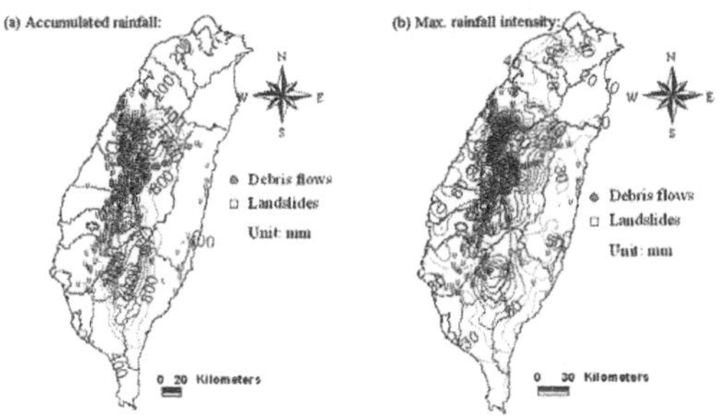

Figure 11. Overlap of landslides and (a) accumulated rainfall, (b) maximum rainfall intensity after Typhoon Mindulle in 2004 (after Chen, 2007).

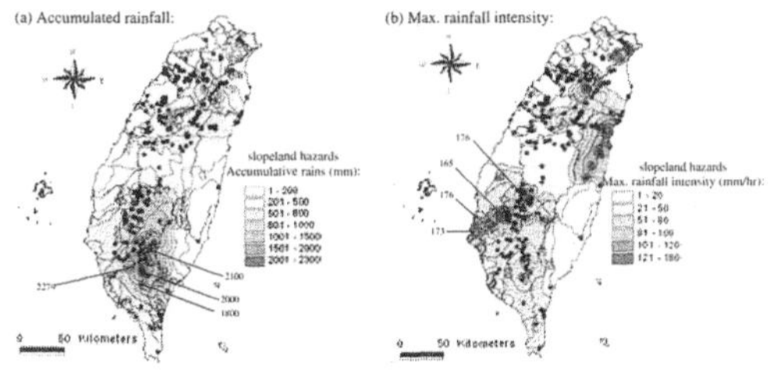

Figure 12. Overlap of slopeland hazard map and (a) total accumulated rain, (b) maximum rainfall intensity during 16:00 15th~16:00 21st of June, 2005, Typhoon Haitang invading Taiwan (source the Water Resources Bureau).

located in northern and eastern Taiwan. The rainfall from typhoons in 2000 attributed to 49 cases of slopeland hazards, 223 cases in 2001, no landslide hazards in 2002 and 2003, and 1,288 landslides in 2004. This trend is obviously correlated to the average annual rainfall shown in Figure 4.

Figure 11 shows the overlap map of the Typhoon Mindulle induced landslide map, and the accumulated rainfall and maximum rainfall intensity isopleth map in 2004. It shows a strong correlation between the accumulated rainfall, the maximum rainfall intensity and the initiation of landslides. Figure 12 plots the overlap map of the rainfall from Typhoon Haitang and the location of landslides. These landslides did not only correlate to the accumulated rainfall and the maximum rainfall intensity but also were affected by post-seismic behaviour in middle Taiwan (Chen et al., 2007a).

6 CONCLUSIONS

The author reviewed the rainfall and landslide characteristics in recent years in Taiwan and an obviously climate change was found. The climate change affected the frequency of typhoons invading Taiwan, their track, their brought rainfall intensity, the accumulated rain, and brought active air currents. The results show that the landslide characteristics in Taiwan were affected by the track of typhoons hitting Taiwan, the following active air currents, the topographic characteristics on the rainfall distribution and the seismic effects in addition to the fragile geologic conditions in Taiwan.

ACKNOWLEDGEMENTS

The author would like to thank the National Science & Technology Center for Disaster Reduction (NCDR), the Central Weather Bureau in Taiwan, the Soil and Water Conservation Bureau, and the Water Resources Bureau for providing valuable materials for this analysis. Financial support from the National Science Council in Taiwan under contract No. NSC 95-2221-E-492-002 is appreciated.

REFERENCES

Chen, C.Y., Chen, L.K., Yu, F.C., Lin, S.C., Lin, Y.C., Lee, C.L., Wang, Y.T. & Cheung, K.W. 2007a. Characteristics analysis for the flash flood-induced debris flows. Submission to Natural Hazards.

Chen, C.Y., Lin, S.C., Yu, F.C. & Cheung, K.W. 2007b. Assessment Model for Debris Flow Hazard Zones. Submission to Proceeding of Geotechnical Engineering.

Chen, C.Y. 2007. Efficiency of Ecological Engineering Following Typhoon Mindulle in Taiwan. Submission to Environmental and Engineering Geoscience.

Chen, H. & Petley, D.N. 2005. The impact of landslides and debris flows triggered by Typhoon Mindulle in Taiwan. Quarterly Journal of Engineering Geology and Hydrogeology: 38, 301–304.

Cheng, J.D., Huang, Y.C., Wu, H.L., Yeh, J.L., Chang, C.H. 2004. Hydrometeorological and landuse attributes of debris flows and debris floods during typhoon Toraji, July 29–30, 2001 in central Taiwan. Journal of Hydrology: 306, 161–173.

Chen, Z.E. & Wang, K.C. 2003. The relationship between the trail of typhoon and environmental safety of a village, Information Resource and Environmental Management Association, 2003 Annul Conference, 12th December, Hualien, Taiwan.

Hsu, M.H., Shi, B.Z., Yu, F.C., 2004. Hazard investigations on typhoon Mindulle and the following storm in 2004, National Science and Technology Center for Disaster Reduction, Report No. NCDR 93-11.

Landslides and Climate Change – McInnes, Jakeways, Fairbank & Mathie (eds)
 ISBN 978-0-415-44318-0

Is the current landslide activity in the Daunia region (Italy) controlled by climate or land use change?

J. Wasowski & D. Casarano
CNR-IRPI, Bari, Italy

C. Lamanna
University of Bari, c/o CNR-IRPI, Bari, Italy

ABSTRACT: The spatial frequency of the active landsliding in the Daunia region (southern Italy) is today 50% higher than in the mid-seventies, even though the precipitation data indicate significant decreasing decadal trends in average annual and winter rainfall. Since 1976 there has also been over 35% increase in the areal extension of sown fields, mainly for cereal cultivation. Thus the higher susceptibility to landsliding can be linked to the land use change and especially to the new ploughing, which has been taking place on the steeper slopes, whereas the above average precipitation in the last few winters (after 2002), with occurrence of intense rainfall, has acted as the causative/triggering factor for the recent slope failures. Thus the assessments of climate change impact on landsliding can be difficult on man-modified slopes, because man is capable of both improving and worsening the stability of slopes at a rate exceeding that of climatic change.

1 INTRODUCTION

The evolution of slopes to stable or unstable forms is typically controlled by numerous geomorphological, physical and, to an increasingly greater extent in recent decades, man-made processes (e.g. Crozier, 1986). Although time-space distributions of slope failures result from an interaction of many variables, landslide case histories demonstrate that the most significant temporal controls are often those related to rainfall-induced processes and their influence on hydrology, groundwater pressures and effective strengths of slope materials (e.g. Wieczorek, 1996). Thus it is generally accepted that changes in climatic conditions can alter significantly the susceptibility of slopes to landsliding (e.g. Schuster, 1996).

It is also known that agricultural and urban changes can influence local and regional slope instability, but it is also recognised that their effects are often minor, poorly understood, and difficult to quantify (e.g. Gostelow & Wasowski, 2004). Nevertheless, there are case studies that demonstrate the importance of distinguishing between man-modified and natural slopes while analysing rainfall-landslide relations for predictive purposes (e.g. Wasowski, 1998). Regarding the Basilicata region of southern Italy, Clarke & Rendell (2000) examined the impact of the farming practices on soil erosion processes, and more recently Piccarreta et al. (2006) considered implications of changes both in precipitation patterns and in land use policy on soil erosion.

In this work we research into the historical variations in the landslide activity occurred in the last three decades in the municipal territory of Rocchetta Sant'Antonio (Daunia region, southern Italy). This predominantly rural area is known for recurrent landslide problems and as such can perhaps be considered a representative portion of the Daunia Mountains. There is evidence that landsliding has increased in recent years and, though detailed studies are lacking, a currently popular view among the inhabitants of the area is that the climate change is to be blamed. However, the investigations into the climate patterns in Italy in the second half of the last century show a general decrease in precipitation (e.g. Brunetti et al., 2004), with, starting from the seventies, a significant lowering of average winter rainfall in southern Italy (e.g. Piccarreta et al., 2004; Polemio & Casarano, 2004). While such trend should lead towards a higher stability of slopes in Daunia, it seems that the pronounced 20th century human alterations of the local environment in the 20th century (Limongelli et al., 2005) have produced the opposite effects.

We recognise that landslide susceptibility is not simple to map and model, even where the controlling physical factors are relatively well known. Therefore, to demonstrate the influence of climate and land use (and land cover) changes on landsliding, we examine the temporal series of landslide frequency and land use maps, as well as the precipitation patterns in the last decades. The occurrence of the above average precipitation in the last winters is the main causative/triggering factor of recent slope failures, but the results also show clear connections between the land-use changes in the last 30 years and the current high landslide activity. We then argue that for hillslope areas profoundly modified by man in the recent decades, the combined effects of land use and rainfall pattern changes need to be considered to avoid a risk of overestimating the relative impact of climate change on landslide activity. We also stress that the analysis of the temporal impacts of climate change on slope stability is difficult, because the historical information on landslide activity is typically episodic and often incomplete.

2 PHYSICAL SETTING, GEOLOGY AND LANDSLIDING

2.1 *General*

The area studied belongs to the municipality of Rocchetta Sant'Antonio, a small hilltop town situated in the southern part of the Daunia Apennine Mountains in Southern Italy (Fig. 1). The moderate relief Daunia region is characterised by gentle hills and low mountains only locally exceeding 1000 m above sea level. In the area studied the elevations range from about 290 to about 680 m.

Only the steepest, highest elevation areas include a significant percentage of arboreous land. Elsewhere, especially where clay-rich units predominate, the vegetation cover is represented mainly by cultivated

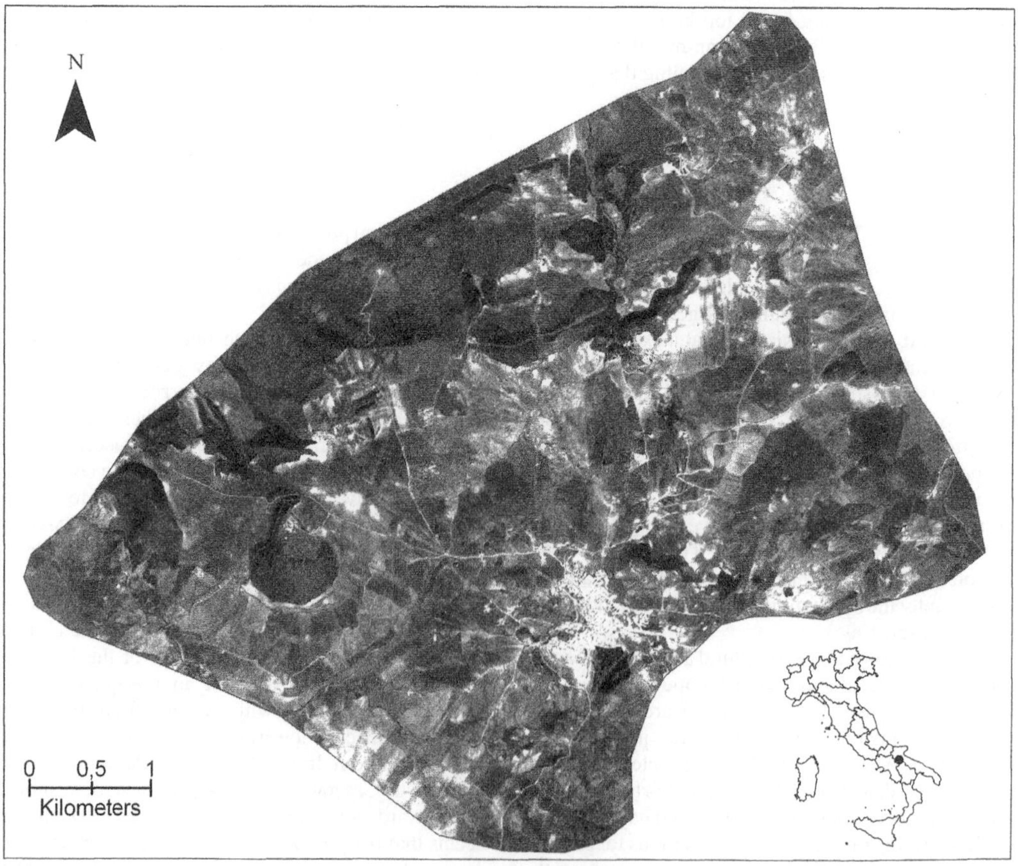

Figure 1. High resolution satellite image of the study area (inset shows location in Italy). Note predominantly agricultural land use.

land (cereals) and locally by grassland. The barren land areas are very limited and correspond to flysch and clay outcrops. The relative density of developed land (urban and rural settlements, roads and other infrastructures) is low.

The Rocchetta territory is characterised by a Mediterranean (sub-apennine) climate with total annual rainfall that typically varies from 600 to 750 mm. Winters are wet and mild, with snow precipitation being very limited. Summers are typically dry and warm.

2.2 *Geology*

The Daunia Apennines belong to the highly deformed transition area between the most advanced frontal thrusts of the Apennine chain and the western-most part of the foredeep (e.g. Dazzaro et al., 1988). The chain units are characterised by a series of tectonically deformed turbiditic (flysch) formations of Paleogene-Miocene age.

The clay-rich flysch units are more prone to landsliding, compared to the formations containing higher proportion of lithoid intercalations (sandstones, limestones, marlstones). The widespread presence of clayey materials with apparently poor geotechnical properties seems to be the underlying cause of landsliding. Furthermore, as a result of the tectonic history of the Apennines, the geological units are intensely deformed and hence have been susceptible to slope movements.

2.3 *Landslides*

In spite of the moderate elevation and relatively modest amount of total annual precipitation, the Daunia Mountains are known for their susceptibility to landsliding (e.g. Cotecchia, 1963). The region has experienced considerable deforestation in the last centuries and this may have prompted widespread landsliding in the rural areas (e.g. Parise & Wasowski, 2000).

Although mass movements are common throughout the entire region (e.g. Zezza et al., 1994), there have been rather few published studies on landslides in the Daunia Mountains (e.g. Iovine et al., 1996, Parise & Wasowski, 2000) and none regarding the Rocchetta Sant'Antonio territory.

There are over 20 hilltop towns in Daunia, most of which have experienced slope instability problems in the past. Although the specific documentation concerning the exact temporal occurrence of slope failures is rarely available, from the information acquired at different municipalities and interviews with local inhabitants, it appears that in recent years there has been an increase in mass movement activity in several urban and peri-urban areas. It is probable that the stability of slopes bordering the hilltop towns has gradually worsened because of residential development and infrastructure growth over the recent decades. This might have led in some cases to re-activations of pre-existing old landslides. Furthermore, the urban expansion onto marginally stable hillslopes has likely resulted in the increases in first-time damaging failures.

Recently Mossa et al. (2005) presented a GIS-based assessment of landsliding regarding the area neighbouring the north-west territory of Rocchetta Sant'Antonio. The over 20% landslide area frequency (both active and old slides) obtained for the overall area (132 km^2) demonstrates the high susceptibility of slopes to failure. The results also showed that the landslide frequency is the highest in 10–15 degree slope class, followed by 5–10 and 15–20 slope classes. This indicates a very low ($\approx$ residual) strength of the slope materials and suggests that a significant portion of failures could represent reactivations of pre-existing landslides.

Indeed, frequent field visits conducted in the last several years in the Daunia Apennines confirm that seasonal remobilisations (mainly in winter and spring time) of pre-existing landslides are common. Nevertheless, also first-time shallow landsliding is widespread in rural areas. In most cases the triggering factors seem to be related to rainfall events.

3 LANDSLIDES AND LAND USE CHANGE

The analysis concerned a 27 km^2 area within the Rocchetta Sant'Antonio municipal territory. A GIS was used to facilitate examination of spatial and temporal relations between landsliding and land use change.

3.1 *Data used to carry out analysis*

The main information and datasets exploited in this work are listed below:

- Topographic base maps at 1:5000, 1:2000 and 1:500 scale, from 2002;
- Digital elevation model (DEM) generated from the above 1:5000 scale maps;
- Black and white aerial photographs in scale 1:25000, from 1976;
- ASTER satellite multi-spectral imagery (using three visible and VNIR bands at 15 m resolution), from 2000;
- IKONOS satellite multi-spectral imagery with 1 m resolution in panchromatic, from 2006

This study also relies on the field inspections conducted in the study area since 1998 and on extensive in situ landslide mapping carried out in 2005–2006.

3.2 *Historical and current landslide distributions*

To investigate the differences between historical (1976) and current (2006) landsliding we compiled two series of landslide activity maps (cf. Soeters &

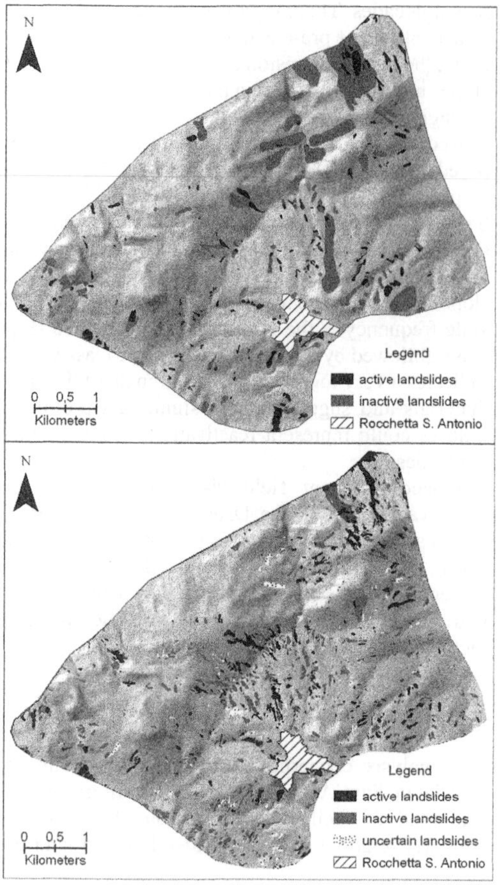

Figure 2. Landslide activity maps draped over a DEM: upper) 1976; lower) 2006.

Van Westen, 1996), derived from aerial photography and satellite image interpretation. For simplicity a distinction was made only between active and inactive landslides. We followed the geomorphic criteria typically adopted for the recognition of landslides and their state of activity from air-borne imagery (e.g. Wieczorek, 1984). Our experience and local knowledge of the study area suggest that the landslides classified as active have moved within the last annual or at most the last two annual seasonal cycles. Clearly, the criteria based on surface expression of geomorphic features are subjective, nevertheless, this work is concerned with the changes in landslide activity based on the variations in aerial frequency of mass movements in time and, therefore, these variations need not reflect the exact absolute values of landsliding.

Firstly, a stereoscopic interpretation of 1976 aerial photos was used to obtain a historical inventory of landslides (Fig. 2). Secondly, a high resolution IKONOS satellite imagery acquired in early spring of 2006 was exploited to obtain the most recent landslide inventory (Fig. 2). To facilitate the interpretation the satellite imagery was draped over a detailed DEM generated from 2002 topographic maps at 1:5000 scale.

Table 1. Landslide statistics (aerial frequency and number per km^2) for 1976 and 2006.

	Active slides		Inactive slides		Uncertain slides*	
Year	%**	No/km^2	%**	No/km^2	%**	No/km^2
1976	2.2	6.5	7.4	3.6	–	–
2006	3.3	17.0	4.1	22.2	0.9	6.1

* Uncertainty mainly due to the lack of stereoscopy in 2006 imagery.
** Percentage referred to the test area (27 km^2).

Table 1 summarises the differences in landsliding. The results are presented in terms of a number of landslides per km^2 and areal frequency of landslides (percentage of the total area occupied by landslides) to facilitate comparisons with other areas affected by landsliding.

We realise that some variation in landslide distributions likely results also from the use of different type (and resolution) of spatial information. For example, several large, inactive landslides with subdued surface features, identified on 1976 aerial photos, were not recognised on the 2006 IKONOS imagery that lacks stereoscopic capacity. Indeed, the 2006 inventory includes a class of uncertain landslides, some of which overlap in part the areas of inactive landslides in the 1976 inventory (Fig. 2). Nevertheless, the differences in the distributions of slope failures classified as certain are pronounced and a clear change in active landsliding can be established. In particular, the comparison of the 1976 and 2006 landslide inventories shows a 50% increase in aerial frequency of active landslides (Table 1). An opposite tendency is seen for the inactive landslides and some of the landslides inactive in 1976 result active in 2006 (Fig. 2).

3.3 *Historical and current land use versus active landsliding*

To quantify variations in land use occurred in the last three decades we used first the same aerial photos to obtain a map of land use in 1976 (Fig. 3). Because the attempts of using automatic classification procedures produced unreliable results, the land use classes were extracted "manually" through photo interpretation. As a short-cut, only two major groupings were distinguished: agricultural land (sown fields with mainly cereal cultivation involving ploughing) and other (including arboreous land, bare land, manmade). This simplification is considered admissible, because the temporal changes in agricultural land,

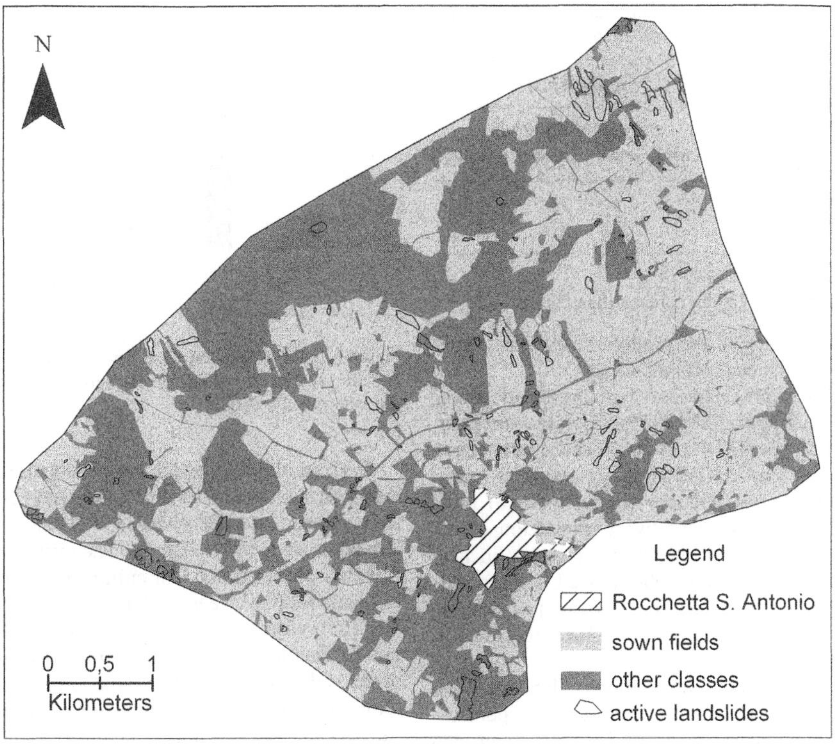

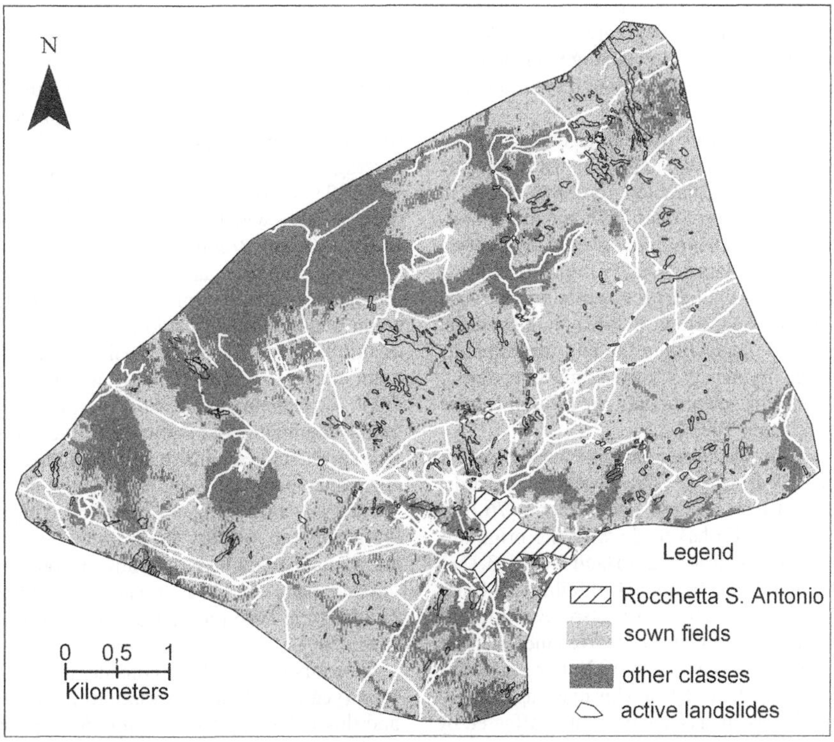

Figure 3. Land cover and distribution of active landslides: upper) 1976; lower) 2006; white lines indicate roads and infrastructures.

Table 2. Land use change and variation in areal frequency of landsliding (in %) from 1976 to 2006.

	Active slides			Inactive slides		
Year	Old* Sown	New** Sown	Other	Old* sown	New** sown	Other
1976	2.0	3.1	2.2	8.4	7.9	5.4
2006	3.3	5.2	1.8	5.0	5.0	1.6

*Areas already sown in 1976.
**Areas sown after 1976 (but not later than 2000).

which represents by far the predominant class; seem to have most influence on landsliding.

A map of recent land use was produced using ASTER imagery from 2000 (Fig. 3). This imagery was preferred over IKONOS data, because the acquisition time of the ASTER imagery resulted more suitable for automatic (supervised procedure) land-use classification. A qualitative examination of the imagery suggested that only minor changes in land use occurred from 2000 to 2006. Indeed, neither catastrophic events (e.g. forest fires) nor major land-use policy changes seem to have affected the study area in the last few years.

To indicate the temporal variations in land use from 1976 to 2000 the map data were expressed as percentage of areal frequency. The results show that about 53% of the land has been used as sown fields in the seventies and that by the year 2000 the percentage increased to over 72%. Such a pronounced change in land use in a mountainous area could have some impact on slope stability.

The influence of the land use change on mass movement activity can be indicated by considering the distributions of active landslides on sown fields in figure 3. A visual comparison of 1976 and 2006 maps indicates that the new land given over to agriculture (new sown fields) is characterised by higher density of active landslides. To highlight the correlations between the change in the areal extent of sown fields and the variations in active landsliding the data were quantified using area frequency statistics and considering separately the fields sown already in 1976 and those sown after that date (Table 2).

In particular, the results demonstrate that the current (2006) very high density (5.2%) of active landslides in the areas that have become sown after 1976 (and no later than by 2000) has to do with the originally high susceptibility of those areas to landsliding (3.1%), exceeding even the areal frequency of failures in fields already sown in 1976 (2.0%). It is also revealing that at the same time the frequency of active landsliding in other areas (not sown neither in 1976 nor currently) has decreased from 2.2 to 1.8%. This may suggest that the stability of those areas have been little affected by the climate change in the last 30 years.

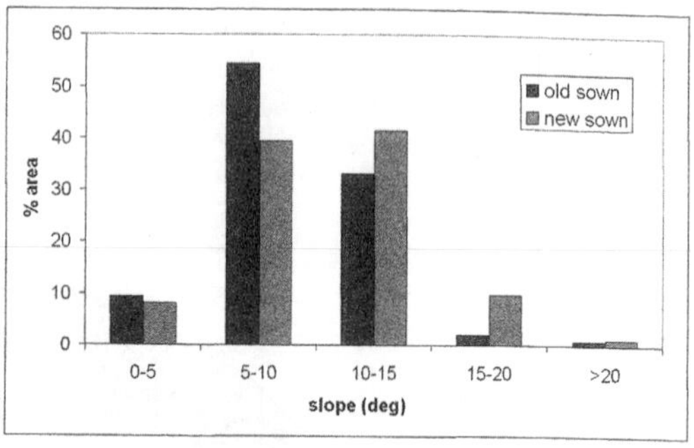

Figure 4. Areal frequency statistics of slope angles for fields already sown in 1976 (old sown) and those sown after 1976 (no later than by 2000).

To further investigate the reasons beyond the higher landslide activity in the fields sown after 1976 we confronted their slope distributions with those of the areas already sown in 1976 (Fig. 4). The results clearly show that the new land used for cultivation is characterised by significantly steeper slopes (most often exceeding 10°), whereas the fields already sown in 1976 exhibit shallower slopes. This strongly suggest that the observed increase in active landsliding have resulted at least in part from the expansion of cultivation onto the steeper and presumably less stable land.

4 PRECIPITATION PATTERNS SINCE 1951

The rainfall data available for the study area come from the Rocchetta Sant'Antonio pluviometric station and span the period 1951–2006 (winter). The average annual precipitation is about 670 mm (range 449–1037 mm). Falls and winters are typically most rainy seasons with the average precipitation of, respectively, 202 mm (range 75–470) and 194 mm (range 64–360). Summers are usually dry with the average precipitation of 104 mm (range 24–266). To reveal some trends in precipitation we analysed time series of rainfall data. Since the available literature on climate change in southern Italy is limited to the year 2000, for comparative reasons we present the general trends in precipitation in the study area until that year. The results indicate that there are significant decreasing trends in annual (−3, 33 mm/yr) and winter (−1, 43 mm/yr) rainfall (Fig. 5). Though relying on the data from only one locality, these trends resemble closely the regional precipitation trends for southern Italy calculated from the data registered in numerous rain gauges (e.g. Polemio & Casarano, 2004; Piccarreta et al., 2004).

Nevertheless, there are considerable short and mid-term (yearly to decadal) variations in the precipitation and this is highlighted in Figure 5. In particular, it is evident that the winter seasons in the last few years

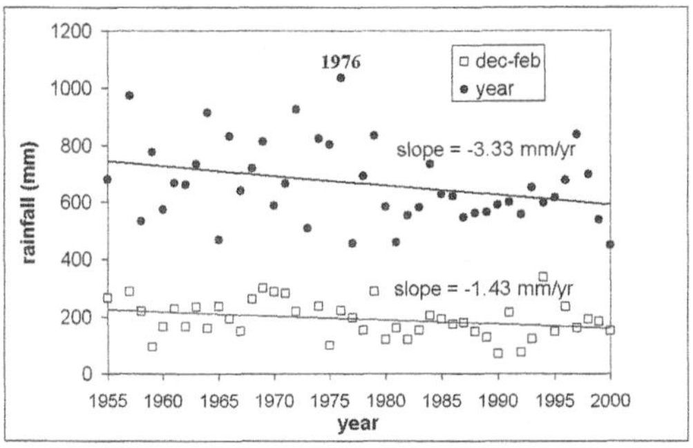

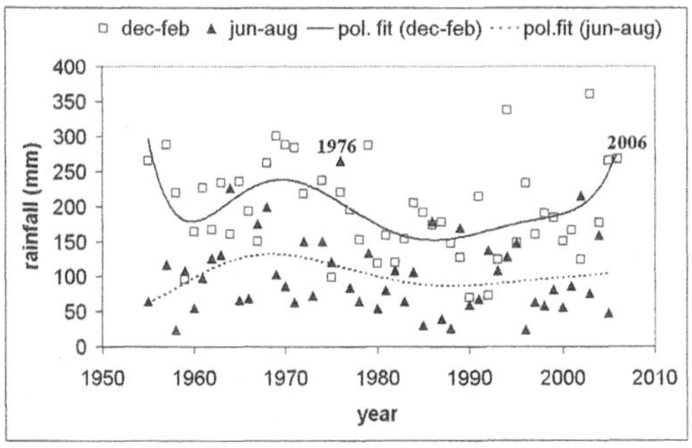

Figure 5. Patterns in average yearly and seasonal precipitation registered in Rocchetta S. Antonio: upper) decreasing yearly and winter (December–February) trends in the period 1951–2000, estimated by least square best linear fit; lower) trends in winter and summer precipitation in the period 1951-winter 2006 (with 6th order polynomial fit).

have been generally very wet. Also the winter 2006, i.e. the period preceding the acquisition of the imagery used for the landslide inventory, have been characterised by the precipitation well above the average. Importantly, the year 1976 and its summer season have witnessed the highest precipitation on the record (respectively 1037 and 266 mm). Furthermore, most of the 1976 summer precipitation occurred in June and July, i.e. shortly before the acquisition of aerial photos used for 1976 landslide inventory.

Therefore, although the imagery used for the two landslide inventories was acquired in two seasons typically marked by very different weather patterns, both 1976 aerial photos and 2006 imagery were acquired shortly after few month wet periods with comparable cumulative rainfall (about 250 mm). This suggests that the slope conditions favourable for landsliding (especially for shallow movements) were to some extent similar.

5 DISCUSSION AND CONCLUSIONS

The results of this work show that the land-use changes in the last 30 years correlate well with the increase in shallow landslide activity in Daunia. It is likely that the 20th century human alterations of the local environment have had negative impact on the stability of local slopes. In particular, deforestation can be historically regarded as one of the main factors that have contributed to landslides in Daunia, as well as in other parts of southern Italy (e.g. Parise & Wasowski, 2000). Following deforestation, the rural areas have been gradually given over to agriculture with fields first worked by hand and then by mechanised means. The latter methods, especially deep ploughing marked another significant geotechnical change and soil strength loss. The original vertically structured soil was generally destroyed, but more importantly a repetitive remoulding (ploughing) introduced a new layer with a more open and aerated structure. This, ploughing along the steep slopes, and especially new ploughing (removing existing natural vegetation), which tends to target the steeper more marginal land for cultivation have resulted in increased erosion and shallow landsliding.

In addition, our recent observations in the Rocchetta area indicate a considerable impact of damaging landslides on the road network. It seems that the situation has worsened in the last few years. This has coincided with the recent (as of 2002), transfer of the road maintenance responsibility from the National Highway Administration to the Regional Government. It is difficult, however, to asses to what extent the increased landslide damage to the road network may result from inadequate maintenance or from natural causes (rainfall), or from both.

Regarding the possible influence of changing weather patterns, the evidence of a general global warming trend appears overwhelming (e.g. Mann et al., 1998). At present, however, the prediction of the exact future climate change scenarios seems to represent an illusive task. Furthermore, at regional level variable changes in climatic conditions can be expected. It follows that a range of possible climate change impacts on slope stability, both negative and positive, need to be considered.

For example, regarding the United Kingdom, the studies on the ongoing climate change and future scenarios (e.g. Hulme et al., 2002) indicate the following general trends in precipitation:

- wetter winters
- drier summers
- higher frequency of heavy winter precipitation
- decreases in snowfall

It is likely that such changes, and in particular increases in winter precipitation, will lead to higher landslide activity (e.g. Forster et al., 2006).

In comparison, a different pattern of climate change is observed in Italy. For instance the analysis of Brunetti et al. (2004) shows a strong decrease in precipitation and higher frequency of extreme rainfall events in the second half of the last century. For southern Italy, the study by Polemio & Casarano (2004) demonstrates that the overall decreasing trend in net rainfall is more relevant for winter months and that the temperatures above the average enhanced the effects of the last two serious droughts (1988–1992 and 2000–2001). Regarding the Basilicata region, which neighbours the study area to the south, Piccarreta et al. (2004) pluviometric analysis indicate the following trends:

- a large decrease in the mean yearly precipitation, starting from the seventies
- a significant decrease in the mean winter precipitation (and lack of significant trends for other seasons
- a higher frequency of severe drought periods

The above changes imply an opposite effect with respect to that predicted for the UK, i.e. they should lead to lower landslide activity in southern Italy. Indeed, in their recent study of 1950–2000 record of extreme rainfall events and landslides in Basilicata, Clarke & Rendell (2006) indicated that an overall reduction in the temporal frequencies of landslides can be expected from the ongoing climate change.

Our results from the Daunia area, however, show that the spatial frequency of active landsliding in the last years is much higher than about 30 years ago. The current increase in landslide activity has coincided with the occurrence of wetter winters starting from 2002. This recent change can perhaps represent only a short term variation within a longer term (decades) general trend of decreasing precipitation. Nevertheless, the in situ observations in Daunia suggest that if similar changes are accompanied also by intense precipitation (e.g. winter 2003), then they may lead to widespread landsliding. The correlation with the land use changes in Daunia in the last 30 years also indicates that the consequences of the variations in rainfall pattern could be especially pronounced on the steep slopes recently modified by man.

The assessments of climate change impact on landsliding can be complex on man-modified slopes, because man is capable of both improving and worsening the stability of slopes at a rate exceeding that of recent climatic or geomorphic change. It seems also that in predictive analysis of climate change impact on landsliding much attention should be given to the period following the occurrence of any major modification of slope setting, both via anthropogenic (e.g. intensive agriculture) and via natural (e.g. extreme rainfall or high-magnitude earthquake) agents. There are indications that following such modifications the thresholds of landslide susceptibility can be significantly altered with respect to those based on previous historical records of precipitation and associated slope failures (e.g. Wasowski, 1998; Parise & Wasowski 1999). Because of this and because regional scale landslide inventories are prepared neither frequently nor on the regular basis, there is a risk of over- or under-estimating the relative importance of climatic variations on slope stability.

ACKNOWLEDGEMENTS

We thank the municipal administration of Rocchetta S. Antonio and in particular Dott. A. Magnotta, Eng. G. Amoruso and Eng. G. Tedeschi of Regione Puglia, and Dott. G. Rampino for providing some of the data used in this study.

REFERENCES

Brunetti, M., Buffoni, L., Mangianti, F., Maugeri, M. & Nanni, T. 2004. Temperature, precipitation and extreme events during the last century in Italy. *Global and Planetary Change* 40: 141–149.

Clarke, M.L. & Rendell, H.M. 2000. The impact of the farming practice of remodelling hillslope topography on badland morphology and soil erosion processes. *Catena* 40: 229–250.

Clarke, M.L. & Rendell, H.M. 2006. Hincasting extreme events: the occurrence and expression of damaging floods and landslides in southern Italy. *Land Degrad. Develop.* 17: 365–380.

Cotecchia, V. 1963. I dissesti franosi del Subappenino Dauno con riguardo alle strade provinciali. *La Capitanata*, 1 (5–6), Foggia.

Crozier, M.J. 1986. *Landslides: Causes, Consequences, and Environment.* London, Croom Helm.

Dazzaro, L., Di Nocera, S., Pescatore, T., Rapisardi, L., Romeo, M., Russo, B., Senatore, M.R. & Torre, M. 1988. Geologia del margine della catena appenninica tra il F. Fortore ed il T. Calaggio (Monti della Daunia – Appennino Meridionale). *Mem. Soc. Geol. It.* 41: 411–422.

Forster, A., Culshaw, M., Wildman, G. & Harrison, M. 2006. *Implications of climate change for urban areas in the UK from an engineering geological perspective.*

Gostelow, P., & Wasowski J. 2004. Ground surface changes detectable by earth observation and their impact on the stability of slopes. *Polish Geol. Inst. Sp. Papers* 15: 43–52.

Hulme, M., Jenkins, G.J., Lu, X., Turnpenny, J.R., Mitchell, T.D., Jones, R.G., Lowe, J., Murphy, J.M., Hassell, D., Boorman, P., Mcdonald, R. & Hill, S. 2002. *Climate change scenarios for the United Kingdom: The UKCIP02 Scientific Report*, Tyndall Centre for Climate Change Research, School of Environmental Sciences, University of East Anglia, Norwich, UK.

Iovine G., Parise, M. & Crescenzi, E. 1996. Analisi della franosità nel settore centrale dell'Appennino Dauno. *Mem. Soc. Geol. It.* 51: 633–641.

Limongelli, L., Uricchio, V., Zurlino, G. (eds.) 2006. *La valutazione ambientale strategica per lo sviluppo sostenibile della puglia: un primo contributo conoscitivo e metodologico.* F.E.S.R. Regione Puglia Assessorato Ecologia.

Mann, M.E., Bradley, R., Hughes, M.K. 1998. Global-scale temperature patterns and climate forcing over the past six centuries. *Nature* 392: 779–787.

Mossa, S., Capolongo, D., Pennetta, L. & Wasowski J. 2005. A GIS-based assessment of landsliding in the Daunia Apennines, southern Italy. *Proc. Intern. Conf. "Mass movement hazard in various environments"* 20–21 October 2005, Cracovia, Polonia.

Parise, M. & Wasowski, J. 1999. Use of landslide activity maps for the evaluation of landslide hazard: three case studies from southern Italy. *Natural Hazards* 20: 159–183.

Parise, M. & Wasowski, J., 2000. Fenomeni di dissesto nell'Appennino Dauno, ed implicazioni per il patrimonio archeologico e storico-culturale. *Atti Convegno Geo-Ben*, Torino.

Piccarreta, M., Capolongo, D. & Boenzi, F. 2004. Trend analysis of precipitation and drought in Basilicata from 1923 to 2000 within a southern Italy context. *International Journal of Climatology* 24: 907–922.

Piccarreta, M., Capolongo, D., Boenzi, F. & Bentivenga, M. 2006. Implications of decadal changes in precipitation and land use policy to soil erosion in Basilicata, Italy. *Catena* 65: 138–151.

Polemio, M. & Casarano, D. 2004. Rainfall and drought in southern Italy (1821–2001), *Proceedings of UNESCO-IAHS-IWHA Symposium "Water Science? A basis for civilization"*, Rome, December 2003, IAHS Publ. 286: 217–227.

Schuster, R. L. 1996. Socioeconomic Significance of Landslides. In A.K. Turner & R.L. Schuster (eds), *Landslides, Investigations and Mitigation, Trans. Res. Board. Spec. Pub. 247*, Nat. Acad. Press., 673 p.

Servizio Idrografico Italiano 1951–2005. *Annali idrologici dei bacini con foce al litorale adriatico e ionico dal Candelaro al Lato*, Sez. Idrogr. di Bari, Fasc. VIII, Ist. Poligr. dello Stato, Roma.

Soeters, R. & Van Westen, C.J. 1996. Slope instability recognition, analysis and zonation, In A.K. Turner & R.L. Schuster (eds), *Landslide investigation and mitigation, NRC, Transportation Research Board Special Report 247*, Washington, 129–177.

Wasowski, J. 1998. Understanding landslide-rainfall relationships in man-modified environments: a case history from Caramanico Terme (Italy). *Environmental Geology* 35: 197–209.

Wieczorek, G.F. 1984. Preparing a detailed landslide-inventory map for hazard evaluation and reduction, *Bull. Int. Ass. Geol.* 3: 337–342.

Wieczorek, G.F. 1996. Landslide Triggering Mechanisms. In A.K. Turner & R.L. Schuster (eds), *Landslides, Investigations and Mitigation Trans. Res. Board. Spec. Pub. 247*, Nat. Acad. Press., 673 p.

Zezza, F., Merenda, L., Bruno, G., Crescenzi, E. & Iovine, G. 1994. Condizioni di instabilità e rischio da frana nei comuni dell'Appennino Dauno Pugliese. *Geologia Applicata e Idrogeologia* 29: 77–141.

Landslides and Climate Changes – McInnes, Jakeways, Fairbank & Mathie (eds)
© 2007 Taylor & Francis Group, London, ISBN 978-0-415-44318-0

Debris-flow activity in the Ritigraben torrent (Valais Alps, Switzerland): Will there be less but bigger events in a future greenhouse climate?

M. Stoffel
Laboratory of Dendrogeomorphology (Dendrolab.ch), Department of Geosciences, Geography, University of Fribourg, Switzerland

ABSTRACT: Tree-ring records and climate proxies suggest that cool summers with frequent snowfalls regularly prevented the release of debris flows at Ritigraben between the 1570s and 1860s, whereas the warming trend in conjunction with greater precipitation totals in summers and falls apparently led to an increase in activity between 1864 and 1895 and in the early 20th century. Simultaneously, the seasonality of events started to shift from June and July to August and September. Given that RCMs project extreme precipitation events to occur less frequently in summer and that wet spells will become more common in spring or fall, it is conceivable that debris flows will not necessarily occur as frequently in the future as they did in the past. But even if the frequency of events is likely to decrease, the magnitude of summertime debris flows and related impacts could be greater than currently.

1 INTRODUCTION

In the recent past, the Swiss Alps have experienced several events of extreme precipitation conditions that have led to widespread flooding and debris-flow activity, severe damage to infrastructure and fatalities on cones or at the mouth of gullies (Rickenmann & Zimmermann 1993, BWG 2002, Schmidli & Frei 2005, Beniston 2006). With the projected changes of the climate and global warming (Christensen & Christensen in press), there is much debate about modifications in the amount, intensity, duration, type and timing of precipitation events as well as about their effect on related flooding, debris-flow and other mass-wasting processes (e.g., Bradzil et al. 2002, Milly et al. 2002, Mudelsee et al. 2003). Consequently, a plethora of climatological and hydrological scenarios have been developed to seize potential impacts of future greenhouse climates, but Goudie (2006) pertinently emphasizes that these scenarios have not been matched for the most part by the development of scenarios of future changes in geomorphological systems.

Previous studies focusing on future changes have mainly been based on data of past debris-flow events covering, at best, parts of the 20th century or on comparably short time series of meteorological records. For instance, Rebetez et al. (1997) compared archival records on debris flows with meteorological data and identified an increase in the occurrence of large debris flows in the Swiss Alps since the 1980s. Marchi & Tecca (2006) identified more records on debris flows in historical documents for the last decades as well. They conclude that the larger number of events could be due both to an increase in the frequency and to a larger availability of information. In the French Alps, Van Steijn (1996) reported high debris-flow activity since the 1980s, but there does not seem to exist a univocal increase in the frequency of debris-flow events as a result of climatic change (Jomelli et al. 2004, in press).

As studies focusing on the role of (changing) climatic conditions on the release of debris flows remain scarce and widely limited by the paucity of data on past events or triggering weather conditions, there is an inherent need to further examine the natural variability of extreme weather events and to gather detailed information on past debris-flow activity before stating any cause-to-effect relationship between the projected global warming and the incidence of geomorphic processes.

On forested cones, data on past debris-flow activity can be considerably improved by means of dendrogeomophological analysis of trees disturbed by former events. In the past, spatial patterns of debris-flow events have been assessed through the coupling of geomorphic with tree-ring data (Stoffel et al. 2006, Bollschweiler & Stoffel in press, Bollschweiler et al. in press). Similarly, growth anomalies in tree-ring records have been used to estimate the magnitude of former events (Strunk 1997, Baumann & Kaiser 1999).

The purpose of this paper is to assess the debris flow activity and process dynamics in an ephemeral torrent originating from a periglacial environment in the Valais Alps (Switzerland). Through the analysis of 2450 tree-ring records obtained from 1204 trees disturbed by past debris flows, we (i) investigate the frequency and seasonality of events and (ii) discuss potential feedbacks of the debris-flow system to the projected changes in precipitation in a future greenhouse climate.

2 STUDY SITE

The analysis of past debris-flow dynamics and growth disturbances in century-old trees was conducted at the Ritigraben torrent (Switzerland, 46° 11' N, 7° 49' E). Figure 1 illustrates the torrent taking its source at approximately 2,600 m a.s.l. In the departure zone, geophysical prospecting indicates the existence of contemporary permafrost (Lugon & Monbaron 1998).

On its downward course to the Mattervispa river, the torrent passes a large forested cone (32 ha) on a structural terrace (1,500–1,800 m a.s.l.), where debris-flow material affects trees within an old-growth stand composed of European larch (*Larix decidua* Mill.), Norway spruce (*Picea abies* (L.) Karst.) and Swiss stone pine (*Pinus cembra* ssp. sibirica). At the confluence of the Ritigraben torrent with the receiving Mattervispa River at 1,080 m a.s.l., depositional forms are lacking and material is immediately eroded.

Debris-flow material consists of heavily disintegrated, weathered metamorphic granites of Permian age (Labhart 2004) and mainly originates from the steep departure zone of the Ritigraben torrent, where an active rock glacier provides material for the initiation of debris-flow events. Further debris is mobilised from the channel, which is continuously recharged with fallen rocks or through lateral channel erosion. While mean rock sizes on the cone surface generally remain well below 2 m in diameter, there is also evidence that boulders with volumes exceeding 10 m^3 have been transported by debris flows in the past.

The high elevation of the source area currently restricts debris-flow activity in the Ritigraben torrent from June to September (Stoffel et al. 2005a). Present-day debris flow activity is initiated primarily by persistent precipitation in fall rather than thunderstorms in summer. The documentation of past events only covers the last two decades (1987–2007) and the "largest event ever" was recorded in 1993 with eleven erosive surges and a volume estimated to 60,000 m^3 (Zimmermann et al. 1997).

Figure 1. The Ritigraben torrent (Swiss Alps) takes its source (S) at 2,600 m a.s.l. and passes through a forested cone (C), before it converges with the Mattervispa river (1,080 m a.s.l.).

3 MATERIAL AND METHODS

On the intermediate debris-flow cone, a majority of the century-old conifers (*Larix decidua*, *Picea abies*, *Pinus cembra*) show visible growth defects related to past debris-flow activity (i.e. tilted stems, partial burying of the trunk, decapitation, destruction or erosion of roots, visible scars). Based on a detailed geomorphic map and on an outer inspection of the stem surface, trees were sampled that have obviously been disturbed by past debris flows.

In the field, at least two cores were extracted per tree using increment borers, one core in the flow direction of past debris flows and another core on the opposite side of the trunk (max. length of cores: 40 cm, Ø 6 mm). In order to gather the greatest amount of data on the growth disturbances caused by past events, increment cores were preferably sampled at the height of the visible damage or within the segment of the stem tilted during past events. In the case of visible scars, further increment cores were extracted from the callus tissue overgrowing the wound.

In addition to the disturbed trees sampled on the cone, we selected undisturbed reference trees from a forest stand located southwest of the cone. For every single reference tree, two cores per tree were extracted perpendicularly to the slope. In contrast to the disturbed trees, increment cores of the reference trees were extracted at breast height (≈130 cm).

In total, 1204 trees were sampled (2450 increment cores): 539 *Larix decidua*, 429 *Picea abies* and 134 *Pinus cembra* trees (2246 cores) from the debris-flow cone as well as 102 trees (204 cores) of the same species from undisturbed reference sites.

In the laboratory, samples were analyzed and data processed following the standard procedures described in Bräker (2002). Single steps of sample analysis included surface preparation, counting of tree rings, skeleton plots as well as ring-width measurements using digital LINTAB positioning tables connected to a Leica stereomicroscope and TSAP 3.0 software (Time Series Analysis and Presentation, Rinntech 2007). Growth curves of the disturbed samples were then crossdated with the corresponding reference chronology constructed from undisturbed trees for each of three conifer species sampled on the cone, in order to separate insect attacks or climatically driven fluctuations in tree growth on the study area from growth disturbances caused by debris flows (Cook & Kairiukstis 1990).

Growth curves were then used to determine the initiation of abrupt growth reduction or recovery (Schweingruber 2001, McAuliffe et al. 2006). In the case of tilted stems, both the appearance of the cells (i.e. structure of the reaction wood cells) and the growth curve data were analyzed (e.g., Braam et al. 1987, Fantucci & Sorriso-Valvo 1999).

Finally, the cores were visually inspected so as to identify further signs of past debris-flow activity in the form of callus tissue overgrowing abrasion scars or tangential rows of traumatic resin ducts formed from cambium damage (Stoffel et al. 2005b, 2006, Perret et al. 2006).

As conifer trees react immediately to damage with the formation of callus tissue or tangential rows of traumatic resin ducts, the intra-annual position of growth disturbances was further used to assess the moment of debris-flow activity in particular years with monthly precision (Stoffel et al. 2005c, Stoffel & Beniston 2006, Stoffel in press). Our tree-ring based data on the intra-seasonal timing of debris flow events were then compared with records from a local meteorological station, operational since 1863 and with archival data on flooding in rivers of the Valais Alps (Lütschg-Lötscher 1926, Röthlisberger 1991).

4 RESULTS

Data on the innermost rings of the 1102 *Larix decidua*, *Picea abies* and *Pinus cembra* trees sampled on the cone varied from AD 1492 to 1962 with 53% of the increment cores showing more than 300 tree rings at sampling height and old trees being quite evenly spread over the cone. The youngest trees, in contrast, are most commonly found near the forest fringe, where anthropogenic interventions influence the age and succession rates of trees (i.e. farming activities, extraction of fire- and construction wood).

Analysis of the disturbed trees allowed reconstruction of 2263 growth disturbances caused by passing debris-flow surges or the deposition of material on the cone. Table 1 shows that signatures of past events were mainly identified on the increment cores via tangential rows of traumatic resin ducts or reaction wood. Abrupt growth recovery or reductions were only occasionally found in the tree-ring series and wounds and overgrowing callus tissue was rarely present on the cores.

In total, dendrogeomorphological analysis of the increment cores allowed reconstruction of 123 debris-flow events covering the last 440 years. The reconstructed frequency of debris flows is given in Figure 2. From the data, it appears that periods of repeated debris-flow activity alter with phases of little or no activity. Such clustering of events is especially obvious in the early 1870s, the 1890s or between the late 1910s and 1935.

In Figure 3, the reconstructed frequency is broken down into 10-yr periods, with bars representing variations from the mean decadal frequency of debris flows for the period 1706–2005, when 3.26 events occurred every ten years. Results illustrate that the frequency of events generally remained well below average during most of the classical "Little Ice Age" (1570–1900; see Grove, 2004) and that periods with increased debris-flow activity only start to emerge after the last "Little Ice Age" glacier advance in the 1860s. This period of increased activity continued well into the early 20th century and culminated between 1916 and 1935. During these 20 years, 14 events were derived from the tree-ring series. Results further illustrate that this episode of important activity was followed by a decrease in debris-flow activity. In a similar way, very

Table 1. Relative number of growth disturbances used to infer past debris-flow activity from increment cores.

Growth disturbance	Absolute number	%
TRD*	987	43.6
Injuries	118	5.2
Callus tissue	22	1.0
Reaction wood	728	32.1
Growth reduction	194	8.6
Growth release	214	9.5
Total	**2263**	**100.0**

* TRD = tangential row of traumatic resin ducts.

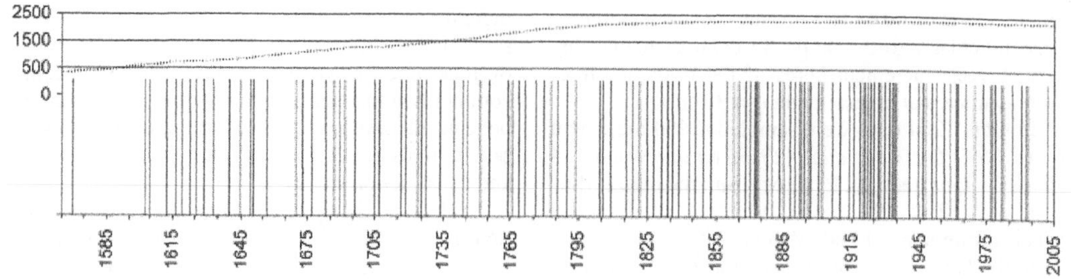

Figure 2. Tree-ring based reconstruction of debris flow activity at Ritigraben between AD 1566 and 2005 containing 123 events. The sample depth gives the number of cores available for analysis at specific years in the past (modified after Stoffel et al, in press).

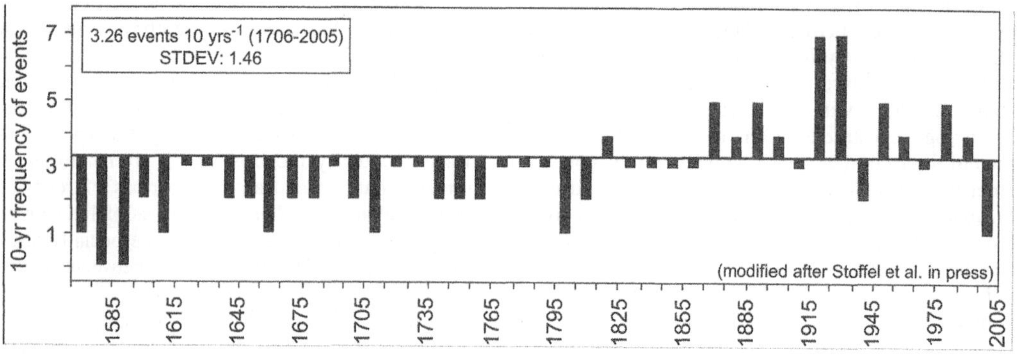

Figure 3. Reconstructed 10-yr frequencies of debris-flow events for the period AD 1566–2005. Results are presented as variations from the mean decadal frequency of debris flows of the last 300 years (AD 1706–2005).

low activity can be observed for the last 10-yr segment (1996–2005) with only one debris-flow event recorded on August 27, 2002. Along with the 10-yr segments of 1706–1715 and 1796–1805, the most recent ten years exhibit the lowest debris-flow activity in the last 300 years.

The seasonality of past events was assessed based on the intra-annual position of tangential rows of traumatic resin ducts in the tree rings, archival data on flooding in rivers of the Valais Alps as well as on meteorological records of the local MeteoSwiss station (1863–2005). Results on the seasonality of debris-flow activity are presented in Figure 4, indicating that events generally occurred much earlier in the summer prior to 1900.

This is especially true for the period 1850–1899, when more than 70% of the reconstructed debris flow events took place in June and July and no incidence in September. In the 20th century, debris-flow activity clearly shifted towards August and September, with not a single event registered for June after AD 1962.

Based on our reconstructions, it also appears that snowfalls and frozen ground apparently inhibit debris entrainment from the starting zone (>2,600 m a.s.l.) during precipitation events between October and May.

5 DISCUSSION

In the study we report here, increment cores extracted from 1102 living *Larix decidua* Mill. *Picea abies* (L.) Karst. and *Pinus cembra* ssp. sibirica trees allowed reconstruction of 2263 growth disturbances belonging to 123 debris-flow events since AD 1566.

On the basis of the evidence presented above, it is possible to characterize climatological as well as meteorological factors driving debris-flow activity in the case-study area. Tree-ring based records of past debris-flow activity suggest that comparably cool summers with frequent snowfalls at higher elevations regularly prevented the release of debris flows most of the time between the 1570s and 1860s. The warming trend in conjunction with greater precipitation totals in summers and falls between 1864 and 1895 did, in contrast, lead to an increase of meteorological conditions favourable for the release of debris flows from the departure zone. Enhanced debris flow activity continued well into the 20th century and the reconstruction exhibits a clustering of events for the period 1916–1935, when warm-wet conditions prevailed during summers in the Swiss Alps (Pfister 1999).

The reconstructed frequency is also in agreement with chronicle data on flooding events in Alpine rivers

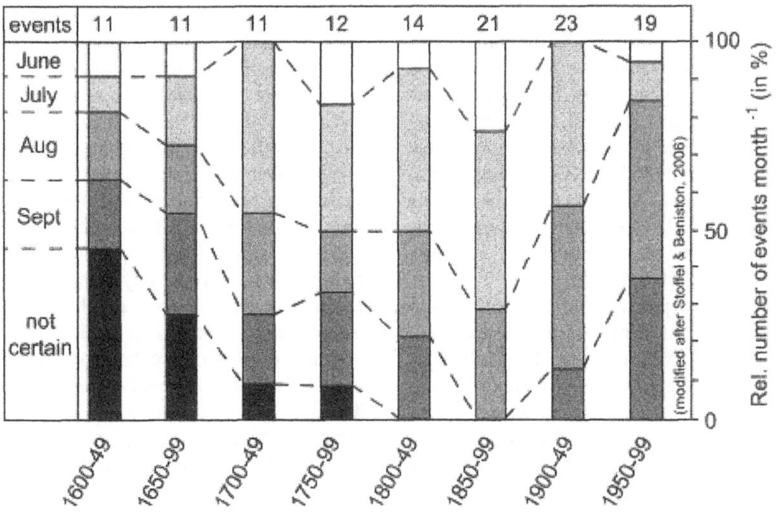

Figure 4. Seasonality of past debris-flow activity as inferred from the intra-annual position of tangential rows of traumatic resin ducts in the tree ring, archival data on flooding as well as meteorological data (1863–2005).

of Switzerland (Lütschg-Lötscher 1926, Röthlisberger 1991), where a scarcity of flooding events are observed for most of the "Little Ice Age" and during the mid-20th century as well. However, it is worthwhile to note that floods in adjacent Alpine rivers became more frequent in the 1830s (Pfister 1999), which is three decades before activity increased in the investigated case-study area. The reasons for this time lag have not been analyzed in details, but it can be assumed that they are due to the comparably cool temperatures that prevailed at high elevation sites during this period.

The seasonality of events underwent changes over the period covered by the reconstruction as well. Based on the tree-ring record, we observe a shift in the debris-flow activity from June and July to August and September over the 20th century, with not a single event registered for June after AD 1962. A comparison of reconstructed debris-flow events with archival data on flooding in adjacent rivers further indicates that convectional rainfalls in summer (i.e. local thunderstorms) would have preferentially triggered debris flows from the 1860s until the 1980s. In contrast, cyclonic rainstorms affecting large parts of the Alps in late summer and early autumn have apparently become more frequent since the 1980s and are responsible for the debris-flow events of 1987, 1993 and 1994. While yet another cyclonic rainstorm caused considerable damage in rivers neighbouring the case-study site in October 2000, frozen ground and snowfalls inhibited excessive runoff and debris entrainment from the starting zone of the Ritigraben located at >2,600 m a.s.l. (Bardou & Delaloye 2004).

The reconstructed shift of debris-flow activity from June and July to August and September can be further explained by the negative trend observed for heavy summer rainfall and the slightly positive trend found in heavy fall precipitation intensities in the study region over the 20th century (Schmidli & Frei 2005).

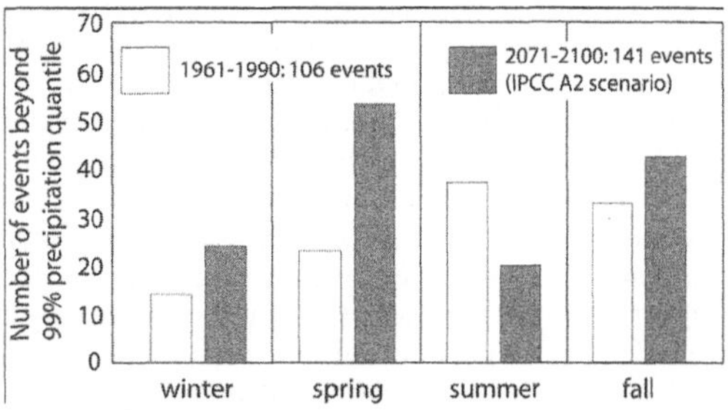

Figure 5. Number of heavy precipitation events beyond the 99% quantile in the Swiss Alps (corresponding to just over 60 mm/day) under current (1961–1990) and a future greenhouse climate (2071–2100) based on the IPCC A2 Scenario (modified after Stoffel & Beniston 2006).

Despite uncertainties related to regional climate simulations of precipitation in complex terrain, recent work by Beniston (2006) based on 4 regional model projections for a "greenhouse climate" by 2100 suggests that mean and extreme precipitation may undergo a seasonal shift, with more spring and fall heavy precipitation events (defined as the 99% quantile values of daily precipitation, which corresponds to just over 60 mm/day) than currently, and less in summer.

Figure 5 illustrates the seasonal shift in the occurrence of heavy precipitation events in the Swiss Alps as discussed by Beniston (2006), for current climate (1961–1990 reference period) and a greenhouse climate for 2071–2100 based on the IPCC (International Panel on Climate Change) A2 greenhouse-gas emissions scenario (Nakićenović et al. 2000). The histograms are based on the HIRHAM regional climate model (Christensen et al. 1998), one of a number of models applied to climatic change studies in Europe in the context of the EUPRUDENCE project (Prudence 2007, Christensen & Christensen in press). This increase in the number of extreme precipitation events (i.e. over 30% between the two periods) supports earlier findings by Frei et al. (1998).

Paradoxically, the impacts associated with future extreme rainfall on debris-flow torrents originating at high elevations may be even reduced. This is because spring and autumn temperatures are suggested to remain 4–7°C degrees below current summer temperatures in a future greenhouse climate, implying lower freezing levels in future springs and autumns as compared to current summers and therefore probably widespread buffering effects of snow on runoff and debris entrainment. Given that mean and extreme precipitation events are projected to occur less frequently in summer and that wet spells will become more common in spring or fall, it is conceivable that debris flows will not necessarily occur as frequently in the future as they did in the past in the case-study area.

However, even if the frequency of summer events is likely to decrease in a future greenhouse climate, the magnitude of future summertime debris flows and related impacts could (theoretically) be greater than currently. This is because warmer temperatures and higher precipitation intensities could result in larger runoff, more important transport capacity of surges and a bigger erosive potential of debris flows.

However, the release of debris-flow surges and the ensuing magnitude of events not only depend on precipitation intensities, their duration and subsequent runoff, but also on sediment availability in the starting zone as well as in the debris-flow channel. After the high-magnitude event in September 1993, when eleven surges mobilized an estimated volume of 60,000 m^3 (Zimmermann et al. 1997), the main channel was almost immediately recharged with debris due to instabilities in and partial collapses of the oversteepened lateral walls, allowing new debris flow activity only one year after the big 1993 event (Bloetzer et al. 1998). Based on observations in the field, we believe that as little as 10% of the debris flow volume reaching the Mattervispa River would actually start directly from the starting zone and that up to 90% of the material would be mobilized from the huge unconsolidated moraine and debris deposits along the flow path of the Ritigraben torrent. It can therefore be concluded that debris is readily available and easily entrained along the debris-flow channel and cannot be considered a limiting factor.

Debris availability and recharge rates could, in contrast, undergo changes in the departure zone due to changes in climatic conditions. Given that temperatures will rise by several degrees in a future greenhouse climate (according to e.g., the IPCC A2 scenario, see Nakićenović et al. 2000), it is conceivable that this will have consequences on the currently prevalent permafrost as well as on the dynamics of the active rock glacier that nowadays feeds the starting zone of debris flows with material. Preliminary results from borehole temperature measurements realised next to the departure zone of the Ritigraben torrent suggest that permafrost is comparably temperate and possibly in an unstable state (Herz et al. 2003). At other locations in the Swiss Alps (Roer et al. 2005, Delaloye pers. comm.), important accelerations have been observed in rock glacier movements ("surges") over the last few decades. It is, thus, possible that rock glacier movements could increase at our study site in the future as well and therefore deliver more debris to the starting zone of debris flows in the Ritigraben torrent. In conjunction with more precipitation events >60 mm/day, these larger amounts of debris could theoretically lead to the entrainment of more material and subsequently to larger debris flows in the system.

On the other hand, should the ice once completely disappear from the rock-glacier body at Ritigraben, one could also imagine that debris would less easily be transported to the starting zone of debris flows and therefore less material be available for the initiation of future events.

The above considerations on potential changes in the seasonality of heavy precipitation events and on potential modifications of rock-glacier dynamics remain highly speculative for the moment. Nonetheless, local authorities should not wait with the planning of appropriate constructive measures so as to (better) protect the buildings located along the currently used channel as well as on the intermediate debris-flow cone from future damage. Regardless of expected changes in the frequency or magnitude of debris-flow events at Ritigraben, they have to be aware that considerably large debris-flow events have repeatedly occurred in the past, and that they will occur in the future as well. As the lateral walls of the currently used channel (that have been incised by the September 1993 event) already started to collapse, it is possible that future debris flows could overtop the channel above 1,650 m a.s.l., reactivate abandoned flow paths and deposit material in the eastern or south-western parts of the cone, theoretically threatening buildings and public infrastructure.

6 CONCLUSIONS

The analysis of 2450 tree-ring sequences from 1102 disturbed conifer trees provided an unusually complete record on past debris-flow activity in a torrent originating from permafrost environments. Reconstructions indicate that debris flow activity remained comparably low during most of the "Little Ice Age" and that high activity culminated in the early 20th century (1916–1935). More recently, debris-flow activity became less frequent and the projected shift of heavy precipitation events (over 60 mm/day) from summer to spring and fall could cause the frequency to remain at this below-average level in the long term. But even if the frequency of summer events is likely to decrease in a future climate, the magnitude of future summertime debris flows and related impacts could probably be greater than currently. This is because warmer temperatures and higher precipitation intensities could result in larger runoff, more important transport capacity of surges and a bigger erosive potential of debris flows. Irrespective of changes in the frequency or magnitude of debris-flow events, local authorities should realize appropriate constructive measures along the channel and on the intermediate cone so as to prevent damage to buildings and other infrastructure.

ACKNOWLEDGEMENTS

This work has been undertaken partly in the context of the FP6 EU-project ENSEMBLES and considerably benefited from the fieldwork and analyses undertake by Mathieu Boéchat, Delphine Conus, Ariane Delachaux, Thierry Falco, Xavier Frainier, Holger W. Gärtner, Michael A. Grichting, Igor Lièvre, Gilles Maître and Mireille Savary. Michelle Boll schweiler is most warmly acknowledged for providing helpful comments on a former draft of the manuscript.

REFERENCES

Baumann, F. & Kaiser, K.F. 1999. The Multetta debris fan, Eastern Swiss Alps: A 500-year debris flow chronology. *Arctic, Antarctic and Alpine Research* 31(2): 128–134.

Bardou, E. & Delaloye, R. 2004. Effects of ground freezing and snow avalanche deposits on debris flows in alpine environments. *Natural Hazards and Earth System Science* 4: 519–530.

Beniston, M. 2006. August 2005 intense rainfall event in Switzerland: not necessarily an analog for strong convective events in a greenhouse climate. *Geophysical Research Letters* 33: L05701.

Bloetzer, W., Egli, T., Petrascheck, A., Sauter, J. & Stoffel, M. 1998. *Klimaänderungen und Naturgefahren in der Raumplanung – Methodische Ansätze und Fallbeispiele*. Vdf Hochschulverlag AG: Zürich.

Bollschweiler, M. & Stoffel, M. in press. Debris flows on forested cones – reconstruction and comparison of frequencies in two catchments in Val Ferret, Switzerland. *Natural Hazards and Earth System Sciences*.

Bollschweiler, M., Stoffel, M., Ehmisch, M. & Monbaron, M. in press. Reconstructing spatio-temporal patterns of debris flow activity using dendrogeomorphological methods. *Geomorphology*.

Braam, R.R., Weiss, E.E.J. & Burrough, P.A. 1987. Spatial and temporal analysis of mass movement using dendrochronology. *Catena* 14: 573–584.

Bradzil, R., Glaser, R., Pfister, C. & Stangl, H. 2002. Floods in Europe – a look into the past. *Science Highlights Pages News* 10.

Bräker, O.U. 2002. Measuring and data processing in tree-ring research – a methodological introduction. *Dendrochronolo gia* 20(1–2): 203–216.

BWG (Bundesamt für Wasser und Geologie) 2002. Hochwasser 2000 – Les crues 2000. *Serie Wasser* 2: 1–248.

Cook, E.R. & Kairiukstis, L.A. 1990. *Methods of dendrochronology – applications in the environmental sciences*. London: Kluwer.

Christensen, J.H. & Christensen, O.B. in press. A summary of the PRUDENCE model projections of changes in European climate by the end of this century. *Climatic Change*.

Christensen, O.B., Christensen, J.H., Machenhauer, B. & Botzet, M. 1998. Very high-resolution regional climate simulations over Scandinavia – Present climate. *Journal of Climate* 11: 3204–3229.

Fantucci, R. & Sorriso-Valvo, M. 1999. Dendrogeomorphological analysis of a slope near Lago, Calabria (Italy). *Geomorphology* 30: 165–174.

Frei, C., Schär, C., Lüthi, D. & Davies, H.C. 1998. Heavy precipitation processes in a warmer climate. *Geophysical Research Letters* 25: 1431–1434.

Goudie, A.S. 2006. Global warming and fluvial geomorphology. *Geomorphology* 79(2): 384–394.

Grove, J.M. 2004. Little Ice Ages: ancient and modern. London: Routledge.

Herz, T., King, L. & Gubler, H. 2003. Microclimate within coarse debris of talus slopes in the alpine periglacial belt and its effects on permafrost. In M. Philipps, S.M. Springman & L.U. Arenson (eds.), *Permafrost. Proc. 8th Int. Conf. on Permafrost, Zurich, 21–25 July 2003*. Rotterdam: Balkema.

Jomelli, V., Brunstein, D., Grancher, D. & Pech P. 2007. Is the response of hill slope debris flows to recent climate change univocal? A case study in the Massif des Ecrins (French Alps). *Climatic Change*: in press.

Jomelli, V., Pech P., Chochillon, C. & Brunstein, D. 2004. Geomorphic variations of debris flows and recent climatic change in the French Alps. *Climatic Change* 64: 77–102.

Labhart, T.P. 2004. *Geologie der Schweiz*. 6th edition. Thun: Ott Verlag.

Lütschg-Lötscher, O. 1926. *Über Niederschlag und Abfluss im Hochgebirge: Sonderdarstellung des Mattmarkgebietes: ein Beitrag zur Fluss- und Gletscherkunde der Schweiz*. Zürich: Schweizerischer Wasserwirtschaftsverband.

Lugon, R. & Monbaron, M. 1998. *Stabilité des terrains meubles en zone de pergélisol et changements climatiques. Deux études de cas en Valais: Le Ritigraben (Mattertal) et la moraine du Dolent (Val Ferret)*. Zürich: vdf Hochschulverlag AG.

Marchi, L. & Tecca P.R. 2006. Some observations on the use of data from historical documents in debris-flow studies. *Natural Hazards* 38: 301–320.

McAuliffe, J.R., Scuderi, L.A., McFadden, L.D. 2006. Tree-ring record of hillslope erosion and valley floor dynamics: landscape responses to climate variation during the last

400 yr in the Colorado Plateau, northeastern Arizona. *Global and Planetary Change* 50(3–4): 184–201.
Milly, P.C.D., Wetherald, R.T., Dunne, K.A. & Delworth, T.L. 2002. Increasing risk of great floods in a changing climate. *Nature* 415: 514–517.
Mudelsee, M., Borngen, M., Tetzlaff G. & Grunewald, U. 2003. No upward trends in the occurrence of extreme floods in central Europe. *Nature* 425: 166–169.
Nakićenović, N., Alcamo, J., Davis, G., de Vries, B., Fenhann, J., Gaffin, S., Gregory, K., Grübler, A., Yong Jung, T., Kram, T., Lebre La Rovere, E., Michaelis, L. Mori, S., Morita, T., Pepper, W., Pitcher, H., Price, L., Riahi, K., Roehrl, A., Rogner, H.H., Sankovski, A., Schlesinger, M., Shukla, P., Smith, S., Swart, R., van Rooijen, S., Victor, N. & Dadi, Z. 2000. *IPCC Special Report on Emissions Scenarios* Cambridge: Cambridge University Press.
Perret, S., Stoffel, M. & Kienholz, H. 2006. Spatial and temporal rockfall activity in a forest stand in the Swiss Prealps – a dendrogeomorphological case study. *Geomorphology* 74(1–3): 219–231.
Pfister, C. 1999. Wetternachhersage. 500 Jahre Klimavariationen und Naturkatastrophen. Bern, Stuttgart, Wien: Paul Haupt.
Prudence 2007. http://prudence.dmi.dk
Rebetez, M., Lugon, R., Baeriswyl, P.A. 1997. Climatic change and debris flows in high mountain regions: The case study of the Ritigraben torrent (Swiss Alps). *Climatic Change* 36: 371–389.
Rickenmann, D. & Zimmermann, M. 1993. The 1987 debris flows in Switzerland: documentation and analysis. *Geomorphology* 8(2–3): 175–189.
Rinntech. 2007. http://www.rinntech.com/Products/Lintab.htm.
Roer, I., Kääb, A. & Dikau, R. 2005. Rockglacier acceleration in the Turtmann valley (Swiss Alps): Probable controls. Norsk Geografisk Tidsskrift – Norwegian Journal of Geography 59, 157–163.
Röthlisberger, G. 1991. Chronik der Unwetterschäden in der Schweiz. *Berichte der Forschungsanstalt für Wald, Schnee und Landschaft* 330: 1–122.
Schmidli, J. & Frei, C. 2005. Trends of heavy precipitation and wet and dry spells in Switzerland during the 20th century, *International Journal of Climatology* 25: 753–771.
Schweingruber, F.H. 2001. *Dendroökologische Holzanatomie.* Bern, Stuttgart, Wien: Paul Haupt.
Stoffel, M. in press. Dating past geomorphic processes with tangential rows of traumatic resin ducts. *Dendrochronologia.*
Stoffel, M. & Beniston, M. 2006: On the incidence of debris flows from the early Little Ice Age to a future greenhouse climate: A case study from the Swiss Alps. *Geophysical Research Letters* 33: L16404.
Stoffel, M., Conus, D., Grichting, M.A., Lièvre, I. & Maître, G. in press. Unraveling the patterns of late Holocene debris flow activity on a cone in the Swiss Alps: chronology, environment and implications for the future. *Global and Planetary Change.*
Stoffel, M., Bollschweiler, M. & Hassler, G.R. 2006: Differentiating events on a cone influenced by debris-flow and snow avalanche activity – a dendrogeomorphological approach. *Earth Surface Processes and Landforms* 31(11): 1424–1437.
Stoffel, M., Lièvre, I., Conus, D., Grichting, M.A., Raetzo, H., ärtner, H.W. & Monbaron, M. 2005a. 400 years of debris-flow activity and triggering weather conditions: Ritigraben, Valais, Switzerland. *Arctic, Antarctic and Alpine Research* 37(3), 387–395.
Stoffel, M., Schneuwly, D., Bollschweiler, M., Lièvre, I., Delaloye, R., Myint, M. & Monbaron, M. 2005b. Analyzing rockfall activity (1600–2002) in a protection forest – a case study using dendrogeomorphology. *Geomorphology* 68(3–4): 224–241.
Stoffel, M., Lièvre, I., Monbaron, M. & Perret, S. 2005c. Seasonal timing of rockfall activity on a forested slope at Täschgufer (Valais, Swiss Alps) – a dendrochronological approach. *Zeitschrift für Geomorphologie* 49(1): 89–106.
Strunk, H. 1997. Dating of geomorphological processes using dendrogeomorphological methods. *Catena* 31, 137–151.
Van Steijn, H. 1996. Debris flow magnitude-frequency relationships for mountainous regions of Central and Northwest Europe. *Geomorphology* 15: 259–273.
Zimmermann, M., Mani, P., Gamma, P., Gsteiger, P., Heiniger, O. & Hunziker, G. 1997. *Murganggefahr und Klimaänderung – ein GIS-basierter Ansatz.* Zürich: vdf Hochschulverlag AG.

Landslides and Climate Change – McInnes, Jakeways, Fairbank & Mathie (eds)
© 2007 Taylor & Francis Group, London, ISBN 978-0-415-44318-0

Coastal zone evolution – evidence of climate change

F. Marabini
ISMAR – Marine Geology Institute – C.N.R., Bologna, Italy

ABSTRACT: The aim of this work is to consider the influence of climate change on the evolution of the coastal zone and the possible present-day implications of human activity on the configuration of the present coastline. The evolution of the Adriatic coastal zone from the Po river delta up to the Ancona promontory is used as an example. In this area there are three distinct physiographic geomorphological units: the Po delta; the sandy coastal zone with the padan plain on the back; and the cliff around Ancona. The results demonstrate the influence and the superimposition of climate cycles at different time-scales, which produce a general regression of the shoreline, but with different values, sometimes even oscillations, dependant upon the different gemorphological characteristics and the influence of human activity during the last fifty years.

1 INTRODUCTION

Around the world, coastal zones and adjacent seas are under increasing environmental pressure associated with global changes, anthropogenic activities and societal trends of rising populations and standards of living.

Initially, the development of the coast was entirely dependant on the interactions between rivers and seas. The most significant natural factor in coastal evolution, development and equilibrium are the fluvial supply of sediments and the river flow regime on one side, and the energy of the coastal sea (wave and currents regime, attenuation of energy etc.) on the other. These factors are strongly influenced by natural global changes in climate (factors such as changes in water and sediment discharge and in modifications of the meteorological-hydrological regime of coastal seas) as well as by variations in sea levels.

The effect of climate change on the evolution of the natural environment is of great importance. Whilst for glaciers rivers, plains and vegetation the effect of climate is easy to recognise, for the coastal zone the situation is more complex.

Utilising dendrochronological curves, it is possible to identify several periodic climatic fluctuations: cod/wet periods alternated with warm/dry periods with a frequency of about a hundred years during the last 3000 years. Within this "large scale" climatic cycle, shorter periods (10–35 years) with the same alternance of climatic conditions can also be identified.

The significance of increasing rainfall, the resulting abundance of sediment yield from the rivers along the coastal zone and the possibility of landslides on the cliffs during the cold/wet periods is evident.

2 CLIMATE CHANGE AND THE COASTAL ZONE

An accurate study of the environmental impact of sea level and shoreline changes due to global warming requires a detailed investigation and interpretation of the events that have occurred during the past 20,000 years. This time interval in fact corresponds to two significant global climatic changes: the last Wurmian glaciation during the Upper Pleistocene; and the warming during the Holocene.

The general increase in temperatures continues today and it is global in extent. This natural increase of temperature, due to CO_2 and green houses gases is likely to become much greater in the next century. This increase of temperature threatens to cause a dangerous increase in sea levels (Fig. 1).

It is known that the general climate trend during the Holocene was towards a general climate warming if we compare it with the last glacial stage (Wurm). Within this general trend during the Holocene some secondary climatic changes (some hundred years length) can be distinguished that have strongly influenced environmental conditions, impacting upon the physical and biological configuration. Within these large-scale climatic cycles, shorter periods (10–30 years) of cold/wet weather interchanged with warm/dry periods are known as "Bruckner cycles" (Bruckner 1890).

Figures 2 and 3 illustrate the cold/warm oscillations with a frequency of some hundred-year time intervals utilising a dendrochronological curve. It is evident that the intensity of climatic variations is not constant.

The final result of this superimposition of different climatic fluctuations at different time scale is that, considering the general trend of the temperature increase

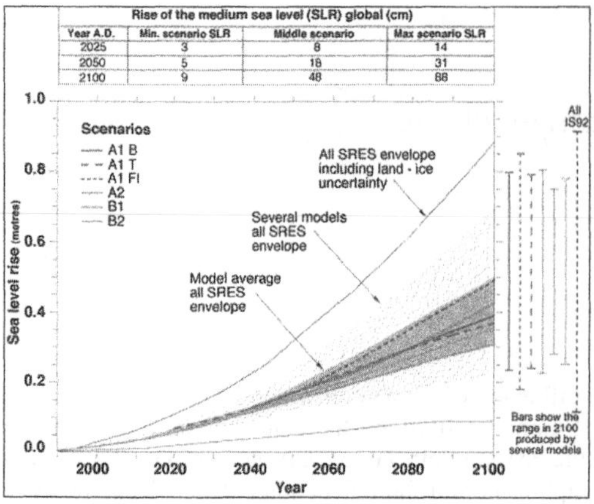

Figure 1. Range of the possible sea level rise up to 2100 with different scenarios (IPCC 2001).

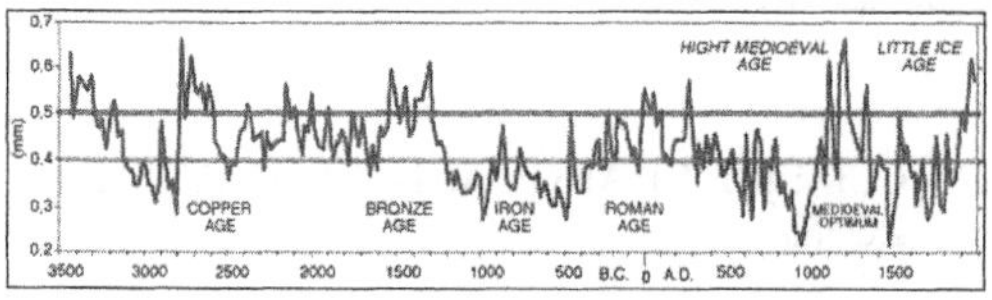

Figure 2. Dendrochronological curve (White Mountains California USA) from 3500 B.C. up to the present day showing: the cold/wet periods and the warm/dry periods corresponding to the oscillations of the extension of the climatic zone of the following Figure 3 (LAMB, 1982).

during the last 10,000 years, the most recent warm period over the past 100 years is warmer than the precedent period.

On the other hand, the most recent cold period is less cold if compared with the precedent cold periods. This produces different effects even during the repetition of the same type of cold or warm periods.

The same effect of gradual increasing of warm characters is evident even during the shorter time fluctuations (Bruckner cycles) of 10–35 year frequencies.

The distribution of the climatic fluctuations at different time scale is important not only to understand the modality of the climatic change, but also even to consider how the effect on the coastal zone evolution may be different.

The superimposition on the general trend of increasing temperatures since the end of the last glacial period on these climatic fluctuations creates some complications to the general regression of the shoreline due to increasing sea levels.

The first evident consequence of the influence of climatic changes in the coastal zone is shown by the deltas progradation during the cold/wet periods, when there is a large amount of sediment yield by floods due to the abundance of rainfall and increasing of landslides in the mountains behind. This event tends

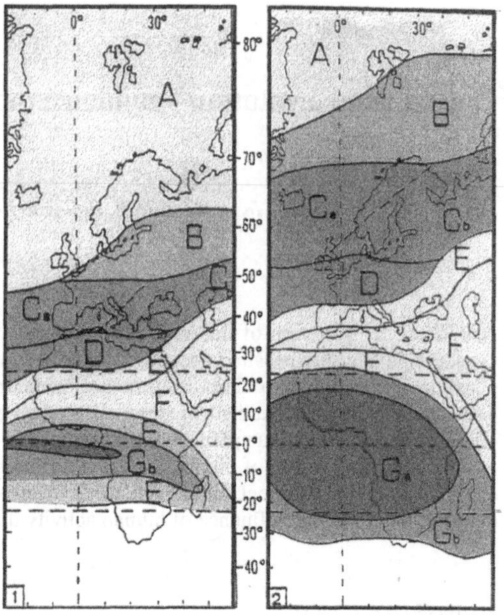

Figure 3. Scheme of the modifications of climatic strips in historical periods. Climatic strips according to the Koppen-Geiger's classification (modified): A = tundra and ice cap polar zone; B = cold-humid sub-artic continental zone; Ca = oceanic humid temperate zone; Cb = cold-semiarid steppe zone; D = Mediterranean temperate zone with dry summer; E = warm-semiarid steppe zone; F = warm-arid desert zone; Ga = humid tropical pluvial forest zone; Gb = savannah tropical zone with dry winter; 1 = climatic strips in the 6th–4th centuries B.C., 5th–8th and 16th–19th centuries A.D. cold-humid periods; 2 = climatic strips during the 2nd–4th and 11th–14th centuries A.D. warm-arid periods.

to mask the general regression of the shoreline due to the increase in temperatures and the consequent sea level rise.

It seems that the mechanism of the climatic changes in three different time intervals (n × 1000 years, n × 100 years, n × 10 years) is clear, even if not always uniform in their repetition.

The significance of the effects of climate change, considering the different time-scales (n × 1000 years, n × 100 years, n × 10 years), of the sequence cold/wet and warm/dry and the variability in duration and intensity of each cycle, in particular of the short time-scale changes, lead to the following conclusions:

- The influence of climate change on the evolution of the coastal zone is of global interest
- The climatic cycles (cold/wet – warm/dry) develop at different time-scales with different order of duration
- Long-term climate change: since the end of the last glacial maximum (10,000 years ago), we have

continued to experience increasing temperatures and a consequent increase in sea levels

- Climate change with a duration of some hundred years: after the "little ice age" (from 1600 A.D. up to 1820), we have today a warm period with an increasing of temperature and a scarce amount of snow and rainfall
- The interference and the superimposition of the climate cycles at different times-scales produces a general regression of the shoreline in the long-term (n × 1000 years) but with oscillation of advancing or regression of the shoreline during the climatic cycles with (n × 100 years) more modest duration
- The variability in the oscillation of advancing or regression is particular evident in the short time-scale (n × 10 years).

3 THE CASE OF THE ADRIATIC COASTAL ZONE

To provide an example of this situation the evolution of the Adriatic coastal zone from the Po River Delta up to Ancona promontory is considered. Figure 4 illustrates the evolution of the Po river delta showing how the maximum advancement (7 Km/century) occurred during the "little ice age" (1600–1820).

The conclusion is that the effect of climatic change on the evolution of the natural environment is of primary importance today as in the past. But it cannot simply be stated that cold/wet climate induces a greater precipitation with a greater sediment supply to the sea and therefore a shoreline advancement. In fact, these weather conditions imply an even greater frequency of strong storms that may cause a shoreline regression.

In a period of cold and wet weather the coastal equilibrium is controlled by the relationship between the large sediment supply by rivers and frequency of strong storms. If the sediment supply is greater than the removal of sediments caused by the wave action, the beach becomes wider. Conversely, if the sediment removal is bigger than the supply, the coastline moves backwards.

In a period of warm and dry weather, the sediment supply to the sea is less and the storms are rare. Therefore, the tendency is that the coastline moves around an equilibrium point or moves slightly seawards.

This suggests that cold and wet weather conditions are significant in coastal evolution. The main influence on the coastal zone from 1600 up until the last century was the fluvial sediment input connected with the "little ice age" climatic phase, which was greater than the output caused by the attack of the waves in storm conditions.

The smaller climatic variations, with a 10–35 year periodicity, have induced a progressive reduction in the coastline advancement, since the beginning of the last century, reaching its full development during the '50s,

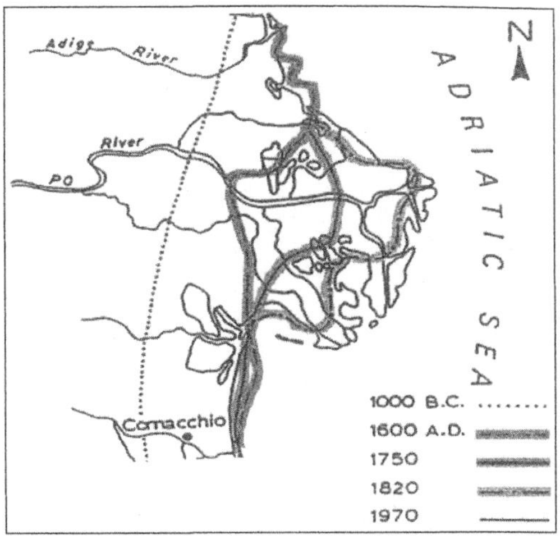

Figure 4. The evolution of the Po river delta (Nelson 1970).

when the sediment supply, typical of cold periods, became inadequate because of human activity on the coastal zone and on the areas lying behind.

In an environment with a lack of sediment supply from rivers, the wave action becomes the controlling parameter causing erosion.

South of the Po River delta there is sandy coastal zone, with the padan plain behind, up to the zone (100 Km south the Po river) where the Apennines Mountains meet the sea.

From here up to the Ancona promontory follows a cliff with a narrow beach with sand and gravel at its toe. These coastal units show, after the end of the Little Ice Age, a general regression of the shoreline due to the diminished sediment yield by the rivers and a prevailing of storm wave action.

Different results may be found if longer time-scales are considered. If shoreline variation due to sea level rise from Boreal-Atlantic up to the present time is considered, in relation to the sandy coast, with Padan plain behind (Fig. 5), only the Boreal-Atlantic shoreline is in open sea, the upper Atlantic and Roman coastline are behind the present shoreline to test the high advancing of the shoreline during periods such as the Little Ice Age (1600–1850 A.D.).

In the south, up to the Ancona promontory, the three ancient shorelines are external to the present one and the effect of sediment yield from the rivers during the Little Ice Age is not sufficient to mask the increasing of the Adriatic sea level. The three ancient shorelines are today at the depth −2 m, −6–8 m, −19–25 m (Fig. 6).

4 CONCLUSION

The influence of climatic changes on coastal zone evolution is of global interest.

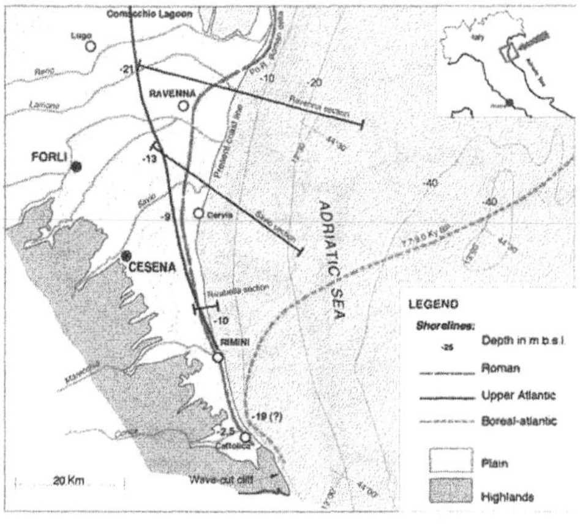

Figure 5. The variations of the Holocene shoreline in the Adriatic Sea, Northern area (ELMI ., 2003).

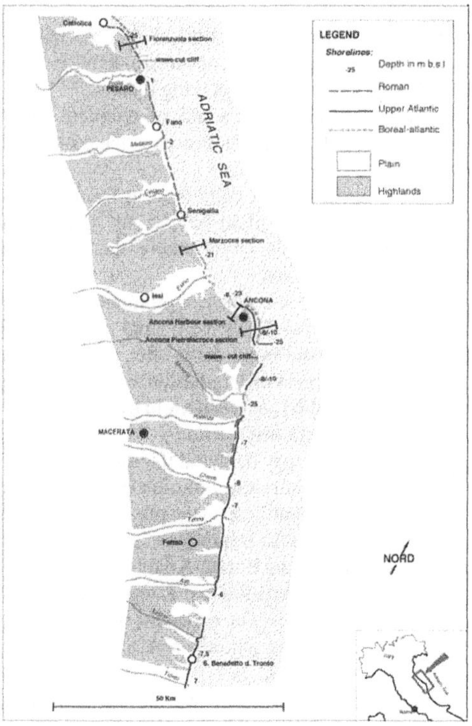

Figure 6. The variations of the Holocene shoreline in the Adriatic Sea, Southern area (ELMI 2003).

Climate cycles develop at different time-scales with different orders of duration (n × 1000 years, n × 100 years, n × 10 years).

The variability in the oscillations of advancing or regression of the shoreline is particular evident at the medium (n × 100 years) and short (n × 10 years) time-scales, whereas the long (n × 1000 years) time-scale is particular important for the final result of sea level rise. This is because it is connected with the continuation of general increase in temperatures since the end of the last glacial era, 10,000 years ago.

In both cases, the only natural and efficient defence is a large amount of sediment yield by rivers to contrast the violence of the storm waves with the following regression of the shoreline and to contrast the sea level rise due to the increase of temperature for the past 10,000 years.

Unfortunately today this is a difficult problem to face as human activity in the coastal zone, with the destruction of the natural coastal environment and insufficient sediment yield to balance the double attack (storm waves action and sea level rise) to the coastal zone.

REFERENCES

Bartholin T.S., (1984) – *Dendrochronology in Sweden.* Instead: Morner N.A. & Karlen W. (Eds.) – Climatic changes on a yearly to millennial basis. D. Reidel P.C., Dordrecht, 261–262.

Bruckner E., (1890) – Klimaschwankungen seit 1700 nebest Bemerkungen über die Klimaschwankungen der Diluvialzeit. Geograph. Abband., IV(2), Wien, 153–184.

Camuffo D., (1990) – Clima e uomo. Garzanti, Milano, 207 pp.

Elmi C., (2003) – La risposta dei processi geomorfologici alle variazioni ambientali nella pianura padana e sulle coste nord e centro adriatiche, Ministero Istruzione, Università e Ricerca. Ed. Glauco Brigati, Genova 2003, 225–259.

Lamb H.H., (1982) – Climate, history and the modem word. Methuen, London.

Marabini F. & Veggiani A., (1990) – L'influenza delle fluttuazioni climatiche sull'evoluzione del delta del Po dal secolo XVI ad oggi. Atti, Conv. sull'Ecologia del delta del Po, Albarella 16–18 settembre 1990, 1–15.

Marabini F. & Veggiani A., (1991) – Evolutional trend of the coastal zone and influence of the climatic fluctuations. Atti, C.O.S.U. II, Long Beach, U.S.A., 2–4 aprile 1991, 459–474.

Marabini F., (1997) – The Po river delta evolution, Geo-Ecomarina 2/1997, Bucaresti 1997, 47–55.

Nelson B.W., (1972) – Mineralogical differentiation of sediments dispersed from the Po delta. Instead: Stanley G.J. (Eds.) – The Mediterranean Sea, a Natural Sedimentation Laboratory, 441–453.

Ortolani F. & Pagliuca S., (1994) – Variazioni climatiche e crisi dell'ambiente antropizzato. Il Quaternario 7(1), 531–536.

Schweingruber F.H., Bartholin T., Schar E. & Briffa K.R., (1988) – Radiodensitometric-dendroclimatological conifer chronologies from Lapland (Scandinavia) and the A/ps (Switzerland). Boreas, 18, 559–566.

Tardy Y., (1986) – Le cycle de l'eau. Climats, paleoclimats et geochimie globale. Masson, Paris, 169–209.

Landslides and Climate Changes – McInnes, Jakeways, Fairbank & Mathie (eds)
© 2007 Taylor & Francis Group, London, ISBN 978-0-415-44318-0

The relationship between rainfall and mass movements on plio-pleistocene sediments in south-western Umbria (central Italy)

D. Valigi & L. Melelli
University of Perugia, Perugia, Italy

L. Faralli, N. Gasparri, R. Piccioni & L.D. Venanti
SGA (Studio Geologi Associati), Perugia, Italy

ABSTRACT: Most of the important historical towns in Umbria are built on Plio-Pleistocene sediments. This lithological complex is affected by landslide hazard (earth slides and earth slumps) as a consequence of its geological structure, in particular where clays percentage is significant. Rainfalls prolonged over time are studied to detect the minimum amounts necessary for generating critical conditions in the subsoil.

1 INTRODUCTION

It is well known that landslides could be clustered in time, following a major hydrometeorological triggering event (severe storm). In this context, a simple view of slope failure triggered by prolonged rainfall is that landslides are probably triggered owing to groundwater seepage/pressure effects. Thus the frequency distributions of the rainfall threshold in landsliding areas need to be exploited to provide a hazards assessment tool in order to foresee and prevent the landslide risk.

These insights into landslides within sediments with medium-low permeability constitute a working hypothesis in which long time series of cumulated rains are evaluated (Galliani et al. 2001, Santaloia et al. 2001, Polemio & Sdao 1999, Pasuto & Silvano 1998, Wasowski 1998, Parise et al. 1997, D'Ecclesis et al. 1991, Capecchi et al. 1988, Cascini & Versace 1988, 1986, Canuti et al. 1985, Govi et al. 1985). To define the "threshold value" of landslide events there is a clear need for a better understanding of the spatial distribution in deposits characterized by medium-high permeability of the shallow landslides induced by extreme and small rain events (Belloni & Martini 1997, Mortara et al. 1994, Cannon & Ellen 1988, Wieczorek & Sarmento 1988, Keefer et al. 1987, Wieczorek 1987, Cancelli & Nova 1985, Moser & Hohensin 1983, Govi & Soriana 1980, Caine, 1980).

The proposed study addresses these goals. In particular, the method of Govi et al. (1985) was used on times series of 60, 90, 120, 180 days prior to landslide events which occurred in the Massa Martana, Todi and Collazzone areas (Umbria, central Italy). These areas belong to the post-orogenic sediments complex of the Tiberino basin (Umbria), which is characterized by marine and lake deposits made up of clay, silty clay, fine and coarse gravel and cobbles, which vary in age from Pliocene to Pleistocene. The Gumbel statistical method was also used, though only for the events occurring in Massa Martana, for the largest daily cumulated rains from one to 180 days preceding the landslide. The cumulated rains probability curves were then evaluated within the associated return times. Results indicate that comparing the curves of cumulate rains of triggered landslide events and evaluating the return times of the cumulate rains shows the rain threshold value for each individual landslide event.

2 GEOLOGIC AND GEOMORPHOLOGIC SETTING

The study area (Fig. 1) is located in the central–southern part of the region of Umbria (central Italy). This area, which forms the middle part of a much vaster intermontane depression that extends northward to the borders of Tuscany and the Marche, is bordered in the east by the western side of Martani ridge, and in the west it reaches as far as the northernmost end of Narnese–Amerina ridge. Its morphologic shape is the result of different tectonic phases. A first compressive period in the Miocene gave origin to the prevalently calcareous mountain chains where formations of the Umbro-Marchigiana Sequence outcrop (Deiana & Pialli 1994). The mountains ridges are anticlines and

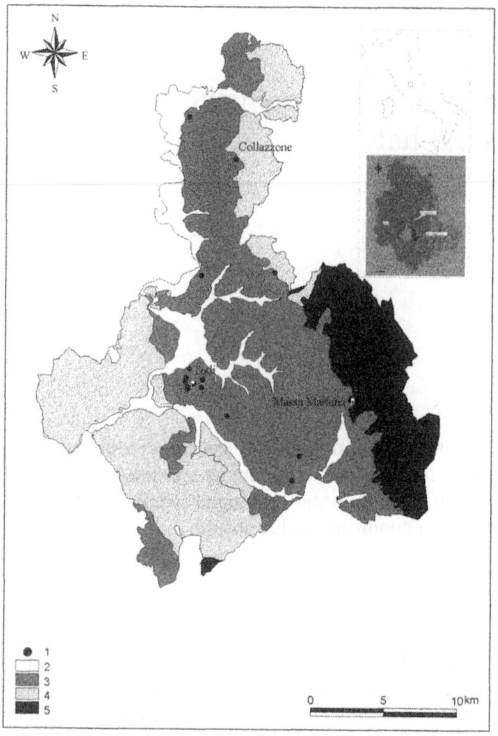

Figure 1. Location map of the study area. 1) Landslide events, 2) Alluvial deposits (Oligocene), 3) Fluvial lacustrine deposits (Piocene-Pleistocene), 3) Flysch bedrock (Miocene-Oligocene), 4) Calcareous bedrock (Lias – Oligocene).

narrow synclines with a NW-SE direction. Several intermontane basins (graben or semigraben) owe their existence to the second extensional tectonic phase that affected the area during the Pliocene-Pleistocene with a general uplift (Calamita et al. 1999). These basins are filled with Plio-Pleistocene fluvial lacustrine deposits and are bounded by normal fault systems (Brozzetti & Lavecchia 1995).

The sequence of the Plio-Quaternary area is made up of four lithostratigraphic units, as shown in the interpretation of Basilici (1993):

- "Fosso Bianco Unit – UFB" (Lower Pliocene): grey clay and sand deposits of lacustrine environment.
- "Ponte Naja Unit – UPN" (Upper Pliocene): in an alluvial fan environment with a thickness of 140 meters.
- "Santa Maria di Ciciliano Unit – USMC" (Lower Pleistocene): in an alluvial environment with lenses of sand and more rarely pebbles (channel deposits) or clay and silt deposits (flood plain environment). The total thickness is around 180 m.
- "Acquasparta Unit – UA" (Lower Pleistocene): mostly calcareous, with a thickness not over 5 meters. The calcareous deposits, which have a silt grain size, are not cemented and sometimes plain parallel laminated.

The USMC and UA units outcrop in the Massa Martana area; in the Todi area the USMC, UPN and UFB units appear; and in the Collazzone area there is the USMC unit. In all the study areas the Plio-Pleistocene units are overlapped by more recent deposits. A series of geo-technical boreholes were drilled in the Todi and Massa Martana areas, allowing the reconstruction of the units and their distribution in different lithofacies.

The vertical and horizontal heterogeneity allows a first order permeability due to the porosity related to the percentage of the smallest grain size. Even if the permeability is low on average, it is still high enough to allow low hydrological circulation in sediments where sands prevail. Moreover, the units with a high degree of cementation (with a travertine matrix or overconsolidated silt and clay sediments) have a secondary permeability due to the open breaks as a consequence of recent tectonic phases. This allows a vertical exchange of groundwater, though quantitatively modest, between the superimposed levels. Several piezometric pipelines, located mainly near the historic centers of Todi and Massa Martana, show the presence of groundwater at different depths below the topographic surface. These groundwaters are currently being drained in order to improve the safety and stability of the buildings in the towns.

3 RAINFALL THRESHOLD EVALUATION

In order to understand the function of rainfall in the triggering mechanism of landslide events occurring in the Tiberino basin Units, the Govi et al. (1985) method was applied. Thirty-nine landslides which took place from 1928 to 2005 in the Todi, Massa Martana and Collazzone areas are taken into account. To apply this method the cumulated rains of 60, 90, 120, 180 days prior to landslides events are collected for each individual landslide event investigated from 1928 to 2000 (Guzzetti et al. 1994) updated to 2005 (Table 1).

Landslide events are widespread on western slopes with a medium-low gradient (less than 32°) with the exception of falls that involve the rock face where the town of Massa Martana is built.

The interstitial pressure, as a consequence of rainfall, is very low because of the medium-low degree of permeability of the sediments involved. This is proved by piezometric data recorded for the groundwater inside the USMC (SGA, 2006). The groundwater level changes and it is always greater than three months, in response to the rainfall regime, the moisture content,

Table 1. List of landslides studied on the Plio-Pleistocene deosits area.

Municipality	Place	Date	Landslide type	Geographic coordinates	
				East	North
Massa Martana	Massa Martana	04/06/1928	Fall	12°31′24,9905″	42°46′35,3257″
Massa Martana	Massa Martana	04/24/1960	Complex	12°31′28,4754″	42°46′35,1094″
Massa Martana	Massa Martana	05/04/1960	Complex	12°31′28,4754″	42°46′35,1094″
Massa Martana	Massa Martana	01/07/1961	Complex	12°31′28,4754″	42°46′35,1094″
Massa Martana	Massa Martana	01/25/1963	Complex	12°31′28,4754″	42°46′35,1094″
Massa Martana	Massa Martana	12/28/1996	Fall	–	–
Massa Martana	Massa Martana	07/1997	Slide	–	–
Massa Martana	Massa Martana	02/04/1999	Fall	–	–
Massa Martana	Massa Martana	07/1999	Slide	–	–
Todi	Rosceto	03/13/1941	Slide	12°28′55,6423″	42°44′03,3738″
Todi	Todi	03/13/1941	Slide	12°24′08,5969″	42°46′52,6219″
Todi	Todi	02/15/1959	Slide	12°24′31,8065″	42°47′03,7805″
Todi	Collevalenza	01/06/1961	Complex	12°29′13,4831″	42°44′52,6497″
Todi	Todi	01/13/1961	Complex	12°24′25,9737″	42°46′54,8637″
Todi	Todi	01/17/1961	Complex	12°24′05,7369″	42°46′48,4077″
Todi	Collevalenza	01/17/1961	Complex	12°29′10,4671″	42°44′48,9204″
Todi	Todi	11/04/1977	Complex	12°24′25,9737″	42°46′54,8637″
Todi	Todi	02/24/1979	Complex	12°24′25,9737″	42°46′54,8637″
Todi	Todi	03/06/1979	Complex	12°24′26,2183″	42°46′42,6808″
Todi	Todi	03/20/1979	Complex	12°24′25,9737″	42°46′54,8637″
Todi	Todi	02/05/1981	Complex	12°24′25,9737″	42°46′54,8637″
Todi	Todi	11/23/1982	Slide	12°25′15,0956″	42°46′50,3657″
Todi	C. Cascinelli	12/28/1982	Slide	12°25′27,7984″	42°46′11,4608″
Todi	Todi	06/23/1984	Complex	12°24′31,7817″	42°47′09,6472″
Todi	Todi	05/01/1985	Slide	12°24′26,6512″	42°47′23,9561″
Todi	Ilci	02/01/1986	Complex	12°24′48,7203″	42°50′27,2159″
Todi	Ilci	02/25/1986	Complex	–	–
Todi	Ilci	03/09/1986	Complex	–	–
Todi	Todi	12/13/1991	Slide	12°24′05,7369″	42°46′48,4077″
Todi	Todi	12/08/1992	Slide	12°24′08,5969″	42°46′52,6219″
Todi	Todi	09/13/1995	Slide	12°24′32,2871″	42°47′31,2801″
Todi	Todi	09/13/1995	Slide	12°24′56,3351″	42°47′36,1676″
Todi	Roseto	01/07/1997	Slide	–	–
Todi	Todi	06/02/1997	Slide	–	–
Todi	Pantalla	10/07/1998	Slide	–	–
Todi	Ripaioli	10/07/1998	Slide	–	–
Todi	Todi (P.ta)	12/20/2004	Slide	–	–
Todi	Todi (Oberdan)	03/15/2005	Slide	–	–
Todi	Todi (Rocca)	04/11/2005	Slide	–	–
Todi	Todi (Europa)	05/08/2005	Slide	–	–
Collazzone	Collepepe	10/30/1928	Slide	12°24′01,5846″	42°55′18,9893″

the vegetation type and density. The rainfall data, used by Govi et al. (1985) are collected in the weather stations closest to the landslide areas (Table 2).

Data cumulated for 60, 90, 120 and 180 days prior to each landslide event were divided by the mean annual rains in order to be normalized. These cumulated rain data were used to draw four envelope curves of the minimum threshold values, one for each normalized cumulated daily rains. Looking at the four curves, it is possible to observes that the minimum threshold values for rainfall prior to the landslide vary depending on the month from which they are counted, moving backward. The trends of the envelope curves for different cumulated rains are substantially similar, especially in the periods of the year when there are the greatest numbers of landslides. There is a constant raising of the threshold from November until December-January. From January to May, in following with the raising of groundwater levels, there is a decrease in the threshold values; these tend to rise from May to September, when the groundwater is in an impoverishment stage. The rare landslide events occurring in the area in this period are of a superficial type and are mostly associated with intense summer

Table 2. Pluviometric stations taking into account for the Govi et alii, 1985 method application.

Pluviometric station	Altitude a.s.l. (m)	Temporal interval	Medium annual rain (mm/y)	Department
Massa Martana	356	1919–1978	1074	Rome Hydrographic department
Massa Martana (2)	340	09/2004–10/2005	–	Umbria Region office
Todi	411	1919–2000	873	Rome Hydrographic department
Todi (2)	316	01/1991–06/2005	800	Umbria Region office
Fratta Todina	214	1919–1978	933	Rome Hydrographic department
Bastardo	330	01/1992–06/2005	801	Umbria Region office

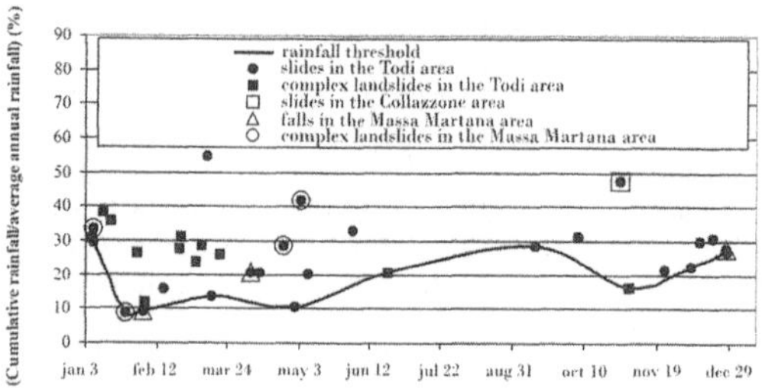

Figure 2. Threshold values of rainfall cumulated for 60 days before the landslide events in Collazione, Todi and Massa Martana area.

rainstorms. It should be kept in mind that the trend of rainfall thresholds in the summer has a lesser statistical significance. Figure 2 shows the trend of rainfall thresholds for normalized cumulated rains 60 days before the landslides.

4 LANDSLIDE EVENTS IN THE TOWN OF MASSA MARTANA

In order to better understand the influence that rainfall events may have had on the landslides, the Gumbel statistical-probabilistic method was applied to the main gravitational movements occurring in the town of Massa Martana from 1928 until the present (Melelli et al. 2001). In applying the method the annual maximum daily cumulated rains for 30 years at the Massa Martana station were used, and the rainfall probability curves for different return times were obtained with the Gumbel method; these latter data were compared to the rain curves for 1, 5, 10, 20, 30, 60, 90, 120, and 180 days recorded prior to the ten landslides occurring in Massa Martana between 1928 and July 1999

Table 3. Date referred to each studied landslide event.

Complex landslides	Fall landslide	Slide landslide
24/04/1960	06/04/1928	07/1997
04/05/1960	29/03/1981	07/1999
07/01/1961	28/12/1996	–
25/01/1963	04/02/1999	–

(Santaloia et al. 2001, Polemio & Sdao 1999, Parise et al. 1997, D'Ecclesiis et al. 1991). Table 3 shows the landslide events taking place in the town of Massa Martana since 1928, which are divided into falls and topple in the upper portion that involves the rock face in the Acquasparta Unit (UA) and slides in the lower portion where there is the Santa Maria di Ciciliano Unit (USMC).

Complex landslides also occurred at the same time as collapses in the UA and rototranslational slides in the USMC. The cumulated rains in the 180 preceding days for Massa Martana make it possible to deduce that:

- for the activation of the four falls approximately 600–700 mm of rainfall were needed, i.e. about 55–67% of the mean annual rainfall (Fig. 3).
- for the activation of the four complex landslides the critical precipitation levels in the 180 days prior to the event are between 840 and 1191 mm, approximately 78–110% of the mean annual rainfall (Fig. 4);

4.1 *Fall type landslides*

The most significant rock fall that involved the travertine rock face upon which stands the town of Massa Martana, took place on the west side of the plate in 1928, the north face in 1981, and the south face in

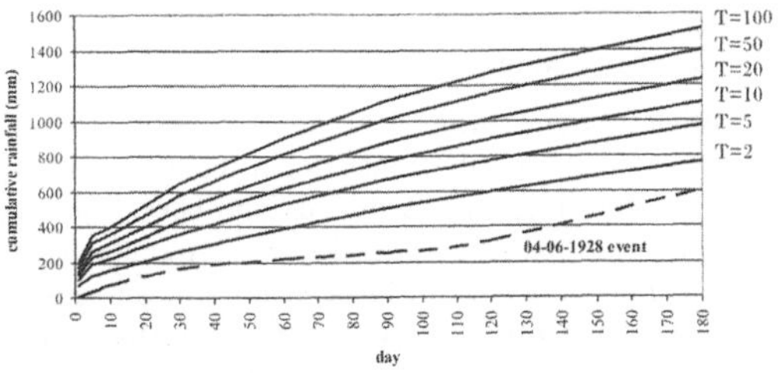

Figure 3. Comparison between rainfall probability curves for peculiar return times and cumulate rains before to falls on April 1928.

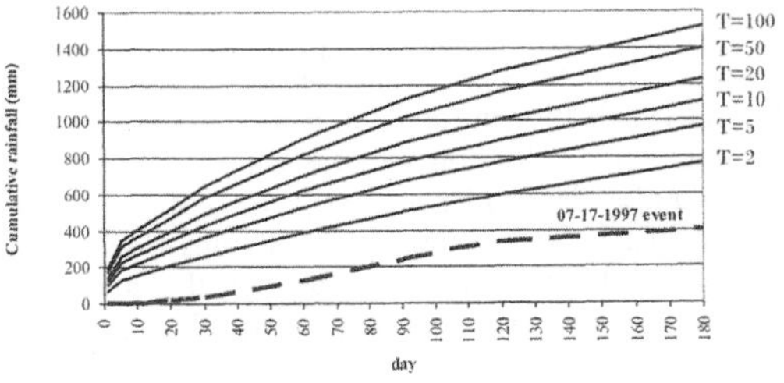

Figure 4. Comparison between rainfall probability curves for peculiar return times and cumulate rains before to slide on July 1997.

1996 and 1999. The falls are in relation to cumulated rains with return times of less than 2 years. Given the type of landslide, rainfall played a supporting role, much less important than other predominant factors (degree and arrangement of fracture systems, rock face morphology, bioturbation of the root systems, physical alteration of the deposits).

4.2 *Slide type landslides*

The slide-type landslides in the base unit (USMC) occurred in July 1997 and 1999 in the area at the foot of the rock face, which is used as farmland. Here also the rains falling prior to the landslides had return times of less than 2 years (Fig. 5).

Along with rain, the underwashing at the base of the slope done by the surface water of the Fosso di Castelrinaldi, the formation of runnels concentrated along the western slope and the agricultural use of the soil after plowing may have influenced the stability of the slopes.

4.3 *Complex landslides*

The complex landslides are the most interesting, as they are the ones that caused the greatest damage. They involve simultaneous collapses in the upper travertine

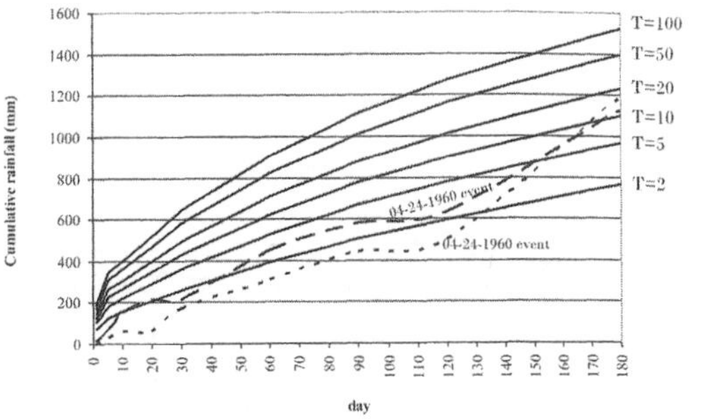

Figure 5. Cumulate rains for complex landslides between 24th of April 1960 and 4th May 1960.

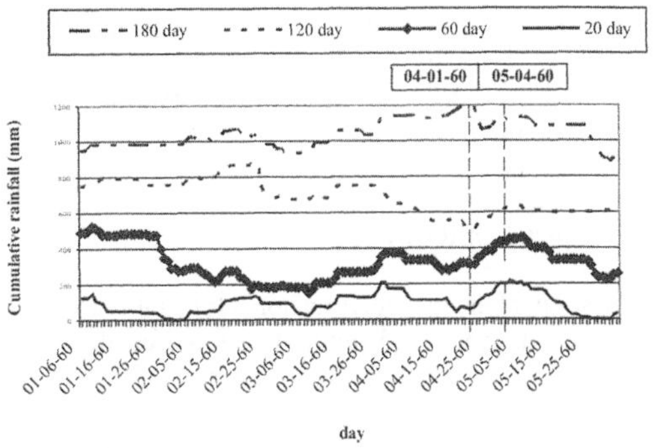

Figure 6. Comparison between rainfall probability curves for peculiar return times and cumulate rains before to 24th of April 1960 and 4th May 1960.

unit (UA) and rototranslational slides in the lower unit (USMC) and they occurred and were reactivated four times in the space of just a few years. In particular, the complex landslide of 24 April 1960 and its reactivation on 4 May 1960, which occurred in the western area of the rock and then propagated to the underlying base unit were among the most serious in the area (Melelli et al. 2001). As can be seen in Figure 5, only the cumulated rains of 180 days had an increasing trend, with an absolute maximum of 1191 mm precisely on the day of the gravitational event and return times of about 20 years (Fig. 6).

In this case the rain that contributed most to the landslide seems to be that which fell in the months of October and December 1959 (approximately 700 mm). Playing an important role in the reactivation of the landslide on May 4th were the heavy rains (150 mm) which continued to fall in the 10 days prior to the first activation and the deterioration of the physical-mechanical characteristics of the soil.

The complex landslides on 7 January 1961 and 25 January 1963 were caused by rains with return times of not more than 10–15 years. In the first case the heaviest

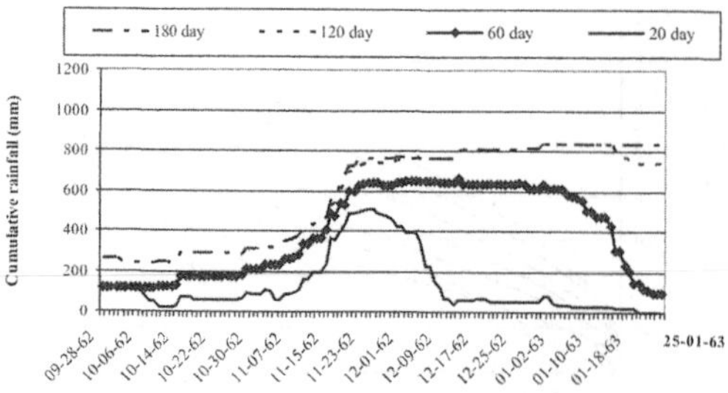

Figure 7. Cumulate rains during the complex landslide occurred on the 25th of January 1963.

rain (about 350 mm) fell in September 1960, whereas for the January 1963 landslide the rain in November (550 mm) was crucial. As regards the landslide of 7 January 1961, the cumulated rains of 120–180 days had an increasing trend in the period prior to the day of activation of the event.

5 CONCLUSIONS

The study of the rainfall thresholds for triggering landslides using the 1985 method of Govi et al. (1985) applied to the units of the Tiber River basin in the municipal areas of Massa Martana, Todi and Collazzone made it possible to obtain the critical rainfall values above which, in the various seasons, there is a greater probability of triggering landslide events: the minimum values are seen in May and November and the maximum values in the summer. Values much greater than the threshold, especially for 180 days, were observed for the four complex landslides which occurred in the town of Massa Martana in the 1960s, especially in the 180 days prior to the landslide. This showed that for Massa Martana the rainfall accumulated in the 180 days prior to the landslide was particularly significant.

The Gumbel hydrological-statistical analysis of the landslides showed that:

- the four collapse-type landslides which occurred in Massa Martana were caused by unexceptional rainfall events with return times of less than 2 years. In these cases rainfall probably did not play a major role, but was only one of many factors that contribute to the instability, such as:
- the morphology of the travertine face, because it has sides with slopes of up to 90°;
- the bioturbation of the root systems of the plant cover near the summit and along the rock faces, as these work their way into the fractures and cavities, facilitating the dislocation of the blocks of stone;
- cryoclastism since the freezing-thawing action causes the cracks to dilate, contributing to the instability of the blocks and cause rapid infiltration of surface water.

What was said earlier about fall-type landslides can be repeated also for the translational slides which took place in the Santa Maria di Ciciliano Unit. The rain that fell prior to these landslides have a trend and values within the norm. In these cases as well rainfall does not have a primary role in the activation of this type landslide and, as mentioned regarding simple collapses, it must be considered one of many factors that bring about this type of landslide, such as:

- the underwashing at the base of the slope done by the surface water (Fosso di Castelrinaldi) and the formation of runnels concentrated along the western slope;
- the agricultural use of the soil on the slopes of the rock, with plowing and the clearing of trees, as they facilitate infiltration and erosion;
- complex landslides in the Massa Martana area are reactivated following heavy rains over a long period. For this type of landslides, the cumulated rains of 180 days prior to the landslide is generally important.

In particular, in the analysis of the cumulated rains 180 days prior to the ten landslides occurring in Massa Martana it was observed that for the activation of the four collapse-type landslides approximately 600–700 mm of rain were needed, i.e. 55–67% of the mean annual rainfall; for the activation of the two slide landslides lesser amounts of rain were needed (approx. 400 mm), and prior to the four complex landslides between 840 and 1191 mm of rain fall, about 78–110% of the mean annual rainfall.

REFERENCES

Basilici, G. 1993. Il Bacino continentale Tiberino (Plio-Pleistocene, Umbria): analisi sedimentologica e stratigrafica. Tesi di Dottorato, V° ciclo, Università degli Studi di Bologna.

Belloni, S., Martini, C. 1997. Clima e dissesti idrogeologici. Acqua & Aria 9: 95–109.

Brozzetti, F. & La vecchia, G. 1995. Evoluzione del campo degli sforzi e storia deformativa nell'area dei Monti Martani (Umbria). Bollettino della Società Geologica Italiana 114: 155–176.

Calamita, F., Coltorti, M., Pieruccini, P., Pizzi, A. 1999. Evoluzione strutturale e morfogenesi plio-quaternaria dell'appennino umbro-marchigiano tra il Preappennino umbro e la costa adriatica. Bollettino della Società Geologica Italiana 118: 125–139.

Caine, N. 1980. The rainfall intensity-duration control of shallow landslides and debris flows. *Geografiska Annaler* 62A: 23–27.

Cancelli, A. & Nova, R. 1985. Landslides in soil and debris cover triggered by rainfall in Valtellina (Central

Alps-Italy). *Proc. IV Int. Conf. & Field Workshop on Landslides, Tokyo* 267–272.
Cannon, S.H. & Ellen, S.D. 1988. Rainfall that resulted in abundant debris flow activity during the storm. In: *"Landslides, floods and marine effects of the storm of Jan 3–5, 1982 in the San Francisco Bay region, California"*, USGS Prof. Papers. 1434: 27–35.
Canuti, P., Focardi, P., Garzonio, C.A. 1985. Correlation between rainfall and landslides. *Proc. 27th I.G.C., Bull. IAEG., Moscow* 32: 49–54.
Capecchi, F., Focardi, P., Garzonio, C.A. 1988. Rainfall and landslides: research into a critical precipitation coefficient in an area of Italy. *Proc. 5th Symp. On Landslides, Lausanne* 1131–1136.
Cascini, L. & Versace, P. 1986. Eventi pluviometrici e movimenti franosi. *Atti XVI Conv. Naz. di Geotecnica* 3: 171–184.
Cascini, L. & Versace, P. 1988. Relantionship between rainfall and landslide in a gneiss cover. *Proc. 5th Int. Symp. On Landslides, Lausanne* 1: 565–570.
D'Ecclesis, G., Grassi, D., Merenda, L., Polemico, M., Sdao, F. 1991. Evoluzione geomorfologia di un'area suburbana di Castronuovo S. Andrea (PZ) ed incidenza delle piogge su alcuni movimenti di massa. Geologia Applicata e Idrogeologia 26: 141–163.
Deiana, G. & Pialli, G. 1994. The structural provinces of the Umbro-Marchean Apennines. Memorie della Società Geologica Italiana 48: 473–484.
Keefer, D.K., Wilson, R.C., Mark, R.K., Brabb, E.E., Brown, W.M., Ellen, S.D., Harp, E.L., Wieczoreck, G.F., Alger, C.S., Zatkin, R.S. 1987. Real-time lanslide warning during heavy rainfall. Science 238: 921–926.
Galliani, G., Pomi, L., Linoni, F., Casagli, N. 2001. Analisi meteoclimatologica e soglie pluviometriche di innesco delle frane nella regione Emilia-Romagna negli anni 1994–1996. Quaderni di Geologia Applicata vol. 1. Le frane della Regione Emilia-Romagna, interventi di protezione civile nel periodo 1994–1999 2131: 74–91.
Govi, M. & Soriana, P.F. 1980. Landslide sudceptibility as a function of critical rainfall amount in Piedmont basins (NW Italy). Studia Geomorphologica Carpatho-Balcanica 14: 43–61.
Govi, M., Mortara, G., Soriana, P.F. 1985. Eventi idrogeologici e frane. Geologia Applicata e Idrogeologia 20(2): 359–375.
Guzzetti, F., Cardinali, M., Reichenbach, P. 1994. The AVI project: A bibliographical and archive inventory of landslide and floods in Italy, Environmental Management 18(4): 623–633.
Melelli, L., Faralli, L., Gasparri, N., Piccioni, R., Venanti, L.D. 2001. Nota preliminare sui fenomeni gravitativi della rupe di Massa Martana. Memorie della Società Geologica Italiana 56: 123–130.
Mortara, G., Cerini, M., Laffi, R., Lattini, C., Beretta, E. 1994. L'evento alluvionale del 22 luglio 1992 nella conca di Bormio in Alta Valtellina. C.N.R. – Quaderno 17 – Suppl. GEAM: 25–36.
Moser, M. & Hohensin, F. 1983. Geotechnical aspects of soil slips in Alpine regions. Engineering Geology 19: 185–211.
Parise, M., Polemio, M., Wasowsky, J. 1997. Rainfall and landslides in the Upper Valleys of Sele and Ofanto Rivers. Int. Symp. Engineering Geology and the Environment, IAEG-IAH, Alba: 955–960.
Pasuto, A. & Silvano, S. 1998. Rainfall as a trigger of shallow mass movements. A casa study in the Dolomites, Italy. Environmental Geology, 35: 184–189.
Polemio, M. & Sdao, F. 1999. The role of rainfall in the landslide hazard: the case of the Avigliano urban area (Southern Appennines, Italy). Engineering Geology 53: 297–309.
Santaloia, F., Cotecchia, F. & Polemio, M. 2001. Mechanics of a tectonized soil slope: influence of boundary conditions and rainfall. Quarterly Journal of Engineering Geology and Hydrogeology, 34: 165–185.
Wasowski, J. 1998. Understanding rainfall-landslide relationships in man-modified environments: a case-history from Caramanico Terme, Italy. Environmental Geology, 35: 197–209.
Wieczorek, G.F. 1987. Effect of rainfall intensity and duration on debris flows in central Santa Cruz Mountains, California. *Geol. Soc. of America, Rev. in Engineering Geology* 7: 93–104.
Wieczorek, G.F. & Sarmiento, J. 1988. Rainfall, piezometric levels and debris flows near La Honda, California, in storms between 1975 and 1983. *In: "Landslides, floods and marine effects of the storm of Jan 3–5, 1982 in the San Francisco Bay region, California", USGS Prof. Papers.*, 1434: 43–63.

Session 2
Facing the challenge – landslides (and other natural hazards) and climate change

In both coastal and inland locations, the occurrence of landslide events can be linked to meteorological conditions. Changes in our climate will have an impact on landslide hazards and vulnerable communities will face increasing risks in future years. It is essential to investigate, asses and evaluate the relationship between landslides (and other natural hazards) and climate change in order to find effective management solutions and improve preparedness for the future.

Part of the Ventnor Undercliff, Isle of Wight, UK Courtesy Wight Light Gallery, Ventnor, Isle of Wight, UK.

Landslides and Climate Changes – McInnes, Jakeways, Fairbank & Mathie (eds)

Radiocarbon dating on landslides in the Northern Apennines (Italy)

G. Bertolini

Emilia-Romagna Regional Authority, Basin Technical Survey, Italy

ABSTRACT: The regional Geological Survey has identified over 70,000 landslide bodies, covering one fifth of the entire mountain area of Emilia-Romagna Region. From a morphological point of view, a good 90% of these are large, ancient earth flows, nourished by shaly and structurally complex formations. The evolutionary stages of past landslide activity are here outlined more precisely thanks to the ^{14}C dating techniques performed on 40 wood remnants collected by means of core borings at different depth inside the landslide bodies. As a paleo-climatic indicator, this data suggests the existence of a wetter period (Sub-boreal, early Sub-Atlantic) within the Holocene in the Northern Apennines.

1 INTRODUCTION

The northern flank of the Apennines, bordering the southern Po Plain, corresponds to the portion of the chain under the administration of Emilia-Romagna regional authority. This 12,000 sq. km. area of land can be considered one of the most landslide-prone areas of Italy.

During the last few decades, this territory has been extensively studied by the regional administration in an effort to reduce major damage, caused by the recurrent reactivation of many landslide bodies.

Studies and investigations have been performed by the regional geological survey (esp. geo-thematic cartography) and by the basin technical services (esp. direct investigations and monitoring).

A detailed knowledge of the landslide distribution throughout the territory is now available thanks to the regional "Landslide Inventory Map" ("*Carta Inventario del Dissesto*", Pizziolo, 1996) and the "Landslide Susceptivity Map" ("*Carta della Pericolosità Relativa di Frana*", Bertolini et al., 2001). Moreover, the use of continuous core boring and topographic and inclinometric monitoring in current practice provides the regional services with a valuable amount of data regarding the dimensions and kinematics of gravitational slope processes.

The evolutionary stages of past landslide activity are outlined more precisely thanks to the ^{14}C dating techniques performed on wood remnants that were buried inside the landslides during the descent and superimposition of minor landslides, usually earth flows, that form the present landslide bodies (Table 1). In many cases, samples of wood of different age were collected by means of core borings at different depth inside the same landslide body, making it possible to outline its evolution over time.

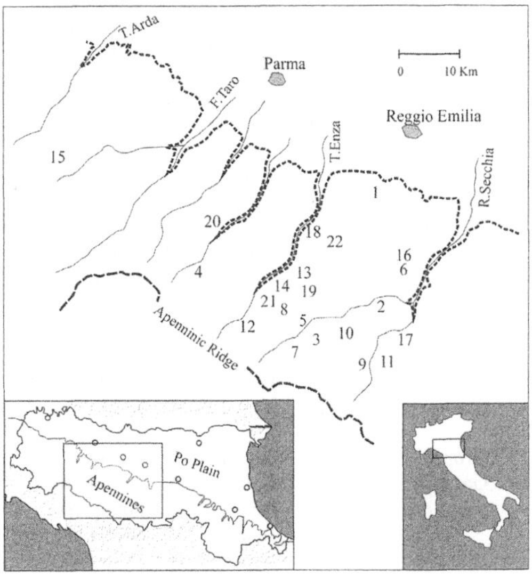

Figure 1. Location map of studied landslides (see Table 1).

2 THE SITUATION

In the Emilia-Romagna Apennines the Regional Geological Survey has identified over 70,000 landslide bodies, covering one fifth of the entire mountain area.

From a morphological point of view, a good 90% of these are large, ancient earth flows.

Ancient dormant earth flows have been areas wrongly judged as suitable for human settlement since ancient times, thanks to the gentle slope of their frontal and mid-accumulation zones, a real trap for many hamlets and villages (e.g. the Corniglio, Morsiano, Talada, Cavola cases, Figures 3, 4, 5 and 6). As a result, 281 inhabited centres (defined as four or more buildings, excluding "scattered houses") lie upon or

Table 1. Ages of wood fragments found inside landslide bodies. A few tree trunks were found complete, as in Sologno, Morsiano, Romanoro, Cinquecerri and Lucola cases. Usually, they were brought to light by fluvial erosion (Figure 9). "Depth" is indicated when the samples were collected by means of continuous core borings. For location see Figure 1.

	Landslide	Depth (m)	Sample (Lab. Code)	Conv. age (yr.BP)	13C/12C ‰	2 sigma(*) Cal. yr BP(**)	Lat. N (geogr.)	Long. E (geogr.)
1	Casoletta	−16	Beta 152644	580±40	−12.1	650–530		
		−6	Beta 123272	100.4±0.6%		modern	44°35′00″	10°32′00″
2	Cavola	−9	Beta 135395	2970±40	−27.8	3255–2990		
		−24	Beta 135394	3250±40	−26.8	3895–3690		
		−32	Beta 137039	3720±40	−28.2	4160–3690	44°24′49″	10°32′49″
		−37	Beta 137040	3660±40	−30	4070–4035/4000–3700		
		−45	Beta 135396	3600±70	−24.1	4090—3700		
3	Sologno	−8	Beta 125331	2350±60	−25	2700–2645/2490–2310 2230–2190		
		−8	Beta 123274	2310±50	−28.6	2365–2300/2250–2170		
		−11	Beta 152652	3900±40	−27.7	4430–4230	44°21′30″	10°23′00″
		−17	Beta 152653	4210±40	−24.9	4850–4800 /4770–4620		
		−24	Beta 147800	4510±40	−24.2	5310–5040		
		−6	Beta 158266	3850±70	−25.6	4430–4080 /4030–4010		
4	Miano	−4	Beta 137041	2360±60	−28	2465–2300/2260–2160	44°29′36″	10°06′03″
		−43	Beta 131904	3940±60	−26.4	4530–4230		
5	Talada	−12	Beta 147801	2260±70	−27.7	2850–2710/2630–2500	44°22′50″	10°21′25″
		−8	Beta 152654	2530±60	−26.7	2760–2370		
6	Magliatica	−3	Beta 167935	9190±60	−27.1	10520–10220	44°28′00″	10°35′10″
7	Cinquecerri	0	Beta 123271	3620±60	−30.3	4090–3815/3785–3730	44°21′14″	10°19′35″
8	Lucola	0	Beta 125334	4380±60	−25	5275–5175 /5070–4840	44°26′33″	10°18′38″
9	Morsiano	−10	Beta 125333	11390±70	−25	13,790−13,670 13,500–13,145	44°18′00″	10°30′00″
		0	Beta 166926	3720±60	−26.3	4240–3900		
10	Minozzo	−10	Beta 152650	960±40	−27.1	950–780	44°22′00″	10°27′00″
11	Romanoro	−10	Beta 123273	2870±60	−27.7	3195—2845	44°18′00″	10°30′30″
12	Succiso	−20	Beta 123270	8380±60	−29.1	9520–9285	44°21′00″	10°14′00″
		−20	LODYC	8300±80				
13	Groppo	−4	Beta 152649	1980±60	−24.8	2060–1820	44°26′10″	10°20′30″
14	Gazzolo	−35	Beta 152648	2530±40	−27.8	2750–2470		
		−9	Beta 152645	8090±70	−27	9240–8770		
		−10	Beta 158267	8000±50	−27.4	9020–8650	44°25′59″	10°17′50″
		−16	Beta 152646	2500±40	−25.9	2740–2370		
		−26	Beta 152647	2480±40	−26.9	2740–2360		
15	Rocca	−11	Beta 152651	2510±40	−26.9	2740–2450/2410–2380		
		−13	Beta 165472	2580±60	−26.9	2780–2690/2660–2480	44°40′00″	09°45′00″
		−2	Beta 144292	119±1.0%		modern		
16	Valestra	−10	Beta125332	3980±60	−26	4570–4260	44°27′30″	10°35′10″
17	Valoria	0	Beta 166925	6840±70	−27.6	7800–7580	44°19′38″	10°32′58″
18	Ienza	−14	Beta 166927	1540±40	−26.9	1530–1330		
		−27	Beta 166928	1200±40	−27.9	1240–1040/1030–1000	44°32′04″	10°22′12″
		−30	Beta 167937	1210±40	−24.8	1250–1050		
19	Cà di Rinaldo	−7	Beta 167936	390±40	−29.5	520–420/400–320	44°26′10″	10°20′30″
20	Signatico	−8	Beta 167938	120±40	−26.1	280–0	44°32′50″	10°07′17″
21	Ramiseto	−6	Beta 165473	2930±40	−27.2	3220–2950	44°24′11″	10°17′02″
22	Vedriano	0	Beta 144293	119.2±0.8%		modern	44°30′00″	10°23′18″

(*)95% probability; (**)Stuiver et Al., 1998 (0 Cal. BP = AD 1950)

are directly affected by active landslides and 1608 by dormant landslides. Dozens of municipal centres are among them.

A good 16% of the total road network is on top of existing landslide bodies, and so is threatened and periodically affected by slope movements.

It has been demonstrated that about the 90% of damage caused by landslide activity in Emilia-Romagna derives by reactivation of existing landslide bodies, mainly ancient earth flows.

The cost of this situation is staggering: in Emilia-Romagna, over the last five years about 390 million Euros has been invested in reconstruction, relocation of villages, consolidation work and monitoring of unstable slopes. Luckily, the number of human casualties caused by reactivation of earth flows in Emilia-Romagna is almost negligible, thanks to the generally slow displacement velocity of this type of landslide.

3 LANDSLIDE SUSCEPTIBILITY IN EMILIA-ROMAGNA APENNINES

According to the regional 1:25.000 *Landslide Susceptibility Map* (Bertolini et al. 2002), geological units show a predisposition to instability that may be quantified on the basis of the "Landslide Density Index".

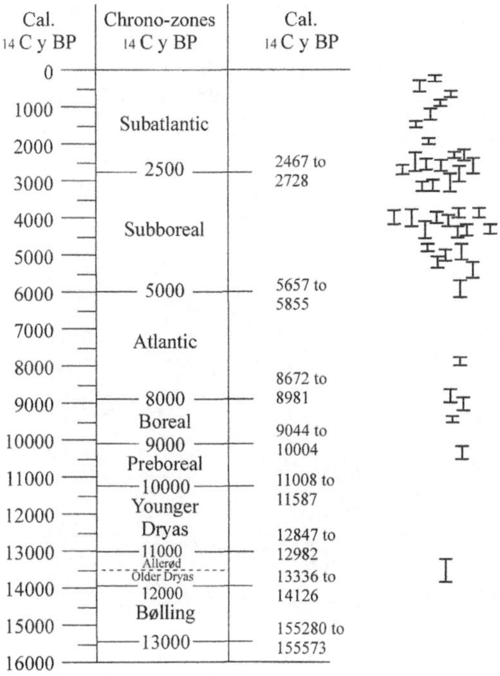

Figure 2. Temporal distribution of the landslide events from ^{14}C dating.

LDI represents the ratio between the sum of landslide areas affecting a given geological unit and its total mapped surface.

The Romagna sector and the Apenninic ridge above 900 m elevation are formed by Oligo-Miocene sandstones (e.g. the "Macigno" and the Marnoso-arenacea Formations,) with a "Landslide Density Index" value (LDI) ranging from 1 to 10% (low susceptibility). Medium susceptibility units (Plio-Pleistocene "Blue Clays" and "Yellow Sands" with an LDI from 10 to 20%) form the Apennine foothills.

In the remaining part of the Emilia Apennines, shaly formations with high susceptibility (LDI from 20 to 40%) prevail: olistostromes, "broken formations" with block-in-matrix structure ("Argille Scagliose" Auctt.), mainly Cretaceous in age.

As regards the relationship between morphology and susceptibility, the majority of landslides are concentrated in areas where the slope gradient usually ranges from about 8 to 11°, which is the usual slope angle of "Argille Scagliose" and similar formations, whereas in slopes whose gradients exceed 25° they are very rare, thanks to the presence of a stronger bedrock.

4 FEATURES AND ORIGIN OF ANCIENT LANDSLIDES

As regards dimensions, it may be calculated that at least 4000 landslides have an amplitude exceeding 10 hectares and 100 exceeding a square kilometre. Several of them exceed 4 km in length. From plan view, they show the typical features of an earth flow: a large crown, a relatively narrower middle "channel" – corresponding to the area of flow – and a wide basal fan reaching the valley floor, with a modest or null slope inclination (es: Figures 3, 4, 5 and 6). Their thickness is usually no more than a few tens of metres. Calculations based on a sample of 46 different landslides (whose movements are monitored by 190 inclinometers) show that the majority of them (52%) reaches a depth ranging from 10 to 30 meters. About 10% of them show a depth exceeding 40 m. The landslide depth has seldom reached the order of magnitude of one hundred metres, as in the Corni-glio case (Larini et al. 2001; Figure 4).

These landslide bodies were -and are- nourished by several different units (often "Argille Scagliose" or other shaly and structurally complex formations) that constitute the bedrock of the whole versant.

Following sedimentological laws, their internal structure shows several superimposed "strata", originating at different moments in the past and from different parts of the versant. Consequently, their lithology is extremely variable and usually characterised by a block-in-matrix structure, with a prevalence of clay matrix produced by the softening of shaly units.

Figure 3. The typical shape of the Large Morsiano Landslide (Reggio Emilia Province). A tree trunk outcropping from the tip has been dated to about 13,500 Cal. Yr BP (Figure 9). Photo by G. Bertolini, 2000.

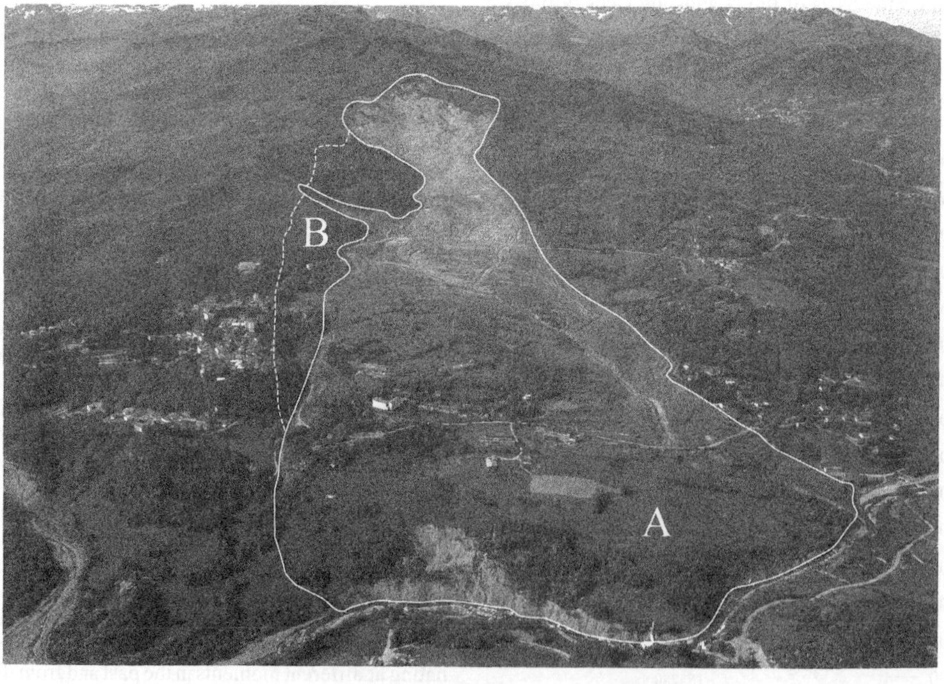

Figure 4. The Corniglio Landslide (Parma Province) is one of the largest in Northern Apennines. It completely reactivated in 1994 after one century of dormancy, destroying 70 buildings. "A" indicates the 100 m thick earth flow called the "lama". "B" indicates parts of the slope affected by movements triggered by the friction of the right flank of the "lama". Photo by G. Bertolini, 2006.

Figure 5. The 2 km long Cavola Landslide (Reggio Emilia province). The arrow indicates the location of the core boring where 5 samples were collected at different depths (see also Figure 7). Photo G. Bertolini, 2005.

Figure 6. The 4 km long Talada landslide. The arrow indicates the location of the drilling where two samples were collected at 8 and 12 m of depth. Photo G. Bertolini, 2005.

In terms of shear strength properties, these materials show a high degree of variability, both spatially and temporally, which is difficult to quantify. The minimum shear strength values are found in argillaceous materials with montmorillonite minerals (for example, *Argille di Viano Formation*; Bertolini & Pellegrini, 2001).

5 THE REACTIVATION

A recurring behaviour pattern can be seen in the majority of reactivations of ancient earth flows that have occurred during the last decade.

In many documented or observed events, when a reactivation occurs, the first movements are large rotational slides in the source area that cause a regression of the main scarp, which is the most unstable part of the versant. The displaced material reaches a liquid state of consistency, thus producing earth/mud-flows moving downward as far as the landslide body's mid-section. The undrained overload induces a sudden increase in porewater pressures, thus triggering a series of imbricate fissures often connected to the base of the whole landslide body. This pattern migrates valleyward, propagating progressive failure along the base of the ancient landslide, which may entirely reactivate first by roto-traslational sliding and

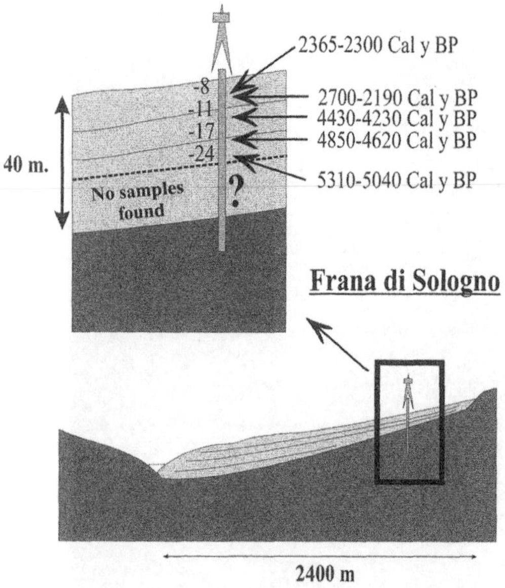

Figure 7. The age of the 14C samples usually grows accordingly with the depth. In the case of the Cavola landslide it was possible to date also the bottom of it, thus allowing to assess the probable age of its origin.

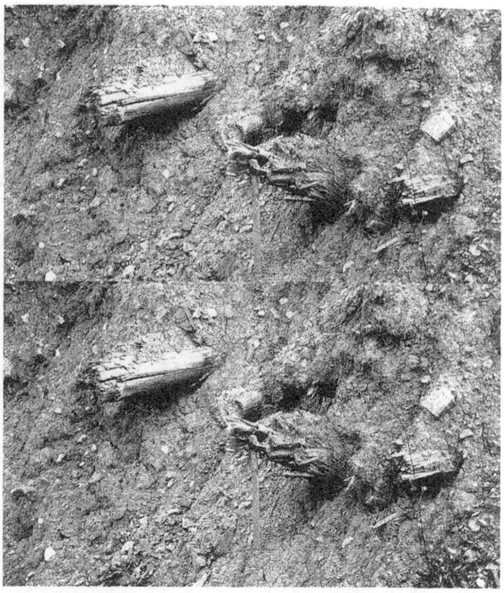

Figure 9. These tree trunks outcrop at the tip of the Morsiano landslide (Figure 3). They date to the Late Glacial period (*circa* 13.500 Cal. Yr BP).

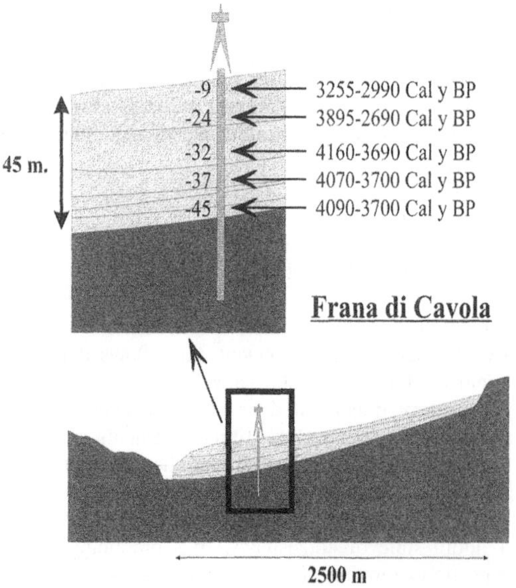

Figure 8. In almost all the cases, with the exception of the Cavola one (Figure 5 and 7), the vast majority of samples were found inside the upper half of the landslide body; in these cases the dated events only represent reactivations of already existing landslide bodies, as in this example.

then by flowing (rare eventuality). This sequence of events, which has been observed in many recent cases, was probably the same in past, historic and pre-historic reactivations.

6 CAUSES OF REACTIVATION

A historical database of Emilia-Romagna Region has been compiled for about one third of its territory and in the near future the regional administration will complete the archive research for the whole territory. The analysis of this data-base, which already contains several thousands records of events, allow us to deduce that long-lasting rainfall plays a major role as a triggering factor in reactivating landslide bodies all year round, while melting snow cover brings to bear particular effect in the months of March and April (Basenghi & Bertolini 2001).

Seismic triggers have seldom been incontrovertibly identified by means of historical investigation on the concomitance between landslides and earthquakes. The definite cases are few and confined to limited areas (Romagna, Reggio Emilia ridge sectors) were strong earthquakes historically occurred. However, from a statistical point of view, it may be stated that, as a triggering factors, the role of earthquakes is not comparable to precipitation.

7 RADIOCARBON DATING RESULTS

^{14}C dating methods add objective data regarding the evolution of landslide activity, both in general terms and in relation to some specific landslides.

The Regional Authority (Bertolini et al., 2004 and 2005) has applied Radiocarbon Dating techniques to wood remnants collected by core boring inside the landslide bodies (Table 1 and Figure 2). All the landslides in question are of the earth-flow type. The landslide's clay matrix allows optimum preservation of the wood remnants (sometime entire tree trunks) that usually appear fresh despite their age (Figure 9).

The thickness of the investigated landslides varies from a minimum of 18 to a maximum of 50 m (30 m on average). Datings were carried out by measuring the ^{14}C with the traditional radiometric method (Liquid Scintillation counting) on large wood samples or by means of mass spectrometry (AMS) on samples weighing a few grams. From the Conventional Radiometric Age, the 2 sigma (95% probability) Calibrated Age was calculated through the Stuiver et al. curve (1998).

These studies, carried out on dozens of cases, demonstrate that these landslide bodies are the result of multi-phase events occurring over a period of thousands of years. They originated as earth-flows after the last glacial maximum and grew during the rainiest periods of the Holocene through the superimposition of new earth flows. Consequently, inside the landslide body, the age of wood remnants usually increases with the increasing depth (Figure 7 and 8). They evolved through phases of rapid growth and others of minor activity, like that which we are experiencing today. Some past accumulation rates describe a landscape dynamism that is unequalled today (e.g.: 4.5 cm/year for a time period of 1000 years in the case of Cavola and ~1 cm/year for 2800 years in the case of Sologno).

As a paleo-climatic indicator, this data suggests the existence of a wetter (Sub-boreal, early Sub-atlantic) period within the Holocene (Figure 2) in the Northern Apennines. However, the number of samples is insufficient to draw conclusions about the EarlyHolocene and pre-Holocene periods of time. In quite a large number of core borings, it has been noted that the frequency of wood remnants found inside the landslide body decreases with the depth (and obviously the age). In other words, the large majority of samples were found inside the upper half of the landslide body; as a result of that, almost all the dated events represent reactivations of existing landslide bodies. Only in one case (Cavola landslide, −45 m depth) the sample was found at the very bottom of the landslide and consequently it may probably be ascribed to its origin.

8 CONCLUSIONS

Radiocarbon dating leads to a better understanding of landslide evolution in the in Northern Apennines during the late Pleistocene and Holocene. The thousands of landslide bodies that presently lie on northern Apennine slopes originated as earth flows probably because of the climate changes that characterized the beginning of the Holocene. In particular, groundwater recharge reached higher values as a combined consequence of permafrost disappearance and increased precipitation. These landslides reactivated and grew in thickness by superimposition of new earth flows during the rainiest periods of the Holocene, with noticeable accumulation rates, thus demonstrating a landscape dynamism unequalled today.

In conclusion, Holocene climatic oscillations played an important role in moulding the Northern Apennine landscape. This was probably not a continuous process, but occurred mostly by impulses depending by different climate phases.

All these hypotheses must be further tested by adding new case records to the present database. Further data may also allow us to comprehensively reconstruct the different phases of Holocene climate oscillation. Several studies, carried out by the regional authority and Parma and Florence Universities (Tellini C., 2004, Bertolini et al., 2004), are presently underway.

REFERENCES

Basenghi, R. & Bertolini, G. 2001. Ricorrenza e caratteristiche delle frane riattivate durante in XX secolo nella Provincia di Reggio Emilia (Appennino Settentrionale). *Quad. Geol. Appl.* 8. Bologna: Pitagora Ed.

Bertolini, G. & Pellegrini, M. 2001. The landslides of the Emilia Apennines (Northern Italy) with reference to those which resumed activity in the 1994–1999 period and required Civil Protection interventions. *Quad. Geol. Appl.* 8. Bologna: Pitagora Ed.

Bertolini, G. & Sartini, G. 2001. La frana del Rio di Sologno: l'evento del febbraio 1996 e il rinvenimento dell' «albero di Sologno» (Comune di Villaminozzo, Provincia di Reggio Emilia). *Quad. Geol. Appl.* 8, Bologna: Pitagora Ed.

Bertolini, G. & Tellini, C. 2001. New radiocarbon dating for landslide occurrences in the Emilia Apennines (Northern Italy). *Trans. Japan. Geom. Un.*, 22 (4), C-23.

Bertolini, G. Canuti, P. Casagli, N. De Nardo, M.T. Egidi, D. Mainetti, M. Pignone, R. & Pizziolo, M. 2002. Landslide Susceptibility Map of the Emilia-Romagna Region. Rome: SystemCart Ed.

Bertolini, G. Casagli, N. Ermini, L. & Malaguti, C. 2004. Radiocarbon Data on Lateglacial and Holocene Landslides in the Northern Apennines. *Natural Hazards* 31:645–662, Kluwer Academic Publishers, Netherlands.

Bertolini, G. De Nardo M.T. Larini, G & Pizziolo, M 2004. Landslides of the Emilia Apennines. *Field Trip Guide Book of the 32nd International Geological Congress,*

August 20–28, 2004, Florence (Italy). Rome: edited by APAT.

Bertolini, G. Guida, . & Pizziolo, M. 2005. Landslides in Emilia-Romagna region (Italy): strategies for hazard assessment and risk management. *Landslides* 2(4) pp 302–312, Springer-Verlag.

Larini, G. Malaguti, C. Pellegrini, M. & Tellini, C. 2001. "La Lama" di Corniglio (Appennino Parmense), riattivata negli anni 1994–1999. *Quad. Geol. Appl.*, 8: 59–114. Bologna: Pitagora Ed.

Pizziolo, M. 1996. Carta Inventario del Dissesto. *Regione Emilia-Romagna*, Bologna.

Stuiver, M. Reimer, P.J. Bard, E. Beck, J.W. Burr, G.S. Hughen, K.A. Kromer, B. McCormac, F.G. v.d. Plicht, J. & Spurk, M. 1998. INTCAL98 Radiocarbon age calibration 24,000 – 0 cal BP. *Radiocarbon*, 40:1127–1083.

Tellini, C. 2004. Le grandi frane dell'Appennino Emiliano quali indicatori geomorfologici di variazioni climatiche. *Rassegna Frignanese*, XXXIII, Accademia del Frignano "Lo Scoltenna" , Pievepelago. Pavullo (Modena): Tipolitografia Benedetti.

Landslides and Climate Change – McInnes, Jakeways, Fairbank & Mathie (eds)
© 2007 Taylor & Francis Group, London, ISBN 978-0-415-44318-0

Developing a strategy for coastal cliff monitoring and management

J.A. Lawrence & R.N. Mortimore
School of the Environment, University of Brighton, Brighton, UK

M. Eade
Brighton & Hove City Council, Brighton, UK

A. Duperret
Laboratoire de Mécanique, Faculté des Sciences at Techniques,
Université du Havre, France

ABSTRACT: The INFORM Project (INformation FOr cliff Recession Management) has been established between the University of Brighton, Civil Engineering and Geology Division and Brighton and Hove City Council, to develop a monitoring and instrumentation network to better inform the planning strategy along the coastal cliff section belonging to Brighton and Hove City Council. This strategy will be partly achieved by a total rock approach to the cliff line geology and investigation of the rock mass properties in order to determine the likely cliff instability hazards. It is intended that these geological and mechanical factors will be used to locate a monitoring network to measure cliff instability processes at critical points and measure the external factors that influence the cliff including extreme climatic events, climate change and sea-level rise. It is hoped that this approach will provide a way of quantifying the evolving hazards and risks and enable more informed decisions to be made about cliff access, public usage and coast protection works.

1 INTRODUCTION

An outcome from two European funded research programmes (ROCC (Risk Of Cliff Collapse) 1999–2001 (Mortimore et al. 2001a) and PROTECT (PRediction Of The Erosion of Cliffed Terrains) 2000–2004 (Busby et al. 2004)) has been the recognition that the chalk coastal cliffs of Europe require:

1. Detailed geological investigation, utilizing a total rock approach.
2. Analysis of physical properties of the materials.
3. Numerical and 3D modeling to investigate and test intuitive and non-intuitive failure mechanisms, and aid location of boreholes and instruments.
4. Constant monitoring to measure the rate and scale of processes, and mechanisms of cliff failure to be integrated with numerical models.

The INFORM Project, (INformation For cliff Recession Management) sets out to build on this previous research by implementing and developing the recommendations of the ROCC and PROTECT Programmes and establishing a cliff monitoring network.

The coastal area between Brighton Marina (TQ 3355 0335) and Saltdean (TQ 3814 0188) is the responsibility of Brighton and Hove City Council, it is a section dominated by chalk cliffs, with a Pleistocene and Holocene Raised Beach (TQ 3355 0335) and a series of dry valleys. The cliff-line and undercliff have high amenity value to the adjacent densely populated urban area. The entire length of cliff between Black Rock to Saltdean is also a Geological Conservation Review (GCR) site and a Site of Special Scientific Interest (SSSI).

In the winter of 2000–2001 several large and numerous small chalk cliff collapses occurred along this cliff section, causing concern to the local residents, businesses and the local council. In response to the cliff collapses two stabilization schemes were implemented to limit further recession. These schemes involved trimming the cliff face, bolting, meshing and soil nailing, as well as limiting access to the foot of the cliffs. As a result of this and the lack of specific information to assist future planning, a programme of research and monitoring has been initiated to better understand the natural processes that affect the cliffs. At the same time the local Shoreline Management Plan (SMP) was being reviewed. That process highlighted the lack of detailed knowledge of cliff evolution making it difficult to set shoreline policy for the next 100 years as required by the SMP review process.

The research has been developed to investigate the mechanisms behind cliff instability in this area and includes: extensive geological and geotechnical fieldwork supported by a novel approach to field instrumentation to monitor cliff movement and laboratory testing to provide physical parameters to evaluate failure mechanisms. Previous research has shown the type of failure in chalk cliffs is dependant on the material properties and the fracture network, this combined with historical data and a site specific approach to monitoring particular sections of the cliff should provide a means to establish:

1. frequency of failures
2. volume of failures
3. type of failures
4. climatic events (general precipitation, storm events, climate change), sea-level change and other controls.

The results will be used to inform local plans for shoreline management and the relocation of coastal infrastructure.

2 GEOLOGY

Three different geologically controlled areas can be identified between Brighton Marina and Saltdean these are:

1. The chalk cliff sections
2. The ancient dry valleys
3. Black Rock Raised Beach and overlying Head deposits

2.1 *The Chalk*

The Chalk from Brighton Marina to Saltdean is entirely composed of the Newhaven Chalk Formation (Mortimore 1986, Bristow et al. 1997 & Rawson et al. 2001). Marl seams stand out as bands of paler chalk at a distance and the predominant conjugate, steeply inclined (60–70° dips) fracture sets create a conspicuous coastal geomorphology and slope failure mechanism. Lithological marker beds are used for correlation within the formations.

Identification of each of the lithological marker beds used for correlation and mapping the chalk through the cliffs (marl seams, flint bands) is supported by key index fossils. Species of fossil sea urchins (echinoids) are especially useful and these include the changing shapes in the genus *Echinocorys* and the abundance of *Offaster pilula* (Mortimore 1997). Using both the biostratigraphy and lithostratigraphy it is possible to establish a detailed stratigraphic cross-section of the chalk cliffs from Brighton to Saltdean (Figure 1).

The beds of chalk rise towards Brighton on the southern limb of the Old Steine Anticline. The Old Nore Marl is in the top of the cliff at Saltdean, it finally disappears out of the succession at Ovingdean. As the Newhaven Chalk is traced westwards in these cliffs it thickens and the individual marl seams become better developed. Exposures of particular stratigraphic importance are present at Black Rock where the Santonian – Campanian stage boundary section is the best developed on this coast. The Brighton Marl and underlying beds are present only at the far western end next to the Pleistocene cliff line at Black Rock (Mortimore 1997, Figure 1)

2.1.1 *The rock mass character of the Newhaven Chalk Formation*

Understanding both the rock mass character and lithology is important in determining the cliff instability issues. Each formation in the chalk has a different rock mass character in terms of style of fracturing and frequency of joints (Mortimore 1993). For the Brighton Marina to Saltdean coast section rock mass data has been collected along the coast using scan-line surveys (Priest & Hudson 1976 & BS 5930 1999). This illustrates the dominant primary discontinuity orientations with a strong north-south set and subsidiary NE-SW and E-W trending sets. These sets are the steeply inclined joints, typical of the Newhaven Chalk Formation (Mortimore et al. 2001a, b & 2004a).

There are both normal and reverse faults present in the cliffs and most of the faults are strongly slickensided, usually with sub-horizontal slickensides, indicating, at least in part, a compressional stress field (Mortimore & Pomerol 1997).

2.2 *The Pleistocene/Holocene*

The Pleistocene and Holocene geology along the coast includes the Brighton Marina, Raised Beach (Black Rock), the formation of dry valleys, the cold climate slope deposits filling the valleys and many periglacial structures including involutions (Figure 2).

2.2.1 *Black Rock, Brighton Marina*

There are three parts to the Black Rock cliffs at Brighton Marina.

(i) The first part is the present cliff-line cut by the sea before construction of the protecting sea wall. This part of the cliff exposes an engineering geology profile which includes a weathered, cryoturbated top section of a few metres that passes down into progressively more competent rock mass containing particular sets of fractures which control cliff stability in the Newhaven Chalk Formation. This

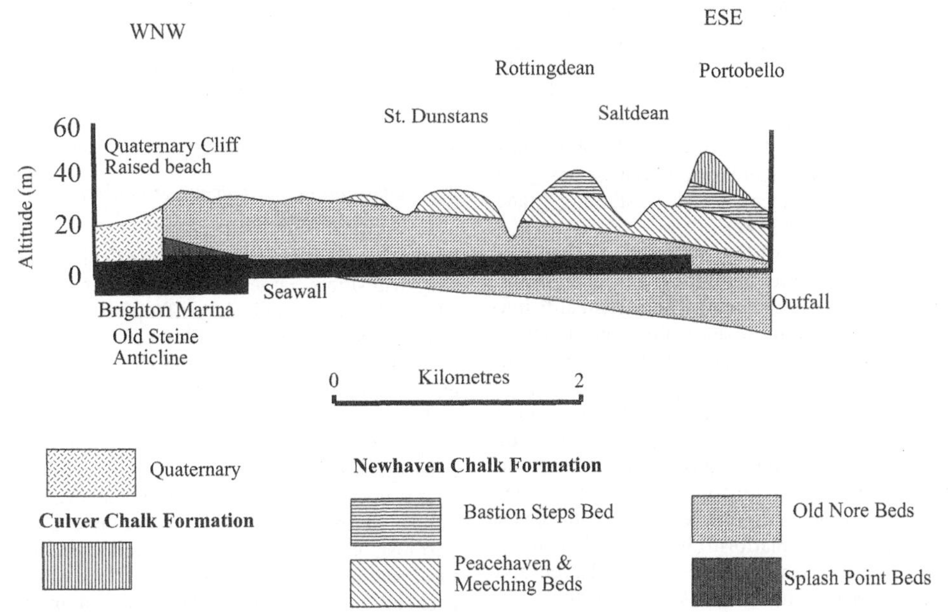

Figure 1. Schematic geological section of the coastal chalk cliffs from Brighton Marina to Portobello cliffs, showing the length of exposure in each of the main divisions of the Newhaven and Culver Chalk Formations. The length of Brighton Marina and the area covered by the seawall is also shown (Adapted from Mortimore et al. 2001).

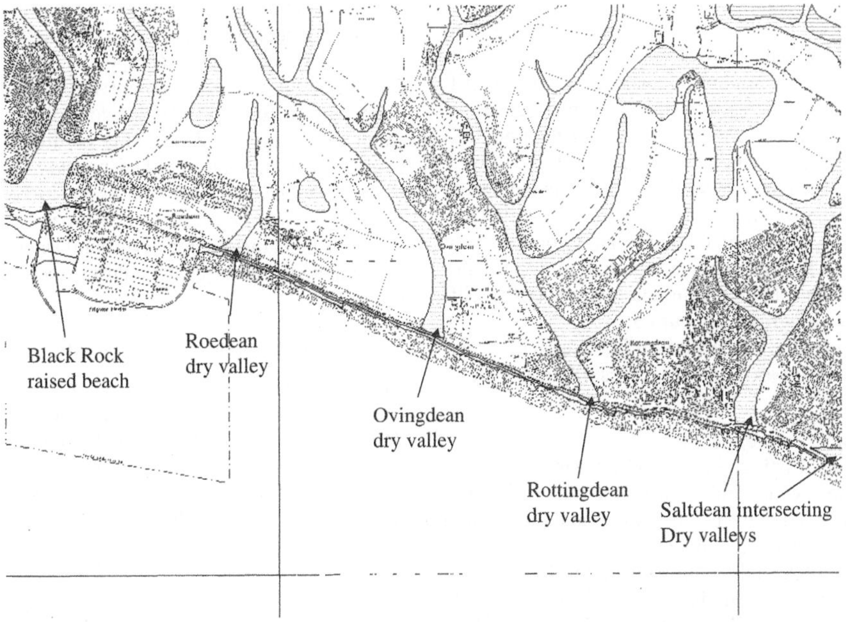

Figure 2. Map of the Palaeogene features from Brighton to Saltdean. Brighton Marina Raised Beach (Black Rock) can be traced inland north of the marina showing the position of the ancient cliff face. There are 4 identifiable dry valley sections. Saltdean dry valleys are the result of two dry valleys which almost intersect at the coast.

profile is well displayed on the access road down the eastern side of Black Rock, Brighton Marina. Periglacial weathering of the chalk during cold (humid) climatic episodes produced an ice churned (cryoturbated involutions) zone covered by various soils below which is a special style of fracturing (horizontally longer blocks). Marl seams (pale bands in the distance) act as weathering grade breaks (Mortimore 1997).

(ii) An ancient chalk cliff-line cut when sea level was higher than the present during an 'interglacial' period (possibly Ipswichian). The ancient cliff-line cuts inland in a north-westerly direction and forms the now buried back face of the cliff buried by Head deposits.

(iii) A mixed recent and ancient chalk cliff line which is only just exposed above the splash wall of the coast protection works incorporated into Brighton Marina. The recent cliff-line erosion has truncated the former cliff line leaving a solid chalk (ancient wave-cut-platform) base upon which the later sediments rest (includes Raised Beach and other sediments, Hutchinson & Millar 1998).

To the west, instead of chalk, the cliff comprises mixed warm and cold-climate, poorly consolidated, Pleistocene sediments resting on a solid chalk wave-cut platform. This part of the cliff is in four parts:

- A Raised Beach of rounded flint-gravel incorporating occasional blocks of chalk fallen from the ancient cliff line.
- Ancient cliff collapses. These pass laterally into poorly cemented sands.
- Scree deposits backed-up against the ancient cliff representing ancient Pleistocene cliff collapses.
- 'Head' deposits formed as scree, solifluction, hill-slope wash or windblown sediment that 'flowed' over the ancient chalk cliff-line, burying the Raised Beach gravels and sands as well as the cliff face. This mixed deposit contains bands or layers of coarse to fine metastable material, and the depositional dips in the sediment reduce as the cliff becomes progressively buried.
- A weathered cryoturbated surface which also contains remnant artificial structures (pipework etc.).

2.2.2 *The dry valleys*

East of Black Rock, the coastal cliffs truncate a series of dry valleys which illustrate a wide range of weathered profiles. Each of the dry valleys and intervening interfluves is identified in figure 2.

The Rottingdean dry valley illustrates the effects of cold wet sediment degradation. The chalk is degraded *in situ* yielding a high percentage of 'fines' (CIRIA Grade D, Lord et al. 2002) Evidence for heave structures is found in the undulating and displaced bedding in the degraded chalk. Such degraded chalk and heave structures probably relate to freeze-thaw cycles in which the 'fines' are displaced by an undulating freezing front, beneath the valleys where primary bedding structures are still recognisable. This degraded chalk is overlain by lobes of silt and flint-gravel which appear to have 'sunk' into the underlying wet chalk.

2.3 *Dissolution, weathering, groundwater and ground conditions*

In the sub-Pleistocene surface, particularly beneath Clay-with-flints, large dissolution pipes are present. Although not present in the cliff top between Brighton Marina and Saltdean, they do occur just to the east at Portobello and are possibly present behind the cliff. Such pipes are typically lined with flint cobbles and contain collapsed, weathered sediments (mostly Clay-with-flints but sometimes with hints of basal Palaeogene sediments).

It is uncertain how much local groundwater conditions either perched or at the 'water table' influences cliff instability. The potential for marl seams to create local perched water pressure in a cliff face was investigated by Duperret et al. 2002. Bedding features such as marl seams and flint bands, particularly sub-horizontal sheet flints, are also horizons of enhanced groundwater flow as seen by the formation of dissolution structures along these beds (Duperret et al. 2001).

Other concentration points for groundwater maybe where fracture sets coalesce (e.g. Mortimore, 1993). Most groundwater movement in the chalk is traditionally considered to have two components, the matrix and the fissures. Most fluid flow is thought to be concentrated in the fissure component. As each chalk formation has its own characteristic style and frequency of fracturing and fissure development, groundwater flow might be expected to be different in each formation. Conjugate inclined primary fracture sets in the Newhaven Chalk are likely to be zones of enhanced groundwater movement. Where the inclined fractures meet then increased groundwater flow might be expected, this is evident during periods of high rainfall.

2.4 *Engineering or hazard domains*

Taking the above aspects of the geology, the coastline can be divided into a number of engineering or hazard domains. These domains reflect the chalk lithologies (in terms of strength/density) and fracture styles within a formation; the depth of weathering (degradation of the chalk); and the depth of slope deposits filling the valleys.

3 GEOHAZARDS IN THE CHALK

The development of a strategy for coastal cliff monitoring and management from Brighton Marina to Saltdean involves a detailed understanding of the:

1. The range of hazards and risks
2. Rock mass properties
3. The potential volumes of cliff collapses
4. The effect and interaction of the coast area and the sea
5. The effects of the sea wall and other coastal protection works
6. The likely effects of climate change and sea level rise in the area under investigation

Each section of cliff contains geohazards specific to that site as well as more general hazards. Specific hazards are generally related to lithology, style of fracturing and cliff height and include types of rock slope failure (Mortimore et al. 2001a). General hazards include continuous spalling of small fragments. The rate at which a hazard develops is partly related to rates of marine erosion at the base of the cliff and partly to weathering and exposure of the face and material capping the cliff. The coast protection between Brighton Marina and Saltdean does slow the rate of hazard development compared to unprotected sites but does not eliminate it.

The sea wall and groynes protect the entire cliff section from Brighton Marina to Saltdean. The under cliff walk is regularly used for recreational purposes (walking, bicycle riding, disabled access), and even small falls are, therefore, a hazard. There are access points at Black Rock, Roedean Ovingdean, Rottingdean and Saltdean. A particular concern is the vulnerability (long-term stability) of the A259 coast road from Brighton to Newhaven. Its closest point to the cliff is at Saltdean, but the road is never far from the cliff edge.

As this is a protected section of coastline major cliff failures are not common. However, during the winter 2000–2001 several collapses occurred in the *in situ* chalk cliffs behind Black Rock, Brighton Marina. In addition, numerous small fragments of chalk and flint from both the *in situ* chalk cliff and the degraded valley-fill material has and continues to fall onto the under cliff pavement and the access pathways. The Brighton Marina failures were related to the structure of the Newhaven Chalk, the length of time the cliffs have been weathering since they were trimmed and the severe wet conditions.

3.1 *Potential volumes of cliff failure*

The scale of failures in the Newhaven Chalk is less than some other formations because of the interconnectivity of the conjugate fracture network. From measurements of previous collapses in the Newhaven Chalk it is possible to estimate the potential size of future collapses in relation to cliff height from Brighton Marina to Saltdean (Figure 3). Although the largest measured collapse has been about 2,000 m^3 figure 3 shows it is possible in this area for a single collapse to be in excess of 10,000 m^3.

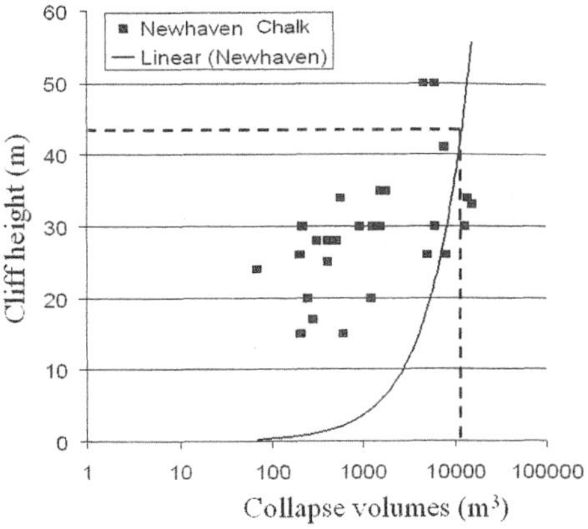

Figure 3. Comparison of cliff height to permissible failures in the Newhaven Chalk Formation. The points show the size of previous collapses in relation to the cliff height. The dashed line shows the maximum cliff height from Brighton Marina to Saltdean (44 m OD) and potential collapse size.

3.2 *Potential areas of collapses*

Areas which are potentially going to fail are those where full face trimming has not been completed, where overhangs are developing and where the main fractures daylight in the cliff at a critical dip angle and strike direction. Sterographic plots of rock mass data collected from scan-lines in the cliffs illustrates which fracture sets daylight in the cliffs and therefore could fail. By plotting the fracture data against the average orientation and dip of the cliff face it is possible to see that (Figures 4 & 5):

1. Wedge failures are possible, as a result of the conjugate style fracturing. This is the most likely type of failure to occur and the fracture network at Saltdean indicates the potential for large wedge type failures (Figure 4, diagonal lines, daylight envelope).
2. Although not as common figure 5 shows the potential for planar failures, large sections of rock could collapse along a plane (Figure 5, hatched daylight envelope).
3. Even though the cliff face has been trimmed back, figure 5 also shows there is a possibility of toppling failures along this cliff section (Figure 5, diagonal lines, daylight envelope).

The information from the sterograms demonstrates that a variety of collapse types are possible in the cliffs from Brighton Marina to Saltdean.

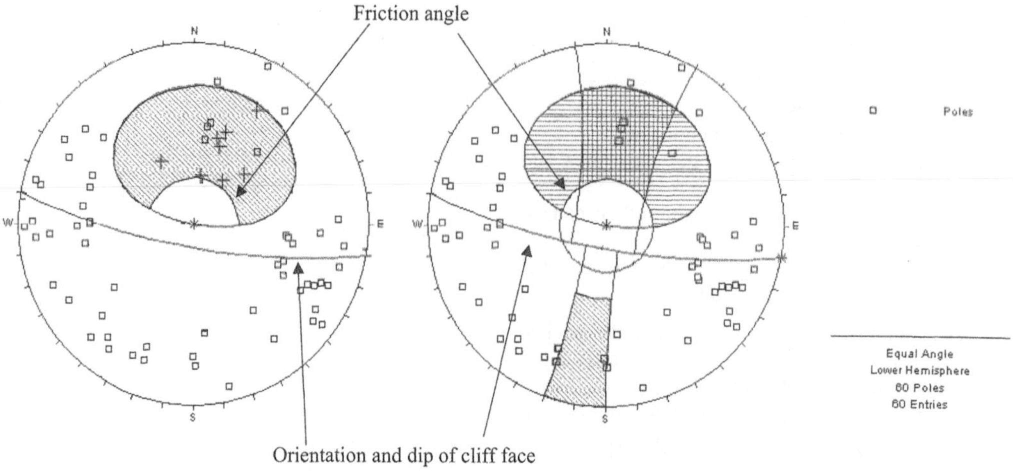

Figure 4. Sterographic projection showing the potential for wedge type failure. Each of the crosses in the diagonally lined daylight envelope reflects poles of the intersection of a conjugate fracture set. When these poles fall into the wedge failure daylight envelope there is the possibility of a wedge type collapse.

Figure 5. Sterographic projection showing the potential for planar and toppling type failures. Each of the poles that fall within the hatched planar failure daylight envelope reflects the possibility of planar failure. Each of the poles that falls into the diagonally lined toppling failure envelope reflects the possibility of toppling failure.

3.3 *Rockmass properties*

An initial set of laboratory tests has been carried out on the chalk from Brighton to Saltdean these involved:

- Saturated Moisture Content tests (SMC)
- Natural Moisture Content tests (NMC)
- Intact Dry Density (IDD)

The results of these tests indicated the Newhaven Chalk is amongst the softest of all chalk with a density ranging from 1.45–1.8 mg/m^3 (Figure 6).

All the samples tested in the cliff face were saturated suggesting the cliff face and the chalk behind it is always at saturation (Figure 6) as a result of rain and wave action. Tests have shown that saturated chalk is softer than dry chalk.

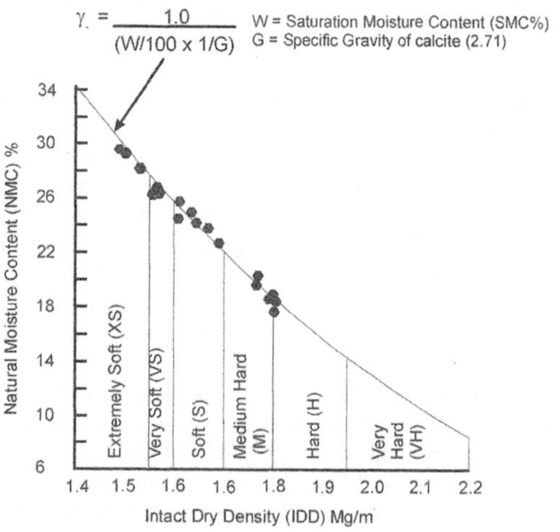

Figure 6. Moisture contents of naturally occurring Newhaven Chalk from cliff faces plotted onto the Intact Dry Density (IDD) – Saturation Moisture Content (SMC) Line (Adapted from Mortimore et al. 2004b).

4 GEOHAZARDS IN THE DRY VALLEYS

The modern cliff line truncates several dry valleys along this section of coast. Each valley displays a unique section of degraded chalk and Pleistocene and Holocene sedimentary structures related to periglaciation. Each valley drained southwards into the English Channel. These valley profiles differ from those seen elsewhere on the coast (e.g. Seven Sisters, Hope Gap, and Portobello) in terms of depth of weathering and overall geological structure.

Within these valleys a different set of geohazards is present. The volume of collapse is likely to be far less because the cliffs are lower and the degraded nature of the material. Spalling is the most likely hazard although small-scale collapses are possible.

5 BLACK ROCK GEOHAZARDS

The west end of Brighton Marina is largely composed of Pleistocene sediments, resting on a chalk platform and backed by a chalk cliff. Each component of this geology, and each type of sediment, has its own geotechnical properties.

The chalk wave-cut platform underneath the Raised Beach is solid, relatively unweathered CIRIA GRADE B/C 2/3 chalk (Lord et al. 2002). The degree of fracturing will affect the ability of this unit to drain freely. The gravels of the raised beach are free-draining.

The Brighton Raised Beach rests on the chalk platform and abuts against the ancient cliff-line. This deposit has a maximum altitude of 11.9 m AoD and is 2–3 m thick. The Raised Beach comprises well-rounded flint gravel (comparable to the present Brighton Beach on the west side of the Marina) with some sand and shell debris. Exposures in this part of the section during construction of the Marina illustrated a thin deposit of wind-blown sand on top of the beach. The flint-gravel beach deposit is locally weakly cemented by material washed down from overlying deposits. In general, the Raised Beach would be expected to be a freely draining unit but would collapse to a natural angle of repose in exposed sections.

Above the Raised Beach are the 'Head' deposits, up to 20 m thick. These have been divided into three units:

1. a lowest unit comprising coarse chalk debris
2. a middle unit of yellowish brown chalk solifluction deposits
3. a coarser solifluction deposit containing Sarsen stones, ironstones and flint with cryoturbation involutions at the very top.

These 'Head' deposits contain a matrix of fine silt (a form of Brickearth) which is likely to be metastable losing shear strength on wetting. This was probably a cause of failures in these deposits during the winter of 2000–2001. Not all the deposits in the Pleistocene section of the Brighton Marina cliffs will be unstable in the same way or to the same degree.

The buried chalk face to the Pleistocene cliff-line may drain into the Head sediments from perched water along marl seams or may follow fracture-controlled fissures and/or a combination of these structures. Artificial drains may also flow through this buried cliff. The interface between the cliff-line and its Pleistocene cover sediments is likely to be a water pathway. This interface is also a point of differential settlement for drains and could be a plane of fracture in drains. Investigation boreholes have, however, so far failed to find any water at this interface.

6 THE EXPECTED IMPACT OF CLIMATE CHANGE

The relationship between rainfall and cliff collapse has been well documented (e.g. Charalambous 2001) indicating that periods of high rainfall lead to increases in the frequency of cliff collapses. Investigations of this phenomenon have been unable to explain the hydrogeological link. Using the largest and most complete rainfall record in the world (dating back to 1875; held by Eastbourne Council) this relationship becomes clearer. The rainfall record combined with historical records of cliff instability (Williams et al. 2004 & Hutchinson 2002) shows a very strong link between years of high rainfall and very large cliff collapses in the chalk of southeast England (Figure 7). It can be assumed that there were many other collapses of varying sizes that went unreported.

The rainfall data also shows that in the last 115 years there have only been 4 years of rainfall over 1100 mm suggesting that these are a 1 in 30 year event (Figure 7). Although 3 out of 4 of the high rainfall events have been in the last 45 years suggesting that these high rainfall events are becoming more frequent, and could be expected as often as every 15 years. Future climate trends expect winter and overall rainfall to increase, (Hutchinson & Bromhead 2002) therefore the frequency of cliff collapse should also increase.

What can also be observed in Figure 7 is that periods of high rainfall are often preceded by an extended dry period. Between 2004–2006 the southeast of England has received below average rainfall so it would not be unreasonable to expect a very wet winter period within the next few years, if previous trends are reproduced.

Observations suggest the final contributing cause of a collapse is temperature. Many collapses seem to occur at either dawn or dusk as the cliff face is warming up or cooling down. This results in a final phase of expansion or contraction that causes blocks to slide and a collapse to occur (e.g. Brighton Marina collapses 2001, Peacehaven collapse 2001).

7 DISCUSSION

A monitoring strategy for the coastal cliff sections requires:

1. A long term plan (e.g. Split into several periods)
2. Understanding of cliff collapse mechanisms through a "total rock approach"
3. Understanding of the range of hazards
   - Volume of collapse
   - Type of collapse
   - Frequency of collapse
   - Mechanisms of collapse
   - Triggers for collapse
   - Hydrogeological effects
4. Understanding of the range of risks (number of people/cost and damage to property)
5. A range of methods is required to identify and reduce the hazards
   - Rockbolts
   - Meshing

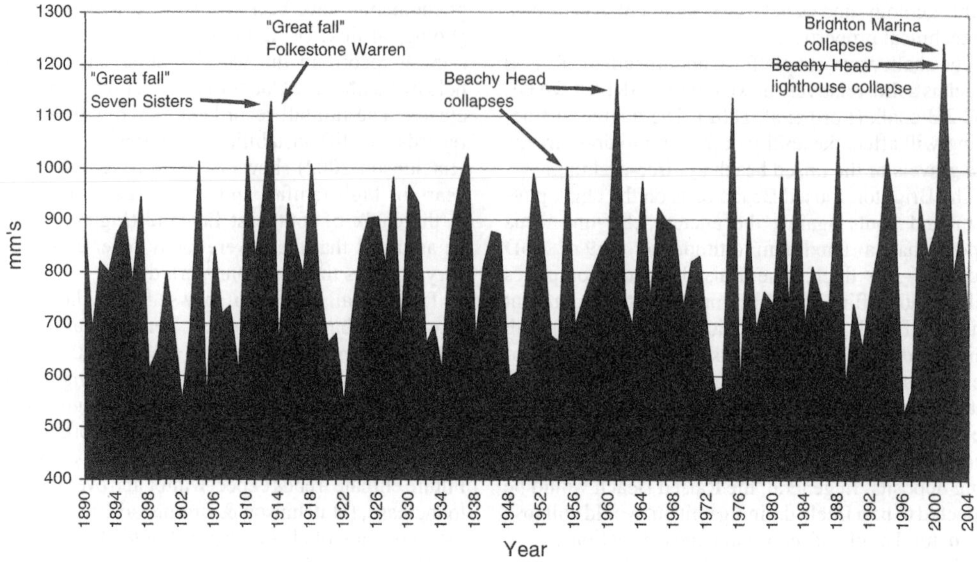

Figure 7. Rainfall record since 1890 from Eastbourne Council. The "year" is calculated from June to June so that the entire winter period can be measured. It shows that periods of high rainfall (>1100 mm per year) coincide with very large historical collapses.

- Shingle banks
- Instrumentation
- Monitoring

6. A warning method to inform local authorities of imminent hazards potentially developing and a plan of action to reduce the risk. Such warning methods would include walk-over surveys as well as automatic recording instrumentation which will alert the relevant authorities. Having raised the warning a predetermined course of action would be implemented (e.g. warning signs, restricted access).

The monitoring system will need continual evaluated to improve the network and the type of data collected. Automatic data logging and real-time results over a period of 4 years or more will test the validity of the instrumentation and the value of the different data sets (e.g. as the data from monitoring increases, the ability to accurately evaluate the cliff face will increase). Part of the investigation will also seek to evaluate the trends in the longer-term impacts of changes in climate and sea level.

Understanding the rock mass properties and taking a total rock approach is the first and possibly most important part of formulating a successful long-term coastal cliff monitoring and management strategy. Even for a short length of coast line such as Brighton Marina to Saltdean the range of geology creates a wide range off cliff collapse hazards each requiring individual investigation, monitoring and management.

The current study of these geological conditions and related hazards is aided by the application of a monitoring strategy. This includes the use of meteorological stations at critical localities to provide site specific data. Meteorological effects have been shown (section 6) to be an important contributing factor to cliff instability.

Site specific meteorological data needs to be compared with site specific evidence of cliff instability to provide:

- A better understanding of the relationship between weather/climate (especially seasonal trends) and cliff instability/collapse.
- A measure of long-term changes in weather patterns and the anticipated effect on cliff instability.

This makes cliff monitoring an indispensable management tool.

A further element of this project will be the combination of historical data with the project data to construct predictive models for periods covering 20, 50 and 100 years.

8 CONCLUSIONS

A new research programme, the INFORM project, seeks to aid the management of the hazards associated with the cliffs from Brighton to Saltdean. The cliffs along this section are divided into 3 broad divisions dependent on rock type, based on lithology and rock mass properties. Each rock type has its own range of hazards. Once the rock types are identified the hazards can be extrapolated further along coastal sections.

Historically large collapses have been associated with periods of high rainfall. It is likely that expected changes in climate and sea level rise will lead to increased instability.

The importance of a "total rock approach" to coastal cliff instability has been identified. It is essential that the lithostratigraphy and rock mass properties be identified early in any investigation of a coastal cliff section. This must be done in conjunction with a study of the external factors that affect stability. Only when this is done can the range of hazards and risks be identified. As climate change occurs and more data becomes available the ability to quantify the hazards will change, and therefore so will any risks creating an iterative process.

The investment now in the INFORM project should provide a basis for predicting trends in cliff instability processes and identifying changing hazards over the next 10 to 100 years.

REFERENCES

BS 5930. 1999. Code of practice for site investigation. British Standards Institution.

Bristow, R. Mortimore, R. N. & Wood, C. J. 1997. Lithostratigraphy for mapping the Chalk of southern England. *Proceedings of the Geologists 'Association*, **109**, 293–315.

Busby, J. P. Lawrence, J. A. Senfaute, G. Mortimore, R. N. Pedersen, S.A.S & Gourry, J. C. 2004. Prediction Of The Erosion of Cliffed Terrains "PROTECT". Technical Report. *British Geological Survey Internal Report. IR/04/142*, 62pp.

Charlambous, A. N. 2001. Brighton Cliff Management Study. Hydrogeological Services International Limited, Under assignment to High Point Rendal. Unpublished.

Duperret, A., Mortimore, R.N., Pomerol, B., Genter, A., Martinez, A. 2002. L'instabilité des falaises de la manche en Haute–Normandie. Analyse couplée de la lithostratigraphie, de la fracturation et des effronrements. *Bulletin d'Information des Géologues du Bassin de Paris*, **39**, No. 1, 5–26.

Duperret, A., Genter, A., Mortimore, R.N., Delacourt, B., & De Pomerai, M. 2001. Coastal cliff erosion by collapse at Puys, France: the role of impervious marl seams within the Chalk of NW Europe. *Journal of Coastal Research*, **18**, 52–61.

Hutchinson, J. N. 2002. Chalk flows from the coastal cliffs of Northwest Europe. IN Evans, S. G. & Degraff, J. V. (Eds.) *Catastrophic landslides: Effects, occurrence, and mechanisms.* Boulder, Colorado, Geological Society of America reviews in Engineering Geology.

Hutchinson, J. N. & Bromhead, E. N. 2002. Keynote paper: Isle of Wight landslides. IN McInnes, R. G. & Jakeways, J. (Eds.) *Instability – Planning and Management.* Isle of Wight, Thomas Telford Ltd.

Hutchinson, J. N. & Millar, D. L. 1998. Survey of the interglacial Chalk cliff and associated debris at Black Rock, Brighton. In *The Pleistocene of Kent and Sussex: Field Guide* (eds. J. B. Murton *et al.*), 135–146. London; Pleistocene Research Association.

Lord, J. A., Clayton C. R. I. & Mortimore, R. N. 2002. *Engineering in chalk.* Construction Industry Research and Information Association. CIRIA, **C574.** 350pp.

Mortimore, R. N. 1986. Stratigraphy of the Upper Cretaceous White Chalk of Sussex. *Proceedings of the Geologists' Association*, **97**, 97–139.

Mortimore, R. N. 1993. Chalk Water and Engineering Geology. In, Jones, G.P., Price, M. & R.A. Downing Eds. *The Hydrogeology of the Chalk of North-West Europe.* Oxford University Press. pp 67–92.

Mortimore, R. N. 1997. *The Chalk of Sussex and Kent.* Geologists' Association. Guide No. 57. 139pp.

Mortimore, R. N. & Pomerol, B. 1997. Upper Cretaceous tectonic phases and end Crétaceous inversion in the Chalk of the Anglo-Paris Basin. *Proceedings of the Geologists' Association*, 108: 231–255.

Mortimore, R. N. Lawrence, J. A. & Pope, D. 2001a. *ROCC (Risk Of Cliff Collapse) Geohazards on the UK Chalk Cliffs of Sussex.* INTERREG II. Espaces Rives-Manche. East Sussex/Seine/Somme. Unpublished.

Mortimore, R. N., Wood, C. J. AND Gallois, R. W. 2001b. *British Upper Cretaceous Stratigraphy.* Geological Conservation Review Series, No. 23, Joint Nature Conservation Committee, Peterborough. 558pp.

Mortimore, R. N. Lawrence, J. Pope, D. Duperret, A & GENTER, A. 2004a. Coastal cliff geohazards in weak rock: the UK Chalk cliffs of Sussex. In Mortimore, R.N. & Duperret, A. (eds). *Coastal Cliff Instability.* Geological Society, London, Engineering Geology Special Publication, **20**, 3–31.

Mortimore, R. N. Stone, K .J. Lawrence, J. & Duperret, A. 2004b. Chalk physical properties and cliff instability. In Mortimore, R.N. & Duperret, A. (eds). *Coastal Cliff Instability.* Geological Society, London, Engineering Geology Special Publication, **20**, 75–88.

Priest, S. D. & Hudson, J. A. 1976. Discontinuity spacing in rocks. *International Journal of Rock Mechanics and Mining Sciences and Geomechanics*, **13**, 135–148.

Rawson, P. F., Allen, P. & Gale, A. S. 2001. The Chalk Group – a revised lithostratigraphy. *Geoscientist*, **11**, 21.

Williams, R. B. G., Robinson, D. A., Dornbusch, U., Foote, Y. L. M., Moses, C. A. & Saddleton, P. R. 2004. A Sturzstrom-like cliff fall on the Chalk coast of Sussex, UK. IN Mortimore, R. N. & Duperret, A. (Eds.) *Coastal Chalk Cliff Instability.* La Harve, Geological Society, London, Engineering Geology Special Publications.

Landslides and Climate Change – McInnes, Jakeways, Fairbank & Mathie (eds)
© 2007 Taylor & Francis Group, London, ISBN 978-0-415-44318-0

Combining slope stability and coast protection at Seagrove Bay on the Isle of Wight

P. Winfield & E. Moses
Royal Haskoning, Haywards Heath, UK

M. Woodruff
Malcolm Woodruff Ltd, UK

ABSTRACT: The instability of the coastal slope at Seagrove Bay is threatening the integrity of the residential properties in Seagrove Bay. After significant rainfall, the factor of safety against slope failure, due to slipping, only remains above unity when there is an adequate beach fronting the seawall. Under storm conditions much of this 'protective' beach can be lost over a single high tide period, whilst often taking some months to recover. The inter-tidal area is heavily designated under the European Habitats Regulations. The impact of climate change, including increased sea level rise and storminess, together with the prospect of higher annual winter rainfall, can be expected to increase the instability and risk of slope failure. This paper describes the detailed geological investigation, slope stability analysis and coastal process analyses which have been carried out and the proposed engineering option to stabilise the slope.

1 PROBLEMS AND CAUSES

1.1 *Ground movement/coastal slope stability*

The coastal slope at Seagrove Bay (Figure 1) has a long history of land slippage (Skempton, 1945). The most recent movement in the north end of the Bay occurred during the winter of 2002/03. At that time old scarps were reactivated and new ones formed in the grounds of the properties, Waters Edge, Bonny Blink and Shorestones, following the particularly wet winter seasons (Figure 1). These properties are located in the central part of the affected frontage. Structural damage to the properties of East and West Rookery, and Seahouse, at the southern end of the frontage, were additionally observed, as was slippage closer to the seawall, within the grounds of Thatched Cottage No. 1 and Montserrat.

Figure 1. New 'scarp' faces formed in the grounds of Bonny Blink were observed following the 2002/03 winter season.

The insitu geology of this area, as indicated in the literature, is confused. Therefore, a geological investigation (Woodruff, 2004) consisting of deep land based boreholes and examination of the exposed deposits on the beach, at a time of extremely low tide, was commissioned by the local residents (King, 2004). This established that the geology of Seagrove Bay is similar to much of the north side of the Isle of Wight and consists of Bembridge Marls over Bembridge Limestone, St Helens Beds, Osborne Beds and Fishbourne Beds over the mixed Sandstones and Limestones of the Nettlestone Beds.

The Bembridge Limestones and Marls are only present at the top of the coastal slope at the north end of the Bay. The St. Helens Beds consist of coarse silt and fine sandy deposits with a thin layer of hard siltstone at its base. The Osborne clays are highly plastic variegated clays showing little structure. The Fishbourne Beds are very highly plastic clays well bedded with very low measured residual shear strengths, of the order of 6 to 7°. The natural water table in the St Helens Beds is high, only about 1 m below ground level near the houses. Water can also be found at the base of the Fishbourne Bed just above the Nettlestone Beds.

The northern and southern parts of Seagrove Bay (Figure 2.) are characterised by ground movements within the coastal slope, in the form of a deep seated slip, or slips, based in the Fishbourne Beds, extending to a depth of some 7 m below beach level at the position of the seawall on the study frontage. A geological cross section through the coastal slope is shown in Figure 4.

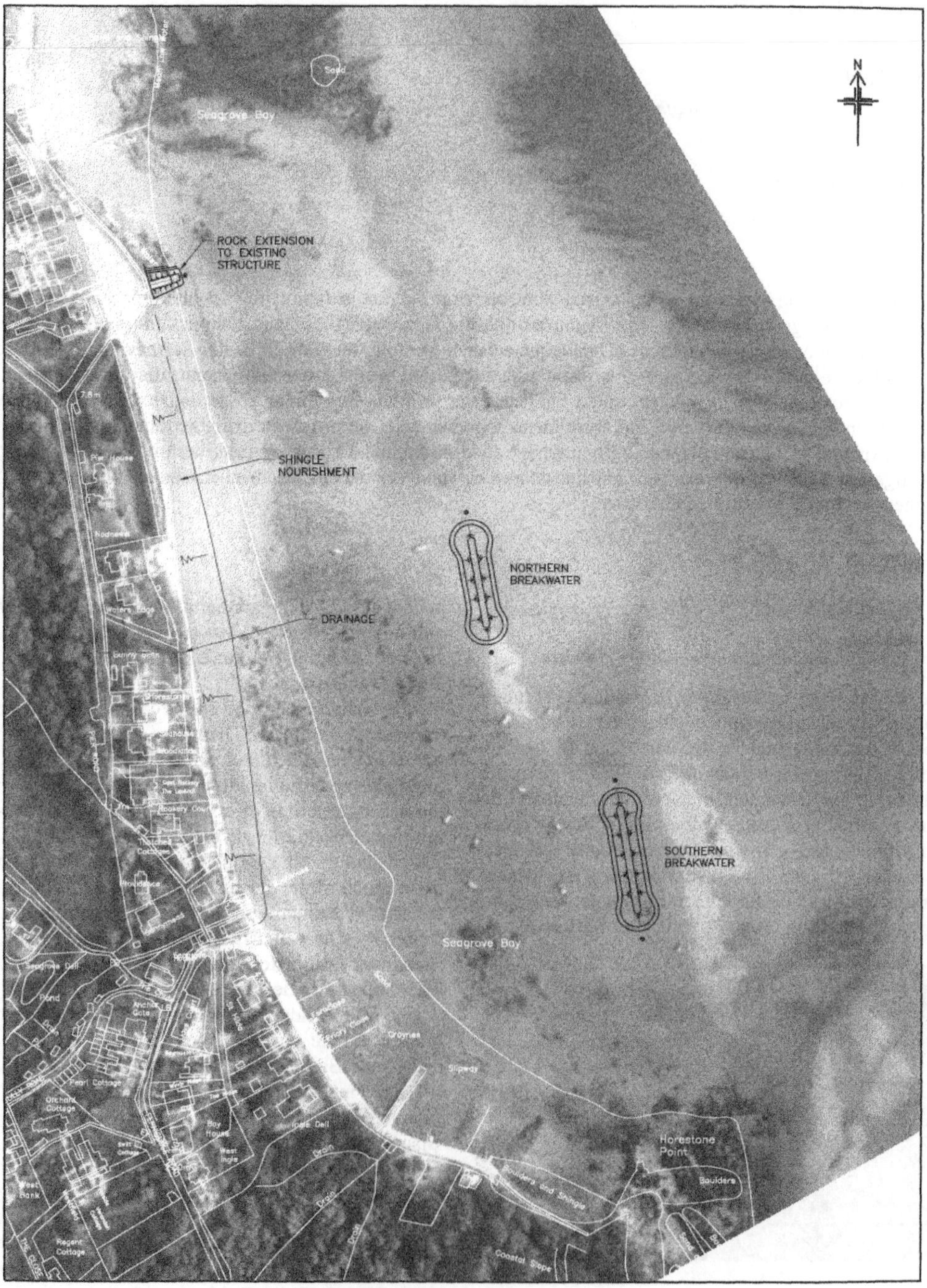

Figure 2. Site plan and scheme layout at Seagrove Bay.

The toe of the slip was identifiable on the beach, through the rotation of the siltstone band from the base of the St. Helens Beds, outcropping at an angle of 45° with the soft clay from the Fishbourne Beds visible immediately to seaward (Figure 3). These rocks are normally covered by mobile beach deposits, but were visible at the time of the recent slip and subsequent survey.

Figure 3. Slip outcrop at Seagrove Bay beach.

The whole slope is critically stable i.e. it has a factor of safety very close to one. Any change of conditions, such as a reduction of toe weight due to a lowering or complete removal of the mobile beach deposits, or an increase in water table level within the slope deposits may lead to failure. The recent slip, which extended below the sea wall and back to the houses, occurred at a time of high antecedent winter rainfall together with a very low beach level. Any mobile beach and the ground immediately behind the seawall serve as toe weighting to the slip surface, and thus help keep the slope in a stable condition. Progressive beach drawdown for extended periods of time over the winter period had reduced the factor of safety against failure of the slope to below one. Complete failure of the slope will lead to collapse or breaching of the seawall, which will then allow the ground behind it to be eroded by wave action, thus reducing the factor of safety against slope failure even further.

Under a 'No Active Intervention' scenario temporary stabilisation of the slope will occur following slope failure/breach of the seawall event due to the slump material from the failure acting as toe weighting to the remaining slope.

A failure cycle will be established, as in time, continued erosion of the slump material will occur and remove the toe weighting and thus reduce the slope stability causing further failures to occur.

The possibility of further movement of these slips, over the winter of 2003/04, led to emergency beach nourishment works being undertaken along the frontage.

1.2 *Lowering beach levels*

A recent trend of lowering beach levels has been observed in the Seagrove Bay area. This anecdotal assessment was studied on behalf of the Isle of Wight

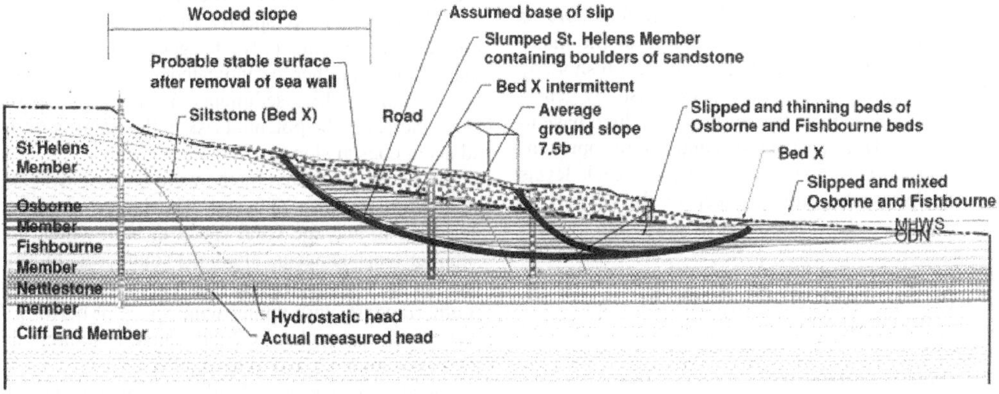

Figure 4. Typical geological section through coastal slope.

Figure 5. Low beach levels allow the seawall to be undermined.

Council. The study concluded that the beach levels in Seagrove Bay are volatile and vary with time possibly in a cyclical manner (Haskoning, 2002). The periodicity of these cycles is unclear and unproven based on the limited beach cross-section data available at present. However, aerial photographic and anecdotal information from long-term local residents indicates that the currently existing offshore bar has not always been present. In spite of the offshore bar, the beach profile data shows that there is currently a trend of falling beach levels due to wave and tidal action.

The low beach levels allow waves with higher energy to come closer inshore and attack the dilapidated seawall. Consequent undermining of the seawall structures has occurred. These waves with higher energy reach the beach from an easterly or south-easterly direction during storm conditions. Waves from these directions exacerbate the problem of lowering beach levels as the reflected wave energy, from hitting the seawall, encourages the mobile beach sediments to be rapidly transported from the upper foreshore, i.e. significant cross-shore transfer of beach material occurs, as well as long shore transport, depending on the incident wave direction. Much of the material is drawn down the beach onto the lower foreshore or into the shallow sub-tidal areas. Some of the material tends to be moved back higher up the foreshore by subsequent lower energy wave action, usually generated by the waves from the south west, which have to diffract around the coast of the Island before approaching Seagrove Bay. Eventually the low beach levels will precipitate the failure/breach of the seawall by undermining of the foundations (Figure 5).

2 CONSIDERATION FOR OPTION DEVELOPMENT

2.1 *Climate change*

It is currently predicted that climate change will result in alterations to our wind and wave regimes; effective rainfall; storm and tidal surge events and mean relative sea level.

Increases in exceptional rainfall magnitudes and periodicities are predicted which could lead to the reactivation of pre-existing coastal landslides within Seagrove Bay.

Rainfall records for Shanklin, which lies some 8km to the south west, reveal a 22% (75 mm) increase in effective antecedent precipitation for September to January, for the period 1839–2000 (Lee, et al. 1315–1321) 4. Note that mean rainfall recorded at Big Meade and Shanklin, averaged 88.8 mm annually (1950–1990), with a slight decline over this period. However, this mean value reflects four periods of exceptionally wet winters (November to January) in 1950–52; 1958–62; 1965–68 and 1987–88. For the eastern part of the Isle of Wight the main periods of greater than average precipitation; between September and January, were 1927–40; 1950–65 and 1978–82 (The rainfall received between October 2000 and February 2001 exceeds any on record for the same period for Shanklin, but may, or may not, be part of another period of excess winter rainfall). Using the Shanklin records in combination with the 1998 UKCIP climate change study (Halcrow, et al. 2001) 5 suggest that mean monthly effective rainfall may increase from 5–6% to 12–25%, with greatest estimated change occurring between December and February. This range of values reflects the 'Low' and 'High' scenarios of climate change employed in the UKCIP study6. The potential recurrence of extreme rainfall intensities is estimated to reduce from 1:200 years at present to 1:45 years by 2080.

Despite the demonstration of strong links between antecedent rainfall, ground movement and landslide events at Luccombe (Lee, et al., 1998) and along the southern Undercliff coast (Lee, et al., 1998), it is difficult to state with certainty that effective rainfall trends will inevitably create more frequent and larger scale slope failures. This means that ongoing event monitoring is important, in order to detect whether any increased instability can be identified.

Rainfall, and more specifically ground water pore pressure, is not the only 'trigger' mechanism; toe unloading is another. As indicated above, marine erosion of any mobile sediments from the toe of the slope increases the potential risk to the stability of the adjacent coastal slope.

Waves are generated by the wind blowing over any significant expanse of water. The largest predicted increases in average wind speed in both winter and summer seasons occurs along the south coast of England. In the winter, stronger winds are expected on the southern coast, with winter wind speed of between 4 and 10 per cent higher by 2080, whilst in summer the increases are predicted to be smaller.

For sites in the eastern English Channel (using data for Shoreham), it is suggested that there will be a small decrease in offshore, but an increase in near-shore, significant wave height. The latter might be in the order of 0.25 cm, for approximately 10% of the year, by 2080.

Such a modest increase in significant wave height needs to be superimposed on the projected rise of mean sea level. In addition, storm surges are likely to increase in both magnitude and frequency. The POL (1996) storm surge model applied to the UK continental shelf, coupled with the Hadley RECM climate change predictions, indicate a 0.1 to 0.2 m increase in sea-level for the 1 in 5 year surge event; and a 0.2–0.4 m increase for a 1 in 50 year recurrence. It is this, in combination with wetter winters, more frequent storm surge events and a generally more variable climate, which is likely to escalate the risk of increased coastal slope instability beyond the mid 21st Century caused by loss of beach sediments due to wave and tidal action.

Therefore, any solution to stabilise the slope, would ideally (particularly in the light of the uncertainty and magnitude of these effects) be adaptable in the future to accommodate climate change.

2.2 *Environment*

Seagrove Bay is either within or in close proximity to a number of conservation designations. These sites of designated conservation importance are listed below:

- Solent and Southampton Water Special Protected Area (SPA)
- Solent and Southampton Water Ramsar Site
- South Wight Maritime Special Area of Conservation (SAC)
- Ryde Sands and Wootton Creek Site of Special Scientific Interest (SSSI)
- Brading Marshes to St Helen's Ledges SSSI
- Priory Woods SSSI
- Solent and Isle of Wight Sensitive Marine Area (SMA)
- Southampton Water and Solent Marshes Important Bird Area (IBA)

The Ryde Sands and Wootton Creek SSSI component of the site is important for wintering and migratory wildfowl and waders and contributes towards the importance of the SPA for waterfowl. The foreshore provides intertidal land where wintering and migratory wetland birds can feed.

The intertidal sandflats are used for feeding by wintering and migratory Dark-bellied Brent Goose, Oystercatcher, Ringed Plover, Grey Plover, Lapwing, Dunlin, Curlew and Redshank, species that contribute to the wintering waterfowl assemblage of the Solent and Southampton Water SPA.

Moreover, the shallow intertidal and subtidal waters provide feeding areas for Annex 1 Divers, Grebes and Mergansers.

The Ryde Sands and Wootton Creek SSSI component of the Ramsar site supports habitats that contribute to the range of habitats of a sheltered channel. The foreshore provides intertidal land where wintering and migratory wetland birds can feed (Isle of Wight Council, 2006).

It was identified that any preferred option must maintain, protect and if possible enhance the existing natural environment (Royal Haskoning, 2005).

One of the government's statutory agencies also requested that consideration be given to the appropriateness of any solution and its possible removal if it was found to be having unanticipated undesirable environmental impacts, i.e. not to be working as intended.

2.3 *Geomorphological and coastal processes*

Summary of the Key issues;

- Seagrove Bay is sheltered from the prevailing south-westerly winds and waves and only exposed to waves from 0° N to 120° N. The largest waves approach the bay from the east and south east.
- Tidal currents within the bay are complicated by the adjacent headlands and local bathymetry. It is likely that tidal flows are in a north-westerly direction throughout much of the tidal cycle.
- Sediment is transported northwards into and out of Seagrove Bay, with localised weak drift reversal in the southern end of the bay. The longshore movement of sediment has resulted in the development of an offshore sand bar across much of the bay.
- It is possible that tidal flows within the bay maintain the channel inshore of the sand bar and interrupt sediment supply from the sand bar to the beach.
- It is likely that the ongoing loss of the beach within Seagrove Bay will continue as any natural response and realignment of the shoreline is prevented by the man-made defences.
- Based on the study of historical OS mapping Horestone Point has eroded by some 20 m in the last 80 years.

3 OPTIONS FOR SLOPE STABILISATION

Several methods of stabilising the slip were considered. Methods included reducing the ground water table through drainage, stabilising the coastal slope and seawall against localised failure through the provision of beach in front of it, increasing the shear strength through piling that intersects the slip planes. All of these methods have been adopted to a greater or lesser extent. In general the drainage and piling stabilisation measures were financed by the residents.

3.1 *Slope drainage and seawall stabilisation works*

A sub-horizontal, shore parallel, drain has been installed in the form of a 6 m deep interceptor drain along Pier Road, which lies on the upslope side of all the affected properties. The water collected from the drain is allowed to freely discharge onto the beach. Much of the seawall in the central section of the study frontage has been stabilised against forward movement or overturning, through the installation of soil nails, but not against undermining. Piling has been used to help stabilise the ground below both some new and existing properties. However, the effect of all of this work on the factor of safety of the slope, although significant, would easily be negated if the beach level were allowed to continue to fall or remain low.

3.2 *Coastal works to provide toe weighting to the coastal slope*

Three basic options for toe weighting and protection of the seawall were initially considered including the following;

i. Beach re-nourishment with annual recycling of sand/shingle together with the reconstruction of the seawall
ii. Timber or rock groynes with recycling of sand/shingle every 2 to 5 years e together with the reconstruction of the seawall
iii. Offshore rock breakwaters with occasional recycling of sand/shingle every 10 to 20 years.

4 MATHEMATICAL MODELLING

A study was undertaken to ensure there was the best understanding of the geomorphological processes at the site based on the data available (Royal Haskoning, 2005). The specific aims of the study were to;

- Review the existing data on the site and develop an understanding of the sediment movement processes within the bay and along the adjacent lengths of coast
- Select an appropriate material for any renourishment of the beach within the Bay
- Consider if structures are required to maintain a beach in position when subject to the ongoing wave and tidal action, and to
- Identify any site surveys and numerical modelling that would be required to develop the design of the coast protection scheme.

The aims and objectives of the further work were numerical modelling together with associated studies were as follows;

- Develop a greater understanding of the marine conditions at the site through a use of existing data in combination with limited site measurements of tidal currents and numerical modelling
- Identify and review the potential options available, to provide a stable beach along the toe of the unstable slope and select the most appropriate option for more detailed studies (Sarker, 2005).
- Develop, with the aid of numerical modelling, a preferred scheme layout (Sarker, 2005).

4.1 *Methodology*

Various 1 and 2 dimension mathematical models were selected to optimise the design of a nourishment scheme that would provide long-term increased loading to the toe of the coastal slope and protection to the seawall foundations to prevent the loss of the properties and infrastructure.

The SWAN model was used to transform offshore waves to an inshore location close to the Bay. The SCATTER model was then used to derive the inshore wave climate needed as input data to the sediment movement model UNIBEST-LT. This model was then calibrated to simulate the present longshore sediment movement mechanism. The cross-shore (on-shore/offshore movement) volatility of a sand or shingle beach was assessed using the UNIBEST-DE and BREAKWAT models respectively. Finally the long-term beach evolution was predicted using the BEACHPLAN model.

Model runs were firstly carried out for a sand nourishment option without any offshore rock breakwater. Studies were then extended to one and two breakwater options. Further model runs were carried out for an alternative shingle nourishment scheme, firstly without breakwater and then with one and two breakwaters.

5 THE PROPOSED SCHEME

To accommodate all the constraints identified during the consultation process (undertaken during the development of the scheme), an innovative combination of solutions was needed to provide a sustainable solution.

All of the options set out in Section 3.2 would provide toe weighting to the slope and also protect the seawall. Due to the wave and tidal action, only the third option would reduce the volatility of the beach sediment volume sufficiently to confidently maintain sufficient toe weighting to the slope and protection of the existing seawall.

The proposed coastal works comprise;

- Re-nourishment of the beach with up to 10,000 m^3 of shingle, graded with a D_{50} between 10 and 20 mm (i.e. the 50% point on a grading curve of the shingle lies between 12 and 30 mm). The shingle would be placed to initially create a 5m wide horizontal berm against the seawall, at a level of +3 m ODN, with a seaward face at approximately 1 in 8 slope.

- Two offshore breakwaters constructed from 3 to 6 tonne rock, which would be placed on a filter layer of rock of 0.3 to 1 tonne. These structures would protect the re-nourished beach from movement and loss due to tidal and wave action. The northern structure would be 50 m long and situated some 160 m offshore from the seawall, the southern structure would be 60 m long and would be situated some 215 m from the seawall.
- Each breakwater would have a crest level of approximately +1.9 m ODN. The gap between the crests of the structures would be some 120 m, see Figure 1. The structures would be visible at low water, just breaking the surface at mean high water and fully submerged at mean high water springs. Navigational marker beacons would be installed at each end of both the breakwaters to warn sea users of the presence of the structures during high water periods.
- A rock spur located at the northern end of the Bay, constructed at the root of the Old Chain Pier, also from 3 to 6 tonne rock around an existing concrete structure, is to limit the loss of mobile sediments from the northern end of the Bay due to littoral action. The rock spur would be approximately 17 m long and stand a maximum of 1 m above the existing concrete structure at the landward end. A navigational marker beacon would also be positioned at the end of the rock spur.

Once the scheme has been completed, the shape of the beach will evolve relatively quickly under wave action, on a daily basis, until it reaches an equilibrium form, after which generally more modest changes would occur depending on the prevailing weather/wave conditions and direction.

This solution was selected because it satisfied all the constraints set out by the various consenting agencies and provided an effective solution in technical terms. The solution has been assessed by the competent authority to have no adverse impacts on the natural environment. The scheme will be of significant benefit to the local community by protecting 16 properties and a sewer serving about 100 other properties. During construction, a series of minor adverse impacts will occur due to temporary disturbance from the presence of construction plant, such as increased noise and disturbance to impact on marine flora and fauna.

Following construction, sand is expected to accrete within the lee of the breakwaters. This is anticipated to increase the area of lower intertidal within the bay, which would be of benefit to walkers and other recreational users of the bay. However, it will have an adverse effect on the use of the boat moorings as it would cause them to dry out for a greater proportion of the tidal cycle than at present.

The offshore breakwater structures and/or rock spur would be easily adjustable in response to climate change, by simply adding, (or possibly removing) rocks to the breakwaters to increase their height. The breakwaters may also need to be removed either when they reach the end of their residual life or if the scheme was found to have unanticipated undesirable environmental impacts. Removing the rock breakwaters, for whatever reason, would be possible and could in practice be undertaken by the reversal of the construction process, using the same type of equipment. The rock could then be re-used elsewhere.

6 CONCLUSIONS

- The environmental constraints on any works undertaken at the site required an innovative amalgamation of techniques to be developed to stabilise the slope in a sustainable manner.
- The slope stability problem to be addressed needed the combined expertise of both geotechnical and coastal specialists.
- The preferred option is easily adaptable for climate change and is removable if unanticipated or undesirable environmental impacts occur.

REFERENCES

Halcrow Maritime. 2001. Universities of Portsmouth and Newcastle and the Meteorological Office Preparing for the Impacts of Climate Change. *Report to SCOPAC*, 101 pp.

Isle of Wight Council. 2006. Appropriate Assessment for Seagrove Bay. Isle of Wight.

King, C. 2004. Geological Study of the Foreshore and Adjacent Coastline at Seagrove Bay. Seaview, Isle of Wight.

Lee, E. M., Moore, R. & McInnes, R. G. 1998. Assessment of the Probability of Landslide Reactivation: Isle of Wight Undercliff, UK. *Proceedings of 8th International Association of Eng. Geologists Conference (Vancouver)*, 1315–1321pp.

Royal Haskoning. 2002. Seagrove Bay Beach Level Investigation.

Royal Haskoning. 2005. Seagrove Bay Coast protection Scheme, *Environmental Statement.*

Royal Haskoning. 2005. Seagrove Bay, Isle of Wight, Development of Nourishment Scheme.

Sarker, M. A. 2005. Appropriate Integration of Numerical Models for Design Optimisation of Coast Protection schemes: *Seagrove Bay case study*.

Sarker, M. A. 2005. Design Optimisation of Coast Protection Schemes: *Seagrove Bay case study.*

Skempton, A. W. 1945. Earth Pressure & the Stability of Slopes. *The Principles and Application of Soil Mechanics.* 31–61pp.

UKIP02. 2002. Climate Change Scenarios for the United Kingdom.

Woodruff, M. 2004. Geotechnical report on the North End of Seagrove bay, Seaview, Isle of Wight.

Landslides and Climate Change – McInnes, Jakeways, Fairbank & Mathie (eds)
© 2007 Taylor & Francis Group, London, ISBN 978-0-415-44318-0

Recent landslide impacts in Scotland: Possible evidence of climate change?

A.J. Mills, R. Moore, J.M. Carey & S.K. Trinder
Halcrow Group Ltd, Birmingham, United Kingdom

ABSTRACT: Landslides and slope failures rarely lead to injuries or fatalities on UK roads although they can lead to significant disruption, damage and other costs including emergency response work and remedial measures. A good example is the landslide impacts on the Scottish road network which were triggered by a prolonged and intense rainstorm in early August 2004. The landslides resulted in blockage of several roads and whilst no one was injured, 57 people had to be airlifted to safety. A similar event occurred in the Shetland Islands in September 2003, following an intense local rainstorm, when several motorists narrowly missed being hit by a fast-moving landslide. The landslides at Channerwick in the Shetlands were unusual in that they originated from blanket peat upland and developed into fast-moving debris flows. The landslides caused the temporary closure of the main road connecting Lerwick, the capital town of the Shetlands, to the international airport at Sumburgh. Clear-up operations and remedial works cost in excess of £1 million and the incident prompted a detailed investigation of the causes and mechanisms of the landslides so that lessons could be learnt and appropriate measures implemented to avoid a repeat event. This paper focuses on the investigations and follow-up actions of the Shetland landslide event from a technical and road authority point of view. The influences of climate change on historical and projected frequency and magnitude of peat landslides is reviewed.

1 INTRODUCTION

Great Britain is not renowned for its landslide hazards, with the most publicised landslide studies and investigations being confined to the upland areas of Scotland and Wales and the coastal cliffs of southern and eastern England (Hutchinson & Bromhead 2002; Lee 1999; Ballantyne 1986 and Miller 1974). The national review of landsliding in Great Britain carried out in the late 1980s (Jones & Lee, 1994) revealed that landslide deposits and surface morphology are widespread, and that far from being a benign landscape, much of Great Britain has a legacy of pre-existing landslides and is subject to contemporary landslide activity in various locations.

Most documented landslides in Scotland belong to one of four categories: non-rotational rock slope failures; rotational rock slope failures; shallow debris flows and debris slides developed in superficial deposits (Ballantyne 1986). The national review of landslides identified 625 reported rock slope failures compared with 115 shallow landslides within superficial deposits. The majority of rock slope failures are ancient features associated with the metamorphic and igneous rocks of the Highland, Hebrides and Midland Valley. The shallow debris flows and debris slides are younger and generally involve failure of glacial sediments or peat blanket and, therefore, the extent of peat deposits and former glacial activity largely controls their distribution. The national review of landslides recognised that the true number of existing landslides may be considerably more than quoted given that many landslides remain unreported, particularly in remote regions.

Two notable multiple landslide events occurred in Scotland following prolonged and intense rainfall in September 2003 and August 2004. The former event occurred at Channerwick on the Shetland Islands, during which several large-scale shallow landslides overran the main trunk road on the island, narrowly missing several vehicles and causing temporary closure of the road. The latter, much publicised event resulted in blockage of several roads over a wider area (Glen Kinglas, Glen Ogle and Cairndow), with 57 people airlifted to safety. Fortunately there was no loss of life in either case.

2 LANDSLIDE IMPACTS IN SCOTLAND

2.1 *Debris flows in central Scotland*

The landslide events of August 2004 had a substantial impact on Scotland's trunk road network, both in terms of disruption experienced by local and tourist traffic, as well as that to goods vehicles. The tourist industry, which reaches its peak in the summer months, was also significantly disrupted.

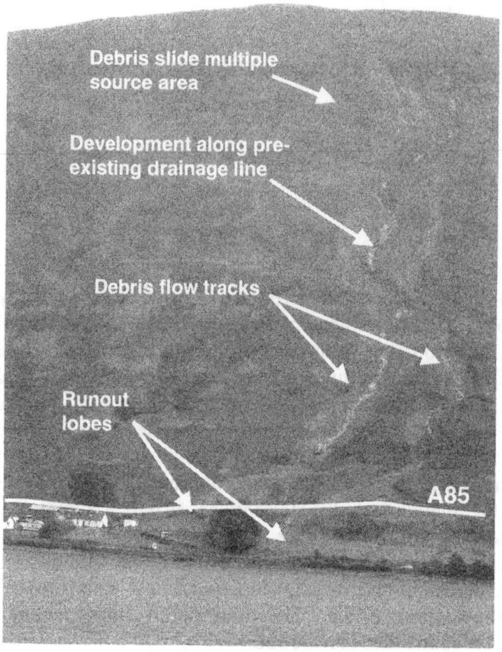

Figure 1. Upland debris slide and flow development causing damage and disruption to the A85 Scotland 2004.

Figure 2. Peat slide debris run-out, Channerwick Burn, Shetland Islands south mainland.

The landslides were triggered by rainfall substantially in excess of the norm. Some areas of Scotland received three times the 30-year average rainfall for August. The rainfall was both intense and long lasting and a large number of landslides, in the form of debris flows, were experienced in Scotland (Fig. 1). A small number of these impacted the trunk road network, notably the A83 between Glen Kinglas and to the north of Cairndow, the A9 to the north of Dunkeld, and the A85 at Glen Ogle. Some 57 people were airlifted to safety after being trapped between two debris flows on the A85 in Glen Ogle (Fig. 1). The A85, carrying up to 5,600 vehicles per day, was closed for four days; the A83, which carries around 5,000 vehicles per day, was closed for two days; and the A9, carrying 13,500 vehicles per day, was closed for two days (Winter et al., 2005).

2.2 *Peat slides in Shetland*

A comparable rainfall induced landslide event occurred in the Shetland Islands on Friday 19 September 2003, when multiple peat slides and debris flows caused temporary closure of the A970 between Cunningsburgh and Levenwick on the south mainland of Shetland (Fig. 2). Up to 30 peat slides occurred within an area of 13 km^2 centred on Channerwick. The largest failures initiated above the A970 and resulted in high velocity run-out of peat debris onto the road, which became impassable. Several cars narrowly missed being hit by the peat slides and debris flows.

Extensive damage was caused to approximately 800 m of crash barriers and fencing, and culverts were blocked by peat debris captured by floodwaters causing widespread flooding. Shetland Islands Council cleared the A970 of debris on the same day but restricted its use over the following weeks whilst regrading and drainage works were carried out on the failed slopes above the road. The total damage costs of the event have been estimated to exceed £1 million in road reconstruction, slope remediation and drainage, site investigation and clear-up operations (Halcrow Group Ltd, 2004).

In common with other recorded peat slides in the UK, the Channerwick failures occurred during an intense rainfall event. The time at which landsliding occurred was estimated to be between 0530 and 0830 in the morning. Site-specific rainfall data was not available and this precluded even coarse estimates of rainfall intensities based on rainfall radar echoes. The storm was noted by residents to be very intense and localised, with locations just a few kilometres away experiencing minimal rainfall. Anecdotal evidence from local residents indicates an average storm intensity of about 33 mm hr^{-1}. The surface water runoff generated by the rainstorm was photographed by a local resident (Fig. 3).

Previously reported landslides in the vicinity of Channerwick occurred in August 2000. Prior to this, landslides had been reported in the Shetlands in the 1990s, 1980s, 1950s and 1935 (The Shetland News 1935; 1950). Accounts of the 1935 and 1950 events cite 'tropical' intensities, and also describe major flooding of the local burns. The only measure of storm intensity is provided for the 1935 failure in Weisdale, in which 90 mm of rain fell in 26 hours.

Figure 3. Excess overland flow during the storm that caused peat slides at Channerwick.

3 LANDSLIDE MECHANISMS

The long term weathering of soils and rocks make upland slopes increasingly susceptible to failure. Such preparatory processes have been described as the 'ripening' of slopes. They are often overlooked as a major cause of landslides given the long timescales over which they operate but they are a fundamental control in the location and timing of upland landslide events.

Steep upland slopes which are mantled by a cover of unconsolidated soils or peat are particularly susceptible to debris slides and debris flows. Debris slides have shorter run-out trails than debris flows but pose a significant hazard where they occur close to transport infrastructure and development, or where their run-out propagates as a debris flow.

Debris slides involve shear failure of unconsolidated material or peat at the interface with underlying weathered rock, which typically varies between 1 to 5 m below ground surface. Rapid increases in porewater pressure along this interface result in significant reduction in effective shear strength, leading to rupture or shear failure along the soil-rock interface. Once mobilised, debris slides remain relatively intact throughout their travel. They are characterised by a

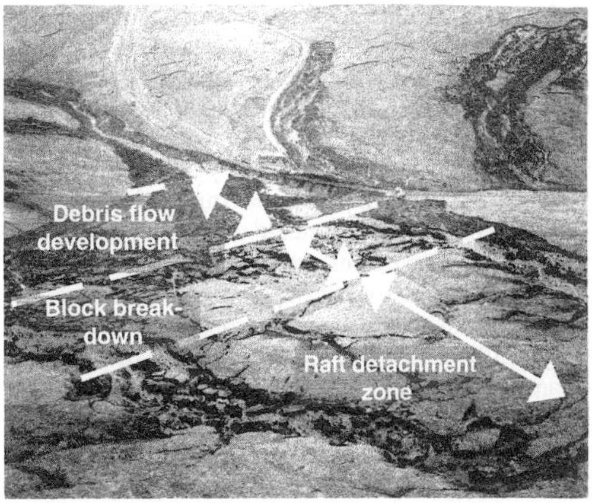

Figure 4. Changes in peat transport mechanisms, Channerwick, Shetland Islands south mainland.

bowl-shaped source area with a debris trail much the same width as the source area. Debris slides have typical length-to-width ratios of between 5:1 and 10:1, or more (Cruden & Varnes 1996; Dikau et al 1997).

Peat slides are a form of debris slide in which the major volume of displaced material is comprised of organic deposits, or peat. Shear failure usually occurs at the interface between the peat deposit and the underlying substrate, usually weathered rock or till (Warburton et al, 2004).

Debris flows are generally characterised by high-velocity events and long run-out and may pose a significant hazard downslope of existing gullies and mountain stream catchments. A debris flow has a distinct morphology comprising a single or multiple bowl-shaped source area, a narrow, often sinuous run-out trail and a lobate debris fan (Fig. 1). They occur on steep hillsides (hillside debris flows) or result from the mobilisation and run-out of debris from slope declivities or stream valleys (channelised debris flows). Hillside debris flows may develop their own channels due to the formation of levées (debris ridges) along the run-out trail (Blinjenberg 1998).

Hillside debris flows typically start as a sliding detachment of material (upland debris slide, peat slide, rock slide etc.), usually initiated during heavy rainfall, which subsequently breaks down into a disaggregated mass in which shear surfaces are short-lived and usually not preserved. Once the slide is in motion and depending on the coherency of the displaced mass, the slide breaks up during movement. The failure develops into a debris flow when the debris comes into contact with surface water, stream flow and on entry into swollen rivers, dramatically increasing the viscosity of the debris-water mix. Figure 4 illustrates the morphological evidence for such transport mechanisms in one of the peat slides at Channerwick.

As a general rule, where the constituent particles of the slide debris cease to be in contact and become supported by fluids, a change in mechanism from debris slide to debris flow takes place. This transition may be very rapid once the slide debris makes contact with surface water or stream flow. In peatlands, multiple subsurface water pathways (pipes) may exist within the peat deposit, supplying water to the moving debris during transport as it passes over intact peat or is supplied by overland flow from upslope.

Shetland Islands Council commissioned Halcrow Group Ltd to investigate the multiple peat slide event at Channerwick and identify the mechanisms and likely causes of failure in order that future landslide potential might be better understood and managed.

4 CHANNERWICK PEAT SLIDE SITE INVESTIGATION

4.1 *Context*

Great Britain and Ireland contain extensive peatlands. Despite this, investigations into the properties of peat have remained limited as human populations have concentrated in areas outside generally removed from its peat uplands. Whilst some progress has been made in the investigation of peat consolidation (e.g. Hobbs 1986 and Adams 1962), the shear strength of peat is still poorly understood. Detailed site investigations in peatlands are notoriously difficult due to the nature of the materials, the relative difficulty of access to the remote uplands in which they are found, and the harsh climate. In such instances, standard geotechnical investigations are often not appropriate and specially designed site investigations are required. Such an investigation was commissioned by Shetland Islands Council following the event at Channerwick and involved a team comprising local highway authority and Halcrow staff. The investigation was undertaken over a 4-week period in May 2004.

4.2 *Geomorphology and hydrogeology*

Site surveys were carried out for the main peat slide failures to evaluate the slope morphology, drainage, and other geomorphological features of interest and to identify sites of potential incipient failure. Observations indicated that all failures initiated as translational peat slides over a range of slope angles with shear surfaces developed at the interface of the base of peat with the underlying weathered schist. Large 'rafts' of peat up to 1.7 m in depth were displaced, many of which broke down into 'block and slurry' peaty debris flows. *In situ* peat surrounding the failures revealed widespread cracking, suggesting wider disturbance of the peat blanket and several areas of potential incipient failure.

Figure 5. Peat pipe in exposed landslide scarp.

The hydrological setting was significant, as extensive pipe networks were observed to discharge into slide scars at the interface at the base of the peat (Fig. 5). In a number of locations, pipe networks could be traced upslope with collapsed pipe ceilings evident on the peat surface. In addition, saturated flush zones were observed above the slide scars. These drainage pathways almost certainly facilitate rapid water transmission from the pool and hummock complexes at the hillslope summits to the peat slides in the mid-slopes. The extent and character of the peat slide and debris flow run-out zones varied for each site, reflecting site-specific run-out propagating factors such as slope angle and depth of failed peat.

4.3 *Site investigation*

Three sites were chosen for detailed investigation (Fig. 6) all of which failed in September 2003. Site 1 was subsequently remediated after the event. Sites 2 and 3 are natural peat slides on the north and south facing slopes above Red Burn. At each site detailed ground investigation comprised dynamic probing, hand augering, trial pitting, peat logging and material sampling. The slopes were also instrumented to monitor groundwater levels, overland flow and ground movement. The logging and classification of peat used a combination of British Standards for soil logging, Troel-Smith (1955) and Von Post Classification (Hobbs 1986).

4.4 *Results of site investigation*

Results of the ground investigation and peat logging identified 4 distinct layers forming the peat blanket, which reached up to 1.7 m in depth in and around the peat slide sites, an upper spongy fibrous peat close to the surface; a mid layer of firm light brown fibrous peat with some organic clay, a dark brown fibrous peat

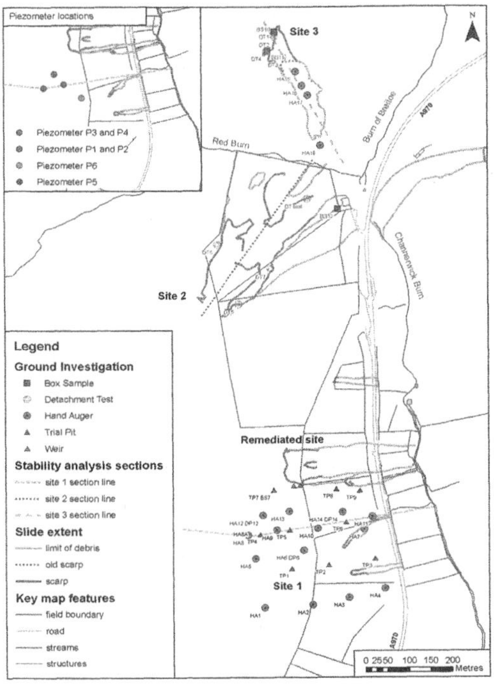

Figure 6. Channerwick ground investigation plan.

and a basal amorphous peat which overlay weathered schist. The organic solids and high water content result in complex mechanical properties of peat, further complicated by the highly fibrous nature of the peat which promotes anisotropy and internal reinforcement (Edil & Wang 2000).

A combination of innovative field strength tests, standard and specialist laboratory tests and infinite slope modelling were conducted to determine the likely mechanisms and sensitivity of parameters from back analysis of failure. *In situ* strength tests were performed on detached blocks of peat (Fig. 7). The tests demonstrated that a rapid change from peak to residual strength could occur at the peat and weathered schist interface, which could explain the rapid acceleration of peat rafts at the onset of failure.

Standard shear box tests were performed on fibrous peat, amorphous peat and weathered schist to determine the effective shear strength parameters for infinite slope stability analysis of the sites. The stability analysis demonstrated that the sites were stable under normal conditions but were sensitive to changes in slope angle and the porewater pressure regime. This confirmed that significant porewater pressures would have been needed to trigger the failures. The limited number of peat slides recorded on the island suggests that only infrequent high-magnitude rain storms have the potential to cause the rise in porewater pressures needed to initiate failure.

Figure 7. Peat detachment testing to define field shear strength of in situ peat blocks.

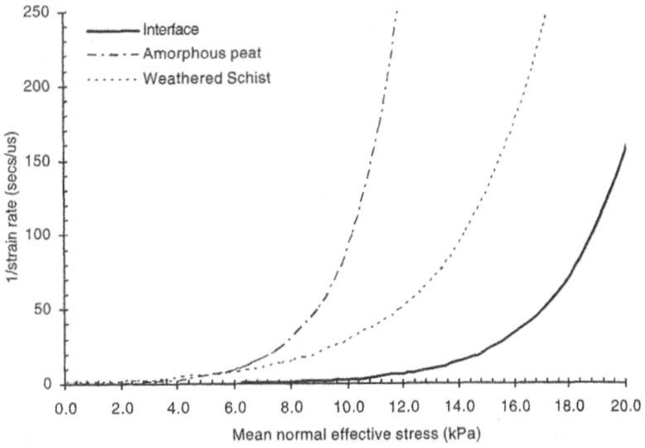

Figure 8. Stress-strain relationship for failed materials tested in back-pressured shear box.

Figure 8 demonstrates that rapid increases in the rate of strain accompanied inflated porewater pressures during testing, and that this effect was most pronounced at the interface between peat and the weathered schist substrate.

5 SHALLOW LANDSLIDE HAZARD POTENTIAL IN SCOTLAND

The importance of understanding landslide mechanisms in upland UK is increasing as political pressure for renewable energy technologies also increases. Wind farms, concentrated in upland and often peat covered areas, are seen by many as the means by which carbon emissions might be reduced, and the UK's reliance on fossil and nuclear fuels might be reduced. The high environmental value afforded to peat uplands has required that the positive impacts of wind farm developments are evaluated against the negative consequences for local peat areas and their often diverse and unique habitats.

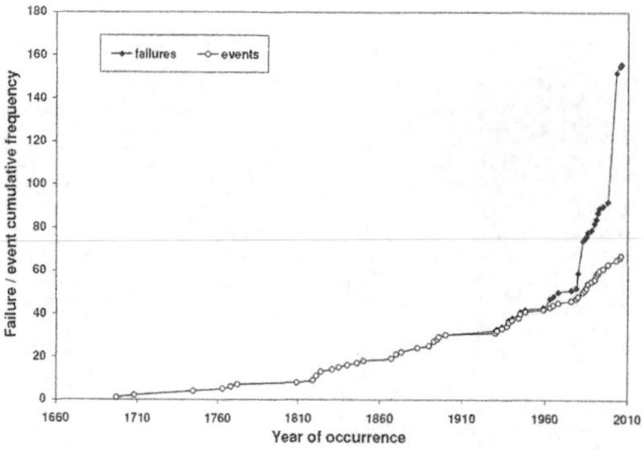

Figure 9. Frequency of peat failures in the UK and Ireland.

Month	Rainfall (millimetres)		
	2002	Mean (1961–1999)	Extreme daily values (1922–1999)
Jan	170	133	43 ('37)
Feb	161	93	47 ('51)
Mar	95	115	47 ('52)
Apr	55	73	37 ('40)
May	70	62	50 ('66)
Jun	63	62	37 ('81)
Jul	57	64	94 ('35)
Aug	55	77	47 ('94)
Sep	44	118	51 ('81)
Oct	94	136	61 ('81)
Nov	144	143	71 ('49)
Dec	120	144	68 ('95)
Year	1128	1220	

Figure 10. Annual (2002), monthly and extreme rainfall for the Shetland Islands.

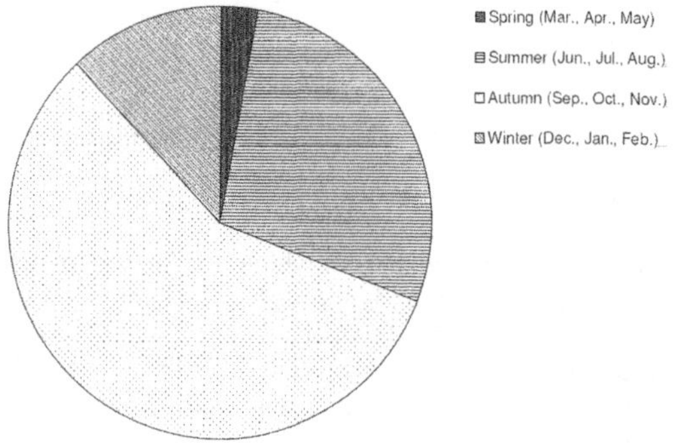

Figure 11. Seasonal distribution of peat landslides in the UK and Ireland.

Just as wind farms and their associated infrastructure may be affected by (or even cause) shallow slides and debris flows, other infrastructure such as road networks, flood defences, drainage and power lines may also be affected. Until recently, the incidence of slides and flows such as those experienced recently in Scotland and Shetland have been limited in number, and few if any measures have been taken to avoid or protect against such events.

Historical records suggest there has been an increase in the frequency and extent of shallow slides and flows in the past several hundred years. In the case of peat slides, this trend has seen a marked upturn in the last few decades (Fig. 9).

Figure 9 illustrates the rapidly increasing frequency of landslides recorded in British and Irish peatland environments and the associated frequency of triggering events, such as intense rainstorms. One severe storm may trigger a single failure in a particular location or trigger multiple failures. Although shallow failures in mineral soils are often quick to recover, peat failures represent a permanent loss of 'soil' cover.

The curve appears to indicate that the rapid increase in landslide frequency in recent years is associated with a growing occurrence of multiple failure events during single severe storm events, rather than an increasing number of individual failures. This indicates that the landslide 'events' in peat are becoming more severe, potentially resulting in greater societal and economic loss per event (depending upon the proximity of population to the failure site). Comparison of this peat landslide record with other landslide datasets would be a useful means of determining if similar patterns exist in non-peatland environments.

Future predicted increases in winter rainfall and storm intensity due to climate change has raised concern that shallow slides and debris flow events may become more frequent in the Shetland Islands.

Figure 10 shows mean monthly rainfall for the Shetland Islands, with extreme daily values and the year in which they occurred (SICEDU, 2003). The values are based on recordings in Lerwick, and in likelihood do not reflect locally intense, orographically influenced rainstorms in the upland areas of the islands.

Figure 11 shows the seasonal distribution of peat landslide events recorded in the UK and Ireland. It demonstrates that a majority of recorded failures have occurred in late summer/autumn months; supporting the notion that landsliding in peat occurs most readily as a response to summer drying, in likelihood followed by intense convective rainfall (for which gauged rainfall intensities are usually unavailable).

UKCIP predictions for regional changes in the United Kingdom do not incorporate the Shetland Islands. However, extrapolating from North Scotland to Shetland under the medium-high emissions scenario allows future peat landslide potential to be gauged in relation to changes in climatic controls. The UKCIP report does not attach probabilities to the emissions scenarios, but the medium-high scenario is considered to be a 'business as usual' evolution of current trends (Galbraith et al 2002). Although annual change in precipitation is expected to remain within 'natural' variability, autumn, winter and spring rainfall are projected to increase by up to 20% (Hulme et al, 2002). Summer rainfall is projected to decline by between 0 and 20%. The tabulated values shown in Figure 10 broadly support this expected trend. Changes in storm intensity, represented by change in number of intense rainfall days per seasons are projected to increase slightly through autumn, winter and spring and remain static in summer. Annual temperature changes for Shetland's maritime climate are projected at approximately 2°C. The implications of these projected trends are that while the incidence of triggering events throughout the year will not change significantly, the pre-conditioning of ground towards failure through drying may be enhanced. The cumulative effect of increased drying in combination with a similar incidence of storminess would be to increase the likelihood of future peat landslide events.

For shallow slides outside peatlands, a trend in rising occurrence has been attributed to a number of factors: a) progressive pedogenesis (podzolisation) which alters the hydrological transmissivity of soils, rendering them more vulnerable to failure during rainstorms (Brooks et al 1995); b) an increase in the frequency and intensity of rainstorms; and c) land management practices, including deforestation, overgrazing or heather burning. These controls apply particularly to landslides in peat environment. The hydrophobic qualities of peat are such that beyond a certain degree of drying, peat deposits are unable to re-wet, and irreversible physical changes occur including loss in density and loosening of structure. The combined effects of this may render artesian pressures more easily attained through increased transmission of surface water to depth, and through loss in mass of the peat blanket, increase buoyancy of the surficial peat deposits.

Under current climate change scenarios, drier summers can be expected to lead to greater drying and shrinkage of soils causing tensile fractures. This tendency is reflected in the clustering of peat slides in the months of July and August, after summer drying and in a period during which high intensity convective storms are more common. In the longer, wetter winter periods, peat slide occurrence is noticeably limited (Mills, 2000).

6 LANDSLIDE HAZARD MANAGEMENT

6.1 *The way forward in Scotland*

The shallow debris slides and flows that occurred in Scotland in 2003 and 2004 caused considerable losses in terms of damage to infrastructure and disruption to traffic. The events serve as an early warning of what could happen if events of this nature were to recur. In both examples cited, many vehicles and their occupants narrowly avoided being hit by the rapid landslides and the consequences could have been far worse.

A number of parallel and integrated approaches are being taken to manage landslide risk in Scotland. In 2002, the Scottish Executive adopted the principles behind the publication Managing Geotechnical Risk (HD22/02) which forms a useful philosophical basis on which to develop a landslide risk reduction strategy. This work is at an early stage of development (Winter et al., 2005), but nevertheless establishes the principles of the risk reduction strategy that will be adopted. The strategy will be informed by a hazard ranking procedure for individual sites so that decisions can be made on the selection of appropriate site mitigation measures according to the likelihood and potential severity of the landslide hazards and consequences.

In 2006, the Scottish Executive made assessment of peat landslide hazard a specific requirement for all wind farm applications over peat covered sites in Scotland. Compilation of criteria for a best practice guide for peat landslide hazard assessments was at least partly based on lessons learned from the Channerwick peat slide investigations. In this guidance, the loss of peat cover represented by an individual landslide event or a landslide cluster is considered not only in terms of potential risks to the viability of the wind farm development and neighbouring land, but also in terms of a net carbon loss from the global terrestrial carbon store.

In 2006, Shetland Islands Council commissioned a second stage of investigation into peat slides in Shetland. This new phase of work is intended to generate planning guidance for areas at risk of future peat slide occurrence. At the same time, a medium-term hydro-meteorological monitoring programme, combined with slope monitoring, is collecting data at the Channerwick site to enable improved understanding of hydrological response times to rainfall events in peat deposits known to be susceptible to failure.

7 CONCLUSIONS

Shallow landslides and debris flows could occur at many locations across the transportation network in Scotland, not least in the many areas of peat-covered upland.

This paper contends that the shallow slides and flows in Scotland are the result of long-term geomorphological processes (slope ripening) and land management practices which result in the gradual reduction of the factor of safety against hillslope failure over time. The triggering event is often very intense rainfall (c50–100 mm in 3–24 hrs). Geomorphological and geotechnical evidence supports a shear failure mechanism fed by high porewater pressures arising from this intense rainfall, and facilitated by desiccation cracking of peat. Increased drying of peatlands and more frequent cracking, coupled with an increase in intensity of rainstorms may give rise to greater occurrence of peat slides in future years.

At Channerwick, a variety of measures are being implemented to mitigate this possibility, including extensive drainage, installation of a hydrometeorological monitoring regime, construction of catch fences, embankments to divert flows away from the road, and a regionally based peat landslide hazard assessment.

ACKNOWLEDGEMENTS

The authors are grateful for the helpful advice and contributions of Shetland Islands Council and Halcrow colleagues, Dr Jeff Warburton, Professor Brunsden, Professor David Petley and Dr Toru Higuchi who assisted with the site investigation and specialist laboratory testing at Channerwick, Shetland Islands.

REFERENCES

Adams 1962. Laboratory Compression Test on Peat. *Ontario Hydro Research News*, Third Quarter, pp. 35–40.

Ballantyne, C.K. 1986. Landslides and slope failures in Scotland: a review. *Scottish Geographical Magazine*. Vol.102 pp.134–150.

Blijenberg, H. 1998. *Rolling Stones? Triggering and Frequency of Hillslope Debris Flows in the Bachelard Valley, Southern French Alps*. August 1998, 223 p.

Cruden, D.M. & Varnes, D.J. 1996. Landslide Types and Processes. Chapter 3, Landslide Investigation and Mitigation, *Transportation Research Board, Special Report* 247, National Research Council, pp. 36–75.

Dikau, R., Brunsden, D., Schrott, L. & Ibsen, M. 1997. *Landslide Recognition*. International Association of Geomorphologists: Publication No. 5, Chichester, UK. John Wiley & Sons. 251 p.

Edil, T.B. & Wang, X. 2000. Shear strength and K0 of Peats and Organic Soils. In *Geotechnics of High Water Content Materials*, ASTM STP 1374 (Edil TB and Fox PJ eds.) American Society for Testing and Materials, West Conshohocken, PA.

Galbraith, R.M., Price, D.J. & Shackman, L. (2005) *Scottish Road Network Climate Change Study*, Scottish Executive.

Halcrow Group Limited 2004. *Shetland A970 Channerwick Peat Slides – Interpretative Report*. Report No. R5917. Shetland Islands Council.

HD22/02 2002. *Managing Geotechnical Risk – Design Manual for Roads and Bridges Volume 4 Geotechnics and Drainage*.

Hobbs, N. 1986. Mire Morphology and the properties and behaviour of some British and foreign peats. *Quarterly Journal of Engineering Geology*, London, Vol. 19. pp7–80.

Hudleston, F. 1930. The cloudbursts on Stainmore, Westmorland, June 18th 1930. *British Rainfall*, 1930, 287–292.

Hulme, M., Jenkins, G.J., Lu X., Turnpenny, J.R., Mitchell, T.D., Jones, R.G., Lowe, J., Murphy, J.M., Hassell, D., Boorman, P., McDonald, R., & Hill, S. 2002 *Climate Change Scenarios for the United Kingdom: The UKCIP02 Scientific Report*. Tyndall Centre for Climate Change Research, School of Environmental Sciences, University of East Anglia, Norwich, UK.

Hutchinson, J.N. & Bromhead, E.N. 2002. Isle of Wight landslides. In *Instability, Planning and Management* (McInnes RG and Jakeways J eds.) Thomas Telford, London, pp. 3–70.

Jones, D.K.C. & Lee, E.M. 1994. *Landsliding in Great Britain*. HMSO, London.

Lee, E.M. 1999. Coastal Planning and Management: The impact of the 1993 Holbeck Hall landslide, Scarborough, *East Midlands Geographer*, 21, pp.78–91.

Lee, E.M., Moore, R. & Clark, A.R. 1996. *Landslide Investigation and Management in Great Britain: A Guide for Planners and Developers*, ISBN 0 11 753180 4, HMSO, London.

Miller, J. 1974. *Aberfan–a disaster and its aftermath*. Constable, London.

Mills, A.J. 2000. *Peat slides: Morphology, Mechanisms and Recovery*. Unpublished PhD thesis, University of Durham. 426 pp.

Moore, R., Lee, E.M. & Palmer, J.S. 2002. A sediment budget approach for estimating debris flow hazard and risk: Lantau, Hong Kong. In McInnes RG and Jakeways J (eds) *Instability, Planning and Management: seeking sustainable solutions to ground movement problems*, Thomas Telford Publishing, 2002.

Moore, R. Lee, E.M. & Clark, A.R. 1995. *The Undercliff of the Isle of Wight: a review of ground behaviour*. South Wight Borough Council.

SICEDU (2003) *Shetland in Statistics*. Shetland Islands Council Economic Development Unit, Lerwick.

Sidle, R.C., Pearce, A.J. & O'Loughlin, C.L. 1985. *Hillslope Stability and Land Use*. American Geophysical Union, Washington D.C.

Troel-Smith, J. 1955. *Characterisation of Unconsolidated Sediments*. Geological Survey of Denmark. IV. Series. Vol. 3.No. 1.

Warburton, J., Holden, J. & Mills, A.J. 2004. Hydrological controls on surficial mass movements in peat, *Earth Science Reviews*, 67 (1–2): 139–156.

Winter, M.G., Macgregor, F. & Shackman, L. 2005. *Scottish Road Network Landslides Study. Summary and Technical Report prepared for the Scottish Executive*.

Landslides and Climate Changes – McInnes, Jakeways, Fairbank & Mathie (eds)
© 2007 Taylor & Francis Group, London, ISBN 978-0-415-44318-0

Geotechnical centrifuge modelling of slope failure induced by ground water table change

S. Timpong, K. Itoh & Y. Toyosawa
Construction Safety Research Group, National Institute of Occupational Safety and Health, Tokyo, Japan

ABSTRACT: This paper describes the development of a new in-flight ground water table control system in the centrifuge, which can be used to investigate the mechanism of slope failure induced by the ground water table change. The failure mechanism of model sandy slopes observed in the centrifuge is similar to that of a rainfall-induced landslide. The variations of pore water pressure and surface settlement measured during the test provide useful information for predicting the occurrence of slope failure in the future.

1 INTRODUCTION

In Japan, the rainfall-induced landslide is one of the most destructive natural disasters that occur frequently in natural and man-made soil slopes, causing significant property damage and loss of life. In general natural slopes are in unsaturated condition and the ground water table is relatively deep. An increase of ground water table due to the infiltration of rainwater into the slope during heavy rainfall periods can cause the generation of positive pore water pressure, increases soil weight and decreases shear strength of soil, which can significantly making the slope more susceptible to failure. Figure 1 shows the variations of the number of landslides and the annual national precipitation in Japan during the period of 1984–2004. The annual national precipitation was calculated from the average annual precipitation of 47 prefectures in Japan.

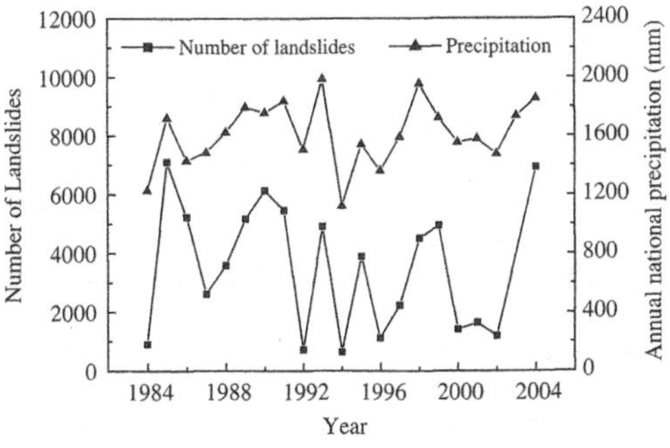

Figure 1. The variations of the number of landslides and the annual precipitation in Japan during 1984–2004.

The variation of landslide event is similar to that of the annual precipitation. It is quantitatively indicated that the landslide events are related to the amount of precipitation.

Global climate change has significant impact on the world environmentally; increasing global temperatures will lead to changes in frequency and intensity of rainfall, typhoon pattern and sea level. Increasing the number and intensity of heavy rainfalls will increase the risk of rainfall-induced landslides in the future. Several case studies illustrated that the frequency and intensity of heavy rainfall events in Japan might change due to the climate change. Iwashima & Yamamoto (1993) demonstrated that the frequency of appearances of extreme daily precipitation was increased in the past 100 years due to the global warming. Higashi & Matsuura (2006) described temporal and regional changes in precipitation in Japan during a period of 1961–2002. Their results indicated that the changes in the heavy rainfall in Japan were related to the El Nino-Southern Oscillation (ENSO). According to the climate change monitoring report of the Japan Meteorological Agency, the mean surface temperature in 2004 was estimated to be +0.99°C above the normal value (1971–2000 average), and it was the second highest after 1990 in the last 107 years. Temperature anomaly had been rising at a rate of about 1°C per 100 years since 1898. In particular, the temperature was rapidly increased since the late 1980s. This tendency was almost the same as that of the worldwide temperatures. The annual national precipitation was 113% of the normal value and the annual precipitations were more than 120% of the normal value at many observation stations from the northern to the western of Japan. Extreme rainfalls in the late of rainy season and a total of ten landfalls of typhoons (the maximum record) caused a large number of landslides in 2004. The main objective of this paper is to purpose a new development of the geotechnical centrifuge modelling technique for slope failure induced by ground water table change during the heavy rainfall periods. The model slope tests were conducted with the aim of increasing the understanding between the variations in pore water pressure and surface settlement during the rainfall events and their effect on slope stability.

2 CENTRIFUGE MODELLING

Although a small-scale model test has been widely used in the study of seepage and rainfall-induced slope failure (Orense et al. 2005), it is exceedingly difficult to simulate the in-situ condition of the slope due to the gravity effect. Realistic self-weight induced stresses in the small-scale model can be conveniently created in the centrifuge and the failure mechanism of the model slope will be more similar to the prototype. In this paper, the model slope tests were performed in the NIIS Mark-II centrifuge (beam type centrifuge) at the centrifuge laboratory of the National Institute of Occupational Safety and Health. The NIIS Mark-II centrifuge has an effective radius of 2.3 m with a maximum acceleration capacity of 100 g. It is capable of simulating the static and dynamic geotechnical problems.

2.1 *Model test container*

A rigid aluminum model test container with acrylic transparent wall on one side was constructed to allow the visual observation of model slope during the centrifuge test. The internal dimensions of the model container were 25 cm wide, 80 cm long and 70 cm high. The model container was divided into three sections; a central portion (70 cm long) for making the model slope, a water supply chamber and a drain chamber (each 5 cm long). These sections were separated by perforated steel walls covered with wire mesh to allow the water flow in and out of the model slope without migration of soil particle. Figure 2 shows the photograph of the model test container used in the present study.

2.2 *In-flight ground water table control system*

Figure 3 shows a schematic view of the in-flight ground water table control system, which is capable of simulating the infiltration from rainfall to bring the ground water table to the desired level. The control system consists of water supply and drain tanks, pressure control system, reversible motors, float switches, two-port solenoid valves, limit micro switches and potentiometer-type displacement transducers. After

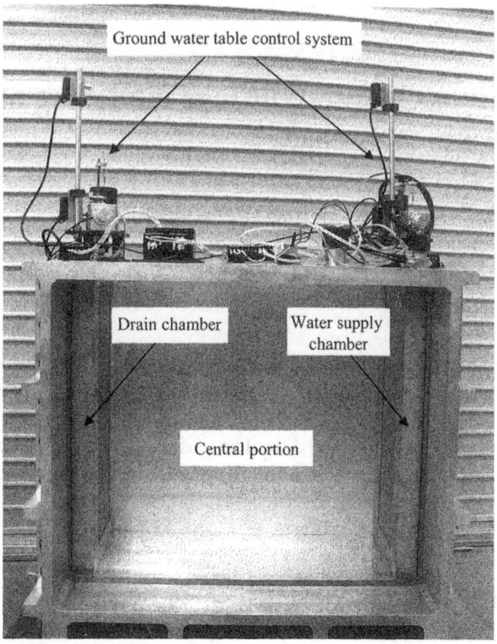

Figure 2. The photograph of the model test container.

the centrifuge acceleration is reached at the intended level, the solenoid valves at the water supply and drain chambers are opened. The water supply tank feeds the water in the water supply chamber and when the water level rises to the trigger point of the float switch, the switch will actuate the solenoid valve to switch off. When the water level in the water supply chamber is lowered due to water seepage through the model slope, the float switch is turned on again and the solenoid value is then opened to allow the water flow in the water supply chamber. Once the water level in the water supply chamber recovers to the prescribed level again, the float switch is turned off to shut down the water supply.

Similar water control system can be described for the drain chamber at the downslope. The water levels in the water supply and the drain chambers are controlled by the potentiometer-type displacement transducers. In order to limit the maximum and minimum position of the reversible motor, the braking system using the limit micro switch is provided for each reversible motor. The water level in the water supply and the drain chambers can be controlled in real-time by the computer during the centrifuge test. The control system was calibrated in the 1 g small-scale model tests of Toyoura sand slope with different relative densities and slope angles. Test results demonstrated that the control system was performing as expected although the test results were not presented in this paper.

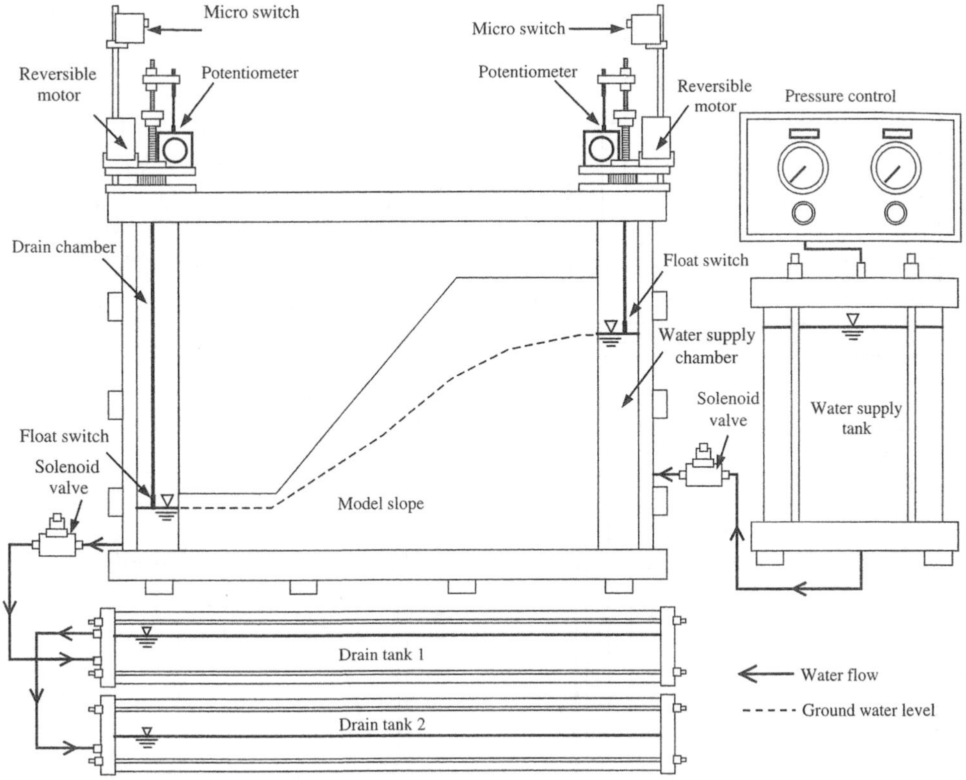

Figure 3. The schematic view of the in-flight ground water table control system.

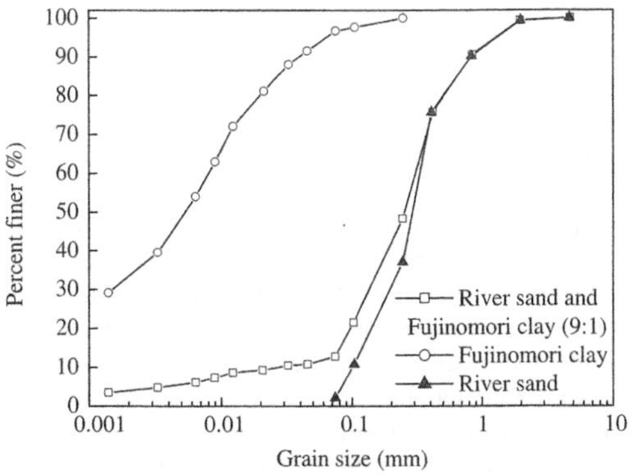

Figure 4. Grain size distribution of the model slope.

2.3 *Preparation of model slope*

The model ground was prepared by mixing the River sand with the Fujinomori clay as a ratio of 9:1 by weight at water content of 15%. The grain size distribution is presented in Figure 4. The physical properties of the soil mixture are specific gravity $G_s = 2.65$, mean particle size $D_{50} = 0.25$ mm, minimum dry density $\gamma_{d(min)} = 1.24$ g/cm^3, maximum dry density $\gamma_{d(max)} = 1.62$ g/cm^3 and permeability $k = 8.55 \times 10^{-4}$ cm/sec. Silicon grease was applied on the model container wall to minimize the friction between the wall and the model slope. Homogenous model ground was constructed in the central portion of the model container by laying out the soil mixture in series of horizontal layers. Each layer was compacted with a hand vibrator to obtain the prescribed height with the relative density of about 80%. Eight pore water pressure transducers (PPT) were embedded into the model ground at specified locations to measure the variation of pore water pressure during the centrifuge flight. After that the resulting block of model ground was trimmed to form the desired model slope geometry. Three linear variable differential transducers (LVDT) were installed on the top of slope to monitor the surface settlement with time. In this paper, two model slope tests were performed to simulate the different patterns of ground water table change. Figures 5 and 6 illustrate schematic views of the model slope geometry and the arrangements of displacement and pore water pressure transducers for the model slope

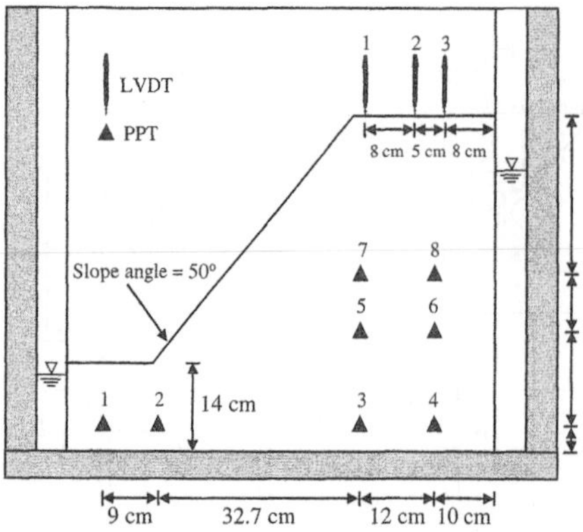

Figure 5. Experimental setup of the model slope test 1.

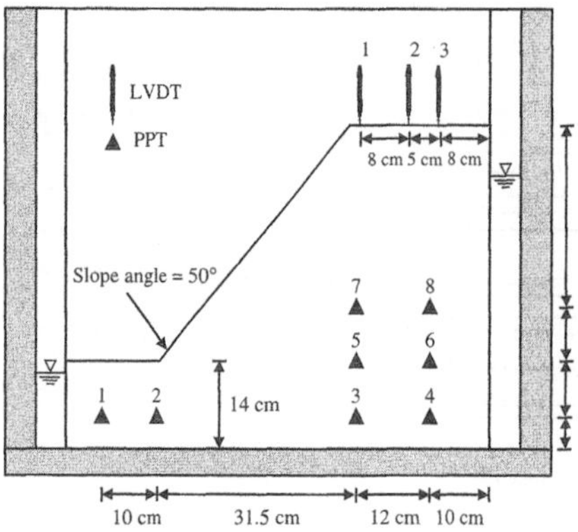

Figure 6. Experimental setup of the model slope test 2.

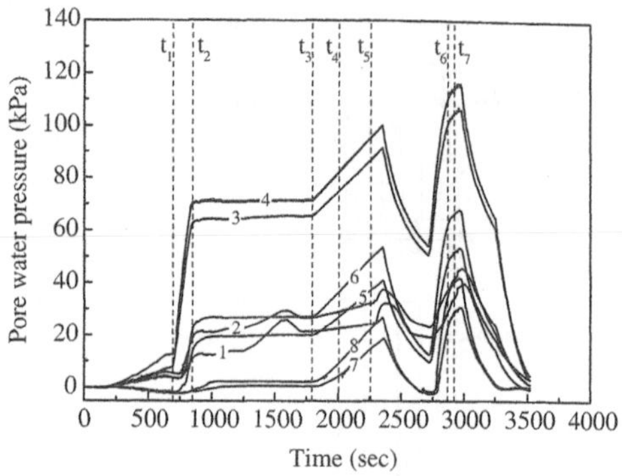

Figure 7. The variations of pore water pressure in the model 1.

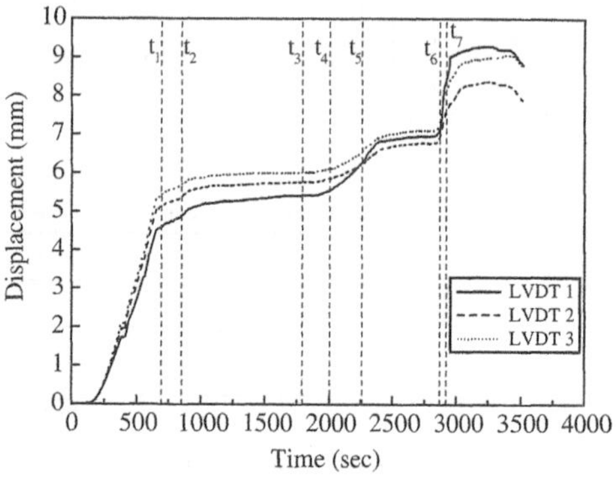

Figure 8. The variations of surface settlement in the model 1.

test 1 and the model slope test 2, respectively. The model container was then loaded on the swinging platform of the centrifuge. The digital video and CCD cameras were installed above the slope surface and in front of the model container to provide the visual observation of the model slope during the centrifuge flight. In addition, the digital video camera was also installed in front of the water supply chamber to monitor the water level. The change of ground water level in the model slope was simulated by the control system and the test was finished when sufficient slope failure was observed.

3 TEST PROCEDURES AND RESULTS

The model slope test 1 was intended to simulate the effect of large fluctuation of ground water table on the slope stability. The centrifuge acceleration level was gradually increased to the acceleration of 28.5 g where the model slope represents an approximately 11 m high slope in the prototype scale. Figures 7 and 8 present the measured pore water pressure and surface settlement, respectively. The number on the line in the Figure 7 represents the corresponding number of the pore water pressure transducer. During the gravity turn on process from 1 g to 28.5 g (t_0 to t_1) the rapid increase in surface settlement can be observed. Upon reaching the testing acceleration, the infiltration was simulated by introducing water into the water supply chamber at the upslope section (at t_1). The build up of positive pore water pressure can be clearly observed in the Figure 7. After that the ground water level was kept constant at the elevation of 30 cm measured from the slope base (t_2 to t_3) and then the ground water level was raised again by moving up the float switch in the water supply chamber. During the increase of the ground water level, partial failure occurred at the toe of slope (t_4 to t_5). The water supply tank was turned off and the ground water level in the water supply chamber was rapidly decreased to simulate the drawdown situation of ground water level during the dry period.

The negative water pressures at the middle of model slope (PPT7 and PPT8) were monitored and

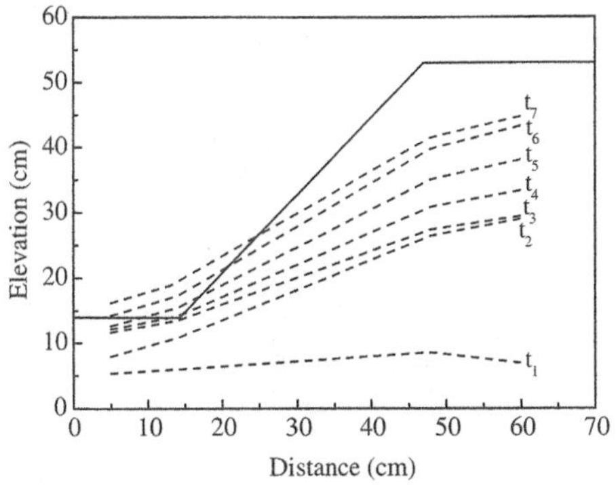

Figure 9. The variations of ground water level in the model 1.

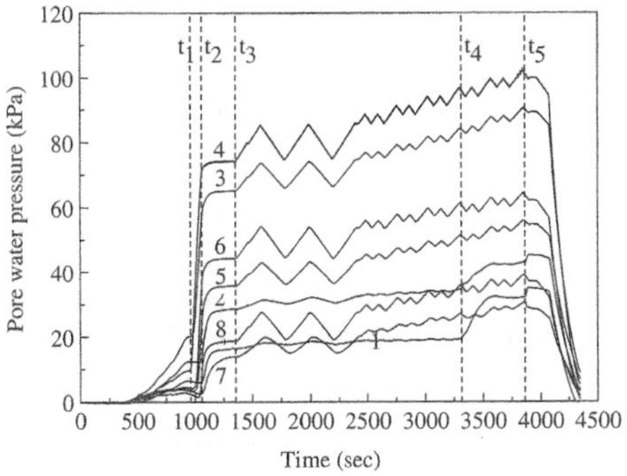

Figure 10. The variations of pore water pressure in the model 2.

significant increases of surface settlement were also observed. The ground water level was rapidly raised again and slope failure from the middle portion to the top of slope was observed during time t_6 to t_7. The rapid increase in the surface settlement was monitored just before the slope failure (Fig. 8), this phenomenon is useful to predict the occurrence of slope failure. Back-calculations were performed using the measured pore water pressure data from the PPT1, PPT2, PPT3 and PPT4 to establish the variations of ground water level in the model slope as shown in Figure 9. It should be noted that after the slope failure occurred the back-calculated ground water level seems to be higher than the slope surface. This is probably due to the fact that after the failure, the water gushed out from the collapsed slope and caused flooding thus the higher ground water level was observed.

The model slope test 2 was conducted to simulate continuous increase of ground water level during the heavy rainfall periods. The test was performed at the acceleration of 30 g, corresponding to the 11 m high slope in the prototype scale as similar to the first model test. Figures 10 and 11 present the monitored time

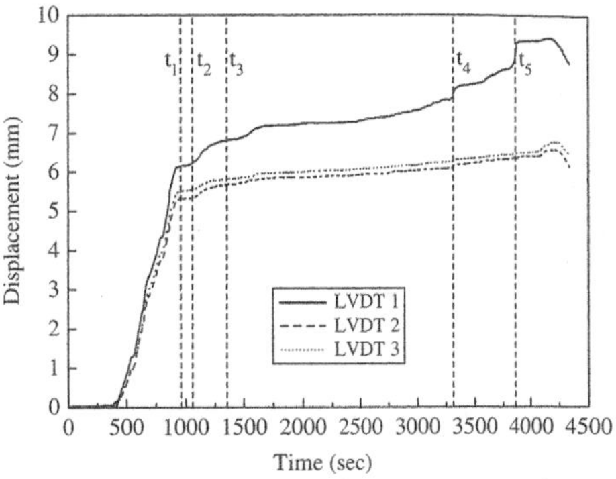

Figure 11. The variations of surface settlement in the model 2.

histories of the pore water pressures and surface settlements during the centrifuge flight, respectively. The rapid increase of surface settlement was observed during the increase of acceleration level form 1 g to 30 g (t_0 to t_1). The ground water level was raised and maintained constant at the level of 30 cm (t_2 to t_3). After that the ground water level was increased in steps with the cyclic pattern. For the first step two cycles of ground water level fluctuation were controlled by moving up and down the float switch at the water supply chamber with the amplitude of 2 cm. Subsequently the ground water level was increased to the next step with the amplitude of 0.5 cm, each step consists of two cycles of ground water fluctuation. The partial slope failure from the toe to the middle of slope was observed at the fifth step of ground water level change (t_4), the rapid build up of pore water pressure was also monitored at the toe slope (PPT1 and PPT2).

The positive pore water pressure continues to build up until the complete slope failure occurred after the sixth step (t_5) of ground water level change and the rapid increase of surface settlement can be monitored near the slope face (LVDT1). Figure 12 shows the back-calculated ground water level from the pore water pressure transducers at the base of slope (PPT1, PPT2, PPT3 and PPT4). It should be noted that flooding was observed near the toe of slope (at t_3) thus the back-calculated ground water level seems to be higher than the slope surface. The higher ground water level observed at the toe of slope after the occurrence of slope failure can be explained similarly to that in the model slope test 1. Figures 13 and 14 illustrate the photograph of slope failure pattern in the model slope test 1 and test 2, respectively. Tension cracks were observed at the top of slope in both of the model tests. The shallow slip surface above the ground water table was also identified in the both model tests. This phenomenon is similar to the failure mechanism of the rainfall-induced slope failure in the natural sandy soil slopes.

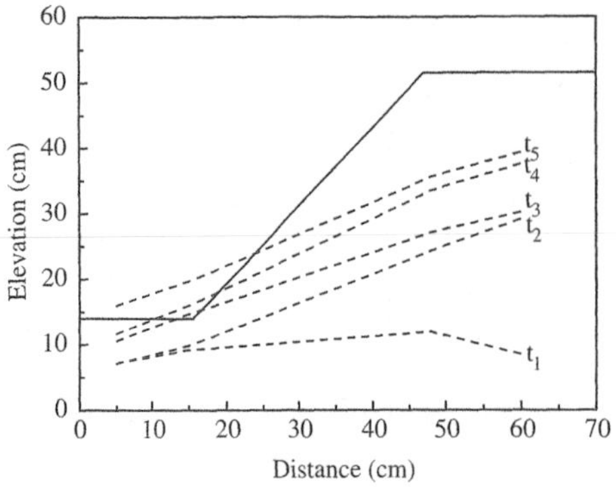

Figure 12. The variations of ground water level in the model 2.

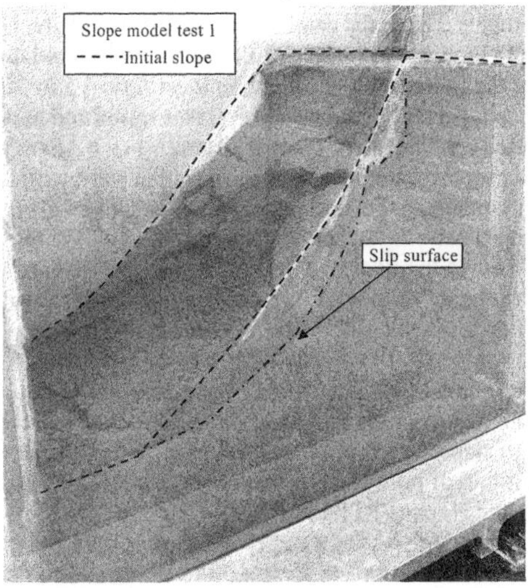

Figure 13. The photograph of failure pattern of the model 1.

4 CONCLUSIONS

The new in-flight ground water table control system was developed in this paper. The performance of the control system was verified by the small-scale model tests of sandy soil slope at 1 g and at centrifuge acceleration level. The mechanism of slope failure observed in the model test was similar to that of the rainfall-induced slope failure. The generation of positive pore water pressure (loss of matric suction) and the increase of saturation of model slope due to the rise in ground water level as a result of rainwater infiltration were sufficient to trigger the slope failure. The movement of wetting front in the model slope can be observed from the variations of pore water pressure and the rapid increase of the surface settlement just before

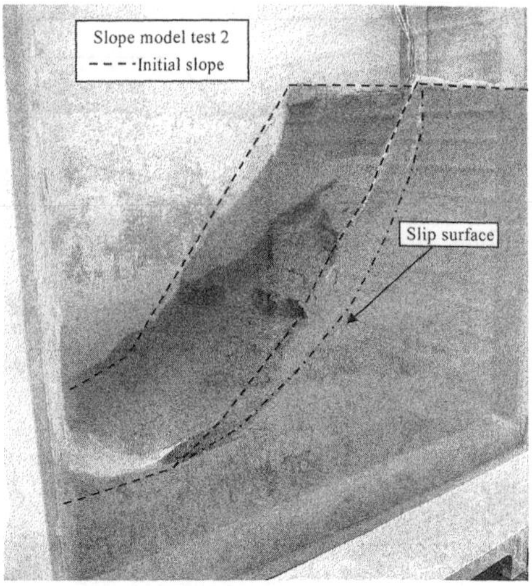

Figure 14. The photograph of the failure pattern of the model 2.

slope failure was also monitored. This information is useful to establish the prediction of the slope failure occurrences in the future. The other factors that influence the slope stability such as the relative density of ground, soil types and the ground water level change patterns should be adequately clarified in the further research in order to gain more understanding about the mechanism of slope failure induced by ground water level change.

ACKNOWLEDGEMENT

The authors would like to thank the Japan Society for the Promotion of Science (JSPS) for providing financial support for this research.

REFERENCES

Higashi, H. & Matsuura, T. 2006. Relationships between changes in annual frequency of heavy precipitation in Japan and ENSO. *The 18th conference on climate variability and change, Georgia, 28 January–2 February 2006*: 1–5.

Iwashima, T. & Yamamoto, R. 1993. A statistical analysis of the extreme events: Long-term trends of heavy daily precipitation. *Journal of the Meteorological Society of Japan* 71: 637–640.

Japan Meteorological Agency. 2005. Climate change monitoring report 2004.

Orense, R.P., Shimoma, S. & Farooq, K. 2005. Real-time prediction of rainfall-induced instability in sandy slopes. *Proceeding of the 16th International Conference on Soil Mechanics and Geotechnical Engineering, Osaka, 12–16 September 2005*: 2559–2562.

Landslides and Climate Changes – McInnes, Jakeways, Fairbank & Mathie (eds)
© 2007 Taylor & Francis Group, London, ISBN 978-0-415-44318-0

Climate change and evolution of landslide hazard at Nefyn Bay, North Wales

G.O. Jenkins & A.D. Gibson
British Geological Survey, Nottingham, England, United Kingdom

A.J. Humpage
British Geological Survey, Cardiff, Wales, United Kingdom

ABSTRACT: An essential tool in the management of present slope instability is an understanding of the climatic history of an area, and how this relates to the deposition and stability of sediments. Late Pleistocene and early Holocene climatic changes have had a significant influence upon sediment deposition and subsequent modification, and this has a direct bearing on Holocene coastal landsliding that continues to the present day. The cliffs of Nefyn Bay, on the northern coast of the Lleyn Peninsula in North Wales, exhibit a complicated succession of sediments deposited as result of changes in geoenvironmental conditions over time. Examination of the cliffs has shown that understanding the complex relationship between sediments, climate and slope stability is essential if we are to successfully manage unstable terrain in the region, given that the frequency of extreme climatic events is forecast to increase in the future.

1 INTRODUCTION

The superficial geology within Nefyn Bay, North Wales, is dominated by a complex sedimentary succession. This succession was deposited under variable geo-environmental conditions, associated with the decay of the Irish Sea Ice Sheet towards the end of the Devensian Ice Age. This environment, the deposits laid down by this environment, and alterations to the area by subsequent environmental changes have defined the nature and mechanisms of the landslide activity and hazard within Nefyn Bay today.

On 2nd January 2001 a series of shallow landslides caused extensive disruption to an access road and car park located mid-cliff in the centre of the bay. One person was fatally injured and a second severely injured when a debris flow pushed their vehicle over the sea wall. Following this event, a programme of research, funded jointly by the British Geological Survey (BGS) and Gwynedd Council, was undertaken to investigate the incident and to assess the landslide hazards in Nefyn Bay. A full account of this hazard assessment was given by Gibson & Humpage (2001a&b). This paper discusses the impact of past and present climate change upon the North Wales coast.

2 LOCATION

Nefyn Bay is a northwest-facing bay on the north coast of the Lleyn Peninsula in North Wales (Fig. 1). Two bedrock headlands enclose the bay – Penrhyn Bodeilas to the northeast and Penrhyn Nefyn to the southwest (Fig. 2).

3 GEOLOGICAL BACKGROUND

The area lies within the British Geological Survey 1:50,000 scale geological map sheet 118 (Nefyn). Bedrock is exposed only at the base of the cliffs on the headlands at each end of the bay (Fig. 3). At Penrhyn Bodeilas, bedrock is composed of a coarse-grained granite of unknown age, whereas the headland at Penrhyn Nefyn is composed of Precambrian metamorphosed granites and Ordovician shales. These metamorphic rocks are also believed to underlie the length of Nefyn Bay but are not exposed.

Superficial deposits exposed in the headlands at either end of the bay differ in character. At Penrhyn Bodeilas, the near vertical, 30 m high sea cliff is dominated by 15–20 m of glacial till, which is divided into

Figure 1. Location of Nefyn Bay, North Wales, United Kingdom.

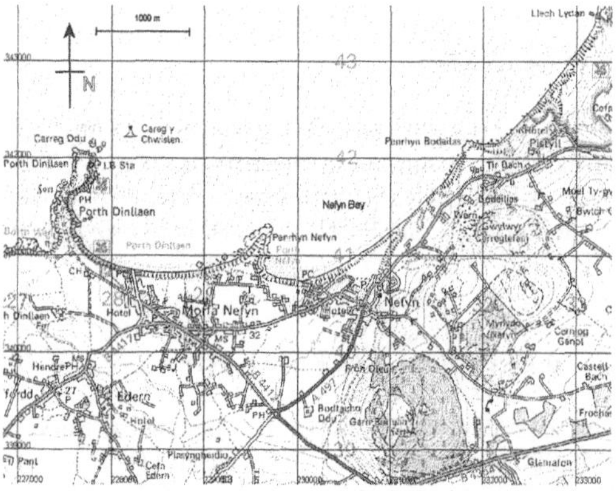

Figure 2. Location of Nefyn Bay, North Wales, United Kingdom. *Ordnance Survey Topography © Crown Copyright. Licence Number: 100017897/2006.*

a lower, structureless, lodgement till containing many clasts (including sandstone, mudstone and metamorphic fragments) ranging in size from gravel to large boulders 3 m in diameter; and an overlying, bedded flow till where clasts show some evidence of flow realignment. The till is overlain by 1–5 m of grey, stiff, laminated clays and soft, massive silts. Above the silt and clay unit, the sea cliffs are capped by 3–15 m of glacio-fluvial outwash deposits varying from gravel to sand and silt. In a few locations the glacio-fluvial deposits are overlain by Head deposits and yellow wind-blown fine silty sand probably of loessic origin.

At Penrhyn Nefyn, in the south west of the bay, the till forms only the lowest metre of the cliff. The remainder of the approximately 25 m high cliff is composed of glacio-fluvial outwash deposits which are predominantly well-bedded, fine- to medium-grained sand and gravel, with individual beds up to 5 m in thickness.

The remainder of Nefyn Bay, between the two headlands, can be divided into two distinctive sedimentary suites. In the southern-central section of the bay, near Penrhyn Nefyn, the cliffs, up to 40 m in height, expose a variable succession of sediments. Rarely exposed at the base of the cliff is a grey, stiff till containing clasts of variable size and shape in a silt/clay matrix. This is overlain by fine to coarse-grained sand with lenses of gravel and isolated lenticular bodies of grey, stiff, laminated silt, up to 10 m in width above which is a persistent, uniform stratigraphy of an erosive based 2 m thick, calcium carbonate cemented, indurated cobble gravel overlain by up to 1 m of soft, grey, laminated clay. The upper third of the cliff in this section of the bay comprises a 1 m thick grey massive silt which grades upwards into a brown sandy silt and silty sand and gravel with lenses of grey-brown, loose sand and gravel.

Within this upper section of the cliff, are large, isolated sedimentary bodies composed predominantly of massive grey silt, within which are rare mudstone clasts up to 100 mm in size and lenticular beds of sand up to 1 m in thickness. The silt bodies appear to be located within basins up to 60 m in diameter and up to 25 m deep. At the top of the cliff, a thin (up to 0.5 m thick) deposit of yellowish brown silt occurs locally. Although not examined in detail, the open texture and uniform grain size indicates that it is likely to be wind-blown loess. Head deposits infilling channels or depressions a few metres deep in the surface topography were also recorded. These consist of fine-grained sediments and angular clasts up to 100 mm in diameter, which show some evidence of sorting and flow.

At the car park, site of the 2001 landslide, syn-sedimentary faulting has resulted in a relative increase in thickness of the succession above the cobble-gravel unit compared to elsewhere in this part of the bay. As a consequence, the upper half of the cliff has a different succession extending laterally approximately 200 m. The upper sand and gravel unit is overlain by an additional grey laminated silt/clay bed, above which is a distinct, discontinuous and occasionally thick unit of stiff brown silt/clay (Bed 7c of Gibson & Humpage, 2001a. Together these form a distinctive terrace within the cliff profile. The upper cliff at this location is

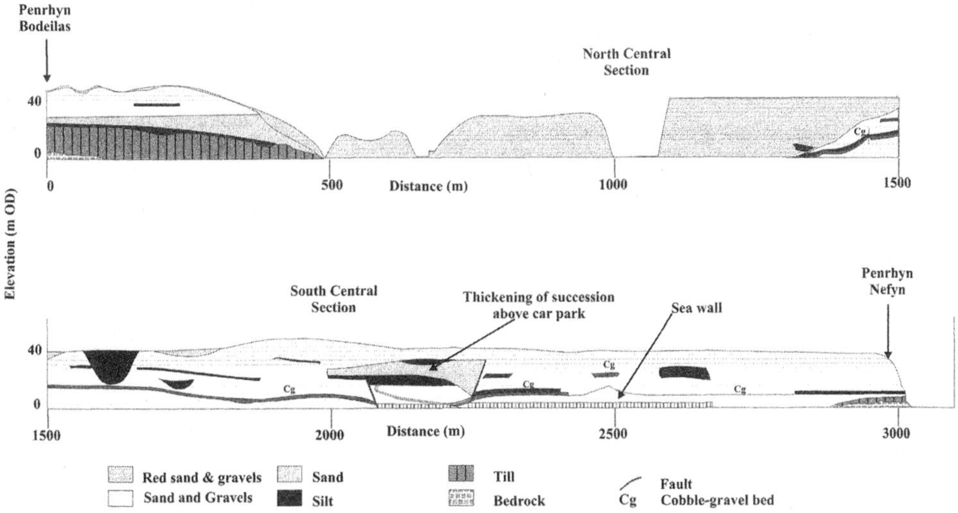

Figure 3. Geological cross section of the cliff section at Nefyn Bay.

dominated entirely by soft, loose, fine- to medium-grained, cross-bedded sand (Bed 7a of Gibson & Humpage, 2001a).

In the northern-central region, near Penrhyn Bodeilas, the sediments are composed of red-brown, well bedded, poorly to moderately sorted, glacio-fluvial outwash sands and gravels, varying in size from fine-grained sand to coarse gravel. Although generally loose, there are isolated gravel-dominated beds cemented by calcareous cement. The red colouring of this succession is probably due to a high proportion of Permo-Triassic sandstone-derived material from the Irish Sea basin. This succession forms the entire cliff section in the middle of the bay but thins both to the northeast, where it forms the upper cliff at Penrhyn Bodeilas, and to the south-west, where it overlies the southern-central deposits described above. This succession appears to be infilling an erosional channel or large basin which has been eroded into the till and older deposits to the south-west (Figure 3).

4 ENVIRONMENTS OF DEPOSITION

The superficial deposits described are interpreted as being laid down by a waning ice sheet. The till at the base of the cliffs was deposited when Nefyn was covered by an ice sheet flowing southwards across the Irish Sea Basin.

As the ice front retreated northwards, the southern-central sedimentary succession (sands, gravels with silt bodies) was deposited in an ice-proximal (proglacial) environment. The lowest of these deposits has been interpreted as glacio-fluvial outwash, with aggrading channel fill deposits laid down in flowing water. The presence of silt-filled hollows within this succession indicates that, there were periods of quiescence, when standing water allowed the deposition of fine sediments. It is possible that these periods occurred when the ice front had retreated just offshore of the present coastline. The idea of a sustained presence of a nearby ice front is further supported by the presence of 'kettle hole' deposits exposed within the cliff section. These formed when blocks of ice fell from the degrading ice front and settled within the glacio-fluvial sediments. When these blocks of ice melted, they left voids into which other deposits could flow. The resultant voids have been filled with silts and fine sands.

Further research is being carried out to determine the depositional environment of deposits in the north-central region of the bay. However, the relatively homogenous nature of the sediments (sands and gravels), and absence of significant silt bodies associated with ice melt structures indicates a more distal glacio-fluvial environment when the ice sheet had retreated farther to the north.

The head deposits, found at the top of the cliff were formed by cold climate conditions following the retreat of the ice sheet, but before significant climatic warming had occurred. Fine grained, uniform deposits, with open structure and uniform texture were found at exposures at the cliff top in a number of locations along the bay. These have been interpreted as wind-blown deposits probably deposited in tundra-like conditions where exposed surface sediment can be easily mobilised by the wind.

5 LATE PLEISTOCENE CLIMATE CHANGE AND EVOLUTION OF LANDSLIDE HAZARD

5.1 *Glacial (>18,000 years BP)*

During the last (Devensian) glaciation, Nefyn Bay was covered by the Irish Sea Ice Sheet that extended southwards from the Irish Sea Basin, across the Lleyn Peninsula and out into Cardigan Bay (McCarroll, 2001). Deposition from this ice front has resulted in the lower, structureless lodgment till observed at Penrhyn Bodeilas in the north of the bay.

From about 18,000 years BP, the climate began to warm, and the Irish Sea Ice retreated in a northerly direction. This resulted in sediment being transported within the glacier being deposited as till. Although there is no direct evidence of surface landforms or mass-movement processes from this period, the exposed till surface was unstable and was able to mobilize and flow into hollows as a flow till. Deposition from this period has resulted in the upper till observed at Penrhyn Bodeilas. Overall, the result of this period was the emplacement of relatively strong, impermeable tills, which outcrop at either end of the bay.

5.2 *Proglacial (c.18,000–c.14,000BP)*

As the ice front retreated, meltwater issuing from the front of the glacier deposited glacio-fluvial sand and gravels and there was periodic ponding of meltwater. In this highly active environment with abundant water and sediment supply it would be expected that any slopes present would be highly unstable and prone to movement. However, there is no visible evidence to suggest any such paraglacial landscape modification within this environment. The greatest impact from this climate was the deposition of a series of sands, gravels and silts in varying degrees of induration.

5.3 *Periglacial (c.14,000–12,000 BP and 11,000–10,000 BP*

As the Irish Sea ice sheet retreated from the Nefyn area, cold climatic conditions continued to persist across the region. Periglacial slope processes will have occurred during two distinct cold phases; at the end of the Devensian glaciation, and again for approximately 1000 years during a period termed the Loch Lomond Stadial.

Although there has been little research on the impact in North Wales of the relatively aggressive tundra-like climate, it is fair to assume that, as elsewhere (McCarroll *et al.*, 2001), this period coincided with significant mass-movement activity. Such processes would have included shallow landsliding, deep landsliding, slope retreat and development of head within all of the sedimentary sequences,

Between these two periods of activity was the 1000 years or so of the Windermere Interstadial. The warm climate during this period generally resulted in less aggressive ground conditions and a reduction in mass-movement activity across Europe. It is likely that the region around Nefyn became vegetated at this time, further reducing the likelihood of mass movement processes.

6 HOLOCENE CLIMATE CHANGE AND EVOLUTION OF LANDSLIDE HAZARD

It is the climatic changes that have occurred during the Holocene, coupled with the legacy of the glacial and proglacial sedimentary environments that have influenced the style and mechanisms of landsliding in Nefyn Bay today. Several studies have utilized peat records and lacustrine oxygen isotope levels to model a climatic record for the Holocene around the Irish Sea Basin (see for instance Barber *et al.*, 1994, Jones *et al.*, 2002, Marshall *et al.*, 2002, Barber *et al.*, 2003). These demonstrate a significant warming at the end of the Loch Lomond Stadial (approximately 10,000 years ago), which, coupled with sea level rise, have been instrumental in the development of the landslide hazard. Even over the last 10,000 years, there have been wetter climatic conditions than those prevalent today (Fig. 4) and short-lived cold periods such as the "Little Ice Age" during the late 17th and 18th centuries AD.

In the upper cliff in the area between the car park and Penrhyn Nefyn, a series of failures, now vegetated, can be observed. These have been interpreted as a series of degraded, non-circular, deep-seated rotational failures. Surface morphology suggests that sliding probably took place along shear zones within the laminated silt and massive clay units above the top of the indurated cobble gravel unit. It is thought that these were the result of the warm, wet climate during the warm Atlantic stage (Chiverell & Innes, 2004). These conditions led to elevated pore-water pressures at impermeable boundaries – the top of the till or indurated beds within the pro-glacial sands and gravels. This situation led to the failure of the upper cliff resulting in a series of horizontal or back-tilted benches upon which debris has subsequently accumulated. The significance of this is discussed below.

During the Holocene, sea level rose rapidly between 9000 – 7000 BP to approximately present day levels

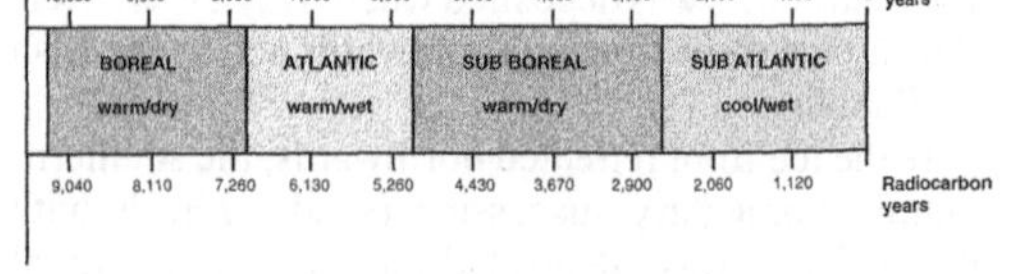

Figure 4. The climatic stages of the Holocene.

(Heyworth & Kidson, 1982). During this period, the sequence of sediments seen at Nefyn was eroded, ultimately resulting in the creation of the modern day coastline.

7 RECENT CLIMATE CHANGE AND EVOLUTION OF LANDSLIDE HAZARD

Over the last 2000 years, the North Wales coast has experienced a relatively temperate climate. There may have been a period of aggressive ground conditions during the Little Ice Age (McCarroll et al., 2001), but there is no clear evidence of this within the survey area. This climate has led to the accumulation of a significant thickness (0.5–10 m) of organic rich debris. Generally the material forms a soft, saturated, deformable layer that typically builds up on the topographic benches formed over the large rotational landslides described above. Virtually the entire surface of the debris accumulation on the various benches seen in cliff profile was soft, wet and covered in tension cracks aligned perpendicular to the slope. Failures within this material were common, tending to occur in a shallow translational or rotational manner depending upon thickness, local hydrology and the gradient of the underlying slope. It was a failure in this surface material, following a heavy rainfall event, which resulted in the 2001 landslide.

Recent and contemporary sea level has resulted in the ongoing removal of material from the toe of the slopes (where undefended). This contributes both to the shallow failures observed at the base of the cliff and to the rockfall activity observed across the bay.

The relatively weak, permeable nature of sediments in the northern-central part of the bay, laid down in a proglacial environment, have formed steeply graded but vegetated slopes. Failures tend to be shallow planar slides (Fig. 5). One example, recorded in 2001, was over 60 m in width but involved a thickness of material no more than 2 m. Elsewhere in the bay, failures observed within the glacio-fluvial sands and gravels were predominantly small slides and flows, generally less than 0.5 m deep and 1–2 m in length. Such failures are seen over the majority of the slope, often associated with small deposits at the base of the cliff that are removed at very high tides.

Failures were also associated with filled kettle holes. Figure 6 shows the largest of these, where recent erosion at the base of the cliff has led to a large mudflow. The failure occurred when part of the kettle basin was breached at its seaward side. The removal of lateral support for the sediments within the kettle basin, caused the fine-grained, cohesionless silt fill to fail rapidly and flow out through the breach leaving behind an empty flow bowl.

Rockfalls have resulted in the presence of a series of fallen blocks, usually of the indurated cobble-gravel unit on the foreshore (Fig. 7). These blocks, some of which were 4–5 m in diameter, had fallen 10–15 m from the cliff, presumably when support was removed by coastal erosion of supporting sediments.

Figure 5. Planar slide at [230950 341150] in the glacio-fluvial sands and gravels.

Figure 6. Kettle hole failure in silts at [230650 340955].

Figure 7. Fallen block of the indurated cobble gravel unit, indicative of a possible rockfall hazard below the cliffs.

8 LANDSLIDE HAZARD AND RISK

The research undertaken by BGS in Nefyn Bay over the past five years was in response to the landslide that occurred in January 2001. The research has highlighted not only the hazards within the study area but also the need to consider the cumulative effects of different depositional and climatic environments when building a picture of landslide hazard, especially where the post-glacial environment has been complex.

Landslide hazards in Nefyn Bay and the risks resulting from them vary over very short distances. At present there is little hazard posed by rotational failures within the cliff. However, the presence of these degraded features has led to the generation of a separate hazard, that of a build up of weak, saturated debris, the behaviour of which is difficult to predict. It was a failure in this type of deposit that occurred in 2001, and this type of deposit remains the greatest hazard within the area today. The risk posed by most of these failures is, in most instances, low, as they will affect only a small area in the immediate vicinity of the failure. Risk is greatly increased when the area is transgressed by the built environment, for instance by the construction of chalets or access roads. Remediation work described by Statham (2002), has mitigated this risk immediately above the car park. However there are still adjacent areas of slope that have the potential to fail in the future, presenting a risk to both properties and life.

A coastal path traverses the top of the cliff, extending from Penrhyn Nefyn to the town of Nefyn before heading inland. Although the risk here of a landslide hazard coinciding with a walker being on the affected section of the path is relatively low, several tension cracks were observed along sections of the path during a visit in October 2006, so a considerable potential landslide hazard and risk is present.

A number of residential properties are located along the cliff top. With a forecast increase in more extreme climatic events, such as heavy rainstorms (DEFRA, 2006), a repeat of the heavy rainfall event such as that in 2001, coupled with continued coastal erosion at the toe of the cliff, presents a significant landslide risk in Nefyn Bay.

9 CONCLUSIONS

Landslide hazard has evolved over time within Nefyn Bay, with this evolution linked intrinsically to long- and short-term climate change. The landslide hazard present today first began to evolve during the last (Devensian) Ice age as the Irish Sea Ice retreated from the northern Lleyn Peninsula and laid down a succession of glacial and proglacial deposits.

Coastal erosion associated with sea level rise has a significant control over slope mechanisms in Nefyn Bay. Failures resulting from coastal erosion vary widely, but include small scale falls of sand and gravel and larger slides in proglacial sediments, rockfalls from bedrock or till, and in some cases, rapid cliff failure due to the cutting of a kettle hole.

Relict non-circular rotational landslides that are stable today, but are presumed to have moved during a wetter climatic period during the Holocene, have indirectly caused the greatest landslide hazard today. These failures have provided a zone of sub-horizontal benches upon which soft, unstable, organic rich debris has accumulated. Recent history has shown that these sediments can be mobilized during rapid onset events such as extreme rainfall episodes.

This research has highlighted that, especially in an area affected by proglacial deposition, the mechanisms, hazards and risk associated with landsliding can be very complex and can vary over short distances. Building such an understanding of slope evolution over time and with changing climate is a necessary step in understanding, and mitigating landslide hazard.

REFERENCES

Barber, K. E., Chambers, F. M. & Maddy, D. 1994. A sensitive high-resolution record of Late Holocene climatic change from a raised bog in Northern England. *The Holocene* 4: 198–205.

Barber, K. E., Chambers, F. M. & Maddy, D. 2003. Holocene palaeoclimates from peat stratigraphy; macrofossil proxy climate records from three oceanic raised bogs in England and Ireland. *Quaternary Science Reviews* 22(5–7): 521–539.

Chiverell, R. C. & Innes, J. B. 2004. Holocene climate change, vegetation and environmental history. In R. C. Chiverell, A. J. Plater & G. S. P. Thomas (eds), *The Quaternary of the Isle of Man and North West England*. QRA Field Guide.

DEFRA, 2006. Climate Change: The UK Programme 2006. The Stationery Office. London.

Gibson, A. D. & Humpage, A. J. 2001a. The Geology and Cliff Stability at Nefyn, North Wales – Interim Report. *British Geological Survey Commissioned Report* CR/01/101, 32p.

Gibson, A. D. & Humpage, A. J. 2001b. The Geology and Cliff Stability of Nefyn Bay, North Wales – Final Report. *British Geological Survey Commissioned Report*, CR/01/267.

Heyworth, A. & Kidson, C. Sea-level changes in southwest England and Wales. *Proceedings of the Geologists Association* 93: 91–111.

Jones, R. T., Marshall, J. D., Crowley, S. F., Bedford, A., Richardson, N., Bloemendal, J. & Oldfield, F. 2002. A high resolution, multi-proxy Lat-Glacial record of climate change and intrasystem responses in northwest England. *Journal of Quaternary Science* 17: 329–340.

Marshall, J. D., Jones, R. T., Crowley, S. F., Oldfield, F., Nash, S. & Bedford, A. 2002. A high resolution Late-Glacial isotopic record from Hawes Water, Northwest England Climatic oscillations: calibration and comparison of palaeotemperature proxies. *Palaeogeography, Palaeoclimatology, Palaeoecology* 185: 25–40.

McCarroll, D. 2001. Deglaciation of the Irish Sea Basin: a critique of the glaciomarine hypothesis. *Journal of Quaternary Science* 16(5): 393–404.

McCarroll, D., Shakesby, R. A. and Matthews, J. A. 2001. Enhanced rockfall activity during the Little Ice Age; further lichenometric evidence from a Norwegian talus. *Permafrost and Periglacial Processes* 12(2): 157–164.

Statham, I. 2002. Failure on the coastal slope above Lon Gam, Nefyn, Llyn Peninsula: mechanism and remedial measures. In D. Nichol, M. G. Bassett & V. K. Deisler (eds), *Landslides and landslide management in North Wales*, 99–105. National Museum of Wales Geological Series No. 22, Cardiff.

Landslides and Climate Changes – McInnes, Jakeways, Fairbank & Mathie (eds)
© 2007 Taylor & Francis Group, London, ISBN 978-0-415-44318-0

Control of soil properties on the Scottish debris flow geohazard and implications of projected climate change

F.D. Milne & M.C.R. Davies
Division of Civil Engineering, University of Dundee, Scotland

ABSTRACT: Debris flows are the most common geohazard effecting areas of high relief in Scotland having impacted upon the road network several times in recent years. The effects of sediment texture, soil strength and permeability on the spatial and temporal distribution of debris flow activity in Scotland are examined at six study sites. Based upon climate change projections made by the Hadley Centre's HadRM3 regional circulation model using UKCIP02 emission scenarios, it is inferred that debris flow activity will increase in Scotland by the 2080s as a consequence of an increase in intense rainfall. Increases in debris flow activity will be most pronounced during winter months, in eastern Scotland and on hillslopes underlain by granite and sandstone lithologies.

1 INTRODUCTION

Debris flows are the most common geohazard effecting areas of high relief in Scotland having impacted upon the road network several times in recent years (Winter *et al*, 2006; Nettleton *et al*, 2006). The primary triggering factor for debris flow generation is high magnitude rainfall which leads to soil saturation, reduced effective stress and thus slope instability. Consequently any changes in precipitation patterns as a result of climatic change can be expected to have an impact on the spatial and temporal distribution of debris flow activity. The importance of intense rainfall as a trigger for debris flow activity is highlighted by the fact that all documented instances of recent debris flow activity in Scotland are associated with high magnitude rainstorms (Common, 1954; Baird & Lewis, 1957; Jenkins *et al*, 1988; Luckman, 1992; Winter *et al*, 2006; Nettleton *et al*, 2006) although rainstorm magnitude cannot be directly related to flow generation due to the importance of antecedent soil moisture conditions as a prerequisite to debris flow activity (Ballantyne, 2002). It is anticipated that future climate change will be characterised increased occurrences in high magnitude rainfall related to a more vigorous hydrological cycle driven by warmer global temperatures and a higher energy atmosphere (Conway, 1998).

Good understanding of the controls on the debris flow process is essential to allow accurate interpretations about the consequences of climatic change on the nature of the geohazard to be made and to allow effective hazard conceptualisation and management. Debris flows occur widely in Scotland where the pre-requisite factors of a mantle of slope material and a sufficient slope angle exist with most recorded debris flows occurring on slopes ranging between 30 and 42° (Innes, 1983). Although a complex and variable suite of contributory factors determine the susceptibility of any given hillslope to debris flow, material properties such as permeability and frictional strength are key determinants on the propensity to debris flow. This is demonstrated by observations of greater densities of debris flow forms in areas underlain with coarse lithologies such as granite and sandstone which yield sandier soils compared to hillslopes underlain with finer grained schistose or extrusive igneous lithologies which commonly develop into soils with a larger silt component (Innes, 1983; Ballantyne, 1986; Curry, 1999). The susceptibility of coarse-grained sediments to debris flow has been attributed to high infiltration rates, which permit a rapid rise in pore-water pressures during rainstorms (Innes, 1983; Ballantyne, 2002).

In this study soils sampled at six study sites around Scotland where debris flow activity is known to occur were investigated to determine the effect of texture, permeability and frictional strength of soils on hillslope propensity to debris flow under present climatic conditions. Inferences were then made on the effect of climatic change on the spatial and temporal distribution of debris flows are using climate projections up to the 2080s derived from the Hadley Centre's HadRM3 regional circulation model.

2 STUDY SITES

Six study sites with differing lithologies were chosen across Scotland for assessment of soil properties on the spatial frequency of debris flows (figure 1). The study sites were chosen where debris flow activity

Figure 1. Location of study sites.

has been recorded in the past (Ballantyne & Eckford, 1984; Jenkins *et al*, 1988; Luckman 1992; Curry 2000; Winter *et al*, 2005) and are situated across Scotland in order to avoid bias towards areas with higher average annual precipitation.

Glamaig is a mountain situated in the Western Red Hills on the Isle of Skye. The study site was focused on the southern and eastern facing slopes of the mountain and is dominated by granite bedrock overlain by a basalt roof pendant (Bell & Williamson, 2002). The basalt bedrock is situated above the debris flow source areas at the eastern side of the study site, but towards the west the basalt lithology underlies the debris flow source areas. Consequently, debris flow activity on both fine-grained basalt and coarse-grained granite can be investigated in close proximity. The Lairig Ghru is a high level glacial breach valley in the Cairngorm Mountains. The study site comprises a 1 kilometre stretch of the pass running from slightly north of Lurchers Crag. The bedrock geology is coarse-grained granite. Glen Ogle is a steep sided valley in the southern Grampian Mountains. The lithology at Glen Ogle is dominated by relatively impermeable upper Dalradian metamorphic rocks, in particular Quartzose-Mica-Schists formed from marine shales during the Caledonian orogeny (*c*.410 Ma BP). The Pass of Drumochter is located in the central Grampian Mountains. The bedrock at the site consists of monotonous sequences of psammitic Schist and localised semipelite derived from shallow marine deposits which were metamorphosed during the Grampian Orogeny (Stephenson & Gould, 1995). The An Teallach study site is centred on the northerly slopes of Glas Mheall Mor, the most northerly peak in the An Teallach massif in the North West Highlands. The site is completely underlain by Torridonian Sandstone (Ballantyne & Eckford, 1984). Mill Glen is situated in the Ochil hills in the Midland Valley of Scotland. The study site is located in the upper reaches of the glen on the hillslopes surrounding the Gannel Burn and is completely underlain with andesitic lavas (Francis *et al*, 1970).

3 METHODOLOGY

At each site the spatial frequency of debris flows per kilometre was determined by dividing the number of observed debris flows at each site by the length of the studied hillslope. A total of 9 debris flows were chosen for sampling across all of the study sites. Soil was sampled from exposed profiles at source landslide scars for investigation of geotechnical properties in the laboratory. All of the chosen source areas had grassland vegetation cover. Analysis was only carried out on particles smaller than 2 mm due to the fact that all the sampled soils are matrix dominated and that permeability and soil strength are largely determined by the soil matrix (Innes, 1983; Ballantyne, 1986; Fannin *et al*, 2005).

Particle size analysis was carried out on the soil samples using a Coulter LS250 particle size analyser to assess the control of soil texture on debris flow susceptibility. Samples taken from debris flow source areas sampled at Glen Ogle 1 and An Teallach 1 were further analysed to determine the shear strength and permeability characteristics of the soils. These soils were chosen as PSD analysis found that these soils had the coarsest (An Teallach 1) and finest (Glen Ogle 1) sediment textures. Effective stress strength parameters c' and ϕ' were obtained using a direct shear box in accordance with BS 1377 (BSI, 1990) under varying normal loads chosen to closely match the effective stresses experienced in the field. The permeability of the soil is investigated using the constant-head permeability test following the procedures outlined in BS 1377 (BSI, 1990) using a specially constructed 51 mm diameter permeameter cell. In both the shear box test and the permeability test, the tested soil specimens were reconstituted from disturbed samples to match the density and water content conditions measured in the field and hand tamped into the apparatus. The *insitu* density of the soil was calculated in the field using a technique devised for this research. A bench was cut into the exposed soil profile and a sample removed using a trowel. Plastic film was then carefully placed in the void made by the removal of the sample. A measured quantity of water was pored into the void to establish the volume of the soil sample and the density of the soil was calculated by dividing the mass of the soil sample by its volume. The moisture content of the soil at the time of sampling and the dry density were also obtained.

In considering the propensity of hillslope to debris flow it is important to acknowledge that slope gradient exerts a critical control on slope stability, with steeper slopes more likely to experience instability. The joint influence of gradient and soil strength on the susceptibility of hillslopes to debris flow generation was investigated by calculating the factor of safety for the source slopes of the An Teallach 1 and Glen Ogle 1 debris flows using equation 1 below:

$$F = \frac{\tan \phi'}{\tan \beta} \tag{1}$$

Where $F =$ factor of safety; $\phi' =$ internal angle of friction; and $\beta =$ source area slope gradient.

4 MATERIAL PROPERTIES

The granitic part of the Glamaig study site had the highest debris flow density at 24 km^{-1}. The Lairig Ghru and the An Teallach study sites both also registered a frequency of debris flow forms in excess of

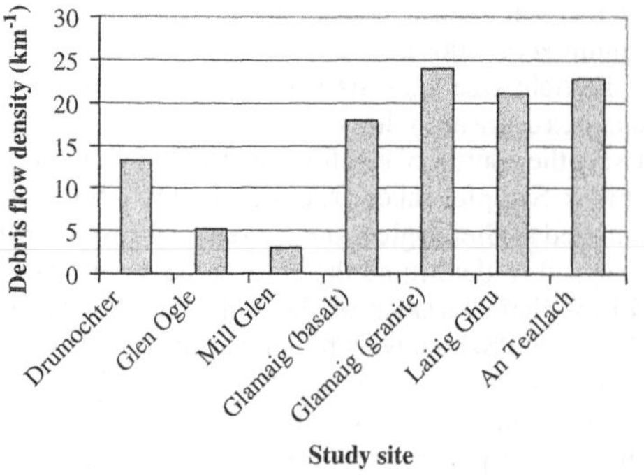

Figure 2. Density of observed debris flows per kilometre for each study site

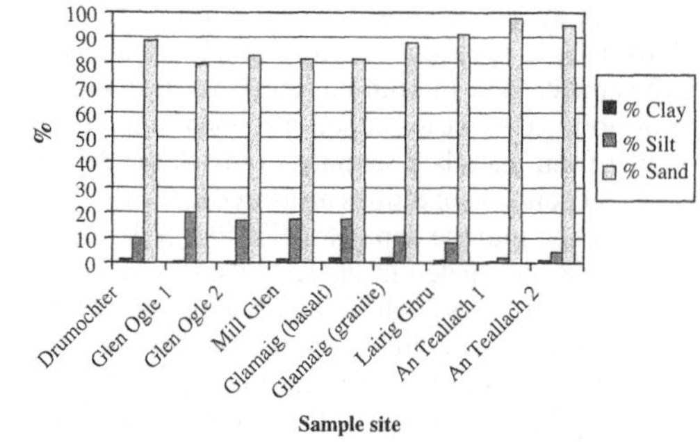

Figure 3. Particle size distribution for soils sampled at debris flow source areas at each study site.

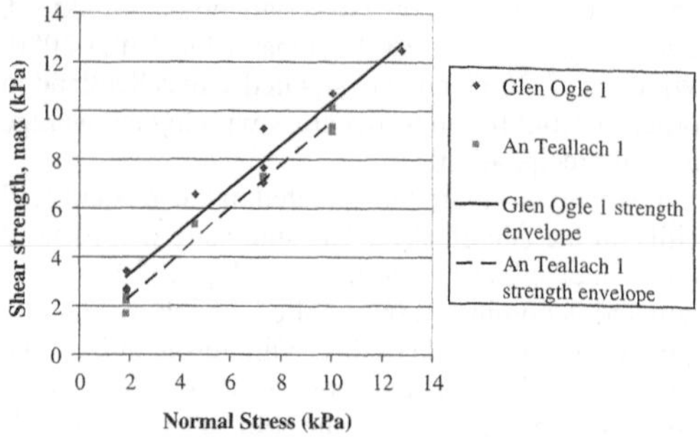

Figure 4. Shear strength envelope for Glen Ogle 1 and An Teallach 1 soils.

$20\,km^{-1}$. Debris flow density at the basaltic part of the Glamaig was found to have a debris flow density of approximately $17.9\,km^{-1}$. The hillslopes at the Pass of Drumochter were calculated to have a debris flow density of $13.3\,km^{-1}$ whereas the study sites at Glen Ogle and Mill Glen were found to have much lower debris flow spatial frequencies with debris flow densities of $5.2\,km^{-1}$ and $3\,km^{-1}$ respectively (figure 2).

Particle size analysis showed that all of the sampled soils were found to be dominated by sand sized particles (>0.06 mm diameter) (figure 3). The debris flows sampled at An Teallach and the Lairig Ghru were found to have the coarsest sediment textures, consisting of over 90% sand sized particles. The debris flows sampled at Glen Ogle, on the basaltic part of Glamaig and at Mill Glen were found to have the finest soil textures with 19.9% and 16.8% silt fractions (0.002–0.06 mm diameter particles) for Glen Ogle 1 and Glen Ogle 2 respectively, 17.1% for Glamaig and 17.3% at Mill Glen. The silt fraction of the soil overlying the granitic part of Glamaig (10.4%) is slightly higher than the silt component of soil developed over granite bedrock at the Lairig Ghru study site (8.1%). The soil sampled at the Pass of Drumochter was found to have a coarser sediment texture than other soils sampled over fine-grained lithologies. This may be due to the fact that the psammitic schist bedrock at the Pass of Drumochter is largely derived from sand dominated shallow marine deposits which disintegrate into coarser particles compared to other schistose lithologies. The clay fraction (<0.002 mm diameter particles) of the sampled soils was very low, registering less than 2% for all soils.

The material tested was recompacted to match the *insitu* densities experienced in the field. Field densities and moisture contents of $1.82\,g\,cm^3$ and 28.3% for Glen Ogle 1 and $1.7\,g\,cm^{-3}$ and 9.4% for An Teallach 1 were measured. The shear strength envelopes for shear box analysis of soil sampled from Glen Ogle 1 and An Teallach 1 are shown in figures 4 and 5. It was found that the soil at Glen Ogle 1 had an internal angle of friction of 41.3° and the sampled soil at An Teallach 42.2° with apparent cohesions (c') of 1.6 kPa for Glen Ogle and 0.5 kPa for An Teallach. The coefficient of permeability (k) was found to be $1.19 \times 10^{-3}\,m\,s^{-1}$ for Glen Ogle 1 and $6.08 \times 10^{-3}\,m\,s^{-1}$ for An Teallach 1. The corresponding dry densities were $1.31\,g\,cm^{-3}$ and $1.54\,g\,cm^{-3}$ for Glen Ogle 1 and An Teallach 1 respectively. The internal angle of friction for the soils was relatively similar. This is indicative of the fact that the sand fraction in both soils is determining the strength of the soil, with the proportion of silt in the Glen Ogle 1 sample not sufficient enough to significantly depress the internal angle of friction. However, the An Teallach 1 soil is approximately 6 times more permeable than the Glen Ogle 1 soil. This indicates that the silt fraction in the Glen Ogle 1 soil contributes to a lowering in permeability although it is important to acknowledge that particle shape, mineralogical composition and the soil fabric would also have influenced the permeability (Head, 1982).

The source area slope gradients of the An Teallach 1 and Glen Ogle 1 debris flows are 36° and 21° respectively. The factor of safety was found to be 1.25 for An Teallach 1 and 2.29 for Glen Ogle 1. The factor of safety for An Teallach is closer to 1, indicating that the slope is less stable here than at Glen Ogle as a result of the greater slope gradient.

5 PROJECTED CLIMATE CHANGE IN SCOTLAND

Under the UKCIP02 scenarios, the average annual temperature in the UK will be between 1° C and 5° C warmer depending on the region and scenario (Hulme, *et al* 2002) with the comparable predictions in Scotland equating to a rise of annual temperature of between 1° C and 4° C (UKCIP, 2006). Higher global temperatures can be expected to lead to a more vigorous hydrological cycle and a consequential increase in the frequency of high magnitude rainfall. According to UKCIP02, by the 2080s the occurrence of 1 in 2 year daily precipitation events in Scotland will increase. This increase could exceed 20% during the winter in eastern Scotland under the medium-high and high emission scenarios. However, in much of western Scotland the winter occurrence of 1 in 2 year daily precipitation events will increase by less than 5% or remain within the limits of natural variability. In the summer, in eastern Scotland the occurrence of 1 in 2 year daily precipitation events will decrease for medium-high and high emission scenarios by between 5 and 25% depending on location (UKCIP, 2006). The intensity of precipitation is also expected to increase over the coming decades. By the 2080s extreme storm event precipitation depths are expected to increase by between 10% and 30% with intensity of winter storms increasing slightly more than this, spring and autumn slightly less and summer extreme rainfall depths are predicted to rise by between 0% and 10% (Winter *et al*, 2005).

Annual and seasonal changes in precipitation for Scotland will vary greatly according to location and scenario. Annual changes are anticipated from 0 to −10% for all scenarios across Scotland, but these changes vary greatly seasonally and regionally (Werritty & Chatterton, 2004). The most extreme seasonal changes for low emissions are in the winter where, by the 2080s, increases in excess of 15% are predicted for some parts of eastern Scotland whilst under high emission scenarios winter precipitation increases in excess of 30% are anticipated for some eastern areas. In the western Highlands precipitation increases are lower remaining lower than 15% for even medium-high to high emission scenarios (UKCIP, 2006). However, it is important to note that in absolute terms the west remains markedly wetter than the east on account of the marked west-east precipitation gradient (Werritty & Chatterton, 2004). Summer precipitation decreases of 0–15% in the north and 15–30% in the south of Scotland are predicted by the 2080s under the low emissions scenario. These become reductions of 15–30% in the north and more than 30–45% in the south under the high emissions scenario (UKCIP, 2006). Despite the increases in winter precipitation, increased temperatures will mean that more precipitation will fall as rain rather than snow (Werritty & Chatterton, 2004).

As a result of the change in climate average soil moisture is likely to increase during winter and decrease during summer months. By the 2080's it is predicted that the average soil moisture content will be between 3% and 5% higher during the winter, and between 10% and 30% lower during the summer and autumn. Although winter precipitation is likely to increase, higher temperatures and reductions in relative humidity mean that evaporation will also increase and under current climate conditions soils are already highly saturated during winter months. This may be the reason why soil moisture levels across Scotland are likely to increase less than may have otherwise been expected, based upon precipitation increases. Changes in autumn soil moisture levels are predicted to be very similar to those in summer. This is due to the long time taken to recover soil moisture levels following hotter and drier summers. The reduction in soil moisture will be most pronounced in southern and eastern Scotland (Galbraith *et al*, 2005).

6 INTERPRETATIONS AND IMPLICATIONS OF CLIMATE CHANGE

Field investigations have shown that there is a greater density of debris flows on hillslopes underlain by coarser grained lithologies with particle size analysis demonstrating that these slopes are generally mantled by soils with a greater percentage of sand sized particles (An Teallach, Lairig Ghru). This finding is in agreement with previous research which has identified a greater frequency of debris flow forms on slopes underlain by sandstone and granite lithologies, which tend to yield abundant coarse sand on weathering and are often mantled by sediment with a coarse-grained sandy matrix, compared to slopes underlain by schists and extrusive lavas which are mantled by soils containing a much higher proportion of fine sand and silt (Ballantyne, 1981; Innes, 1983; Curry, 1999; Ballantyne, 2002). Permeability data for the coarsest (An Teallach 1) and finest (Glen Ogle 1) sampled soils has shown that the soil with a larger sand content has a greater permeability. Since the internal angle of friction is similar for both soils the data suggests that the permeability of soils determined by sediment texture could be an essential geotechnical control on debris flow susceptibility.

However, it is important to acknowledge that factors other than soil properties can influence the susceptibility of a hillslope to debris flow activity. The influence of such factors are perhaps apparent on the basaltic part of Glamaig where a high spatial frequency of debris flows (17.9 debris flows per kilometre) is supported by soils with a high silt fraction (17.1%). This suggests that factors other than sediment texture such as slope gradient, vegetation cover and prevailing rainfall conditions heavily determine propensity to debris flow at this study site. Stability analysis of the source slopes of the An Teallach 1 and Glen Ogle 1 debris flows, in which the steeper slope at An Teallach was shown to be less stable than that at Glen Ogle, highlights the key control slope gradient has on debris flow susceptibility. The lower factor of safety at An Teallach means a smaller increase in water table is sufficient to trigger debris flow activity at An Teallach compared to Glen Ogle. Therefore, the greater density of debris flows at the An Teallach site compared to Glen Ogle can be shown to be as a result of a combination of more permeable soils and greater slope gradients.

The primary triggering factor for debris flow generation around the world is high magnitude rainfall and therefore increases in intense rainfall can be expected to result in an increase in debris flow activity across Scotland. This will result in a more potent debris flow geohazard with a greater risk towards transport communications and buildings in areas of high relief due to a higher frequency of debris flow events. The anticipated increases in rainfall in the east can be expected to result in a more substantial increase in debris flow activity on hillslopes in eastern Scotland compared to western Scotland where smaller projected increases in rainfall will perhaps result in a more subtle increase in debris flow activity. Decreases in summer precipitation as well as hotter drier weather will lead to reductions in summer soil moisture of between 10 and 30% by the 2080s. As antecedent soil moisture conditions are a key determinant of debris flow activity it can be anticipated that debris flow will be less prevalent in summer months under projected climatic conditions. Summer reductions in soil moisture will also lead to increases in slope strength due to lowering groundwater levels and increased soil suctions in the slope material as a result of negative pore pressures. However, cracking of the soil during dry summer months will provide pathways for water to infiltrate into the slope and saturate slope material when a period of heavy rainfall follows the dry weather. This means that the greatest risk to slope stability may be when the wetter weather begins to commence in the autumn. Increases in winter precipitation can be expected to lead to a greater frequency of debris flow during winter months facilitated by increases in winter soil moisture contents. Although the number of debris flows in winter can be expected to increase due to rises in rainfall, the number triggered due to snowmelt will decrease as a result of a reduction in snowfall. Accordingly, under UKCIP02 scenarios, meteorologically driven seasonal episodes of slope instability may occur by the 2080s as a result of enhanced autumn infiltration through cracks in the soil mantle and winter percolation saturation of the soil.

Inaccuracies in climate change projections made by the HadRM3 RCM will compromise the accuracy of any associated inferences on the future frequency and pattern of slope failure on Scottish hillslopes. It is important to note that climate models generally predict averages and that error limits may be substantial. Models are also generally considered to be incapable of predicting localised summer storms. Such storms are often responsible for triggering debris flow, and consequently climate data may not give a full picture of the relationship between precipitation and landslides. (Winter *et al*, 2005).

7 CONCLUSION

Field investigations have shown that there is a greater density of debris flows on hillslopes underlain by coarser grained lithologies with particle size analysis demonstrating that these slopes are generally mantled by soils with a greater percentage of sand sized particles (An Teallach, Lairig Ghru). Permeability data for the coarsest (An Teallach 1) and finest (Glen Ogle 1) sampled soils has shown that the soil with the higher sand content has a greater permeability. Due to the fact that the internal angle of friction is similar for both soils the permeability of soils determined by sediment texture can be demonstrated to be a critical geotechnical control on the susceptibility of a given hillslope to debris flow activity.

Climate change projections made by the Hadley Centre's HadRM3 Regional Circulation Model using the UKCIP02 emission scenarios suggest that the climate of Scotland will experience increases in the frequency of high magnitude rainfall by the 2080s (UKCIP, 2006). It can be expected that this will lead to a greater frequency of debris flow activity in areas of high relief in Scotland by the 2080s with the critical sediment texture/permeability control on debris flow density ensuring that the greatest frequency of debris flows will occur on slopes mantled with sediment with coarse sandy matrixes yielded from granite and sandstone lithologies. This will result in a more potent debris flow geohazard with a greater risk towards transport communications and other infrastructure in areas of high relief due to a higher frequency of debris flow events. Anticipated increases in rainfall in the east can be expected to result in a consequential upturn in debris flow activity on hillslopes in eastern Scotland compared to western Scotland where smaller projected rises in rainfall will perhaps result in a more subtle increase in debris flow activity. The temporal distribution of the debris flow geohazard may be determined by meteorologically driven seasonal episodes of slope instability by the 2080s as a result of enhanced autumn infiltration through cracks in the soil mantle formed during proceeding dry summer months, and winter percolation saturation of the soil. The seasonal intensification of debris flow activity will be greatest in winter as a result of increased winter rainfall, whereas hotter, drier summers and the consequential decreases in summer soil moisture can be expected to result in a reduction in the frequency of summer debris flows.

REFERENCES

BSI. 1990. *Methods of Testing Soil for Civil Engineering Purposes*. B 1377. BSI. London.

Baird, P.D. & Lewis, W.V. 1957. The Cairngorm floods: summer solifluction and distributary formations. *Scottish Geographical Magazine*.73. 91–100.

Ballantyne, C.K. 1986. Landslides and slope failures in Scotland: a review. *Scottish Geographical Magazine*. 102. 134–150.

Ballantyne, C.K. 2002. Debris flow activity in the Scottish Highlands: temporal trends and wider implications for dating. *Studia Geomorphologica Carpatho-Balcanica*. 36. 7–27.

Ballantyne, C.K. & Eckford, J.D. 1984. Characteristics and evolution of two relict talus slopes in Scotland. *Scottish Geographical Magazine*. 100. 20–33.

Bell, B.R. & Williamson, I.T. 2002. Tertiary igneous activity. In: Trewin, N.H. (ed). *The Geology of Scotland*. The Geological Society. London.

Common, R. 1954. A report on the Lochaber, Appin and Benderloch floods, May, 1953. *Scottish Geographical Magazine*. 70. 6–20.

Conway, D. 1998. Recent climate variability and future climate change scenarios for Great Britain. *Progress in Physical Geography*. 22. 350–374.

Curry, A.M. 1999. *Paraglacial modification of drift-mantled hillslopes*. Unpublished PhD Thesis. University of St Andrews. 350pp.

Curry, A.M. 2000. Holocene reworking of drift-mantled hillslopes in the Scottish Highlands. Journal *of Quaternary Science*. 15(5). 529–541.

Fannin, R.J., Eliadorani, A. & Wilkinson, J.M.T. 2005. Shear strength of cohesionless soils at low stress. *Geotechnique*. 55. 467–478.

Francis, E.H., Forsyth, I.H., Read, W.A. & Armstrong, M. 1970. *The Geology of the Stirling District*. Memoir of the British Geological Survey, Sheet 39, Scotland. HMSO. London.

Galbraith, R.M., Price, D.J., Shackman. L., Easteal, R., Motion, A. & Barnett, C. 2005. *Scottish Road Network Climate Change Study*. Scottish Executive. Edinburgh.

Head, K.H. 1982. *Manual of Soil Laboratory Testing: vol 2, Permeability, Shear Strength and Compressibility Tests*. Pentech. London.

Hulme, M., Jenkins, G.J., Lu, X., Turnpenny, J.R., Mitchell, T.D., Jones, R.G., Lowe, J., Murphy, J.M., Hassell, D., Boorman, P., McDonald, R. & Hill, S. 2002. *Climate Change Scenarios for the United Kingdom: The UKCIP02 Scientific Report*. Tyndall Centre for Climate Change Research. Norwich.

Innes, J.L. 1983. Debris Flows. *Progress in Physical Geography*. 7. 469–501.

Jenkins, A. Ashworth, P.J. Ferguson, R.I. Grieve, I.C. Rowling, P. & Stott, T.A. 1988. Slope failures in the Ochil Hills, Scotland, November 1984. *Earth Surface Processes and Landforms*. 13. 69–76.

Luckman, B.H. 1992. Debris flows and snow avalanche landforms in the Lairig Ghru, Cairngorm Mountains, Scotland. *Geografiska Annaler*. 74. 109–121.

Nettleton, I.M., Tonks, D.M., Low, B., MacNaughton, S. & Winter, M.G. 2005. Debris flows from the perspective of the Scottish Highlands. In: Sennesest, K., Flaate, K. & Larsen, J.O. (eds). *Landslides and Avalanches*. ICFL 2005 Norway. Taylor & Francis Group. London. 271–277.

Stephenson, D. & Gould, D. (1995). *British Regional Geology: the Grampian Highlands*. 4th edn. HMSO. London.

Werritty, A. & Chatterton, J. 2004. *Foresight Future Flooding Scotland*. Department of Trade and Industry.

Winter, M.G., MacGregor, F. & Shackman, L. 2005. Introduction to Landslide Hazards. In: Winter, M.G., MacGregor, F. & Shackman, L. (Eds). *Scottish Road Network Landslides Study*. Scottish Executive. Edinburgh.

Winter, M.G., Heald, A.P., Parsons, J.A., Shackman, L. & MacGregor, F. 2006. Scottish Debris Flow Events of August 2004. Quarterly Journal of Engineering Geology and Hydrogeology. 39(2).

UK Climate Impacts Programme (UKCIP). 2006. http://www.ukcip.org.uk.

Session 3
(Double Session): Advances in hazard modelling and prediction (short- and long-term)

One of the most important ways of reducing risks arising from natural events is to continue to develop a 'culture of prevention' (ISDR, 2002). Hazard monitoring and modelling, as well as prediction of hazard events, is a rapidly advancing field of science and there are many new techniques being developed as well as challenges of data management. Improving prediction of natural hazards is essential in order to reduce the impact of these events on vulnerable communities.

Blackgang, Isle of Wight, UK
Courtesy Wight Light Gallery, Ventnor, Isle of Wight, UK.

Landslides and Climate Change – McInnes, Jakeways, Fairbank & Mathie (eds)
© 2007 Taylor & Francis Group, London, ISBN 978-0-415-44318-0

Hydromechanical modelling of a large landslide considering climate change conditions

L. Tacher
Engineering and Environmental Geology Laboratory, Swiss Federal Institute of Technology, Lausanne

Ch. Bonnard
Soil Mechanics Laboratory, Swiss Federal Institute of Technology, Lausanne

ABSTRACT: In order to define the safety conditions and the sustainable development of the large Triesenberg landslide (5 km^2), the Principality of Liechtenstein has undertaken a modelling research to determine both the critical hydraulic heads and the potential influence of climate changes. The main objectives of this research focused on critical groundwater levels and related movements, in order to supply significant mid and long-term information for land planning. First a geological model was built to define the parameter fields; then a transient unsaturated hydrogeological model computed the groundwater pressure fields that were introduced in the geomechanical model. The results of this three stage 3-D modelling process show a moderate variation of the slide movements when considering climate change scenarios, as the progressive snowmelt, due to more rain events in winter, will absorb the present critical pressure peaks.

1 INTRODUCTION

The question of the possible relations between landslide activity and effect of climate change has been raised for many years (IPCC, 1996). Many pessimistic predictions have been formulated that may have a certain justification for shallow slides or debris flows. However, in the perspective of large landslides, covering several km^2, an appropriate and general answer is not easy to express, for several basic reasons:

- The relation between rainfall events or episodes and movements is complex and depends on several characteristics of the landslide (rainfall pattern, geological structure, permeability field, geomechanical parameters) (Tacher et al., 2005).
- Very few extensive investigations have been carried out at a large spatial and temporal scale to assess the long-term behaviour of very large landslides (Noverraz et al., 1998), so that local and specific observations are often unduly extrapolated.
- The detailed expected evolution of the climatic conditions in typical landslide-prone areas like the Swiss Alps has only been published recently (OcCC, 2004) and is not accounted for in many general reports on disasters and climate change (Bader & Kunz, 1998).

Indeed the problem of the potential evolution of landslide movements with new climatic conditions can be dealt with in two different perspectives:

- A first approach leads to determine the real evolution of the movements of a set of large landslides, on which landmarks have been surveyed for about a century or more, with the corresponding evolution of the annual rainfall conditions in the area, considering that only a major trend in the precipitation amount may cause a significant increase in the average velocities of the landslide mass. It has been showed that a slight acceleration can be observed for slides covering 1 km^2 or less in the regions where a notable rainfall increase has been recorded in the second half of the XXth century; however, for larger slides (with an area of several km^2 and up to 45 km^2), the acceleration is minimal when a slight long-term rainfall increase is observed to nil when no rainfall increase is recorded (Noverraz et al., 1998).
- A second approach aims at modelling the hydrogeological and geomechanical behaviour of the landslide during exceptional rainfall events, which are the main causes of acceleration phases. Then the future evolution of such critical meteorological scenarios are anticipated, on the basis of the foreseen

trends of climate change within the concerned region, so as to simulate the behaviour of the landslide in such conditions and then quantify the expected future crises.

This is the general approach that is presented here through an application to a very large landslide on the East border of Switzerland, the Triesenberg Landslide.

2 PRESENTATION OF THE LANDSLIDE AND OF THE RESEARCH

The Triesenberg landslide extends over a significant part (i.e. 5 km^2) of the Principality of Liechtenstein (160 km^2), located to the East of Switzerland (Fig. 1). It also includes two villages, Triesen at its toe and Triesenberg at mid-slope, the infrastructures of which incur occasional damage, in particular during crisis episodes.

The movements of this landslide are quite ancient and date back to the end of the Wurmian period; presently they are generally slow (i.e. some mm/year to cm/year) in normal conditions and locally may reach velocities of a few dozens of cm/year during severe crisis periods. As the slide displays a relatively slow movement, many buildings have spread on the slope in particular during these last decades, due to the real estate development.

The research carried out deals with the numerical modelling of groundwater flow in the slope and the geomechanical modelling of the slide movements; it has been led in close partnership with the consulting office Bernasconi in Sargans, Switzerland, which has been collecting relevant data for many years, as well as with the authorities of Liechtenstein (Office of Public

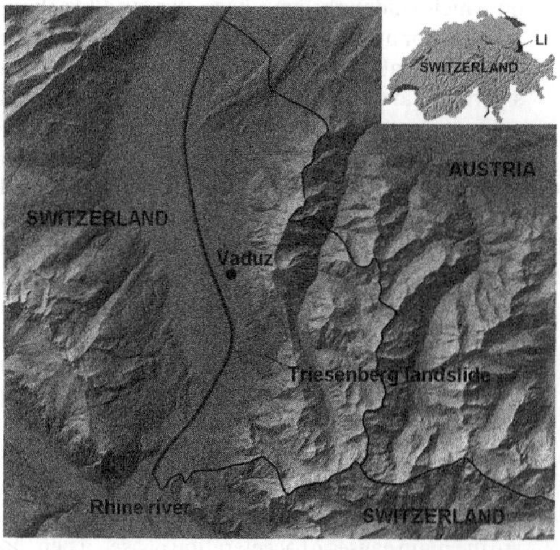

Figure 1. Location of the Principality of Liechtenstein and of the Triesenberg landslide.

Works, TBA). The objectives of the research are the following:

- To determine the critical hydrogeological conditions that cause an acceleration of the slide, that may expose the population and assets to danger, and that justify the triggering of an alarm system.
- To foresee the behaviour of the slide under the possible effect of climate change, so as to establish bases for the sustainable development of the slope.

The first aim of the models developed does not consist in determining the possibilities of stabilizing the overall slope, as it can be expected that such works would by far pass the planned investments by the authorities of the Principality. What is aimed at is to live with the slide, and not to slow it down.

Three iterative phases have been developed in this research:

- 3-D structural geological modelling of the area, in order to determine the geometrical limits of the numerical models and the parameter fields.
- 3-D transient groundwater numerical modelling, considering unsaturated conditions.
- 2-D and 3-D geomechanical numerical modelling, in order to compute the displacement fields, by using the hydraulic heads computed in the previous step.

A similar analysis for another large landslide in Switzerland had been previously carried out, namely the la Frasse landslide, which extends over 1 km^2 (Tacher et al., 2005). In comparison to that case the specific difficulties presented by the Triesenberg landslide mainly refer to its large area, to its essentially unsaturated hydrogeological conditions and to the slow movement velocities. The models developed have been calibrated thanks to numerous hydrogeological data (piezometer readings, spring rates), as well as geomechanical information (GPS and inclinometer data, laboratory triaxial tests), in particular for the reference year 2000. Furthermore, the computed displacements could be compared with the displacement fields established with IN-SAR (Synthetic Aperture Radar Interferometry) (Colesanti & Wasowski, 2004).

After calibrating the parameter models with respect to the crisis of 2000, the impact of climate change has been analysed by modifying the boundary conditions of the hydrogeological model, on the basis of the relevant climatic scenarios, as set up by the Swiss Commission for the assessment of climate change (OcCC). Then the respective computed groundwater pressures have been introduced in the geomechanical model, as it was done for the year 2000. All the numerical simulations were carried out using the finite element codes Feflow® (Version 5.1) (Feflow groundwater modelling software, Wasy AG, Berlin) and Z_SOIL® 2D and 3D (Version 6.24) (Z_SOIL_PC,

Zace Services Ltd, Lausanne), as well as the 3-D geological model using the Geoshape software developed at the GEOLP/EPFL.

3 MAIN FEATURES OF THE LANDSLIDE

3.1 *Morphology and geology*

The slope is oriented from North-East (up) to South-West (down). It presents some small undulations but is generally fairly regular. Based on a digital terrain model, the mean slope is 24° (Figs 2, 3). Three parts are distinguished:

- In the upper part, deep-seated slope movements occurred, probably at the end of the Wurmian glacial retreat (14,000 years); they are now underlined by a terrace in the topography at the top of the slope (Fig. 3) and were triggered by a deep landslide, the socalled Prehistoric landslide. The deep-seated slide is composed of Triassic "Buntsandstein" sandstones, shales and limestones of the Lechtal nape and vanishes at the top of the lower active slide. This upper zone, largely inactive, is not considered to cause a driving force on the slope. Approximately, it covers 1.7 km^2 with a volume of 74 million m^3.
- The prehistoric landslide is known by some boreholes. It is more than 80 m deep and is made of flysch (clayey shales of the Austro-alpine nape). A comparison of inclinometer and geodetic data shows that this zone is today stabilised; moreover, no movement has been observed at the toe of the landslide, where it lies under the Rhine river alluvia (gravels).
- The active landslide (Table 1) covers the prehistoric one. It is also composed of flysch and takes place on a slip surface located at an average depth of 10–20 m. According to the inclinometer data, available from 1995 to 2002, the slip surface is approximately one meter thick (Fig. 4).

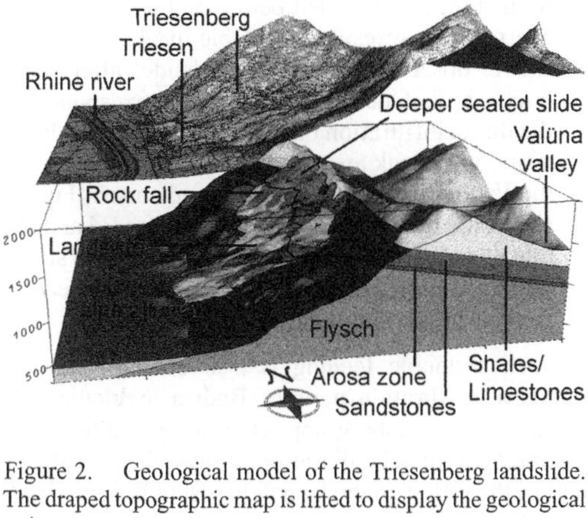

Figure 2. Geological model of the Triesenberg landslide. The draped topographic map is lifted to display the geological units.

The analysis of the observed intensities and directions of the movements (Fig. 5) showed that the area is indeed composed of three instability zones that can be considered as independent.

This is confirmed by the reduced depth of the slip surface close to the assessed boundaries of the three areas. Furthermore, the damage to infrastructures and buildings is mainly concentrated along these same boundaries, indicating differential movements, and additional movement monitored by GPS measurements completed in 2004 (Frommelt AG) gives the same results in terms of intensity and distribution.

The estimated range of velocity of the observed movements is given in Table 2.

The practical consequence of the decomposition of the landslide into three distinct systems is to allow defining three different modelling areas for the 3D mechanical modelling, supposing negligible cinematic interactions between them (the hydrogeological model involves a single regional area). Processing three parts has the advantage of decreasing the finite element size of each individual mesh and, therefore, of increasing the accuracy of the 3D computations (François et al., 2007).

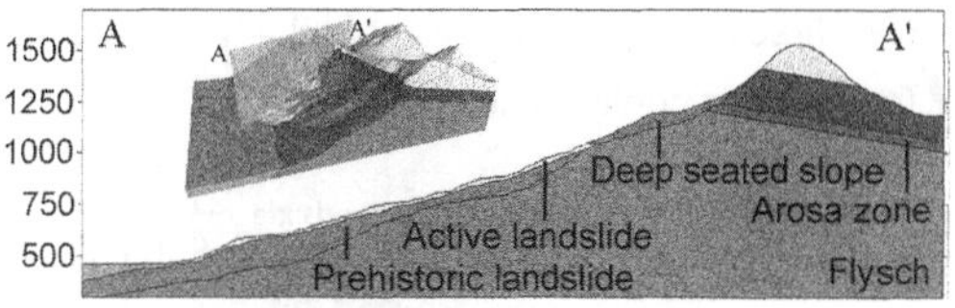

Figure 3. Geological vertical cross section.

Table 1. Main features of the Triesenberg active landslide.

Aspect	Characteristics
Area	3.1 km^2
Altitudes	min. 460 m, max. 1500 m a.s.l.
Length	2300 m
Width	1500–3200 m
Mean depth	10–20 m
Volume	37 millions m^3
Mean slope	24°
Mean velocity	0 to 3 cm/year
Soil	Flysch (clayey shales) including elements of limestone and sandstone
Vegetation	Pasture land and some wooded zones
Investigations	Hydrogeology, boreholes with inclinometers, GPS, RMT geophysical methods, laboratory tests, modelling
Possible damage	Infrastructures of two villages

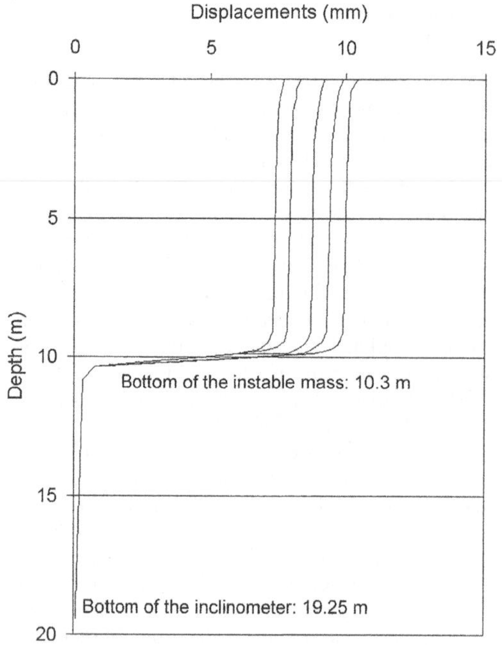

Figure 4. Displacement data in inclinometer KL1A from December 1996 to March 2000 (location close to piezometer B4, see Fig. 6).

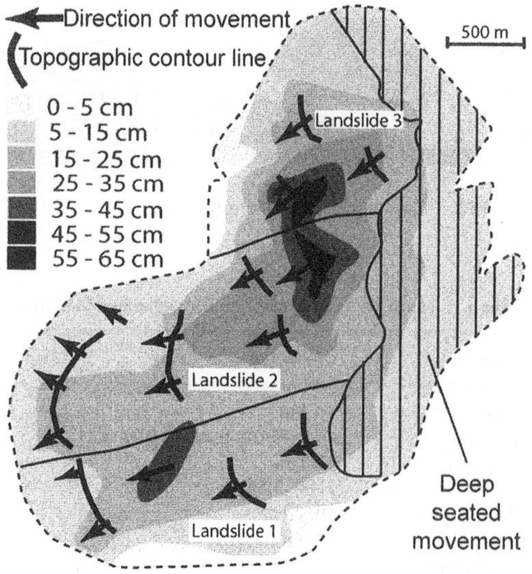

Figure 5. Total displacements between 1976/1981 and 1996/1997 (i.e. during some 20 years) in the whole instability area, and boundaries of the three landslides.

3.2 *Hydrogeology*

Besides supplying geomechanical models with hydraulic pressures, the hydrogeological analysis and modelling has the aim of better understanding the particular hydrogeological behaviour of the slope.

Table 2. Order of magnitude of the observed average velocity of the landslide movements.

Term	Considered period	Velocity
Medium term	>20 years	0.5 to 3 cm/year
Short term	±1 year	0.1 to 4 cm/year
Exceptionally	<1 month	6 cm/year

The yearly observation of displacements shows a close dependence of the movements on the seasons. A reactivation is generally perceived in the spring, which corresponds to the snowmelt period. This indicates that the main driving force of the movements is the variation of pore water pressure in the slope. However, a reactivation may also occur following a storm event.

As stated above, the Triesenberg landslide develops into the Austro-Alpine flysch, mainly composed of clayey shales. At the top of the slope, the "Buntsandstein" sandstones, shales and limestones are much more pervious (ca. 5×10^{-3} m/s). Between these two units, the tectonic Arosa zone (Fig. 3) is a very important feature of the hydrogeological system due to its low permeability: a part of the Valüna Valley groundwater (Fig. 2) flows on the Arosa zone and feeds the basal surface of the landslide, causing the Triesenberg groundwater basin to be much larger than its topographic watershed. This mechanism is proven by three observations:

- Tracer tests (Bernasconi, 2002a) performed in 1999 in the upper Valüna Valley gave positive results in several springs on the landslide.
- As the mean altitude of the Valüna Valley is higher than that of the Triesenberg slope, snowmelt occurs later and leads to a delayed increase of piezometric head and spring rates in the landslide, with respect to the normal snowmelt period on this slope.
- Storm events result in double peaks in piezometric observations in the landslide; after some hours, a first peak occurs which corresponds to the direct infiltration on the slope; some days later, a smoother peak appears, which is the inflow from the Valüna Valley through the sandstones. (Fig. 6) shows the hydraulic head in piezometer B4 and the rate of spring Q16 during the August 6, 2000 storm period; at these two observation points, the second peak occurred about ten days after the event.

Such a double feeding is also effective outside intensive infiltration periods. Both a hydraulic balance of the Triesenberg slope (Bernasconi, 2002b) and a numerical model calibration suggest that about one half of the inflow in the landslide is supplied by a base flow from the Valüna Valley through the sandstones

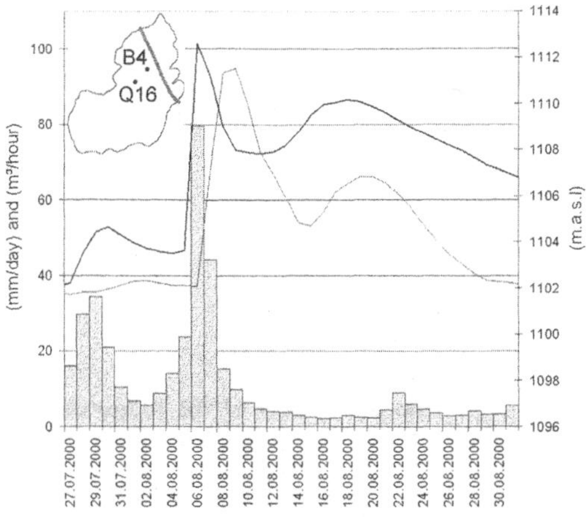

Figure 6. Rate of spring Q16 and hydraulic head in piezometer B4 in August 2000. In the small location map, the thick line is the trace of the Arosa zone at the base of the landslide.

Table 3. Parameters of the three intact samples tested in the laboratory. The identification parameters of the sample No 5966-9 have not been determined.

Sample	Liquid limit %	Plasticity index %	Fine particles <0.02 mm	Classification USCS
5966-1	26	13.1	50%	Clay CL
5966-2	22.2	10.6	37.6%	Clayey sand SC
5966-9	–	–	–	Clayey sand SC

Sample	Young* modulus MPa	Confining pressure kPa	Friction angle °	Cohesion kPa
5966-1	97	300	25	0
	134	500		
	223	800		
5966-2	128	560	30	11
5966-9	276	300	30	17
	330	500		
	415	800		

*The Young modulus is determined in unloading conditions starting from 2% of axial strain under a confining pressure mentioned in the adjacent column.

covering the Arosa zone (ca. 9 mio m^3/year). Indeed, at the top of the slope, the piezometric behaviour is directly influenced by the inflow from the Valüna Valley. The hydraulic head in piezometer B4 is rather constant, close to 1101 m a.s.l, a value that is kept steady by the base flow due to the capacitive function of the sandstones. In the lower part of the slide, the inflow peaks from the Arosa zone are smoothed by the landslide aquifer. The piezometric peaks in borehole B8 are less sharp and do not exceed 5 m (Fig. 13). Groundwater discharge occurs through some one hundred springs distributed over the landslide (Fig. 8), as well as at its toe, in the Rhine River alluvia. The water table is located about 20 to 30 m below the soil surface at the top of the landslide, whereas at the bottom, it almost reaches the ground surface.

3.3 *Materials*

From several soil samples, extracted at different locations, three main categories of soil with three quite different mechanical properties have been identified:(i) the loose soil forming the landslide mass, (ii) the slip surface material and (iii) the bedrock (i.e. the compacted prehistoric landslide mass). Geotechnical tests (triaxial shear tests under different confining pressures and oedometer compression tests) were carried out at LMS/EPFL on three particular samples, leading to the determination of the material parameters given in Table 3.

The hydraulic properties of the material are discussed in section 4.

4 HYDROGEOLOGICAL MODELLING OF THE YEAR 2000

The year 2000 was chosen to perform the modelling; during this year, a critical phase with a reactivation of the movements was observed, showing a good correlation with the snowmelt phase in April. A violent thunderstorm also occurred in August. This year has a return period of the annual rainfall of 42 years, according the Gumbel law (Fig. 9). Another reason of this choice is the availability of calibration data for both hydrogeological and geomechanical models. As the slide is very thin, the unsaturated zone is of relatively high importance, which justifies computing groundwater flows in unsaturated regime, i.e. the flows are governed by Richard's equation (Hillel, 1980).

4.1 *Parameters*

Spatially, the 3-D FE mesh extension encompasses the entire landslide area (slide plus old deep-seated slide zone). 51,772 triangular prismatic linear elements are distributed over four layers of equal thickness. The bottom of the model is the slip surface. The old, deep-seated slide zone has little importance in the model since it is almost completely unsaturated. It is discretised only to allow for the introduction of flux boundary conditions at the bottom of the mesh along the Arosa zone (Fig. 8).

Parameters are assigned to both top layers with respect to data obtained by RTM geophysical investigations (Fig. 7), while for both bottom ones, homogenous values are considered ($k = 2 \times 10^{-4}$ m/s,

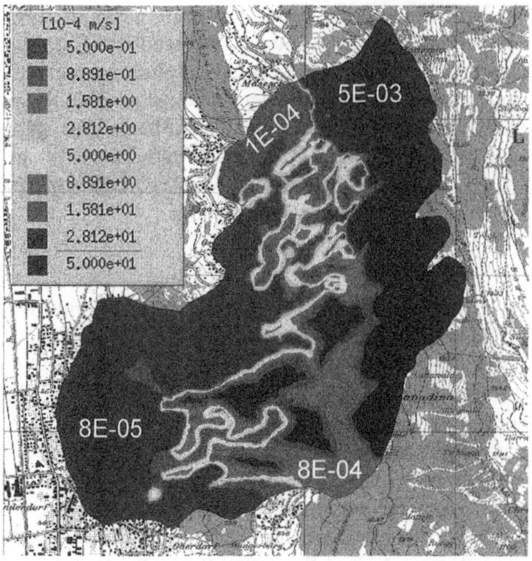

Figure 7. Hydraulic conductivity (k) map of both upper layers of the hydrogeological model.

$n = 0.15$). In all four layers, the high values to the East correspond to the old deep-seated slide zone ($k = 5\times10^{-3}$ m/s). k is everywhere considered isotropic. The porosities n are linearly traced on hydraulic conductivities, leading to values ranging from 0.15 to 0.25. In the old deep-seated slide, $n = 0.15$.

Assigning homogeneous properties to both bottom layers, rather than continuing the surface heterogeneities down to the slip surface, is justified by avoiding giving too much weight to structures that are probably very superficial. Furthermore, with depth, the heterogeneity of k has a significant impact only if high permeability heterogeneities are connected. If not, they behave as isolated pockets, which is the assumption made.

Unsaturated parameters linking permeability to saturation and suction to saturation are not documented on the site. They were assigned in a homogeneous way to represent coarse materials, i.e. a sharp transition between saturated and unsaturated zones (linear model with a fringe pressure $= -4.1$ m and residual saturation $= 0.1$).

4.2 *Boundary conditions*

4.2.1 *At the surface*

Hydraulic head conditions ($h =$ elevation z m) with constraint of outgoing rate were imposed at the location of springs (Fig. 8). At the toe of the slope, a $h = z - 2$ m condition was imposed at the contact with Rhine Valley alluvia (as for all four other interfaces).

Direct infiltration was computed using the COUP model (Jansson & Karlberg, 2001) for five elevation

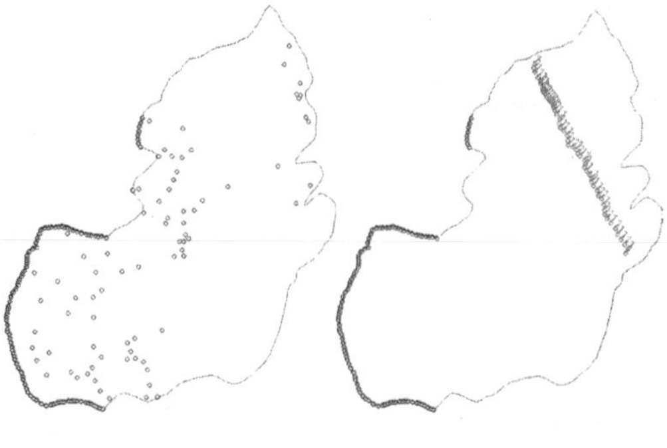

Figure 8. Boundary conditions of the hydrogeological model. Left: Upper interface of the mesh with hydraulic head conditions at the location of the springs and along the Rhine river valley. Right: Bottom interface of the mesh with rate conditions along the Arosa zone.

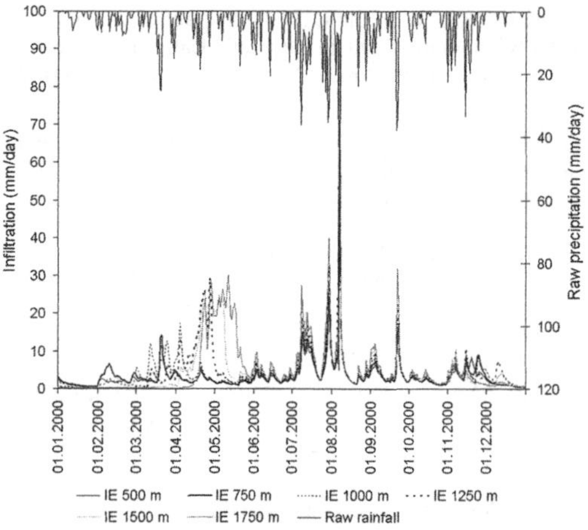

Figure 9. Top curve: Raw precipitations (rainfall and snow converted in water height) at Balischguad gauge station (elevation ~ 1000 m a.s.l.). Down: Computed infiltration (IE) for the six elevation classes.

classes, i.e. the zones included between the isolines 625, 875, 1125, 1375 and 1625 m a.s.l. (Fig. 9).

4.2.2 *At the bottom*

The infiltration class for $z > 1625$ m a.s.l. is not used as direct infiltration on the slope (maximum elevation ~1500 m), but to compute the feeding by the Arosa zone as rate conditions. The calibration of the COUP model is based on snow height measurements that are available in the village of Malbun, located in the Valüna valley (Fig. 10).

To simulate both conductive and capacitive effects of the sandstones, infiltration from the Valüna Valley was reduced and a base flow was assumed as follows (Fig. 11) :

$$Q_{Arosa} = \text{Infiltration}_{1625} * S_{Arosa} * p$$

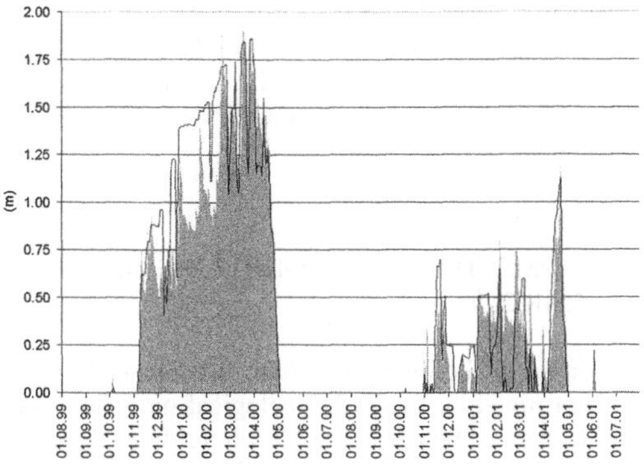

Figure 10. Computed (line) and measured (grey surface) snow height in Malbun.

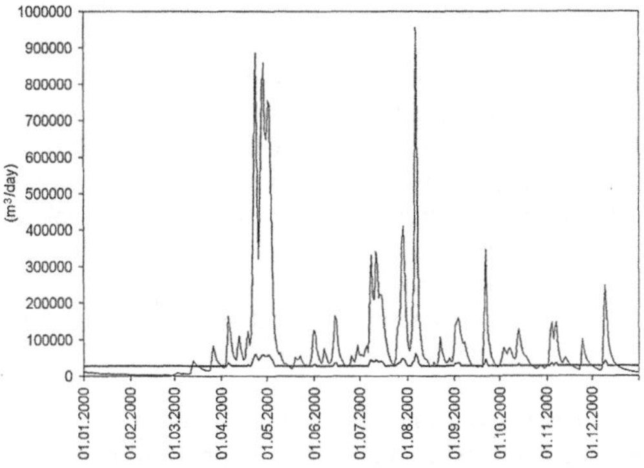

Figure 11. Thin line: Total infiltration on the Valüna watershed. Thick line: Feeding of the Triesenberg slope via the Arosa zone; this rate is distributed at the bottom of the model at the nodes situated along the Arosa zone.

Q_{Arosa}: rate feeding the landslide through the sandstones above the Arosa zone in m^3/day.

$Infiltration_{1625}$: infiltration computed by the COUP model in the Valüna Valley (m^3/day. m^2)

S_{Arosa}: area of the Valüna basin (ca. 12 km^2)

p: percentage of water of the Valüna basin directed towards the landslide through the limestones, estimated to 25%.

Then, when applying this relation:

if $Q_{Arosa} < 27{,}000\ m^3$/day: $Q_{Arosa} = 27{,}000\ m^3$/day. This represents the base flow.

if $Q_{Arosa} > 27{,}000\ m^3$/day: $Q_{Arosa} = Sqr(Q_{Arosa} - 27{,}000)*10\ m^3$/day. This calibration represents the absorption of infiltration events.

Except for the rate conditions along the Arosa zone and head conditions at the toe, no boundary conditions were set up at the bottom interface, which means that the hydraulic relations of the landslide with the underlying flyschs were neglected. Indeed, intact flyschs are assumed to have a very low permeability.

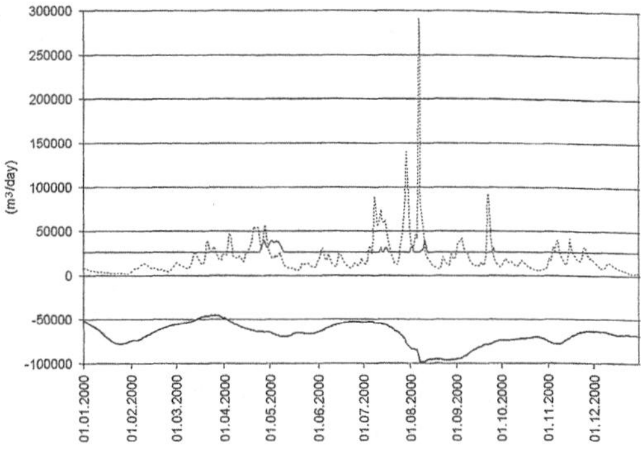

Figure 12. Hydraulic balance of the model for the year 2000. Thin line: Feeding by the Arosa zone. Dot line: Direct infiltration on the slope. Thick line: Outgoing rates through springs and the Rhine river valley.

4.3 *Results*

From the model results, in terms of volumes, the direct infiltration reached 7.52 mio m^3 in 2000, while the inflow through the Arosa zone was about 9.86 mio m^3 (Fig. 12). The cumulated rate of the springs reached 1.06 mio m^3, which represents only a few percent of the total outflow; the balance flow seeps in the Rhine river alluvia.

The outflow curve is smoother than the inflow events, due to the capacitive function of the landslide. Typically, the August 6th storm response was absorbed and delayed. In May, the snowmelt episode did not lead to spectacular changes in the hydraulic balance because, due to the slope topography, the melting occurred progressively from the bottom to the top. For example, when the snowmelt occurred in the Valüna Valley, it had already finished on the landslide several days to weeks earlier.

More relevant from the geomechanical point of view is the piezometric behaviour. It is illustrated by piezometer B8 (Fig. 13). The respective calibration was carried out by comparing the water table data with the hydraulic head computed at the bottom node at this site, i.e. at the slip surface. Both main events of the year 2000 led to a peak more than 2 m high. Just after these peaks, the head decrease was slower in the model than in the reality. This can be explained by the relative smoothing of the parameter field, mainly over depth.

Heterogeneities are also responsible for another observation: during the snowmelt event, the model reacted with a delay of some days with respect to the monitoring data. Local pervious heterogeneities that were not considered in the model accelerated the piezometer response to inflows in the Valüna Valley. Such a delay did not occur at the beginning of August since inflows concern both the Triesenberg and Valüna basins.

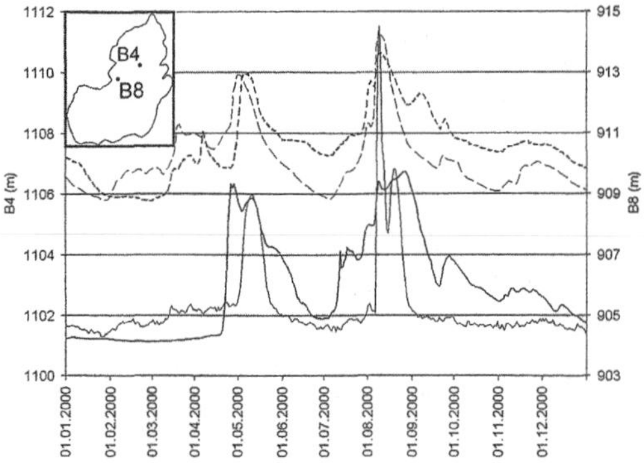

Figure 13. Measured and computed hydraulic heads in piezometers B4 and B8. Thin line: B4 measured. Thick line: B4 computed. Thin dotted line: B8 measured. Thick dotted line: B8 computed.

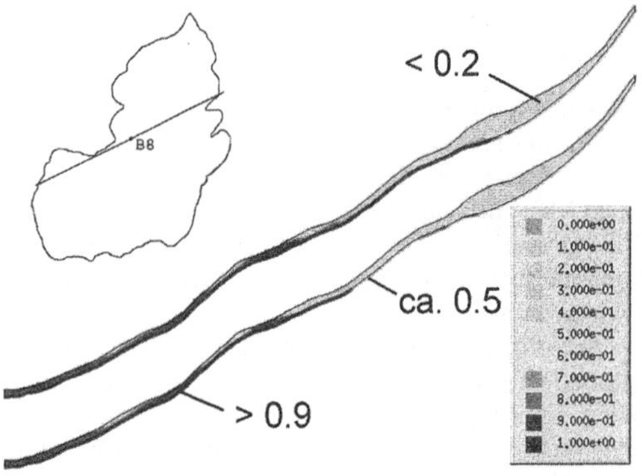

Figure 14. Computed saturation profiles. Top: April 23, wet conditions; Bottom : June 11, drier conditions.

Spatially, (Fig. 13) also shows that according to the observation, the head peaks are smoother and less high downward the slope, due to the capacitive function of the landslide.

At the slope scale, saturation profiles (Fig. 14) show seasonal conditions, mainly in the middle of the slope. In the old, deep-seated slide zone, inflow from the Valüna Valley is not sufficient to saturate the medium, neither in winter (dry) nor under wet conditions (summer), in accordance with observations. At the toe of the slope, saturation up to the subsurface is permanent.

The feeding role of the Valüna Valley through the sandstones covering the Arosa zone is confirmed by the model, since any calibration that did not consider it led to a strong hydraulic head deficit during winter and to the impossibility to calibrate the flat behaviour in piezometers located just below the Arosa zone.

The numerical results suggest that the model globally fits reality, despite a simplification of the parameter fields, a rough estimation of the unsaturated parameters and a minimal knowledge of the real hydraulic balance. Computed hydraulic pressures are thus suitable as an input in the hydro-mechanical models in order to describe the direct causes of the movements during crises.

5 GEOMECHANICAL MODELLING OF THE YEAR 2000

The effect of the hydraulic head variation with time, as determined by the hydrogeological modelling, on the mechanical behaviour of the whole slide, has been modelled by a FE code, Z-SOIL (2-D and 3-D), using a Biot-type formulation, implying the conservation of mass and momentum of both fluid and solid phases (François et al., 2007).

Two soil constitutive laws were used for the landslide mass, first an elastic model with high rigidities for the upper loose soil, and then a modified Cam-Clay elasto-plastic model for the material composing the slip surface, the depth of which is assessed to be 1 m. The displacements of the landslide mass are thus induced by pore pressure changes with time, applied at the lower part of the slip surface layer, as derived from the hydrogeological model.

The geotechnical characteristics of the slip surface layer were determined by undrained triaxial tests, which lead to define not only shear strength parameters (like $\phi' = 25$ to $30°$), but also the elasto-plastic parameters (elastic and plastic compressibilities, dilatancy angle, initial critical pressure) that are required for the geomechanical modelling.

In the 2-D model, the results obtained exhibit two main active zones, one on top of the slope and the other one in the middle of the slide profile. The value of the maximum displacement, reaching 19 mm for the 2000 crisis, near the village ofTriesenberg, is relatively close to the monitored data in this zone (continuous inclinometer readings at some distance), but the movement pattern with time shows various steps corresponding to both critical hydrogeological episodes (May and August), without presenting a continuous base movement, probably related to the cyclic behaviour of the landslide mass (that is not completely considered in the model). In the 3-D model, the maximum displacement values are in general slightly lower, but they appear within zones where damage has been reported (Fig. 15).

Parametric studies have also been carried out to evaluate the effect of the selected friction angles (between 30 and 21°) and of the range of water pressure variation (the computed data through the hydrogeological model were multiplied by 1.25 and 1.5 respectively).

Both simulations display nearly linear variations and prove that, even in extreme conditions, it is not

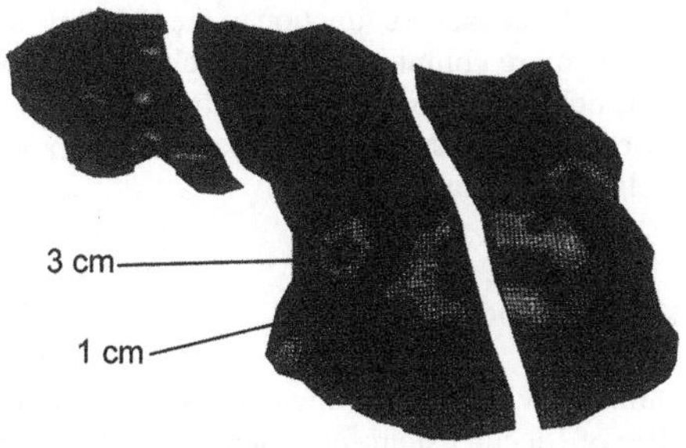

Figure 15. Distribution of the obtained displacements after 291 days of simulation, from January 1 to October 18, 2000 (water pressure data multiplied by 1.25) and location of the more active zone in the central slide (3 cm).

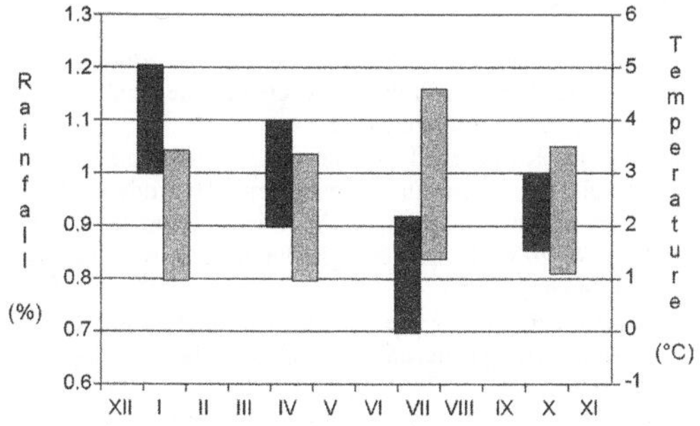

Figure 16. Climatic scenario for 2050 for the North of the Alps, after OcCC. Horizontal axis: months of the year. Black bars: Rainfall change in % with respect to present average values. Grey bars: Temperature change in °C.

expected that the movements will lead to a catastrophic behaviour of the whole slope.

6 MODELLING OF CLIMATE CHANGE IMPACT

According to (IPCC, 1996), the air temperature should increase in the medium term, especially in summer, and the rainfall should increase in winter, but decrease in summer. The climatic scenario for 2050 used in this study is issued from the Swiss "Organe consultatif sur les Changements Climatiques" (OcCC, 2004), more specific to the North of the Swiss Alp context (Fig. 16).

According to this scenario, it can thus be expected that:

- In winter, the total infiltration would increase and rain would partly replace snow accumulation. On the other hand, snowmelt at the beginning of the spring would be less important.
- In summer, the storm events would remain similar, if not slightly worse, but the total infiltration would be smaller than today because of higher evapotranspiration.

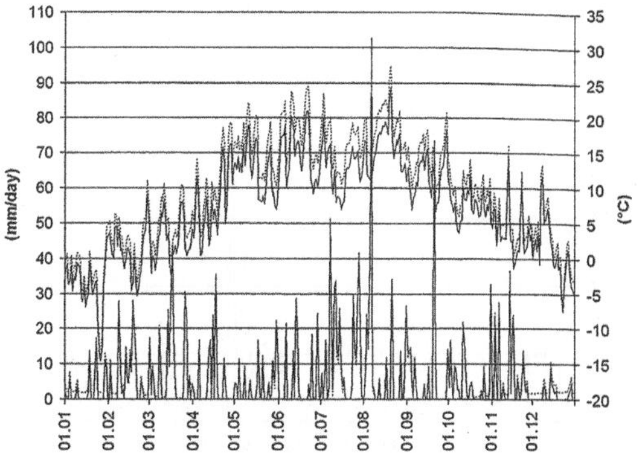

Figure 17. Climatic scenario for 2050 in the Valüna valley. Upper curves: Temperatures in 2000 (solid line) and in 2050 (dotted line). Lower curves: Raw rainfall in 2000 (solid line) and in 2050 (dotted line).

In this study, the rainfall in 2050 is considered to increase by 2 mm/day in winter and to decrease by 2 mm/day in summer. Those values are added as a one year sinusoidal transformation to the records for the year 2000. Similarly, the temperature curve for 2050 is obtained by adding a one year sinusoidal function to the records of the year 2000, considering a warming of 1.5 °C in winter and 3.5 °C in summer (Fig. 17).

Considering the impact on landslides, such a scenario is not obviously more severe, mainly for the landslide zones in altitude. Indeed, besides the total infiltration, the groundwater pressure fluctuations have a major effect on the movements. By diminishing the rather massive infiltration period of snowmelt, the 2050 scenario smoothes out the groundwater head curve at spring time. In particular in the Valüna valley, the fast snowmelt at the beginning of May 2000 might be replaced by a succession of less important episodes of rain, falling on a thin accumulation of snow.

The target of the models is here to consider the most unfavourable scenario as far as the landslide movements are concerned. Thus these worst case infiltration conditions for 2050 are as follows, even if they are not the most plausible:

- No consideration of the decreasing of raw rainfall in summer. The infiltration curve is left intact from May 1st.
- Keeping the snowmelt event of the end of April.
- In winter, adding infiltration periods without decreasing the accumulated snow height.

In practice, the 2050 infiltration scenario implies to add infiltration days between January 1 and April 20 to the year 2000 conditions. For all altitude classes of

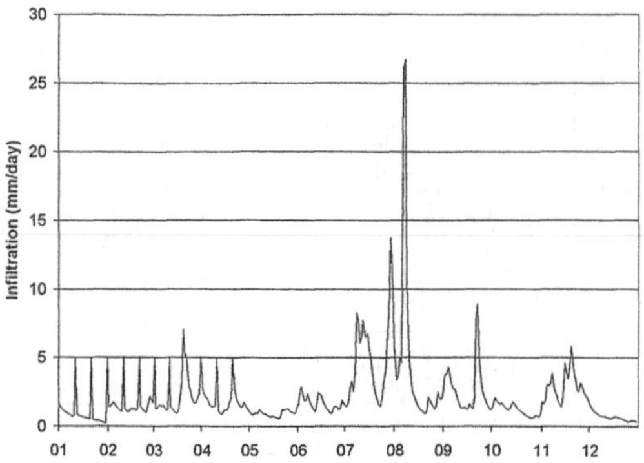

Figure 18. Scenario 2050. Infiltration conditions for altitudes below 625 m.

infiltration (see §4.2), a 5 mm/day event is introduced each ten days (Fig. 18). This represents an additional infiltration of 55 mm/year.

The results of both hydrogeological and geomechanical models with such modified boundary conditions are very similar to those obtained for the year 2000. Typically, the hydraulic heads in piezometer B8 (Fig. 13) are changed by some centimetres only.

7 CONCLUSIONS

A detailed hydrogeological and geomechanical modelling as it was recently applied at the Triesenberg and La Frasse landslides allows a significant modelling of large landslide movements during crises, provided sufficient information is available. The application of predicted climatological conditions in the future then supplies quantitative values of possible movements, considering appropriate scenarios. However, extremely rare conditions with a very remote probability cannot be modelled reliably, as the boundary conditions may significantly differ from the ones considered in the original model.

The analysis of several large landslides in other contexts (Bonnard et al., 2004) also shows that the effect of climate change on landslides within the next 50 years or so must not be overemphasized. Indeed, as shown here, the progressive snowmelt that will begin earlier than before tends to reduce the occurrence of critical situations in the spring or summer. On the contrary, it is clear that the expected increase of storm intensity, as foreseen by some climatologists, may produce more violent and frequent small slides and debris flows; but this specific prediction is not relevant for large landslides and cannot justify a development of more disasters related to this type of phenomena. Indeed, due to the heterogeneity of the material at a large scale, to the increased range of altitude where infiltration occurs, to the capacitive function of the landslide mass and to the more complex hydraulic relationships with the bedrock, the response to climatic events may be significantly smoothed and delayed, which explains this relatively optimistic vision.

ACKNOWLEDGMENTS

The authors wish to thank the authorities of the Principality of Liechtenstein for supporting this research, Dr Riccardo Bernasconi geological office for supplying data and advice, as well as all the colleagues who participated in the modelling.

REFERENCES

Bader, S. & Kunz, P. 1998. *Climat et risques naturels – La Suisse en mouvement.* Rapport scientifique final PNR31. Zürich: V/D/F éditeur, 312 p.

Bernasconi, R. 2002a. Tiefbauamt des Fürstentums Liechtenstein. Hangsanierung Triesenberg – Hydrogeologische Überwachung – Ergebnisse der Markierversuche Valünatal 1999/2000, Hydrogeologischer Bericht Nr. 1124–04, unpublished report.

Bernasconi, R. 2002b. Tiefbauamt des Fürstentums Liechtenstein – Hangsanierung Triesenberg – Hydrogeologische Überwachung – Messdaten 1999–2001, Hydrogeologischer Bericht Nr. 1124–03, unpublished report.

Bernasconi, R. 2003. Tiefbauamt des Fürstentums Liechtenstein – Hangsanierung Triesenberg – Geophysikalische Messungen und Geologische Sondierungen, Hydrogeologischer Bericht Nr. 1124–02, unpublished report.

Bonnard, C., Forlati, F. & Scavia, C. 2004. *Identification and mitigation of large landslide risks in Europe : advances in risk assessment.* IMIRILAND Project. 317 p. Leiden: Balkema, ISBN 90 5809 598 3.

Colesanti, C. & Wasowski, J. 2004. Satellite SAR interferometry for wide-area slope hazard detection and site-specific monitoring of slow landslides. *Proc. 9th Int. Symp. on Landslides.* 795–802. London: Taylor & Francis.

EPFL 2006. Lehrstühle für Ingenieur – und Umweltgeologie (GEOLEP) und für Bodenmechanik (LMS). Tiefbauamt des Fürstentums Liechtenstein. Geologische, hydrogeologische und geomechanische Modellierung des Erdrutsches von Triesenberg (Fürstentum Liechtenstein), Abschlussbericht, Lausanne, unpublished report.

François, B., Tacher, L., Bonnard, Ch., Laloui, L. & Triguero, V. 2007. Numerical modelling of the hydrogeological and geomechanical behaviour of a large slope movement: The Triesenberg landslide (Liechtenstein), *submitted to Canadian Geotechnical Journal.*

Hillel, D. 1980. *Fundamentals of soil physics.* New York: Academic Press.

IPCC, 1996. *Climate change 1995 – The second Assessment Report.* Contributions of working Group I, 572 pp.; Group II, 878 pp.; Group III, 448 p. Cambridge University Press.

Jansson, P.-E. & Karlberg, L. 2001. *Coupled heat and mass transfer model for soil-plant-atmosphere systems.* Royal Institute of Technology, Department of Civil and Environmental Engineering, Stockholm, 325 p.

Noverraz, Fr., Bonnard, Ch., Dupraz, H. & Huguenin, L. 1998. *Grands glissements de versants et climat*. Rapport final PNR 31. Zürich: V/D/F éditeur. 314 p. ISBN 3 7281 2612 8.

OcCC, 2004. Die Klimazukunft der Schweiz – Eine probabilistische Projektion, Christoph Frei, Institut für Atmosphäre und Klima, ETH Zürich.

Tacher, L., Bonnard, Ch., Laloui, L. & Parriaux, A. 2005. Modelling the behaviour of a large landslide with respect to hydrogeological and geomechanical parameter heterogeneity, *Landslides Journal*, 2(1): 3–14.

Landslides and Climate Change – McInnes, Jakeways, Fairbank & Mathie (eds)
© 2007 Taylor & Francis Group, London, ISBN 978-0-415-44318-0

Using artificial intelligence in an integrated risk management programme for a large alpine landslide

F. Oboni
Oboni Riskope Associates Inc., Vancouver, Canada

C. Angelino
Riskope Italia, Torino, Italy

J. Moreno
AILandslide, Lausanne, Switzerland

ABSTRACT: This paper demonstrates the use of artificial intelligence (AI) for the forecast of sliding velocities (AILandslide) within the frame of an integrated risk management program for a large alpine landslide named Cassass (AILandslide). The slide and the forecasting of the velocity through various classic methods have already been described in several referenced prior papers.

The Cassass Landslide is located in the NW Italy Piedmont region and impinges on the main access corridor to 2006 Winter Olympics, the international Frejus railroad line, the Frejus Tunnel Access Highway, hydro-electrical facilities as well as the village of Salbertrand. The slide underwent a paroxysm (i.e. a sudden acceleration and generalized failure) in 1957. This paper summarizes the slide's major features and the results brought in by prior studies, including a long term monitoring program implemented by the Highway Operator and financed by the Regional Government.

In recent years the area has been afflicted by rather extreme meteorological patterns. The impact of these events on the instrumentation, the evolution of the monitoring program and the risk mitigation measures undertaken are discussed. Results yielded by the probabilistic modelling of this slope are discussed in terms of the periodic risk re-evaluation, the influence on the mitigation program and decision-making, and, finally the impact on future monitoring. Artificial intelligence is also used to predict velocities as a function of prior rainfall. The velocity-forecasting tool (www.ailandslide.com) is integrated with the slope analysis thus leading to an integrated evaluation tool that can be maintained online (real time).

A final chapter discusses the stabilization work undertaken (drainage tunnel) as well as the latest developments of the situation and, of course, expected future developments.

1 INTRODUCTION

The Cassas Landslide is located in the NW Italy Piedmont region and impinges on a corridor encompassing main transportation lines, hydro-electrical facilities and a large village. The slide, or more correctly the slide system, covers an area spanning a length of 1.4 km by 0.6 km and has an active sliding surface approximately 50 m deep in the area of an inclinometer known as I4. Attention is focused on a subset of this system, which underwent a paroxysm (i.e. a sudden acceleration and generalized failure) in 1957 before returning to a "normal behaviour", characterized by velocities ranging between 20 mm/yr and 150 mm/yr as a function of their location within the slide, long term meteorology.

The slide has been the object of monitoring for more than a decade by various agencies and its behaviour with respect to antecedent rainfall studied in detail with classic methodologies (Oboni, 2005). The landslide impinges on an international transportation corridor (Fréjus tunnel railroad and highway), a large rest area with restaurants and gas stations, as well as on several private and public infrastructures. Models were developed to predict how a future catastrophic paroxysm would interact with the valley floor, the river and various structures/potential targets (Fig. 1), leading to the formulation of appropriate emergency plans.

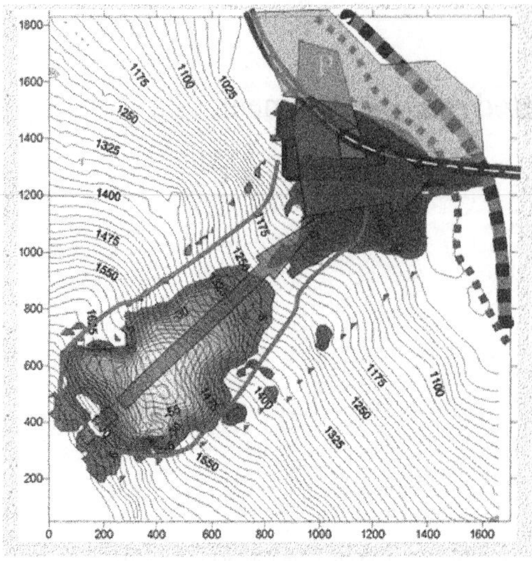

Figure 1. Study of the areas potentially invaded by a 10 Mm^3 potential paroxysm of the Cassass Landslide.

1.1 *Initial risk assessment*

Furthermore a formal quantitative risk assessment (QRA) was performed (Roberds, 2001, Cheung et al., 2001, IUGS, 1997, Fell, 1994) and updated in several occasions over the last decade meanwhile related monitoring programs were launched (CTM, 2002–2004; Polithema & Oboni Associates, 2003).

Within the RA the slope was modelled by using the Oboni & Bourdeau probabilistic slope stability analysis method (Oboni & Bourdeau, 1984, Oboni et al. 1984) as a tool to quantify paroxysm initiation probabilities and mobilized lengths within the active sliding body. Data for this approach were derived from pre-existing studies. The initial model was developed after careful evaluation of all the available data. The main results of the Oboni & Bourdeau analyses can be summarized as follows:

- The slope would behave as a series of "independent" bodies where the uphill one would reactivate, slide down to take support on the prior, downhill one, causing its sliding, slow down and repeat the cycle unless a major heave of the water table would create the conditions for a massive reactivation.
- The slope was not prone to sudden (within days or hours) reactivations, but could feature paroxysms lasting various weeks in case of particularly unfavourable meteorological conditions.
- It was predicted that a heave of 6–8 m of the water table in certain areas, monitored by piezometers, would most likely cause a significant acceleration of the sliding velocity in that area.

1.2 *Monitoring and complementary analysis/monitoring approach*

The monitoring system has undergone several reconstruction and technical evolution phases over the last decade. Beside data acquisition stations and classic inclinometers the site is nowadays also monitored with motorized optical instruments, which report via GSM (digital telecom) to a central monitoring centre (CTM, 2002).

Five level velocity-alert criteria have been established for the Cassas landslide (Polithema & Oboni, 2003). These criteria drive the alert status, changes in the frequency of monitoring, and, of course, can trigger the emergency plan, which encompasses several reactivation scenarios. The Regional Civil Protection Centre can trigger emergency plans specifically designed for various types of reactivation that could occur within the sliding mass, i.e. volumes going from a few hundreds of thousands of cubic meters to the largest considered potential reactivation phenomenon (Regione Piemonte, 2004).

The landslide went in a pre-alert level in the period following the year 2000 flooding which was captured by the monitoring program implemented within the risk assessment study.

The complementary analysis/monitoring approach yielded interesting predictive/ observational results, which drove stabilization actions summarized at the end of this paper. Indeed, the integration of predictive probabilistic analyses with appropriate monitoring methods, followed by an appropriate period of observation and calibration lead to a good understanding of the parameters that influence the Cassas landslide behaviour. Among these the main one is the antecedent rain, net of evapotranspiration. A parametric study indicated that antecedent precipitation for periods of up to 300 days (ten months) displayed the strongest correlation with inclinometers velocity (Oboni, 2005).

The observed strong correlation made it possible to propose a simple relationship between the net antecedent rain mentioned before and the velocity at a given topographic point. Of course these results are and will remain valid within the landslide, provided global conditions do not change over time, and cannot be transferred to another landslide without a similar step-by-step, carefully designed approach. However, they constituted the formalization of a generally understood behavioural characteristic of large landslides, i.e. that these phenomena respond to long term cumulated antecedent rains rather than isolated, intense rainy events.

This paper illustrates how the research was pursued by using Artificial Intelligence systems capable of learning from past experience (measured rain-velocity data) and then predicting future behaviour (we will refer to the AI system applied to landslides as AILandslide). AILandslide makes it possible to develop on-line applications to yield spot analyses to be performed on landslides that are equipped with online instruments. This will enable Civil Protection Command Centres to update their hazard evaluation as situations unfold (Regione Piemonte, 2004).

2 USING ARTIFICIAL INTELLIGENCE TO MODEL THE RAIN-VELOCITY RELATIONSHIP

The need for a predictive analysis of monitoring velocity/displacement data (inclinometers, extensometers, etc. versus pluviometric data) of landslides arises from the need to trigger alerts, organize public safety actions, civil protection in areas where accelerations of the impinging sliding movements may generate high consequences. The same need arises when alert status has to be removed, and evacuated people are to be allowed back to their residences/work places.

AI systems are capable of predicting performances in many fields and have been used in missiles guidance systems, environmental engineering, commerce and stock exchanges, mechanical and maintenance engineering. The application to natural hazards, namely landslides, which we refer to as AILandslide is an important evolution in the civil protection/geohazard field. Thus Artificial Intelligence (AI) has been used to analyse monitoring data response with respect to antecedent rain.

The AILandslide system "learns" from the past and based on its cumulated experience makes predictions that become more and more precise as the experience on a specific landslide widens. Before the learning cycles begin, the model has to be custom tailored for any given monitoring point. AI allows reliable predictions based on past performances, significantly reducing false alerts, thus avoiding many costly errors.

AILandslide has been successfully deployed on various alpine landslides in Europe, with very significant results. The application to the Cassass landslide, object of this paper, demonstrates the outstanding predictive capabilities when using past rainfall to predict future movements/velocity.

2.1 *Customization*

Like a child, AILandslide demands a learning phase during which it analyzes the input parameters and adjustments are made. Each monitoring point needs a specific learning/customisation phase.

The predictive results are significantly influenced by the quality of the inputs. Quality of the inputs is measured by accuracy, duration, and continuity.

2.2 *Data to collect*

For a given landslide the required data are, on top of usual geological, geotechnical, geographic and climatologic data, the following:

- Rainfall Data: if possible daily precipitation, covering at least the monitoring span and continuing in the future insofar as predictions are requested.
- Temperature and Solar Radiation: if possible daily averages, to allow a precise evaluation of evapotranspiration. In case these data are not available, literature formulae can be used to yield approximation of this parameter (see Cassass analysis below).
- Movements History: Inclinometric (or other instruments) monitoring data over a sufficient time span
- History of mitigation activities/human activities on the landslide: this is important because it may lead to the preparation of two models, i.e. one before the implementation of the mitigative works, and one afterwards.

As new deformation measures and pluviometric data are inputted, AILandslide will generate new predictions. The quality of the predictions decays, of course with the range: short terms predictions are better than long term ones. The required frequency of these predictions is a "client's parameter" which will depend on the general environment (geographic, risks, prevailing meteorology) of the landslide. It is possible at any time to simulate evolution scenarios by inputting rain scenarios, thus answering questions like: what will be the deformation in the next six months if it rains, from this date on, like last year? What if the rain is double?

3 APPLICATION TO CASSASS LANDSLIDE

3.1 *Rainfall data, temperature and solar radiation*

As mentioned above these parameters constitute the basis of any AILandslide application. The first step is to evaluate the net antecedent rain, i.e. the rain minus the evapotranspiration. In the Cassass study there were no local detailed records on temperature and solar radiation, so the evapotranspiration was estimated using literature (Allen et al., 1998).

As the inclinometer readings were performed discretely at a rate of 4 measures per 12 months, it was necessary to generate intermediary velocity points by interpolation (dotted points in the measured velocity in Figure 3). Modern monitoring with automatic online readings (inclinometers or surface instruments) would allow a significant increase of the accuracy of the predictions.

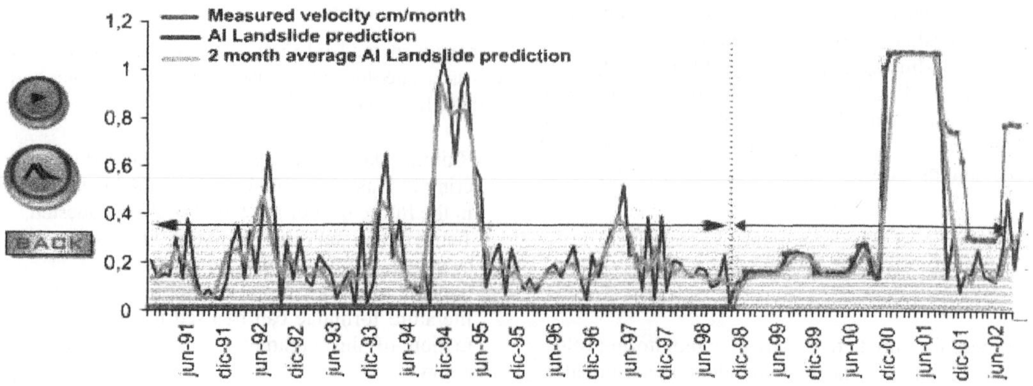

Figure 2. Precipitation, Estimated Evapotranspiration and resulting net precipitation for the period going from January 1998 to July 2002 at Cassass Landslide, based on neighbouring pluviometer stations.

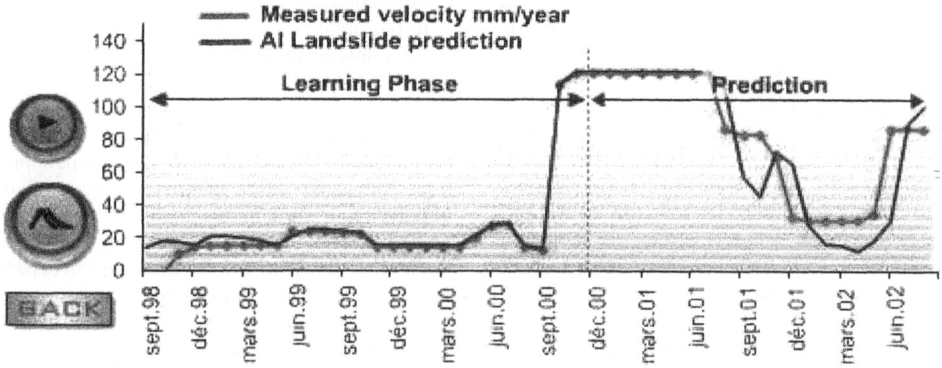

Figure 3. Measured velocity of Inclinometer I4 at Cassass Landslide as compared with predictions during the Learning Phase and a Predictive Phase.

Figure 3 depicts the measured velocity of one specific instrument at Cassass Landslide (Inclinometer I4) together with the AILandslide prediction during the Learning Phase (September 1998 to December 2000) and a first true Predictive Phase, from December 2000 to July 2002.

The Learning Phase was chosen to include an acceleration period resulting from a particularly severe rainy period (over a season), culminating with the October 13th–16th 2000 flooding in Regione Piemonte (Regione Piemonte, 2000). As it can be seen in Figure 3, AILandslide was then able to mimic with success the slowing down of the movements and the acceleration that ensued in late summer-fall 2002.

Figure 4 depicts another analysis that was performed using the AILandslide system. As it can be seen the Learning Phase described in Figure 3 was used to evaluate the velocities of the topographic point where Inclinometer I4 is installed during the years that actually preceded its first installation (i.e. before December 1998).

The analysis depicted in Figure 4 shows that the flooding in the region prior to the one of fall 2000 (Arpa Piemonte, 2006), i.e. November 1994, provoked, following AILandslide "back-prediction" an acceleration similar, but lower in intensity and duration, than the last one.

The integration of the velocity plot allowed the evaluation of the total displacement occurred between Dec 1990–Dec 1998 (total estimated: 21.1 cm) and between Dec 1998 and Jul–Aug. 2002 (total: 20.5 cm): the long term average velocities almost doubled in the second period when compared with the prior one.

4 MITIGATION

Several alternative stabilisation techniques were studied, taking into account their life expectation,

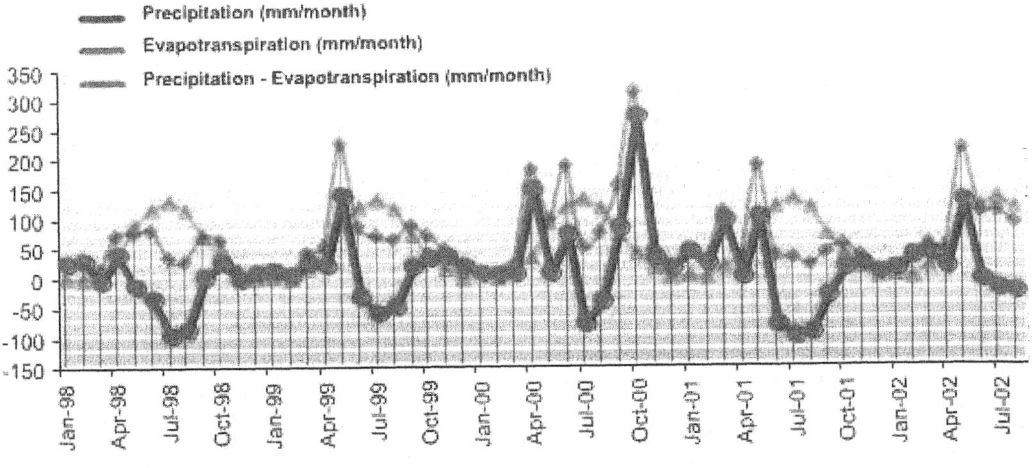

Figure 4. Use of AILandslide to evaluate the velocities of a specific surface location (Inclinometer I4 location) during a time frame preceding the inclinometer installation (before Dec. 1998).

maintenance criteria, environmental impact (the slope is in a National Park), costs, and, of course residual risks. Risk Management has to be clearly differentiated from Hazard Management and generally leads to more sustainable choices (Oboni, 2003, IUGS, 1997, Einstein, 1988).

The main three design candidates were the ones listed below with some of their main pros/cons:

1. A deep drainage by vertical shafts equipped with submerged pumps.
   - Low cost.
   - Need for regular reconstruction, at least at the beginning of the drainage action.
   - Low environmental impact.
2. A 600 m long tunnel in "stable" ground, reaching underneath the slide from a side, equipped with ascending drainage boreholes at its end.
   - High costs.
   - Long to build.
   - High environmental impact-needs a road in stable forested areas.
   - Low maintenance.
3. A 150 m long tunnel within the sliding mass, parallel to the movement vectors, equipped at its end with sub-horizontal drains reaching the sliding surface.
   - Intermediate cost.
   - Short building time.
   - Low impact because access runs mostly through ancient landslides devastated areas.
   - May require heavy maintenance in the future.

Finally, the 150 m long tunnel alternative was chosen and it is now almost completed. The excavation of the tunnel, 3 m×3 m, was performed with light engines and no explosives, under an umbrella of sub-horizontal micropiles to stabilize the ceiling. At each stage an exploration drill was performed at the point of excavation to gain information on the next 30 m of terrain.

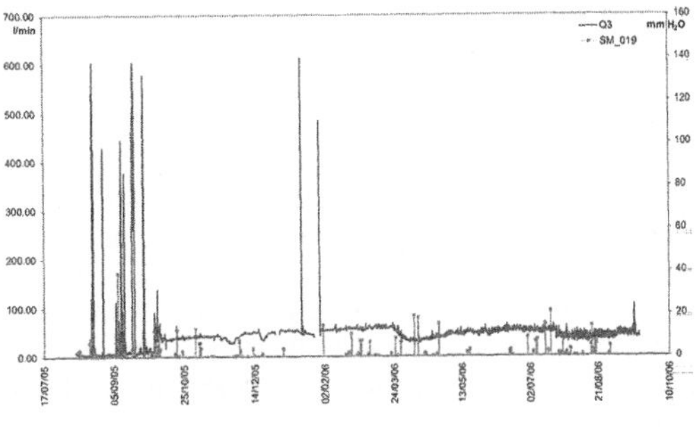

Figure 5. Outcoming flows from the semi-completed drainage tunnel between July 2005 and August 2006.

Figure 5 displays the flows drained by the semi-completed tunnel from July 2005 to August 2006. Unfortunately the measuring station was the object of vandalism and there are no more data after August 2006. The peaks in the plot correspond to measurement errors and should be discarded.

As it can be seen the average drainage is in the order of 30 l/min, or 1300 m^3 per month, with a remarkable constant flow. As the pluviometry of the last few years has been below average, the drainage acts, up to date, on the water present at proximity of the sliding surface. Only once the pluviometry would be such as to recharge the water table within the sliding mass the drainage tunnel would see the flow increase. The inclinometers display average annualised velocities in the range of 10 m/yr to 20 mm/yr in the last five months.

5 CONCLUSIONS

After years of attempts to define a predictive instrument for the velocity of medium to large active Alpine landslides that exhibit periodic paroxysms (acceleration and generalized failures) after particularly unfavourable meteorological cycles, the use of Artificial Intelligence has shown very promising results.

Using prior classic multiparameter correlation studies as a guide, the AILandslide application has been built linking antecedent rain (over a span of several months) to inclinometric velocities.

After testing the application on various landslides, the Cassass landslide was chosen as a full-scale pilot application. The application was used to formulate predictions as well as to estimate cumulative displacements of the landslide in the past, when monitoring data were not present.

The AILandslide application has been integrated with success into a complex framework that includes:

- Monitoring.
- Probabilistic analysis.
- Quantitative risk.
- Alert levels.
- Catastrophe Emergency planning.

In the future it is expected that AILandslide applications will allow real time prediction of velocities of large landslides under various sets of rain scenarios.

This will allow Civil Protection to deploy in a reasonable and sustainable way their assets and deliver protection to the population exposed to natural hazards.

The integration of probabilistic predictive analysis and AILandslide will bring observational approaches in landslide engineering to a new level of sophistication where, finally, all the monitoring investments will produce results that are fully used and interpreted.

REFERENCES

Arpa, (2006), *Evento Alluvionale del 5–6/11/1994*, Regione Piemonte.

Allen, R.G., Pereira, L.S., Raes, D. & Smith, M. (1998), *Crop evapotranspiration – Guidelines for computing crop water requirements* – FAO Irrigation and drainage paper 56, FAO – Food and Agriculture Organization of the United Nations Rome.

Cheung, W.M., Shiu, Y.K. & Pang, P.L.R. (2001), Assessment of global landslide risk posed by old man-made slopes in Honk-Kong, *International Conference on Landslides, Causes, Impacts and Counter measures*, pp. 497–505, 17–21 June 2001, Davos, Switzerland.

CTM, (2002–2004), *Monitoring System of the Landslide* (in Italian), unpublished report to SITAF, 2002, 2003, 2004.

Einstein, H.H. (1988), Special Lecture: *Landslide Risk Assessment Procedure*, Proceedings 5th ISL, Lausanne.

IUGS, Working Group on Landslides (1997), Committee on Risk Assessment, Quantitative Risk Assessment for Slopes and Landslides; *The State of the Art, IUGS Proceedings*, Honolulu, Balkema.

Fell, R. (1994), Landslide Risk Analysis and Accidental Risk, *Canadian Geotechnical Journal*, V31, p. 261.

Oboni, F. & Bourdeau, P.L. (1983), Determination of the Critical Slip Surface in Stability Problems – *Proc. of IVth Int. Conf. on Application of Statistics and Probability in Soil and Structural Engineering,* Florence. Universistà di Firenze (Italy) 1983, Pitagora Editrice, pp. 1413–1424.

Oboni, F., Bourdeau, P. L. & Russo, F. (1984), Utilisation des Processus Markoviens dans les Analyses de Stabilité des Pentes – *2ème Séminaire sur l'Utilisation des Méthodes Probabilistes en Géotechnique*, LMS, EPFL, Lausanne.

Oboni, F. (2003), Ten Years Experience in Linear Facilities Risk Assessment (LFRA), *International Conference on Application of Statistics and Probabilities,* San Francisco ICASP9 2003.

Oboni, F. (2005), Velocity-Rain Relationship at the Cassass Landslide, *Landslide Risk Management Conference, Vancouver.*

Polithema & Oboni Associates, (2003), *Landslide Emergency Plan* (in Italian),Unpublished Report to SITAF.

Regione Piemonte, *Alluvione in PiemonteRapporto sull'evento alluvionale del 13–16 ottobre 2000.* www.regione.piemonte.it/alluvione/rapp.htm

Regione Piemonte, Protezione Civile, (2004) The New Command Center, *pamphlets & other advertisment material (in Italian).*

Roberds William J. (2001), Quantitative landslide risk assessment and management, *International Conference on Landslides, Causes, Impacts and Counter measures*, pp. 585–595, 17–21, Davos, Switzerland.

Landslides and Climate Change – McInnes, Jakeways, Fairbank & Mathie (eds)

Case studies of landslide risk due to climate change in Sweden

C. Hultén, Y. Andersson-Sköld, E. Ottosson, T. Edstam & Å. Johansson
Swedish Geotechnical Institute (SGI)

ABSTRACT: Based on global climate change scenarios, SRES A2 and B2 IPCC emission scenario, the climate change in Sweden through to 2100 has been simulated by the Rossby Centre (SMHI, 2005). According to these scenarios the annual temperature will increase by 2.5–4.5°C over the next century and the sea level will rise by up to 0.8 m along the coast of southern Sweden. The change will be most pronounced in winter, with shorter periods of snow and ice as well as shorter periods with frost in the ground. The climate borders will move north. In most of Sweden the annual precipitation will increase by 5–25% over the next century. In the south-east, however, the annual precipitation will be the same or even less than today. Throughout the whole of Sweden, more extremes are to be expected, i.e. more frequent storms and more heavy rainfall (e.g. IPCC, 2001, SMHI, 2005).

Rough calculations have been made, as an initial estimate of the impact of increased precipitation on slope slide risks in two case studies located in the south-west and north of Sweden. In the case studies the impact on landslide risk of changes in precipitation, erosion and water levels has been studied. Reduced safety for all types of slopes included in this investigation is to be expected. Additional risks, such as higher frequency of mudflows due to increased precipitation, are expected.

1 INTRODUCTION

Several signs indicate ongoing global warming and the main cause is most likely anthropogenic emissions of greenhouse gases, e.g. (IPCC, 2001, EEA, 2004, WHO, 2002). Global scenarios show that the warming will continue for at least the next 50 to 100 years, e.g. (IPCC, 2001, SMHI, 2005). Based on global climate change scenarios, ECHAM4/ OPYC3 and SRES B2 IPCC emission scenario, climate change in Sweden through to 2100 has been simulated by RCA3 at the Rossby Centre (SMHI, 2005). According to those scenarios, the annual temperature in Sweden will increase by 2.5–4.5°C over the next century. The change will be most pronounced in winter, with shorter periods of snow and ice as well as shorter periods with frost in the ground. The climate borders will move north. The simulated changes in precipitation for Sweden over the next century can be seen in Figure 2. In most of Sweden the annual precipitation will increase by 5–25% over the next century. In the south-east, however, the annual precipitation will not change and there could even be an average decrease each year compared with today. Throughout the whole of Sweden, more extremes are to be expected, i.e. more frequent storms and more heavy rainfall, e.g. (IPCC, 2001, SMHI, 2005).

There are well-known correlations between geotechnical conditions and climate, but there are, to our knowledge, no studies where, for example, the relationship between slope stability (landslides and avalanches) and the influence of an increase in precipitation or more extreme rainfall are described quantitatively for conditions in Northern Europe. In Sweden, the Swedish Geotechnical Institute (SGI) has just recently initiated a study to investigate whether the climate change will lead to an increase in natural hazards and accidents, such as erosion and landslides, and if this will have a significant impact on the mobility and movement pattern of contaminants in soil and groundwater. In this paper an investigation of the possible impact of changes in precipitation on landslide events is described, with a focus on pore water pressure, the groundwater level and erosion for selected case studies.

2 CASE STUDY AREAS

Two case study areas, the Göta älv (Göta river) valley in south-west Sweden and a typical river valley, a steep, sandy river bank, Krokvåg, in the north of Sweden, have been chosen for further calculations and investigations. The two locations are close to Göteborg and Härnösand and are shown on the map in Figure 1. The map also includes the most frequent landslide areas in Sweden. The precipitation changes according to

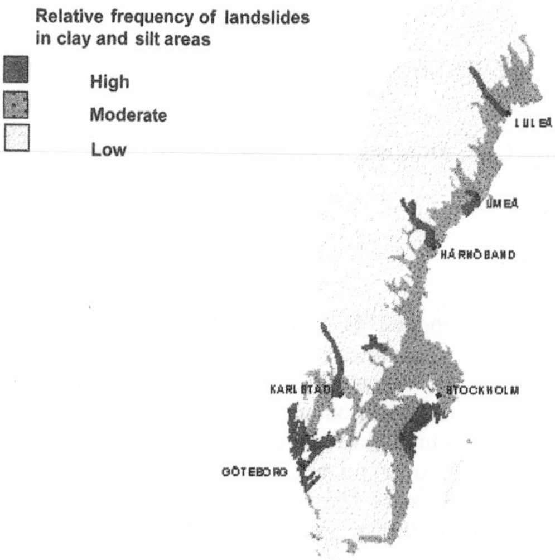

Figure 1. Map of Sweden showing the freqency of landslides and ravines (www.sgu.se, *2005*).

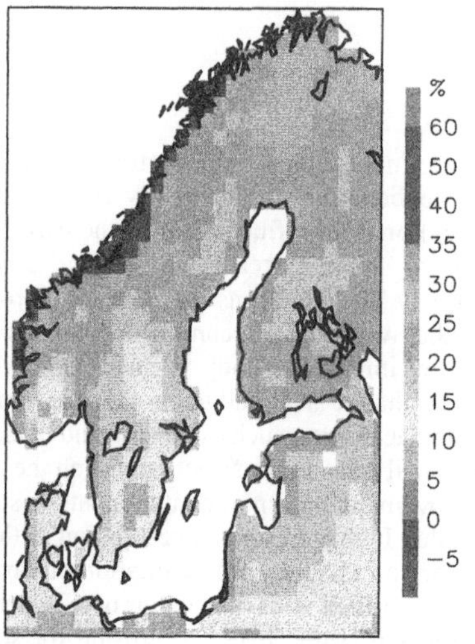

Figure 2. Simulated precipitation changes, year 2100 compared with the period 1961–1990 (SMHI Rossby Centre, 2005).

climate change scenarios in Sweden are shown in Figure 2 (SMHI, 2005). The case study areas are located in areas with frequent landslide events and where a significant increase in precipitation is expected.

2.1 *South-west Sweden – Göta älv valley*

The Göta älv valley runs from Lake Vänern in the north to Göteborg in the south (Figure 3) and is one of the most frequent landslide valleys in Sweden, see also Figure 1. A number of landslides occur every year but in general they are fairly small, shallow and caused by erosion. Limited slides of the, occasionally, very steep slopes under the water table also occur. An aerial photo scanning survey conducted in 1982 revealed 150 slide scars in the river valley (Viberg, 1982). Larger landslides have also occurred in the river valley (e.g. Alén et al., 2000). The documented larger slides in the area are shown in Figure 3. The first documented landslide occurred in 1150, [1] in Figure 3, and among the more recent are Surte, 1950, [12] in Figure 3, and Göta, 1957, [14] in Figure 3 (Alén et al., 2000). In all of these, and in many of the other larger landslide events, one or more people were killed (e.g. Hultén et al., 2006). The high frequency of landslides in the river valley is due to the geological history, which has resulted in deep clay layers. Most of the clays have been deposited in a marine environment and quick clay is therefore common in the area. Quick clay is a soil where the skeleton collapses due to vibrations, or stirring, and the clay thereafter behaves as a liquid. When a landslide occurs in a quick clay area the process is very rapid and the extent of the slide can be very large, with major consequences.

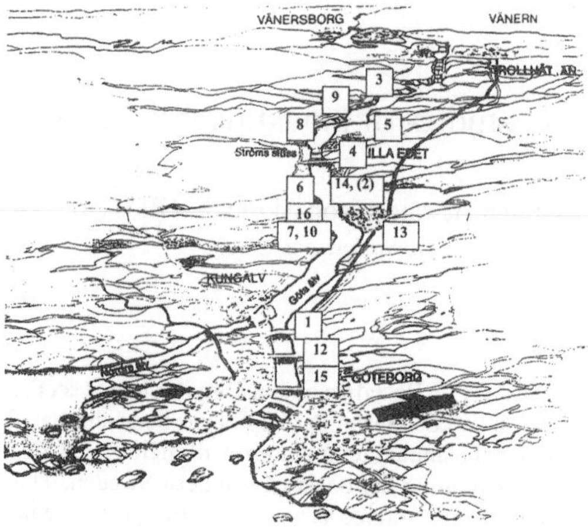

Figure 3. Map of documented landslides in the Göta älv valley since 1150 (Alén et al., 2000).

In the northern part of the Göta älv valley the soil depths are moderate. Bedrock is common together with and mixed with sediment areas. The slopes towards the river are in general more silty, high and steep. The clays are firm and often contain water-permeable layers. The pore pressure variations related to precipitation will therefore be larger than for homogenous clays, which will also result in a greater impact on slope stability. (Alén, et al., 2000).

Towards the south of the valley the clays are soft and homogenous and the clay and soil depths increase.

In the southern part, riverside shelves and slopes under the water table often occur between the shoreline and the deepest part of the river. The stability of the slopes close to the river depends to a large extent on the topography of the riverside shelves and slopes under the water table. (Alén, et al., 2000).

In 1938, a regulation of Lake Vänern, and thereby also the river (Göta älv) was done to optimise the use of the power stations along the river. The water flow in the river prior the regulation varied between 200 and $850\,m^3/s$ and the deviation between normal water flow and high water flow was moderate. Lake Vänern acted as a natural buffer, thereby minimising the variations in the water flow from day to day. After the regulation both higher and lower water flows have been permitted. Water flows higher than $1000\,m^3/s$ occur more and more frequently after the regulation. This may have a great impact on the erosion processes and the of slope stability conditions along the river (Sundberg et al., 1963; Göta älvs vattenvårdsförbund, 1996).

2.2 *North of Sweden – Krokvåg*

In Figure 4 a schematic profile of a common valley of Sweden is shown (the Indalsälven valley). The case study area, a steep, sandy river bank in Krokvåg, is a typical example of such a valley. It is located 140 km north-west of Härnösand at 150 a.s.l. and the soil layers here are (starting at the bottom): rock, moraine, gravel and sand (glacial fluvial deposits), varved clay (glacial fluvial deposits), clay and silt (post-glacial fluvial deposits), sand and silt with clay layers and nearest the ground surface gravel and sand (post-glacial fluvial deposits) (Lundqvist 1969). The layers are not always completely developed thus contributing to erosion in the top layers or variations in the deposition conditions in the lower layers. The soil layers are silt in the river slope part, sand on the plane just above the steep, sandy river bank, and nearest the mountain slope silt and clay (Lundqvist, 1969). The mountain slope consists mainly of rock and moraine although glacial fluvial deposits may also occur in the area.

Above the water table, in the unsaturated zone, there is an attracting force between the soil particles due to the surface tension of the liquid layer around the gas bubbles increasing the effective stress and thereby making a positive contribution to the strength of the soil. This phenomenon is called false cohesion and is common in silty and sandy soils. The pore pressure in general is positive below the water table (in the saturated zone) and negative or zero above the water table (the unsaturated zone). When designing, preparing geotechnical constructions or analysing and interpreting test results in silty soils, it is crucial to consider the impact of the negative pore pressures. In 1994, pore pressure and water level measurements were conducted in the area (Fallsvik, 1994). The groundwater

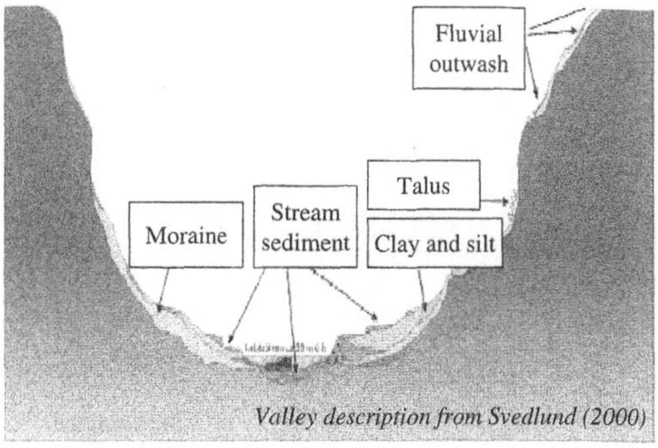

Figure 4. Slope north of Sweden, Krokvåg. The scales differ on the horizontal and vertical axes. (Lundqvist 1969).

level was in October 1993 approximately 2 m below the surface of the plane just above the steep, sandy riverbank and negative pore pressures were found (Fallsvik 1994).

2.3 *General Swedish landslide aspects*

Many rivers are regulated and thus have the same water flow variations as the Göta älv. There are a number of rivers in Sweden which already have a pronounced frequency of landslide events as shown in Figure 1. Most of these river valleys are, like the two case studies, located in the western and northern parts of Sweden where the precipitation is expected to increase by up to 25% through to 2100 (Figure 2). The geological conditions in these river valleys make the stability conditions sensitive to changes in groundwater levels, pore pressure changes and erosion. The case studies presented below will give some indication of the impact on the slope stability of changes in erosion, groundwater levels and pore pressures resulting from expected climate changes in the areas.

Not taken into account in quantitative estimates, but still of great importance, are mud flow events. During occasions with heavy rainfall and in areas with steep slopes, with water in narrow streams such as a ravine, mud flow events may occur. Mud flows are rapid soil movements in which large volumes of water and soil mixtures flow down a slope. The masses are heavy and the high kinetic energy and the intense erosion are very destructive. Often further soil masses, but also trees and stones, follow the mud flow, increasing the mud flow volume downhill. Most often the mud flow follows existing routes although new paths can also be formed. Along less steep parts of the path, soil masses, trees and bushes from small landslides can come to rest and pile up. Occasions with high water flows may cause new movements of the soil masses and thereby new mud flows that will continue down the hill. Restarted mud flows are common on hillsides with moraine landslides.

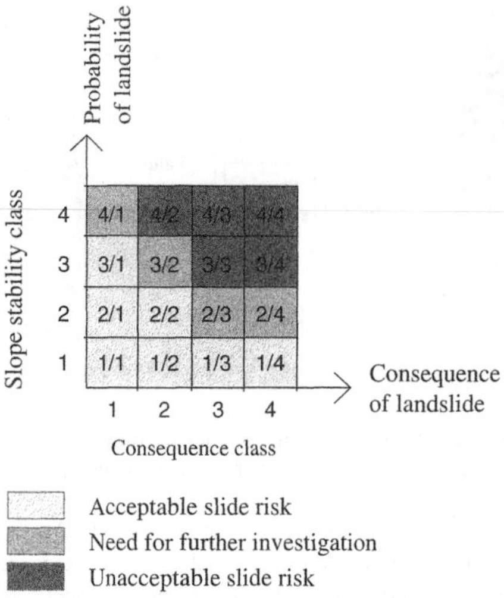

Figure 5. Slope stability classes, consequence classes and slide risk levels.

3 RISK ANALYSIS METHODOLOGY

One method of estimating the landslide risk is a probabilistic approach, (risk = probability × consequence), which has also has been used in studies of the stability conditions in some areas along the Göta älv (Alén et al., 2000, Hultén et al., 2006).

Here the probability of a slide has been estimated by statistical analysis and calibrations of the safety factors. Based on the statistical analysis, one of four stability classes has been set for the part of the river valley in question. The four classes run from negligible probability (stability class 1) for a landslide event to occur to a very significant probability (class 4) for a landslide event to occur (vertical axis in Figure 5).

The consequences, i.e. the damage to life, buildings, the infrastructure and the environment, of a landslide also are divided into four classes. The four classes run from very limited damage (consequence class 1) to catastrophic (class 4) (horizontal axis in Figure 5). The risk is a result of 16 combinations of the classes as described in the risk matrix in Figure 5.

To simplify the use of the matrix, three major risk groups have been defined: acceptable risk level, need for further investigation, and a non-acceptable risk level as shown in the matrix in Figure 5. (Alén et al., 2000). This three-level classification is based on experience from previous stability investigations in the river valley area and also elsewhere in the south of Sweden. Stability classifications, expected consequences and the geometrical extent of slides (both calculated and observed) have been accommodated when constructing the matrix.

4 EXPECTED CLIMATE CHANGE EFFECTS

Based on the climate change scenario simulations, the precipitation is expected to increase by up to 25% in the western and northern parts of Sweden and thus also in the two case study areas (Figure 1). The increase in precipitation will cause increased water flow, run-off and increased groundwater levels. The global temperature increase will also result in a rising water level along the coasts. The impact of the sea water level will only be of importance in the southern parts of Sweden as it will be counteracted by land upheaval, which is more pronounced further north.

Increased groundwater flow and run-off may cause erosion in watercourse areas. Steep sand and silt slopes are most sensitive and the probability of erosion is increased if the rain follows a period of drought. Estimates based on the regional climate change scenarios show that there will be increased water flows, especially in autumn and winter, while the spring water flows will be reduced, apart from in the very north of Sweden (Andréasson, 2004). Both the erosion and the precipitation itself may contribute to landslide events.

4.1 *Göta älv*

Precipitation influences the groundwater level. A recent study by the Swedish Geological Survey (SGU) (Engdahl, 2005) was conducted at three SGU stations in the Göteborg region. At the stations the groundwater table in the lower aquifer has been measured since 1971. During the period 1998–2002, the precipitation during September to April was 40% higher than normal. According to this SGU investigation, the 40% precipitation increase resulted in groundwater level increases of 0–0.9 m varying between the stations. The groundwater variation and precipitation has a long-term impact on the pore pressure fluctuations. There are, to our knowledge, no studies that quantitatively show the relationships between precipitation and the risk of slides under geological conditions as in Scandinavia. There are some studies which show that the longer the continuous precipitation period the greater the risk of landslides and the lower the soil permeability the longer before the impact on the soil stability may be observed (Tsaparas et al., 2002).

Here, some first attempts to investigate the possible impact of increased precipitation on slope stability have been made. The investigation is based on tests and sensitivity analysis using methods developed for past and present climatic conditions. The investigation is further based on a scenario where the groundwater level may rise 1 m in the clay layer due to increased precipitation. According to the study by SGU this may be realistic for a precipitation increase of ≥40%, which also may occur under present climatic conditions (Engdahl, 2005). This assumption has been used for different conditions described in Table 1.

Table 1. Case study scenarios to investigate the possible impact of increased precipitation and water flow in the Göta älv valley.

	Conditions	*Stability class*	*Calculated safety factor, F*	*Changes in safety factor, F*
Figure 7	Cross-section in the south-west of the Göta älv (Bäckebol, Göteborg)			
1	Present conditions	3	1.23	
2	Change (increase) in groundwater level (1 m) Erosion as described in the text above.	4	1.13	8–10% reduced risk margin
Figure 8	Cross-section in the north-east of the Göta älv (Lilla Edet)			
1	Present conditions	4	1.04	
2	Change (increase) in groundwater level (1 m). Erosion as described in text above	4	0.95	9% reduced risk margin, the slope can not be regarded safe.
3	The river water level is assumed to decrease 0.5 m.	4	1.02	2–3% reduced risk margin
4	Change (increase) in the groundwater level (1 m). Erosion as described in text above. The river water level is assumed to decrease 0.5 m.	4	0.93	11% reduced risk margin, the slope can not be regarded as safe

As part of a national investigation of climate change vulnerability and adaptation capacity the Swedish meteorological and hydrological Institute (SMHI) has investigated two realistic tapping scenarios for Lake Vänern (SMHI, 2006). Of the two climate change scenarios, one corresponds to the maximum flow observed today at the north of the Göta älv (Vargön power station), close to Lake Vänern), which is $1030\,m^3/s$ and the other scenario is an increased maximum flow to $1400\,m^3/s$. Both scenarios include an expected rise in the water level in the ocean (Göta älv outlet) in the region of 0.5 m. The $1400\,m^3/s$ flow corresponds to a situation where the water level in Lake Vänern is unchanged. A maximum flow of $1030\,m^3/s$ will most likely result in an increased water level in Lake Vänern (Bergström et al., 2006). In this paper these two scenarios are considered but also a base case scenario, corresponding to the conditions today, with the maximum permitted tapping of $1030\,m^3/s$ at the Vargön power station and no change in the sea level (± 0 m), see Figure 6.

Increased precipitation and tapping in the river will result in an increased groundwater level and pore pressure in the ground. The increased water flow will also result in increased erosion. Furthermore, an increased need for variations in water flows will result in larger variations and fluctuations in the water level, which will create situations with high pore water pressures in combination with a low water level in the river. All these factors individually contribute to lower stability and in addition some are co-working to reduce the stability.

Another part of the national investigation of climate change vulnerability and adaptation capacity has involved the impact on erosion due to changes in tapping and precipitation by Lund University (Larson et al., 2006). According to Larsson et al. (2006) the erosion along the river shorelines may be as much as 2 to 3 metres during the next century, while the bottom erosion may be 0.5 to 1.5 m. Under present conditions there is also ongoing erosion in the river, although the changes in tapping, including an increase in the maximum flow by 20%, is expected to result in a 50% increase in the erosion rate. The real erosion magnitude depends on local conditions, such as turbulence, the local water flow, erosion protection etc. and large variations (both higher and lower erosion rates) are to be expected. In this investigation the erosion is set at 2.5 m on each side of the river and the bottom erosion is set at approximately 1 m in the sections that are not protected against erosion.

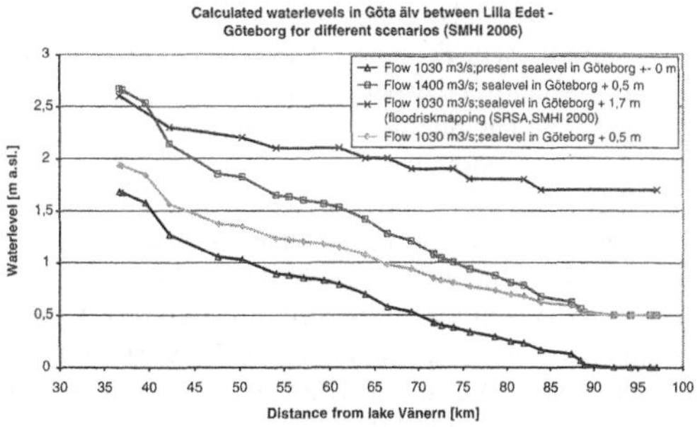

Figure 6. Calculated water levels in the Göta älv between Lilla Edet and Göteborg for different flow scenarios (SMHI 2006).

4.2 *Krokvåg*

The stability in this slope is likely to due, in part at least, to false cohesion in the slope. As mentioned previously, the groundwater level in October 1993 was measured at approximately 2 m below the surface of the plane just above the steep, sandy riverbank and negative pore pressures were found (Fallsvik 1994). According to climate change scenarios the precipitation will increase by up to 30% in this area during the period 1961–2100 (SMHI 2005). The increase in precipitation will most likely result in a groundwater level in the upper magazine that will be near the surface for most of the year and the surface run-off will increase, which together with groundwater leaching in the steep, sandy river bank will result in increased erosion of the slope (Engdahl, M. et al., 2005).

Long, dry periods will cause the water to evaporate from the ground in silty soils, thereby reducing the pore pressures. In the same way, the pore pressure will increase during rainy periods. Under certain extreme conditions the unsaturated zone will become saturated and the pore pressures will change from negative to positive. If the groundwater level is increased during rainy periods the false cohesion will disappear and the slope stability will decrease. The false cohesion may also disappear after dry periods since the water surface tension of the gas bubbles then disappears due to the water being evaporated.

In areas with silty soils, where the groundwater is only a few metres below the surface, the capillary zone almost reaches the lower edge of the dry crust (i.e. the top part of the soil where the soil has been affected by drought, frost in the ground and weathering). In these cases almost all the soil profile is water-saturated and precipitation will only affect the groundwater conditions in the dry crust. There is, however, only very limited water needed to infiltrate before the groundwater reaches the bottom of the dry crust (or even higher up in the ground), which can drastically affect the soil-bearing capacity.

5 CHANGES IN THE LANDSLIDE RISK

Here some first attempts to investigate the possible impact of increased precipitation on slope stability will be presented. The investigation is based on a scenario where the groundwater level may rise 1 m due to increased precipitation. According to the study by SGU (Engdahl, M. et al., 2005) this may be realistic for a precipitation increase ≥40%, which also may occur under present climatic conditions.

5.1 *Göta älv valley*

Two representative cross-sections along the Göta älv are shown in Figure 7 and Figure 8. The stability

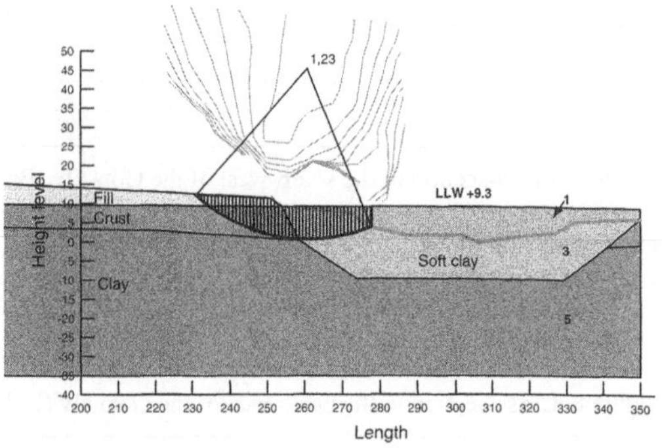

Figure 7. Calculated safety factor under present conditions for a cross-section in the south-west of the Göta älv valley (Bäckebol) (stability class 3 under present conditions).

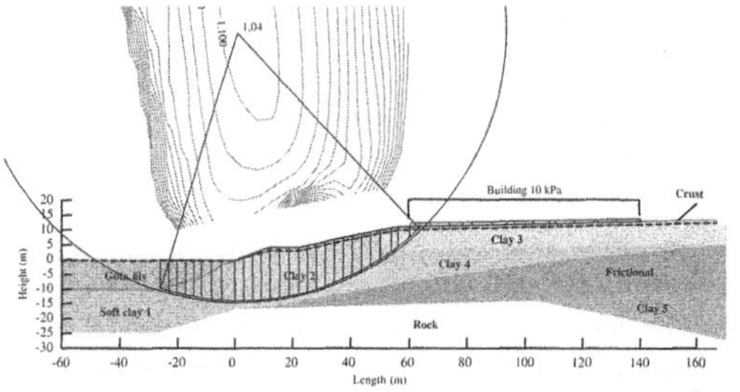

Figure 8. Calculated safety factor under present conditions for a cross-section in the north-east of the Göta älv valley (Lilla Edet) (stability class 4).

analyses for the cross-sections indicate that the stability factor is about 1.2 under present conditions, i.e. corresponding to stability class 3 in the south-west of the Göta älv valley (Figure 7) and 1.04 in the north-east cross-section, corresponding to stability class 4 (Figure 8). The calculations of increased erosion, 2.5 m on each side of the river and the bottom erosion of 1 m, indicate that the expected erosion will change the stability class from 3 to 4, i.e. a significant probability of a landslide occurring.

In Table 1 all scenarios for the two cross-sections studied are shown together with the calculated change in the stability factor of safety.

As can be seen in Table 1, the stability classes will increase 0.5–1 class during the next century in the river valley due to increased erosion, pore water pressures and groundwater levels. In principle, this implies an increase in the probability of landslides to occur in all soft clay areas along the river. The conclusion is that in all probability landslides will occur at cross-sections that are currently on the limit but are still regarded as safe, due to the expected precipitation changes.

Based on the results presented in Table 1, a survey along the river valley was conducted in order to identify present and future stability classes. Furthermore,

Table 2. Estimated stability classes (present and future) and estimated requirement of additional measures along the river valley due to expected climate change and changes in the river regulation.

Present stability class	Estimated future stability class	*Estimated need for reinforcements due to expected tapping and water flows in the river in the future.*	
		Flow 1,030 m³/s	*Flow 1,400 m³/s*
4	>4	Approx. 30–50% of the area	Approx. 50–70% of the area
3	3–4	Approx. 20–40% of the area	Approx. 40–60% of the area
2	2–3	Approx. 5–15% of the area	Approx. 10–20% of the area
1	1–2	Less/little reinforcement to increase the slope stability is needed	Less/little reinforcement to increase the slope stability is needed

measures needed to reduce the geotechnical consequences of expected water flow changes and changes in precipitation patterns to an acceptable risk level were assessed.

This survey mainly includes areas along the river where slide risk analyses or general stability surveys have been done previously (approximately 40 km of shore line). The stability consequences of flood protection embankments were also analysed in the survey.

Based on this survey, the consequences of the two flow scenarios (1,030 and 1,400 m^3/s) have been estimated. It should be noted that already under present conditions there is a need to achieve an acceptable landslide risk in some areas (stability classes 3 and 4). However, the study presented here only includes additional measures required due to the expected climate change and the two tapping scenarios. The present risk-reducing measures to estimate future stability classes are given in Table 2. The table also includes an estimate of the extent of the additional measures required with areas with different stability classes. As can be seen, there is a significant number of areas where measures are needed in order to obtain an acceptable landslide risk.

5.2 *Krokvåg*

A representative steep, sandy river bank in the north of Sweden is shown in a section at Krokvåg in Figure 9. The present stability conditions, i.e. the calculated present safety factor is 1.22.

An increase in precipitation will result in an increase in the groundwater level and the water level in the river may also increase. The increase in the groundwater level will result in water nearer the surface and the negative pore water pressures that were found in previous field measurements at the site will be reduced or disappear. Increased precipitation and heavier rainfall will also result in higher groundwater leaching and surface run-off, which will in turn cause increased

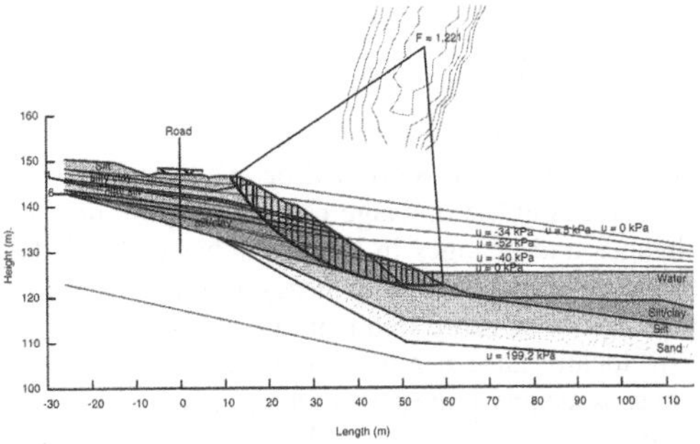

Figure 9. Calculated safety factor under present conditions for a steep, sandy river bank at Krokvåg in the north of Sweden.

slope erosion. Calculations have been conducted for one scenario with only an increase in water levels and one scenario with increased erosion as described in Table 3 below.

In the scenario where only the negative pore pressures disappear, as is likely to occur and is assumed in the calculations when increasing the groundwater level by 1 m and the river water level by 0.5 m, the safety factor is reduced to almost 1 and the slope cannot be regarded as having an acceptable landslide risk. If the increased erosion is also taken into account, the safety factor will be less than one and the probability of a landslide occurring thus becomes high. In addition, the increased water flow may cause fine-grain soil to be transported by the groundwater entering the river at the slope base, thereby increasing the erosion. This type of erosion may cause the creation of large holes in the ground.

This type of slope is common and typical of the northern parts of Sweden and in general an increase in the landslide risk in watercourse areas is to be expected due to climate change. Present infrastructure and buildings near the slopes will be affected and climate change must be taken into account in the planning process.

Table 3. Case study scenarios to investigate the possible impact of increased precipitation at Krokvåg.

	Change in conditions	*Calculated safety factor, F*	*Changes in safety factor, F*
1	Present conditions	1.22	
2	Change (increase) in groundwater level (1 m). The river water level is assumed to increase 0.5 m.	1.03	The slope cannot be regarded as safe
3	Increase in groundwater level (1 m). The river water level is assumed to increase 0.5 m. Increased erosion up to 1.0 m by the slope base.	0.98	The slope cannot be regarded as safe. Increased precipitation results in a groundwater level near the surface. The surface water run-off will thus increase which, combined with increased flow and percolation in the permeable layers, will cause increased slope erosion.

5.3 *Some examples of other forms of impact of climate change in Sweden*

5.3.1 *Firmly layered clay areas*

Firmly layered clay soils will be even more affected by changes in the drainage conditions, pore pressure changes and changes in the groundwater level than soft clays. Occasions with heavy rainfall, which are expected to be more frequent due to climate change, may locally have a more rapid impact than in soft clays, especially if the safety factor under present conditions is already near one or if there are high external loads. In these soils the pore pressures may remain high in the soil while the water in the watercourse has already dropped after the rain, which will contribute to reducing the slope stability.

5.3.2 *Mud flows*

In the climate scenarios for the Swedish mountain area an increase of up to 25% in annual precipitation is expected and the heavy rain occasions are expected to increase. The result will be that the heavy rains, described as 100-year rains, will occur much more frequently than today. The mud flow risk in the mountain areas will thus increase and areas not yet affected by mud flows will be affected in the future.

6 CONCLUSIONS

The expected climate change will result in increased precipitation and heavy rainfall, which will increase erosion and the probability of mud flows and landslides. The geographical areas that will be most affected are those that already have a pronounced number of landslides (Figures 1 and 2) and in the future the probability, and perhaps also the consequences, will be more severe. The most pronounced effects are expected in the northern and western parts of Sweden.

The most sensitive are steep slopes in erosion-sensitive watercourse areas. Slopes with layers of permeable soils will be affected more rapidly by changes in groundwater and pore pressures. In areas with soft, deep clay, the effects will be less pronounced as the pore pressure changes will be slower.

In many parts of Sweden, the slope stability in built-up watercourse areas will be reduced due to climate change. Many areas will require measures in order to compensate for increased erosion and landslide risks. In the investigated parts of the Göta älv valley, climate change and tapping conditions will cause an increase in the stability class by 0.5–1 for large parts of the investigated areas. The higher the stability class, the higher the probability of a landslide occurring. For the present non-acceptable class (class 4) 50–70% will become significantly less stable due to climate change and increased maximum river water flows. For class 3, 40–60% of the investigated areas will fall into class 4.

Many areas are today reinforced, but to cope only with present conditions. Increased erosion and increased water levels and pore pressures will result in risks that are not acceptable today. In those areas re-reinforcement or reconstructions are also needed for the same acceptable risk level to be maintained.

Many of the slopes today are the result of several forces and processes, including erosion, and consequently they are often on the edge of being stable. Especially in watercourse fine-grain soil areas the slide risk of such slopes can be expected to increase significantly. In built-up areas this may have negative effects, such as clogged water flows and increased mudding in the water systems.

Reduced slope stability in quick clay areas may result in landslides with a large extent and consequences for society and the infrastructure.

Increased precipitation will increase the water flows, which may result in mud flows and may in turn cause erosion, flooding, landslides and damage to buildings and the infrastructure.

Not considered specifically here are the south-east parts of Sweden. The climate scenarios show that the annual precipitation in those areas will decrease and the soil evaporation will increase, resulting in a lack of water. The water levels will be reduced and thus the supporting forces. Landslides are, however, not common in those areas at present, the pore pressures are not likely to increase and the landslide risks are not expected to change significantly compared with the present situation.

There are several examples of an increase in risk of climate change due to the increase in precipitation in the Nordic countries. The increased precipitation will lead to increased surface run-off, causing higher water flows and water levels in rivers and other watercourses. This may lead to increased erosion and pore water pressures. An increase in inundation and flooding is also to be expected. Landslide events are to be expected as erosion increases. In addition, water penetration of constructions in the ground will lead to reduced soil-bearing capacity and an insufficient drainage system capacity. The precipitation changes may cause reduced slope stability and bearing capacity for roads and railways. Water, gas and electrical systems and dams may be affected as well as contaminated soils and shorelines due to erosion and hydrological changes. It is of great importance to further deepen the investigations in order to identify critical areas and to be able to act in time and cost efficiently in order to reduce the consequences for society. (Hultén et al., 2005).

REFERENCES

Alén, C., Bengtsson, P-E., Berggren, B., Johansson, L., Johansson, Å. 2000. Land slide risk analysis of the Göta älv valley – Methodology (Skredriskanalys i Göta älvdalen- Metodbeskrivning). SGI Rapport 58, Statens geotekniska institut, Linköping. (In Swedish)

Andréasson, J., S. Bergström, Carlsson, B., Graham, L.P., Lindström, G. 2004. Hydrological Change – Climate Change Impact Simulations for Sweden. Ambio 33(4–5): 228–234.

EEA. 2004. Mapping the impacts of recent natural disasters and technological accidents in Europe. Environmental issue report No 35, European Environment Agency, Copenhagen, Denmark.

Engdahl, M. 2005. SGU, Sveriges geologiska undersökning, Private communication.

Fallsvik, J. 1994. National road Z87 at Krokväg, Ragunda municipality, Jämtland. Report 2. Geotechnical conditions, stability analysis and suggested measures. (Riksväg Z87 vid Krokvåg, Ragunda kommun, Jämtlands län. Delrapport 2. Geotekniska förhållanden, stabilitetsanalyser och förstärkningsförslag.) SGI, Statens geotekniska institut. Dnr: 2-9309-439. (In Swedish)

Hultén, C., Edstam, T., Arvidsson, O., Nilsson, G. 2006. Increased tapping from lake Vänern and related geotechnical conditions in the Göta älv valley. (Geotekniska förutsättningar för ökad tappning från Vänern till Göta älv. Underlag till klimat och sårbarhetsutredningen.) SGI, Statens geotekniska institut 2006. (In Swedish)

Hultén, C., Olsson, M., Rankka, K., Svahn, V., Odén, K., Engdahl, M. 2005. Släntstabilitet i jord- konsekvenser av ett förändrat klimat.(Slope stability – consequences of climate change) Delrapport inom uppdraget Jordskred och Ras i Klimatförändringens Spår. SGI Varia 560-1, Statens geotekniska institut, Linköping. (In Swedish)

IPCC. 2001. Climate change 2001: Impacts, adaptation and vulnerability. IPCC WGII report, Cambridge University Press, Cambridge, UK.

Larsson, M., Hansson, H. 2006. Sediment transport and erosion in Göta älv, impact of climate change. (Sedimenttransport och erosion i Göta älv, inverkan av klimatförändringar.) Institutionen för teknisk vattenresurslära, Lunds tekniska högskola, Lund. (In Swedish)

Lundqvist, J., 1969. Desriptions to the soil map of Jämtland (Beskrivning till Jordartskarta över Jämtlands län.) Sveriges geologiska undersökning (In Swedish).

SMHI. 2005. Climate scenarios from 2005 (Klimatscenarier från 2005, R&D, Rossby Centre, SMHI, www.smhi.se/

Sundborg, Å., Norrman, J. 1963. Göta älv – Hydrology and morfology with specific regard to erosion processes. (Göta älv – Hydrologi och morfologi med särskild hänsyn till erosionsprocesserna.) SGU avhandlingar och uppsatser i 4:0, Serie CA nr 43. (In Swedish)

Tsaparas, I., Rahardjo, H., Toll, D.G., och Leong, E.C. 2002. Controlling parameters for rainfall-induced landslides. Computers and Geotechnics, Vol. 29, s. 1–27.

Viberg L, 1982. Mapping and classification of slope stability conditions in clay areas. (Kartering och klassificering av lerområdens stabilitetsförutsättningar.) SGI Rapport nr 15, Statens geotekniska institut, Linköping. (In Swedish)

WHO. 2002. Floods: Climate change and adaptation strategies for human health. Report WHO meeting, London, UK, 30 June-2 July 2002. EUR/02/5036813.

Landslides and Climate Change – McInnes, Jakeways, Fairbank & Mathie (eds)
© 2007 Taylor & Francis Group, London, ISBN 978-0-415-44318-0

Overview mapping of landslide and flooding hazards using LIDAR monitoring and GIS-processing

J. Fallsvik & K. Lundström
Swedish Geotechnical Institute

ABSTRACT: The expected climate change will cause higher precipitation and hence higher runoff, erosion and more frequent landslides in parts of Sweden. Overview mapping of landslide hazards using LIDAR monitoring and GIS-Processing facilitates digital databases, which can be used as a planning tool to minimizing the increased risks.

1 SCOPE

1.1 *Objective of the paper*

The objective of the paper is to describe the advantages of using advanced measuring technique for achieving topographical data for performing overview mapping of landslide hazards, and also the increased landslide hazard due to flooding of lakes and river shores. In Scandinavia, flooding will be more common and far-reaching in the future because of the expected climate change.

1.2 *The climate change*

Scenarios for the climate change until 2100 have been simulated for the Nordic region by Rossby Centre (see for instance SMHI, 2005). Besides higher temperatures, the annual precipitation in most parts of the Nordic countries is expected to increase indicating more extremes, i.e. more frequent storms and more heavy rainfall. Therefore, flooding, erosion, landslides, debris flows, etc. will be more common and/or extreme. Furthermore, for the prevailing climate, GIS-based hazard mapping on extreme flooding along some Swedish rivers has been performed. This will increase the runoff; hence what we call extreme flooding today will be more common tomorrow.

1.3 *Overview landslide hazard mapping*

To find areas with prerequisites for landslides in areas covered by fine-grained soils, in for example Norway and Sweden, nation-wide survey hazard mapping is performed. In Sweden this mapping is based on overview information on the slope inclinations and the soil conditions in the mapped areas, roughly the existence of clay and silt layers in the terrain coinciding with slope inclinations greater than a certain angle, primarily 1:10. Especially riverbank slopes can be unstable, and in quite many cases prone for devastating landslides also impacting built-up areas.

The mapping is carried out manually; hence it is time consuming. In order to make this mapping more efficient, an ArcGIS-based method, called NAKASE, to perform the mapping has been developed Viberg et al, (2001). An ArcGIS algorithm was improved and used for performing calculations based on the local soil and topographical conditions, which performs overview assessments on the landslide prerequisites based on empirical criteria. The criteria can be adjusted to different soil regimes; e.g. areas predominated by soil layers comprised by soft clay, silt, sand, etc. Following the described method, different sites in Sweden have been mapped.

1.4 *Influence on the stability conditions from flooding*

During the first period of time after flooding, the landslide hazard in the riverbank slopes is increased due to remaining high pore water pressures in the soil layers. Therefore, a further GIS-based evaluation of the hazard mapping method is included in the EU project called LESSLOSS (see Fallsvik, 2006a), where flooding and landslide hazard digital maps has been combined to gathering a map indicating sub-areas with prerequisites for landslides, which also can be flooded, hence emphasising the remaining higher level of hazard. Further, the input data is based on a digital soil map database and detailed topography information, i.e. laser scanning on land and echo sounding of the river bottom.

Also taking into account the increased flooding hazard due to the climate change, the method can be used in future overview mapping of the stability conditions in built-up areas. The method can also easily be used as a tool when planning new buildings and infrastructure. In addition, as the output data is in a digital shape, it can easily be used as input overlay data in further GIS-processing for planning purposes. In this paper the GIS-based hazard mapping method is described and the results of the mapping in selected sites are presented.

2 THE CLIMATE CHANGE

Today, there is an almost complete international agreement, that a global climate change will occur, mainly because of the man made increase of the compound of the greenhouse gas carbon dioxide in the atmosphere. Globally, the Earth's average temperature will increase around 1 degree until 2100. However regionally, according to scenarios, outlined by the Rossby Centre of the Swedish Meteorological and Hydrological Institute (SMHI 2005), the average temperature in Scandinavia will increase as much as 2–6 degrees during the coming century.

2.1 *Changes in precipitation in Northern Europe*

Due to the climate change, in most parts of Northern Europe also the annual precipitation is expected to increase SMHI (2005), see Figure 1.

In Sweden, the increase of the precipitation is expected to differ widely between different parts of the country. Some parts will get up to 30% more precipitation, but in other parts the amount of precipitation will be unchanged.

Further, also the evaporation is expected to increase because of the higher temperatures. Therefore, as an example, in southeast Sweden, the runoff is expected to remain unchanged or slightly decreased despite higher precipitation, Figure 2.

Also the pattern of the precipitation will change. The amount of precipitation, which will fall, as snow will decrease, whereas the amount, which will fall as rain, will increase. The precipitation caused by frontal heave of air will be almost unchanged, whereas the amount of rain due to convection cells in the atmosphere will increase. Thereby, the amount of precipitation, which will fall as intensive rains, will increase. Heave due to convection is more common during the summer season. Therefore, in view of the fact that the summers will be longer, the amount of precipitation in as convectional rain will increase.

The precipitation conditions will also be more variable, e.g. periods with intensive rains, which will alter with drought periods. More intensive storms will create strong winds, and hence high waves and flooding of coastlines.

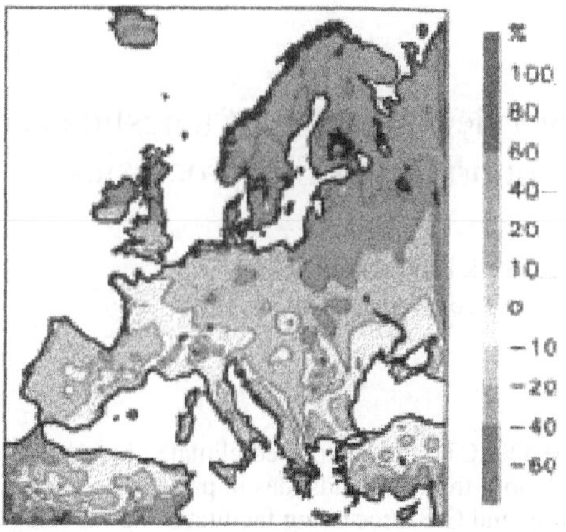

Figure 1. The climate change in Europe. Increase of the annual precipitation until 2071–2100, SMHI (2006). (The map is presented in colour on the conference CD.)

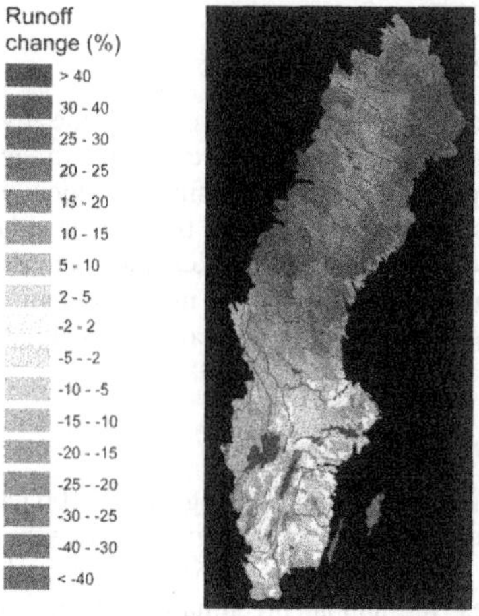

Figure 2. The climate change in Sweden. Increase of the annual runoff until 2071–2100, (SMHI, 2006). (The map is presented in colour on the conference CD.)

Built up areas adjacent to rivers, lakes and seashores will be exposed to flooding, waves and erosion related to the climate changes, in their turn triggering landslides. When future both more common and dramatic weather phenomena occur, our cities, towns and infrastructure, will be vulnerable.

The landslide hazard will be changed due to the climate change. On commission by the Swedish Climate and Vulnerability Investigation, Fallvik et al, (2006) has performed a coarse estimation how the landslide hazards will change during the coming 70–80 years, Figure 3. The estimation is based on climate change so called index maps, SMHI (2006).

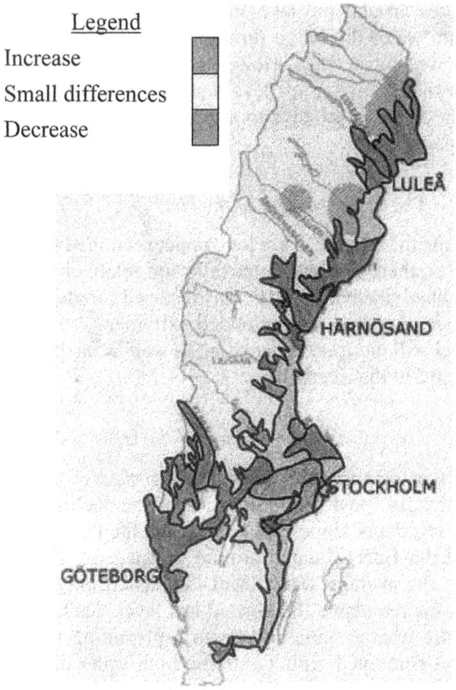

Figure 3. Landslide hazard change in Sweden due to the expected climate change during the coming 70–80 years, Fallsvik et al (2006). The map is performed by overlay between two index maps indicating the change in expected runoff and erosion respectively.

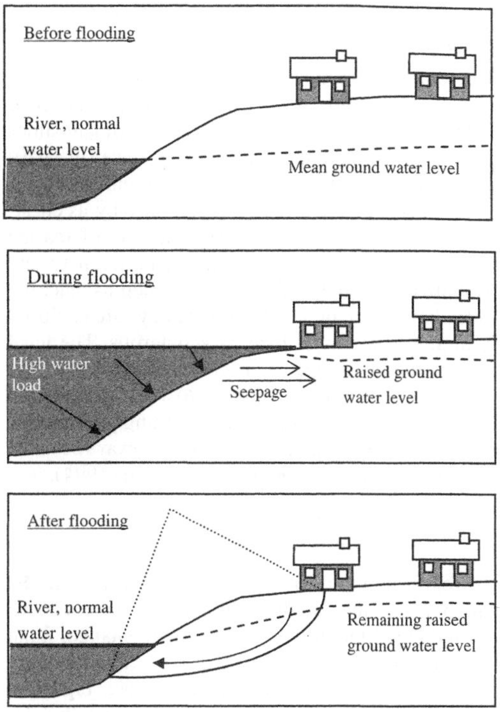

Figure 4A–C. The geo-hydrological situation before, during and after flooding, after Ottosson (2001).

2.2 *Influence on river environment*

Simplified, from a geomorphologic point of view, the topographical structure of a river, e.g. the natural riverbed including its flanking riverine areas, is adopted to the current climate conditions of today – the prevailing conditions generated by today's "normal" variation of the water discharge. During hundreds of years, the water has eroded, transported and accumulated sediments, forming the watercourse to its present shape. However, a changed water discharge due to the feared new precipitation pattern may lead to repeated severe flooding, and to changed topography due to increased erosion, landslides in the river banks, etc.

Particularly, landslides may be triggered due to shifting between extreme high and low water levels in the river. Figure 4A–C.

During flooding, the extreme high water level leads to infiltration of water into the soil layers in the river banks, hence generating a higher ground water level and higher pore pressures (higher water pressures in the pores of the soil).

High pore pressure leads to lower soil strength. However, during flooding, the high water level acts as a counterweight towards the slope, so normally a landslide will not be triggered as long as the flooding prevails. Later, when the water level sinks, often quite rapidly, and hence the counterweight decreases, the increased pore pressure, will not decrease in the same pace, particularly in fine-grained low-permeable soil layers as clay and silt. Therefore, due to the lower soil strength compared to the situation before the flooding, a landslide may be triggered, provided the slope inclination of the riverbank is steep enough.

Off coarse, also higher precipitation and hence higher ground surface runoff leads to increased ground water infiltration and higher pore pressures.

3 THE GÖTA RIVER

Measurements have been performed by using airborne LIDAR on land and multi beam echo sounding on the river bottom to achieve dense and high quality topographical three dimensional information in the landslide problem area Lilla Edet on the Göta River on the Swedish west coast. The topographical information was used as in-data for the NAKASE algorithm, see chapter 1.3 above.

3.1 *The Göta River Valley – a geotechnical problem area*

The southwest region of Sweden around and north of the Gothenburg area – especially the Göta River Valley – is well known for a large number of landslides in clay, for example the landslides in Surte (in 1950) and Göta (1957) attracting particular attention.

The region in question is characterised by its coastline and long valleys with deep deposits of marine soft clays. Clay layer depths varying between 15 and 40 m are common, but also clay layer depths reaching 100 m have been found. The long valleys are enclosed by high formations of bare rock outcrops. The level of the ground surface often varies widely, and clay slopes have been formed next to erosive watercourses. The geological processes have accordingly given rise to slope stability problems, which must always be taken into consideration, Ahlberg and Ottosson (1994).

3.2 *Future influence by the climate change*

The relatively high dense populated areas around the flat shores of the Swedish "inland sea" lake Vänern are in danger due to the future climate change. Lake Vänern is drained by the Göta River, having its estuary in Gothenburg on the Swedish west coast, Figure 5. The water discharge in the river is controlled by hydropower dams. Allowed minimum and maximum extremes are restricted by water control directives.

Even in the current climate situation, there is a fragile hydrological balance because of the mutual dependence between the water level in Lake Vänern, the water discharge in the Göta River and the water level at the river estuary, e.g. the prevailing water level in the sea. Expected higher precipitation due to the climate change will generate flooding of the flat shores around the lake Vänern, provided the maximum water discharge through the Göta River is not allowed to increase. However, a higher discharge in the river a higher sea level lead to flooding and erosion of the riverbanks to a higher degree than today, Hultén (2006).

Figure 5. The position of the Lilla Edet test site. The Lilla Edet town is situated on the banks of the Göta River, which is the largest river in Scandinavia, draining the "inland sea" Lake Vänern.

3.3 *The upper section of the river*

Along the around 35 km long upper section of the Göta River, there are two water falls and relatively steeply inclined riverbank slopes. An increased variation of the water discharge and extended extremes of the water level will increase the erosion, as well as the landslide hazard in the riverbanks.

3.4 *The downstream section of the river*

Further downstream, the bottom profile of the river is flat. In most sections, parts of the inclination of the riverbank slopes are flat. Below the last waterfall at Lilla Edet, situated around 60 km from the estuary, the average water level is situated only around 0.6 metres above the normal sea level. Therefore, in future weather situations with a prevailing high sea level combined with a extreme high water discharge in the river, the water discharge will be hold back, generating severe flooding of the areas around the flat river bank slopes.

Compared with the situation of today, the climate change scenarios outline more common situations with high water levels in the sea in the future. This is despite the post-glacial isostatical up-lift of the Scandinavian ground surface, e.g. 2 mm yearly on the Swedish west coast.

3.5 *The Lilla Edet test site*

The Lilla Edet test site is situated on the west side of the Göta River in the municipality of Lilla Edet. In a general inventory performed by the Swedish Geotechnical Institute (SGI) in the municipalities along the Göta River Valley, the frequency of previous landslides in different municipalities in the area was studied, Inganäs & Viberg (1979). The number of previous landslides appeared to be 130 in the area of the Lilla Edet municipality, which was the outstanding highest frequency of all the studied municipalities.

In the urbanised areas of the Lilla Edet village (4000 inhabitants), dwellings, schools, service areas, factories, a major lock, a water power plant and other constructions are situated close to the riverbanks. The soil layers in the area contain clay including layers of quick clay, and hence many landslides have occurred in the area and its vicinity, both during prehistoric and modern time. Some of these landslides have affected built up areas, hence causing loss of human life and property.

In 1957, a major landslide occurred in the southern rim of Lilla Edet, however mainly in the neighbouring village Göta. Three employed were killed and three were seriously injured in the completely destroyed paper mill in Göta.

4 TOPOGRAPHICAL MEASUREMENTS

In the so-called LESSLOSS EU-project, Fallsvik (2006a), laser scanning has been performed in an 8 km^2 large area covering the Lilla Edet situated on both sides of the Göta River, Figure 6. In the river, multi-beam echo sounding of the bottom topography has been performed. The two methods are described in this chapter.

4.1 *Laser scanning of the topography*

Infrared laser scanning was performed by using the TopEye™ airborne topographic survey system, to capture topography and high-resolution digital images with high precision and in near real time, based on scanned laser and digital images.

The flight was performed at approximately 400 m above ground, and the raw point density was between 7–10 points per m^2. The helicopter survey routes are illustrated in Figure 6. Systematic shifts between the survey routes are verified by checking data from overlapping flight lines.

As an independent check on the processed laser point clouds, an ortho-imagery was done in order to verify the laser data and imagery compared to known points and surfaces. However, these images also accomplished an illustrative basis when constructing maps based on the laser scanning.

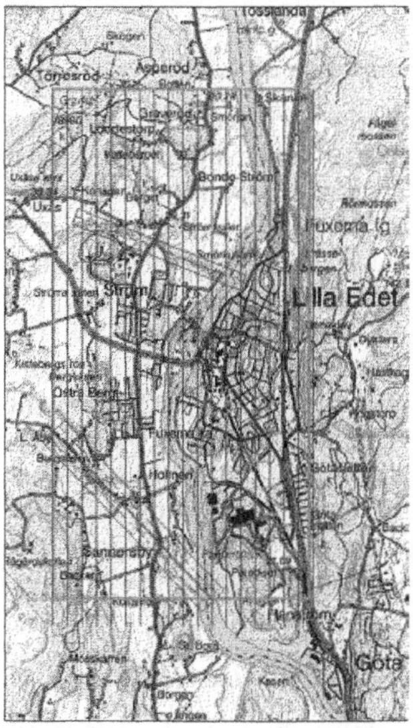

Figure 6. Lilla Edet test site. The lines indicate the pattern of the helicopter routes for performing the TopEye laser scanning, covering 8 km^2. (The map is presented in colour on the conference CD.)

4.1.1 *Estimation of the true ground surface*

The laser scanning achieves echoes from the ground surface as well as vegetation, and buildings, etc. To avoid the echoes from the latter objects hiding the real ground surface, the achieved data was processed by using algorithms developed to identifying the typical geometrical shapes of obstacles. Where obstacles are found (trees, bushes, houses, etc.), the algorithm "neutralises" them, by replacing the obstacles with a virtual ground surface normalised to the neighbouring ground surface, see Figure 7.

4.1.2 *Accuracy in x-, y- and z-directions*

The accuracy in estimation of heights is ±0.1 m. In x- and y-direction the accuracy is achieved by the point density, which is between 7–10 points per m^2.

4.1.3 *Limitations*

The TopEye laser equipment is based on infrared light. The laser scanning provides detailed information on the topography on land.

However, infrared light cannot penetrate water if it is not clear enough. The river water is muddy. Therefore, laser scanning of the river bottom topography could not be performed.

4.2 *Multi-beam echo sounding of the bottom topography of the Göta River*

In the task to analyse the slope stability conditions, detailed information on the bottom topography in the river is essential.

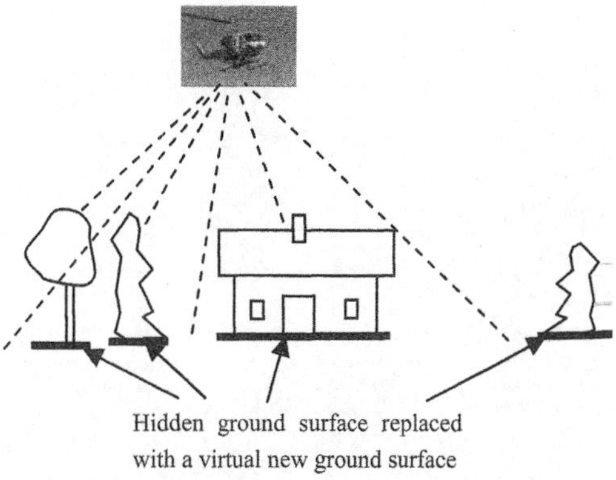
Hidden ground surface replaced
with a virtual new ground surface

Figure 7. Where obstacles are found (trees, houses, etc), that hide the ground surface, an algorithm "neutralises" them, by replacing the obstacles with a virtual ground surface normalised to the neighbouring ground surface.

For creating a detailed terrain model of the bottom topography of selected sections of the Göta River, bathymetric measurements in an earlier project had been performed by Marin Mätteknik AB (2004). These measurements were commissioned by SGI and financed by the Swedish Road Administration, Banverket (the Swedish National Railway Administration) and the municipalities of Lilla Edet and Ale.

The measurements were carried out by using multibeam echo sounding, which meets (gives around the same) accuracy of the laser scanning on land.

The multi-beam echo sounding could only be performed where the water depth exceeds 1 m under the keel of the measuring vessel. Therefore, the river bottom topography could not be measured within a narrow zone with shallow depth close to the shores.

A base station transceiver was established on a fixed base point, which sends corrections to the receiver onboard the vessel. This system increases the positioning accuracy to ±0.015–0.020 m horizontally and ±0.020–0.30 m vertically.

The multi-beam echo sounder, Simrad EM 3000, performs marine measurements, which provides depth-data within a width on the bottom, which is 7–8 times the water depth, Figure 8. The multi-beam echo sounder is mounted on the vessel, and the survey can be performed in a water depth exceeding 2 m. Measurements can be carried out close to the river banks, and the measured data will be gathered in a 3-dimentional model. The results are reported as maps, in digital models or as sets of XYZ co-ordinate data points.

The multi-beam echo sounder works with 254 beams, which simultaneously collects the water depth data.

Figure 8. Multi-beam echo sounder (Simrad EM3000), principle. (The figure is presented in colour on the conference CD.)

Within the area to be measured, collected high-resolution data was covering the major part of the river bottom. However, the shallowest zone close to the river shores could not be measured because of deep drawing of the measuring vessel. The pattern of the vessel routs during measuring was held parallel. The water depth and the bottom topography determine the distance between the survey lines, so a guaranteed 25% overlap of measured data will was acquired. This practice ensured that all sectors of the river bottom was covered, however, it also gained a control of the precision of the data collection.

In the Göta River the measurements have been performed along 3–4 parallel survey lines, depending on the local water depth conditions. The measurements were carried out according to the demands stated by the Swedish Maritime Administration.

As a summary, after corrections due to position, vessel movements, the sound velocity through water, and checking the actual water stand, the accuracy of the measurements of the water depths typically was ±0.1 m.

5 SOIL LAYERS

The digital quaternary soil map produced by the Swedish Geological Survey, SGU, has been used as one information layer in the GIS-processing of the soil condition on land. For the soil condition on the river bottom results from side scanning sonar has been used as another GIS-layer.

5.1 *The soil layers on land*

The digital soil map is based on stereo interpretation of aerial photos combined with field verification. The field verification was performed using a hand-held probe, auger, and shovel. Complementary information on the soil layer stratigraphy and thickness has been collected from open cuts and by borings performed by the SGU and others.

On general SGU soil maps, the soil conditions 0–0.5 m below the ground surface are shown. However, in the LESSLOSS-project, Fallsvik (2006a), also the existence of more profound clay layers on a level deeper than 0.5 m below the ground surface are shown on the map, even if coarser superficial soils cover the clay, such as washed deposits, fluvial deposits or peat.

5.2 *The soil layers on the river bottom*

The soil layers on the Göta River bottom have been scanned using side-scanning sonar. The river fairway was used as a reference line for mapping. The resulting sonar image looks like a greyscale aerial photo. Generally, darker shades on the sonar image indicate coarse-grained soils, whereas lighter shades indicate fine-grained soils, such as clay. The resulting soil assessment based on the sonar images was verified by studying seismic profiles, and by probing.

6 OVERVIEW LANDSLIDE HAZARD MAPPING (CLAY AND SILT SLOPES)

The Swedish method for nation wide overview landslide hazard mapping for areas with slopes in silt and clay has been developed in cooperation between SGI, the Swedish Rescue Services Agency and Chalmers University. The method has been adopted by the Swedish Rescue Services Agency, to form the first step of the Swedish national programme for landslide risk reduction. The mapping is carried out nationwide, however comprises only existing built up areas. (A related method has been developed for overview landslide hazard mapping of slopes in coarse soils.)

The objective of the mapping is to find slopes in clay or silt soil layers in urbanised areas being steep enough indicating that the stability conditions must be further investigated. In the former glaciated regions in Norway, Sweden and Canada, landslides have not occurred in natural clay slopes where the slope inclination is roughly below 1:10, Inganäs & Viberg, (1979). Therefore, the inclination 1:10 is chosen as the upper limit criteria for clay slopes.

Following the method, the mapped areas are divided into three so-called Stability Zones, describing the prerequisites for initial slope failure. Areas where landslide hazard could not be neglected are divided into the two Stability Zones, I and II, whereas the Stability Zone III comprises land with other soils than clay and silt.

The Stability Zone I comprises land where the slope inclination exceeds the criteria 1:10 indicating prerequisites for spontaneous or proceeding landslides in slopes containing clay or silt soil layers, e.g. areas which may be primarily affected by an initial slide or slip. Many slopes in Zone I could have a satisfactory stability; however, the stability conditions first have to be further investigated and calculated.

The Stability Zone II comprises areas containing clay or silt soil, which have no prerequisites for initial slope failure, but may be affected secondarily by landslides in Zone I acting backwards or forwards. After changes of the conditions by human activities (e.g. construction work, land filling, excavation, and change of ground water conditions), and after landslides in adjacent Zone I areas, the stability conditions in Zone II could change. In these cases, the stability may have to be investigated.

The Stability Zone III comprises areas with bedrock outcrops, or where the soil layers only contain coarse soils. Before activities as blasting, piling or other vibration creating activities and change of the ground water conditions, are performed within Zone III, their influence on the stability conditions within adjacent Zone I or II areas must be investigated.

The division of the mapped area is performed by time-consuming manual measurements on the elevation lines topographical maps combined with studies of printed soil maps. The method has been described by Fallsvik and Viberg (1998).

6.1 *Prototype for national digital map data base on landslide prerequisites in clay and silt areas in Sweden*

The national survey investigation of the landslide hazard in built up areas in clay and silt soils has been going on in Sweden since the nineteen eighties. However, there is also a demand for landslide hazard maps outside built up areas as a tool for city and infrastructure planning, planning and executing of rescue actions, and judgement of landslide hazard for existing constructions outside built up areas.

Therefore, in 2000–2001, SGI was commissioned by the Swedish Ministry of Environment to develop a prototype landslide hazard map produced by GIS-technique. This work was carried out in co-operation with SGU, Lantmäteriet (the Swedish land surveying authority) and the Swedish Rescue Services Agency. The project "National digital map data base on landslide prerequisites in clay and silt areas in Sweden" is abbreviated the NAKASE-project, Viberg et al, (2001).

For the prototype, which used a database built on the iso-lines of a topographical map, an algorithm was designed to process the data and classify the terrain into the Stability Zones I, II and III.

7 COMBINATION OF DATA

The data from the laser scanning and the multi-beam echo sounding was combined in a Digital Terrain Model (DTM), Fallsvik (2006a).

As a summary, overview landslide hazard mapping in Lilla Edet was performed by using ArcGIS as described in Table 1.

The resulting digital database of the LESSLOSS GIS-performed overview landslide hazard mapping along the Göta River through Lilla Edet, based on laser scanning and multi beam echo sounding, is shown in Figure 9.

The LESSLOSS result has been compared with the result of earlier manually performed landslide hazard mapping. The LESSLOSS result gave a higher level of detail and is also influenced by the achieved information on the topographical conditions on the river bottom.

Table 1. Summary of landslide hazard mapping in Lilla Edet in the LESSLOSS-project sub task, Fallsvik (2006a).

Effort	Method
The topography on land	Helicopter borne laser scanning (LIDAR)
The soil conditions on land	Overlay achieved from the SGU Digital soil map, revised with reference underlying sited fine grained soils
The Göta River bottom topography	Multi-beam Echo Sounding
The soil conditions below the river bottom	Overlay achieved from the SGU's classification based on information from side scanning sonar images
The stability zones	GIS processing by overlay technique involving the information above and by using the GIS-algorithm revised and extended from the Swedish NAKASE-project
The field control	Evaluation of the laser scanned digital terrain model with respect to analysis results of field check. Studies of the detailed digital topographical information.

8 FURTHER GIS-PROCESSING

Both the topographical database based on the laser scanning and multi beam echo sounding as well as the GIS-produced digital landslide hazard database could be reused for further processing for multi purpose planning of the society. In the following two examples are given.

8.1 *Detailed stability calculations*

In the LESSLOSS project, stability calculations based on the detailed digital topographical database were performed in a selected test section in Lilla Edet, Fallsvik (2006b). The section is situated on a slope flanking the Göta River.

In two earlier performed stability investigations, the geometry of the section had been measured manually. In the first overview investigation the geometry was estimated roughly by regard to the elevation iso-lines on a map. In the second investigation the geometry was measured by levelling, which is the level of detail normally used for detailed stability investigations in Sweden. In these two earlier investigations the geometry of the river bottom was estimated by manual plumbing from a small boat.

The geometries of the section, achieved by respectively the two earlier investigations and by the new detailed digital topographical database, are compared in Figure 10.

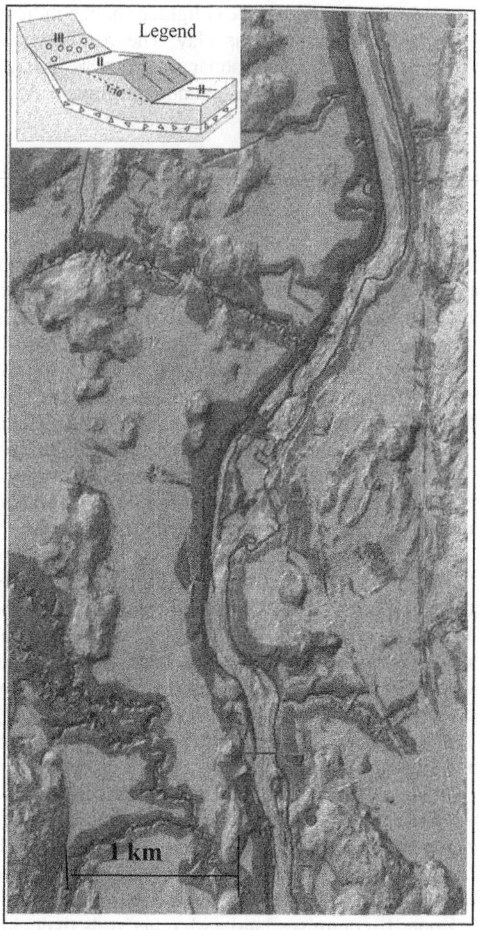

Stability Zone I (red): Areas with soil layers compiled by clay and/or silt. Prerequisites for initial landslides.
Stability Zone II (yellow): Areas with soil layers compiled by clay and/or silt. No prerequisites for initial landslides in clay or silt.
Stability Zone III (green): Areas with outcrops of firm rock or coarser soil layers, i.e. not compiled by clay and/or silt. No prerequisites for initial landslides in clay or silt.

Figure 9. Map presenting the resulting digital database of the GIS processed overview landslide hazard mapping along the Göta River through Lilla Edet, Fallsvik (2006a). The processing was based on topography information achieved by laser scanning on land and multi beam echo sounding on the river bottom. The GIS processing was based on the NAKASE algorithm. The very high level of detail and the GIS hill shade presentation function, gives the image the three-dimensional impression. (The map is presented in colour on the conference CD.)

The results of detailed stability calculations based on each of the different geometries are listed in Table 2.

The computer programme Slope/Windows, Version GeoStudio (2004) was used for the stability

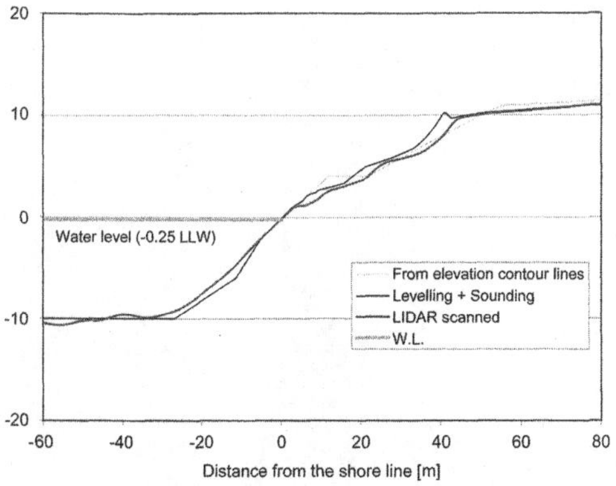

Figure 10. Comparison between geometries of a section to the Göta River achieved manually by regard to the elevation iso-lines on a map, levelling and plumbing, respectively, digitally by laser scanning and multi beam echo sounding, Fallsvik (2006b). (The figure is presented in colour on the conference CD.)

Table 2. Results of detailed stability calculations based on geometries achieved by measurements with three different level of detail, Fallsvik (2006b).

Used method for achieving the geometry of the section	Calculated factor of safety, combined analysis $F_{combined}$
Measuring on the elevation contour lines on a local topographical map on land and manual plumbing of the river bottom	1.03
Levelling on land and manual plumbing of the river bottom	1.00
Laser scanning on land and multi beam echo sounding of the river bottom	1.07 (see Figure 11)

calculations. The programme calculates all circular sliding surfaces satisfying a selected grid of centre points and a number of selected radius tangent lines placed in the soil layers. The rigorous stability calculation theory according to Morgenstern/Price was chosen to be followed; a theory commonly used in Sweden for slopes bordering to shores. Where the calculated sliding surfaces pass through cohesional soil layers, both undrained analysis and combined analysis were performed in all slices. Here only the results of the combined analysis are reported.

In the combined analysis, in each slice the shear strength is estimated to the lowest value of the drained and the undrained shear strength respectively:

$$s = \min[c' + \sigma' \tan(\varphi'); c_u] \quad (1)$$

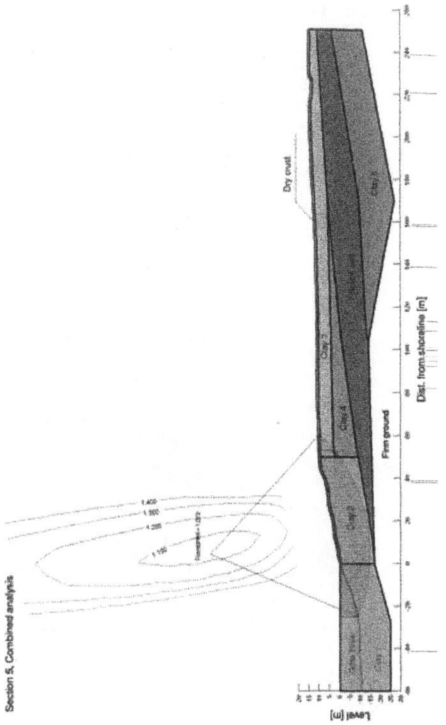

Figure 11. Test section in Lilla Edet, Fallsvik (2006b). Stability calculation performed by combined analysis. The topography was achieved by laser scanning on land and by multi beam echo sounding on the river bottom. (The figure is presented in colour on the conference CD.)

where:

s = shear strength [kPa]
φ' = effective angle of internal friction [°]
c_u = undrained shear strength [kPa]
c' = effective cohesion [kPa]
$\sigma' = (\sigma_n - u)$ = effective normal stress [kPa]
σ_n = total normal stress [kPa]
u = pore-water pressure [kPa]

Perpendicular to a calculated section, the stability conditions differ along a slope predominantly depending on the topography. The soil layer conditions however do not differ to the same degree. To demonstrate how the stability conditions differ along the same slope, ten sections were drawn parallel to the section. The internal distances between each parallel section are 10 m. Stability calculations were performed in each of these parallel sections (see Figure 12.)

The laser scanned and multi-beam sounded topographical database facilitated the achievement of the geometry of these parallel sections. In comparison, ordinary levelling and manual plumbing of these ten extra sections had been far too expensive to carry out in a typical ordinary stability investigation.

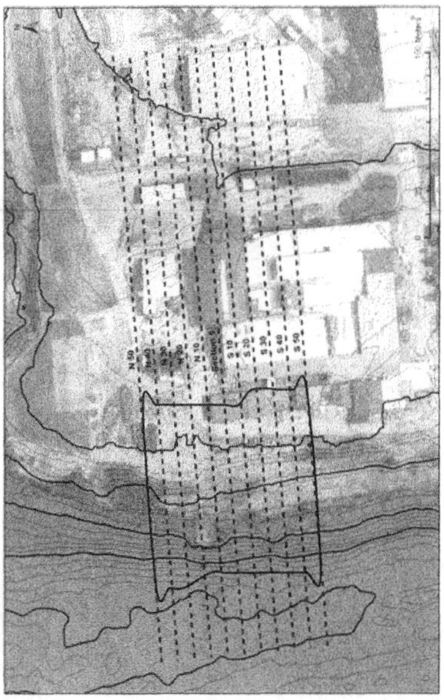

Section	*Fcomb*	Section	*Fcomb*
N50	0.99	S10	1.07
N40	0.95	S20	1.08
N30	0.97	S30	1.09
N20	1.02	S40	1.10
N10	1.06	S50	1.11
The test section	1.07		

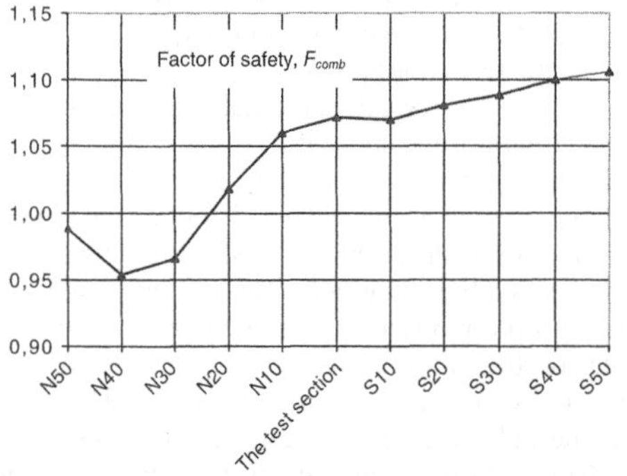

Figure 12. Position of the test section and ten selected parallel sections. The lowest F-factors calculated of the sliding surfaces in each of the parallel sections are reported in the table and diagram. The shaded area shows the position in each section of the end- and centre-points of the sliding surfaces with the lowest calculated F-factor according to combined analysis.

Area, which will be flooded within Zone I	
Area, which will not be flooded within Zone I	
Area, which will be flooded within Zone II	
Area, which will not be flooded within Zone II	
Zone III	

Figure 13. The LESSLOSS digital landslide hazard database combined with a flooding scenario for the Göta River developed by the Swedish Rescue Services Agency (2006). The GIS-overlay indicates landslide prone areas, which also has a potential to be flooded. (The map is presented in colour on the conference CD.)

8.2 *Overlay between landslide and flooding hazard maps*

By using GIS-based so called overlay technique, the LESSLOSS digital landslide hazard database was combined with a flooding scenario for the Göta River was developed in the Swedish Climate and Vulnerability Investigation (2006), also available as a digital database. By using the overlay technique sub areas could be indicated having the potential both to be flooded and to be influenced by landslides, Figure 13.

REFERENCES

Ahlberg, P., Ottosson, E., (1994), Planning and localisation of built-up areas and infrastructures based on landslide risk analysis, International conference on landslides, slope

stability and safety of infra-structures, Kuala Lumpur, Malaysia, Proceedings, p 9–16
Fallsvik, J., (2006a), Application of laser scanning digital terrain model (LS DTM) in landslide hazard zonation, Zonation and landslide hazard by means of LS DTM, Deliverable Report 7, Risk Mitigation for Earthquakes and Landslides, LESSLOSS, EU Integrated Project, Sixth Framework Programme, Global Change and Ecosystems, Landslide monitoring and warning systems, In-situ and remote monitoring techniques
Fallsvik, J., (2006b), LIDAR data for slope stability analyses, Deliverable Report 6, Risk Mitigation for Earthquakes and Landslides, LESSLOSS, EU Integrated Project, Sixth Framework Programme, Global Change and Ecosystems, Landslide monitoring and warning systems, In-situ and remote monitoring techniques
Fallsvik, J., Alexandersson, H., Edsgård, S., Hågeryd, A.-C., Lind, B., Löfling, P., Nordlander, H., Tunholm, B., (2006), Overview estimation of soil layer movements due to a changed climate, Published in Swedish, "Översiktlig bedömning av jordrörelser vid förändrat klimat", Klimat och sårbarhetsutredningen, the Swedish Climate and Vulnerability Investigation, the Swedish Geotechnical Institute (SGI), SGI reg. No. 1-0611-0652, Linköping, Sweden
Fallsvik, J., Viberg, L., (1998), Early stage landslide and erosion risk assessment – A method for a national survey in Sweden, Erdwissenschaftliche Aspekte des Umweltschutzes. Arbeitstagung des Bereiches Umwelt, 4, Wien, April, 1998, Tagungsband, pp 151–153, Vienna, Austria
Hultén, C., Edstam, T., Arvidsson, O. & Nilsson, G. (2006), Geotechnical prerequisites for increased draining through the Göta River, basis for the Swedish Climate and vulnerability investigation, published in Swedish: Geotekniska förutsättningar för ökad tappning från Vänern till Göta älv", Swedish geotechnical institute, SGI, Linköping
Slope/Windows, Version GeoStudio (2004), Computer programme for slope stability analyses, Geo-Slope International Ltd., Calgary, Alberta, Canada
Swedish Rescue Services Agency (2006), Overview mapping of flooding hazards in Göta River, GIS-data for overview mapping of Göta River
Viberg, L., Inganäs, J., (1979), Inventory of landslides in clay in Sweden, Published in Swedish 'Inventering av lerskred i Sverige', Swedish Geotechnical Institute, Varia 13, 8 p, Linköping
Viberg, L., Hågeryd, A.-C., Fallsvik, J., Ottosson, E., Fredén, C., Grånäs, K., Edsgård, S., Jonsson, H., Johansson, K., Johnsson, A., (2001), NAKASE, National digital map data base on landslide prerequisites in clay and silt areas, Development of database prototype and proposal for production, (in Swedish), I and II., Swedish Ministry for Environmental Protection, Swedish Geotechnical Institute in co-operation with Swedish Geological Survey, Lantmäteriverket, LMV and Swedish Rescue Services Agency, 2001, 1. 41 p; 2. 11 p + map, Linköping, Sweden
SMHI (2006), Climate scenarios, R&D, Rossby Centre, 2006-10-24, www.smhi.se/sgn0106/if/rc/clmscen05.htm

Landslides and Climate Change – McInnes, Jakeways, Fairbank & Mathie (eds)
© 2007 Taylor & Francis Group, London, ISBN 978-0-415-44318-0

GPS and GIS based technique for a full-system of registration, management and publishing information of slopes along highways in JKR (PWD) Malaysia

R. Majid, S. Shaharom, A. Mohamad & R. Ahmad
Slope Engineering Branch, PWD Malaysia

ABSTRACT: Over the years, landslide hazard assessment has played an important role in developing land utilization regulations aimed at minimizing the loss of lives and property. A variety of approaches have been used in landslide assessment in JKR (PWD) and these can be classified into heuristic approach, statistical approach and deterministic approach etc. However, there is little work on the satisfactory integration of these models with GIS and GPS to support slope management and landslide hazard mitigation.

This paper described the method of registering slopes along the highway in Malaysia followed by the assessment using the indirect and the direct method, based on GIS and GPS technology to produce landslide hazard map. Previously, only the indirect method was used. The technique is by filling in a special form of the slope parameters taken through visual inspection. Slope parameters include slope morphology, landslide activity, material, drainage, activities nearby etc the data then entered onto the PC for analysis based on multi criteria evaluation technique (indirect method).

With the introducing of the new technique, slope along the road were digitised (delineated) and inventoried and slope parameter were taken using PDA (GPS ready). The data then downloaded electronically to the server/PC and the assessment of slope hazard can be done with both the indirect method and the direct method. In the direct method, detailed geomorphology and geology was carried out, using uniquely coded polygons, which were evaluated one by one by an expert to assess the stability of the slope.

With all the data collected visually and spatially, gathered and analysed quickly, will be published through the Internet in the form of GIS map, so that it will become easier and understandable by the slope custodian as slope manager and the public as the road user.

1 INTRODUCTION

Malaysia has a history of tragic landslides. Malaysia highway industry is facing significant challenges due to heavy rains and landslides resulting in unpleasant incidents in the recent past. These incidents caused large numbers of casualties and huge economic losses. Economic loss from landslides is great and there is evidence that it is on the increase as development takes place, under the pressures of expanding road networks on unstable hillsides areas and also couple with the global climate change. Landslides destroy or damage residential and industrial developments, agricultural and forestlands, have negative impacts on water quality in rivers and streams and can ultimately lead to loss of life.

In recent years the use of GIS for landslide hazard modelling and assessment has increased because of the development of commercial system, such as Arc/Info (ESRI products) and the quick access to data obtained through Global Positioning System (GPS) and remote sensing.

Figure 1. Landslide May, 06.

The continued efforts to tackle the slope safety problems in Malaysia have resulted in the progressive development of good practice and state-of the-art

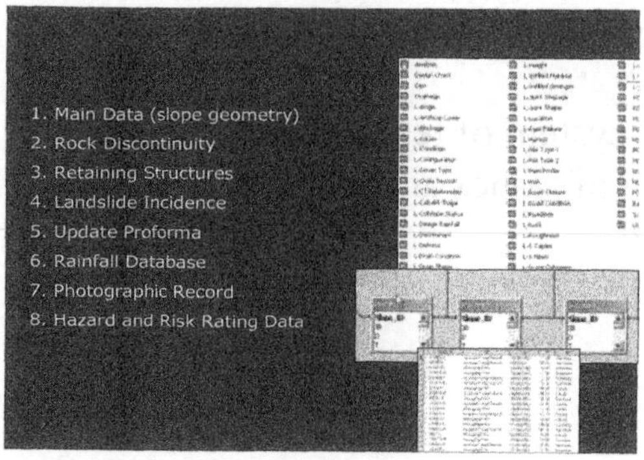

Figure 2. Non-Spatial Database.

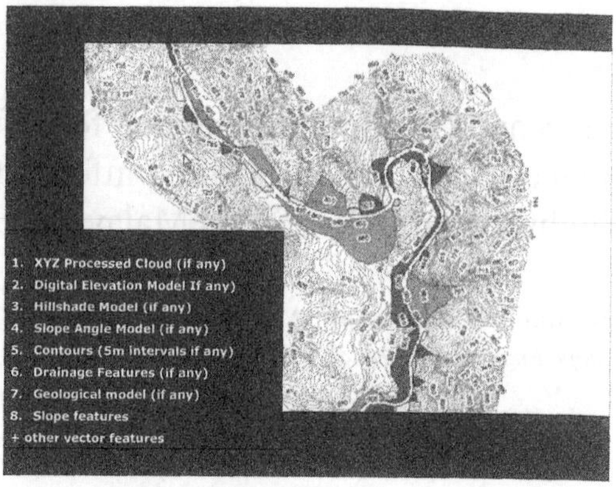

Figure 3. Spatial Database.

advances in slope engineering. Recent advances in the management of landslide risk in Malaysia also include other non-technical developments such as improvements to the slope management system with the engagement of the GIS and GPS system.

These are now accessible to the geotechnical profession, and are applicable to slope engineering in terms of achievable resolution, capability, time and cost.

2 METHODOLOGY

2.1 *Provide road slope geodatabase*

Non-spatial database was design to store and access the data efficiently from the Desktop GIS. This database has all the spatial attribute information for each slope, which has been extracted from PDA.

The spatial database design contains the database components, spatial information and data updating. This database development involves the data extracting from the aerial information and produce to GIS layer using GIS software (ArcGIS). Figure 3 below briefly shows the data model for GIS application within the develop system.

2.2 *Software customisation*

This part is divided into three components as shown in Figure 4.

2.3 *Management component*

This part explained the utilizing of the customized ArcGIS Desktop. Programming code were wrote to provide

- *Tools* to facilitate users to transfer data to PDA for field work
- *Tools* to facilitate users to update the data collected in the PDA back to the database
- *Code* to transfer customized forms to PDA
- *Tools* for slope database management activity

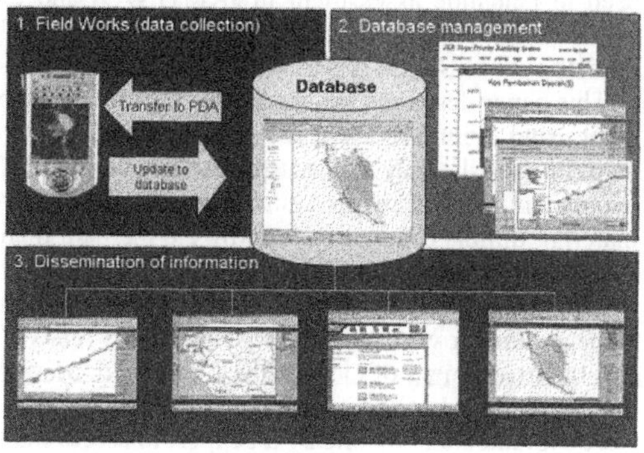

Figure 4. Software customisation.

- *Tools* to facilitate users to locate area of interest immediately
- Code for calculating the slope risk score automatically
- *Tools* for creating maps according to JKR standard.
- *Tools* for creating reports and graphs
- *Tools* for searching pictures and sketches automatically (customized hotlink search)
- *Tools* for retrieving slope historical data.

A few GIS layers were also included in the system namely:

- State boundary
- District boundary
- Cities and Towns Location
- Federal and State Roads
- Geological Formation
- Various Thematic maps derived from DEM (if any)
- Kilometre Post Location etc.

2.4 *Data collection component using PDA*

ArcPad software is customized in the PDA device, which is also integrated with GPS receiver to facilitate

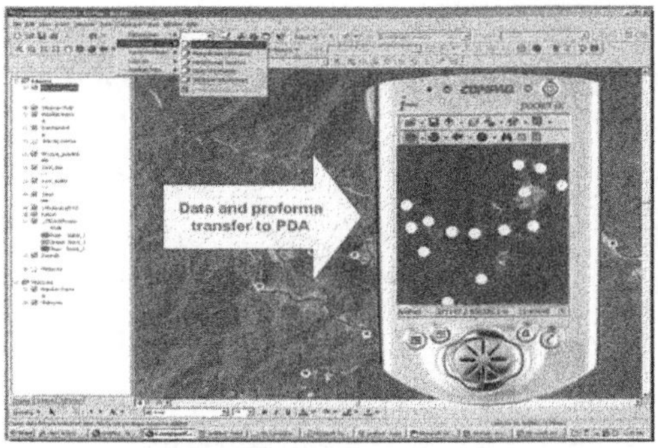

Figure 5. PDA with integrated GPS.

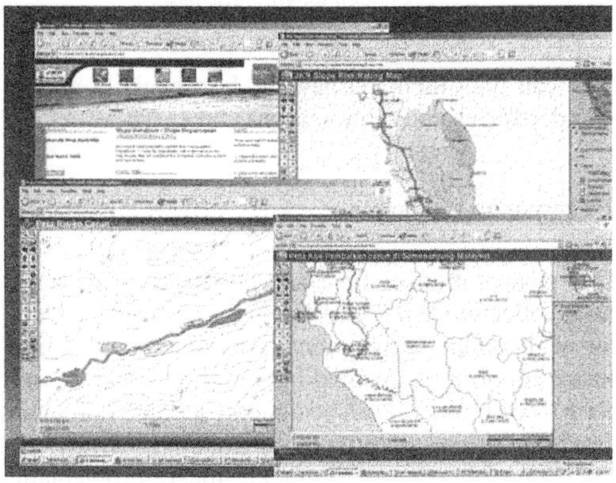

Figure 6. Data Publication.

slope data collection in the field. Together with the kilometre post location layer, it will help the process of collecting data.

2.5 *Data publication component using the web*

ArcIMS is being utilized and customized in order to publish all the information that had been collected and process. Publication is done through the internet. End user is able to view and used all the slope information and risk maps through Internet Explorer or Arc Explorer.

3 ANALYSIS AND DISCUSSION

3.1 *Data collection*

Conventional road slope data were collected by visual inspection during sunny day only. This is because of the usage of the paper proforma, which were specifically design for filling in the data. Special equipments such as clinometer, range finder, compass, measuring tape etc were used to assist in the works. Before the physical data of the slope were taken, the person needs to delineate the length and perimeter of the slope in order to inventories the slope. This needs an experience person to do it especially for slope of natural and embankment type. But with the new system develop (digital field mapping with the aid of GPS), additional information such as ortho-images, contours, and other geo-reference data were supplied to the data collector prior to the task. This information will be installed in the PDA and operated in the ArcPad format. With this information, the process of demarcating of a road slope is making easier and simpler. All the slope data and physical properties were filling in the PDA by just clicking the pull down menu. This will reduce a lot in terms of human error and time consume. Furthermore, the data can be taken at all type of weather condition, 'on and off' the vehicle.

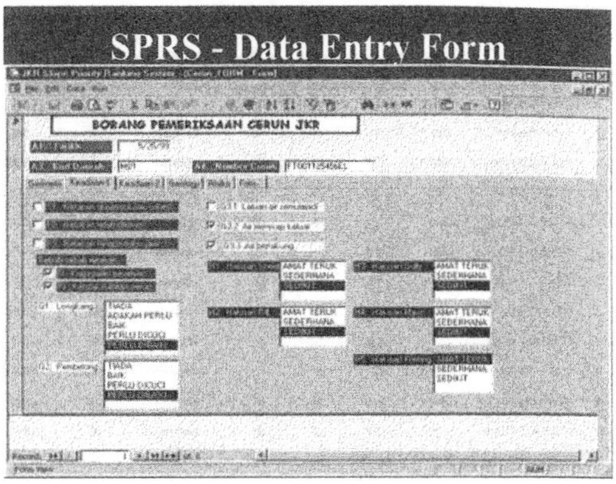

Figure 7. Previous System (Paper Form).

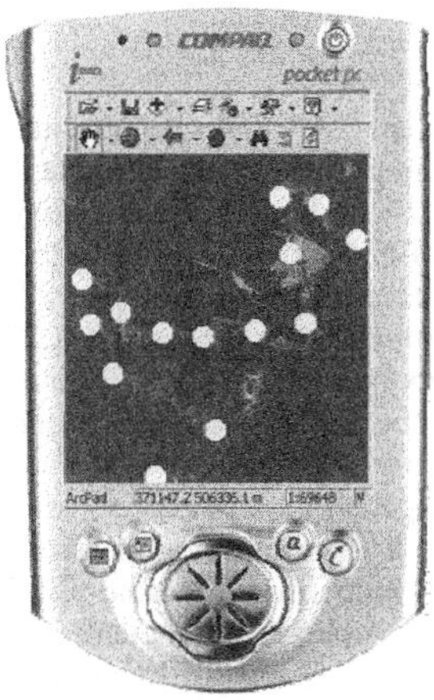

Figure 8. New System (Digital Form).

3.2 *Analysis*

From the previous system, data captured from the field were then entered manually into a system developed

JKR Slope Priority Ranking System *sorted by Risk Rating*

No.	Feature No.	Daerah	panjang	tinggi	sudut	RISK RATING	Score	KOS RM juta
1	FT055-031-350CL	B06	0.030	30	60	Moderate	52	0.081
2	FT055-035-960CL	C07	0.310	13	70	Moderate	45	0.809
3	FT055-023-700CL	B06	0.030	20	60	Moderate	44	0.054
4	FT055-021-700CL	B06	0.030	20	60	Moderate	44	0.054
5	FT055-014-050CR	B06	0.007	24	65	Moderate	40	0.014
6	FT055-005-710CR	B06	0.013	27	75	Moderate	40	0.028
7	FT055-005-480CR	B06	0.022	25	65	Moderate	40	0.047
8	FT055-005-140CR	B06	0.032	25	75	Moderate	40	0.065
9	FT055-005-120CR	B06	0.012	25	65	Moderate	40	0.026
10	FT055-026-300CR	B06	0.028	29	60	Moderate	36	0.073
11	FT055-026-650CR	B06	0.016	36	60	Moderate	38	0.052
12	FT055-017-530CR	B06	0.017	22	65	Moderate	36	0.031
13	FT055-023-750CR	B06	0.038	30	60	Moderate	36	0.103
14	FT055-016-550CR	B06	0.011	13	70	Moderate	36	0.012

Figure 9. Old system publication format.

using MS Access 2000. Entering data manually is time consuming and prone to error. The data then analysed and will give slope hazard classification and display in the form of text format only. With the new system in place, data transfer is done digitally from PDA to a desktop computer with just one click. The data then analysed to give the hazard classification of the slope. The result can be displayed in the form of text format or graphical format. Combine with other geocoded map of geology, vegetation, slope aspect, curvature profile, etc with give a more meaningful assessment of the slope stability. Furthermore, if ortho-photos or satellite images are available, the risk elements then can be identified and a more quantified risk assessment can be done thus producing risk maps.

3.3 *Publish information*

Traditionally, the outputs of the slope analysis were used solely by JKR for slope maintenance and management. But now, with the output being displayed in digital and graphical format, the information can be shared with the public and road users so that they can plan their journey safely especially during thunderstorms and rainy seasons.

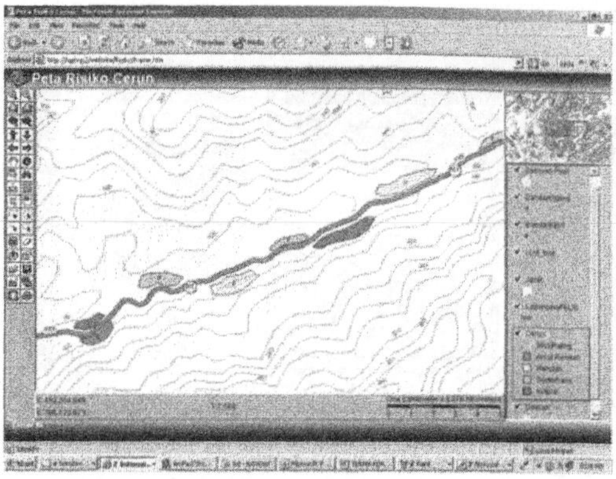

Figure 10. New system publication format.

4 CONCLUSIONS

This system reduces time and human error in slope data collection. GPS positioning assist data collector in the slope registration/mapping and the digital forms simplify the method of capturing slope parameters. Data collected can be quickly updated into main database.

GIS enable experts to quickly locate problematic slopes visually for further action. GIS layers like geology, aerial photos, hydrology, contours and vegetation complement existing slope engineering method in judging slope stability.

Continuous efforts by the geotechnical and ICT profession have significantly reduced the risk of landslides and improved slope engineering practice in Malaysia. Much remains to be learnt and developed, and further knowledge-based and technology driven advances in slope engineering practice in Malaysia are anticipated in near future especially with the formation of Slope Engineering Branch, PWD back in 2004.

Landslides and Climate Change – McInnes, Jakeways, Fairbank & Mathie (eds)
 ISBN 978-0-415-44318-0

Process-based empirical prediction of landslides in weakly lithified coastal cliffs, San Francisco, California, USA

B.D. Collins & R. Kayen
United States Geological Survey, Western Coastal and Marine Geology, Menlo Park, California, USA

N. Sitar
University of California, Dept. of Civil and Envir. Engineering, Berkeley, California, USA

ABSTRACT: Coastal landslides in weakly lithified sediment are a common occurrence in many parts of the world, including the west coast of the United States. Here, geologically young (Quaternary), marine terrace deposits form steep, near vertical cliff exposures up to 30 m in height and extend for many kilometres along the coastlines of the states of California, Oregon, and Washington. A comprehensive research study begun in 2001 documented and monitored the effects of winter storms on several sections of cliff south of San Francisco, California, USA. We present the results of five seasons (2001–2006) of weekly observations of these cliffs, documenting failure occurrences, modes, and mechanisms along a 1.5 km stretch of coast, and correlate these with storm event precipitation totals and storm-induced ocean wave run-up heights. We utilize the results of the correlations to outline thresholds for the likelihood of cliff failure in order to form a process-based, short-term, methodology for landslide prediction. The methodology is generalized for long-term (decadal) predictions of landslide occurrence based on rates of sea-level rise, and possible changes in precipitation.

1 INTRODUCTION

Landslide and erosional failure of weakly lithified coastal cliffs (coastal bluffs or seacliffs) are a common occurrence along many of the world's coastlines. Whether composed of weakly cemented sands, nominally cohesive clays, or weaker bedrock members such as chalk, these lithologies form dynamic geomorphologic landscapes, especially when exposed to a suite of coastal and storm processes. Increasing development along these landscapes has been an inevitability in many parts of the world, including the west coast of the United States. In some locations, such as near the city of San Francisco, California, coastal cliffs and bluffs composed of weakly to moderately cemented sand have undergone continuous residential development for the past 50 years (Fig. 1). In many of these locations, initial coastal setback limits have either not been adequate or were never planned for, and residences and public utilities have been lost to coastal cliff retreat.

Given the continuous force of winter storms along this stretch of coast and climatic predictions of global warming and resulting sea-level rise (CCCC, 2006), cliff failures will be an inevitability for the foreseeable future, as will the associated catastrophic loss of property and potentially lives. There is therefore a need to predict future occurrences of cliff failures given projected sea-level, wave action, and precipitation trends.

Figure 1. Typical weakly lithified coastal cliffs with nearby residential development, Pacifica, California, USA.

In this paper, we develop empirical correlations for predicting cliff failure along a 1.5 km stretch of coastal cliff located in the city of Pacifica, California. While much research has focused on these cliffs due to their high failure rate (Clough et al. 1981, Hampton & Dingler, 1998, Lajoie & Mathieson, 1998, Hampton,

2002, Sallenger et al. 2002, Sitar, 1983, Snell et al. 2002), this study presents the first work aimed at systematically establishing the wave and precipitation scenarios that lead to cliff failure. The methodology relies on five years of direct observations of coastal cliff failures, and pairs these observations with high-resolution records of coastal tides, open ocean wave heights, and precipitation records. The empirical correlations are modelled after existing methodologies for landslide occurrence dependent on cumulative seasonal rainfall (e.g. Nilsen et al. 1976) and rainfall intensity-duration thresholds (e.g. Caine, 1980; Cannon, S.H., 1988). While the effect of precipitation on seepage driven failures in moderately cemented coastal cliffs is handled in much the same way as existing correlations, we introduce a new methodology for identifying these thresholds for the effect of wave action on more weakly cemented cliffs.

2 REGIONAL SETTING

The Pacifica coastline extends for several kilometres along a tectonically active portion of the California coast, south of the city of San Francisco (Fig. 2). Located immediately south of the Mussel Rock splay of the San Andreas Fault zone, the geology of the area consists of elevated marine terraces composed of variably lithified Holocene beach and dune deposits and Pleistocene alluvial deposits. These terrace deposits overlay older, Mesozoic Franciscan bedrock members that are intermittently exposed at beach level and form small headlands between the currently eroding marine terrace deposits. Deposition of the terrace deposits occurred prior to the relative sea level fall in the area approximately 5000 years ago and which resulted in the uplifted, steep exposures of coastal cliff now present (Cleveland, 1975). Today, vertical movement of the San Andreas Fault system continues to uplift this landscape.

While tectonics has played an important part of the genesis of these cliffs, the effects of winter storms far outpace periodic seismic events when considering decadal geomorphologic change. Observations of coastal cliff landslides have been made during both large and small earthquakes in 1906 (Lawson, 1908), 1957 (Bonilla, 1959), 1989 (Plant & Griggs, 1990, Sitar, 1991) and 2002 (Collins & Sitar, 2002), however, there has been far greater recorded crest retreat and landslide frequency as a result of winter storms (Collins, 2004).

The wave and storm climate of the area is typical of the northern Pacific coast of the United States and generally consists of a Mediterranean-type climate. Following an extensive dry period during the summer and autumn months, a very active storm season begins in November with increased swells from the northwest followed typically by a series of large storms that descend from the far-north Pacific (Storlazzi & Wingfield, 2005). The result is a very wet winter and spring with an average of nearly 470 mm of precipitation falling between the beginning of November and the end of May (NWS, 2006). The resulting connection between precipitation and landslides in the San Francisco area has been well documented (e.g. Ellen & Wieczorek, 1988; Godt, 1999) and the Pacifica area is no exception.

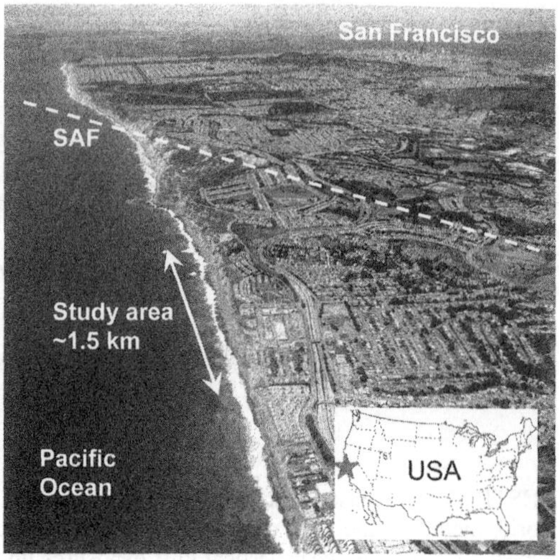

Figure 2. The Pacifica, California study area, located south of the San Andreas Fault zone (SAF).

Winter storms also bring periods of high wave action with offshore significant wave heights reaching up to 7.6 meters compared to an annual mean value of 1.8 meters (NOAA/NDBC, 2006). The effect of these storms is to remove built-up sand from the beaches, by up to two meters in some cases (Collins, 2004). This generally exposes the toe of the coastal cliffs to much more severe and more frequent wave action, especially when coupled with high spring-tides (Fig. 1). Wave contact with the cliffs generally occurs at one of the two high tides per day, although during periods of high storm activity, waves may reach the toe even at low tides.

3 SITE GEOLOGY AND FAILURE MODES

The present study focuses on a 1.5-kilometer length of coast consisting of seven individual cliffs (Fig. 3) reaching an average height of 24 m. Geologic mapping of the area has shown that the northern half of the study area (Cliffs N4, N3, N2, N1 and S1-North) consists of weakly cemented sands while the southern half (Cliffs S1-South and S3) are composed of moderately cemented sands (Collins, 2004). The sand units can be distinguished from one another

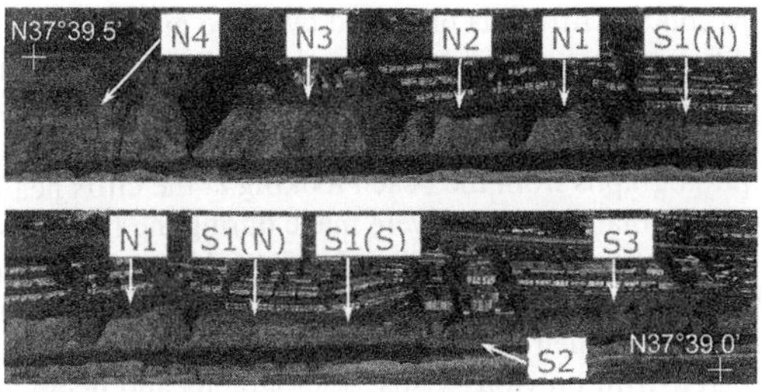

Figure 3. Oblique view of the Pacifica study area showing locations of observed cliffs. Cliff S2 was not observed for failures. Photo courtesy of the California Coastal Records Project.

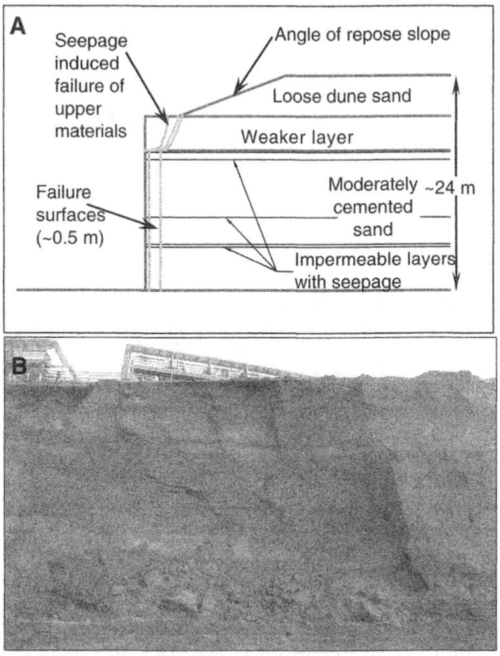

Figure 4. Schematic diagram (A) and photo (B) of weakly cemented coastal cliff failure mode. The failure surface is typically inclined at 65° to the horizontal.

qualitatively by a field test of slope inclination or quantitatively by unconfined compressive strength (UCS). In general, units with slopes of more than 70° are moderately cemented, whereas slopes less than this are weakly cemented. Geotechnical testing indicates that moderately cemented materials exhibit UCS of up to 400 kPa, with weakly cemented units falling in the 5 to 30 kPa range, consistent with categorical divisions for cemented sands developed by Shafii-Rad & Clough (1982).

Our detailed observations have also shown the predominant failure mode (translative shear failure or tensile-induced topples) as well as the failure

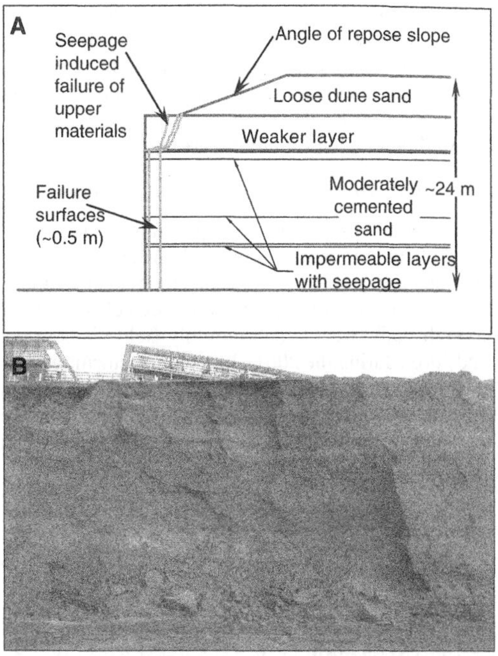

Figure 5. Schematic diagram (A) and photo (B) of moderately cemented coastal cliff failure mode. The failure surface is typically near-vertical.

mechanism (wave action or precipitation induced seepage) are almost wholly dependent on the lithology of the cliff. In the weakly cemented cliffs, the primary mode of failure is through geometrically constrained, translational shear as a result of wave action (Fig. 4). Early season winter storms with long period energy remove sand from the fronting beach and a combination of high tide and wave run-up allows sea levels to contact the cliff toe. Wave erosion leads to the development of vertical cliff profiles up to several meters in height at the toe, which become unstable from the cliff materials low shear strength (Collins & Sitar, 2005).

In the moderately cemented cliffs, the failure mode is through tensile exfoliation of the outer cliff face from precipitation induced seepage and wetting (Fig. 5). In these materials, the tensile strength is reduced by wetting and subsequent saturation and the cliff face, existing in a state of quasi-stable tension, collapses (Collins & Sitar, 2005). Our observations, laboratory testing, and analyses show that a near-complete state of saturation is necessary for failure, and given the low permeability of the more heavily cemented sand units, that saturation requires a period of significant wetting.

4 METHODOLOGY

We combined a program of regimented site specific monitoring for new cliff failures with high frequency tide, offshore wave height, and precipitation data to identify triggering conditions of coastal cliff failures. We then developed empirical relationships based on a general implementation of precipitation and wave climate proxies for determining intensity and duration thresholds for each failure mechanism.

4.1 *Monitoring and observations of failures*

The coastal cliff monitoring program consisted of direct observations and digital, oblique photography of the study area cliffs. Observations for evidence of failures were made of the seven study area cliffs (Fig. 3) through daily, weekly and monthly field visits to the study area during the 2001–2006 winter storm seasons (November to May). Cliff S1 was divided into north and south sections due to its length and the change in material properties that occurs moving from north to south. Cliff S2 is protected by slope grading and a 4.5-meter tall riprap seawall and was not observed for native failure mechanisms.

Observations were made during low tide to take photographs from the beach looking at the cliffs head on. In total, 128 visits were made during the five winter seasons from 2001–02 to 2005–06. Visits consisted of walking the length of the beach in front of the cliffs and along their crest to document any changes compared with previous visits. Photographs and detailed descriptions of failures were noted on each occasion. When a failure was observed, the probable failure mechanism was determined based directly on field observations of the failure mode and the wave and climate conditions that had occurred since the last field visit. Observed failures for each of the five winter seasons investigated are plotted individually in Figure 6.

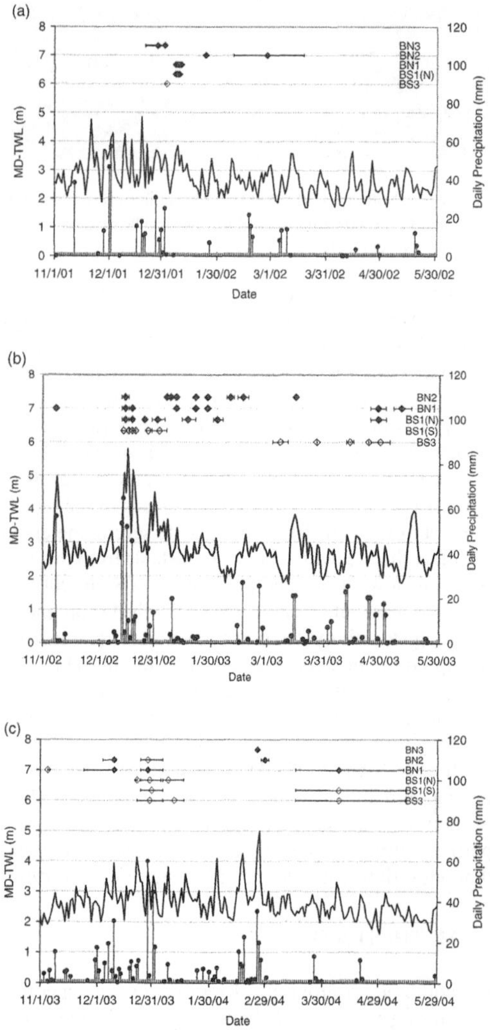

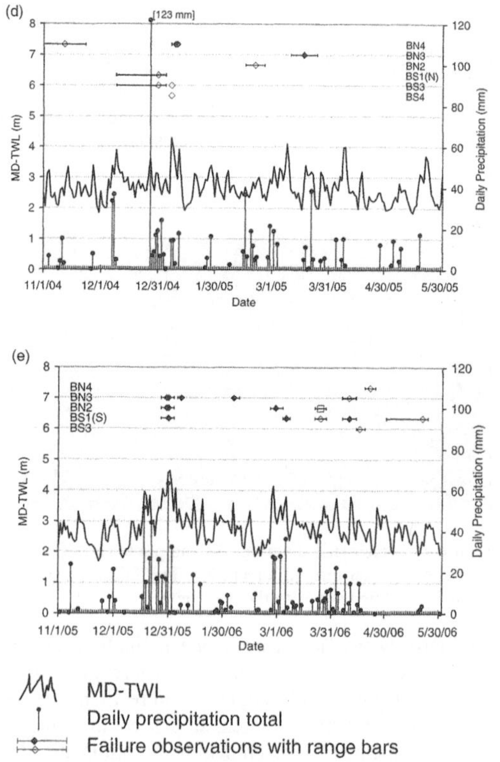

Figure 6. Failure observations, maximum daily total water level, and daily precipitation for the (a) 2001–02, (b) 2002–03, (c) 2003–04, (d) 2004–05, and (e) 2005–06 winter storm seasons. Failure observations with a closed symbol are wave action induced failures; open symbols are precipitation induced failures. Location of symbols within range bars indicate the best approximation of the time of failure within the time period in which failure may have occurred.

While an attempt was made to document the occurrence of all failures, the primary effort was directed to identifying those failures that produced crest retreat. A majority of the cliff dataset is available for viewing at http://eriksson.gisc.berkeley.edu/bluff.

4.2 *Failure event indicators*

Statistical correlations of observed failures were developed through processed-based indicators of wave action and precipitation. To estimate levels of relative wave climate, we utilized the methodology outlined by Komar (1998) for calculating the relative sea-level elevation on the beach from tides and wave run-up. This provided a measure of the maximum wave run-up level during each of the winter seasons when failures were observed.

The methodology sums the recorded tide level with a measure of wave run-up. We utilized local data from the nearest open-ocean tide gauge located in Point Reyes, California, to the north of the city of Pacifica (NOAA/NOS, 2006). This tide level (η) takes into account both astronomical tide changes and storm surge related anomalies. Wave run-up was measured indirectly, through an empirical correlation developed by Ruggiero et al (2001) for similarly dissipative beaches along the Oregon, USA coast, and which links the deep water significant wave height (H_s) with the wave run-up component. Ruggiero et al's formulation predicts the vertical level exceeded by 2% of the run-up heights ($R_{2\%}$), which gives a measure of the maximum exceedence, suitable for investigating wave conditions leading to cliff failure. The complete wave run-up equation defining the total water level (R) is:

$$R = \eta + R_{2\%} \quad (1)$$

where

$$R_{2\%} = 0.5H_s - 0.22 \quad (2)$$

Since both tide and wave height data were available on an hourly basis, we calculate the maximum daily total water level (MD-TWL) as an indication of wave conditions capable of leading to cliff failure. The resulting maximum daily total water level is plotted for each of the five winter seasons in Figure 6 and seasonal maximums and averages are reported in Table 1. Complete details of the specific methodology used in developing these data sets is available in Collins (2004).

In the moderately cemented cliffs, we observed that high levels of precipitation typically occurred immediately prior to a failure event. Therefore, we utilized daily precipitation records as an indicator of failure conditions. Daily precipitation data were obtained from a nearby rain gauge located 4 km from the study area (NWS, 2006) and are plotted for each of the winter seasons in Figure 6. Seasonal cumulative totals and the maximum total 48-hour precipitation events are reported in Table 1.

4.3 *Failure event correlations*

From the data presented in Figure 6, parameter sets were developed that capture the main characteristics of each failure mechanism. For wave action, our understanding of the failure mode indicates that failure is dependent on the occurrence of a high relative sea-level on the beach (R) and a preliminary early season time period of high wave energy. Thus, the MD-TWL data were used directly as an indicator of waves sufficiently strong to make contact and erode the cliff toe: a measure of wave action intensity.

Our observations also indicated that the beach must be at a lowered elevation to allow high tides and strong wave action to make cliff contact. Since the beach elevation is lowered in the early season (November–December), and builds back up in the late season (April–May), we developed a wave interaction duration parameter to act as a proxy for season-wide wave effects. This parameter, the cumulative difference between the MD-TWL and the average MD-TWL ($\Sigma_{\Delta MD-TWL}$) for an individual season is defined as:

$$\Sigma_{\Delta MD-TWL} = \sum_{i=1}^{Nov-May} \left[(MD\ TWL)_i - \left(\frac{\sum_{i=1}^{Nov-May}}{\#days} \right) \right] \quad (3)$$

This parameter provides a trend that begins low or negative, quickly increases to higher positive values in the early-middle portion of the storm season, and then slowly declines in the middle to late season (Fig. 7). In developing this parameter, each season's average and cumulative exceedence is calculated separately since beach and wave climate vary from year to year, and cliff failure conditions are dependent on the individual season climate. The data in Figure 7 show the difference between a winter season with a very active

Table 1. Seasonal comparison of wave climate and precipitation data.

Winter season	TWL season average (m)	Maximum season TWL (m)	Cumulative season precip. (mm)	Maximum 48-hr precip. event (mm)
2001–02	2.7	4.9	416	104
2002–03	2.8	5.8	729	118
2003–04	2.6	5.0	540	68
2004–05	2.7	4.3	728	130
2005–06	2.8	4.6	846	80

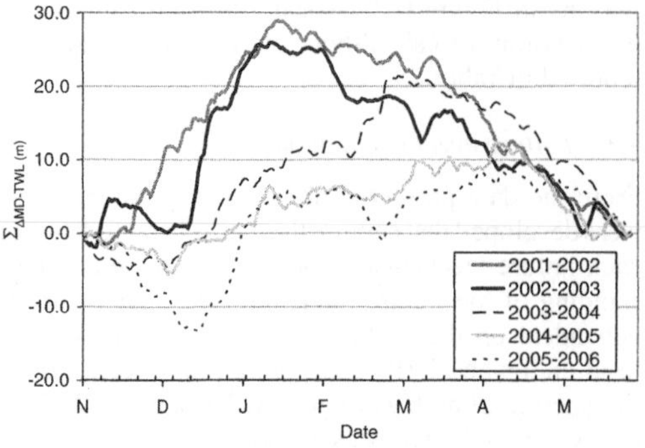

Figure 7. Plot of cumulative difference of maximum daily total water level ($\Sigma_{\Delta MD-TWL}$) from the season average for each of the 2001–2006 winter seasons.

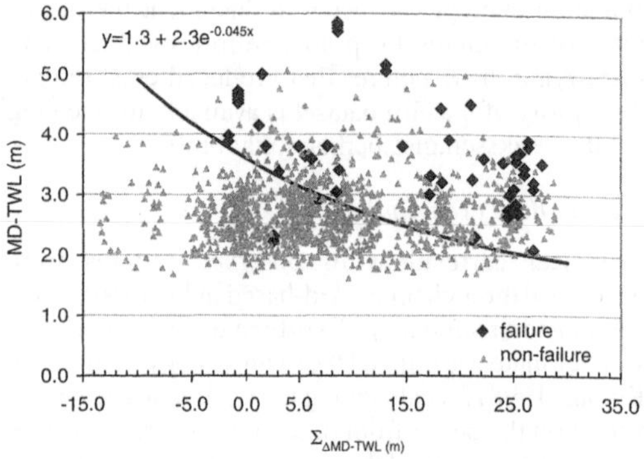

Figure 8. Failure correlation of weakly cemented cliffs with wave action. Failures are predicted to occur in the area above the trend line.

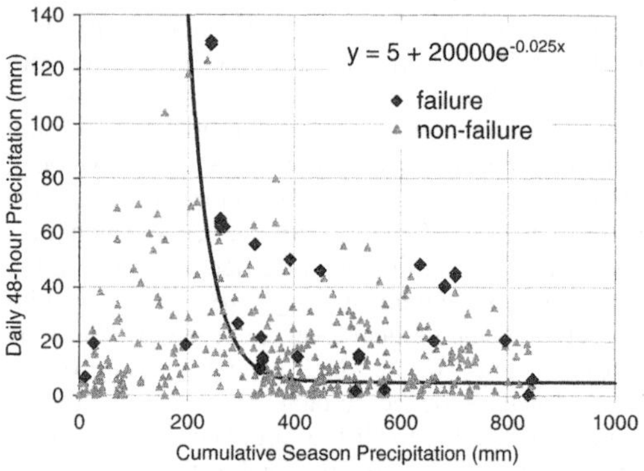

Figure 9. Failure correlation of moderately cemented cliffs with precipitation. Failures are predicted to occur in the area above the trend line.

early to middle season (2002–2003) and a season with a less active early season, and only minimally active mid-season (2004–2005).

Parameter sets developed for identifying proxies for precipitation-induced failures were more straightforward. We utilized the daily, 48-hour precipitation total as a measure of precipitation intensity, summing the daily readings from consecutive 24-hour periods. Likewise, we used the cumulative season precipitation coinciding with each 48-hour total as a measure of season-wide duration.

5 RESULTS

5.1 *Wave action failure mechanism*

Plots of each days MD-TWL and cumulative MD-TWL difference from the season average ($\Sigma_{\Delta MD-TWL}$) for all five storm seasons are shown in Figure 8 along with the associated values for each observed failure outlined in Figure 6. Of note, is the large number of points spread about the mean cumulative difference value of 0.0, indicating that most daily values do not significantly exceed the mean. When the mean is exceeded, it is for a short period of time, as verified by the lower density of points in the upper values (greater than 15 m). It is in this region that many of the days with associated failure events lie.

Due to the observable decrease in the intensity parameter of the failure events with increasing $\Sigma_{\Delta MD-TWL}$, an exponential function was best fit directly to the data to establish a threshold for failure. The equation:

$$y = 1.3 + 2.3e^{-0.045x} \quad (4)$$

is plotted in Figure 8 and captures almost 95% of the failure events, each located above the threshold. The threshold indicates that periods of wave climate exhibiting extremely high MD-TWL fail under any beach condition, but that lower levels of MD-TWL require a sufficiently lowered beach condition, quantified by the $\Sigma_{\Delta MD-TWL}$ parameter.

5.2 *Precipitation failure mechanism*

The daily 48-hour precipitation total, along with the associated cumulative season precipitation for each day with a non-zero 48-hour total is plotted in Figure 9. An exponential equation was fit to the data to delineate the precipitation parameter sets that result in failure from those that do not. The equation of this curve is:

$$y = 5 + 20000e^{-0.025x} \quad (5)$$

where y is the daily 48 hour precipitation total and x is the cumulative season precipitation coinciding with the specified 48-hour total. We selected an exponential function to model the physical reality inherent in the data set. That is, with increasing cumulative precipitation, failures occur with much lower 48-hour precipitation totals although a minimum threshold of 5 to 20 mm is still required. Likewise, the function also captures the behaviour and requirement of cliff saturation for failure. An antecedent rainfall threshold is indicated by the correlation, requiring at least 200 to 250 mm of seasonal rainfall before failures occur. While some failure data points do not fit the model equation, this is expected given the inherent variability and multi-dependent conditions under which coastal cliff failures occur. However the correlation equation does fit 82% of the failure data points above the model threshold.

6 DISCUSSION

The data and results presented in Figures 6–9 allow us to make both short- and long-term predictions of the likelihood of coastal cliff landslides within the study area. Short-term (1 to 5 year) predictions are based on the assumption that the current wave and storm climate, along with the frequency of failures, will follow similar patterns as that observed during the 2001–2006 time period. We base our long-term (decadal) predictions, on the other hand, on the results of state-of-the-art climate model predictions developed specifically for the California coastline.

6.1 *Short-term failure predictions – wave action*

Short-term predictions of the continuing effects of wave action on the weakly cemented cliffs in the Pacifica study area are made using a single-parameter threshold and a dual-parameter, intensity-duration threshold. We find from the data presented in Figure 6 that the average MD-TWL for all wave action induced failures to be 3.7 m, which is significantly above the average MD-TWL of 2.7 m for all seasons, and also well above the 1σ level ($2.7 + 0.6 = 3.3$ m). Therefore, we establish a 3.7 m MD-TWL predictive failure threshold, with a strong likelihood of cliff failure when the effects of tide and wave action reach this level.

We move one step further by utilizing the results presented in Figure 8, which show that given a two-parameter estimate of measures of wave action intensity and early season beach erosion duration, that failures are likely when a specific day's parameters plot above the curve defined by Equation 4. For the five-season time period from 2001–2006, 434 MD-TWL values exceeded this threshold (Eq. 4) out of a possible 1061 days or approximately 40.9% of the time. Likewise, failures occurred on 52 of these days, which represents 12.0% of the exceedence days or 4.9% of the entire season. We therefore restate these values as predictions: wave action induced failures occur in the study area approximately 5% of the time during the winter storm season, and 12% of the time in which the MD-TWL exceeds the Equation 4 threshold.

6.2 *Short-term failure predictions – Precipitation*

Short-term predictions of failure of the moderately cemented cliffs from a precipitation mechanism were developed following the same methodology as for the wave action failure mechanism in the weakly cemented cliffs. Here, we used the daily total precipitation to form a single parameter prediction of failure conditions. From Figure 6, the average 48-hour precipitation event total for all failures resulting from precipitation is 36 mm. This is roughly double the average precipitation event of 15 mm for all five seasons and just above the 1σ level ($15 + 18 = 33$ mm). We therefore define a 40 mm precipitation event as a threshold level for precipitation induced failures in the moderately cemented cliffs.

The results from Figure 9 allow development of a two-parameter prediction based on the proxies of rainfall intensity and duration as measured by the 48-hour cumulative event total and the coinciding seasonal cumulative total. Here, we define a high likelihood of failure when these parameters plot above the threshold defined by Equation 5. Note that values for the season only represent those days with a 48-hour precipitation value, days without rain are not plotted. During the five seasons investigated, there were 181 days out of a total of 444 days, with precipitation parameters plotting above the threshold. Of these, 27 resulted in failure. Thus, the possibility of failure occurred 40.8% of the time during the season's precipitation days, and failure occurred 6.1% of this time. Of the occurrences plotting above the threshold, failure occurred on 14.9% of these days. We therefore state a short-term prediction of these precipitation induced failures as occurring on roughly 6% of the winter season days with a non-zero 48 hour precipitation value, and approximately 15% of the time that the precipitation parameters plot above the failure threshold defined by Equation 5.

6.3 *Implications of global warming on climate in California*

Long-term predictions of continued cliff failure along the California coast depend on projected climate conditions during the next 100 years. We utilized several recently published reports by the California Climate Change Centre (CCCC, 2006a, b, c) which was tasked by the California state government to perform and publish research on the implications of global warming on California's climate. The CCCC used three state-of-the-art global climate models to provide estimates of climate effects for three greenhouse gas emission scenarios. The greenhouse gas scenarios were: B1 – low emissions with high economic growth and stable population but less dependence on fossil fuels,

A2 – medium-high emissions with uneven economic growth and continuous population growth with triple the CO_2 concentration relative to pre-industrial levels, and A1fi – high emissions with high fossil fuel dependent economic growth, a peaking – then declining population, and with more than triple the CO_2 concentration compared to pre-industrial levels (CCCC, 2006a).

Our long-term predictions for coastal cliff landsliding are based on the reported results from this body of research and used their estimates of sea-level rise, increased activity in oceanic wave climate, and new precipitation scenarios to model these effects on the observed failure mechanisms of the Pacifica study area cliffs.

6.4 *Long-term failure predictions – wave action*

Predictions of the long-term effect of climate change on coastal cliffs affected by wave action were investigated by incorporating results from the CCCC reports on relative changes in future sea level that may occur from greenhouse gas emission driven global warming (CCCC, 2006b). We incorporated only one of the major findings of the report detailing the probable local rise of sea level along the California coast. While the CCCC report also makes clear predictions of a more active wave climate in the northern Pacific Ocean as a result of global warming and an increase in extreme wave heights from a positive shift of the El Niño Southern Oscillation (ENSO) index, we did not incorporate those finding here. However, we do point out that an increase in wave action and extreme sea level events will unquestionably result in a greater quantity and higher frequency of cliff failures.

During the next 70 to 100 years, the CCCC model simulations project that sea level will rise, relative to the year 2000 sea level, by 11–54 cm using the B1 greenhouse gas emission scenario, by 14–61 cm using the A2 scenario, and by 17–72 cm using the A1fi scenario (CCCC, 2006b). For comparison, the report shows that the historic sea-level rise for the past century has been 19.3 cm/century in the San Francisco, California region. Thus, the average and higher bounds of the simulated ranges are a significant increase over recent historic levels.

The effect of a rising sea level trend on coastal cliff landslides was investigated by directly imposing a uniform sea level trend to the 2001–2006 winter season data sets and calculating the number of days that would result in cliff failure. We calculated this value assuming that the ratio between the number of failures and the number of points exceeding the failure threshold remains constant at 12% based on the 2001–2006 data:

$$\frac{\#_{days\ with\ failures}}{\#_{days\ exceeding\ threshold}} = \text{constant} = 12\% \quad (6)$$

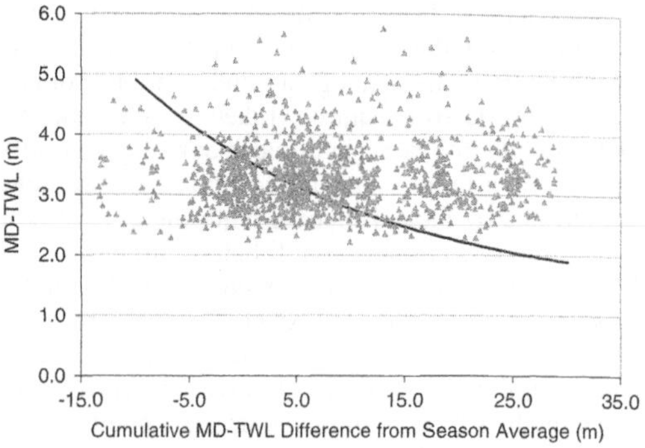

Figure 10. Prediction of wave parameters for a 5 year time period in 100 years with sea-level rise trend of 60 cm imposed on 2001–2006 season data set. Equation 4 failure threshold is plotted for comparison.

Table 2. Prediction of failure events for a 5-year period in 100 years based only on sea level trend.

Sea level trend (m/100yr)	# Points exceeding threshold	# Days of predicted failure	% increase of failure	Global emissions scenario
0	434	52	0	sea level decline
0.2	506	61	17	present rate
0.4	594	71	37	A2 average
0.6	675	81	56	A2 maximum
0.72	719	86	65	A1fi maximum

This assumption takes into account the inherent variability of the timing and occurrence of events while also allowing future projected wave action parameters to be evaluated using the established threshold (Eq. 4). As an example, for a projected sea level rise trend of 60 cm in Figure 10, the wave parameters are shifted upwards and to the right compared with Figure 8. The Equation 4 threshold is exceeded on 675 occasions and the number of days where failure is likely to occur increases to approximately 30 additional events from 52 to 81 events. (Table 2). Predictions of failure events for various additional trends in sea level coinciding with the CCCC greenhouse gas scenarios are outlined in Table 2. With an average level of cliff retreat during a single failure event on the order of 0.5 to 1.5 meters (Collins, 2004), we can make clear predictions that the failure rate and cliff retreat rate will increase in the future given any trend of sea level rise.

6.5 *Long-term failure predictions – precipitation*

According to the CCCC report on Climate Scenarios for California (CCCC, 2006c) the climate model scenarios do not predict any clear trend for the long-term effects of emissions-based climate change on precipitation. In general, the report states that California is likely to maintain its present Mediterranean-style climate, with winter storm seasons bringing the majority of the annual precipitation. The report states that little change in annual precipitation totals are predicted, with only slightly wetter winters and reciprocal drier spring conditions.

The effect of this scenario on the long-term failure rate of the Pacifica coastal cliffs is therefore evident. Based on the CCCC's reports conclusions, we do not expect precipitation-induced failures to either increase or decrease during the next 100 years.

It is potentially important, however, to understand how our predictive formulation would behave under wetter or dryer precipitation scenarios. Using Figure 9 as a baseline, under a wetter climate scenario, the cumulative season precipitation level will increase, which will locate more of the season-wide data points to the right within the figure. In this case, the delineating trend line for failure/no-failure remains in the same position, and the likelihood of more days being located above the threshold will increase. With higher single day precipitation totals, the effect will be similar, with the parameters for more days moving upwards in the figure, and likewise a higher chance for failure with more days plotted above the failure threshold. Under a dryer climate scenario, the reverse will be true for each of these parameters: the data points will move down and to the left in Figure 9, and there will be less likelihood for failure from this mechanism.

7 CONCLUSIONS

Coastal cliff failures in weakly lithified sediments are an inevitability in most coastal settings around the world and we have shown that the cliffs located south of San Francisco, California are no exception. Detailed observations and wave and precipitation data sets collected over a five season time period from 2001–2006 have shown that empirical correlations that predict distinct failure thresholds can be identified (Eqs. 4–5). We identified these thresholds for the two predominant failure modes that occur in the study area: wave action induced translative shear failure in weakly cemented sand cliffs and precipitation driven seepage and tensile failure in moderately cemented sand cliffs. The thresholds were calculated as exponential functions, consistent with a processed based understanding of higher failure thresholds when intensity parameters are high and duration parameters are low, and lower failure thresholds with the reverse case.

The correlation equations also provide starting points for carrying out forecasting analyses aimed at understanding the impact of future trends in sea-level, wave action, and precipitation events. Given the current understanding of the effect of increased greenhouse gas emissions to global warming and state-of-the-art climate model predictions, we project an increase in wave action driven failures of the weakly cemented cliffs, potentially reaching over 165% of the current number of failures per year. This represents an increase from 52 failures per 5-year period measured during the 2001–2006 winter seasons to 86 failures per 5-year period for a sea level trend consistent with a high greenhouse gas emissions (A1fi) scenario. Obviously, other scenarios are also possible. On the other hand, we predict a constant level of precipitation-induced failures in the moderately cemented cliffs of the Pacifica study area consistent with current and projected levels of precipitation.

Future work in this area will be aimed at incorporating the effect of sea level extremes and wave height anomalies into our future predictions of wave driven failures in the weakly cemented cliffs in the study area. We also intend to model wave action more rigorously by incorporating the effects of short and long period waves on either removing or accreting sand to the beach – an issue tied directly into the proper modelling of the failure threshold duration parameter.

ACKNOWLEDGEMENTS

Funding for this research was provided by the U.S. Geological Survey, Mendenhall Postdoctoral Research Program, the U.S. Geological Survey, Western Coastal and Marine Geology Program, and the University of California, Coastal Environmental Quality Initiative. Pamela Patrick and Ann Marie Puzio, former graduate students at the University of California, Berkeley were responsible for several years of the data collection phase of this work and their diligence with site visits and observations are gratefully acknowledged. The authors wish to thank Monte Hampton for valuable insights and observations of the study area and Jonathan Allen at Oregon DOGAMI for helpful suggestions to the wave run-up analyses. The photo in Figure 2 is used with permission from Cotton, Shires and Associates and the photos in Figure 3 are used with permission of the California Coastal Records Project.

REFERENCES

Bonilla, M.G., 1959, Geologic observation in the epicentral area of the San Francisco earthquake of March 22, 1957., G.B. Oakenshott, ed., *Spec. Rep. 57, California Division of Mines*, pp25–37.

Caine, N., 1980, The rainfall intensity-duration control of shallow landslides and debris flows: *Geografiska Annaler*, Vol. 62A, pp23–27.

Cannon, S.H., 1988, Regional rainfall-threshold conditions for abundant debris-flow activity, *in* Ellen, S.D. and Wieczorek, G.F., eds., Landslides, floods, and marine effects of the storm of January 3-5, 1982, in the San Francisco Bay region, California: *U.S. Geological Survey Professional Paper No. 1434*, pp35–42.

CCCC, 2006a, Our Changing Climate: Assessing the Risks to California, *California Climate Change Center, General Publication, Report No. CEC-500-2006-077*, Prepared by Luers, A.L., Cayan, D., Franco, G., Hanemann, M., and Croes, B. http://www.climatechange.ca.gov/ biennial_reports/ 2006report/index.html, 64p.

CCCC, 2006b, Projecting Future Sea Level, *California Climate Change Center White Paper, Report No. CEC-500-2005-202-SF*, Prepared by Cayan, D., Bromirski, P, Hayhoe, K, Tyree, M, Dettinger, M, and Flick, R., http://www.climatechange.ca.gov/bienial_reports/ 2006 report/index.html, 64p.

CCCC, 2006c, Climate Scenarios for California, California Climate Change Center White Paper, Report No. CEC-500-2005-203-SF, Prepared by Cayan, D., Maurer, E., Dettinger, M., Tyree, M., Hayhoe, K., Bonfils, C, Duffy, P., and Santer, B., http://www.climatechange.ca.gov/biennial_reports/2006report/index.html, 64p.

Cleveland, G.B., 1975, Landsliding in Marine Terrace Terrain, California: *Special Report 119 of the California Division of Mines and Geology*, 24p.

Clough, G.W., Sitar, N., Bachus, R.C. & Shaffii-Rad, N., 1981, Cemented sands under static loading, *ASCE Journal of Geotechnical Engineering*, Vol. 107(GT6), June 1981, pp. 799–817.

Collins, B.D., 2004. *Failure Mechanics of Weakly Lithified Sand Coastal Bluff Deposits*: Doctoral dissertation, University of California, Berkeley, 278p.

Collins, B.D. & Sitar, N., 2002. Geotechnical Observations of Recent Coastal Bluff Failures, Pacifica, California, *Geological Society of America – 98th Cordilleran Section Annual Meeting, Corvallis, Oregon, May 13–15, 2002.*

Collins, B.D. & Sitar, N., 2005. Failure mode identification and hazard quantification for coastal bluff landslides, *Proceedings 2005 International Conference on Landslide Risk Management*, Eds. O. Hungr, R. Fell, R. Couture, & E. Eberhardt, Vancouver, BC, Canada, June 2005, pp. 487–496.

Ellen, S.D. & Wieczorek, G.F., 1988. Landslides, floods, and marine effects of the storm of January 3–5, 1982 in the San Francisco Bay region. *U.S. Geological Survey Professional Paper 1434*: 310p.

Godt, J.W., 1999. Maps showing locations of damaging landslides caused by El Niño rainstorms, winter season 1997–98, San Francisco Bay region, California, J.W. Godt, ed., *U.S. Geological Survey Miscellaneous Field Studies Maps MF-2325-A-J.*

Hampton, M., 2002, Gravitational failure of sea cliffs in weakly lithified sediment: *Environmental and Engineering Geoscience*, Vol. 8(3), pp175–192.

Hampton, M. & Dingler, J., 1998, Short term evolution of three coastal cliffs in San Mateo County, California: *Shore and Beach*, Vol. 66(4), pp. 24–30.

Komar, P.D. (1998). Beach Processes and Sedimentation, Prentice Hall, New Jersey.

Lajoie, K.R. & Mathieson, S.A., 1998, 1982–83 El Niño Coastal Erosion, San Mateo County, CA: *U.S. Geologic Survey Open-File Report 98-041*, 61p.

Lawson, A.C. (1908). The California earthquake of April 18, 1906, *Publication No. 87, Report of the State Earthquake Investigation Commission*. Vol. 1, 451p.

Nilsen, T.H., Taylor, F.A. & Dean, R.M., 1976, Natural conditions that control landsliding in the San Francisco Bay region – an analysis based on data from the 1968–69 and 1972–73 rainy seasons: *U.S. Geological Survey Bulletin 1424*, 35p.

NOAA/NDBC, 2006, National Oceanic and Atmospheric Association, National Data Buoy Center (online), <http://www.ndbc.noaa.gov/>, Site #46026, San Francisco, California.

NOAA/NOS, 2006, National Oceanic and Atmospheric Association, National Ocean Service Observed Water Levels and Associated Ancillary Data (online), http://co-ops.nos.noaa.gov/data res.html, Site #9415020, Point Reyes, California.

NWS, 2006, National Weather Service Forecast Office, San Francisco Bay Area/Monterey (online), <http://www.wrh.noaa.gov/Monterey/rtp02/>, Site # Pacifica 2S, Pacifica, California (PCAC1).

Plant, N. & Griggs, G.B., 1990, Coastal landslides caused by the October 17, 1989 earthquake, *California Geology*, Vol.43(4), p75–84.

Ruggiero, P., Komar, P. D., McDougal, W. G., Marra, J. J. & Beach, R. A. (2001). "Wave Runup, Extreme Water Levels and the Erosion of Properties Backing Beaches." *Journal of Coastal Research*, 17(2), 407–419.

Sallenger, A.H., Krabill, W., Brock, J., Swift, R., Manizade, S. & Stockdon, H., 2002, Sea-cliff erosion as a function of beach changes and extreme wave run-up during the 1997-1998 El Niño: *Marine Geology*, Vol. 187, pp. 279–297.

Shafii-Rad, N. & Clough, G.W. 1982, The Influence of Cementation on the Static and Dynamic Behavior of Sands: *John A. Blume Earthquake Engineering Center, Report No. 59, Stanford University, California, December 1982*, 315p.

Sitar, N., 1983. Slope stability in coarse sediments, *ASCE Special Pub.: Geologic Environment and Soil Properties*, R.N. Yong, ed., pp. 82–98.

Sitar, N., 1991, Earthquake-induced landslides in coastal bluffs and marine terrace deposits, *Spec. Pub. No. 1, Loma Prieta Earthquake, Ass. of Eng. Geology*, pp67–82.

Snell, C.B., Lajoie, K.R. & Medley, E.W., 2000, Sea-Cliff Erosion at Pacifica, California Caused by the 1997/98 El Niño Storms: *Slope Stability 2000, ASCE Geot. Spec. Pub. No. 101, Proc. of Geo-Denver, Denver, Colorado*, August 5–8, 2000, pp. 294–308.

Storlazzi, C.D. & Wingfield, D.K, 2005, Spatial and temporal variations in oceanographic and meteorologic forcing along Central California:1980–2002: *U.S. Geological Survey Scientific Investigations Report 2005-5085*, 30 pp.

Landslides and Climate Change – McInnes, Jakeways, Fairbank & Mathie (eds)
© 2007 Taylor & Francis Group, London, ISBN 978-0-415-44318-0

A comparison between analytic approaches to model rainfall-induced development of shallow landslides in the central Apennine of Italy

D. Salciarini & P. Conversini
Department of Civil and Environmental Engineering, University of Perugia, Italy

ABSTRACT: Several predictive approaches have been proposed in the scientific literature to assess landslide susceptibility. To model the rainfall-induced development of shallow landslides in a study area in the eastern Umbria Region of central Italy, we compare the results from three different approaches. The TRIGRS (Transient Rainfall Infiltration and Grid-based Slope-stability) model couples an infinite-slope stability analysis with a one-dimensional analytical solution for transient pore-pressure response to rainfall infiltration. The SHALSTAB code is a widely used model that couples an infinite-slope stability analysis with a steady-state hydrological model. The TM model is an approach based only on the topographic information. We show that the last two methods produce a greater over-prediction of the unstable hillslopes in the study area compared to TRIGRS results. Differences in the distribution of predicted slope failures between the models are primarily a function of the spatially varying soil depth and spatially distributed mechanical and hydrological properties that can be considered in TRIGRS simulations. Next, we show that TRIGRS is able to account for the influence of expected rainfall duration and depth on shallow landslides initiation. We model the climate-depending instabilities on a scenario basis, using precipitation-duration-frequency curves for the expected rainfalls.

1 INTRODUCTION

Shallow landslides and debris flows are natural processes that have the potential to bring about damage, loss or other adverse effects to the built environment (Crozier & Glade 2005). Such phenomena are frequent on steep slopes mantled with loose soils. Typically, intense and prolonged rainfall is the major cause of the reduction of the soil strength and consequent slope failure. In the central Apennines of Italy, four surficial environments have been identified as prone to shallow landsliding: landslide deposits, highly fractured rocks, scree or talus deposits, and glacial deposits (Guzzetti & Cardinali 1991).

Landslide susceptibility is defined as the propensity of a certain area to undergo landsliding (Crozier & Glade 2005). To assess landslide susceptibility, several predictive approaches have been proposed in the scientific literature, such as: statistical models based on logistic regression techniques; physically based, spatially distributed models that are based on the physics of landslide processes; and "topographical" models based on the definition of a topographic index that depends on the slope gradient. Statistical models rely on the correlation between landslide locations and various geological, gemorphological, and land-use characteristics to quantify landslide susceptibility (e.g. Coe et al. 2004). Coupled physical models can be used to examine shallow landslide and debris flow occurrence at the scale of a river basin or region and as a function of physical properties of the soil mantle and climate (i.e. rainfall) (e.g. Montgomery & Dietrich 1994, Iverson 2000, Baum et al. 2002, Savage et al. 2003). These models allow the quantification of the stability of each unit (grid cell or topographic element in which the study area is subdivided). Two important codes that implement physically based theories are SHALSTAB (Dietrich & Montgomery 1998) and TRIGRS (Baum et al. 2002). Topographical Models (TM) are subset of the statistical models and partition the landscape into susceptibility categories based only on topographic attributes (e.g. Salciarini et al. 2006b).

In the first analysis, this paper discusses the results obtained from the SHALSTAB code, TRIGRS code and TM method for slope stability modelling, by applying each model to a study area in central Italy and comparing the results to the landslide data from the Landslide Inventory Map of Umbria Region (Guzzetti & Cardinali 1989, 1990). Because TRIGRS implements a transient hydrological solution, it is the only model suitable for evaluating stability conditions as function of time and depth. SHALSTAB (Montgomery & Dietrich 1994, Dietrich & Montgomery 1998) is a quantitative time-independent approach that has a strong dependence of shallow landslide and debris flow initiation on the topography.

Figure 1. Map showing the location of the study area.

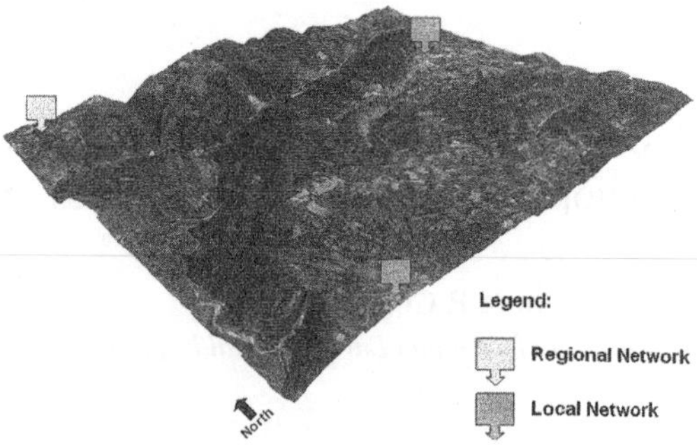

Figure 2. Perspective view showing the location of the rain gauges within the study area.

Unlike TRIGRS (Baum et al. 2002, Savage et al. 2003), it cannot represent the spatial variation and distribution of the soil physical parameters. Both of the approaches assume an understanding of the physical processes that govern the phenomenon initiation. Finally, Topographical Models are very easy to use, but much less accurate and can be subject to bias in the interpretation of the results.

In the second analysis, we show how TRIGRS may be used to assess regional shallow landslide instabilities that depend on climate. We derived the climatic factors for the study area from the probabilistic distribution of historic rainfalls and we used them for modelling the influence of different rainfall scenarios on shallow landslide initiation.

2 STUDY AREA

The study area is located in the Apennine Mountains, in the eastern part of the Umbria region of central Italy (Fig. 1). Developed areas on alluvial fans and transportation corridors that cross the toes of unstable slopes have been periodically exposed to hazards from shallow landslides and debris flows.

In the study area shallow landslides and debris flow events commonly occur in small basins (smaller than 3 km^2), in a hilly environment (with an average elevation of 800 m above sea level). Average slope gradient, within the watersheds varies between 25° and 30° and fan deposits of mobilized sediment can vary from 3,000 to 16,000 cubic meters (Conversini et al. 2005; Salciarini et al. 2006a). Shallow landslides that quickly develop into debris flows exclusively involve colluvium and talus underlain by calcareous or marl bedrock. The initiation is mostly due to the rapid infiltration through the permeable soil mantle. The rapid seepage is then detained from the lower less permeable substrate, and this causes a transient rising water table (Salciarini et al. 2006a).

In the central Apennines the bedrock is a marl and limestone sequence characterized by an increasing marl fraction in the upper part of the series (Calamita & Deiana 1986). Erosion and weathering products from these sedimentary formations mantle slopes in the study area. These surficial deposits typically have varied sorting and thicknesses. Vegetation on the hillslopes is typically dense deciduous forest.

The climate of the Umbria Region is typical of the central Apennine, characterized by hot dry summers and a mild winter. It is mainly derived from the interaction between the mountain chain and local atmospheric circulation (Corradini & Melone 1988). Within the study area there is one rain gauge at Forsivo, 818 m above sea level, that is managed by the Regional Hydrological Service and hourly rainfall information is available from 1992 to the present (Fig. 2). In addition, there are two rain gauges that belong to the local network: at Norcia (450 m above the sea level) and Poggiodomo (606 m above sea level) (Fig. 2). Average annual rainfall ranges from 830 mm to 944 mm. The heaviest rainfall, typically caused by frontal systems from the northwest, occurs in November and April, while the minimum is in July. The average monthly rainfall varies between about 70 and 80 mm and the average of the maximum daily rainfall varies between about 23 and 28 mm.

The main triggering factors for shallow landslides and consequent debris-flow occurrence are short period storms (varying from less than an hour up to four hours) of very intense rainfall over limited areas (Salciarini et al. 2006a). Analyses of previous events reveals that daily cumulative rainfall that triggers shallow landslides and consequent debris flows frequently exceeds average peak daily values (>28 mm/day) (Salciarini et al. 2006b). A detailed historical record of the relation between rainfall, shallow landslides, and debris flows is still not available for the study area.

2.1 *Data collection*

The grid-based digital elevation model with a 5-m cell size used as input to SHALSTAB and TRIGRS was derived from digital topographic contour maps at scale 1:5000, using the Topogrid function in ArcGis.

As rainfall test, we consider a real rainstorm recorded at the Forsivo rain gauge (the National Network rain gauge showed in Fig. 2) characterized by a significant concentration of rainfall over a small area (the Poggiodomo rain gauge, about 10 km away, recorded no rainfall). During this storm about 41 mm of rain fell in 16 hours and caused the initiation of both debris flows and shallow landslides.

As test data set we use a landslide inventory map produced by the Italian National Research Council (CNR), based on aerial photo interpretation and field surveys. In this map, the source areas of shallow landslides and debris flows are displayed with a circle.

3 THEORETICAL BASIS OF THE MODELS

3.1 *The SHALSTAB code*

The SHALSTAB approach (Dietrich & Montgomery 1998) combines an infinite slope stability model with a steady state hydrological model. The modelled basin is divided into topographic elements defined by the intersection of contours and flow tube boundaries, as sketched in Figure 3.

The hydrological model reduces to a calculation of wetness W, expressed as:

$$W = \frac{I \cdot A}{b \cdot T \cdot \sin\alpha} \tag{1}$$

where I is the net rainfall rate, A the upslope drainage area, b the outflow boundary length, T is soil transmisivity which is depth integrated saturated hydraulic conductivity, and α the local slope.

Combining this hydrological model with an infinite slope stability model for soil with cohesion, c, the stability equation implemented in the code is the following (Montgomery et al. 1998):

$$\frac{I}{T} = \frac{\sin\alpha}{(A/b)}\left[\frac{c}{\rho_w g z \cos^2\alpha \tan\varphi}\right] + \frac{\gamma_s}{\gamma_w}\left(1 - \frac{\tan\alpha}{\tan\varphi}\right) \tag{2}$$

where z is the soil depth, g is gravitational acceleration and γ_w is the bulk density of water. The model has three topographic terms that are derived from the digital elevation model: drainage area, A, outflow boundary length, b, and hillslope angle, α. The material properties that need to be assigned to apply this

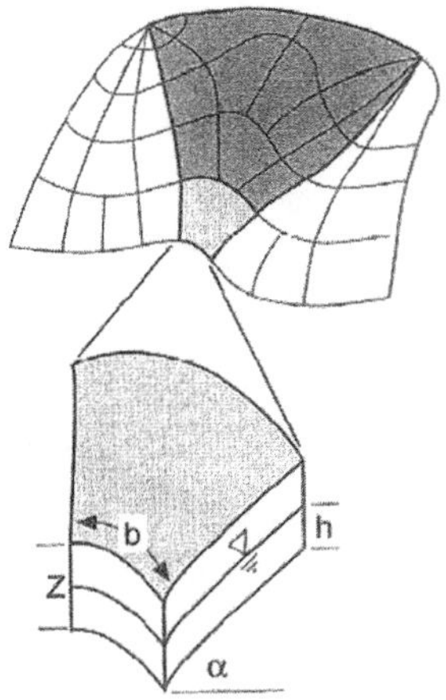

Figure 3. Sketch of the catchments subdivision into topographic elements (after Montgomery & Dietrich 1994).

model are: the soil bulk density, γ_s, the angle of internal friction of the soil, ϕ, the effective soil cohesion, c, and the soil transmissivity, T (defined as K_{sat}/soil-depth).

SHALSTAB produces a stability index (SI) based on the safety factor calculations, ranging from 0 to 1.5. This stability index is defined as the probability that a location is stable assuming uniform distribution of the soil parameters over their range of values (Morissey et al. 2001). Based on the stability index value the territory is classified into 6 categories. The first three categories (1–3) are for regions that should not fail with the most conservative parameters within a specified range. For classes 4 and 5 the probability of failure is less than and greater than 50% respectively. The sixth class is for regions defined as "unconditionally unstable" that should fail even with the less conservative parameters in the specified range.

SHALSTAB can be used in a "deterministic" manner assuming that the lower boundary and the upper boundary of the parameters variability range are identical. It's important to notice that a single set of parameters can only allow us to identify areas with equal topographic control on landslide initiation.

3.2 *The TRIGRS code*

TRIGRS (Transient Rainfall Infiltration and Grid-based Regional Slope-stability) calculates transient pore pressure response and the attendant changes in the safety factor, due to rainfall infiltration over digital topography. The code extends Iverson's method (Iverson 2000, Baum et al. 2002) by implementing the solution for complex storms, a solution for an impervious lower boundary at finite depth, and a simple runoff-routing scheme. Infiltration, hydraulic properties, and slope stability input parameters are allowed

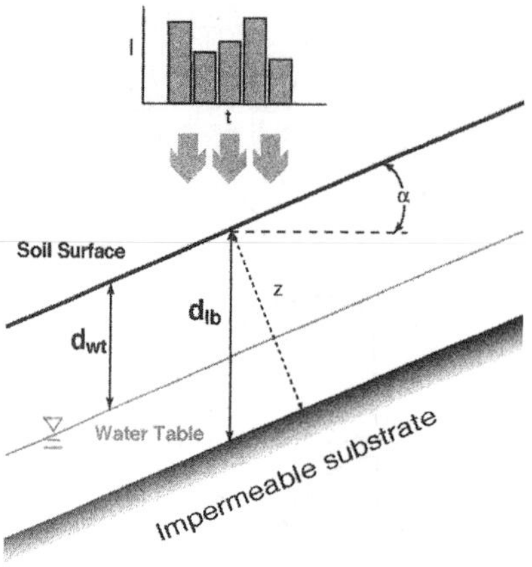

Figure 4. Conceptual sketch of the hydrological model in TRIGRS (after Godt 2004).

to vary over the grid areas thus making possible to analyse complex storm sequences over geologically complex terrain.

Slope stability is calculated using an infinite-slope model. The factor of safety, *FS*, is defined as the ratio of the resisting and driving forces and is calculated at a depth *Z* by:

$$FS = \frac{\tan\phi'}{\tan\alpha} + \frac{c' - \psi(Z,t)\gamma_W \tan\phi'}{\gamma_S d_{lb} \sin\alpha\cos\alpha} \tag{3}$$

where ϕ' is the soil friction angle for effective stress, c' is the effective cohesion, ψ is the pressure head as a function of depth, Z, and time, t, d_{lb} is the depth of the impervious lower boundary, and γ_w, and γ_s are the unit weights of water and soil, respectively. The infinite slope is stable when $FS > 1$, in a state of limiting equilibrium when $FS = 1$ and $FS < 1$ denotes unstable conditions. Thus the depth Z where FS first reaches one will be the depth of landslide triggering at time t.

The hydrological model implemented in TRIGRS is based on solutions to a linearised form of the Richards equation (Iverson 2000, Baum et al. 2002, Savage et al. 2003;). This solution is appropriate for initial conditions where the hillslope is saturated, tension saturated, or nearly saturated. Figure 4 is a conceptual sketch showing a hillslope inclined at an angle α subject to time-varying surface infiltration, I, with a tension-saturated zone above a water table at a depth, d_{wt}, vertically below the ground surface.

The water table overlies an impermeable boundary at a depth, d_{lb}, below the ground surface. Figure 4 also shows the slope-normal coordinate z, and $Z = z/cosa$ is defined as the vertical coordinate.

Table 1. Parameters used for SHALSTAB.CO application.

c (Pa)	ϕ (°)	γ_s (N/m^3)	K_{sat} (m/s)
5,000	32	18,000	1×10^{-4}

3.3 *The topographical model*

To apply the topographic model and obtain slope categories for the study area, the topographic slope at each inventoried shallow landslide location is determined using the digital elevation model. Next, a cumulative distribution of slope angles at these locations is constructed. Finally, we determine the range of slope angles that include a given percentage of shallow landslides in the inventory (i.e. 40%, 50%, 60%, etc.). In this way, it is possible to compute the percentage of the map area that falls within those ranges of slope angles.

4 APPLICATION OF THE MODELS

4.1 *Application of SHALSTAB*

The study area was divided into topographical elements based on the 5-m digital elevation model. We used an extended version of SHALSTAB, called SHALSTAB.CO, which takes into account the effects of soil and root cohesion on the soil strength (Montgomery et al. 1998). Therefore, besides the digital surface information, the other data required for the calculation are: soil cohesion, internal friction angle, transmissivity and soil weight. For the purpose of the comparison between deterministic approaches, we elect to use the parameters with their exact value, shown in Table 1, rather than with their probabilistic variability. The soil thickness is assumed to be uniform with a value of 2 m.

In Figure 5 the result provided by SHALSTAB.CO is shown.

The response of the hillslope to the test rainfall, recorded at the Forsivo rain gauge, is a widespread failure for a large part of the region. Note that, being SHALSTAB.CO a steady-state model, the rainfall is fed into the code as a steady-state rainfall of mean intensity equal to 2.64 mm/h.

4.2 *Application of TRIGRS*

The study area was divided into a regular 5-m grid based on the digital elevation model. The parameters used for the analysis are shown in Table 2 and are the same of the SHALSTAB.CO simulation.

In addition, for the TRIGRS analysis we assumed that the soil thickness, d_{lb}, varies exponentially as function of the slope angle α by the following relation (Salciarini et al. 2006b):

$$d_{lb} = 14\,exp(-0.0693 \cdot \alpha) \tag{4}$$

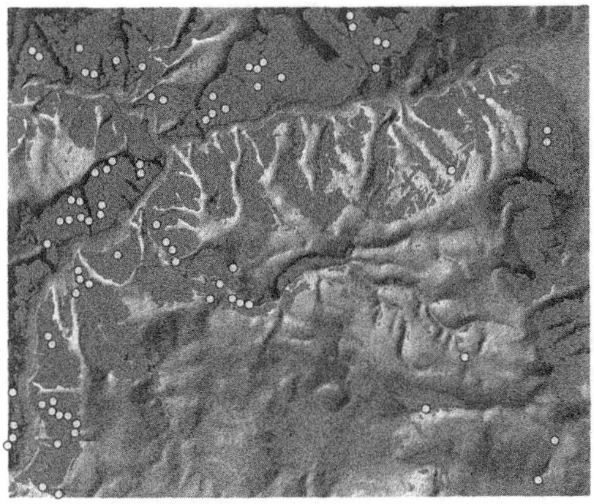

Figure 5. Failure prediction provided by SHALSTAB.CO application. White circles indicate the location of shallow landslides (Guzzetti & Cardinali, 1989).

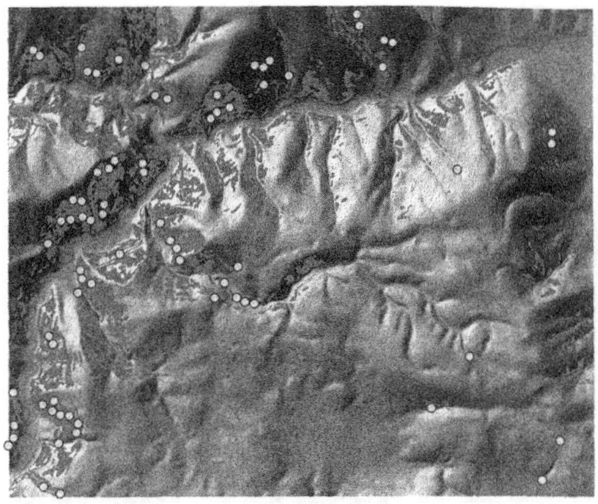

Figure 6. Potential instability prediction from TRIGRS, assuming a uniform distribution of the soil physical parameters for the entire study area. White circles indicate the location of shallow landslides (Guzzetti & Cardinali, 1989).

Table 2. Parameters used for TRIGRS application.

c'(Pa)	ϕ(°)	γ_s(N/m^3)	K_{sat}(m/s)	D$_0$(m^2/s)
5,000	32	18,000	1×10^{-4}	5×10^{-4}

Figure 6 shows the results provided by TRIGRS at the end of the rainfall input. The red areas display where the safety factor is less than 1, denoting instability. According to the TRIGRS prediction, 11.7% of the study area is potentially unstable. Thus, the area that TRIGRS predicts to be potentially unstable is about 1/3 less than SHALSTAB.CO. The difference in these results can be attributed to the spatially varying soil thickness in the TRIGRS application.

The assumption of uniformity for the physical parameters over the entire study area can be further improved using TRIGRS if we subdivide the study area into 5 zones, based on surficial geology. These zones are:

- Soils (surficial, loose, coarse grained, variably sorted materials of varying thickness) that are primarily products of erosion and weathering of bedrock.
- Well-consolidated ancient landslide deposits (inactive) within the marl-clay formations.
- Layered rocks dominated by marl-clay formations (Scaglia Cinerea Formation, Marne a Fucoidi Formation and Bisciaro Formation).
- Layered calcareous rocks (Calcare Massiccio, Maiolica, Calcari Diasprigni and the older rocks of the Scaglia Series).
- Competent and massive rocks.

Table 3. Parameters used for TRIGRS application, subdividing the study area into 5 zones.

ZONE	c'(Pa)	ϕ(°)	γ_s(N/m^3)	K_{sat}(m/s)	D$_0$ (m/s^2)
1	10,000	20	19,000	1×10^{-4}	5×10^{-4}
2	5,000	32	18,000	1×10^{-4}	5×10^{-4}
3	30,000	30	20,000	1×10^{-8}	5×10^{-8}
4	50,000	40	21,000	1×10^{-6}	5×10^{-6}
5	100,000	40	21,000	1×10^{-6}	5×10^{-6}

The range of variability of the physical parameters for each surficial material has been evaluated, and parametric studies were carried out to assess the strength and hydraulic parameters for each zone in a previous paper (Salciarini et al. 2006b). Values based on this work are shown in Table 3.

The results provided by TRIGRS when the study area is subdivided into five zones are shown in Figure 7.

The code predicts only about 4.5% of territory as potentially unstable. The area predicted to be unstable is about 2 times smaller than the predicted unstable area in the previous simulation with spatially uniform physical parameters. We see that the grid cells characterized by high strengths that were predicted as unstable in the previous simulation (where the topographic control prevailed) are now predicted to be stable, owing to the subdivision of the study area into 5 different soil property zones.

4.3 *Application of the TM model*

Performing the slope classification, as described in section 3.3, we obtain the result shown in Table 4.

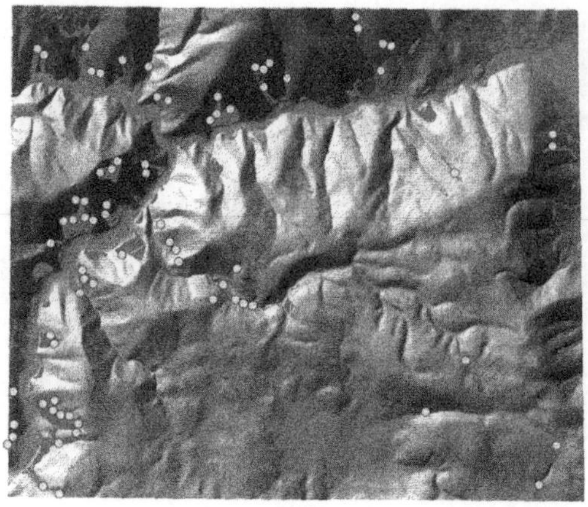

Figure 7. Failure prediction provided by TRIGRS assuming a different distribution of the physical parameter, depending on the surficial geology. White circles indicate the location of shallow landslides (Guzzetti & Cardinali, 1989).

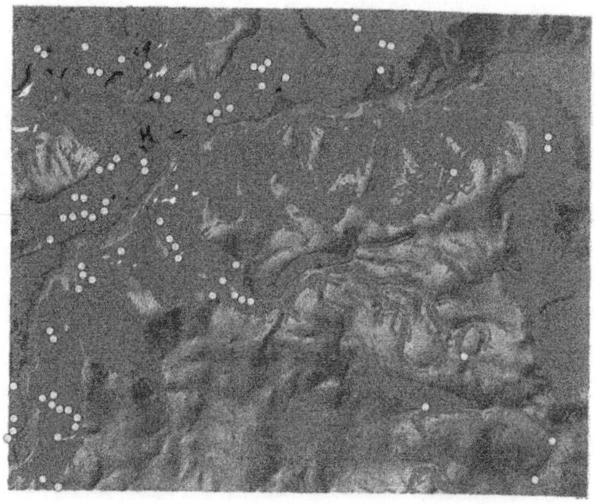

Figure 8. Spatial distribution of the V° slope-categories shown in Table 4. White circles indicate the location of shallow landslides (Guzzetti & Cardinali, 1989).

Table 4. Creation of the slope categories, using the TM model.

Category	Slope range	Percentage of the study area predicted to be unstable (%)	Percentage of agreement between the model predictions and the inventoried landslides (%)
I	30°–35°	15.25	28.5
II	27.5°–37.5°	28.40	60.0
III	25°–40°	38.00	78.6
IV	22.5°–42.5°	44.40	85.7
V	20°–45°	50.0	90.5

Increasing the range of slope angles beyond that for category V improves the percentage of agreement between the model prediction and landslide inventory only slightly.

The spatial distribution of slopes that fall in the Category V range is shown in Figure 8.

5 COMPARISON OF THE RESULTS AND DISCUSSION

To quantify the agreement of the different methods results with the inventoried location of shallow landslides by CNR, we conducted an analysis of the total area predicted to be unstable encompassed within the white circles which represent landslide source areas. We assume that the model prediction agrees with the landslide inventory when at least 6 predicted unstable cells (each cell is 5 m × 5 m) or topographical elements lie within the land-slide source circle. Table 5 lists the total area predicted to be unstable (first column) and the percentage of agreement between the prediction and the landslide inventory (second column) for each model.

Table 5. Summary of the area predicted to be unstable and the agreement between the prediction and the landslide inventory.

	Percentage of predicted instability over the territory (%)	Percentage of agreement btween the model predictions and the inventoried shallow landslides (%)
TM method	50.0	90.5
SHALSTAB.CO	34.2	88.0
TRIGRS 1 zone	11.7	82.0
TRIGRS 5 zones	4.5	80.0

An ideal landslide susceptibility map should maximize the agreement between known and predicted landslide locations and minimize the over-predictions. There are two distinct types of prediction errors: prediction of a landslide where none has occurred ("false positives") and no prediction of a landslide where one has occurred ("false negatives"). To evaluate the total error in the prediction we compute the sum of these two errors for each of the model results (Table 6). The optimal map will have the lowest value of the total error since it both maximizes the agreement of the landslide inventory with the area predicted to be unstable and minimizes the over-predictions.

We can state that false positive and false negative represent the over-prediction and the under-prediction,

Table 6. Summary of the errors produced by the codes.

	False Positives (%)	False Negatives (%)	Total error (%)
TM method	48.0	9.5	57.5
SHALSTAB.CO	33.0	12.0	45.0
TRIGRS 1 zone	11.2	18.0	29.2
TRIGRS 5 zone	4.3	20.0	24.3

respectively. They have different implications for practical applications. Decision makers might give a different importance to each of them in terms of land management. The error that we have defined "false negatives" is crucial, because it indicates that the model is not able to accurately reproduce the actual triggering condition. The error defined as "false positives" could instead associate to the degree of caution that we need to have in the land management. Decision makers might decide to prefer a "conservative" model, that is over-predictive, rather than a model that misses the prediction of the effective landslides, or vice versa, anyway this is an administrative decision.

From a scientific point of view, each of these types of error represents an inaccuracy of the model in landslide prediction. For this reason we compared the different models results in terms of the minimization of the sum of these two errors, without giving different weights to each of them.

The comparison show that results from the TM and SHALSTAB.CO models capture the greatest percentage of landslides in the inventory, however, as shown in Figures 5 and 8 these two methods also predict that a large part of the study area is potentially unstable. The two TRIGRS results produce much smaller total errors, particularly the simulation that accounts for the subdivision of the physical properties into 5 zones.

Predictive models like SHALSTAB are very efficient in numerous study cases (Montgomery et al. 1997, Guimaraes et al. 2003, Dietrich et al. 1998) but some assumptions on the morphologic and geologic conditions have to be verified for their application. For instance, the lithological characters and the spatial distribution of the soil mantle should be homogenous over the study area to obtain a good agreement with the real cases. In the central Apennine, this hypothesis is not always valid, because the outcropping formations are very variable and we can find the rapid alternation of cliffs and loose soil in the same hillslope. The possibility of considering the study area subdivided into zones with different hydraulic and mechanical properties make a significant improvement in the results because local variation in geology and soils strongly influence patterns of landsliding.

However, as noted by (Morissey et al. 2001) the choice of a model depends, primarily, on the available data for the study area. For instance, most of the successful TRIGRS applications are for cases with abundant input data. In the central Italy applications, the TRIGRS code gives correct prevision only after an accurate calibration of the model (Salciarini et al. 2006b).

6 INSTABILITY PREDICTIONS THAT DEPEND ON CLIMATIC FACTORS

Climate-landslide models are used to model the landslide movements, in different climate scenarios. We introduce the climate information (rainfall) in TRIGRS model using the historical climate data as input. Therefore, TRIGRS is able to simulate the occurrence of shallow landslides as function of different climatic scenarios, using a probabilistic distribution for the expected rainfalls.

The climatic factors for the study area were developed by the Hydrological Service of Umbria Region, using 50 years of rainfall histories recorded at the national network rain gauges. Since both rainfall intensity and duration play a key role in hillslope hydrology and lead to the conditions for slope failure, we perform a series of analyses considering the extreme expected rainfalls provided by the PDF (Precipitation Duration Frequency) curves, varying the recurrence time, and assuming a uniform distribution of rainfall during the event. The PDF curves for the study area are estimated from the regional rainfall climatic factors provided by the Pluviometric Atlas of the Umbria Region (Regione Umbria, 1991).

The PDF curve is described by the following general law:

$$h_T(d) = m_1 \cdot (1 + V \cdot K_T) \cdot d^n \qquad (5)$$

where the rainfall climatic factors given by the atlas are: $n = 0.37$, $m_1 = 26.08$, $V = 0.36$.

In Figure 9 we display the PDF curves, calculated from equation (5), for six recurrence periods ($T = 2, 5, 10, 25, 50$, and 100 years) and for five rainfall durations ($d = 1, 3, 6, 12$, and 24 hours).

Using the hydraulic and geotechnical parameters shown in the previous section, we ran TRIGRS simulations using rainfall from the PDF curves for various return periods (Salciarini et al., submitted).

Table 7 shows the percentage of predicted unstable territory (where $FS < 1$) by TRIGRS for different rainfall duration and depth.

For each recurrence interval and for each rainfall duration, TRIGRS predicts different failure scenarios. The results show the rainfall-duration robust control on hillslope stability. For prolonged rainfall the predicted unstable territory increases significantly. This is due to the effect of the progressive water table increase

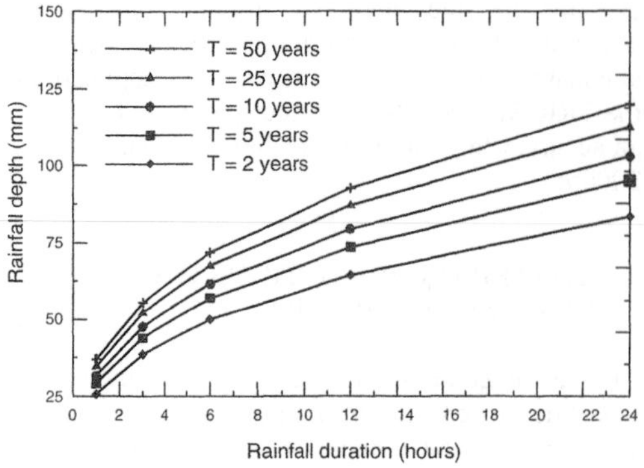

Figure 9. PDF curves for the study area.

Table 7. Results provided by TRIGRS simulations using different rainfall duration and depth.

T	Rainfall duration				
(years)	1 hour	3 hours	6 hours	12 hours	24 hours
2	—	—	—	—	1.46%
5	—	—	—	0.76%	1.91%
10	—	—	—	1.23%	2.11%
25	—	—	—	1.58%	2.30%
50	—	—	0.28%	1.80%	2.44%

and consequent increase of pore pressure. The recurrence interval of rainfall of a given duration also has an effect on the total area predicted to be unstable. For example, the area predicted to be unstable by TRIGRS for rainfall of 12 hours duration increases by about a factor of 2.4 when the return period is increased from 5 to 50 years.

The model results were compared with the observations in the study area, that reveal that the main triggering factors for debris flow and shallow landslides occurrence are prolonged rainstorms (varying from 12 up to 24 hours) with a short peak of very intense rainfalls (varying from less than 1 hour up to 4 hours), which impact a very localized area (Salciarini et al. 2006a). We find that the model results are able to correctly identify the rainfall pattern (prolonged, high-frequency rainfalls) that induces slope failure. However, the model predictions generally overestimated the total triggering rainfall, and consequently under-predict the recurrence time of failures. We attribute this to the assumption of a constant rainfall rate for a storm (Salciarini et al., submitted).

This paper addresses the possibility of the TRIGRS model to simulate different climate scenarios. The model can be coupled with a historical variations of historic climate and can be used to assess the effects on slope stability for different hillslopes.

For each recurrence interval and for each rainfall duration TRIGRS predicts different failure scenarios. The results show the rainfall-duration robust control on hillslope stability. For prolonged rainfall the predicted unstable territory increases significantly. This is due to the effect of the progressive water table increase and consequent increase of pore pressure. The recurrence interval of rainfall of a given duration also has an effect on the total area predicted to be unstable. For example, the area predicted to be unstable by TRIGRS for rainfall of 12 hours duration increases by about a factor of 2.4 when the return period is increased from 5 to 50 years.

The model results were compared with the observations in the study area, that reveal that the main triggering factors for debris flow and shallow landslides occurrence are prolonged rainstorms (varying from 12 up to 24 hours) with a short peak of very intense rainfalls (varying from less than 1 hour up to 4 hours), which impact a very localized area (Salciarini et al. 2006a). We find that the model results are able to correctly identify the rainfall pattern (prolonged, high-frequency rainfalls) that induces slope failure. However, the model predictions generally overestimated the total triggering rainfall, and consequently under-predicted the recurrence time of failures. We attribute this to the assumption of a constant rainfall rate for a storm (Salciarini et al., submitted).

This paper addresses the possibility of the TRIGRS model to simulate different climate scenarios. The model can be coupled with historical variations of historic climate and can be used to assess the effects on slope stability for different hillslopes.

7 CONCLUDING DISCUSSION

We have presented a comparison between two analytic approaches and a method based on slope angle, to estimate the potential susceptibility to shallow landslides and consequent debris flows in a study area in the Umbria Region of central Italy. The physically-based models (TRIGRS and SHALTAB.CO) couple solutions for groundwater flow with infinite slope-stability models. The TM model only considers the effect of topographic slope on the location of shallow landslides and debris flows initiation. Assuming that a certain amount of uncertainty will always remain in hazard and susceptibility analysis, the performance of a model can be assessed by estimating the error that is produced in a prediction. Results from each of these models were evaluated in terms of agreement with the landslide inventory data for the study area.

The analysis has revealed that the minimum error in the prediction has been provided by the TRIGRS approach when the study area is subdivided into different zones characterized by differing values for the soil properties. Thus, the accuracy improves when variable soil thicknesses and soil parameters are considered over the study area.

Then, different rainfalls are used as input into the model to examine the variation of the hillslope response to rainfall return period and duration. The input rainfalls are obtained by a statistical analysis of historical climate information. To effectively assess the effects of climate variability on landslide occurrence the TRIGRS code can be coupled with a model for estimating future rainfall patterns. Indeed, the results show that the model can be successfully combined with the climate information to produce forecasting of the hillslope stability.

REFERENCES

Baum, R.L., Savage, W.Z. & Godt, J.W. 2002. TRIGRS – A Fortran Program for Transient Rainfall Infiltration and Grid-Based Regional Slope-Stability Analysis. U.S. Geological Survey Open-File Report 02- York, 764 p.

Calamita, F. & Deiana, G. 1986. Geodinamica dell'Appennino umbro-marchigiano (In italian): Memorie della Società Geologica Italiana. v. 35, p. 311–316.

Carslaw, & Jaeger 1959. Conduction of heat in solids. Oxford University Press, Oxford, 510 p.

Coe, J.A., Michael, J.A., Crovelli, R.A., Savage, W.Z., Laprade, W.T. & Nashem, W.D. 2004. Probabilistic assessment of precipitation-triggered landslides using historical records of landslide occurrence, Seattle, Washington. Environmental Engineering and Geoscience, v. 10, no. 2, p. 103–122.

Conversini, P., Salciarini, D., Felicioni, G. & Boscherini, A. 2005. The debris flow hazard in the Lagarelle Creek in the eastern Umbria region, central Italy. NHESS, v. 5, p. 275–283.

Corradini, C. & Melone, F. 1988. Spatial distribution of pre-warm front rainfall in the Mediterranean area. Nordic Hydrology, v. 19, p. 53–64.

Crozier, M.J. & Glade, T. 2005. Landslide hazard and risk: issues, concepts and approach. In: Landslide hazard and risk. Glade, Anderson and Crozier Eds. Wiley, 824 p.

Dietrich, W.E. & Montgomery, D.R. 1998. SHALSTAB – A digital terrain model for mapping shallow landslide potential. Technical report by NCASI.

Godt, J.W. 2004. Observed and modeled conditions for shallow landsliding in the Seattle, Washington, area. Boulder, University of Colorado, Ph.D. dissertation, 151 p., 1 pl., 32 figs.

Guimaraes, R.F., Montgomery, D.R., Greenberg, H.M., Fernandes, N.F., Trancoso Gomes, R.A. & Carvalho, O.A. 2003. Parametrization of soil properties for a model of topographic controls on shallow landsliding: application to Rio de Janeiro. Engineerin Geology, v. 69, p. 99–108.

Guzzetti, F. & Cardinali, M. 1989. Carta inventario dei fenomeni franosi della regione Umbria e aree limitrofe. G.N.D.C.I., pub. n. 204, map at 1:100.000 scale.

Guzzetti, F. & Cardinali, M. 1990. Landslide inventory map of the Umbria region, Central Italy. 6th ICFL-ALPS 90, Milan, Italy, p. 273–284.

Guzzetti, F. & Cardinali, M. 1991. Debris-flow phenomena in the Central Appenines of Italy. Terra Nova, v. 3, p. 619–627.

Iverson, R. 2000. Landslide triggering by rain infiltration. Water Resources Research, v. 36, p. 1897–1910.

Morissey, M.M., Wieczorek, G.F. & Morgan, B.A. 2001. A comparative analysis of hazard models for predicting debris flow in Madison County, Virginia. USGS – OFR 01-67.

Montgomery, D.R. & Dietrich, W.E. 1994. A physically-based model for the topographic control on shallow landsliding. Water Resources Research, v. 30, p. 1153–1171.

Montgomery, D.R., Sullivan, K. & Greenberg, H.M. 1998. Regional test of a model for shallow landsliding. Hydrological Processes, v. 12, p. 943–955.

Reid EM 1997. Slope instability caused by small variations in hydraulic conductivity. Journal of Geotechnical and Geoenvironmental Engineering 123 (8):717–725.

Salciarini, D., Conversini, P., & Godt, J.W. 2006a. Characteristics of debris flow events in eastern Umbria, central Italy. In: Proceeding of IAEG2006, London, UK.

Salciarini, D., Godt, J.W., Savage, W.Z., Conversini, P., Baum, R.L. & Michael, J.A. 2006b. Modeling regional initiation of rainfall-induced shallow landslides in the eastern Umbria region of central Italy. Landslide, v. 3 n. 3.

Salciarini, D., Godt, J.W., Savage, W.Z., Conversini, P., Baum, R.L. 2006. Modeling land-slides recurrence in Seattle, Washington (submitted).

Savage, W.Z., Godt, J.W. & Baum, R.L. 2003. A model for spatially and temporally distributed shallow landslide initiation by rainfall infiltration. Proceedings, 3rd International Conference on Debris Flow Hazards Mitigation: Mechanics, Prediction, and Assessment, September 10–12, 2003, Davos, Switzerland, p. 179–187.

Savage, W.Z., Godt, J.W. & Baum, R.L. 2004. Modeling time-dependent slope stability. Proceedings 9th International Symposium on Landslides, Rio de Janeiro, Brazil, June 27–July 2, 2004, p. 23–38.

Landslides and Climate Change – McInnes, Jakeways, Fairbank & Mathie (eds)
© 2007 Taylor & Francis Group, London, ISBN 978-0-415-44318-0

Assessing the influence of climate change on the activity of landslides in the Ubaye Valley

J.-P. Malet, A. Remaître & O. Maquaire
CNRS, UMR 6554, University of Caen-Basse-Normandie, Caen, France

Y. Durand, P. Etchevers & G. Guyomarc'h
Centre National de Recherche Météorologique, Météo-France, CNRS URA 1357, Grenoble, France

M. Déqué
Centre National de Recherche Météorologique, Météo-France, CNRS URA 1357, Toulouse, France

L.P.H. van Beek
Utrecht University, Faculty of Geosciences, Utrecht, The Netherlands

ABSTRACT: Effects of past, present and future climate characteristics on landslide activity can be assessed by historical information, geomorphological evidences, and modelling of slope hydrology and mechanics. However there are a series of problems related to that approach. First, quantitative assessments of the spatial and temporal relationships between climate and landslide occurrences are often hampered by the inaccuracy and uncertainty of the historical records, and by the scarcity of climate data in mountain areas. Second, the climate-landslide coupling is complex, because climate is related to landslides via the nonlinear soil water system. Process-based models have been proposed to understand this complex interaction. However, landslide triggering systems show complex responses in relation to geotechnical, hydrological, and climate properties. Third, the uncertainty in future climate parameters is high, especially if the time context is greater than weather records or because of the low-resolution of the downscaled climate time series. The objectives of this paper is to present some results on the climate-landslide relationships for two landslide types observed in the Ubaye Valley, and to propose a method for assessing the impacts of climate change on landslide frequency.

1 INTRODUCTION

Many factors can affect landslide incidences including climate characteristics, seismic activity, human activity or compound factors. As a consequence, the causation between climate and landslides have never been simple, and is regarded by many researchers as problematical (Corominas, 2000), especially in terms of spatial and temporal scales. Climate acts as a complex agent on the magnitude and frequency of landslides via the nonlinear soil water system (Iverson, 2000; Bogaard & van Asch, 2002). Consequently, combining regional and local analyses is necessary to minimize the effects of predisposing factors such as geo(morpho)logic conditions and landscape history, to elucidate the relationships between climatic factors and landslides. Moreover, understanding landslide hydrology and landslide mechanics is the basis for quantitative assessments of the possible influence of climatic changes on slope stability.

This work deals with the understanding of changes in the activity of rainfall-controlled landslides in the Ubaye Valley (South East France) in the last century by combining archive analysis, field investigation, hillslope monitoring and model simulations. The work presents a method for assessing the climate change impacts on landslide frequency by combining climate time series predictions and hydrological-slope stability modelling. The main hypotheses of the work are:

1) The analysis of the climate-landslide relationships in a specific climatic sub-zone and eco-region (20 × 30 km) characterized by a mountainous Mediterranean climate.
2) The analysis of the climate-landslide relationships for the recent time (e.g. last 50 years of the 20th century); influences of climate variation since the last deglaciation are not considered. Climate records since the late 1900s indicate a limited magnitude of climatic changes: the variability of 10-years

average parameters (rainfall amounts, mean temperatures) is quite small though higher variability is observed on smaller time scales.

3) The analysis of the temporal pattern of landslides observed in a geologically homogeneous lithology (black marls). The work focuses on the characterization of two rainfall-controlled landslide types representative of the study area: rotational or translational slides occurring in moraine deposits and at the contact with the black marls bedrock, and deep-seated large mudslides occurring in weathered and reworked black marls.
4) The identification of the hydrological and geomechanical mechanisms controlling the behaviour of the landslide types, and the possibility to reproduce their -observed- temporal pattern with a simple process-based slope model.
5) The introduction of meteorological parameters, characterising a climate change scenario, as boundary conditions to the slope hydrology and stability models (Buma & Dehn, 1998). The 'changed' meteorological parameters (e.g. time series) are issued from the downscaling of General Circulation Models (GCMs) at the local scale of the study area.

Considering the above mentioned hypotheses and the small magnitude of climatic variations during the interval (1950–2005) in which landslide data have been analysed, it is possible to consider modern climatic records as representative of the actual climate-landslide relationships, and to further apply climate change scenarios to model landslide frequencies. Besides the proposed methodological framework to assess impacts of climate changes on landslide activity, the paper discusses the differences in landslide frequencies observed on the period 1950–2005 and simulated for the period 2069–2099.

2 CHARACTERISTICS OF THE STUDY AREA

The Ubaye Valley is representative of climatic, lithological, geomorphological and land-use conditions observed in the South French Alps, and is highly affected by landslide hazards (Flageollet et al., 1999). Within the valley, the Barcelonnette Basin (Fig. 1) extends over an area of 200 km^2, from 1100 m to 3100 m a.s.l, where Quaternary glaciers and about 26 torrents have carved out a large tectonic window in autochthonous Jurassic marls very sensitive to weathering. Black marls are overlaid by flyschs and limestones of allochtonous sheet thrusts, and mostly covered by moraine and deposits. The slope gradients range from 20° to 50°.

The study area is located in the dry intra-Alpine zone, characterized by a mountainous Mediterranean

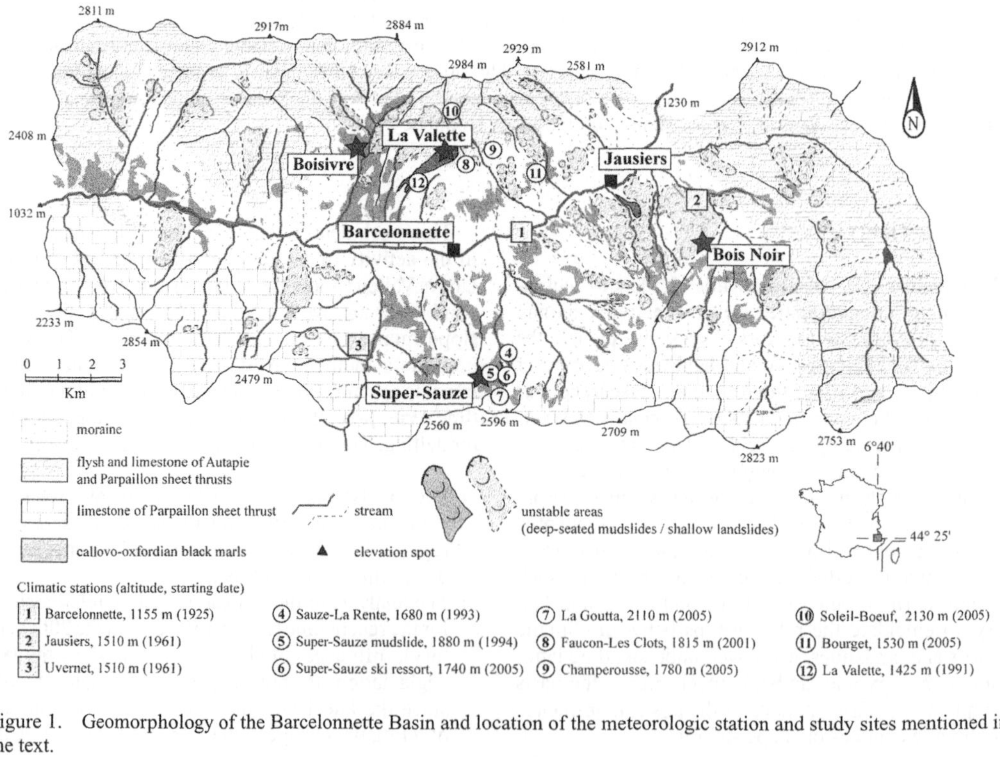

Figure 1. Geomorphology of the Barcelonnette Basin and location of the meteorologic station and study sites mentioned in the text.

climate with a high inter-annual rainfall variability (735 ± 400 mm over the period 1928–2005), a mean annual temperature of ca. 7.5°, and the presence of a snow pack on the upper slopes for 4 to 6 months. Melting of the snow cover adds to the effects of rain in the triggering of landslides in spring. Locally, summer rainstorms can be intense, yielding sometimes intensities of more than 40 mm.h^{-1}.

These predisposing geomorphologic and climatic factors explain the occurrence of several landslide types in the area. The La Valette, Poche and Super-Sauze mudslides are significant examples of large and active landslides affecting black marls slopes for several decades. The moraine slopes are highly affected by shallow rotational and translational slides triggered at the contact with the lithic bedrock (Thiery et al., submitted). Finally, the torrents are experiencing the occurrence of muddy debris flows since more than 100 years (Remaître et al., 2005).

3 CLIMATE-LANDSLIDE RELATIONSHIPS

The climate-landslide relationships have been analysed for the recent time (period 1950–2005) at several time intervals in order to attempt to determine climate conditions (and possibly rainfall thresholds) able to trigger or reactivate landslide in the study area.

The long-term rainfall pattern indicates cyclic variations of 10–15 years of the annual rainfall amount (Fig. 2). These variations comprised periods of 5–7 years of annual rainfall amounts in excess to the mean annual rainfall observed for the period 1900–2004. For example, two large mudslides were triggered or reactivated during periods of excess cumulative rain: the Super-Sauze mudslide had a major reactivation in the end of the 1970s and the La Valette mudslide occurred in 1982 (Fig. 2). As well, the Bois Noir translational slide occurred during the most recent period of excess cumulative rain at the beginning of the 1990s (Fig. 2). However, some relatively dry periods (like the period 1982–1990) are also characterized by the occurrence of landslides, testifying of the complexity of the landslide-climate relationships. On longer time scale, datings of eccentricities of tree rings in some slopes of the study area demonstrated landslide frequencies of ca. 4 ± 2 years (Verhaagen, 1988; Buma, 2000), indicating climate characteristics favourable to a continuous activity of slope processes. However, as frequently observed elsewhere, rainfall is one of the factors that accelerate or triggers landslides, together with other factors such as land-use or tectonic activity (Corominas, 2000). At the monthly scale, no systematic correlation has been made evident between rainfall amounts and landslide frequency, though the spring (deep mudslide) and summer seasons (shallow slump) are the most favourable periods for landslides. At the daily scale, on the basis of about 20 events for which an exact date of occurrence is available, two types of climate situation have to be considered (Fig. 3):

1) Type A situation is characterized by heavy daily rainfall following a 30-days dry period. For most of the observed landslide events, this climate situation corresponds to violent summer storms. The triggering of shallow slumps, translational slides, hillslope mudflows and/or channel debris flows is usually associated with this type. Although, the correlation between landslide events and the rainfall of the triggering date is quite good, no threshold can be established due to the inaccurate knowledge on rainfall variability, especially in the upper part of the hillslopes. For these reasons, several meteorologic stations have been installed around landslide areas since 1993 (like in the Sauze and Faucon catchments; Fig. 3, Fig. 1).
2) Type B situation is characterized by heavy cumulative rainfall distributed over a 30-days very humid period. This climate situation characterizes either

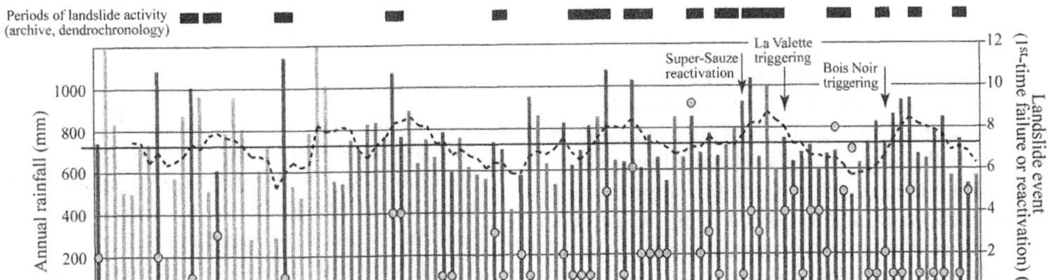

Figure 2. Annual rainfall and references of landslide activity (archives investigation e.g. RTM dataset, dendrochronology analyses).

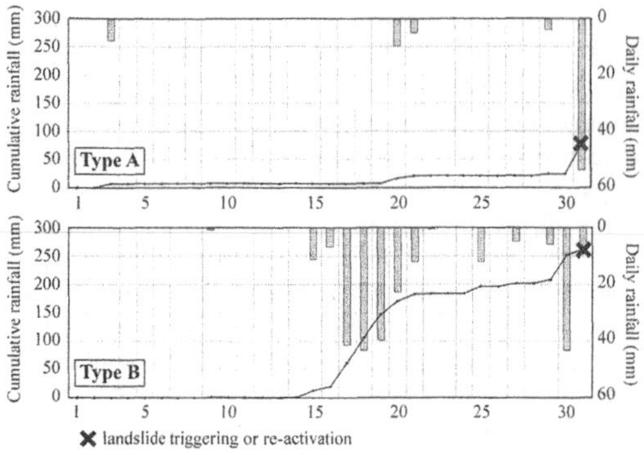

Figure 3. Schematic climate situations of landslide triggering or reactivation in the Barcelonnette Basin. Example of a landlside observed on 15 September 1960 (Type A), and of a landslide observed on 16 November 1963 (Type B). The X-coordinate represents 'day' from the onset of failure.

the progressive saturation of the topsoil, the rising of a permanent groundwater table and the build-up of positive pore pressures. Cumulative rainfall over a short period is often enough to trigger or reactivate a landslide, though little rainfall is observed at the date of occurrence. Triggering or reactivation of deep-seated mudslides are frequently associated with this type (Malet et al., 2005).

Thus, as often observed elsewhere, rainfall is one of the most important elements to be considered in landslide triggering/reactivation. The complexity of the landslide-climate relationships suggests the use of modelling techniques, validated on well-documented sites, to gain knowledge and identify possible trends in landslide frequency.

4 METHODOLOGY TO INVESTIGATE THE EFFECTS OF CLIMATE CHANGE ON LANDSLIDE FREQUENCY

The proposed methodology simulates the effects of climate change on landslide frequency with process-based models of slope hydrology and slope stability. Climate change scenarios (e.g. climate variables) of General Circulation Models (GCMs) are used as input conditions for the slope models. These climate scenarios are within the range for which the slope models have been elaborated and validated. (Fig. 4) depicts the general methodology of combined downscaling of GCMs and slope stability modelling.

The models are applied on two unstable slopes (Fig. 1) for which detailed information on slope geometry, hydrology and activity is available:

1) The Super-Sauze mudslide is a clay-rich flow-like landslide developed in weathered black marls (thickness varying from 10 m to 20 m), and characterized by a complex vertical structure (Malet & Maquaire, 2003). Groundwater fluctuations are controlled by water infiltration both in the soil matrix and in large fractures as well as recharge from the torrents bordering the landslide (Malet et al., 2005; Montety et al., 2007). The mudslide is persistently active since the late 1960s, with velocities comprised between 0.005 to 0.3 m.day^{-1}. Acceleration periods (not exhaustive) have been observed in 1978–1982, 1995, 1999, 2000, 2006.
2) The Boisivre rotational landslide is a slump characterized by a top morainic layer (thickness 1.5 m), underlain by a weathered and unsaturated black marl layer (thickness 5 to 6 m) which overlies the bedrock of unweathered marl (Mulder, 1991). Landslide activity is a consequence of high groundwater levels in the weathered marl layer and the temporal occurrence of a perched water table in the top moraine deposits (Caris & van Asch, 1991). The landslide has been active for decades with a return-period for reactivation of ca. 3–4 years.

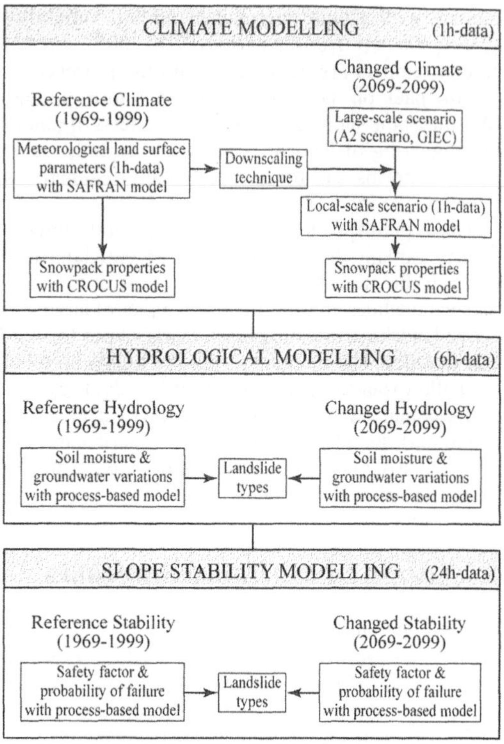

Figure 4. Chain of modelling steps for local assessment of the effects of climate change on slope stability and landslide frequency.

4.1 *Slope Hydrology Model*

The Slope Hydrology Model (van Beek, 2002; Malet et al., 2005) uses net precipitation, temperature and net radiation, as well as geometrical and hydrological characteristics, in order to simulate water flows within the slope. The hydrological model consists of three permeable reservoirs and the underlying impervious bedrock. It describes saturated and unsaturated transient flows in the vertical and lateral directions assuming freely drainable water. Storage (antecedent soil moisture condition) and fluxes (precipitation P, infiltration I, evaporation Ep, surficial runoff R, percolation in the unsaturated zone Pe, saturated lateral flow Q_{sat}) are considered in the different reservoirs and define the hydrological balance of the system (Fig. 5a). Groundwater generation is simulated by imposing boundary conditions: the lower boundary condition is state-controlled and is specified as fixed value for the matric suction; the upper boundary condition is flux-controlled and account for the climate inputs at the surface. A complete mathematical description of the model can be found in van Beek (2002).

The dynamic model is implemented in the PCRASTER GIS package (Wesseling et al., 1996). This approach provides a unified theoretical description of most of the water fluxes observed within a landslide, and allows the easy integration of complex geometry as the effects of topography can readily be incorporated and GIS routine functions can be used to define flow paths in each soil layer.

4.2 *Slope Stability Model*

The Slope Stability Model is a limit equilibrium model loosely coupled to the hydrological model (e.g. the simulated groundwater levels define the pore pressures introduced in the slope stability model). The model calculates the safety factor of the slope, which is the ratio between the available total resisting forces and the total driving forces. The resisting forces decreases with increasing pore pressure and when the safety factor drops below unity, failure is predicted. The limit equilibrium approach only considers yielding by plastic failure and uses the Mohr-Coulomb failure criterion. The stability of the slope is calculated with the Janbu force diagram (Janbu, 1957). This allows to take into account non-uniformly distributed forces throughout the soil mass and the effects of inter-slice forces. Slope stability is therefore dependent, not of the local cell attributes only, but of several adjacent cell attributes (Fig. 5b). This is only possible because detailed geometries (e.g. slip surface depth) are available for both landslides.

Quantification of slope stability is deterministic and requires the inputs of the slip surface depth and the geotechnical properties. Probability of failure is determined by the uncertainty in the shearing

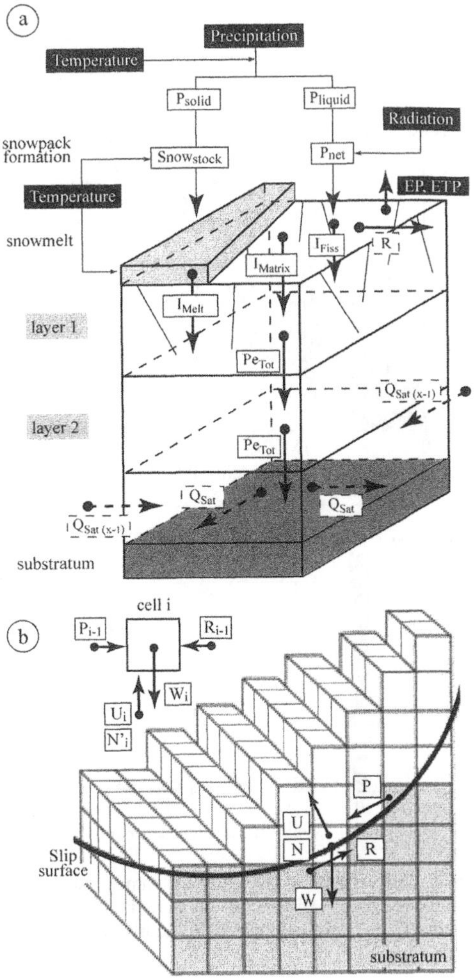

Figure 5. Architecture of the hydrology (5a) and slope stability (5b) model. (a) Schematic representation of the storages and fluxes simulated by the hydrological model, and relation between the calculation cells; (b) Schematic representation of the force diagram used in the slope stability model for each calculation cell i. W is the total weight, U is the pore pressure force at the base, N is the effective normal force, P is the pressure term, R is the resisting shear force.

resistance estimated from the observed distribution of the geotechnical parameters (Maquaire et al., 2003).

4.3 *Climate modelling for the 20th & 21st centuries in the Ubaye Valley*

The climate modelling part of this study comprises the simulation of meteorological surface parameters for the last 30 years of the 20th and 21st centuries:

1) Several meteorological surface parameters have been computed at an hourly timestep for the period

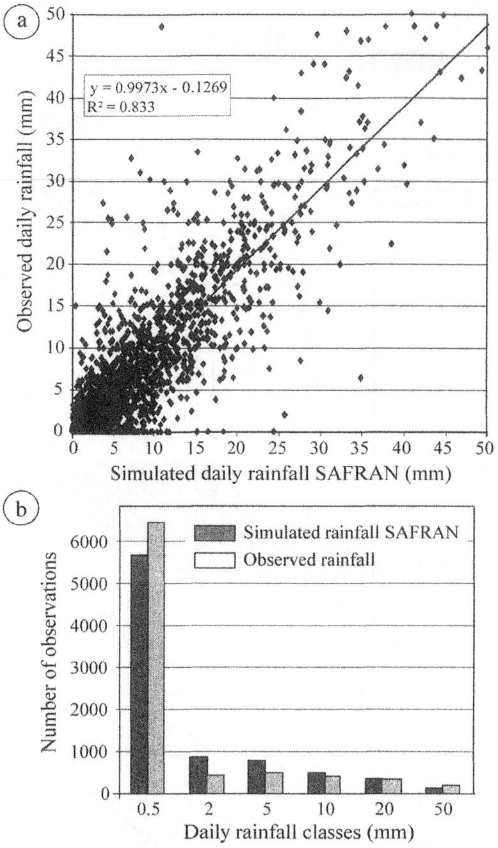

Figure 6. Meteorological land surface parameters observed and simulated with SAFRAN for the period 1969–1999 at Barcelonnette. (a) Rainfall amounts; (b) Rainfall frequencies.

1969–1999 with the SAFRAN model (Durand et al., 1993; 1999). SAFRAN is a meteorological application used for more than 10 years in real-time avalanche hazard forecast. SAFRAN is able to spatialise meteorological parameters for several slope configurations (elevation, aspect) by combining observed meteorological surface data, preliminary estimations of general circulation fields (ERA-40 reanalyses) through appropriate downscaling operators, and climate classifications (synoptic weather types). The reanalyses of the ERA-40 outputs (ECMWF, 2004) have been used at a spatial scale of ca. 125 km for the last 30 years.

Time series of air temperature, air humidity, wind speed, rain and snow precipitation, long-wave radiation, direct and scattered solar radiation, infra-red atmospheric radiation and cloudiness were simulated. Quality of the regionalization operators has been tested at Barcelonnette site, without using the observed meteorological dataset in the SAFRAN simulations.

Comparison of the datasets indicates good agreement of the simulated climate with the observation at a daily time scale (Fig. 6). The simulated time series were then re-calculated in order to take into account local topographical settings (elevation, aspect, slope gradients, orographic mask); this has allowed to correct the time series from the effects of the relief on the initial radiation, temperature and air humidity fields. The SAFRAN downscaling operators are integrating composite large-scale data fields of GCMs, classification of weather types, and site observations. These operators are used in routine for snow avalanches hazard forecast.

2) The disaggregation procedure used to simulate the meteorological parameters over the period 1969–1999 has then be applied to the outputs of the climate change scenario A2 of GIECC over the Alps for the period 2069–2099.

As the ARPEGE-IFS GCM is a simplification of the atmosphere system, the simulated daily variability is quite different from the observations (Gibelin & Déqué, 2003; Déqué et al., 2006) with an underestimation of the extreme values. Therefore, the local climate is only characterized from a statistical viewpoint. A perturbation method has been used in order to downscale the parameters to the local scale with the constraint to keep constant the density function of the ARPEGE-IFS GCM simulation for the period 2069–2099 for each SAFRAN parameters. The method, which presents a high temporal coherence, does not take into account climate evolution in terms of frequency of weather types or modifi-cation in the frequency of extreme events. The average values of the perturbations associated to the A2 climate change scenario are presented in (Fig 7). The main characteristics of the changed climate for Southeast France are: a) higher temperatures in summer, b) more rainy winters, c) drier summers, d) a decrease in soil water content except for winter freezing areas.

A regional operator, based on the minimisation of the XY and Z distances between the study sites and the ARPEGE-IFS GCM grid outputs has been used. Hence, it was possible to compare the Probability Density Functions (PDFs; Fig. 8a), and the corresponding centiles (Fig. 8b), obtained with the disaggregation of the reference (1969–1999) and changed (2069–2099) climate. The differences observed are important, e.g. the average annual temperature rises from 2.4°C to 6.2°C over one century, and the distribution is shifted towards the extreme values. A climate change operator was defined from the analysis of the reference outputs (1969–1999) and the changed outputs (2069–2099) (Fig. 8b). For an X-coordinate on Figure 8b, which represents a centile of the temperature distribution, the probability to obtain a temperature equal or lower than the reference climate or the

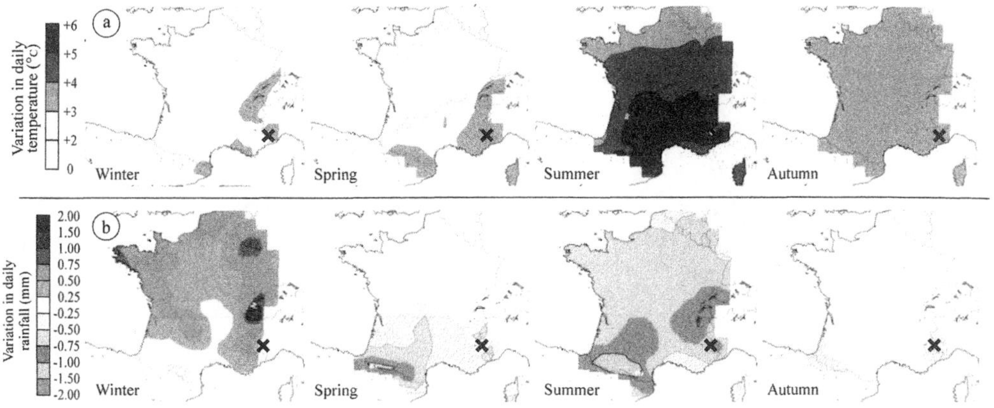

Figure 7. Impacts of climate change with the ARPEGE-IFS experiment based on the A2 scenario of GIECC. (a) Temperature; (b) Daily rainfall.

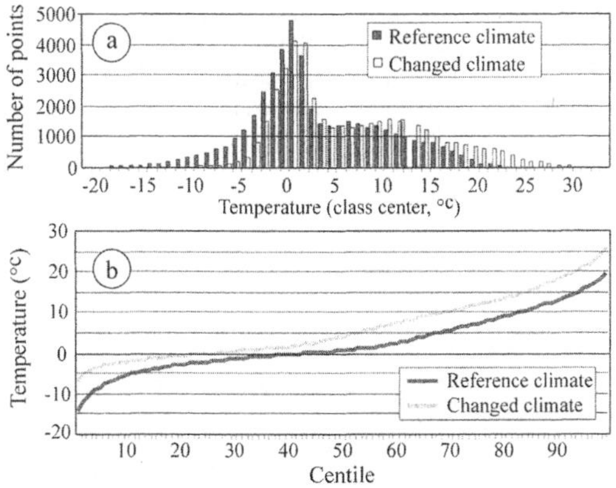

Figure 8. Impacts of climate change at Super-Sauze (1800 m). (a) Probability Density Functions of air temperature for the reference climate (1969–1999) and the changed climate (2069–2099). (b) Centiles of air temperature derived from Figure 8a.

changed climate is equal. Figure 8b indicates that the increment between two populations varies according to the reference temperature.

Then, the operator has been applied to the reference dataset computed by SAFRAN in order to create a 'climate change' dataset which fits the statistical distribution of the changed climate. The average values of the variations of several meteorological parameters are presented in Table 1.

4.4 *Snowpack modelling*

As snow cover has a critical influence on landslide activity in the study area (Malet et al., 2005), a detailed simulation of the snowpack properties has been realised. The CROCUS model (Brun et al., 1989; 1992) is able to simulate all the mass and energy transfers within the snowpack, as well as the snow characteristics, including the presence of liquid and melting waters.

Figure 9 represents the average daily values of snow depths for the two climates at 1800 m. a.s.l. The impact of climate scenario A2 is particularly dramatic for snow depth at this elevation.

Table 1. Average impacts (30 years) of the climate change scenario A2 on several meteorological parameters downscaled on local sites of the Barcelonnette Basin, at respectively 1800 m and 2100 m a.s.l. T is air temperature (°C), RH is relative air humidity (%), P_L is liquid precipitation (mm), P_S is solid precipitation (mm), N is nebulosity ($1/100^e$), W is wind velocity ($m.s^{-1}$), Q is total surface energy potential ($W.m^{-2}$).

	T	RH	P_L	P_S	N	W	Q
Z = 1800 m	+3.8	−2.4	+0.11	−0.5	−4	−0.04	+23
Z = 2100 m	+3.8	−2.3	+0.47	−0.6	−4	−0.02	+26

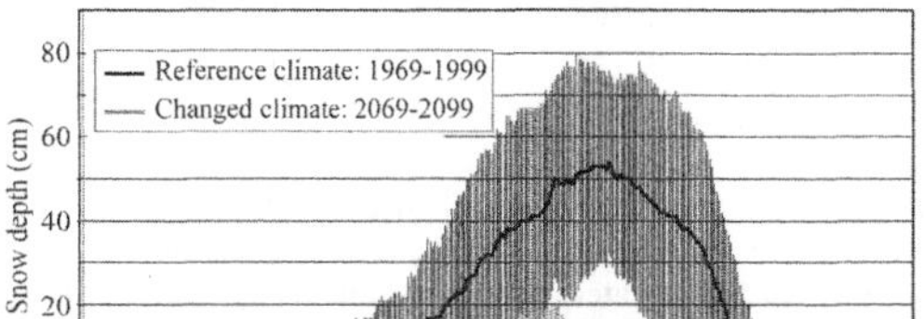

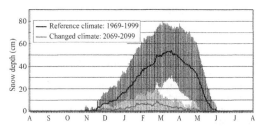

Figure 9. Impacts of climate change at Super-Sauze (1800 m) in for the daily average snow depths. The X-coordinate represents the months (from August to July); the vertical lines express snow depth variability.

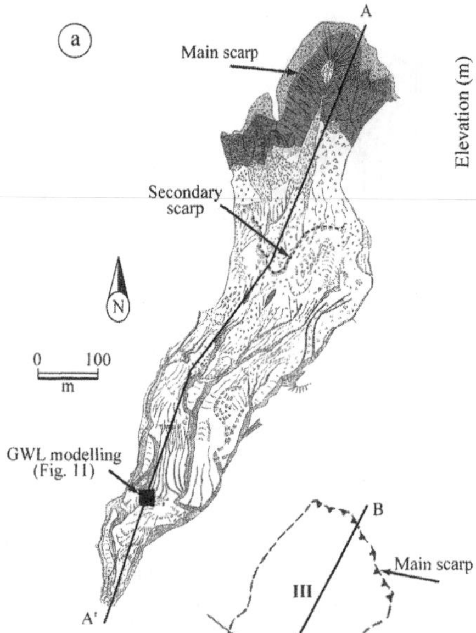

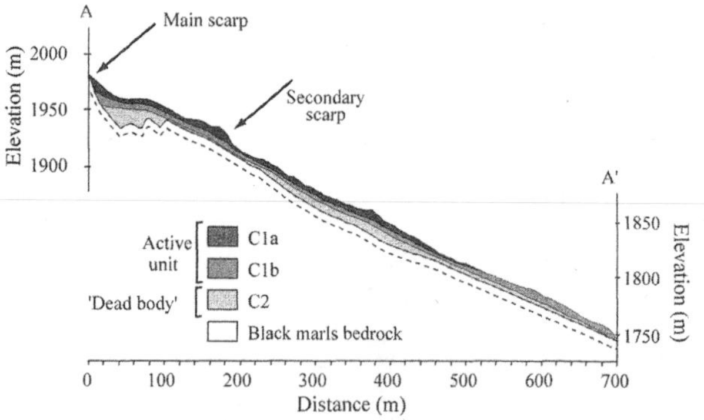

Hydrological parameters	Model input values		Geotechnical parameters	Model input values	
	Sauze	Bois.		Sauze	Bois.
Kv_1 (cm.d^{-1})	0.40	0.40	γ_d (kN.m^{-3})	15	16
Kv_2 (cm.d^{-1})	0.30	0.30	γ_w (kN.m^{-3})	17	19
Kv_3 (cm.d^{-1})	0.30	/	ϕ_1 (°)	28	32
Rat_K (-)	0.01	0.01	ϕ_2 (°)	28	25
θsat_1 (-)	0.38	0.38	ϕ_3 (°)	24	/
θsat_2 (-)	0.24	0.24	c'_1 (kPa)	2	2
θsat_3 (-)	0.21	/	c'_2 (kPa)	2	1
$\|h\|_{BC}$ (m)	2.98	2.98	c'_3 (kPa)	1	/

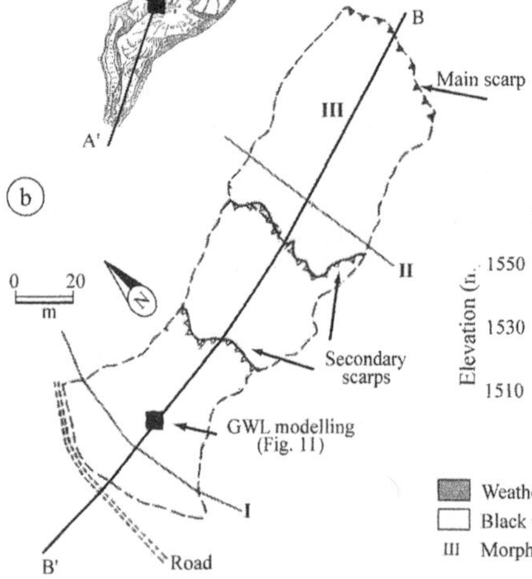

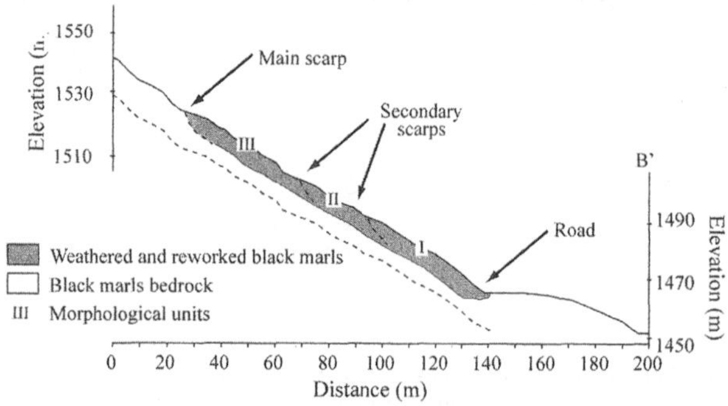

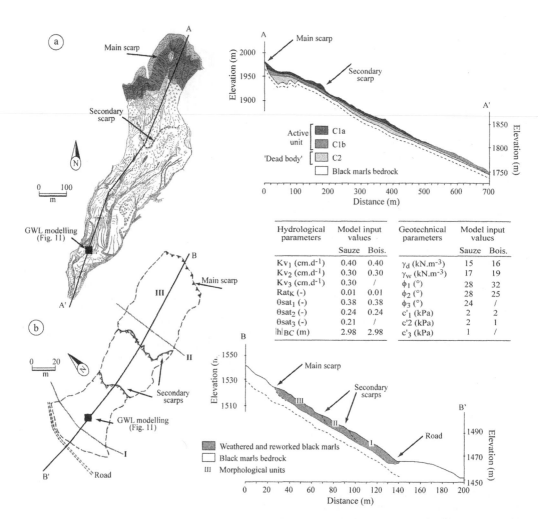

Figure 10. Geomorphological maps of the investigated sites, cross-section and hydrological and geotechnical properties used in the model runs. (a): Super-Sauze mudslide; (b): Boisivre translational slide (modified from Caris & van Asch, 1991). K_v is Saturated vertical conductivity; Rat_K is Ratio Lateral/Vertical conductivity; θsat is effective porosity; $|h|_{BC}$ is matric suction at bedrock; γ_d is dry bulk density; γ_w is wet bulk density; ϕ is angle of internal friction; c' is effective cohesion. The lettering in index represents the soil layers. The parameter values are average values of laboratory tests or field tests.

5 APPLICATION TO THE METHODOLOGY TO THE SELECTED SITES

5.1 *Model implementation*

The slope models have been applied to the sites assuming a compromise between the complex topography of the landslides and the distribution of soil properties. In both cases, a cell resolution of 5 m has been used. Geometry of the layers is derived from detailed geotechnical and geophysical investigations (Malet & Maquaire, 2003; Caris & van Asch, 1991). Depth of the slip surface is known for both sites, and is therefore fixed in the model. Hydrological and geotechnical properties (Fig. 10) are derived from both laboratory and soil characterization, and further optimisation of the model performance on the basis of multi-source field observations (Malet et al., 2005). The assigned parameters values are indicated on Figure 10. The timestep resolution has been set at 6 hours for the groundwater modelling, and 24 hours for the slope stability modelling. At Super-Sauze, model performance has been extensively validated on several piezometers and monitoring periods (Malet et al., 2005); at Boisivre, the hydrological model has

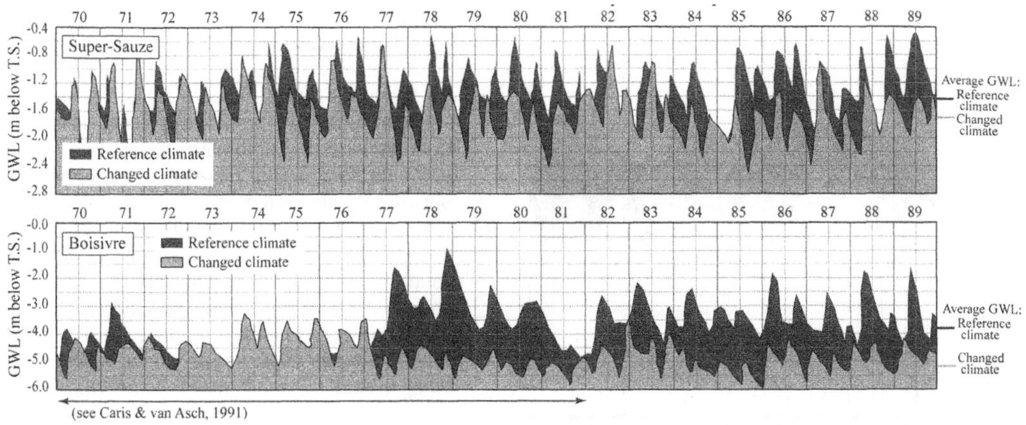

Figure 11. Simulated hydrology of the landslides for the reference climate and the changed climate. Simulations were carried out with the entire set of meteorological land surface data for the period 1969–1999 and 2069–2099. The model outputs for the period 1970–1989 and 2070–2089 are presented. The model outputs consist in the average groundwater levels of 100 model runs.

been tuned against the groundwater simulations of Caris & van Asch (1991). Both experiments show good agreement with the observations.

5.2 *Simulated hydrology and probability of failure for both sites and climate*

Figure 11 shows the results of the groundwater modelling for the reference climate and the changed climate for both sites. Uncertainties were considered in the simulation by performing 100 model runs with many possibilities of model input values (e.g. average hydrological and geomechanical parameter values and their variation range). Therefore, the model outputs in Figure 11 represent the average groundwater levels (GWL) of 100 model runs.

The hydrological model simulates a general decrease in both GWLs and amounts of water storage in the unsaturated soil in the next century. This is particularly true for the Boisivre landslide with a decrease in the average GWL of ca. −2.3 m in comparison to the reference climate (Fig. 11). GWL lowering is explained by a marked decrease in soil effective porosity (e.g. soil moisture in the unsaturated topsoil) with the changed climate, resulting in less vertical percolation to the saturated zone.

This result has to be related to higher potential evapotranspiration in the changed climate than in the reference climate for this landslide situated on the south-facing hillslope of the Ubaye Valley.

At Super-Sauze, the average GWL is lowered of ca. −0.6 m with the climate change scenario. This result may be explained by relatively small variation in the yearly amounts of rainfall on the north-facing hillslope in the changed climate; in terms of water supply, the decrease in snow depths (Fig. 9) seems to be compensated by higher amounts of liquid rainfalls in the winter season which does not change the hydrological behaviour of the landslide.

As a consequence, the slope stability model simulates a general decrease in landslide activity which is particularly marked for the Boisivre landslide. The frequency of unstable areas (Fig. 12) is calculated according to the total area of the landslide (e.g. a frequency of 0.5 indicates that 50% of the calculation cells have a safety factor value lower than 1.1). The most interesting result of this exercise is that parts of the landslides will still fail or be reactivated in the next century, though a decrease in unstable cells of 10 to 20% is simulated.

5.3 *Discussion*

Results of the impact scenario must be interpreted cautiously. First, our approach presupposes that the geometry of the landslides does not vary over time (e.g. a static safety factor is evaluated at each time step for a fixed calculation grid). Therefore, dynamic deformation models should be used to increase the reliability of the impact assessments.

Second, only the meteorological inputs are considered in the slope models; many other factors like land-use changes, material availability or vegetation-feedback mechanisms are also controlling landslide frequency. Finally, given the large uncertainties in our knowledge of the landslide mechanisms, in the simplification of our models and in the forecast of future climate time series, probabilistic analyses should be considered in the impact assessments. Several model runs should be performed with PDFs of the meteorological, hydrological and geomechanical parameters in order to compute PDFs of groundwater levels and

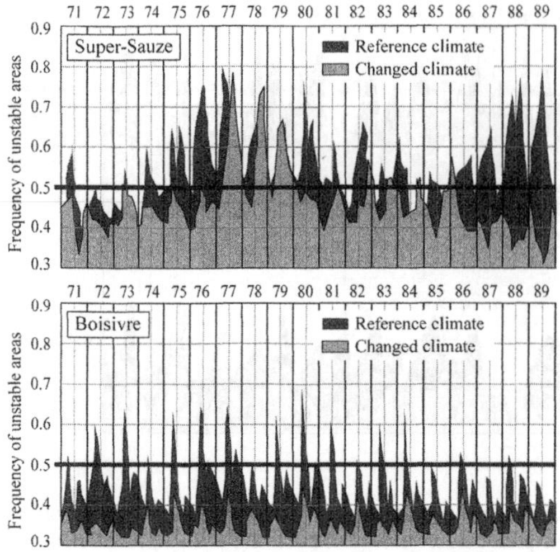

Figure 12. Frequency of unstable areas calculated with the pore pressures simulated by the hydrological model, for the reference climate and the changed climate. The model outputs for the period 1970–1989 and 2070–2089 are presented.

safety factor values. This was not within the scope of this paper, but should be tested.

6 CONCLUSIONS

This paper describes a methodological framework to assess the influence of climate change scenarios on slope hydrology and landslide frequency by combining climate modelling, groundwater modelling and slope stability modelling. A scheme to simulate high-resolution time series of meteorological land surface parameters is proposed. Climate modelling associates a disaggregation procedure to downscale the climate general circulation parameters at the site scale taking the A2 climate change scenario of GIECC as a guess for the future climate, and two meteorological process-based models to spatialise the dataset and evaluate snow properties.

The simulated meteorological parameters are used as boundary conditions in a process-based groundwater model able to reproduce spatially variable ground water flows. The time-dependent hydrological behaviour of two unstable slopes are evaluated, and assessed against monitoring data. A slope stability model is then applied in order to evaluate a time-dependent safety factor in relation to pore pressure variations.

For the climate change scenario hypothesized, and given the uncertainties associated to the climate parameter modelling and the very simple concepts used in the groundwater and slope stability models, the impact simulations indicate:

(1) On the south-facing slopes, a drastic reduction of slope instability for rotational slides, associated to an increase in evapotranspiration and a consequent decrease in soil moisture and effective soil porosity in the unsaturated topsoil.
(2) On the north-facing slopes, influence of climate change for mudslides is limited with a quite small reduction in slope instability. This situation is explained by a small variation in the total yearly rainfall amounts (e.g. the critical decrease in snow depths observed in winter seems to be compensated by an increase in liquid rainfalls).

Although the process-based models used in this study do not claim to simulate all behaviour, they do establish some interesting trends in impact assessments of climate change on slope stability. This study indicates also that more understanding about site-specific landslide activity and mechanisms is very important to forecast 'reliable' scenario. Long-term monitoring and modelling of selected pilot study sites (e.g. hydrology and kinematics) is necessary to regionalize the information and to reduce the uncertainties in our simulations.

ACKNOWLEDGMENTS

This paper is part of the ACI-FNS Programme '*Aléas et Changements Globaux*': Project GACH2C '*Glissements Alpins à Contrôle Hydrologique et Changement Climatique*' (2005–2007) financially supported by the French Ministry of Research. Coordinators: O. Maquaire & J.-P. Malet.

REFERENCES

Bogaard, T.A & van Asch Th.W.J. 2002. The role of the soil moisture balance in the unsaturated zone on movement and stability of the Beline landslide, France. *Earth Surface Processes and Landforms* 27: 1177–1188.

Brun, E., David, P., Sudul, M. & Brugnot, G. 1992. A numerical model to simulate snow cover stratigraphy for operational avalanche forecasting. *Journal of Glaciology* 38(128), 13–22.

Brun, E., Martin, E., Simon, V., Gendre, C. & Coléou, C. 1989. An energy and mass model of snow cover suitable for operational avalanche forecasting. *Journal of Glaciology* 35(121): 333–342.

Buma, J. & Dehn, M. 1998. A method for predicting the impact of climate change on slope stability. *Environmental Geology* 35(2–3): 190–196.

Buma, J. 2000. Finding the most suitable slope stability model for the assessment of the impact of climate change on a landlside in Southeast France. *Earth Surface Processes & Landforms* 25: 565–582.

Caris, J. & van Asch Th.W.J. 1991. Geophysical, geotechnical and hydrological investigations of a small landslide in the French Alps. *Engineering Geology* 31:249–276.

Corominas, J. 2000. Landslides and climate. In: Bromhead, E.N., Dixon, N. & Ibsen, M.-L. (eds), Landslides in Research, Theory and Practice; *Proc. 8th intern. symp. on landslides*, Cardiff, 26–30th June 2000. London: Thomas Telford, London. Cd-Rom.

De Montety, V., Marc, V., Emblanch, C., Malet, J.-P., Bertrand, C., Maquaire, O. & Bogaard, T.A. 2007. Identifying origin of groundwater and flow processes in complex landslides affecting black marls: insights from an hydrochemistry survey. *Earth Surface Processes and Landforms* 32: 32–48.

Déqué, M., Jones, R.G., Wild, M., Giorgi, F., Christensen, J.H., Hassell, D.C., Vidale, P.L., Rockel, B., Jacob, D., Kjellström, E., de Castro, M., Kucharski, F. & van den Hurk, B. 2005. Global high resolution versus Limited Area Model climate change projections over Europe: quantifying confidence level from PRUDENCE results. *Climate Dynamic* 25(6): 653–670.

Durand, Y., Brun, E., Mérindol, L., Guyomarc'h, G., Lesaffre, B. & Martin, E. 1993. A meteorological estimation of relevant parameters for snow models. *Annals of Glaciology* 18: 65–71.

Durand, Y., Giraud, G., Brun, E., Mérindol, L. & Martin, E. 1999. A computer based system simulating snow pack structures as a tool for regional avalanche forecast. *Journal of Glaciology* 45(151): 469–485.

ECMWF. 2004. *ERA-40: ECMWF 45-years reanalysis of the global atmosphere and surface conditions 1957–2002.* ECMWF Newsletter 101. Reading: ECMWF.

Flageollet, J.-C., Maquaire, O., Martin, B. & Weber, D. 1999. Landslides and climatic conditions in the Barcelonnette and Vars basins (Southern French Alps, France). *Geomorphology* 30: 65–78.

Gibelin, A.L. & Déqué, M. 2003. Anthropogenic climate change over the Mediterranean region simulated by a global variable resolution model. *Climate Dynamic* 20: 327–339.

Iverson, R.M. 2000. Landslide triggering by rain infiltration. *Water Resources Research* 36(7): 1897–1910.

Janbu, N. 1957. Earth pressures and bearing capacity calculations by generalized procedure of slices. In: Soil Mechanics and Foundation Engineering; *Proc. intern. conf.*, London.

Malet, J.-P. & Maquaire, O. 2003. Black marl earthflows mobility and long-term seasonal dynamic in southeastern France. In Picarelli, L. (ed), Fast Slope Movements: Prediction and Prevention for Risk Mitigation; *Proc. intern. conf.*, Napoli, 11–13 May 2003. Bologna: Patron Editore.

Malet, J.-P., van Asch, Th.W.J., van Beek, L.P.H. & Maquaire, O. 2005. Forecasting the behaviour of complex landslides with a spatially distributed hydrological model. *Natural Hazards and Earth System Sciences* 2005-5: 1–15.

Maquaire, O., Malet, J.-P., Remaître, A., Locat, J., Klotz, S. & Guillon, J. 2003. Instability conditions of marly hillslopes, towards landsliding or gullying? The case of the Barcelonnette Basin, South East France. *Engineering Geology* 70(1–2): 109–130.

Mulder, H.F.H.M. 1991. Assessment of landslide hazard. Doctoral Thesis. Utrecht University: Utrecht.

Remaître, A., Malet, J.-P. & Maquaire, O. 2005. Morphology and sedimentology of a complex debris flow in a clay-shale basin. *Earth Surface Processes and Landforms* 30: 339–348.

Thiery, Y., Malet, J.-P., Sterlacchini, S., Puissant, A., Maquaire, O. in press. Landslide susceptibility assessment by bivariate methods at large scales. Application to a complex mountainous environment. *Geomorphology* (in press).

van Beek, L.P.H. 2002. The impact of land use and climatic change on slope stability in the Alcoy region, Spain. Doctoral Thesis. Utrecht University: Utrecht.

Verhaagen, P. 1988. Dendrogeomorphological investigations on a landslide in the Riou-Bourdoux Valley. Internal Report. Utrecht University: Utrecht.

Wesseling, C.G., Karssenberg, D., van Deursen, W.P.A. & Burrough, P.A. 1996. Integrating dynamic environmental models in GIS: the development of a dynamic modelling language. *Transactions in GIS* 1: 40–48.

Landslides and Climate Change – McInnes, Jakeways, Fairbank & Mathie (eds)
© 2007 Taylor & Francis Group, London, ISBN 978-0-415-44318-0

Pre-failure behaviour of slope materials and their significance in the progressive failure of landslides

J.M. Carey & R. Moore
Halcrow Group Ltd, Birmingham, UK

D.N. Petley
University of Durham, Department of Geography, UK

H.J. Siddle
Halcrow Group Ltd, Cardiff, UK

ABSTRACT: The concept of pre-failure behaviour within a landslide mass was first established by Terzaghi (1950) who noted that, for landslide failure to occur, a progressive reduction in shear strength was required at the shear surface. Similarly, observations by Varnes (1978) noted that deforming materials undergo pre-failure creep as a result of deformation of the material. These observations have been developed by others into methods for predicting slope failures and landslides. Despite these prediction methods, the mechanisms by which pre-failure creep occurs remains poorly understood. This paper develops a predictive pre-failure model through detailed analysis of surface movement patterns of first-time and pre-existing landslides. The analysis includes specialist pore-pressure re-inflation tests where elevated pore pressures are simulated under a constant deviatoric stress. The results demonstrate how subtle variations in pre-failure creep are related to material deformation and local porewater pressure conditions at the shear zone. This new research contributes to the development of site-specific ground behaviour and failure prediction models for complex landslides undergoing first-time and reactivation failures.

1 INTRODUCTION

It has long been established that a detailed understanding of the material physical, hydrological and geotechnical properties are essential to determine the potential of landslide initiation and reactivation (e.g. Varnes, 1978 Hutchinson, 1967; 1984; 2001). These factors have been recognised as key in controlling the material deformation and resultant movement characteristics of a slope (Petley et al., 2004; Petley et al., 2005a). However, progress in this field has been surprisingly limited over the last 35 years (Petley et al., 2005b).

The concept of understanding pre-failure behaviour within a landslide mass was first recognised by Terzaghi (1950), who identified the connection between material creep and landslides. Terzaghi noted that landslides, which appear to occur with no prior warning, could not take place unless the ratio between the average shearing stress on the potential surface of sliding had previously decreased from an initial value greater than one to unity at the instant of the landslide event. This highlights the requirement for a preceding gradual decrease of the ratio, which in turn, involves a progressive deformation of the slice of the material located at the landslide shear surface. Field observation of the opening of tension cracks, toe bulging and other site features are a sign of a slope in distress. In general, eyewitnesses to landslides that appear to occur as a surprise probably fail to detect these pre-failure strain indicators (Terzaghi, 1950).

Later, Varnes (1978) noted that a deforming material would progress through phases of strain development (Figure 1). This is characterised by a primary phase of accelerated creep related to a period of elastic deformation followed by a secondary creep phase of creep plastic deformation as the material continues to deform and weaken. The displacement curve then becomes concave upward as a tertiary phase of brittle deformation occurs. At this stage the material experiences a drop from peak to residual strength and as a consequence accelerates to final failure. However recent damage mechanics literature (e.g. Martel, 2003) suggests that this initial creep phase may not be elastic but the result of the contrasting process of strain hardening and strain weakening. During primary creep the initial increase in deformation is related to

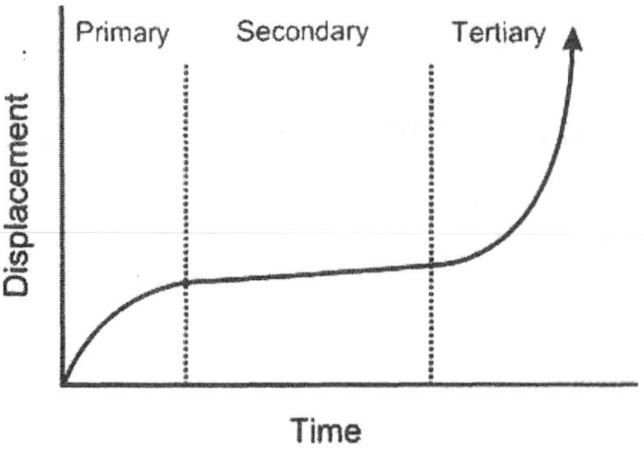

Figure 1. Three phrase creep model of deforming materials (after Varnes, 1978).

the domination of strain hardening forces. During secondary creep, strain hardening and weakening forces are both occurring almost in balance and therefore the rate of deformation is slower and almost constant. These strain hardening forces during this stage begin to slowly give way to the dominance of weakening forces. As a consequence deformation begins to increase, which may explain the inflection point often observed in many creep deformation patterns. During the tertiary phase the weakening forces clearly dominate and eventually deformation results in final failure.

Based on these observations, approaches for monitoring the pre-failure surface movements to forecast the time of failure has been established. This approach has been successfully adopted by a number of authors (e.g. Fukuzono, 1990; Main, 2000; Main et al., 1999) following the creep curve methods of Saito (1965). These methods were based on the concept that the time of failure of a slope can be predicted from creep rupture. This is based on the observation that the increment of the logarithm of acceleration was proportional to the logarithm of velocity of the surface displacement immediately before catastrophic failure (Fukuzono, 1990). Fukuzono proposed a simple method for predicting the time of failure using the reciprocal of mean velocity, which forms a negative linear trend immediately before catastrophic failure. Final failure can be calculated at the time when 1/v equates to zero and therefore the acceleration of the slope is infinite. Successful estimates have been made of landslide failure time based on the reciprocal of movement velocity against time as proposed by Fukuzono. However the success of these methods has been limited by uncertainty in the estimates of failure times because the fundamental physics controlling the nature and shape of the creep curves have yet to be elucidated (Hutchinson, 2001). Therefore to develop and understand the uses of these methods in interpreting landslide behaviour a more detailed understanding of the mechanics of progressive landslide development and in particular the progressive development of shear surfaces is essential. This must be critically dependent on the deformation properties of the failing materials, the groundwater and hydro-climatic regime of the landslide, and the dynamic interaction of these fundamental properties.

A number of previous literature has analysed the patterns of movement in landslides and their relationship to triggering factors such earthquakes, intense rainfall (e.g Allison & Brunsden, 1990, Warren and Palmer, 2000) and pore pressure alteration (e.g. Cooper et al., 1998). Only recently however have these movement patterns been successfully linked the deformation properties of the basal shear surface (Petley et al., 2005a) through the analysis of the 'Saito approach' which allows the prediction of progressive first time failures (Voight, 1988; Fukuzono, 1990). Recent research has highlighted that these methodologies may determine whether the basal surface is ductile or brittle which will have important implications on the mechanisms, cause and magnitude of failure.

2 PROGRESSIVE LANDSLIDE DEVELOPMENT MODEL

The concept of progressive failure in landslide systems has been used for over 35 years since its formulation by Bjerrum (1967). Surprisingly, the detailed mechanisms through which this process occurs remain poorly understood. One particular weakness is the lack of correlation between observations of movement in real landslide systems and the understanding of the deformation processes occurring at the basal shear zone (Martel, 2003). However, recent studies have analysed the deformation mechanisms and movement type together with novel pore pressure reinflation testing using a triaxial cell (Petley et al., 2005a). This research has been used to propose a new conceptual model for progressive failure (Petley et al., 2005a; 2005b).

The model demonstrates that a slope with no initial signs of instability (zero displacement) will undergo micro-cracking when pore pressures are elevated to a point where the effective strength of the material is sufficiently reduced. This process would cause irreversible deformation of materials forming the shear surface and as a consequence displacement will be observed. When the pore pressure rises beyond a key threshold, the effective strength is again lowered to a point where further micro-cracking will take place and further strain will develop. This continual deformation through raised pore pressure may occur multiple times over long timescales. For example, in some slopes the 1 in 100 year return interval rainfall event may be required to promote micro-cracking. The progressive weakening of a material to the point of failure, therefore, may occur over timescales that range from

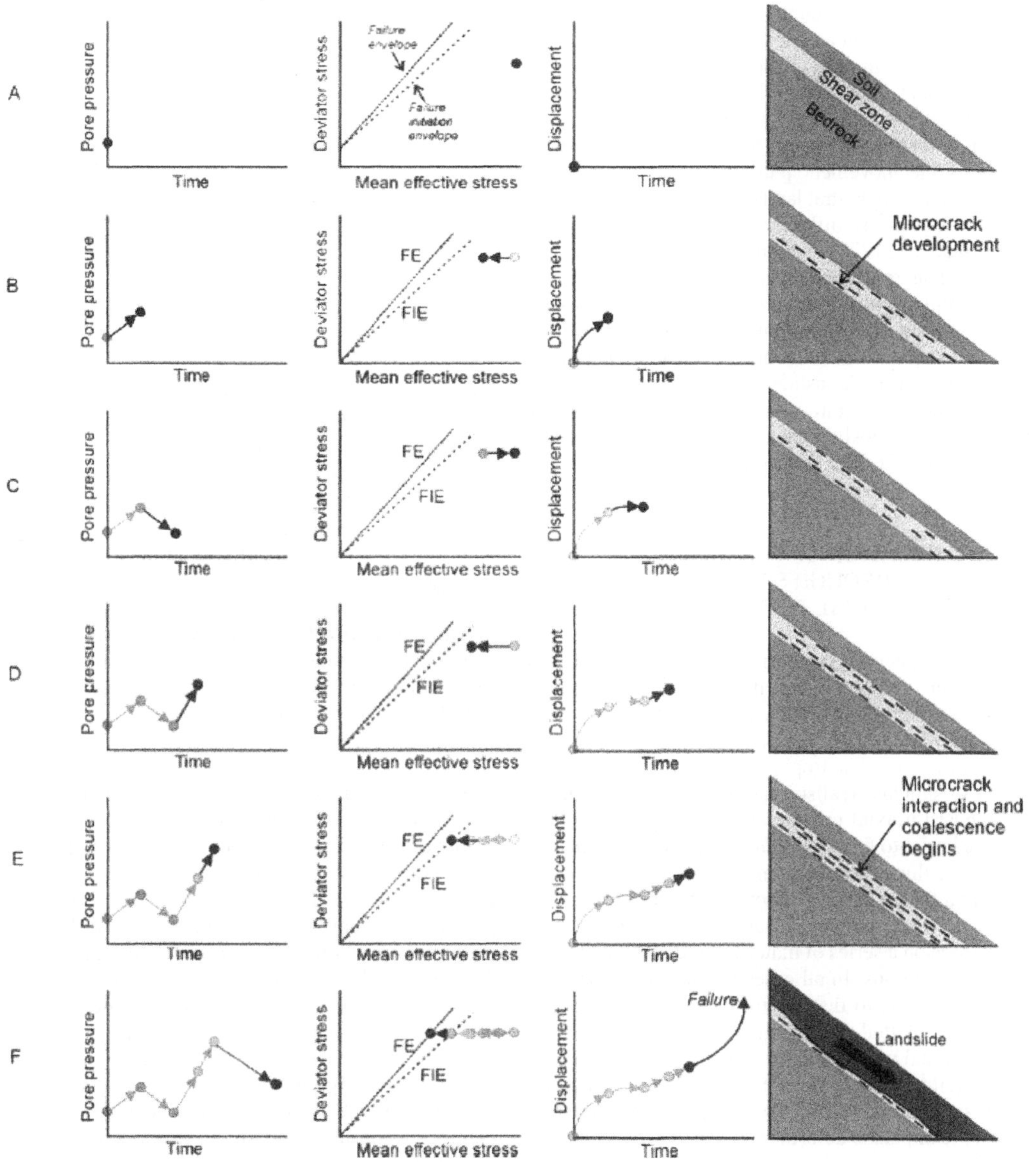

Figure 2. Progressive landslide development model proposed by Petley et al., (2005b).

seconds to millennia. The continued process will therefore reduce the operating shear strength at the shear zone. Final failure begins when the failure initiation line is reached (Petley et al., 2005b). At this point the micro-crack density within the sample becomes sufficiently high that interaction between them can occur, leading to an increase in the stress intensity at the tip of the adjacent micro-cracks, causing them to lengthen without a need for an associated change in the bulk stress state (Petley et al., 2005b). The lengthening of micro-cracks in turn increases the stress intensity at the crack tips of adjacent micro-cracks with each step, and thus the processes accelerates catastrophically leading to a hyperbolic increase in strain rate. Eventually the micro-cracks coalesce and a continuous shear surface is developed. At this point displacement is controlled by other factors such as the deviator stress acting upon the mass, the frictional resistance of the shear surface and the drop from peak to residual strength in the material. In all cases the landslide will fail but the rate of strain during runout is likely to be controlled by these other influences.

The model initially proposed by Petley et al., (2005a), and further developed by Petley et al., (2005b), explains the linearity in 1/v – t space as proposed by Saito (1965) and Voight (1988). It further explains the movement patterns observed in cohesive materials in natural landslides and correlates well with pore pressure inflation testing results (e.g. Zhu and Anderson, 1998; Tsukomoto, 2002; Petley et al., 2005a). This model requires further validation using real, complex landslide systems, especially those associated with landslide reactivation and landslides occurring in materials with little or no cohesion. Therefore, further testing of landslides moving on pre-existing shear surfaces is required to develop real time methods for assessing landslide behaviour and forecasting the potential and likelihood of future failure.

3 JUSTIFICATION OF THE LANDSLIDE MODEL THROUGH SPECIALIST GEOTECHNICAL TESTING

A series of specialist geotechnical tests have been performed on a series of different materials to determine pre-failure behaviour (Petley et al., 2005b, Petley and Allison 1996). Tests have been conducted in a triaxial cell. In these tests isotropic consolidation was undertaken to simulate realistic landslide failure surfaces. A standard triaxial cell was then performed but terminated prior to failure at a pre-determined deviatoric stress. At this point deviatoric stress was held constant and failure initiated through inflating pore water pressures in the sample. A number of tests have been performed on a series of materials and under different failure conditions. In all cases the deforming sample has illustrated two distinct forms of failure when 1/velocity is plotted through time (1/v – t space). First, a linear trend which is in all circumstances related to brittle failures where failure occurs along a singular shear surface, and second, an asymptotic trend related to ductile or plastic deformation which occurs along a plastic zone.

A series of standard consolidated drained and undrained triaxial experiments were carried out on both undisturbed and remoulded samples in order to define the failure envelope and the steady state envelope for this material. These results were used to undertake reinflation tests. Pore reinflation was undertaken at a nominal rate of 10 kPa per hour. The first experiment used an initial mean effective stress of 1000 kPa. The sample was subjected to undrained shear at a strain rate of 0.004 mm per minute to generate a deviatoric stress of c. 105 kPa. After allowing the sample to stabilise, reinflation of pore pressure was undertaken. The development of strain for the final

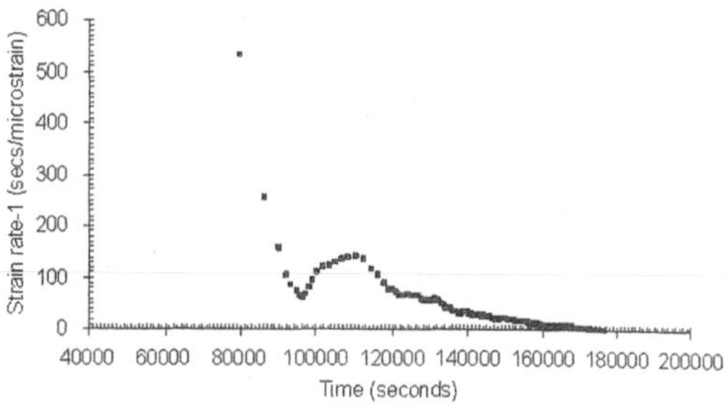

Figure 3. Full Saito plot for Pore pressure reinflation test at 1000 kPa (After Petley & Allison, 2006).

portion of the reinflation component illustrated a distinct linear trend in 1/v – t space with a very rapid rate of deformation at final failure.

A sample subjected to similar conditions but in a remoulded state showed a very different pattern of deformation. In this instance the sample shows clear asymptotic behaviour with strain accumulating throughout the majority of the sample and producing a final barrel shape.

Thus these two laboratory tests confirm linearity is associated to shear surface generation during brittle deformation. When brittle failure cannot occur the sample displays the asymptotic trend in 1/v – t space (Petley et al., 2005).

Further investigations of these relationships have been carried out where samples have been consolidated along a k0 uniaxial stress path to produce a realistic stress state in the sample. A series of monotonic drained compression tests were performed at a displacement rate of 0.05 mm per hour to determine the peak and residual failure envelopes. Four further pore pressure reinflation experiments have been undertaken from a pre-selected initial mean effective stress state until failure occurred. In all cases results demonstrate brittle failure along a singular shear surface plotting a general linear trend in 1/v – t space. The shapes of the 1/ velocity plots are informative. Analysis of the reinflation tests undertaken at 1000 kPa shows a complex trend in 1/v – t space. In this test the early stages of deformation show a rapid increase in strain rate followed by a period of strain rate decline. Strain rate begins to increase again at approximately 110 00 seconds displaying a curved trend through time. At approximately 166 000 the trend becomes linear but with small fluctuations from the linear trend (Figure 3).

When the same data were plotted against mean effective stress, (Figure 4) measurable deformation begins shortly after the residual strength failure envelope. Following this point increasing strain rate was continually recorded but its pattern was not linear until the peak strength failure envelope was reached. This linearity was then maintained to failure but with cyclic fluctuations in strain rate.

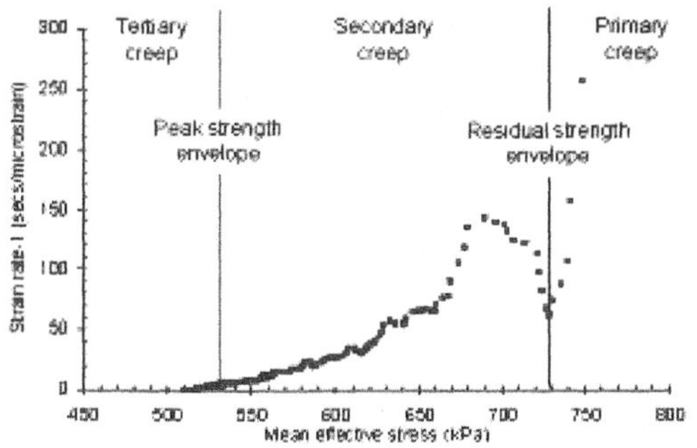

Figure 4. Interpretation of three phase creep behaviour of the 1000 kPa sample (After Petley & Allison, 2006).

These results appear to confirm that the three phase creep model is the result of brittle cracking processes, as suggested by Petley et al., (2002) and Kilburn and Petley (2003) and further illustrate the significance of the progressive failure model proposed by Petley et al., (2005a; 2005b)

4 CONTEMPORARY GROUND MOVEMENT CHARACTERISTICS VENTNOR ISLE OF WIGHT

A review of landsliding in Great Britain (DoE, 1994) identified the Ventnor Undercliff as the largest urban development affected by coastal landsliding in Britain. The Undercliff is situated on the south coast of the Isle of Wight between Blackgang and Luccombe, extending 12 km along the coastline and up to 1 km inland. The Undercliff is an ancient landslide complex of considerable age (Moore et al., 2007). The majority of the coastal landslides that form the Undercliff have been relatively stable in historical times, which has allowed occupation and development of the area. The area became popular during the Victorian period when much of the development seen today took place. The Undercliff is currently inhabited by a population in excess of 6,000 located in Ventnor, Bonchurch, St Lawrence and Niton.

The impacts of coastal instability and ground movement on urban development within the Undercliff have long been recognised (Lee & Moore, 1991). Evidence given by Aubrey Strahan of the Geological Survey, at the Royal Commission on Coast Erosion and Afforestation in 1906 highlighted the slow creep experienced in many parts of Ventnor. Contemporary movements within the town have been slight, however, the cumulative cost of damage to roads, buildings and services has been significant. Over the last 100 years about 50 houses and hotels have had to be demolished in Ventnor because of ground movement (Moore et al., 1995).

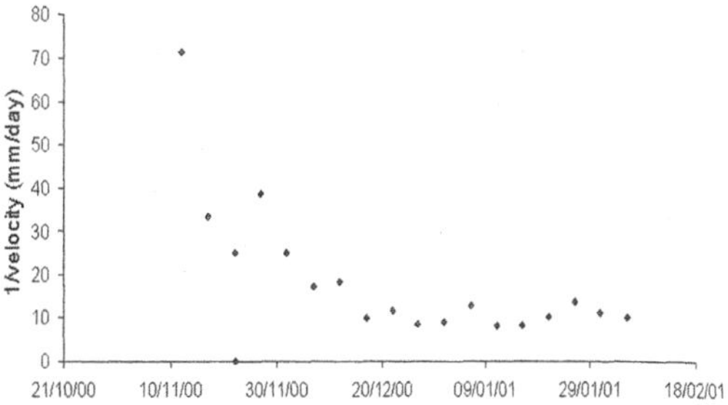

Figure 5. Average 1/v – t for the 2000–01 Acceleration of the Ventnor Landslide; recorded in crackmeter 1 at the Lowtherville graben.

Whilst the occurrences and damage to buildings is well documented and the relationships between landslide events and rainfall have been established (Moore et al., 2007) only recently have continuous, automated data of ground movement, rainfall, and pore water pressure at varying depths become available for Ventnor Town. Critically these patterns of movement and their relationship with the hydrogeological regime of the landslide system provide valuable evidence for the nature of the basal shear zone, how it is activated during intense and prolonged rainfall and whether rapid accelerations of the landslide system may be possible under future climate change scenarios. This information is critical if credible forecasting methodologies are to be established and sustainable remedial measures adopted.

Historical records over the past 200 years indicated that the Ventnor area has been continuously affected by ground movement and occasional landslide events (Moore et al., 1995; Lee and Moore, 1991). Ground movement in Upper Ventnor was first after 1954, suggesting that the Lowtherville graben is a relatively new feature. In recent years, detailed data have been collected through automated piezometers, crack meters, inclinometers and settlement cells located at strategic locations throughout Ventnor Town (Figure 5). These data can be correlated with automatic climate data collected in the town at Ventnor Park to analyse ground movement and the relationship with the landslide porewater pressure conditions.

Ground movement data collected from inclinometers (Moore et al., 2007) shows evidence of movement of approximately 30 mm from September 2002 to 16 August 2004 at a depth 97–98 m below ground level Movement is occurring along a pre-existing shear surface at the lower Gault Clay / Carstone transition zone. This is supported by shear surfaces recorded in boreholes from previous ground investigation (Halcrow, 2006).

Two automatic crack meters were installed at the Lowtherville graben, Newport Road in December 1995, recording displacement every three hours. Initial

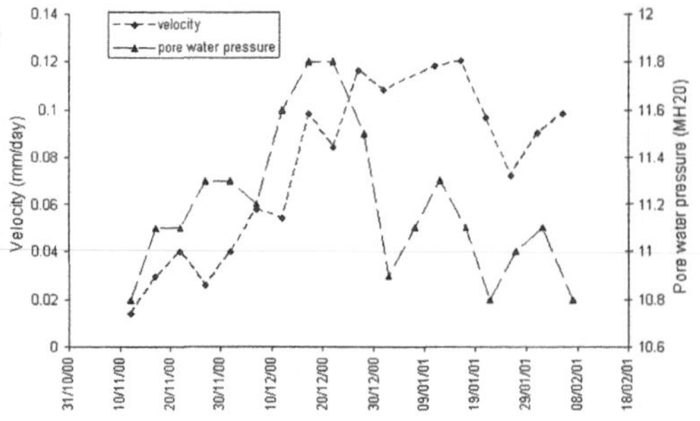

Figure 6. Relationship between acceleration at the Lowtherville graben and porewater pressure conditions recorded at the Winter Gardens.

analysis of this data set shows that movement has occurred at a relatively constant rate of approximately (6.64 mm/yr) from 1994 to 2000, followed by a period of accelerated movement from November 2000 to February 2001 when approximately 6.4 mm displacement recorded over a 4 month period (Moore et al., 2007).

Analysing this period of acceleration in terms of 1/v – t space shows an asymptotic curve (Figure 5), which has been linked to ductile deformation or movement along a predefined shear surface (Petley et al., 2002). Crack meter data have been collated with pore water pressures recorded at the Winter Gardens. Figure 6 suggests that the Ventnor landslide crept as a result of raised pore pressures and then returned to it initial displacement rate as pore pressures dissipated. This behaviour is consistent with previously observed patterns of movement (e.g. Allison and Brunsden, 1990; Moore and Brunsden, 1996) where deformation is occurring along a ductile shear zone.

5 FIRST TIME FAILURE – NEW TREDEGAR LANDLSIDE SOUTH WALES

The New Tredegar landslide is situated on the west facing side of the Rhymney Valley in Caerphilly County Borough, South Wales. The valley at this location has a depth of c.350 m and a width of just over 1 km and the landslide occupies an area of 25 ha. The site can be divided into three units; (1) an old landside of unknown age, (2) A landslide that occurred in 1905, (3) a destructive landslide which occurred in 1930 (Bentley, & Siddle, 2000). The morphology of the area of the 1930 landslide indicates that failure occurred along a curvilinear surface within the Rhondda Beds. A sub-vertical joint set in the Brithdir sandstones formed the upper part of the failure surface. Aerial photography, taken prior to 1960, reveals that mudslides had taken place from the toe area of the rotated block (Halcrow, 1989). There have been no subsurface investigations of the landslide, although the general geological sequence has been established from shaft sections in the Elliot pits approximately 2 km from the site (Halcrow, 1989). The sequence comprises Brithdir sandstone overlying Rhondda Beds. The seam crops at the base of the sandstone and has been extensively mined in the uphill margin of the landslide. The rocks generally dip at about 2° SW, obliquely out of the hillside. The first evidence of movements to affect the Colliery occurred as a result of railway excavations into older landslide deposits. The area continued to suffer gradual movements damaging a road and railway embankment, but following extreme rainfall a large landslide was initiated in March 1905. This landslide displaced the road laterally up to 50 feet and heaved the railway line by up to 4 feet (Knox, 1927).

Following remedial works the colliery reopened but had to be closed again in 1927, reopening in 1930. However, development of new fissures caused so much concern that the colliery commissioned daily of pegs on the slope above the colliery (Bentley & Siddle, 2000). These records provide the only example known of monitored displacements prior to a major landslide event in South Wales.

Figure 7 illustrates how this model may be applied to the New Tredegar landslide. The initial movement occurs as pore pressures are elevated to a point where deformation begins. At this stage strain hardening processes are elevated above the strain weakening processes, expressed in 1/v – t space as an asymptotic trend that represents initial creep (Figure 7a). This period is followed by secondary creep, characterised by a period of slower deformation related to the increase in weakening processes that act to balance the strain hardening processes. This movement is characterised by continued accelerated creep phases during periods in which pore pressures are inflated sufficiently to reduce mean effective stresses and develop further micro-cracking. At this stage micro-cracking is still spread out throughout a large area of the material and therefore movements are asymptotic to reflect the apparent overall ductile behaviour of the landslide (Figure 7b). During these periods the overall material shear strength of the shear zone is reducing, and so it is likely that the mean effective stress state required at which movement can occur is continually increasing. Eventually, the landslide passes into a state of tertiary creep (Figure 7c), which in this case occurs at an elevated but not exceptional groundwater level. This causes a transition in the behaviour of the shear surface similar to phase E in the Petley et al., (2005b) model where the material strength properties reduce, and the micro-crack density within the shear zone is sufficiently high that interaction between them occurs. At this stage strain hardening processes have been dramatically reduced and strain weakening processes are dominant as a singular shear surface develops.

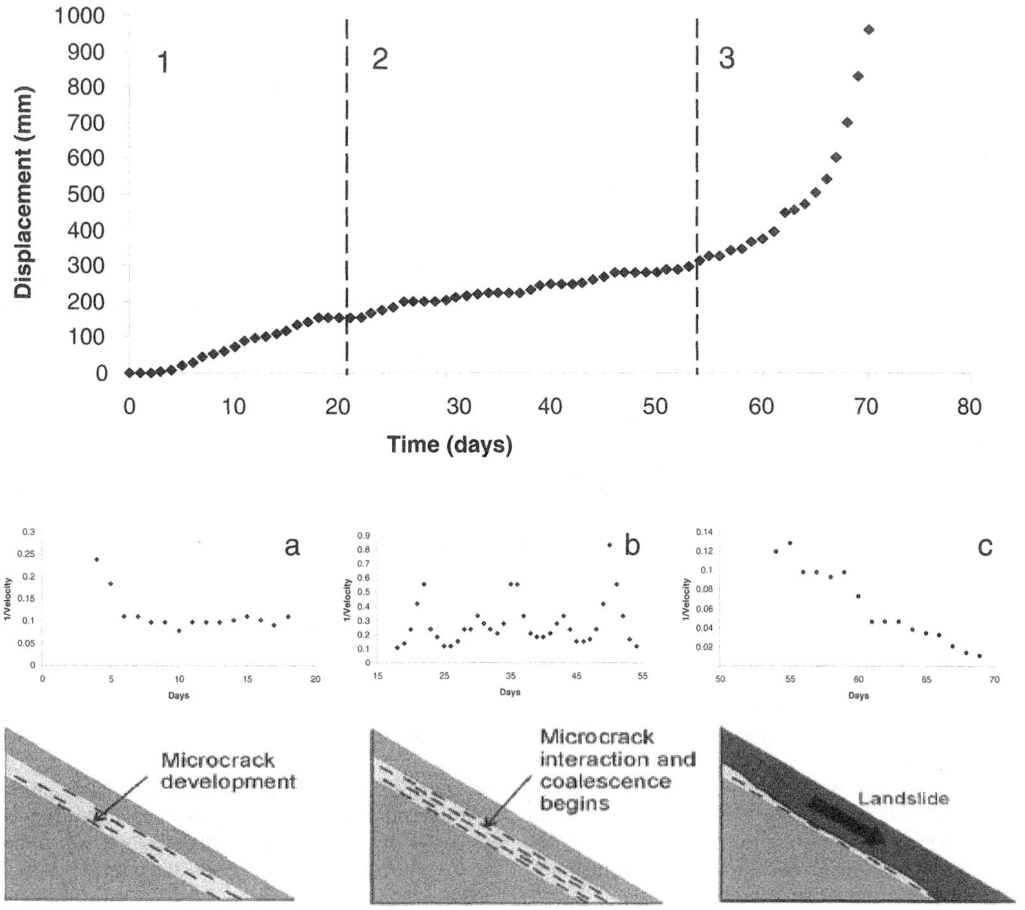

Figure 7. Progressive development of the New Tredegar landslide (Daily displacement records after Bentley and Siddle, 2000).

It is also noted that the deformation patterns in the landslide begin to become divorced from the rainfall characteristics as the reduction of rainfall at approximately day 53 is not sufficient to prevent the landslides progression to the linear trend. It is clear that at this stage the landslide is no longer capable of slowing down. This loss of relationship between movement and groundwater conditions was also observed at Vaiont where attempts to reduce lake water levels were not sufficient to prevent the final catastrophic acceleration (Petley and Petley, 2005).

Therefore at the final stage, the landslide continues to accelerate as the micro-cracks coalesce and lengthen. The shear stresses are now concentrated at the micro-crack tips. The stress concentration in the unsheared material under a constant shear load allows the rate deformation to accelerate geometrically. This period of final shear surface development is depicted by the linear trend in $1/v - t$ space, and the fluctuations in rainfall during this period have limited impact on the rate deformation. It is clear therefore at this stage that the shear surface development is now strain controlled until a singular shear surface is formed. At this point the landslide fails as the theoretical displacement rate becomes infinite and other processes start to reduce the strain rate. During the final stages of deformation a transition point between apparent ductile/ plastic to purely brittle behaviour is observed in $1/v - t$ space (Figure 7b–c). This pattern of deformation is similar to that demonstrated at Vaiont prior to the catastrophic 1963 disaster (Petley & Petley, 2005).

The 1930 New Tredegar landslide event supports the proposed models of progressive development of landslide failure proposed by Petley et al., (2005a; 2005b). These data further support the recent analysis of the Vaiont dam landslide (Petley & Petley, 2005) illustrating the early ductile movement patterns followed by a transitional trend from ductile deformation to brittle deformation. It further supports the existence of 'Saito' linearity during the final phase of failure (Fukuzono, 1990). The development of the linear trend in 1/v – t is recognisable at approximately day 56 which further highlights the potential use of this method for landslide forecasting and landslide hazard warning.

6 DISCUSSION AND IMPLICATIONS OF CLIMATE CHANGE ON FUTURE LANDSLIDING

Four climate change scenarios have been estimated by UKCIP based on up to date IPCC global climate models. These scenarios estimate an increase in the UK temperature of between 2 and 3.5 degrees with a 20–30% increase in winter rainfall and possibly a 50% increase in summer rainfall. Sea level changes on the South Coast could rise by between 26 and 86 cm with extreme sea levels becoming 10 to 20 times more frequent (Hulme et al., 2002).

The progressive landslide development model illustrates the connectivity between porewater pressures and shear surface development in both first time failures and continued movement in existing landslide complexes. The porewater pressures in landslides are controlled by rainfall events, as shown at both Ventnor and New Tredegar, and therefore the UKCIP predicted increases in rainfall are likely to increase the likelihood of future landslide events. This could increase the frequency of first time failures and the frequency and magnitude of reactivation failures.

The correlation between landslide behaviour where ductile deformation is occurring or landslides are moving along pre-exiting shear surfaces suggests that their could be a direct relationship between porewater pressures and ground acceleration once the material residual strength is reached. Further rigorous laboratory and landslide monitoring, however, are required to establish this relationship more thoroughly. From the evidence presented, however, it is possible to hypothesize that continued ground movements will occur in landslide complexes where the mean effective stress state in the landslide is reduced to the residual strength of the shear surface material. Once porewater pressures are sufficiently high to create these conditions it is likely that the rate and degree of movement will be controlled by the porewater pressure characteristics acting at the shear surface.

In the case of first time failures the increase in storm frequency and prolonged rainfall events would be expected to create an increased risk of both pre-failure creep and first time failures. When porewater pressures are elevated to a point where the material reaches the residual strength envelope creep will occur. These movements are likely to increase and decrease in magnitude and frequency as the pore pressures fluctuate. During the secondary creep phase it is probable that the porewater pressures required to cause these creep movements will continually reduce as the degree of micro-cracking across the shear surface progressively reduces the overall effective stress state of the material. When the pore pressure conditions and combined reduction in effective stress reach the peak strength failure envelope a shear surface will develop and catastrophic landslide movement can be expected. Under the climate change scenarios the increases in rainfall will exacerbate antecedent groundwater conditions. This will increase the potential of elevated porewater pressures being experienced more frequently in marginally stable slopes and therefore the potential for landsliding will be increased.

7 CONCLUSIONS

The results demonstrate the relationship between porewater pressure conditions and anticipated ground movements in both pre-existing landslide complexes and first time failures of marginally stable slopes. The movement patterns recorded at New Tredegar demonstrate the secondary creep phase of movement in a slope progressing to first time failure. Analysis in 1/v – t space reveals that the slope progresses through a series of asymptotic ground movements before the negative linear trend is established up to final failure. These results support the progressive landslide model proposed by Petley et al., (2005) and are supported by the pore pressure reinflation tests carried out on intact samples.

The ground acceleration characteristics at Ventnor support the anticipated post failure behaviour of a landslide moving on a previously deformed shear surface. The results suggest that movement occurs above a specific pore pressure level above which acceleration and deceleration is directly related to the pore water pressure characteristics experienced at the basal shear zone.

When these results are considered in relation to future climate change predictions it is evident that both first time failures and landslide reactivations could be expected to increase in the future. Further research is required to establish stronger links between ground movements and pore water pressures which will involve both ongoing specialist testing of material behaviour under differing elevated porewater pressure scenarios and further collecting and collation of ground movement data from different landslide sites. This paper however demonstrates that effective monitoring of landslide movement patterns, groundwater and rainfall conditions combined with an understanding of material behaviour under elevated porewater pressures may offer a real time method for forecasting potential future hazardous ground movements.

ACKNOWLEDGEMENTS

The research presented in this paper is part of an ongoing PhD project being undertaken by collaboration between Halcrow Group Ltd and the International Landslide Centre, University of Durham. The monitoring data has been collected by the Isle of Wight Council Centre for the Coastal Environment as part of the Landslide Management Strategy.

REFERENCES

Allison, R.J, and Brunsden, D. (1990). Some mudslide movement patterns. Earth Surface Processes and Landforms, 15, (4), 297–311.

Bentley, S.P, and Siddle, H.J. (2000) New Tredegar Landslide, Rhymney Valley, *In: Landslides and landslide management in South Wales*.

Bjerrum, L. (1967) Progressive failure in slopes of overconsolidated plastic clay and clay shales, *Journal of the Soil Mechanics Foundation Division of the American Society of Civil Engineers*, 93, 1–49.

Cooper, M.R; Bromhead, E.N; Petley, D.J and Grant, D.I. (1998). Selbourne cutting stability experiment. Geotechnique. 48, 83–101.

Fukuzono, T. (1990) Recent studies on the time prediction of slope failures, *Landslide News*, 4, 9–12.

Halcrow (2006). Ventnor Undercliff, Isle of Wight: Coastal Instability Risk, Interpretative Report & Quantitative Risk Analysis, Isle of Wight Council.

Halcrow. (1989) Landslide and Undermining Project, Final Report, Department of the Environment, Welsh Office.

Hulme M., Jenkins G.J., Lu X., Turnpenny J.R., Mitchell T.D., Jones R.G., Lowe J., Murphy J.M., Hassell D., Boorman P., McDonald R., & Hill S. 2002 Climate Change Scenarios for the United Kingdom: The UKCIP02 Scientific Report. Tyndall Centre for Climate Change Research, School of Environmental Sciences, University of East Anglia, Norwich, UK.

Hutchinson, J.N. (1967) Mechanisms of producing large displacements in landslides on pre-existing shears, *Memoir of the Geological Society of China*, 9, 175–200.

Hutchinson, J.N. (1984) Landslides in Britain and their counter measures, *Journal of Japan Landslide Society*. 12, 1–25.

Hutchinson, J.N. (2001) Landslide risk- to know to foresee, to prevent, Geologia Tecnica and Anbientale, 9, 3–24.

Kilburn, C.J, and Petley, D.N. (2003). Forecasting giant, catastrophic slope collapse: lessons from Vajont, Northern Italy. *Geomorphology*, 54, (1–2), 21–32.

Knox, G. (1927) Landslides in South Wales Valleys, *Proceedings of the South Wales Institute of Engineers*, 43

Lee, E.M. and Moore, R. (1991) Coastal Landslip Potential Assessment: *Isle of Wight Undercliff, Ventnor, Technical Report*. Department of the Environment.

Main, I. (2000). A damage mechanics model for power-law creep and earthquake aftershock and foreshock sequences. *Geophysical Journal International*, 142 (1), 151–161.

Main, I.G. (1999). A damage mechanics model for power law creep and earthquake aftershock and foreshock sequences, *Geophysical Journal International*, 139, F1–F6.

Martel, S.J. (2003) Mechanics of landslide initiation as a shear fracture phenomenon, *Marine Geology*, 203, 319–339

Moore, R, and Brunsden, D. (1996). A physio-mechanism of seasonal mudsliding. *Geotechnique*. vol. XLVI (2), 259–278.

Moore R; Carey, J.M; McInnes, R and Houghton, J. (2007). Climate change, so what? Implications for ground movement and landslide event frequency in the Ventnor Undercliff, Isle of Wight. This Conference.

Moore, R ; Lee, M, and Clark, A.R. (1995). The Undercliff of the Isle of Wight: A review of behaviour. Cross Publishing.

Petley, D.N, and Allison, R.J. (2006). 'On the movement of landslides,' *International Conference on Slopes, Malasyia, Special lecture*, 115–137.

Petley, D.N, and Allison R.J. (1997). The mechanics of deep-seated landslides. Earth Surface *Processes and Landforms*, 22, 747–722.

Petley D.N; Bulmer, M.H.K, and Murphy, W. (2002) Patterns of movement in rotational and translational landslides, *Geology*, 30, 719–722.

Petley, D.N; Higuchi, T; Petley, D.J; Bulmer, M.H; and Carey, J. (2005a). The development of progressive landslide failure in cohesive materials. *Geology*, 33, (3), 201–204.

Petley, D.N; Higuchi, T; Dunning, S; Rosser, N.J; Petley, D.J; Bulmer, M.H, and Carey, J. (2005b). A new model for the development of movement in progressive landslides, In: Hungr, O; Fell, R; Couture, R and Eberhardt,E. *Landslide Risk Management*, A.T. Balkema, Amsterdam.

Petley and Petley (2005). On the Initiation of large Rockslides: Perspectives from a new analysis of the Vaiont movement record, In: Evans, S. (de.) *Large Rock Slope Failures*, Kulwer: Rotterdam (NATO science series).

Saito, M. (1965). Forecasting the time and occurrence of a slope failure. *Proceedings of the 6th International Conference on Soil Mechanics and Foundation Engineering*, 2, 537–541.

Terzaghi, K. (1950). 'Mechanisms of landslides,' *Geological Society of America*, Berkley volume, 83–123.

Tsukomoto (2002). Determination of shear strength of Hawaiian residual soil subjected to rainfall induced landslides. *Geotechnique*, 52, 143–144.

Varnes, D.J. (1978). Slope movement types and processes, in Schuster, R.L; Krizek, R.J. (eds), *Landslides-Analysis and Control*. National Academy of Sciences Transportation Research Board Special Report, 176, 12–33.

Voight, B. (1988) A relation to describe rate dependent material failure, *Science*, 243, 200–203.

Warren, C.D., & Palmer, M.J. 2000. Observations on the nature of landslipped strata, Folkestone Warren, United Kingdom. *Landslides in research, theory and practice, Proc. 8th Intern. Symp on Landslides, 3, Cardiff,* 1551–1556.

Landslides and Climate Change – McInnes, Jakeways, Fairbank & Mathie (eds)
 ISBN 978-0-415-44318-0

Physical and numerical modelling of chalk slopes

K.J.L Stone
University of Brighton, Brighton UK

V. Merrien-Soukatchoff
Ecole des Mines, Nancy, France

ABSTRACT: The results of a centrifuge model test of a chalk slope are presented to demonstrate a novel modelling technique where the prototype rock mass is modelled using a naturally occurring micro-fractured chalk. This approach offers the ability to replicate at reduced scale the characteristics of the prototype rock mass. Consequently parametric studies can be carried out to investigate the influence of climate change, such as increased precipitation and wave action on the stability of coastal chalk cliffs. Furthermore a numerical analysis of the centrifuge test is reported to illustrate the use of the centrifuge modelling technique as a tool in the validation and verification of numerical modelling techniques and procedures.

1 INTRODUCTION

The impact of climate change over the next few decades is likely to have a significant effect on the erosion and weathering of the UK coastline. Rising sea levels, combined with increased precipitation and more frequent and violent storms will have a significant impact on the mechanisms that drive erosive processes in the coastal environment. In particular, the soft chalk cliffs of south and south-east England are particularly vulnerable to these changing climatic factors and increased recession rates as the result of more frequent and extensive collapses are predicted. It is thus important that the mechanisms that govern these collapses, and the influence of climatic changes on these mechanisms, are well understood such that effective management programmes can be created and implemented.

The stability of chalk cliffs, and the parameters that influence that stability, can be investigated at field scale using monitoring and back analysis of actual failures including the application of advanced numerical modelling techniques, or at reduced scale using physical model tests. This paper presents a method for replicating a fractured chalk mass at reduced scale for testing in a geotechnical centrifuge. The results from such physical model tests can be used to validate and verify numerical approaches as well to investigate specific prototype scenarios, and in particular the role of climatic factors on stability.

1.1 *Modelling a rock mass*

To date a significant obstacle in the development and validation of models of the behaviour of jointed rock masses has been the difficulty of creating realistic physical models of the jointed rock mass for study at the laboratory scale.

Past research has mainly utilised synthetic rock materials, usually mixtures of cementatious materials (gypsum or Portland cement) and fine aggregates such as sand or kaolin. This approach has proved very successful in modelling homogenous weak rock materials for example, Johnston and Choi (1986), Jewell et al (1992), Indraratna (1990). For a comprehensive review of the use of synthetic rock modelling materials reference is made to Stimpson (1970). However, while these methods are particularly useful for producing models of homogenous rock masses, they are of limited use for replicating a jointed rock mass. This is primarily due to the inherent problem of correctly replicating the joint features of the rock mass, for example the joint strength and stiffness characteristics, and the random nature of the joint sets.

Attempts to replicate a jointed rock mass by construction of 'brick' models (Wold, 1984 and 1985) have provided useful qualitative data but these are of limited value since they do not correctly model the joint characteristics. A similar approach was also used to generate models with inclined joint sets by casting thin plate elements for toppling studies (Stewart et al., 1994). A different approach involving the cutting of joint sets into a setting mixture (Rowlands, 1992), was successfully used in conjunction with a base friction apparatus to replicate self-weight effects (Bray and Goodman 1981). Stone et al. (1992) constructed layered gypsum models with planes of weakness to represent horizontal bedding sets for use in centrifuge model tests.

In this paper a new approach to modelling a fractured rock mass is demonstrated where a model slope is formed from naturally occurring micro-fractured chalk. The material is found in the peri-glacial zone of the upper chalk formation (Mortimore 1979). Using this material for the model has the significant advantage that the random nature of the joint sets and associated joint characteristics of the prototype material are preserved. Neither of these characteristics can be reproduced in synthetic rock models.

2 CENTRIFUGE MODELING

Geomechanical materials, such as soil and rock, are dependent on stress level. In conventional small scale model tests, performed in the earths gravitational field, it is not always possible to maintain similarity with prototype situations, and to ensure that stress levels in areas of interest reach prototype values. A geotechnical centrifuge can subject small models to centripetal accelerations which are many times the earth's gravitational acceleration. By selecting a suitable acceleration level the unit weight of the material being tested can be increased by the same proportion by which the model dimensions have been reduced. Thus stresses at geometrically similar points in the model and prototype will be the same. In other words, if a 1/N scale model of a prototype is spun at Ng on the centrifuge, then the model's behaviour is thought to be similar to the prototype's behaviour.

The centrifuge model test reported here was carried out on the 1.8 m balanced beam centrifuge at City University Geotechnical Centrifuge Centre.

2.1 *Scaling relationships*

In order for a model to correctly represent a corresponding prototype, it is necessary that the physical phenomena that control the prototype response are correctly represented in the model. In order to achieve this, scaling relationships are derived to maintain the proportional influence of prototype parameters at reduced scale. The most effective method for deriving scaling relationships is through the use of dimensional analysis. Derivation of scaling relationships applicable to rock mechanics problems have been presented in detail elsewhere, and in particular for centrifuge model testing the reader is referred to Hoek (1965) and Clark 1981. A summary of some basic scaling relationships for centrifuge testing of models is presented below in Table 1.

Table 1. Centrifuge scaling relationships.

Parameter	Scaling factor
Acceleration	N
Seepage velocity	N
Length	1/N
Stress	1
Strain	1
Force	$1/N^2$
Time (diffusion processes)	$1/N^2$
Time (creep)	1
Energy	N^3

Figure 1. In-situ micro-fractured chalk (fracture spacing 1–5 cm).

2.2 *Model preparation*

The centrifuge model was constructed from a block of micro-fractured chalk obtained from an exposure at Newhaven near Brighton. This material is periglacial chalk and is usually encountered in a narrow band (4–5 m thick) immediately beneath the topsoil. The chalk is highly fractured with a typical fracture spacing of between 2–5 cm, see Figure 1. Downward percolation has resulted in the transport and deposition of fine grained material within the fractures of the chalk. The periglacial soils associated with the chalk are typically clayey silts with liquid and plastic limits of 30 and 20% respectively (Fookes and Best 1969).

A block of material (approximately 180 mm wide × 400 mm long × 300 mm high) was cut-out of the exposed periglacial formation and transported to the laboratory. After trimming the block, it was noted

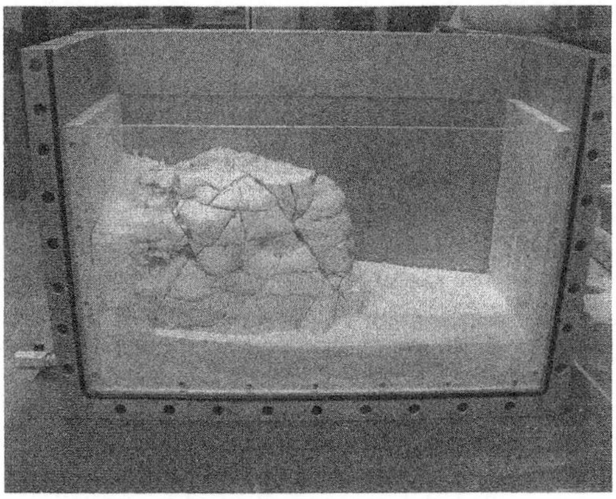

Figure 2. Completed model prior to centrifuge test.

that the sample contained a joint structure that would be suitable for a simple slope stability study without any further cutting or reshaping.

Figure 2 shows the completed model prior to transportation to City University for testing. To prevent damage to the model during transport, it was covered by cling-film and the box filled with sand, such that the model was fully restrained.

2.3 *Centrifuge test procedure*

The model was placed in a centrifuge strong box, one side of which contained a window through which the model could be viewed. Two displacement transducers were set up to monitor displacements of the model surface. The model was also monitored by video cameras. One camera viewed the model via the side window and was set up so that still images could be captured and logged throughout the test.

The test plan was to induce failure of the model slope by increasing the g-level. Failure would result if slippage occurred on the clay filled joints developing a failure mechanism, or as the result of breakage of intact rock triggering a slope failure. To capture the model response video images were saved every second as the centrifuge acceleration increased. Data from the displacement transducers was also logged.

2.4 *Results*

It was anticipated that failure of the model would result from either sliding along clay filled joints or a compression failure of the material leading to a rock fall, or perhaps a combination of both.

Since the clay filled joints were clearly dry prior to the test, the model was soaked for about 40 minutes in an attempt to generate an undrained rather than a drained or frictional response, which would of course be independent of g-level. This was partially successful and some slippage occurred as the g-level

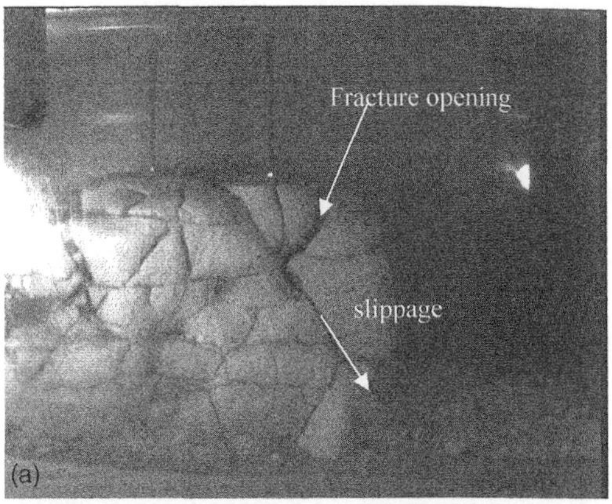

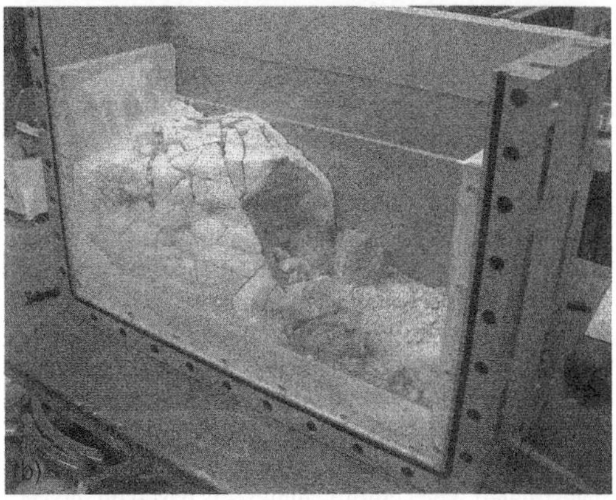

Figure 3. (a) Model in-flight on the centrifuge showing significant slippage on main joint and (b) post-test photograph showing failed slope and debris.

was increased as indicated in Figure 3a. However, the clay was not sufficiently softened to cause a complete failure mechanism to form.

In order to trigger a failure, material was removed such that the slope toe was partially undermined. During the subsequent centrifuge run a failure of the slope was observed at an acceleration of 78 g. The failure mechanism was due to fracturing of material over the undermined slope toe leading to further collapse and sliding of the blocks above. The failed model is shown in Figure 3b where the wedge shaped failure scar is clearly visible together with the fallen blocks.

3 NUMERICAL MODELLING

A simple analysis of the centrifuge model was carried out using the block generator RESOBLOK developed at the Ecole des Mines, Nancy. This program has a downstream module for Block Stability Analysis (BSA) which allows studying the stability of blocks

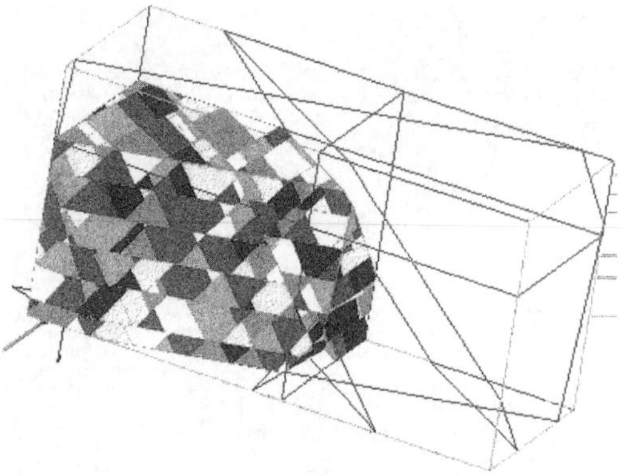

Figure 4. RESOBLOK model used for limiting equilibrium analysis.

next to an excavation or a topographic surface by simple computations based on limit equilibrium (Asof, 1991). The possible movements considered are free fall, sliding (along one or several planes) and rotation. The blocks are assumed to be non-deformable. The density of the blocks as well as joint properties, cohesion and friction angle, are input parameters required to perform the limiting equilibrium stability analysis. The computation is iterative: in the first step BSA checks if isolated blocks are able to detach. The unstable blocks are removed and the boundaries of the model re-defined to take into account the new shape. In following steps the stability of the remaining blocks are investigated. The stability analysis is based on the vectorial Warburton method (Baroudi et al., 1990) and follows two stages: a geometrical evaluation and a mechanical computation.

In order to simulate the effect of enhanced gravity and self weight forces in the centrifuge test, the density of the block material was increased by multiplying it by the acceleration level. The RESOBLOK model generated for the analysis is shown in Figure 4.

The results of the BSA analysis indicated that no blocks were unstable before removing material, in the form of a small block, from the toe of the slope, a process which simulates the undercutting of the model toe carried out in the centrifuge test. This is entirely consistent with the centrifuge model where no instability is apparent until undercutting of the toe is carried out. However after removing the small block, the model is very unstable and very weak, and collapse results.

4 DISCUSSION

From the results presented it is apparent that the behaviour of the model, as a fractured rock mass, correlates well to the expected behaviour of an equivalent prototype. For example, the slippage observed on the clay filled joints is a typical failure mechanism observed in the field during periods of prolonged precipitation. The final observed failure mechanism triggered by undercutting at the cliff base is a significant factor responsible for triggering failures in the chalk cliffs of Southern England. It is clear from the test results, that once collapse has been triggered, failure of the rock mass is dictated by the joint geometry. This observation is significant, in that the results of the model are not necessarily specific to chalk if the fracture geometry is the controlling factor. However, if the joint properties and intact rock properties are the controlling factors that determine the response of the model, then the test result is specific to the model material properties (i.e. the intact chalk strength and joint characteristics). This is likely to be the case when the stresses in the rock mass are large (i.e. corresponding to significant depths).

The limiting equilibrium analysis demonstrates that the physical model can indeed be used as a calibration tool for numerical models. Whilst the numerical approach reported here is of a fairly basic nature it clearly illustrates that if the geometry of the rock mass is well characterized then an appropriate solution can be achieved. Removal of 'blocks' to simulate erosion and initiate collapse is clearly an important tool to predict cliff collapse along vulnerable sections of coastline.

5 CONCLUSIONS

From the numerical and physical modelling procedures reported in this paper the following conclusions can be drawn.

- reduced scale models of a fractured rock mass can be made using naturally occurring micro-fractured chalk
- the behaviour of the micro-fractured chalk model adequately captures the behaviour of an equivalent prototype rock mass
- the numerical analysis of the centrifuge model was able to capture the observed failure mechanism and predict the response of the rock mass in terms of the extent and severity of the collapse.

ACKNOWLEDGEMENTS

The authors would like to express their thanks to Prof. Neil Taylor of City University Geotechnical Centrifuge Centre for his assistance in performing the centrifuge test.

REFERENCES

Asof M (1991) Etude du comportement mécanique des massifs rocheux fractures en blocs (méthode à l'équilibre

limite) : réalisation et application. *Ph.D. thesis*, LAEGO, Ecole des Mines, INPL, France, 142 p.
Baroudi H; Piguet JP.; Chambon C.; Asof M. (1990) Utilization of the block generator "Resoblok" to complex geologic conditions in an open pit mine, *Proceedings of the International conference on Mechanics of jointed and faulted rock*, Vienna, Austria, April 18–20, 1990, Edited by A. A. Balkema. pp 529–535.
Bray JW and RE Goodman (1981) "The Theory of Base Friction Models." Int. Journal of Rock Mech. and Min. Sci., Vol. 18, pp.453–468.
Clark GB (1981) "Geotechnical centrifuges for model studies and physical property testing of rock and rock structures." Colorado School of Mines Quarterly, Vol. 76, No. 4.
Fookes P and R Best (1969) "Consolidation characteristics of some late Pleistocene periglacial metastable soils of east Kent." Q. Jl. Engng. Geol. Vol.2. pp.103–128.
Hoek E (1965) "The design of a centrifuge for the simulation of gravitational force fields in mine models." Journal of S African Inst. of Mining and Metallurgy. Vol. 65 No. 9, pp.455–487.
Indraranta B (1990) "Development and applications of a synthetic material to simulate soft sedimentary rocks." Geotechnique, 40 (2), 189–200.
Jewell RJ, Stone KJL and D Adhikary (1992) "Modelling the stability of rock slopes." Western Australian Conf. on Min. Geomech., Szwedzicki, Baird, Little (eds), Kalgoorlie, WA.
Johnston IW and SK Choi (1986) "A synthetic soft rock for laboratory model studies." Geotechnique, 36 (2), 251–263.
Mortimore RN (1979) "The relationship of stratigraphy and teconofacies to the physical properties of the White Chalk of Sussex." PhD Thesis, Brighton.
Rowlands D (1992) "Surface subsidence associated with partial extraction: An Australian case study." Effects of Geomechanics on Mine Design, Kidybinski and Dubinski (eds), pp.27–45, Balkema, Rotterdam.
Stewart DP, Adhikary D and RJ Jewell (1994) "Studies on the stability of rock slopes." Centrifuge'94, Leung, Lee and Tan (eds), Balkema, Rotterdam, ISBN 9054103523
Stimpson B (1970) "Modelling material for engineering rock mechanics." Int. Jour. Rock Mechs. Min. Sci., 7, 77–121.
Stone KJL, Jewell RJ and I Misich (1992) "Surface and subsurface subsidence over coal mines." Proceedings, 6th Australian-New Zealand Conference on Geomechnics, Christchurch, New Zealand, pp 269–273.
Wold MB (1984) "Physical model study of caving under massive sandstone roof conditions at Moura, Queensland." CSIRO Aust., Div. Geomech., Geomechanics of coal mining report No. 57.
Wold MB (1985) "A blocky physical model of longwall caving under strong roof conditions." Proc. 26th US Symposium on Rock Mechanics, pp 1007–1014.

Landslides and Climate Change – McInnes, Jakeways, Fairbank & Mathie (eds)
© 2007 Taylor & Francis Group, London, ISBN 978-0-415-44318-0

PAGeR – a cost-effective approach to geotechnical risk assessment and remediation for flood defence embankments

D.T. Shilston
Atkins Ltd, Epsom, UK

T.A. Cash, D. Norman, J. Appleby & A. Daykin
Atkins Ltd, Leeds & Warrington, UK

ABSTRACT: The performance and overall stability of flood defence embankments present particular challenges to organisations responsible for their maintenance, especially in response to changes in climate, in river or flood regimes and in public expectations. A phased approach to this problem is described. The approach, known as PAGeR, has been developed and used in investigations for the Environment Agency on rivers in the north-east of England. It comprises three phases:

Phase 1 – Geotechnical Screening and Qualitative Risk Assessment of sub-catchments or lengths of coast;
Phase 2 – Geotechnical Option Assessment of selected high and very high risk areas, using Phase 1 data;
Phase 3 – Geotechnical Outline Design of selected high and very high risk areas, with additional ground investigation data (e.g. for the Environment Agency's Project Appraisal Report stage).

This paper concentrates on PAGeR Phase 1 as its methodology includes newly-developed approaches to data acquisition, data processing and to assessing the stability risk of flood embankments.

1 INTRODUCTION

This paper describes an approach to geotechnical risk management and stability measures that has been developed and successfully applied to flood embankments on rivers in north-eastern England. PAGeR (the Phased Approach to Geotechnical Risk) is relevant to the broad themes of the conference in a number of ways:

- Landslides (or slope instability) are one of the major causes of failure for flood defences. Although flood embankments are man-made, the assessment of their stability uses the same approaches as those applied to natural slopes and to other types of man-made cut and embankment slopes. Also, of course, flood embankments are founded on natural ground with all its inherent strengths and weaknesses.
- Climate change and other changes to the environment are having a major effect on the design and performance criteria adopted for flood embankments – rivers are experiencing increased flood heights, and coasts are subject to sea-level rise and an increasingly stormy wave climate. The PAGeR methodology was developed to accommodate these changes as simple variations to the input to the geotechnical and risk models and to subsequent remedial works.
- Public awareness of the causes and consequences of flooding means there are growing societal expectations and pressures for improvements in flood defences; yet budgets are tight and tightening. PAGeR was developed to help provide a cost-effective and auditable technical and managerial solution to these challenges.

2 THE PROBLEM OF FLOOD DEFENCE EMBANKMENTS

Coastal and river flood defence embankments can be very long, and their construction and likely performance in flood conditions may be unknown or only poorly understood (see Figures 1 and 2). There is therefore an obvious need for simple, cost-effective and auditable ways of:

1. assessing the likelihood of geotechnical failure of flood defence embankments;
2. assessing the potential impacts associated with geotechnical failure; and
3. identifying options and making decisions about maintaining embankments or up-grading them to provide increased standards of flood protection to land and property.

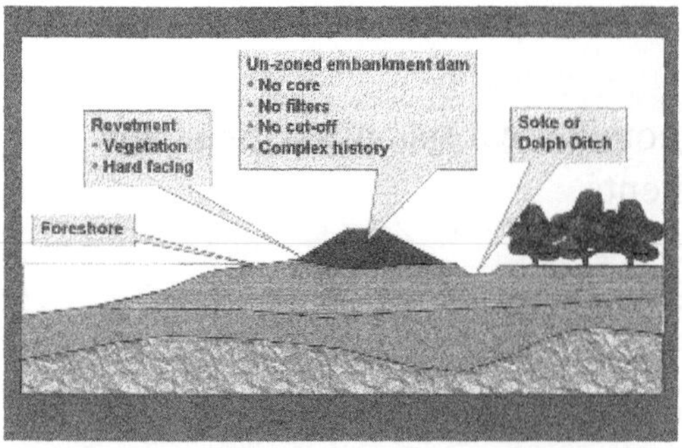

Figure 1. Typical features of UK flood embankments.

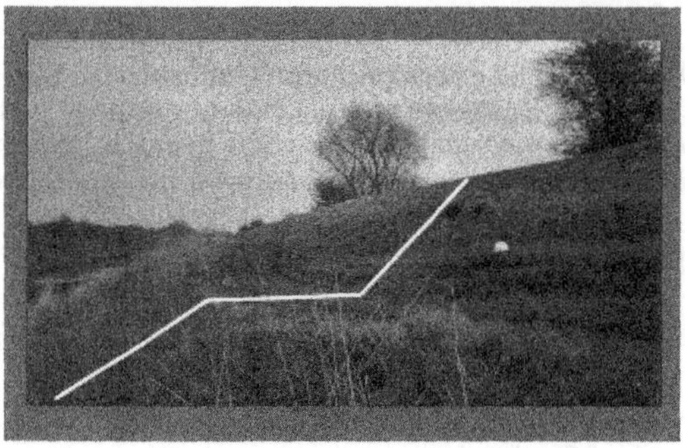

Figure 2. Flood embankment with small slope failure, River Don.

In response to these challenges and with the active encouragement of the Environment Agency, Atkins' water & geotechnical specialists have developed *PAGeR* - a phased approach to assessing and remediating the geotechnical stability of substantial lengths of flood embankments. PAGeR has been applied to rivers in the Environment Agency's North East Region (a total of more than 80 km of the Rivers Aire, Don, Ouse and the Dutch River).

3 OVERVIEW OF PAGER

PAGeR is a straightforward, systematic and repeatable procedure for assessing and managing the condition of flood defences from a geological and geotechnical perspective. It brings together site investigation methods, analytical processes and decision-making tools in a phased and cost-effective way. It helps the managers of flood defence embankments make auditable decisions about where to focus their maintenance or up-grade efforts (and budgets). And it then helps in the selection and design of the work that is to be done. The PAGeR methodology has three phases:

Phase 1 – Geotechnical Screening and Qualitative Risk Assessment of sub-catchments or lengths of coast;

Phase 2 – Development of Geotechnical Options for selected high and very high risk areas, using Phase 1 data;

Phase 3 – Geotechnical Outline Design of selected high and very high risk areas, with additional ground investigation data (e.g. for the Environment Agency's Project Appraisal Report stage).

There is a decision point or 'gate' at the end of Phases 1 and 2 at which results are reviewed and decisions are made about moving on to the next stage. This paper concentrates on Phase 1 as its methodology includes newly-developed approaches to data acquisition, data processing and to assessing the stability risk of flood embankments.

4 PAGeR PHASE 1

PAGeR Phase 1 considers possible factors and events that could lead to the geotechnical stability failure of flood embankments within a particular river sub-catchment or length of coast due to:

Bank erosion: stability failure is induced by narrowing of the foreshore and oversteepening of the flood defences.

Through-seepage and under-seepage: Stability failure is induced by elevated porewater pressures in the flood defences and their foundations (natural strata) during the flood and afterwards (rapid drawdown).

'Failure' in this context is defined as the breach of the embankment. Small failures that do not threaten its overall stability are not considered.

A desk-study and walk-over survey of the flood embankment are the initial part of Phase 1. Standard techniques are used to gather and assess information on:

- Geological background, records of previous ground investigations and history of the embankments
- Construction, performance and maintenance
- Present condition and indicators of geotechnical characteristics

This information is drawn together to provide an initial conceptual model of the condition and stability of the flood embankment and to design a simple ground investigation. As well as determining the technical objectives the desk study and walk-over survey needs to determine the access and logistical constraints for the ground investigation – poor access and vegetation are common problems which need to be overcome.

Knowledge on the geometry of the flood embankments and adjacent river bed is obtained from LiDAR (or FlyMap) and bathymetric surveys. These surveys are carried out for other purposes and are usually already available from the Environment Agency.

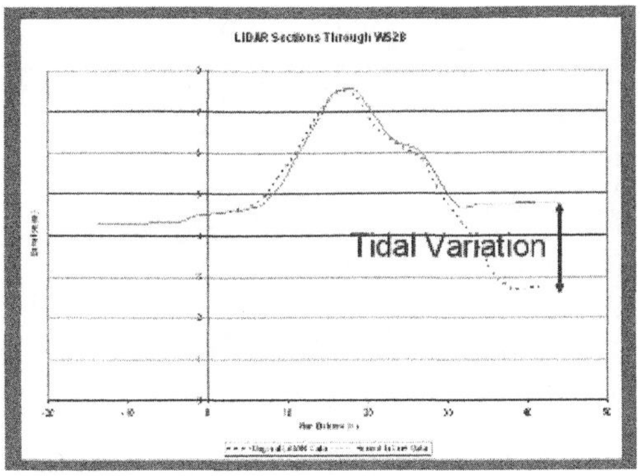

Figure 3. Cross-section of a flood defence embankment compiled from LiDAR data.

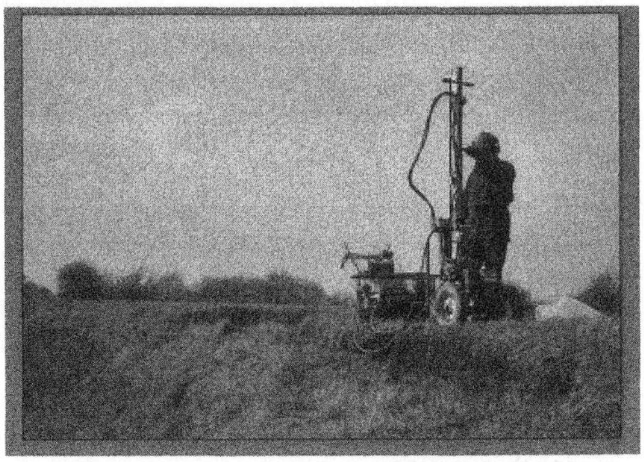

Figure 4. Window sampler in use in crest of an embankment with small slope failure.

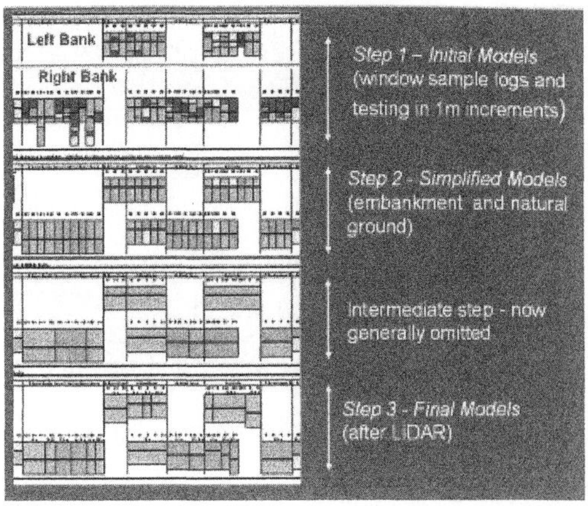

Figure 5a. Development of ground models – PAGeR Phase 1.

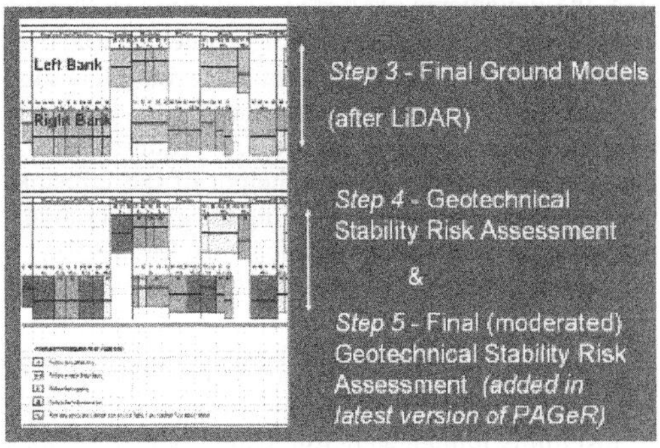

Figure 5b. Development of geotechnical stability risk models - PAGeR Phase 1 (cont.).

PAGeR thus uses existing data and does not usually incur additional costs. Bathymetric survey information is important as LiDAR cannot provide information below the water-level (see Figure 3). The survey information is 'ground-truthed' (verified) and supplemented by simple field observations and measurements taken during the walk-over survey.

The third part of data gathering during PAGeR Phase 1 is the ground investigation. Its objective is to obtain information by using simple portable equipment that overcomes the problems of access at low cost. For the PAGeR assessment we require information on material types and stratigraphy. Much reliance is placed on careful description of the samples and arisings from the exploratory holes. Some simple laboratory testing is done to help determine material type where classification is marginal, but very little information is collected directly on soil strength or permeability.

The Phase 1 investigation work undertaken in the Environment Agency's North East Region has used a 2-man *'window sampler'* to put down exploratory holes to depths of up to about 10 m. Geophysical survey techniques have not yet been used within a PAGeR project due to access and other constraints. Recently, however, Atkins has been using geophysical methods to investigate seepage through dam embankments and we are looking forward to the opportunities for bringing the technical advantages of geophysics into the PAGeR process.

Having completed the data gathering, the Phase 1 of the PAGeR methodology moves on to the development of 'ground models' in five steps, as shown on Figures 5a and 5b. The models subdivide the flood defence embankments into c. 100 m lengths based on the exploratory holes, desk study, walk-over and survey data. Each model is a summary of the ground conditions encountered in the window samples (or other exploratory holes), with the ground being classified as sand, silt, clay, peat or sand & gravel. Initially the models break the exploratory hole results into 1m thick increments extending up and down from the estimated original ground surface beneath the

Calculated Factor of Safety against Stability Failure	Qualitative likelihood of failure and Qualitative risk of failure
>1.2	Low
1.1 - 1.2	Moderate
1.0 - 1.1	High
<1.0	Very High

Assumes that impact (or consequence) of a flood or rapid drawdown event has a constant value (*i.e.* does not vary over space or time).

Figure 6. Slope stability – factors of safety and qualitative risk assessment.

embankment (shown by the thick horizontal black line in the models in Figures 5a and 5b. In the second step, the increments are further simplified so that the embankment is modelled as being of one material over a foundation of a second material (for example a sand embankment over a clay foundation, again separated by a thick black line in Figure 5a). Addition of the LiDAR data enables the true height of the embankment to be determined, this giving the ground models ready for geotechnical assessment (see Figure 5b).

Four modes of failure are considered for each of the Step 3 ground models:

1. Instability during flood conditions (leading to breach of the embankment)
2. Instability immediately following flood conditions (rapid draw-down, leading to breach)
3. Failure through piping (potentially developing into a breach)
4. Failure through bank erosion (potentially developing into a breach)

Failure risks from Modes 1 and 2 are assessed for each ground model through conventional slope stability analysis of a series of prescribed bund geometries based on the survey data and field observations. The computed factors of safety are then converted into qualitative risk of failure using the relationship shown in Figure 6 and illustrated in Figure 5b. An example of the application of the risk categories for ground models along a section of the Upper River Aire is shown on Figure 7.

The final (Step 5) part of PAGeR Phase 1 was introduced in a recent project. Atkins' team realised that, although the ground models had proved to be a useful predictive tool, more use could be made from information from the walk-over survey and maintenance records. Step 5 achieved this by explicitly using this 'ground-truth' information to moderate the modelling and analytical results, as shown by the risk classification scheme in Figure 8.

Ground Model (model number, embankment height, embankment composition / underlying strata)	Flood conditions		Rapid drawdown (slope stability)	Initial assess'nt of overall risk
	Slope stability	Piping		
10 (2m clay / clay)	*L*	*L*	*L*	*L*
11 (2m clay / silt)	*M*	*L*	*L*	M
12 (2m clay / sand)	*L*	*M*	*L*	*M*
16 (2m sand / clay)	*L*	*H*	*L*	*H*
18 (2m sand / sand)	*L*	*VH*	*L*	*VH*
25 (3m sand / clay)	*H*	*H*	*L*	*H*
28 (4m clay / clay)	*VH*	*L*	*L*	*VH*

Figure 7. Example of initial assessment of geotechnical stability risk (Upr. River Aire).

		Initial overall risk class			
		Low	Moderate	High	Very High
Mass movement observed in walk-over	None	L	M	H	VH
	Few	M	H	VH	VH
	Many	VH	VH	VH	VH

Ground-truthing moderation yields Final Overall Geotechnical Risk Class (in colour)

Figure 8. Moderated risk classes as a result of incorporating ground truth information.

Soil type	Alignment of river bank	Vulnerability rating
Clay	Straight to gently curved	*Low*
	Sharply curved	*Moderate*
Silt	Straight to gently curved	*Moderate*
	Sharply curved	*High*
Sand	Straight to gently curved	*Moderate - High*
	Sharply curved	*Very High*

Figure 9. Example of the qualitative assessment of vulnerability to river bank erosion.

The potential for failure of a flood defence through piping (Mode 3 in the list above) and by reduction in foreshore width and/or undermining (Mode 4) are assessed using tables that qualitatively relate ground conditions to vulnerability. Figure 9 is an example for the assessment of vulnerability to river bank erosion. Such tables such as this are developed by fluvial geomorphologists from knowledge of the river's shape and the materials of which its bank are formed.

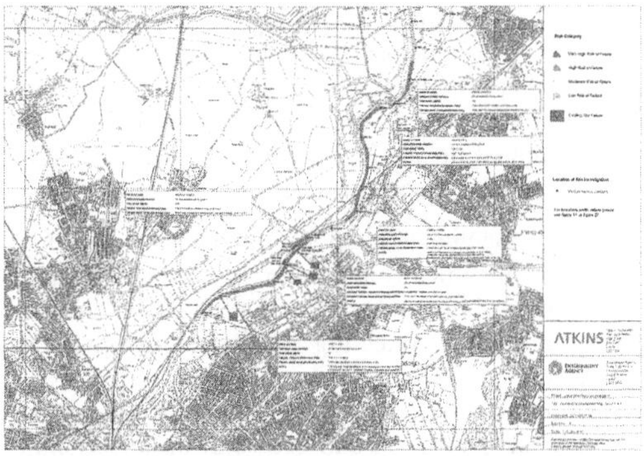

Figure 10. PAGeR Phase 1 geotechnical risk map – an example from the River Don.

We have found that PAGeR Phase 1 results are most effectively summarised and presented as maps that can be handled using Geographic Information System (GIS) software. An example from the River Don is illustrated in Figure 10.

A further example is some work done on a short (3.3 km) length of flood defence embankments on the Upper River Aire. Twenty six window sample holes and seven dynamic probe holes were undertaken. The final outcome was that some 50% of the embankment length was assessed to have a high risk of geotechnical failure and some 20% to have a low risk.

Detailed results of the PAGeR Phase 1 work are taken as input to the overall flood risk assessment for the river. Flood modelling and an assessment of impact of flooding are used to evaluate flooding scenarios for various river level elevations and durations of flooding. Technical and cost-benefit analyses are used to identify lengths of flood defence embankment that could be upgraded or remediated, and to establish a scale of priorities for the work.

PAGeR Phase 1 has some limitations that need to be recognised. Interestingly it can be deceptive in that it is a preliminary product, but can look like the end product – it is not, as Phases 2 and 3 need to follow.

The wide spacing of sub-surface investigation points may miss sharp changes in ground conditions, and it assumes standard soil strength parameters based on index properties. Phase 1 therefore requires experience and judgement to reduce data and to produce the simplified ground and risk models. These limitations are addressed in Phases 2 and 3.

As a further development, Atkins is intending to introduce a more formal assessment of uplift based on a method developed by the US Army Corps of Engineers. We have recently applied this method on a flood defence project on the River Eden in north-west England. Atkins is also evaluating the use of geophysical survey techniques to supplement and extend the PAGeR methodology.

We suggest that the advantages of Phase 1 of the PAGeR methodology are considerable:

- Uses standard LiDAR data or good topographic data, walk-over and maintenance information, and very cheap forms of ground investigation.
- Concentrates on maximising knowledge in key areas – material types, stratigraphy and past performance.
- Provides a logical and repeatable process for assessing risk, based on simple tried and tested technology.
- Doesn't assume areas of weakness – assesses all areas equally; but results then moderated by walk-over and maintenance information.
- Can compare different sites directly to obtain relative risk and priority scales.
- Compatible with standard or customised GIS database systems.
- Fits budget constraints commonly applied to early (screening) stage of flood defence projects.
- Provides appropriate basis for subsequent phases of work.

5 PAGeR PHASE 2

The second Phase of the PAGeR methodology involves the evaluation of options for up-grading or remediating specific lengths of embankment identified by the geotechnical work, flood modelling, risk assessment and cost-benefit analyses undertaken during Phase 1. Phase 2 is undertaken by examining the Phase 1 data on a site-specific basis. Typically these are the lengths identified as high and very high risk at Phase 1.

The simplifications made in Phase 1 are replaced in Phase 2 by details from specific exploratory holes, site-specific rather than generalised cross sections, and (sometimes) by improved estimates of river water levels. No new information is required, and Phase 2 can therefore be undertaken rapidly, leading to the production of an Options Report.

6 PAGeR PHASE 3

PAGeR Phase 3 is the geotechnical outline design of specific sites within the high and very high risk areas identified and assessed in Phases 1 and 2. It requires additional site-specific ground investigation data and takes the work up to the Environment Agency's Project Appraisal Report stage.

A combination of boreholes and cone penetration testing is usually employed, together with more detailed walk-over survey and geomorphological mapping, topographic and (sometimes) bathymetric surveying. This work provides a detailed understanding of the stability conditions of specific lengths of embankment, including the production of geotechnical models and cross sections that are modelled using conventional slope stability software.

For example, on a scheme in Goole boreholes were put down and inclinometers installed in two areas of active instability. A similar approach involving detailed investigation and long-term monitoring has been adopted at Reedness, where river bank and flood defence instability have been a problem for many years.

Evaluation of up-grade or remedial options must be done collaboratively between the various specialist disciplines and, of course, with the client (such as the Environment Agency).

7 CONCLUSION

One of the attractive aspects of PAGeR is the simple way in which the results are presented at the early screening or assessment stages. This is most important as it enables non-specialists or specialists in non-geotechnical disciplines (such as hydraulic modellers and risk managers) to understand and use the findings in their work. PAGeR's other key advantages include:

- Phases 1 and 2:
  - Use and integrate standard topographic survey, desk study, walk-over and maintenance information, and very cheap forms of ground investigation;
  - provide a logical and repeatable process for assessing risk, based on simple tried and tested technology;
  - can compare different sites directly to obtain relative risk and priority scales – it does not assume areas of weakness; and
  - fit the severe budget constraints commonly applied to early / screening stage of flood defence projects;
- Taken together, Phases 1, 2 and 3 provide a logical and technically auditable approach to evaluating the risk of embankment failure and to identifying options and to selecting and designing a preferred option.
- PAGeR can be attractive to geotechnical specialists because geological and geotechnical uncertainties are explicitly included and decisions at one phase are challenged and validated in the next.
- From an owner's or manager's perspective, PAGeR offers many advantages, including its speed, low cost per kilometre of embankment and its repeatable and auditable results.

ACKNOWLEDGEMENTS

Atkins' authors thank their colleagues who have contributed to the development of the PAGeR methodology and its implementation on projects. They also thank the Environment Agency for encouragement during the work and for permission to publish this paper.

Landslides and Climate Change – McInnes, Jakeways, Fairbank & Mathie (eds)
 ISBN 978-0-415-44318-0

Development of a pore pressure prediction model

H. Persson
Swedish Geotechnical Institute/ Chalmers University of Technology, Gothenburg, Sweden

C.G. Alén
Chalmers University of Technology, Gothenburg, Sweden

B.B. Lind
Swedish Geotechnical Institute, Gothenburg, Sweden

ABSTRACT: To improve the estimations of maximum pore pressures in slope stability investigations, and to enable prediction of how the slope stability will be affected by climate change, development of a model based on climate parameters and variations in ground water levels is in progress. The model aims at incorporating prognoses of climate change in predictions of pore pressures in fine sediment slopes and is based on existing models, which are modified for the purpose of pore pressure prediction. The main models are the hydrological HBV model from the Swedish Meteorological and Hydrological Institute and a statistical model for ground water level prediction, developed at Chalmers University of Technology. In a case study simulations of ground water levels, together with predictions of maximum ground water levels, have been carried out. Simulations have been done both for recent climate conditions and with a climate change considered. In the present study no thorough analyses of the results have been done; the results however indicate future possibilities with the used method, which will be further developed in the continued research.

1 INTRODUCTION

In consequence of climate change the weather in most of Sweden is expected to become wetter during the next 100 years. Increases in precipitation and runoff with about 30% (but with large regional variations) are expected in most of the country (Rossby Centre, 2006). A preliminary Swedish study found that during a period with precipitation 40% above mean the ground water levels in the studied area rose by up to 0.9 m (Hultén et. al., 2005). High ground water levels, and thus high pore pressures, decrease effective stresses and consequently also decreases slope stability. Stability calculations of the clay slopes along the Göta älv river in Sweden indicate that the safety factor may decrease as the pore pressure increases due to increase in precipitation (Lind et al 2006; Hultén et al 2006).

The impact of climate change on slope stability has now been identified in many parts of the world (e.g. Buma, 2000; Dehn et. al., 2000).

To improve the estimations of maximum pore pressures in slope stability investigations, and to enable prediction of how the slope stability will be affected by climate change, development of a model based on climate parameters (e.g. precipitation and temperature) and variations in ground water levels is in progress.

Although the climate change issue has actualised the subject for the present study, predictions of pore water pressures are essential for reliable slope stability calculations also without effects of climate change.

This paper outlines the modelling strategy and presents results from a case study. The present study has been done within a newly started PhD project that will continue the research within the field of pore pressure prediction.

2 COMMON METHODS FOR PORE PRESSURE PREDICTIONS

For slope stability calculations in drained conditions maximum pore pressure that can be expected to occur within a certain design period has to be estimated. There is however no established standard for how this is best done. A common method is to calculate what pore pressure conditions that are required for the slope to fail; compare these calculated pressures with pore pressure observations from the slope and consider whether the calculated pressures are reasonable. If the pressures can be expected within the design period the slope is likely to fail and if the pressures are considered unreasonable a lower level has to be estimated from experience. Another estimation method is to add a "safety margin" (from experience) to the observed pore pressures.

Both these methods are commonly used in Sweden (Johansson, 2006) and discussions with geotechnical engineers in Canada and Norway indicate that also in these countries the described methods are common.

3 METHODOLOGY

The new model is based on existing models, which are modified and will be combined for the purpose of pore pressure prediction. The main models are the hydrological HBV model developed at SMHI[1] (Bergström, 1976, 1992; Lindström et al., 1996, 1997) and a statistical model for maximum ground water level prediction developed at Chalmers Technical University (Svensson, 1984, Svensson and Sällfors, 1985), referred to as the Chalmers model.

An earlier study had the aim to compare the HBV and the Chalmers models in terms of ground water prediction (Rosén, 1991). However, the study never finished.

3.1 *The HBV model*

The HBV model is a well-established conceptual rainfall-runoff model that has been used for calculations in around 40 countries (SMHI, 2006). The model was originally developed for runoff simulations and hydrological forecasting, but has since then been used for an increasingly number of different applications, e.g. prediction of ground water levels (Sandberg, 1982; Bergström and Sandberg, 1983; Lindström et al., 2002). The HBV model has also been used for investigations of the influence from climate change on runoff and soil moisture (Andréasson et. al., 2004).

The main input parameters to the model are observations of precipitation and temperature (see Figure 1). Additional inputs are e.g. estimates of potential evapotranspiration, topography and ground surface information of the catchments. The principle of the model is that precipitation increases the soil moisture, which causes percolation to the ground water zone. The ground water zone is divided into two separate reservoirs, upper and lower, which are the origin of the quick and slow components of the runoff hydrograph.

Output from the model is normally outflow from the ground water reservoirs. In the present study however the water stored in the ground water reservoirs is in focus. Ground water levels are calculated from the ground water storage by considering "fictive porosity", since the real porosity not is known, together with a reference level (according to eq. 1).

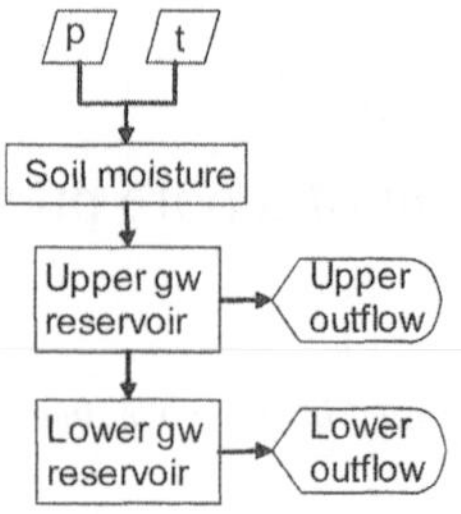

Figure 1. General structure of the HBV model. [p is short for precipitation and t for temperature.].

$$z_{gw} = \frac{1}{c_1} \cdot LZ + c_2 \tag{1}$$

where

z_{gw} = ground water level
LZ = ground water storage in lower reservoir
c_1 = constant ("fictive porosity")
c_2 = constant (adjustment of reference level)

To calibrate the model, hydrological soil parameters as well as percolation and runoff coefficients are modified.

3.2 *The Chalmers model*

The Chalmers model is a statistical model for calculating maximum ground water pressures, with a defined return period, that can be expected at a specific site. The model is recommended in Sweden for predictions of maximum pore pressures in slope stability calculations (RSAES, 1995). However, it is not commonly used and other less reliable methods are dominating.

The model is based on an assumption of correlation between the ground water levels in similar aquifers. To make a prediction of the ground water level with a defined return period for a specific site, ground water measurements over a short period from a ground water pipe[2] in the site can be compared with a long ground water observation series from a nearby reference area. It has been found that the correlation of these levels is good, provided that the long time measurements are from an area that is geologically and climatologically similar to the prediction site. The local topographical position also is relevant since the ground water level generally fluctuate more in a *high* position than in a *low*[3]. The ground water level with a certain return

[1] Swedish Meteorological and Hydrological Institute.

[2] The notation *ground water pipe* is used to describe a ground water measuring station (or *well*) with an open system.

[3] A high position is located close to a water divide whereas a low position is in the bottom of a valley.

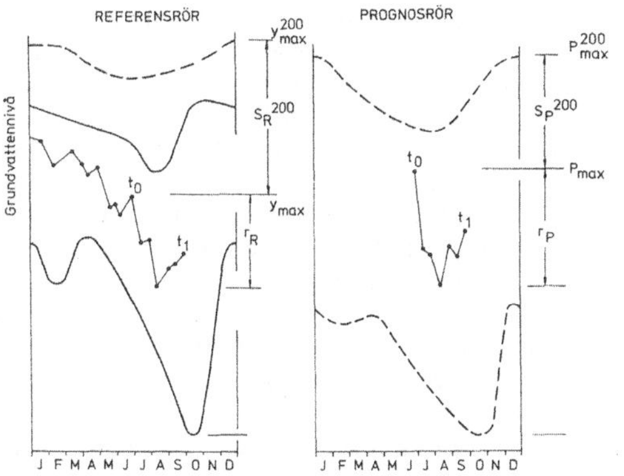

Figure 2. Maximum and minimum ground water levels (for each month in a year) with a return period of 200 years calculated with the Chalmers model. [*Referensrör* (left) means reference pipe and *Prognosrör* (right) means prediction pipe. *Grundvattennivå* (on the y-axis) means ground water level.] [From Svensson and Sällfors (1985).].

period in the reference area is calculated by assuming that the maximum ground water levels from every hydrological year[4] (for which there are observations) follow a normal distribution. The estimate of a ground water level with 200 years return period is given by eq. 2 (Svensson and Sällfors, 1985).

$$P^{200}_{\max} = P_{\max} - S_R^{200} \cdot \frac{r_P}{r_R} \qquad (2)$$

where

$P_{max}(200)$ = maximum level in the prediction pipe with return period of 200 years

P_{max} = maximum level in the prediction pipe during the observation period

$S_R(200) = |y_{max}(200) - y_{max}|$

$y_{max}(200)$ = maximum level for the reference pipe with return period of 200 years

y_{max} = maximum level for the reference pipe during the observation period

r_P = variation of the ground water level in the prediction pipe during the observation period

r_R = variation of the ground water level in the reference pipe during the observation period

The method for predictions of maximum (and minimum) ground water levels (for each month in a year) with a return period of 200 years are shown in Figure 2.

The model was originally developed for prediction of ground water pressures in confined aquifers. However, it has been shown that the pore pressures in clay have variations proportional to the variations in the confined aquifers (Berntsson, 1983). Thus the model is recommended also to be used for pore pressure calculations (RSAES, 1995).

[4] October 1st – September 30th.

4 CASE STUDY

A case study has been carried out to test and illustrate the possibilities of combining the two models.

4.1 *Studied site*

The studied site is called Harestad, and is located on the Swedish west coast about 50 km north of Gothenburg.

The Harestad area is an approximately 30 km^2 large, very flat, valley. The valley bottom consists of clay deposits on a thin (generally less than 0.5 m) sand layer that lies directly on bedrock. The clay thickness is more than 30 m in the middle of the valley, decreases towards the sides and disappears completely in the low and rocky surrounding hills. Agricultural fields cover the entire valley bottom whereas the surrounding hills have mixed forest and meadows.

The hydrogeology in the area has been studied earlier by Bergström et. al. (1982) and Svensson and Sällfors (1985).

4.2 *Data*

Ground water observations in the Harestad area have been made by the Geological Survey of Sweden since the early 1970's. There are several observation pipes in different geological positions that generally are registered twice a month. In total there are 16 pipes of which eight are in use today. The ground water pipes that have been used for analyses are called *pipe 1*, *2* and *10*.

Data with precipitation and temperature for the area have been received from SMHI, including both observations and predicted future values considering a climate change.

4.3 *Modelled period*

Modelling has been done for two different time periods; one with recent climate conditions and one with a climate change considered. The simulated periods are 1990–2005 and 2086–2100 respectively.

4.4 *Scenarios*

Two of the used ground water pipes, *pipe 2* and *10* are located on the sides of the valley with about 5–10 m clay on top of a few decimetres of sand. *Pipe 1* however, is placed in the middle of the valley in a 30 m layer of clay on top of a thin layer of friction material.

Prediction of ground water levels has been done using the HBV model for three sub-areas in which the pipes *1*, *2* and *10* are located. To adapt the model to local conditions, the calculated levels are calibrated against observed ground water levels from each pipe respectively. For calibration of the HBV model it is recommended to use observations from at least 5–10 years (Lindström et. al., 1996). Simulations have been done using SMHI's research version of the model (in Fortran), in which modifications of the original model easily can be made.

Using the Chalmers model, maximum ground water levels with return periods of 200 and 10 years respectively have been calculated. The predictions were made for *pipe 10* with *pipe 2* as a reference pipe.

The calculations using the Chalmers model were carried out with a simple program by Bengtsson and Svensson (1994).

5 RESULTS AND DISCUSSION

The results must be seen as an indication of possible outcomes from predicting ground water levels and pore pressures according to climate change scenarios. In the present study only rough calibrations and no thorough analyses of the results have been done.

5.1 *The HBV model*

The results from calibration can be seen in Figure as simulated ground water levels together with observations from *pipe 1*, *2* and *10*. None of the calibrations is perfect and in certain periods the simulations differ significantly from the observations. However, there are also most certainly errors in the observations due to e.g. iced pipes. For *pipe 2* the observations ended in 1998 and since then there is no data available for calibration. Hence, the calculated levels from 1998 to 2005 illustrate the forecasting aspect of the model.

The HBV model can easily get overcalibrated so that the fit of simulated values to observations becomes very good whereas the model reacts unrealistically when used with other input data. This problem is especially large when forecasting using data predictions considering climate change (including changed temperature and precipitation patterns). Therefore climate change simulations have to be carefully analysed.

Figure 4 shows simulations of a future situation with a climate change taken into account. The predicted ground water levels appear to be slightly higher than today's observed levels for *pipe 1* but very similar to observations for *pipe 2* and *10*. This is due to different calibrations for each site, where *pipe 2* and *10* have a higher surface run off coefficient than *pipe 1*. High surface run off has the effect that, even though the precipitation is higher in the climate change simulation, the ground water recharge is not affected to any large extent. To be able to interpret these results, and to improve the calibrations, more thorough analyses are required.

5.2 *The Chalmers model*

Calculated maximum ground water levels in *pipe 10*, with 10 and 200 years return period respectively, are shown in Figure 5. Different length of observations in the prediction pipe results in different predicted levels, where longer observation series give a more reliable result. As can be seen in Figure 5 the observed ground water levels (for the case with 1 year of reference observations) exceed the level with 10 years return period a few times in the shown period,

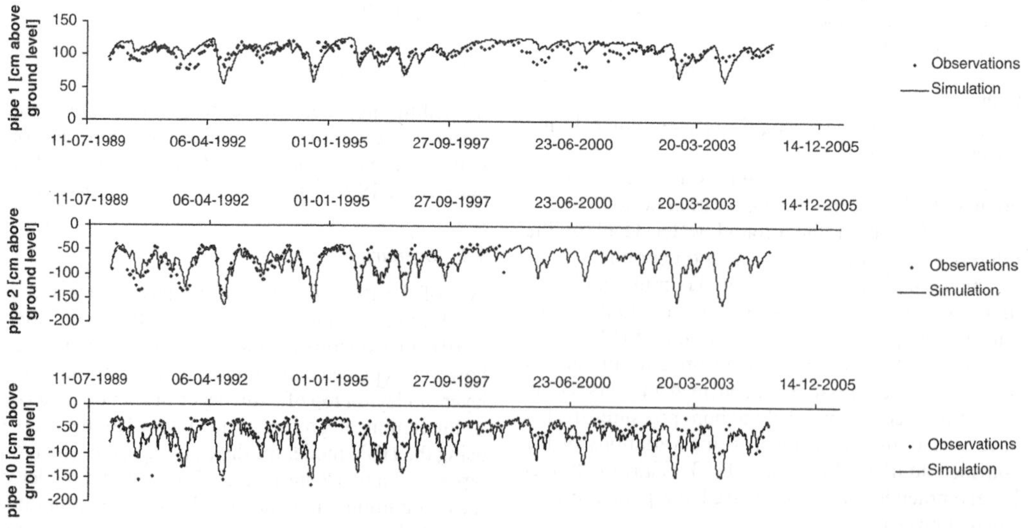

Figure 3. Simulated and observed ground water levels for *pipe 1, 2* and *10* during the calibration period 1990–2005. In *pipe 1* there is an artesian pressure whereas in *pipe 2* and *10* the ground water level is below the ground level.

while the level with 200 years return period is not exceeded.

Calculations with the Chalmers model carried out by Swedish consultants have sometimes resulted in obviously unreasonable levels (Engström, 2006 and Larsson, 2006). The reasons for these anomalies have not been studied, but a well-known problem with the Chalmers model is to find a suitable reference area with long time ground water records. As an alternative to observations from reference areas the authors suggest that simulations of ground water levels can be made with the HBV model.

5.3 *Comparison of the results*

The maximum ground water levels for *pipe 10* with a return period of 200 years are calculated to be less than 5 cm below ground level (see Figure 5). Simulations with the HBV model for the same pipe including climate change effects indicate that the water level will not rise above ~20 cm below ground level (see Figure 4). The difference may derive mainly from the fact that the prediction with the HBV model is not made for a situation with a 200-year return period. Anomalies in the models are however also contributing factors.

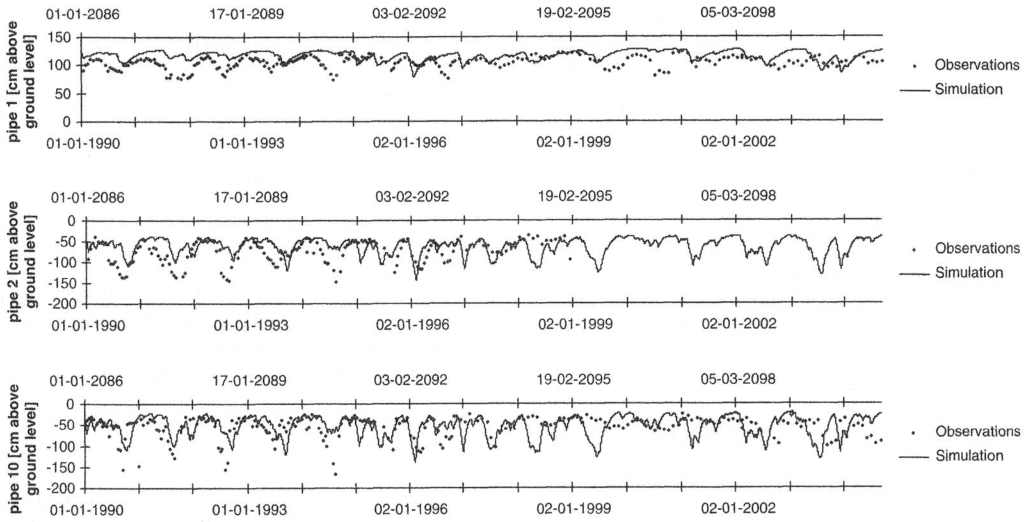

Figure 4. Predicted ground water levels for *pipe 1, 2* and *10* during the period 2086–2100 with a climate change considered. The ground water observations shown are from the calibrated period 1990–2005 and are here displayed only as reference levels and no correlation should be expected with predicted levels.

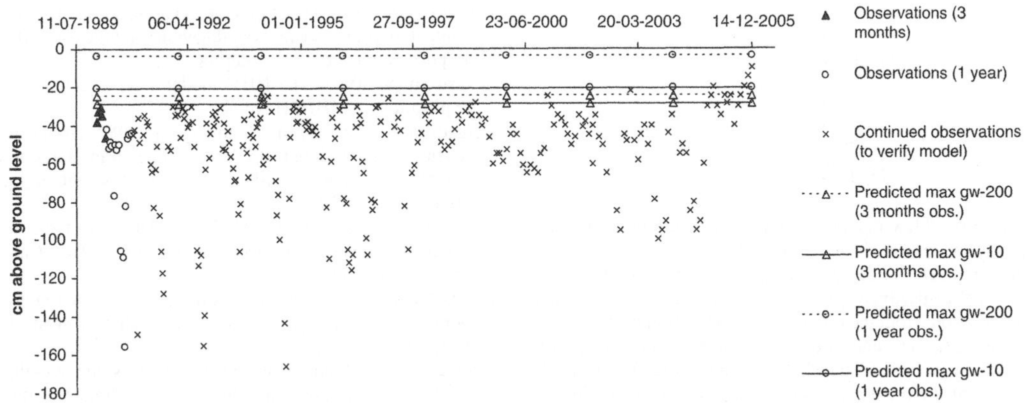

Figure 5. Calculated maximum ground water levels for *pipe 10* with return periods of 10 and 200 years respectively. Calculations have been done both with 3 months and 1 year of observations in the prediction pipe. Continued observations have been done in the prediction pipe and are displayed as comparison.

6 CONCLUSION

It has been shown that it is possible to simulate ground water levels using the HBV model. Predictions of maximum ground water levels with a certain return period using the Chalmers model have also been done. The ground water reference network in Sweden is sparse and do not represent all geological areas and climate regions for which predictions are made, which limits the usefulness of the Chalmers model. To improve the quality of these predictions, ground water levels simulated with the HBV model for a relevant area can be used as reference instead of observed levels.

Simulations of ground water levels using the HBV model have also been done for a future situation with a climate change considered. These simulations require a physically realistic calibration to give reasonable results and more thorough analyses are required to interpret the results. However, the simulations show on the possibilities of considering climate change in pore pressure predictions.

6.1 *Combined analyses*

The next step of development will include continued analyses of the HBV simulation results and improvements in model structure and calibration. Also results from the Chalmers model will be further analysed with possible model improvements as result.

The models will then be combined and used together, so that simulated ground water levels from the HBV model are used as reference levels in the Chalmers model. The structure of the intended pore pressure prediction model for slope stability calculations is illustrated in Figure 6. Simulations can then be carried out for many different areas to create reference data for calculations using the Chalmers model.

In a future climate situation the distribution function for ground water levels probably will be different from today's distribution. Thus, to enable predictions of maximum pore pressures with consideration to a changing climate, using the Chalmers model, observations from the prediction pipe need to be transformed to fit the expected future ground water situation as simulated (see Figure 6).

To get more data for calibration, two field stations will be established. These stations will be installed in areas with thick clay deposits and measure pore pressures on different levels in the clay as well as in the underlying friction material. One of the stations is planned to be located on a slope along the Göta älv river and the other in a flat area.

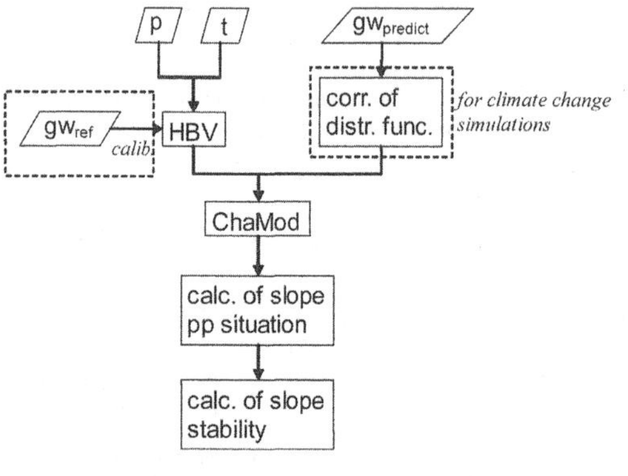

Figure 6. Flow diagram of the parts in the pore pressure prediction model. [*HBV* is the HBV model, *ChaMod* is the Chalmers model, *p* and *t* are precipitation and temperature respectively, gw_{ref} and $qw_{predict}$ are ground water observations from the reference and prediction pipes respectively (gw_{ref} are only used for calibration purposes), *corr. of distr. func.* is a (planned) correction of the observed values from the prediction pipe to make predictions with the Chalmers model including climate change considerations possible, *calc. of slope pp situation* refers to a local slope pore pressure estimation and *calc. of slope stability* is the final stability calculation.].

ACKNOWLEDGEMENTS

We would like to thank the Swedish Rescue Services Agency, the Swedish Research Council Formas, the Swedish Road Administration, Banverket and the Swedish Geotechnical Institute for financial support.

REFERENCES

Andréasson, J., Bergström, S., Carlsson, B., Graham, L. P., Lindström, G., 2004. Hydrological Change – Climate Change Impact Simulations for Sweden. Ambio vol. 33 (no. 4–5): 228–233.

Bengtsson, M-L. and Svensson, C., 1994. Beräkning av grundvattennivåers extremvärden och årstidsvariationer. Datorprogram i excel. Rapportnr. B404.

Bergström, S., 1976. Development and application of a conceptual runoff model for Scandinavian catchments. SMHI Reports RHO, No. 7, Norrköping.

Bergström, S., 1992. The HBV model – its structure and applications. SMHI RH No. 4

Bergström, S., Persson, M., Sandberg, G., 1982. Grundvattensimulering i sluten akvifer, exempel från Harestadsområdet. Vannet i Norden 4, Nordic Association for Hydrology: 13–29.

Bergström, S., Sandberg, G., 1983. Simulations of Groundwater Response by Conceptual Models – Tree Case Studies. Nordic Hydrology 14 (2): 71–84.

Berntsson, J. A., 1983 Portrycksvariationer i leror i Göteborgsregionen. Rapport No. 20, Swedish Geotechnical Institute, Linköping.

Buma, J., 2000. Finding the most suitable slope stability model for the assessment of the impact of climate change on a landslide in southeast France. Earth surface processes and landforms 25(6): 565–582, june 2000.

Dehn, M., Buerger, G., Buma, J., Gasparetto, P., 2000. Impact of climate change on slope stability using

expanded downscaling. Engineering Geology, vol. 55, no. 3, pp. 193–204, Feb 2000.
Engström, P., 2006. Personal communication, Geotechnical engineer, Sweco, Stockholm, Sweden.
Hultén, C., Olsson, M., Rankka, K., Svahn, V., Odén, K., Engdahl, M., 2005. Släntstabilitet i jord, Varia 560:1. Swedish Geotechnical Institute.
Johansson, Å., 2006. Personal communication. Authority examiner of geotechnical investigations in Sweden, Swedish Geotechnical Institute.
Larsson, P-G., 2006. Personal communication, Geotechnical engineer, Bohusgeo, Uddevalla, Sweden.
Lind, B.B., Andersson-Sköld, Y., Hultén, C., Rankka, K., Nilsson, G. 2006: Safe roads in times of changing climate. Proceedings, Full Papers, Transport Research Arena, TRA, Europe 2006, Göteborg, Sweden June 12th–15th 2006.
Lindström, G., Gardelin, M, Johansson, B., Persson, M., Bergström, S., 1996. HBV-96 – En areellt fördelad modell för vattenkrafthydrologin, SMHI Report RH 12, Norrköping
Lindström, G., Johansson, B., Persson, M., Gardelin, M. & Bergström, S., 1997. Development and test of the distributed HBV-96 hydrological model. Journal of Hydrology, Vol. 201, 272–288.
Lindström, G., Bishop, K., Ottosson Löfvenius, M., 2002. Soil frost and runoff at Svartberget, northern Sweden – measurement and model analyses. Hydrological processes 16: 3379–3393.
Rosén, B., 1991. Prognoser av grundvattennivåer/portryck – Etapp 1: Jämförande beräkningsexempel med Lathunden och HBV-modellen. SGI Varia 320, Swedish Geotechnical Institute, Linköping.
Rossby Centre, 2006. Rossby Centre, Swedish Meteorological and Hydrological Institute. www.smhi.se, 2006-10-25.
RSAES, 1995. Guideline for slope stability investigations, Commission on slope stability, Royal Swedish Academy of Engineering sciences. Swedish Geotechnical Institute, Linköping.
Sandberg, G., 1982. Utvärdering och modellsimulering av grundvattenmätningarna i Ångermanälvens övre tillrinningsområde. Hydrologiska/Oceano-grafiska avdelningen, SMHI (Sveriges Meteorologiska och Hydrologiska Institut).
SMHI, 2006. Kortfattad beskrivning av HBV-96-modellen. PDF document received from Göran Lindström at SMHI in spring 2006.
Svensson, 1984. Analysis and use of ground-water level observations. PhD thesis, Chalmers Technical University, Geological institution.
Svensson, C. and Sällfors, G., 1985. Beräkning av dimensionerande grundvattentryck – 1. Göteborgsregionen. Meddelande nr 78, Geohydrologiska forskningsgruppen, Chalmers Technical University, Gothenburg.

Landslides and Climate Change – McInnes, Jakeways, Fairbank & Mathie (eds)
© 2007 Taylor & Francis Group, London, ISBN 978-0-415-44318-0

Impact of climate change on rock slope stability: Monitoring and modelling

G. Senfaute, V. Merrien-Soukatchoff, C. Clement, F. Laouafa, C. Dünner & G. Pfeifle
INERIS – LAEGO, Ecole des Mines de Nancy. Parc de Saurupt, Nancy Cedex, France

Y. Guglielmi
Géosciences-Azur, CNRS, Valbonne, France

H. Lançon
Société SITES, Ecully, France

J. Mudry
Université de Franche-Comté, département de Géosciences, Besançon cedex, France

F. Darve, F. Donzé & J. Duriez
Laboratoire Sols Solides Structures Domaine universitaire, France

A. Pouya & P. Bemani
LCPC, Toulouse, France

M. Gasc & J. Wassermann
LRPC, Toulouse, France

ABSTRACT: The "Rochers de Valabres" is a fractured rock slope located in the French Alps and affected by rockfall. This slope has been chosen as an experimental site for improving knowledge on mechanisms leading to rockfall. Since 2002 scientific investigations (field observations, monitoring and numerical modelling) have been carried out on the site. A seismic network has recorded a significant seismic activity correlated with existing discontinuities. Mechanical measurements using *tiltmeters* monitor the slope movements in order to improve the knowledge on thermo-hydromechanical process. Different geometrical and mechanical models have been performed. In 2005 the site was chosen to concentrate research on understanding the effect of climate change on rock slope stability in the framework of a French national program named STABROCK. This paper presents the results of field measurements and numerical modelling carried out on the experimental site and the complementary investigations will start within the new research. The aim is to study more accurately the impact of environmental factors on rock slope instability mechanisms and to assess the possible effect of the climate change. Technical results will be integrated in a process of risk management.

1 INTRODUCTION

Global warming due to the greenhouse effect and specially changes in precipitation patterns and air temperature might therefore have influences on future slope stability (Dehn et al 2000). Several research studies have demonstrated the relation between climate change and landslide activity (Buma and Dhen, 1998, Moreiras, 2005). Climate change scenarios for SE England suggest an increase in both precipitation and temperature over next century (Rowntree et al 1993). This change is likely to have a number of potentially contradictory impacts on different types of rock mass movement processes. (Collison et al 2000).

The literature contains many examples which demonstrate the importance of climate conditions on the initiation of landslides, rockfalls, rockslides. Many authors claim that rockfalls are most frequent when the temperature fluctuates across the freezing point and that summer rockfall activity mainly is associated with heavy rainfall. However, many rockfalls are not directly correlated with weather factors; additional processes must be active, for example mechanical or chemical weathering not clearly

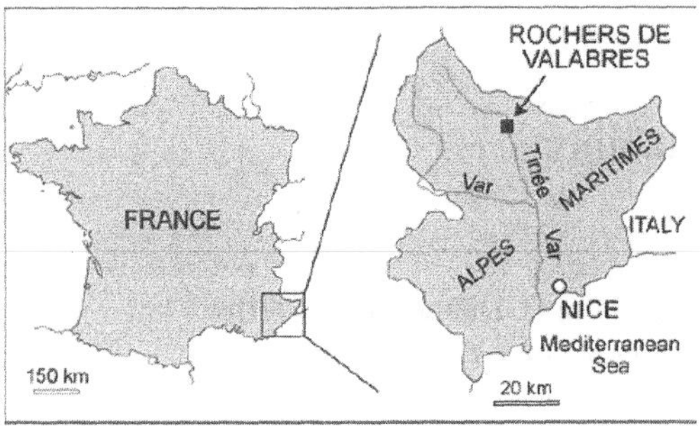

Figure 1. Location of the "Rochers de Valabres" experimental site in the Southern French Alps.

directly depend on weather conditions (Sandersen et al 1996). The challenge is to study the possible effect of meteorological factors (thermal, hydrological...) on rock slope instability mechanisms. The relationship between mechanical, hydromechanical, hydrogeochemical and climatic factors are to be investigated. This research will be achieved by multidisciplinary approaches including: field observations, on site and laboratory experiments and numerical modelling.

Scientific investigations started 2002 in a large-scale fractured rock slope, located in the French Alps and named "Rochers de Valabres" experimental site (figure 1). The recent history of this site is marked by two important rockfalls, which occurred in May 2000 and October 2004. In May 2000 the rockfall involved about 2,000 m^3 of material. This rockfall caused total isolation of the upper valley for a few weeks due to road traffic interruption, along with major economic and social impacts.

The first investigations have consisted of field observations and measurements, numerical modelling and monitoring techniques (Gunzburger et al 2005, 2004, Merrien-Soukatchoff 2004). At present, the monitoring system is composed of microseismic, geodetic, meteorological and mechanical devices. In 2005 the site has been chosen to concentrate monitoring effort to understand the effect of climate changes on rock slope stability in the framework of a French national program named STABROCK supported by the ministry of Equipment and Transport. This paper presents the first results of field measurements and numerical modelling carried out on the "Rochers de Valabres" experimental site since 2002 and the complementary investigations, which will go on within STABROCK research program.

2 GEOLOGICAL SETTING AND ROCKFALL CAUSES

Due to the glacial erosion, the upper Tinée Valley is generally relatively wide with abundant alluvial filling. Yet, at the place known as the "Gorges de Valabres" the Tinée River flows through a deep defile and cut the hard gneissic rocks of the Hercynian substratum, with slopes of several hundred meters on both banks. The rock mass is affected by a metamorphic foliation and three main families of fractures. Two of them are parallel to the slope side, one dipping toward the inside of the rock mass, another dipping towards the outside. The third family is subvertical. This structure leads having vertical wedges that can slide towards the valley by plane sliding if mechanical conditions are combined for sliding.

The majority of rock slope movements results from predisposition, preparatory and trigger factors (Gunzburger 2005). The causes of medium to long-term changes in resistance or disturbing forces are generally known as preparatory factors: in order to be efficient, their small (and almost imperceptible) effects must be cumulated up until rupture. Causes related to changes over the short-to-medium term are often referred to as triggers (or trigger factors) and constitute the most direct causes of failure. This distinction must not obviously be construed as a dichotomy, due to the existence of a continuous transition between preparatory and trigger factors. The key focus inherent in this model is to essentially convey the notion that rockfalls result from numerous, complex and interacting causes that may act over widely varying time frames; hence, it should never be considered that rockfalls must have been generated by just the most recent and apparent changes. Moreover, preparatory and trigger factors do not act identically on all slopes, given that initial conditions are not the same. Some slopes are in fact more conducive than others to rockfall activity, due to background factors such as topography (height and steepness of valley walls), vegetation, lithological parameters, and fracture geometry and density. These elements will be referred to as predisposition factors. Unlike preparatory and trigger factors that are defined by their action time frame, predisposition factors do not exhibit any evolution over time; instead, they serve to define the general framework of the slope that incites, to varying degrees, the onset of instabilities. The most frequently documented factors have been summarised in Table 1. Distinction has been made by their physical nature: mechanical, hydrological, thermal or geochemical phenomena.

3 STATE OF THE RESEARCHES CARRIED OUT ON THE EXPERIMENTAL SITE

A synthesis of scientific investigations carried out on the "Rochers de Valabres" site since 2002 is presented below.

Table 1. Classification of the most widespread rockfall causes (After Gunzburger 2005).

	Predisposition factors	Preparatory factors	Trigger factors
Mechanical	Steep side valley	Rise solpe steepnees due to the valley incision	High magnitu earthquake
	Fracture network	Seismic activity	Freezing and thawing of water
	Neotectonic stress	Damage process Fatigue	
Hydrological and meteorological	High precipitation	Regular rainfall	Heavy rain fall episode
Thermal	Temperatures contrasts	Daily a seasonal temperature oscillations	?
Geo-chemical	Mineralogical content of the rocks prone to weathering	Progressive weathering	?

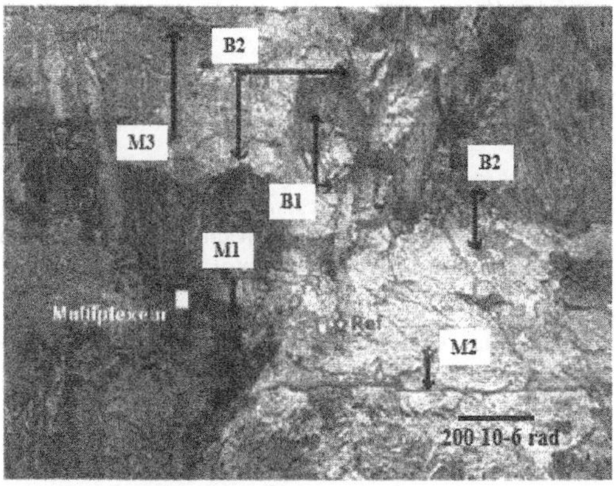

Figure 2. Location of 6 tiltmeters set on the slope surface (vectors express tilt magnitude in 10^{-6} radians and direction) using to measure the daily variations.

3.1 *Experimental set-up*

3.1.1 *Mechanical measurements*

The objective is to use tiltmeter monitoring to improve the knowledge of Thermo-Hydromechanical processes that induce multi-temporal elastic and non-elastic slope surface movements. Six short-base tiltmeters (3 bidirectional and 3 monodirectional being Applied Geomechanics sensors 755 series, having an accuracy of 1 μrad) were fixed at several locations of the slope surface (figure 2). Measurements are continuously registered with a 15 minutes sampling-rate interval, with a Gantner data station. Station is located close to the sensors and all devices were wrapped with thermo-isolating materials, so that temperature effects on connecting cables were negligible. Rock surface temperature, air temperature and rainfall are also monitored close to each tiltmeter.

Synchronously to the large Valabres experiment, this approach is tested on a smaller rock slope where changes in fluid pressures and deformations are simultaneously monitored at the discontinuities and in the rock matrix, using short-base extensometers and pressuremeters, as well as tiltmeters fixed at the slope surface. Tiltmeter monitoring of slope surface movements combined with thermo-hydromechanical numerical modelling appears to be an efficient method (Guglielmi et al 2006):

- to localize the highly deformable and permeable discontinuities that potentially cause slope movements. Indeed, as in the case of monitoring fluid displacements in reservoir production, tiltmeters appear capable of identifying or tracking the fast emptying of highly deformable discontinuities in a slope;
- to roughly estimate the hydromechanical properties of those discontinuities and the evolution of their properties over time.

The estimation relies on the direct inversion of infinitesimal tilt oscillations linked to free-surface oscillations within the slope aquifer. Furthermore, we are also currently studying using the sun's daily thermal loading at the slope surface. A change in the tilt signal over time could be a good indication of a change in the slope Thermo-Hydromechanical properties linked to progressive failure propagation using Mohr-Coulomb model, surface tilt clearly appears to be sensitive to failure located along short segments of the discontinuities. Thus, permanent monitoring of infinitesimal land surface rotations might be a good way to follow progressive failure in the slope.

3.1.2 *Microseismic monitoring*

In August 2003, a seismic network was installed on the experimental site. The seismic network is constituted of five stations: four uni-directional geophones (frequency range between 40 Hz and 1 kHz) and one high-frequency tri-directional accelerometer (frequency range between 1 and 10 kHz). The five sensors were laid out around a potentially unstable zone,

Figure 3. Microseismic monitoring network installed on the "Rochers de Valabres" experimental site.

affected by the rockfall of May 2000 (figure 3). The objectives of this instrumentation were:

- to detect seismic precursory events before rockfalls. During the stable period, microseismic activity has to be analyzed (waveform and frequency analysis, energy calculations...) in order to be able to identify unusual activity, that could be precursors of instability;
- to correlated microseismic activity with meteorological and mechanical data;
- to identify potential high-risk zones using the seismic source location and the determination of focal mechanism.

The microseismic system has been operational since August 2003. Two mainly types of seismic events have been recorded:

- type 1: seismic events associated with small earthquakes. The French national network has also recorded these events with magnitudes between 1 and 3.5. These signals are low frequency, less than 60 Hz (figure 4);
- type 2: seismic events significantly different from those associated with small earthquakes. These signals are high frequency (greater than 100 Hz) and short duration (figure 4).

Between August 2003 and July 2006 any rockfall was detected. Nevertheless, the system recorded 1934

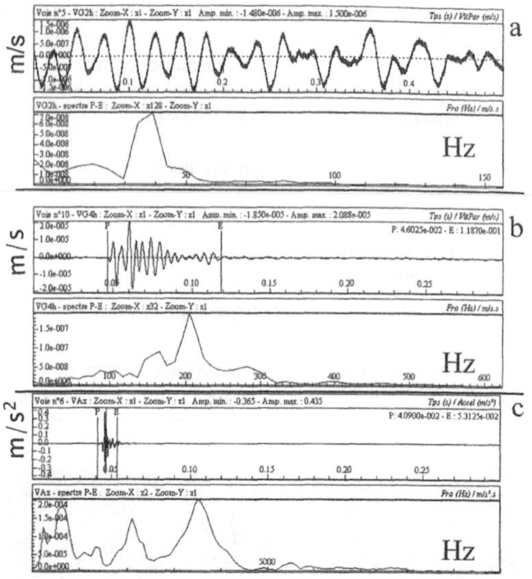

Figure 4. (a) Signal of a small earthquake; (bc) Signals of a seismic event induced by local microruptures recorded by a geophone and an accelerometer respectively. Signals in time (top) and frequency spectrum (bottom).

microseismic events of type 2 which could be associated with local fractured zone. Microseismic events are generally detected by one or two stations and seldom by the entire network. Consequently, the signal is affected by a strong attenuation due to the high fracture rate of rock masses (4.5 meters are corresponding to 65% in amplitude decrease). The data spatial-temporal evolution during this three years is not regular and several microseismic crises have occurred. The application of statistical techniques (Principal Components Analysis and Discriminant Factor Analysis) showed the contribution of the climatic variations on the crisis occurrence. Specially, temperature variations seem to have an influence on seismic events detected by the accelerometer (high frequency signals). However, such potential correlations would be confirmed.

The location of seismic events recorded was carried out using two kinds of methods:

- seismic wave rotation technique for location of seismic events detected by only one 3D accelerometer sensor (Vidal 1986, Cichobiz 1993, Abdul-Wahed et al 2001);
- location using the iterative nonlinear inverse technique called CHEAP and manual picking of the P-wave first arrivals time (Tarantola & Valette, 1982). This technique is applied to signals recorded by more than 4 sensors (geophones).

These locations methods are only applied to signals recorded with signal-to-noise ratio sufficiently high. Consequently, less than 100 events were located. The

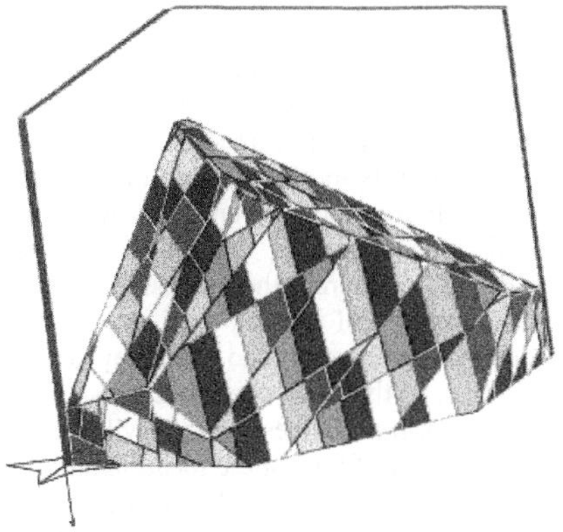

Figure 5. Simplified representation of the Rochers de Valabres topography and statistical distribution of discontinuities.

source localization appears to be associated with existing and visible discontinuities. Thus, the major fracture and the local small fracturing located around the 3D accelerometer sensor seem to produce microseismic activity.

3.2 *Numerical modelling*

Different geometrical and mechanical models have been performed on the "Rochers de Valabres" site. They aim to represent topographic, geological and structural features of the site. Two types of geo-models were employed: a simplified but statistical representation of the site using RESOBLOK software (LAEGO) and a more precise and deterministic model using gOcad geomodeller – Earth Decision Sciences, LIAD-ENSG (Mallet 2002). This modelling step is essential for evaluation of further thermo hydro-mechanical consequences. It is the transcription of predisposition factors that are essential to understand the site behaviour, a mentioned in section 2.

The RESOBLOK representation (see figure 5) has allowed confirming that the principal failure mechanism was plane slide on valley-dipping discontinuities, of blocks delimited by subvertical and inner-dipping discontinuities. The gOcad modelling (see figure 6) allows to accurately modelled the geological and structural aspects of the site. From both RESOBLOK and gOcad (see figure 6) modelling, it is possible to extract cross sections useful for further mechanical modelling.

2D thermo-mechanical computations using distinct element method (Gunzburger et al 2005) have confirmed the observed short-term reversible effect of temperature changes in the Valabres site configuration, with the observed temperature variation. The predictive model of large temperature variation with low mechanical characteristics shows the possibility of large plastic displacement or of failure. Yet at present computed displacements are broadly lower than the observed ones. Different assumptions can be to explain the difference:

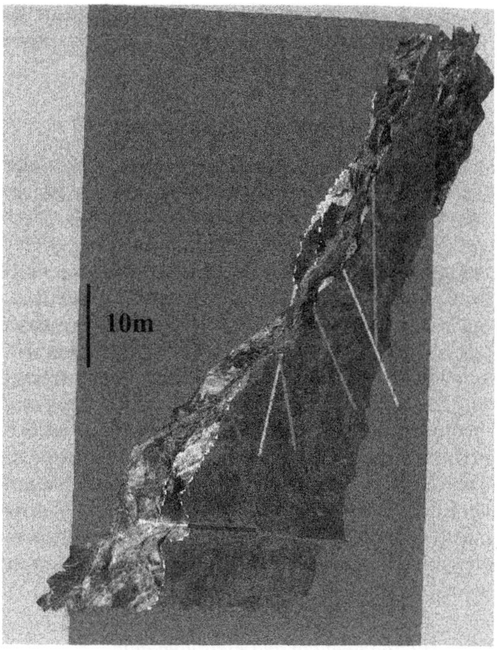

Figure 6. gOcad representation of the "Rochers de Valabres" topography and discontinuities and intersect of fracture in a vertical cross section.

- the inaccuracy of topographic measurement. There is a need of alternative measurement due to the unsuitability of topographic measurements to detect thermal effect;
- the thermal boundary conditions used on the free surface of the slope (imposing the same temperature history both on rock surface and on air) are not satisfactory. There is a need to take into account the radiation and/or convective effects;
- the 3D effect;
- the simple discontinuities behaviour employed for this first modelling (elastic perfectly plastic); and
- the need for parameter calibration, due to the insufficient in situ measurements.

4 A NEW RESEARCH PROGRAM "STABROCK"

STABROCK research program is complementary of investigations carried out on the "Rochers de Valabres" experimental site. The objective is to study more accurately the impact of metrological factors on mechanisms leading rock slope instabilities and evaluated the effects of the future climate change. The scientific objective and methodological approach suggest within this program is presented below.

4.1 *Innovative mechanical measurement*

Measurements will be performed in small boreholes that intersect a major fracture zone where a microseismic activity was detected in the rock slope. Measurement devices consist of one-meter long SOFO optic fibre extensometers fixed to the bore-hole walls by two anchors located on both sides of the fracture zone. High frequency (>100 Hz) deformations normal and tangential to the fault zone direction will be monitored for varying time periods and correlated to microseismic activity. Such a high-frequency in-situ monitoring of a natural rock fracture remains seldom performed although it can allow characterizing the localized fracture dynamic thermo-hydromechanical deformation changes (Cappa et al 2005). Consequently, this local measurement will be complementary to the tiltmeters array that monitors larger scale effects.

4.2 *Hydro-chemical and hydrodynamic measurement*

In the outcropping part of the fissured hard rock massif, water flows occur in the called unsaturated zone, before reaching the saturated aquifer. Recent results obtained in karst, as well in the Mediterranean area as in the Jura mountains, demonstrate that this "unsaturated" zone can be a medium with a large residence time, and therefore a medium for water storage.

As this water flows through thin fissures, with a low storativity, high hydraulic heads develop there during intense effective rainfall or generalized snowmelt episodes. The "Rochers de Valabres" site, which offers both possibilities to measure water input in the fissures (a gallery crosses the rock massif) and at the natural outlet of the slope (Luicetta spring), is designed to quantify the water volumes which are displaced by heavy effective rainfall episodes (pressure transfers). The assessment of these volumes will use the output of natural fluorescence of humic substances, which trace the actual transit of waters infiltrated through the soil. This methodology is implemented in parallel in karstic field areas, in the Jura and Vaucluse limestones.

The aim of the water chemistry and content in humic substances monitoring is the detection of the major infiltration episodes, in order to assess their contingent impact in terms of stability. STABROCK research program will enable to evaluate the relevance of the approach, and will be the prelude to the use of the "Rochers de Valabres" area as a long-term observatory area of the changes in the water balance which could be related to the present global climate change.

4.3 *Microseimic monitoring*

In order to perform the microseismic network existing on "Rochers de Valabres" site more sensors will be installed on the slope. The tree-years of microseismic monitoring help us to characterize the rock mass as well as the microseismic events that we can expect (see 3.1.2). Consequently, within the framework of "STABROCK" research, the extension of the microseismic network will include four new sensors. The exact design of the network to be deployed is actually in discussion. In the same way, processing and analysis will be continued. Investigations will carry out on the identification of the source location and the focal mechanisms which will give information on the source process leading rock fracturing. Microseismic activity will be correlated with mechanical, hydrological, meteorological measurements to qualify the capacity of microseismic method to identify the effect of environmental phenomena on the rock slope instability.

4.4 *Acoustic emission in laboratory*

An acoustic emission (AE) is defined as transient elastic wave generated by the rapid release of energy within a material (Lockner 1993). The development of faults and shear fracture systems for a variety of rock types involves the growth and interaction of microcracks. Acoustic emission (AE), which is produced by rapid microcracks growth, is a phenomenon associated with brittle fracture and provides interesting information about failure process in rock.

The rock mass of "Rochers de Valabres" site is strongly fractured and faulted. The laboratory study is an approach for understanding the failure process, damage precursors and order to help the analysis of observed and measured phenomena on experimental site. Two kinds of tests will be carried out:

- unixial compression tests using acoustic emission and mechanical measurements. Figure 7 shows the experimental procedure prepared to study the localisation of microcracks with 8 piezoelectric transducers and 4 strain gauges (2 horizontal, 2 vertical);
- shear tests on discontinuities using acoustic emission and mechanical measurements. Theses tests are very innovative and will allow to identify and to qualify the acoustic signals associated with shearing discontinuities.

4.5 *New approaches of numerical model*

The aim of numerical modelling in the STABROCK project is to better understand the mechanical phenomena governing the rock mass movement and to determine the external perturbations which have a key role on its instability. This aim should be achieved by performing a relevant numerical-mechanical model

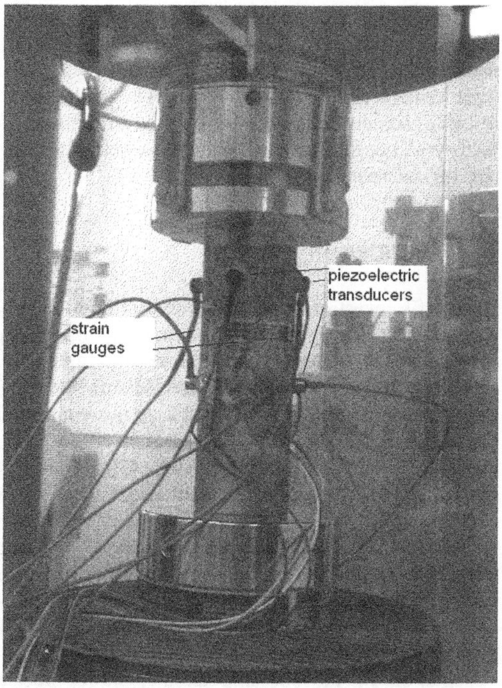

Figure 7. Uniaxial test using acoustic and mechanical mesurements on a gneiss sample of the "Rochers de Valabres" experimental site.

able to explain the rock mass movement and the strain-stress field in the rock mass. Comparing analytical, numerical, and experimental results will lead to calibration. This task will carry out by federating several numerical methods: distinct element method assuming strong discontinuities and finite elements method with non-linear joint elements. In fact, due to the importance of discontinuity features on the mechanical behaviour of the rock mass, they are to be introduced in the modelling process.

Numerical modelling associated with the in situ measurement can be use as a help for grading the different phenomena describe in Table 1. At present the relative and couple role of the different predisposition, preparatory and trigger factors is not well grasp. We do not know for example if a heavy rain is more awkward than a heatwave or a freezing period. The numerical modelling compared to the in situ measurement and data statistical analysis will help us to make a hierarchical classification of the climatic phenomena. The preliminary modelling has shown the importance of more realistic thermal boundary conditions and the necessity to take into account the complex 3D geometry.

4.5.1 *Distinct element modelling*

To understand phenomena several 2D cross section are available. The external perturbations are principally the thermal and hydraulic ones...(see table 1). They will be introduced in the model in accordance with the in situ data measures foreseen, as described in the previous section. The consequences of these inputs are strain and angle variation, displacement, possible rupture that will also be measured in the framework of the program. The comparison between model output and measured consequences will lead to model calibration. The distinct element model will be used with simple constitutive relations to assess the weight of the different input data.

More refined fracture interface relations (belonging to the class of "incrementally non-linear relations", C. Lambert et al 2004) will be introduced in order to take into account a material instability criterion (Darve et al 2004). This criterion, which is based on the sign of stress-strain second order works computed on every fracture, allows to determine which fractures are unstable according to local negative values of second order works and to exhibit a possible failure mechanism. If this criterion is integrated along all the existing fractures, the so-called "global value of second order work" can be used as an indicator of effective failure when it is vanishing along a given loading program applied to the cliff. When this global value of second order work is nil, a perturbation applied to the cliff will induce a failure mechanism as exhibited by the unstable fractures detected by the local nil values of the second order work. These computations will be at first performed on 2D cross section, using UDEC code, in order to test the influence of the different parameters. Yet, there is also a need to run 3D distinct element model using 3DEC computer code.

4.5.2 *Finite element modelling*

CESAR-LCPC (Humbert et al 2005) is a general finite element computation code dedicated primarily to civil engineering problems. This code is developed by the "Laboratoire Central des Ponts et Chaussées" (LCPC) which have a long experience in the modelling of geomaterials constitutive behaviour and geotechnical structures for research as well as for industrial projects. Recently Goodman (1968) joint element has been implanted in CESAR-LCPC (Elmi et al 2006). This joint element is a simple rectangular two-dimensional element with eight degrees of freedom which uses relative displacements as the independent degrees of freedom.

The constitutive models already available in the code for this joint element are elastic model and Mohr-Coulomb elastic-perfectly plastic model. A research work is in progress in LCPC for implantation the more realist models of rock joint in the code. One of these models is the constitutive mode for rock joints proposed by Jing et al. (1993). This model takes into account the roughness of the surfaces of the joint due to presence of the asperities, anisotropy in the morphology of the fracture surface, degradation of this morphology during plastic shear displacement, dilation, pre- and post-peak evolution of the shear resistance and also changes of normal and shear stiffness as a function of the normal stress. Another attempt is the development and implementation of a model able to take into account the damage of the joint when it is under normal traction stress. In this model, beyond the damage threshold, the normal stiffness decreases and tends toward zero for the large displacements. Extending of these models to hydro-mechanical phenomenon which is able to modelling effects of fluid on the joint behaviour and finally hydro-mechanical interaction of the fluid and the joint walls is a perspective of this work.

4.6 *Risk assessment and management*

The expected outcomes of the "technical" part of this research program, especially indicators enhancing risk monitoring processes and efficiency of early warning systems, are designated to be integrated in the larger process of risk management from analysis of the phenomenon and the vulnerable stakes to strategic actions of risk reduction.

Hence, the usefulness of any novel tool assisting risk managers depends on its adaptation to contingencies. So, at least two series of crucial issues should be considered in order to better understand – and reduce – different types of uncertainty (Rowe 1994): the complexity of the phenomenon, and the complexity of the societal context. While the complexity of the phenomenon is of rather "technical" nature and corresponding to the "hazard analysis" step in the process, the second is not.

Defining and carrying out any action of mitigation or attenuation needs a contextual analysis first: identification of societal or contextual factors and identification of actors, stakes and liabilities in the actual risk management processes' structure. Identified actors are invited to express their needs regarding the tools to be developed. This will help integrate operational aspects in the indicator design stage of the STABROCK project.

Designing the risk management process in the light of those insights, then, should draw on normative as well as positive considerations to fix an acceptable risk level, to conceive and assess potential strategies of action, to plan and realize those. Out-comes, finally, should be monitored and evaluated in order to enhance progressively, and guarantee over the longer term, the processes' efficiency. To help concerned actors – governmental agencies, local elected, or other stakeholders – going through the whole process, a book of guidelines to efficient hazard management will be formalized, insisting especially on the importance of risk communication and the respect of participatory principles.

Considering as well societal as well as technical parameters while addressing the whole process of risk management in the dynamic context of ongoing climate change, the STABROCK project should allow to build indicators suitable for the aim of enhancing the overall coping capacity of those concerned by the natural hazard at stake.

5 CONCLUSIONS

Scientific investigations carried out on the "Rochers de Valabres" site allowed studying the behaviour of rock slope stability using field measurements and numerical modelling. A microseismic network installed on the site has recorded a significant local seismic activity. Some of these seismic events have been correlated with existing visible discontinuities on the slope. Mechanical measurements have been carried out by using *tiltmeters* to monitor the slope surface movements. Rock surface temperature, air temperature and rainfall are also monitored close to each *tiltmeter*. Different geometrical and mechanical models have been performed. Two types of geomodels are employed: a statistical fracture representation of the site and another more precise with deterministic representation. These modellings are essential for further thermo hydro-mechanical evaluations.

The STABROCK research program aims to carry out complementary investigations on the "Rochers de Valabres" site. The objective is to study more accurately the impact of metereological factors on mechanisms leading to rock slope instabilities and to evaluate the possible effect of the future climate change. Scientific steps of this research are the following:

- to develop innovative sensors based on optical fibres in order to be able to measure temperature and deformations of the rockmass on a wide range of scale (from centimetre to kilometre);
- to use hydro-chemical and hydrodynamic measurements and recording: water pressure, natural fluorescence, turbidity. The aim is the detection of the major infiltration episodes in order to assess the impact on the slope stability;
- to perform the microsismic network in order to improve the analyses of seismic activity recorded. Microseismic activity will be correlated with mechanical, hydrological, meteorological measurements to qualify the capacity of this method to identify the effect of environmental phenomenon on the rock slope instability;
- to use, at laboratory scale, the acoustic emission for understanding the failure process, damage precursors for helping the analysis of measured phenomenon;

- to correlate in situ and laboratory data with numerical modelling in order to understand the mechanical phenomena governing the rockmass movement and to determine the environmental phenomenon which have a key role on slope instability;
- to integrate technical results in a process of risk management.

ACKNOWLEDGEMENTS

This research is funded by the « Ministère des Transport, de l'Equipement, du Tourisme et de la Mer » and the « Ministère de l'Environnement et du Developpement Durable ». We also thank EDF (Electricité de France) for providing the authorization of using their site where the Rochers de Valabres slope is located and the Mercontour National Park.

REFERENCES

Arno Z., Wagner F., Dresen G. 1996. Acoustic emission, microstructure, and damage model of dry and wet sandstone to failure. Journal of Geophysical Research, Vol. 101, No B8, pp 507–521.

Abdul Wahed M., Senfaute G., Piguet J.-P. 2001. Source location estimation using three-component seismic station. ISRM European Symposium EUROCK, Espoo, Finland.

Buma J., Dehn M. 1998. A method for predicting the impact of climate change on slope stability. Environ. Geol. 35, pp 190–196.

Cappa F., Guglielmi Y., Gaffet S., Lançon H., Lamarque I. 2006. Use of In Situ Fiber Optic Sensors to Characterize Highly Heterogeneous Elastic Displacement Fields in Fractured Rocks. International Journal of Rock Mechanics and Mining Sciences 43, pp 647 – 654.

Collison A., Wade S., Griffiths J., Dehn M. 2000. Modelling the impact of predicted climate change on landslide frequency and magnitude in SE England. Engineering Geology , 55, pp 205–218.

Clement C., Merrien-Soukatchoff V., Dünner C., Sausse J. 2006. Ecoute Microsismique appliquée aux versants rocheux instables. Exemple des Rochers de Valabres (06), JNGG 2006, Lyon 27–29 juin 2006.

Darve F., Servant G., Laouafa F., Khoa H.D.V. 2004. Failure in geomaterials. Continuous and discrete analyses. Comp. Meth. Appl. Mech. and Eng., Vol. 193, n° 27–29, pp 3057–3085.

Dehn M., Bürg G., Buma J., Gasparetto P. 2000. Impact of climate change on slope stability using expanded downscale. Engineering Geology , 55, pp 193–204.

Elmi F., Bourgeois E., Pouya A., Rospars C. 2006. Elastoplastic joint element for the FE analysis of the Hochstetten sheet pile wall", NUMGE06, Sixth European Conference on Numerical Methods in Geotechnical Engineering, Graz, Austria, 6–8 September 2006.

Goodman R. E., Taylor R. L., et Brekke T. L., 1968. A model for the mechanics of jointed rock, ASCE, Journal of the Soil Mechanics and Foundations Division, Vol. 94 (SM 3), pp 637–659.

Guglielmi Y., Cappa F., Rutqvist J., Tsang TC., Thoraval A. 2006. Coupled hydromechanical behaviour of a multi-permeability fractured rock slope subjected to a free-water surface movement: Field and numerical investigations. "Coupled thermo-hydro-mechanical-chemical processes in geo-systems – Fundamentals, Modelling, Experiments and Applications", Elsevier Geo-Engineering Book Series, Edited by J.A. Hudson and C-F. Tsang, Elsevier Ltd., Oxford, UK, 2006).

Gunzburger Y., Merrien-Soukatchoff V., Senfaute G., Guglielmi Y. 2004. Field investigations, monitoring and modeling in the identification of rockfall causes. Landslides: Evaluation and Stabilization, Proceeding of the Ninth International Symposium on Landslides, Rio de Janeiro, pp 557–563.

Gunzburger Y., Merrien-Soukatchoff V., Guglielmi Y. 2005. Influence of daily surface temperature fluctuations on rock slope stability: Case study of the Rochers de Valabres slope (France) International Journal of Rock Mechanics and Mining Sciences 42: in press.

Humbert P., Dubouchet A., Fezans G., Remaud. 2005. CESAR-LCPC: A computation software package dedicated to civil engineering uses, Bulletin des Laboratoires des Ponts et Chaussées (BLPC) N° 256–257 July – August – September 2005.

Jing L., Stephansson O., Nordlund E. 1993. Study of rock joints under cyclic loading conditions. Rock Mech Rock Eng, 26, pp 215–32.

Lambert C., Darve F., Nicot F. 2004. Rock slope stability from microscale to macroscale level, in Numerical Models in Geomechanics, Pande and Pietruszczak eds, Balkema publ., pp 85–90.

Lockner D. 1993. The role of acoustic emission in the study of rock fracture. Int. J. Rock Mech. Min. Sci. & Geomech. Abst. Vol 30, No 7, pp 883 – 899.

Mallet J. L. 2002. Geomodeling. Applied Geostatistics. Oxford University Press.

Merrien-Soukatchoff V., C. Clément, G. Senfaute & Y. Gunzburger, 2005, Monitoring of a potential rockfall zone: The case of "Rochers de Valabres" site, International Conference on Landslide Risk Management. 18th Vancouver Geotechnical Society Symposium, May 31–June 3, 2005.

Moreiras S. 2005. Climatic effect of ENSO associated with landslide occurrence in the Central Andes, Mendoza Province, Argentine. Landslides. pp 53–59.

Rowntree P., Murphy J., Mitchell J. 1993. Climate change and future rainfall predictions. J. Inst. Water Environ. Management, 7, pp 464–470.

Rowe D. 1994. "Understanding Uncertainty", Risk Analysis Volume 14, Issue 5, pp. 743–750).

Sandersen S., Bakkehoi S., Hestnes E., Lied K. 1996. The influence of meteorological factors on the initiation of debris flows, rockfall, rockslides and rockmass stability. Landslides, Senneset ed. Balkema, Rotterdam. pp 97–113.

Senfaute G., Amitrano D., Lenhard F., Morel J. 2005. Laboratory study of chalk rocks damaging by acoustic methods and correlation with in situ results. Revue Francaise de Géotechnique No. 110, pp 9–18.

Session 4

Experience of landslide hazard and risk management and better practices for the future

A coordinated approach to landslide management (whether rapid events or slow-moving phenomena) is essential and successful landslide hazard management should inform the planning and political processes. A review of landslide management in the context of climate change impacts examines how managers are incorporating climate change data into risk management strategies. For those responsible for managing risk in vulnerable locations, a critical issue is to develop solutions for improved management. It is also crucial to effectively communicate information clearly to both technical and non-technical audiences. There are many examples of good practice solutions to hazard management and it is important to exchange ideas and share expertise on these issues.

Seaview Duver, Isle of Wight, UK
Courtesy Wight Light Gallery, Ventnor, Isle of Wight, UK

Landslides and Climate Change – McInnes, Jakeways, Fairbank & Mathie (eds)
© 2007 Taylor & Francis Group, London, ISBN 978-0-415-44318-0

Landslide management and mitigation on the Scottish road network

M.G. Winter
TRL Limited, Edinburgh, United Kingdom

L. Shackman
Transport Scotland, Glasgow, United Kingdom

F. Macgregor
Consultant to Transport Scotland, Glasgow, United Kingdom

ABSTRACT: In August 2004 a series of landslides in the form of debris flows occurred in Scotland. Critically, the A83, A9 and A85 routes, which form important parts of the major road network were all affected by such events. While debris flows occur with some frequency in Scotland, they affect the major road network only relatively rarely. However, when they do impact on roads the degree of damage, in terms of the infrastructure and the loss of utility to road users, can have a major detrimental effect on both economic and social aspects of the use of the asset. Following these events the Scottish Minister for Transport acted swiftly by commissioning a study to determine a way forward for dealing with such events in the future. An integral part of the study commissioned is the development of management and mitigation options and these are based upon the delineation of four categories of hazard ranking. The rankings are based upon the results of an initial GIS assessment backed up by more detailed site specific assessments. The two highest categories of hazard ranking are intended for potential further action by means of either exposure or hazard reduction. This paper focuses on the context of the study and the development of the inclusive approach taken to the execution of the work and also describes both the study itself and the potential effects of climate change on debris flow frequency and severity in Scotland.

1 INTRODUCTION

In August 2004 a series of landslides in the form of rainfall-induced debris flows occurred in Scotland. Critically, some of these affected important parts of the major road network, linking not only cities but also smaller, remote communities.

While debris flows occur with some frequency in Scotland, they have, in the past, affected major communications links relatively rarely. However, when they do impact on roads the degree of damage, in terms of the infrastructure and the loss of utility to road users, can have a major detrimental effect on both economic and social aspects of the use of the asset. Additionally, there is a high potential for such events to cause serious injury and even loss of life although, fortuitously, such consequences have been limited to date.

The impacts of such events can be particularly serious during the summer months due to the major contribution that tourism makes to Scotland's economy. Nevertheless, the impacts of any debris flow event occurring during the winter months should not be underestimated. Not surprisingly, the debris flow events of 2004 created a high awareness of the effects of landslide activity in the media in addition to being seen as a key issue by politicians at both the local and national level.

Rainfall in August 2004 was substantially in excess of the norm; some areas received over 300% of the 30-year monthly average and in Perth and Kinross 250% to 300% was typical. While this percentage was less towards the west, parts of Stirling and Argyll & Bute received between 200% and 250% of the monthly average (Source: www.metoffice.com).

The rainfall was both intense and long lasting. Anecdotal but informed evidence suggests that the storms that followed the extended antecedent rainfall period were uncharacteristically intense. A large number of debris flows occurred in the hills of Scotland. A small number of these intersected with the major road network, notably the A83 between Glen Kinglas and to the north of Cairndow (9 August), the A9 to the north of Dunkeld (11 August), and the A85 at Glen Ogle (18 August).

Figure 1. Road users are airlifted to safety in Glen Ogle (© Perthshire Picture Agency: www.ppapix.co.uk).

While there were no major injuries to those caught up in the events, some 57 people were taken to safety by helicopter after being trapped between the two main debris flows on the A85 in Glen Ogle (Figure 1). The A85, carrying up to 5,600 vehicles per day (all vehicles two-way, 24 hour annual average daily traffic), was closed for four days. The A83, which carries around 5,000 vehicles per day, was closed for slightly in excess of a day and the A9, carrying 13,500 vehicles per day, was closed for two days prior to reopening, initially with single lane working under convoy. These traffic flow figures are for the most highly trafficked month of the year for each of the roads, and this occurs in either July or August on these routes. Minimum flows occur in either January or February and are roughly half those of the maxima. The figures reflect the importance of tourism and related seasonal industries to Scotland's economy. The events of August 2004 are described by Winter *et al.* (2006a; 2006b).

This paper describes the response by Transport Scotland to these events. This response is intended to allow the systematic and effective assessment, ranking, management and mitigation of the potential hazards from such events on the Scottish Road Network in the future. The context of the commissioning and execution of the study is also discussed as are the potential effects of climate change on debris flow frequency and severity in Scotland.

2 RESPONSE TO EVENTS

2.1 *Initial study*

Following the landslide events of August 2004 the Scottish Executive, and its successor organisation Transport Scotland, recognised the need to ensure that in the future it has a system in place for assessing the hazards and associated risks posed by debris flows. (Note that Transport Scotland was launched as an agency of Scottish Executive in January 2006.) The then Scottish Minister for Transport commissioned two studies. The first was to determine a way forward for dealing with such landslide events in the future (Winter *et al.*, 2005a; 2005b); an integral part of this work was the development of management and mitigation options. A second parallel study (not considered further here) was to examine broader issues, other than landslides, in respect to climate change (Galbraith *et al.*, 2005a; 2005b).

The landslides study comprises two parts of which Part 1 dealt with the following activities:

- Considering the options for undertaking a detailed review of side slopes adjacent to the trunk road network and recommending a course of action.
- Outlining possible mitigation measures and management strategies that might be adopted.
- Undertaking an initial review to identify obvious areas that have the greatest potential for similar events in the future.

This part of the study was, from the outset, intended to lead to a second part (Part 2) which would include the development of a system to allow a detailed review of the network. This would identify the locations of greatest hazard and allow those hazards to be ranked and appropriate mitigation and/or management measures to be selected.

The overall purpose of the landslides study is thus to ensure that Transport Scotland has a system in place for assessing the hazards posed by debris flows. In addition, the system will be capable of ranking the hazards in terms of their potential relative effects on road users. This will allow the future effects of debris flow events to be managed and mitigated as appropriate and as budgets permit. This will ensure that the exposure of road users to the consequences of future debris flows is minimised whilst acknowledging that it is not possible to prevent the occurrence of such events.

2.2 *Participation and procurement*

It was clear that a consistent, repeatable and reproducible system was required. This was especially important as a variety of consultants was likely to be involved in the data gathering, analysis and interpretation process. Inevitably each would have a different, but nonetheless valid, approach when operating independently. Such a situation would make any comparison between individual consultant's results and recommendations impossible for the purpose of, for example, allocating funds on a priority basis across the network. It was apparent at the outset that a unified system acceptable to all of the major players in the industry was required.

Transport Scotland has long-standing and close involvement with organisations and specialists in many disciplines in the course of its management and upgrading of the Scottish trunk (major) road network. It was appreciated at the outset that many of these organisations and specific individuals within them could make valuable contributions to what would be a complex project involving this nationally important issue. In addition, it was appreciated that the organisations and individuals would want to be involved in such a high profile project, the outputs form which would be to the benefit of Scotland in the broadest sense.

Thus, rather than opting to take a 'single consultant' approach, Transport Scotland saw the best way forward as managing the project from within, with an external, high profile facilitating project manager; the management team for the project was formed by the authors of this paper. The skills of a wide range of respected individuals from various key organisations could then be brought in, on a reimbursable basis, at relevant points in the lifetime of the project to carry out specific tasks.

It is this innovative approach which united the expert community in the work and was a key contributor to the success of the project. This opportunity was realised at an early stage by the key decision-makers in Transport Scotland, and the positive effects of their support and the enthusiasm for the project shown by the then Minister of Transport should not be underestimated.

The input of a full range of experts and stakeholders was required at the inception stage in order for the studies to be defined and undertaken successfully.

Accordingly a Working Group of individuals was assembled and a Project Workshop was held on 28 September 2004 in order to capture the knowledge vested with individual experts. The Project Workshop was facilitated by Professor Malcolm Horner of the University of Dundee and comprised presentations given by acknowledged experts followed by focused discussion sessions designed to open out the knowledge base and determine the way forward with the project. Following the Project Workshop the authors of this paper assigned tasks to members of the Working Group (including to themselves) in terms of the preparation of a report. The report was produced under the editorial control of the authors of this paper and the input of individuals was demonstrated by the authorship of sections within the report.

Members of the Working Group were selected to enable the individuals most suited to the various tasks to bring their knowledge, expertise and experience to bear on the relevant issues.

The work has been funded through a variety of existing contracts with the close and active involvement and support of Transport Scotland engineers as key members of the team as, indeed, exemplified by the authorship of this paper. The involvement of TRL is in providing the facilitating project manager to lead the experts drawn from Scotland's geotechnical community as well as making specific and substantial technical contributions. TRL has also been responsible for sub-contracting expertise from the British Geological Survey, Donaldson Associates, EDGE Consultants and Arup. The Independent Geotechnical Checker was represented by Halcrow and the (then) two Operating Companies, Amey and BEAR, by W.A. Fairhurst & Partners and Jacobs Babtie respectively.

The foregoing refers to the organisations involved in the project. However, the Working Group, including the editors of this report, comprised individuals each of whom was selected on the basis of their knowledge, expertise and experience and, indeed, their suitability to bring those characteristics to bear on the issues at hand. Appointments to the Working Group were based on the knowledge and experience, of individuals rather than on the appointment of the organisations who employ them. Individual members of the Working Group did, however, employ the services of colleagues as appropriate.

The Working Group comprised:

- Alan Forster, British Geological Survey.
- Andrew Heald and Julie Parsons, Jacobs Babtie/BEAR.
- Forbes Macgregor, Transport Scotland (now consultant to Transport Scotland).
- Stewart Martin, Halcrow (assisted by Dr Steve Hencher and Dr Roger Moore).
- Paul McMillan, W. A. Fairhurst & Partners/Amey (assisted by Dr David Brown).
- Ian Nettleton, EDGE Consultants.
- Julie Parsons, Jacobs Babtie/BEAR.
- Lawrence Shackman, Transport Scotland.
- Andy Sloan, Donaldson Associates.
- Matt Willis, Arup.
- Dr Mike Winter, TRL Limited (facilitating project manager).

2.3 *Reporting*

The work to produce the Part 1 study report was completed in time for a launch seminar at the Royal Museum in Edinburgh on 14 June 2005. The Minister for Transport introduced the event and almost 150 key players from local government, consultants, contractors, academia and other interested bodies attended.

The chapters of the report (Winter *et al.*, 2005a) are as follows:

1. Introduction to landslide hazards.
2. Background to Scottish landslides and debris flows.
3. Debris flow information sources.
4. Debris flow types and mechanisms.
5. Key contributory factors to debris flows.
6. Proposed methodology for debris flow assessment.
7. High hazard areas and early opportunities in Scotland.

8. Debris flow management and mitigation options.
9. Summary and recommendations for debris flows in Scotland.

In addition to this technical report a summary report (Winter *et al.*, 2005b) was prepared. It was intended to inform a wider audience of Transport Scotland's actions both since the events of August 2004 and planned for the future. The reports are available from both the Scottish Executive (www.scotland.gov.uk) and Transport Scotland (www.transportscotland.gov.uk) websites by simply searching "Landslides" to find the documents which are available for free download. (The climate change documents referred to earlier are available in similar fashion.)

2.4 *The broader context of the study*

When events such as those experienced in Scotland in August 2004 occur the attention of the media and thus the general public is almost guaranteed. However, a funded, strategic response is by no means assured, as discussions between the first author and colleagues from a number of other countries have highlighted. Why a funded, high level response from the Scottish Ministers was forthcoming on this occasion may be attributable to some extent to the following:

- Timing of events: The events occurred in August one of the two months, along with July, with peak tourist visitor numbers. Tourism is one of Scotland's largest business sectors, providing direct employment for 200,000 people and generating visitor spending of more than £4 billion a year (source: www.scotland.gov.uk).
- Location of events: In recent memory such events have affected only relatively remote parts of the road network. But in this case critical routes were affected. This is especially the case for the A9, with high levels of traffic travelling between the heavily populated Central Belt in the south and Scotland's most northerly city, Inverness.
- The Scottish Parliament: The Scottish Parliament was formed on 1 July 1999 with a wide range of powers devolved from the United Kingdom Government at Westminster. This has created a positive atmosphere in which decisions are made to resolve Scottish issues when the need exists, with funding appropriate to the need.
- Decision-making: The powers devolved to the Scottish Parliament allow a short chain of communication between Ministers and executive decision-makers.

3 DEVELOPMENT AND IMPLEMENTATION

The first part of the landslides study (see Section 2.3) might properly be described as a scoping study, while the second part is intended to develop and implement the work.

The methodology for the assessment of hazard and exposure is being developed and this will provide a hazard ranking. Hazard ranking has been used as an analogue for risk as not all aspects of risk are being assessed and the nature of a regional assessment such as this means that many aspects of the assessment are semi-quantitative or even semi-qualitative. The hazard ranking allows the selection of an appropriate management. Following that, the second stage is to test the methodology before applying it more widely to the trunk road network.

Figure 2 presents a flowchart of the work to be undertaken. This stage of the work is divided into four elements and can be summarised as follows:

- Development of a debris flow hazard and exposure assessment system to provide a hazard ranking of 'at-risk' areas of the road network.
- Undertaking a computer-based GIS assessment as a first stage in the hazard assessment process.
- Undertaking site specific hazard and exposure assessments of areas identified by the GIS as being of higher hazard.
- The identification and development of appropriate management processes for each category of hazard ranking.

Importantly, formal communication with Local Authorities, who are responsible for the local road network, has been established for this phase of the work via regular update meetings with the Society of Chief Officers of Transportation Scotland (SCOTS). SCOTS have been represented by personnel from Perth & Kinross Council, South Lanarkshire Council, Moray Council and Highland Council. Dave Spence (Highland Council) has also been added as an active member of the Working Group for Part 2 of the study.

During Part 1 of the project a number of lengths of perceived high hazard were identified. These lengths of road involved are in excess of 160 km (Winter *et al.* 2005a). It was thus considered unrealistic to undertake suitably prioritised further evaluations in advance of a GIS-based assessment as described below.

The GIS-based assessment is used as a first stage in the hazard assessment process. This enables site specific assessments to be targeted in order to obtain better value from such relatively resource-intensive activities. It also allows the elimination of large areas of the network having minimal hazard. Indications are that, in addition to serving its primary function of elimination, the GIS-based assessment has not only confirmed the areas of high hazard identified during Part 1 of the study but also identified additional areas of similar character.

It is particularly important to note that the site-specific assessments, that follow the GIS-based

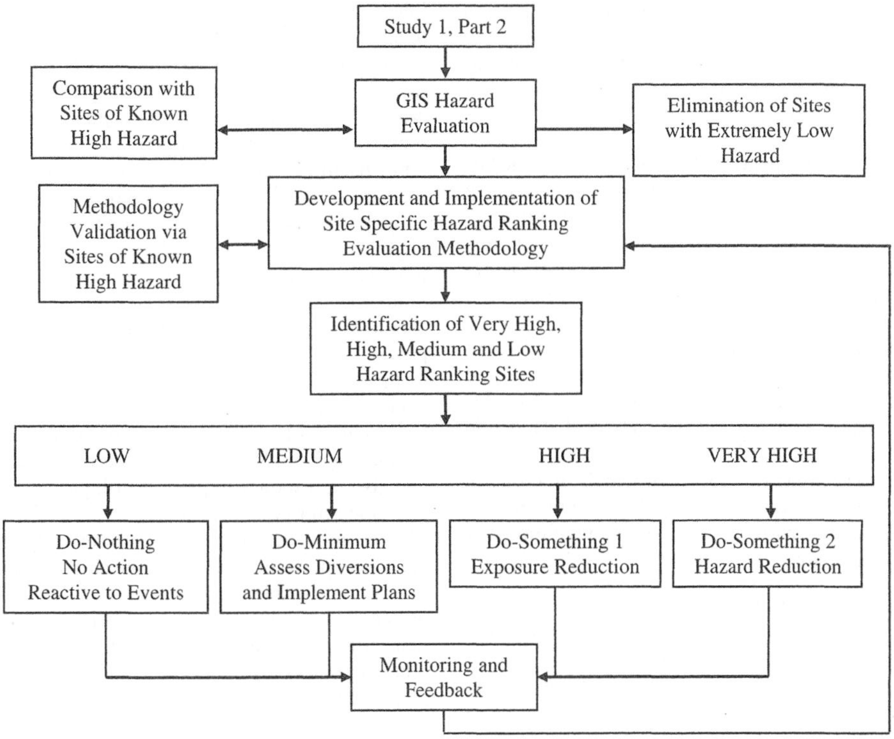

Figure 2. Structure of the work and management and mitigation options.

assessment, will not be a 'drive-by' survey; they will require a highly specialised detailed site examination using an overall consistent approach. Prior to undertaking any site surveys it is important that the system for consistently describing and identifying hazards and the associated exposure is established. Some of the factors that will need to be incorporated in such a system, such as slope angle and the broad nature of the geology are incorporated into the GIS-based assessment. Other, more detailed, factors such as the effects of forestation will need to be incorporated into the site-based survey. Once a hazard assessment has been completed it may be combined with an assessment of the exposure of the road user to that hazard to give a hazard ranking. This will allow, in-turn, an appropriate management option to be selected from the range of options being developed.

There are a number of potential options which could be applied to the management of debris flows (Figure 2). These are addressed in the following paragraphs.

The 'Do-Nothing' approach is intended to be applied to sites of low hazard ranking for which substantial expenditure is inappropriate. For such sites, whilst it is not possible to eliminate the chance of a debris flow event affecting such areas it is seen as an unlikely, largely unforeseeable event and/or the exposure is less serious than at other locations where resources may be better expended.

The 'Do-Minimum' option, with the potential to mitigate the impacts of debris flows to some extent involves simply ensuring that forward plans are in place to ensure that diversion routes are available and may be exploited in an expedient and well organised manner. Diversion route maps and contingency plans are currently held for many areas of the trunk road network.

Whilst it is not possible to eliminate the chance of a debris flow event affecting such areas any occurrence is seen as unlikely and largely unforeseeable. Any residual exposure cannot readily be quantified and is unlikely to justify the commitment of additional resources which may be better expended at other locations.

'Do-Something 1' is the first management option where site specific action is contemplated. Such action essentially involves the reduction of the exposure of road users to hazards by managing the access to and/or actions of the road-using public on the network at times either when events occur or precursor rainfall has indicated a high likelihood of debris flows occurring.

The reduction of exposure lends itself to the use of a simple and memorable three-part management tool (Winter *et al.*, 2005c), as follows:

- Detection: The identification of either the occurrence of an event, by instrumentation/monitoring (e.g. tilt meters or acoustic sensors) or observation (e.g. Closed-Circuit Television or visual patrols during high likelihood periods), or by the measurement and/or forecast of precursor conditions (e.g. rainfall).
- Notification: The dissemination of detected event information by for example variable message and static signs, media announcements (radio, TV, traffic guidance systems and the web) and 'landslide patrols' in marked vehicles.
- Action: The proactive process by which intervention reduces the exposure of the road user to the hazard, by for example road closure, convoying of traffic or traffic diversion.

In the short-term to medium-term this Detection-Notification-Action (or DNA) approach to mitigation must be reactive to debris flow events. There may be a case for reacting to extremely heavy rainfall events. However, a caveat to this is the need to consider carefully at what levels the triggers should be set, in so far as the relationship between rainfall and landslides/debris flows in Scotland is by no means fully understood.

In the longer-term, the detection of precursor triggering conditions (i.e. rainfall) may enable both the notification and action phases to be taken prior to the occurrence of major events (see also Section 4). However, an extensively enhanced rainfall detection network will be required across Scotland. Even once this is in place it is fully expected that it will take some considerable time and effort to ensure that sufficient data has been obtained and analysed so as to be able to introduce a reliable warning system. Even then it must be expected that atypical events, which are not the subject of warnings, and false alarms may be expected. A programme of public and media education and awareness-raising is likely to be desirable to minimise any potential adverse reaction to such scenarios.

'Do-Something 2' involves more major works in order to achieve hazard reduction (as opposed to exposure reduction in the 'Do-Something 1' case). The approaches involved entail physical measures such as the protection of the road, reduction of the opportunity for a debris flow to occur or realignment of the road away from the area of high hazard. Such options need to be considered in the context of the policy governing Transport Scotland's overall trunk road maintenance and construction programme. In general, these are likely to be of high cost necessitating their restriction to a very few areas of the highest hazard ranking.

The challenge with hazard reduction is in identifying locations that are of sufficiently high hazard ranking to warrant spending significant sums of money on engineering works. The costs associated with installing remedial works over long lengths of road are almost certainly both unaffordable and unjustifiable. Moreover the environmental impact of such engineering work should not be underestimated, having a lasting visual impact at the least and potentially more serious impacts. It is considered that such works should be limited to locations where their worth can be proven.

In addition, simple measures such as ensuring that channels and gullies are kept open can be effective in terms of hazard reduction. This requires that the maintenance regime is fully effective both in routine terms and also in response to periods of high rainfall, flood and slope movement. It is also important that maintenance and construction projects currently in design take the opportunity to limit any hazards by incorporating, where suitable, measures such as higher capacity or better forms of drainage, or debris traps. In particular, critical review of the alignment of culverts and other conduits close to the road should be carried out as part of any planned maintenance or construction activities.

Typically, the reduction in hazard will entail physical engineering works to change the nature of a slope or road to reduce the potential for either initiation and/or the potential for a debris flow to reach the road once initiated. Debris flows are dynamic in nature and are quite often initiated some distance above the road; when they reach the road they are relatively fast moving, high energy flows. The energy of these systems has a significant impact on the nature of the engineering works that can be used to reduce the hazard to the road and its user. Hence, there are three broad approaches to selection of hazard reduction works:

- Accept that debris flows will occur and protect the road.
- Carry out engineering works to reduce the opportunity for a debris flow to occur.
- Realign the road.

In relation to the first option there are not many examples of such engineering works in Scotland or the rest of the UK, but in some upland areas of mainland Europe such engineering is relatively commonplace. The energy of the debris flow is such that a rigid barrier constructed to protect the road would have to be designed for very high loads. A debris flow has significant momentum; to bring it to a sudden stop, as is the case with a rigid barrier, requires the near instantaneous dissipation of a lot of energy, imparting very high loads.

4 CLIMATIC INFLUENCES

4.1 *Rainfall patterns and landslides*

Landslides are often cited as being caused by storm rainfall and the link between high intensity rainfall and debris flows has been documented in Japan (Fukuoka, 1980), New Zealand (Selby, 1976) and Brazil (Jones, 1973) amongst other places. However, the influence of antecedent rainfall prior to storm events was clear from the events experienced in Scotland in August 2004.

In a study based in the Santa Cruz Mountains of California, Wieczorek (1987) noted that no debris flows were triggered before 28 cm of rainfall had accumulated in each season. This clearly acknowledges the importance of pre-storm, or antecedent, rainfall, a factor that has also been recognised in studies in Southern California (Campbell, 1975), New Zealand (Eyles, 1979) and Alaska (Sidle and Swanson, 1982). Wieczorek (1987) also notes that in the case of high permeability soils such as those found in Hong Kong (Brand *et al.*, 1984), the period of antecedent rainfall may be short or the amount of necessary antecedent rainfall may be supplied by the early part of the storm event.

4.2 *Scotland's rainfall climate*

The climate of Scotland in terms of its rainfall may be very broadly divided into east and west. Data presented by the Meteorological Office (Anon, 1989) indicates that in the east rainfall generally peaks in August while in the west the maximum rainfall levels are reached during the wider period September to January (Figure 3). Although rainfall levels in the west are relatively low in August they do increase from a low point in May. Both scenarios indicate that the soil may be undergoing a transition from a dry to a wetter state at or around August, indicating an increased potential for debris flow and other forms of landslide activity. The central area, as represented by Pitlochry in Figure 3, has a mix between the rainfall characteristics of the 'east' and the 'west'. The rainfall peak is both lower and shorter (December and January) than in the west, but there are also small sub-peaks in August and October. A broadly similar pattern is found for Perth.

Clearly, the soil water conditions necessary for debris flows may be generated by long periods of rainfall or by shorter intense storms. It is however widely accepted that Scottish debris flow events are usually preceded by both extended periods of heavy rainfall (otherwise known as antecedent rainfall) and intense storms.

4.3 *Potential climate change*

The UKCIP (UK Climate Impacts Programme) report considers three periods: the 2020s, the 2050s and the 2080s. In general terms small changes are noted in the predictions for the 2020s. These changes increase slightly for the 2050s and slightly further still for the predictions for the 2080s, reflecting the temporal trends in temperature and precipitation. Whilst climate models generally predict averages and the associated error limits can be substantial, it is also important to note that inter-annual variability is predicted to increase for many climate factors. This means that average changes, as discussed above, may mask more important variability effects.

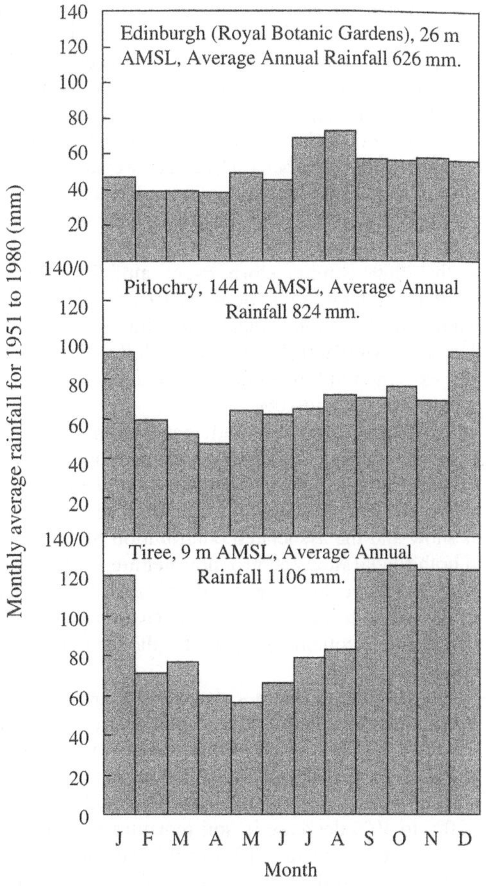

Figure 3. Average rainfall patterns for selected locations in Scotland. Edinburgh is in the East of Scotland, Pitlochry in the centre and Tiree in the West.

Climate change models for Scotland in the 2080s (www.ukcip.org.uk and Galbraith *et al.*, 2005a) indicate that, while overall precipitation levels will decrease, in the summer precipitation will decrease but that they will increase in the winter. However the models are generally considered to be incapable of predicting localised summer storms. These storms are believed to be at least partially responsible for triggering the events of August 2004, and climate data may not give a full picture of the relationship between precipitation and landslides. Furthermore, it is important to note that climate models generally predict averages and that the error limits can be substantial. Predicted changes in the number of 'intense' wet days generally indicate a net increase of less than one day per annum by the 2080s, with slightly fewer intense wet days in the summer and more in the winter. However, by the 2080s extreme storm event rainfall depths are predicted to increase by between 10% and 30%, with intense winter rainfall increasing slightly more than this, and spring/autumn rainfall by slightly less. Summer extreme rainfall depths are predicted to increase by between 0% and 10%.

Peak fluvial flows are anticipated to increase progressively during the twenty-first century. Eastern Scotland is expected to experience larger increases than north-west Scotland for example. The occurrence of snow and the associated contribution of snowmelt to both fluvial flow and groundwater are, on the other hand, predicted to decrease. Reductions in snowfall are predicted to be greater for the eastern and southern parts of Scotland and least for the central upland areas.

Changes in the factors discussed above coupled with increased potential evapotranspiration, particularly in the summer, and a longer growing season, leading to increased root uptake, are expected to have substantial effects on soil moisture. The models predict a 10% to 30% decrease in soil moisture for summer/autumn and an increase of 3% to 5% in the winter. The winter figures reflect the fact that soils can only contain a finite amount of water and most Scottish soils are already close to saturation in the winter.

Reduced soil moisture during the summer and autumn months may mean that the short-term stability of some slopes formed from granular materials is enhanced by suction pressures (often described as negative pore water pressures). Soils under high levels of suction are vulnerable to rapid inundation, and a consequent reduction in the stabilising suction pressures, under precisely the conditions that tend to be created by such as short duration, localised summer storms. In addition, non-granular soils may form low permeability crusts during extended dry periods as a result of desiccation. Providing that these do not experience excessive cracking due to shrinkage, then they may increase runoff to areas of vulnerable granular deposits. Such actions could lead to the rapid development of instabilities in soil deposits, potentially creating conditions for the formation of debris flows. The complicating factors are the potential inability of current climate models to resolve storm events and the precise nature of the localised failure mechanisms that will lead to the initiation of an individual debris flow. It is highly unlikely that the measurement of soil suction could provide a practical and reliable means of debris flow forecast.

The importance of the potential effects of climate change impacts on slope stability is exemplified by the existence of an Engineering and Physical Sciences Research Council (EPSRC) Network: Climate change impact forecasting for slopes (CLIFFS) (Dixon *et al.*, 2006). This is funded to provide a 'talking-shop' for such issues and to develop collaborative working arrangements to study such impacts and to develop coping strategies.

4.4 *Mechanics of unsaturated slope failure*

That rainfall can cause landslides was vividly demonstrated in February 2005 when catastrophic landslides occurred during intense rainfall in both California in the U.S. and British Columbia in Canada. Property destruction and tragic loss of life were the results of the various landslides. Over approximately a seven-month period, the Malibu area of California received over 585 mm (23 inches) of cumulative precipitation. Then in February 2005 the area received an additional 228 mm (9 inches) over a period of about four days, at which time the landsides occurred (GeoSlope, 2005).

Analyses by GeoSlope, replicating the rainfall conditions experienced in California and British Columbia in February 2005 yielded some interesting results. The analysis confirmed that a typical model slope remained stable for seven months during which 585 mm of cumulative rainfall fell but became unstable after a further 228 mm over a period of four days. Typically the failure could not be attributed to increased positive pore water pressures as the failure surface did not penetrate below the water table. GeoSlope attributed the failure to decreases in suction. This type of behaviour corresponds well with that predicted from unsaturated soil mechanics theory (Wheeler *et al.*, 2003) and the broad style of this type of failure mechanism is supported by experiment (Springman *et al.*, 2003).

5 CONCLUSIONS

In August 2004 parts of the Scottish road network were adversely affected by a series of landslide events in the form of rainfall-induced debris flows. While such events are relatively common, such a significant degree of interaction with the transport infrastructure is unusual.

Transport Scotland has initiated and led a rapid, proportionate and structured response to these events. The response to these events has been described in some detail. An inclusive approach has been adopted to the involvement of a wide range of acknowledged experts each of whom has a vested interest in producing an appropriate system for the future assessment, ranking, management and mitigation of debris flow hazards on the Scottish road network. The range of skills required to deliver the work required goes somewhat beyond the usual geomorphological, geological and geotechnical disciplines associated with landslides; therefore those familiar with the management of roads formed a key part of the team.

When events such as these occur the attention of the media and thus the general public is almost guaranteed. However, a funded, strategic response is by no means assured, as discussions between the first author and colleagues from a number of countries have highlighted. The reasons for the proactive response from Government are postulated as being due to the timing and location of the events, and the ability and willingness of decision-makers from both the political and executive branches of government to take decisions at an appropriate level (a fact not unrelated to the existence of the relatively new Scottish Parliament).

The way forward in terms of the development of a systematic approach to the assessment and ranking of debris flow hazards has been described. This will provide a systematic means of ranking hazards and thus allow the allocation of funds on a priority basis across the network. Four levels of hazard ranking are described and two of these are intended to attract site-specific actions.

At sites of high hazard ranking, exposure reduction is anticipated; this is based upon the logical sequence of Detection-Notification-Action (DNA). In the short-term the DNA sequence will be applied in a manner so as to react to debris flow events. In the longer-term an enhanced network of rainfall gauges that will allow the detection of precursor conditions such that notification can be made and action taken in advance of debris flows occurring is planned.

At sites of very high hazard ranking, hazard reduction is anticipated; such actions will, in the main, involve the expenditure of significant sums of money on remedial or other engineering works. The costs associated with installing such works over long lengths of road are almost certainly both unaffordable and unjustifiable. Moreover the environmental impact of such engineering work should not be underestimated, having a lasting visual impact at the least and potentially more serious impacts. Such works will be limited to locations where their worth can be proven.

Clearly many of the mitigation techniques and activities described will require a heightened level of awareness of the issues surrounding landslides in general and debris flows in particular from engineers, the public and the media. As such partnership, education and knowledge dissemination will be a key part of the ongoing work.

It seems clear that the system as described will provide significant benefits in terms of the reduction of the impacts of future debris flow events on the Scottish road network.

The approach to forecasting rainfall-induced landslide events is considered in the context of both the current climate and of potential future climate scenarios. Currently long-term antecedent rainfall followed by short intense storm rainfall is believed to characterise many debris flow events in Scotland. Climate change models would appear to indicate that Scotland's climate will become generally drier, but that the frequency, intensity and duration of storm events may increase. It may thus be inferred that the frequency and severity of debris flow events in Scotland are most likely to increase as a result of climate change.

In developing an understanding debris flow events in a drier summer climate there appears to be a clear role for the application of unsaturated soil mechanics, whether such understanding is to be at the conceptual level or at the analytical level.

ACKNOWLEDGEMENTS

The authors wish to acknowledge the input to this work of their colleagues from the Scottish Road Network Landslides Study Working Group as detailed in this paper and also to the British Geological Survey staff who undertook the GIS-based assessment in close collaboration with members of the Working Group.

REFERENCES

Anon. 1989. *The climate of Scotland – some facts and figures*. London: The Stationery Office.

Brand, E.W., Premchitt, J. & Phillipson, H.B. 1984. Relationship between rainfall and landslides in Hong Kong. *Proceedings, IV International Symposium on Landslides*, 1, 377–384. Toronto: Canadian Geotechnical Society.

Campbell, R.H. 1975. Soil slips, debris flows, and rainstorms in the Santa Monica Mountains and vicinity, southern California. *US Geological Survey Professional Paper 851*, 51p.

Dixon, N., Dijkstra, T., Forster, A., & Connell, R. 2006. Climate change impact forecasting for slopes (CLIFFS) in the built environment. *Engineering Geology for Tomorrow's Cities: Proceedings, 10th International Association of Engineering Geology Congress*, p. 43 and DVD-Rom. London: The Geological Society.

Eyles, R.J. 1979. Slip-triggering rainfalls in Wellington City, New Zealand. *New Zealand Journal of Science*, 22(2), 117–122.

Fukuoka, M. 1980. Landslides associated with rainfall. *Geotechnical Engineering*, 11, 1–29.

Galbraith, R.M., Price, D.J. & Shackman, L. (Eds). 2005a. *Scottish road network climate change study*. 100p. Scottish Executive, Edinburgh.

Galbraith, R.M., Price, D.J. & Shackman, L. 2005b. *Scottish road network climate change summary report*. 31p. Scottish Executive, Edinburgh.

GeoSlope. 2005. Why do slopes become unstable after rainfall events? http://www.geo-slope.com/res/Whydoslopesbe-comeunstableafterrainfallevents.pdf. *Direct Contact*, April 2005.

Jones, F.O. 1973. Landslides of Rio de Janeiro and the Serra das Araras Escarpment, Brazil. *US Geological Survey Professional Paper 697*, 42p.

Selby, M.J. 1976. Slope erosion due to extreme rainfall: a case study form New Zealand. *Geografiska Annaler*, 58A, 131–138.

Sidle, R.C. & Swanson, D.N. 1982. Analysis of a small debris slide in coastal Alaska. *Canadian Geotechnical Journal*, 19(2), 167–174.

Springman, S.M., Jommi, C. & Teysseire, P. 2003. Instabilities on moraine slopes induced by loss of suction: a case history. *Géotechnique*, 53(1), 3–10.

Wheeler, S.J., Sharma, R.S. & Buisson, M.S.R. 2003. Coupling of hydraulic hysteresis and stress-strain behaviour in unsaturated soils. Géotechnique, 53(1), 41–54.

Wieczorek, G.F. 1987. Effect of rainfall intensity and duration on debris flows in central Santa Cruz Mountains, California. In: *Debris Flow/Avalanches: Process, Recognition and Mitigation* (Eds: Costa, J. E. & Wieczorek, G. F.). Reviews in Engineering Geology, VII, 93–104. Boulder, CO: Geological Society of America.

Winter, M.G., Macgregor, F. & Shackman, L. (Eds). 2005a. *Scottish road network landslides study*. 119p. Scottish Executive, Edinburgh.

Winter, M.G., Macgregor, F. & Shackman, L. 2005b. *Scottish road network landslides summary report*. 27p. Scottish Executive, Edinburgh.

Winter, M.G., Macgregor, F. & Shackman, L. 2005c. Introduction to landslide hazards. In: *Scottish Road Network Landslides Study* (Eds: Winter, M. G, Macgregor, F. & Shackman, L.), 9–11. Edinburgh: The Scottish Executive.

Winter, M.G., Heald, A., Parsons, J., Shackman, L. & Macgregor, F. 2006a. Scottish debris flow events of August 2004. *Quarterly Journal of Engineering Geology and Hydrogeology*. 39(1), 73–78.

Winter, M.G., Macgregor, F. & Shackman, L. 2006b. A structured response to the Scottish landslide events of August 2004. *Engineering Geology for Tomorrow's Cities: Proceedings, 10th International Association of Engineering Geology Congress*, p. 125 and DVD-Rom. London: The Geological Society.

Landslides and Climate Change – McInnes, Jakeways, Fairbank & Mathie (eds)

Climate change impacts on landslide mechanisms and hazard in southern UK

A. Bracegirdle & C.O. Menkiti
Geotechnical Consulting Group

A.R. Clark
High Point Rendel

ABSTRACT: The UK Climate Impacts Programme (UKCIP) has published a review of precipitation data, identified trends of increased precipitation, and published predictions of precipitation for the UK up to 2080 based on four different emissions scenarios (UKCIP02). All the scenarios examined indicate a continuing trend of increased precipitation for winter months and drier summer months. A pattern of increased landslide activity is becoming clear in the southern UK, and elsewhere, which appears to be consistent with recent climate trends. This paper examines several examples of recent UK landslides in their historical context. It is suggested that changes in landslide mechanisms that may accompany higher rainfall intensity. Designers are faced by uncertainty over future groundwater conditions and should now question the applicability of historical precedent. In particular, designers should consider the effects of the increased saturation of steep slopes, where rapid failure mechanisms such as debris flows are possible.

1 INTRODUCTION

1.1 *Climate change*

It has been estimated that an increase in average annual temperatures of between 2.0 and 3.5 CE can be expected over the next 80 years (Hulme et al, 2002). This is likely to be accompanied by changes in precipitation, storminess, and sea level change. The largest change in winter precipitation is expected in the south and east, where it is predicted to increase by between 15 and 30%. Rainstorm intensity is also expected to increase considerably, with the current 100-year rainfall intensity occurring on about a 10-year return period. Drier summer months are expected, with up to 50% less precipitation in the south of the UK than present.

The predictions of storm surges in sea levels, wave heights and wind directions are subject to considerable uncertainty at present. These factors are fundamental to determining coastal attack and flooding. Although sea level changes have been of little significance in the last century, they will assume greater importance in the future. Recently, there has been considerable interest in coastal flooding. Hall et al (2006) estimate that the annual cost of coastal flooding in the UK will increase from the present £0.5B to between £1B and £13.5B, depending on the emissions scenario adopted. In a more extreme, but nevertheless plausible scenario, Tol et al (2006) point out that a sea level rise of up to 5 m is possible in the event of the collapse of the West-Antarctic Ice Sheet.

By contrast, until present, the effects of changes in precipitation on landsliding and landslide risk have received relatively little public attention.

1.2 *Recent experiences with new and first-time landslides*

As outlined by Vaughan et al (2002), slopes in which the permeability of the soil is less than average rates of infiltration are generally saturated and are unlikely to be substantially affected by increases in precipitation. The stability of intermediate-permeability partially saturated slopes is however highly sensitive to rainfall intensity. The water level rise of water levels in short-duration, high intensity rainfall in partially saturated fissured clay or rock, for example, is typically eight times the depth of rainfall. Such slopes can be expected to be sensitive to climate change and rainfall intensity.

A substantial increase in landslide activity was seen in the south of the UK between January and March 2001. This period, was characterised by a prolonged wet period with a number of extremely high intensity rainstorms.

Dramatic chalk falls were seen on the Isle of Wight, Brighton and in Mupe Bay in Dorset. Debris flows were seen in steep country in the UK and Channel Islands driven by unusually intense rainfall. On the Isle of Wight, for example, a damaging and potentially very dangerous talus failure occurred that ultimately led to the removal of much of the talus in Shanklin in order to manage the risk posed by these slopes.

The talus failure at Shanklin draws into perspective the potential threat posed by relatively steep slopes of sandy material, such as found elsewhere on the Greensand escarpments in the south of England. An examination of the distribution of landslides in the Weald District shows there are at least 18 known landslides along the Greensand escarpment in the Weald and these typically comprise lobes of head driven by talus and water from the escarpment. Many of these have become exceptionally active since 2001 although, to date, the landslides have been slow moving and incremental. This behaviour does not preclude the possibility of rapid debris flows at the known landslide sites or new landslides on the escarpment.

The very large landslide complexes at East Cliff Lyme Regis and Fairlight have been subject to mapping and observation over many years. An examination of historical data available for these sites shows that in each case there has been a marked acceleration in landsliding. The rates of regression of these landslides at these sites now outstrip the rate of toe erosion. Mudsliding has on both sites become highly aggressive leading to concerns over loss of property and infra-structure.

The case histories provided in the following sections illustrate failure mechanisms that can be linked to high levels of precipitation. This paper seeks to provide some guidance to designers and to highlight particular areas of concern.

2 STEEP SLOPES

2.1 *Chalk falls*

A number of chalk falls have occurred recently in natural slopes standing at angles in excess of 70°. The winter of 2001 saw the highest precipitation in southern England since records began in 1727. More than 500 landslides were recorded over this period, many of which were large and potentially dangerous chalk falls.

Examples include the landslide in Mupe Bay in Dorset in March 2001. The volume of the slide was of the order of 50,000 m^3 and led to a substantial landward recession of the cliff top. Comparison of the 1891 Ordnance Survey map and recent aerial photographs shows there had been no substantial regression of the cliff line or the foreshore between 1891 and 2001.

Massive chalk falls also took place in Dover in January 2001 and Beachy Head and Brighton Marina in

Figure 1. Bedding plane failure, Whitecliff Bay, Isle of Wight.

April 2001(Palmer et al, 2002). Very large cliff failures have been seen in the Lias at Charmouth and the Bridport Sands at Bridport which hade been subject to aggressive coastal erosion.

Conceptual models have been available for the degradation of chalk cliffs and the formation of scree for some time (for example Hutchinson (2001, 1983)). While the factors leading to cliff failure are often complex, the landslide events of 2001 were evidently triggered by extreme precipitation. Given the trend of increased ratio of winter/summer precipitation since 1980 (Hulme et al, 2002) and recent future climate predictions, it seems likely that these events mark the onset of changes in "medium-term" behaviour as defined by Brunsden and Lee (2000).

2.2 *St Helier, Jersey*

Also in March 2001, a large toppling failure was observed in Westmount Quarry in St. Helier, Jersey. Although the quarry slope had been subject to occasional small rock falls, the slope had been generally stable for a period of about 100 years. The toppling failure involved about 5,000 m^3 of weathered rock which destroyed the parish depot (see Fig. 2) and left a large graben feature at the head of the slope and tension cracks extending into the adjacent property.

Catch fencing had been installed prior to and was destroyed by the toppling failure. The event has been attributed to high groundwater levels within the rock

Figure 2. Westmount Quarry failure.

Figure 3. The carpark, March 2001 *courtesy of Peter Marsden.*

behind a vertically inclined zone of clay gouge within and parallel to the line of the slope. The zone of clay gouge, which extends into the adjacent property, had not been identified prior to the collapse.

Some two months earlier the slopes on the adjacent property slope had been subject to high surface water flows down the slope which deteriorated into a rapid debris flow. An apartment block on the site sustained damage as a result of the debris flow. The slope had been steepened three years earlier, at which time rock bolting, anchors and netting were installed to stabilise the slope. There were no reports of surface water flows within the recent history of the site, and the drainage measures installed at the top of the slope did not prevent surface from water running down the slope.

The records of precipitation for the winter of 2000/2001 in Jersey show this to have been the wettest period within the 100 years of recorded data. Antecedent rainfall corresponds to an extrapolated 10,000 year return period event. At both sites, work on the slopes had been carried out without a detailed borehole investigation which might have identified the presence of the clay gouge zone. As is often the case, design relied on observation and logging of the rock faces. There appears to be no historical precedent for the failures that took place.

The case is of interest because the failures are in part attributable to a geological feature that had not been identified and also to extreme precipitation. In undertaking work on steep slopes it is now evident that designers can not necessarily rely on historical precedent and the logging of surface exposures. At the very least, designers have a duty to warn clients of the potential hazards and uncertainties where it is impractical or there is no budget to carry out detailed ground investigations. The case also reinforces the need to carefully consider surface water and extreme rainfall scenarios when assessing the future behaviour of slopes.

2.3 *Talus slopes, sandown*

Coastal erosion of the Lower Greensand led to the formation of sea cliffs in Sandown Bay which, since Victorian times, have been progressively protected by reclamation and coastal defences between Sandown and Shanklin. Until recently, large accumulations of talus had formed at the foot of the protected cliffs in Shanklin.

The talus stood at angles of 30° to 40° to the horizontal, to heights of up to 15 m. The talus material, derived form the Lower Greensand, contained up to 20% of silt and clay sizes. In the 1980s, cliff falls had triggered localised undrained failures of the talus (Barton (1984)), which increased in severity in 1988, causing damage to property and raising concern over safety.

In response to these concerns, the Isle of Wight Council commissioned their consultants to undertake a study of the cliff and talus, ranking the risks of landslides within the study area (McInnes (1996)). The study, described by Clark et al (1993), identified sections of cliff and talus deposits as being potentially unstable. Works to both the cliff and talus deposits were carried out within the funding available to manage the risk.

Notwithstanding the risk mitigation works, a large section of talus failed on 21 March 2001 during intense rainfall and easterly winds. The volume of the failure was about 5,000 m^3; it extended over a length of 80 m with a run-out angle of about 10°. The run-out angle is consistent with debris slides of comparable volume (for example, AGS(2000) and Wong and Ho (2000)). The slope failure was very rapid and was mostly dissipated within the car park area of a beach-front hotel (Fig. 3) although broke through into the rear of the hotel (Fig. 4). Remarkably, there were no fatalities.

With some notable exceptions, such failures have until recently been rare in the UK. While the risk assessment of the cliff line made in the late 1980s did recognise the possibility of extensive run-out of the talus, other areas of cliff line had been correctly assigned higher relative risk. In spite of the protection afforded by talus to the base of the cliffs, much of the talus was removed following the talus failure in 2001.

The case is of interest because it illustrates that rapid debris flows normally associated with mountainous or high-rainfall regions are a reality in the south of England at the onset of a new regime of increased winter precipitation. It is also of interest to examine the other areas of similar geology in the south east that might be similarly affected in the future.

2.4 *The Weald greensand escarpment*

Much has been learnt from studies of landslides emanating from the greensand escarpment (for example, Weeks (1969), Skempton et al (1967, 1976) and Bromhead et al (1998)). At least 18 landslide sites are

Figure 4. Hotel interior, Shanklin, March 2001 *courtesy of Peter. Marsden.*

shown on geological maps and form and arc around the interior of the Weald, as shown on Fig. 5. Typically, these landslides comprise extensive lobes of sand and clay sliding on residual shear surfaces in the Atherfield Clay and Weald Clay. Movements tend to be incremental and are often triggered by rainfall and infiltration of groundwater through the greensand from the higher ground behind the escarpment.

Since 2001 there has been increased landslide activity of some of these sites. Landslides at Holmebury Hill have closed footpaths and large movements at Leith Hill led to a lengthy closure of the Abinger Road in 2001/2002, which is showing signs of continuing distress.

As discussed by Corominas (2000), large landslide masses tend to react to prolonged periods of wetness, while shallow events are often triggered by short-term high intensity rainfall. This general observation applies when comparing the four-month antecedent rainfall that appears to drive the large landslide complexes in the Undercliff on the Isle of Wight and the 1 to 2 day rainfall duration that is associated with the Shanklin talus failure and the St Helier debris flow. In both of the latter cases, however, intense rainfall also corresponded to high seasonal rainfall.

Given the present scenario of increasing precipitation in winter months and more intense precipitation, it is perhaps not surprising that the first sign of change would be the increased activity of existing residual landslides. The effect of this is to draw the colluvial material down-slope from the escarpments, potentially leading to steepening of the relatively more stable talus slopes. Many of the talus slopes along the escarpment are currently standing to heights of 15 m or more and at

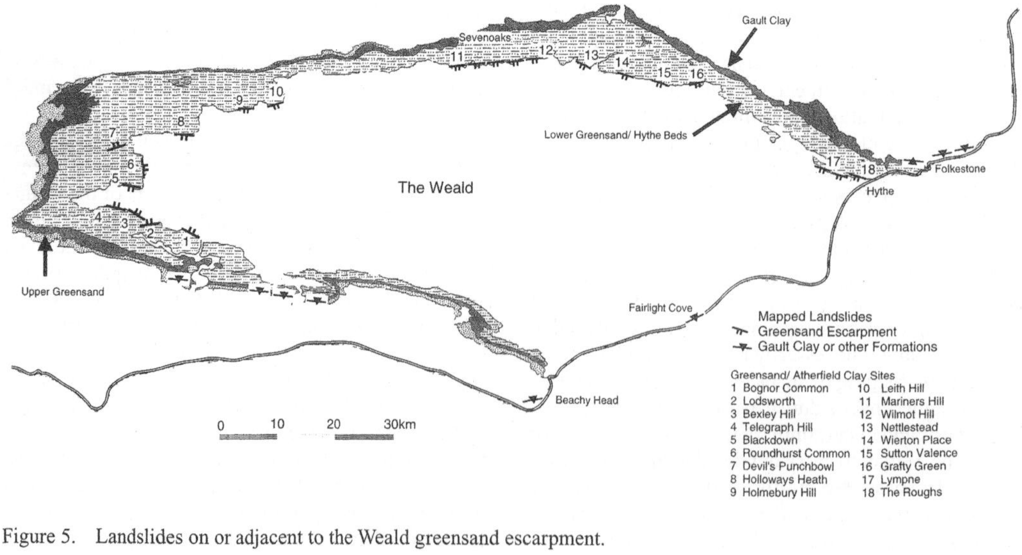

Figure 5. Landslides on or adjacent to the Weald greensand escarpment.

angles in excess of 30° . A similar situation is found at some locations along the Undercliff between Ventnor and Castlehaven on the Isle of Wight.

Collison et al (2000) consider the response of soil-moisture deficit of a 4 km section of the escarpment near Hythe to changes in climate. They conclude that the frequency of large landslides will remain unchanged while the frequency of smaller landslides will reduce. While their conclusions may well be true in the long-term, the period of adjustment and change is likely to bring about medium-term changes in landslide behaviour. As will be discussed further below, there is evidence of more aggressive activity of large landslides in recent years.

As seen in Shanklin, there is also a risk of large debris flows in steep talus slopes that have been relatively stable in the past. While the greensand talus slopes remain relatively dry, they will continue to enlarge, showing only modest signs of instability. If, however, the talus becomes saturated due to higher intensity rainfall superimposed on wetter winters, increased groundwater flow through the talus and surface water flow, then rapid debris flows are likely to occur. The speed and run-out distance of these failures is cause for concern, as is their effect on the large residual landslides that are found extensively below the escarpment.

3 LARGE RESIDUAL LANDSLIDES

3.1 *Fairlight Cove*

Fairlight Cove is located on the Weald coast (see Fig. 5). A large landslide at Rockmead Road, Fairlight, seen in Fig 6, has caused substantial loss property in recent years.

The landslide is on three tiers, the lowest of which is a large lobe that in recent years has advanced seawards. A schematic section through the landslide is shown on Fig. 7. Geomorphological mapping of the landslide suggests the basal shear surfaces are parallel to the bedding at two levels in the Fairlight Clays. The shear surface at the base of the upper tier is seen in Fig 8 below. High water pressures are present within permeable layers of the clay sequence at the rear of the landslide. In addition, there is active ponding of surface water and very active mudsliding in all three tiers.

Between 2002 and 2006, mudsliding has been progressively removing material from the upper two tiers and as a consequence there has been a substantial regression of the cliff line.

Historical mapping and aerial photographs have been used to assess the rate of landward movement of the toe and the cliff top. As can be seen from Fig 9. average regression rates of 0.46 and 0.06 m/yr at the toe and cliff top were experienced between 1875 and 1995 on a section taken through the centre of the landslide. The cumulative regression over this period has resulted in a substantial steepening of the cliff profile. Since the mid 1990s however, the overall slope of the cliff has returned to about that of 1875. It is likely that the form of the cliff is responding cyclicly by episodic flattening followed by long periods of toe erosion and slope steepening. It is nevertheless of interest that the present activity corresponds to a period of increased

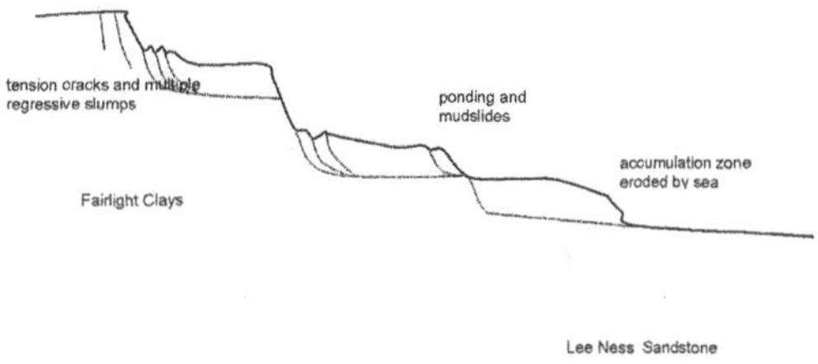

Figure 7. Schematic section through Rockmead Road landslide.

Figure 6. Aerial photograph, Rockmead Road Landslide.

Figure 8. View uphill from the second tier.

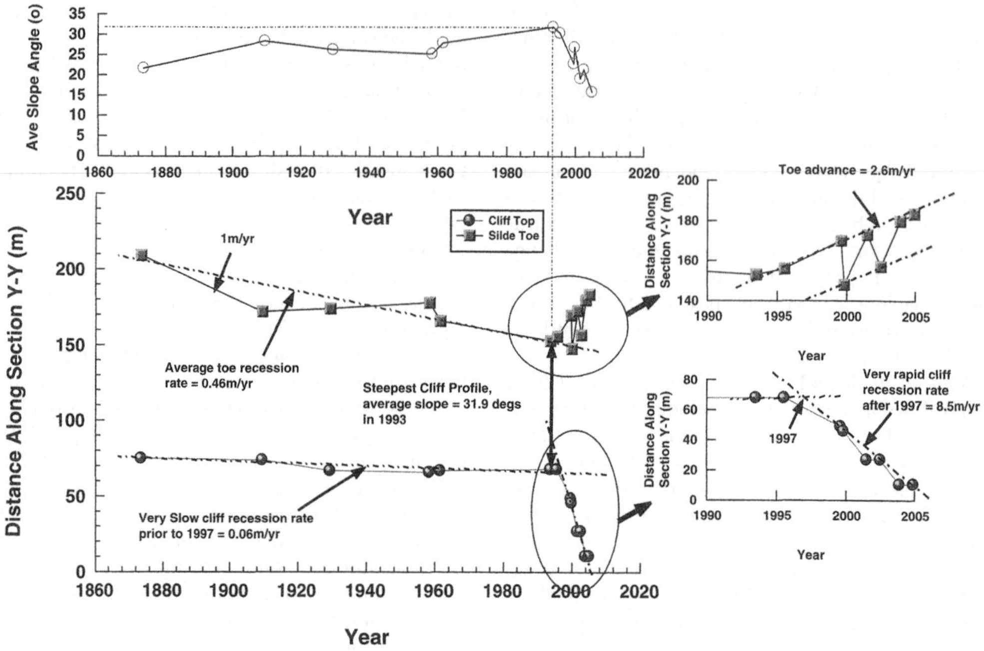

Figure 9. Cliff and toe regression, Rockmead Road landslide.

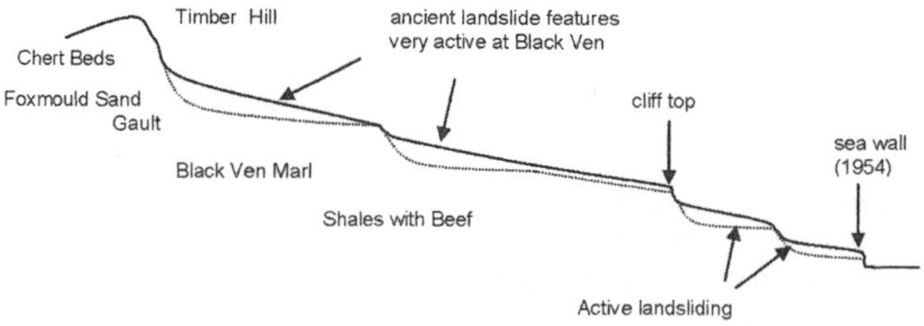

Figure 10. Schematic cross section of landsliding at East Cliff, Lyme Regis.

winter precipitation. Estimates for regression of the cliff line over the next 80 years, based on the present regime vary between 1.0 and 1.5 m per year in the absence of landslide mitigation works.

In 2006, DEFRA approved funding to construct a rock berm at the toe of the landslide and carry out re-grading and drainage works.

3.2 *East Cliff, Lyme Regis*

East Cliff, Lyme Regis has a long history of coastal erosion and landsliding. The area is famous for the Black Ven Landslide, 1km to the east of East Cliff (Brunsden and Chandler (1996)). The geology comprises Cretaceous Greensand and Gault unconformably over calcareous mudstones with interbedded limestones of the Lower Lias, as is the case in many other coastal landslides, such as Fairlight and the Undercliff of the south coast of the Isle of Wight, shear surfaces are active at different levels, giving a stepped or benched ground profile. This is shown schematically in Fig. 10. The width of the complex is approximately 1 kilometre at East Cliff and the most active landsliding is in the lower two tiers, within the "Shales with Beef" and "Disturbed Lias".The Upper two tiers within the Black Ven Marl have shown little movement until very recently.

Mudslides have been very active in the last 10 years particularly in the lower tiers and up to 10 m of lateral displacement of ground markers has been recorded within the last 5 years. In 2004 a series of large diameter (1050 mm (4 no.) and 750 mm (50 no.)) bored piles were installed above the second tier as a temporary measure to prevent regression of the "cliff top" shown in Fig. 10 into a car parking area. There is concern that continued regression and renewed activity at Timber

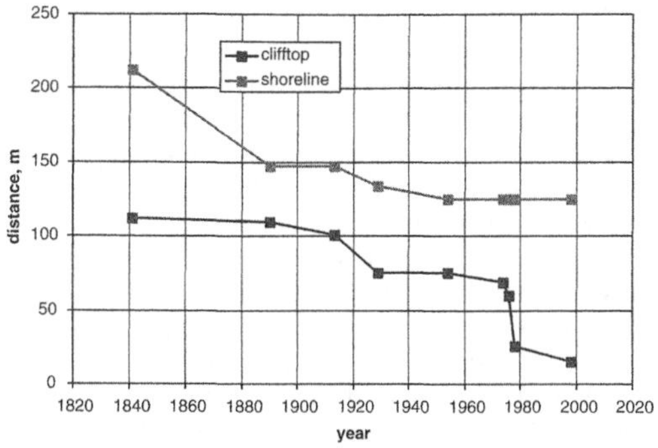

Figure 11. Regression rates, East Cliff, Lyme Regis.

Hill as Black Ven extends westwards, will threaten property and road access to Lyme Regis.

Studies are currently being undertaken by West Dorset District Council and High Point Rendel to examine methods of stabilising both East Cliff and Chuch Cliffs, further to the west. The design work will build on the experiences gained on the Lyme Regis stabilisation works described by Fort et al (2007).

Research has been carried out on the landslide at East Cliff to determine the rates of regression of the "cliff top", using Ordnance mapping and aerial photography. The regression, expressed as a distance from a fixed point, is shown on Fig. 11, with data extending back to 1841. The 1841 data points are based on a tythe map, which is likely to be less accurate than the later ordnance mapping. Ignoring the tythe map data, an average rate of recession of 0.5 m per year is seen for the shoreline up until 1954 when a sea wall was constructed. The rate of recession of the cliff top between 1880 and 1976 is also 0.5 m per year. Major failures took place between 1976 and 1980, and very aggressive mudsliding has taken place since then. The mudslides, which have moved tens of meters since 1980, have progressively emptied the lower tiers of the landslide with the discharge of mudslide material over the sea wall.

The rate of the cliff top regression at East Cliff shows a substantial increase in the mid-late 1970's, followed by mudsliding. This and the recently detected movement of the upper tier are consistent with wetter conditions that have been experienced in recent years and the onset of climate change.

4 SOME LESSONS FROM RECENT LANDSLIDE BEHAVIOUR IN SOUTHERN ENGLAND

4.1 *Steep rock slopes*

As discussed earlier, a large number of rock falls occurred during the wet winter period of 2001. The mechanics of such falls are reasonably well understood. The frequency of such falls is likely to continue to increase given the climate change scenarios available. Local authorities have a duty to warn the public of the increased level of hazard and to set reasonable restraints on developments close to steep slopes.

Figure 12. Debris flow, housing estate, 2003.

4.2 *Debris flows*

As shown by the case in Shanklin in 2001, short duration, high-intensity rainfall can lead to large debris flows. Such flows occur where low plasticity materials become saturated. The likelihood of such failures taking place is increased by rising groundwater levels in response to greater winter precipitation. The Greensand escarpment in the Weald is identified as a feature where vulnerable slopes may exist.

Similar slopes exist elsewhere. Fig. 12 for example shows the aftermath of a sudden debris flow that occurred in December 2003 in a 30 year old engineered fill constructed as part of a housing development. The slope, comprising a clayey sand had been standing at an angle of about 30° and became saturated by increased groundwater flow and heavy rainfall. The run out temporarily dammed a small stream and was lost downstream when the dam collapsed a few hours later. Deep drains were installed following the failure and the slope buttressed with a gabion wall. While damage to property was limited and there was no injury or loss of life, the failure and subsequent damming of the stream was potentially very dangerous.

Experience shows that vulnerable slopes are typically steeper than 30° and comprise fine sands or silty sands. The permeability of these materials is probably within the range 10^{-4} to 10^{-7} m/s. Each situation must however be considered in its own right and the consequences of the run-out carefully considered. Surface water drainage and erosion is clearly extremely important; designers can no longer rely on historical precedent and observations of groundwater and surface water run off in the assessment of slopes without considering extreme rainfall scenarios and climate change.

4.3 *Landslide regression*

Studies at Fairlight and Lyme Regis have shown sudden accelerations of ground movement and recession from the mid 1970's onwards. This behaviour may be part of a medium-term cyclic behaviour pattern or an adjustment to increased precipitation.

Very aggressive mudsliding has been observed at these and other sites in recent years. The effect is to progressively empty colluvial bowls, leading to the steepening and undermining of rear scarps, and increased rates of recession.

The mechanisms of ground movement are well understood, although there is increased uncertainty over future groundwater conditions and landslide regression rates. As with steep slopes, local authorities and planners have an on-going duty to warn the public and sensibly restrict development. A cautious approach is needed in view of the present uncertainties.

4.4 *Comments on the assessment and stabilisation of slopes*

In times of climate change, designers and developers can take little comfort from the history of slopes; the fact that a given slope may be showing little sign of instability does not preclude the possibility of future instability. There is a need for an awareness of the potential vulnerability of slopes to precipitation and groundwater, particularly where the materials are susceptible to the formation of debris flows, and the consequences of instability.

Slopes are often particularly vulnerable to surface water when vegetation dies or is removed. Higher summer temperatures and other changes may lead to the decline of slope vegetation in some areas; where vegetation is to be established on slopes, the species should be selected on its ability to survive greater climatic extremes.

There is an increasing need for designers to carry out either a simple assessment of the infiltration of groundwater and groundwater flow or to assess a worst credible groundwater condition. Fort et al (2007) describe such an approach in their design for the Lyme Regis coast protection and slope stabilisation scheme. In this case, a factor of improvement of 1.05 was required for the worst credible groundwater condition, allowing for the functioning of only sustainable drainage works. In this case, sustainable refers to drainage that is not vulnerable to slope movement, clogging or blockage.

When carrying slope stability assessments, it is often necessary to make assessments of the operational bulk field strength of the soils present. Such strengths may be influenced by structure, bedding or other discontinuities, weathering and cementing. Care must be taken that the sensitivity of the slope is not over or under estimated due to the selection of the strength parameters. A parametric study using a range of plausible bulk strengths and groundwater conditions is always sensible in such situations.

When stabilising slopes, it is rare that the ground investigation will provide a detailed understanding of all aspects of soil behaviour and groundwater. In the absence of a complete understanding, designers should use as wide a range of different methods as possible. Ideally, reliance on a single stabilising technique, such as piles or drainage, should be avoided.

Wherever possible, designers should observe the principle of avoiding the progressive collapse. In this case, structural elements should be capable of sustaining overloading without brittle failure. Increased load factors should be applied where brittle failure modes cannot be avoided. This principle should also apply to the overall stability of structures. For example, sliding modes of failure of retaining walls are preferable to anchorage failure or toppling modes.

Flexibility during construction works on slopes should be retained wherever possible. Conditions are often revealed during construction that can be favourably utilised, or alternatively render designed works un-buildable.

Designers are now faced with greater uncertainty than ever before. As has been discussed above, there is a need for designers to recognise and allow for greater uncertainty in their work. An awareness of climate change and it consequences is essential for designers to meet this challenge.

REFERENCES

AGS (2000) Landslide Risk Management Concepts and Guideleines. *Australian Geomechanics Society, sub-committee on Landslide Risk Management*, Appendix D, after Findlay et al (1999), 83–84.

Barton, M.E. The mechanics of cliff failure at Shanklin, Isle of Wight. Proc. 4th Intl. Conf. on Landslides, Toronto, **1**, 605–609.

Brunsden, D. and Lee, E.M. (2000) Understanding the behaviour of coastal landslide systems: an inter-disciplinary view. In *Landslides: Research, theory and Practice* (eds Bromhead, Dixon, and Ibsen) Thomas Telford Keynote Papers CDROM (avail. from: e.bromhead@kingston.ac.uk).

Brunsden, D. and Chandler, J.H. (1996) Development of an episodic landform change model based upon the Black Ven mudslide, 1946–1995. In MG Anderson and SM Brooks (eds) *Advances in Hillsope Processes*, Vol. 2, John Wiley.

Corominas, J. (2000) Landslides and climate. In *Landslides: Research, theory and Practice* (eds Bromhead, Dixon, and Ibsen) Thomas Telford Keynote Papers CD ROM.

Fort, D.S., Martin, P.L., Clark, A.R. and Davis, G.M. (2006) Lyme Regis Phase II Coast Protection and Slope Stabilisation Scheme, Dorset, UK- Designing for Climate Change. In

Hall, J.W., Sayers, P.B., Walken, M.J.A and Panzeri, I. (2006) Impacts of climate change on coastal flood risk in England and Wales: 2030–2100. *Phil. Trans. Royal Soc., Math. Physical and Eng. Sci.* 364 (1841), 1027–1049.

Hulme, M., et al (2002) Climate Change Scenarios of the United Kingdom: The UKCIP02 Scietific Report, Centre for Climate Change Research, School of Environmental Sciences, University of East Anglia, Norwick, UK.

Hutchinson, J.N. (2001) Reading the ground: morphology and geology in site appraisal. The Forth Glossop Lecture. *Quarterly Journal of Engineering Geology and Hydrogeology*, **34**, 7–50.

Hutchinson, J.N. (1983) *Engineering in a landscape. Inaugural Lecture, 9 October 1979.* Imperial College of Science and Technology, University of London.

McInnes, R.G. (1996) A review of coastal landslide management on the Isle of Wight, UK. Landslides. ed. Senneset, Publ. Balkema, 301–307.

Palmer, J.S., Clark, A.R., Cliffe, D. and Eade, M. 2002.The management of risk on Chalk Cliffs at Brighton, UK. In: Instability, Planning and Management. pub. Thomas Telford. pp. 355–362.

Skempton , A.W., and Petley, D.J. (1967) The strength along structural discontinuities in stiff clays. Proc ICSMFE, 2, 29–46.

Skempton, A.W. and Weeks, A.G. (1976) The quaternary history of the lower greensand escarpamnt and Weald Clay vale near Sevenoaks, Kent. Phil. Trans. R. Soc., London, A283, 493–526.

Weeks, A.G. (1969) The stability of natural slopes in south-east England as affected by periglacial activity, Quarterly Journal of Engineering Geology, 2, 49–62.

To, R.S.J. et al (2006) Adaptation to five metres of sea level rise. *Journal of Risk Research*, **9**, 467–482.

Vaughan, P.R., Kovacevic, N. and Ridley, A.M. (2002) The influence of climate and climate change on the stability of embankment dam slopes. Proc. 12th Connf. of the British Dam Soc., *Reservoirs in a Changing World* Ed. Paul Tedd, publ. Thoams Telford, 337–352.

Wong, H.N. and Ho, K.K.S (2000) Learning from slope failures in Hong Kong. In *Landslides : Research, theory and Practice* (eds Bromhead, Dixon, and Ibsen) Thomas Telford Keynote Papers CDROM.

Landslides and Climate Change – McInnes, Jakeways, Fairbank & Mathie (eds)

Monitoring systems using chemical sensors to protect railroad operation from landslide disasters

H. Sakai
Railway Technical Research Institute, Japan Railways

T. Yamada
West Japan Railway

ABSTRACT: The climate has changed in Japan in recent years to cause local heavy rainfalls in short lengths, especially during typical rainy seasons before and after summer as well as just in summer. Such heavy and localised downpours attack specific places to trigger ground disasters including landslides resulted from subsequent increases in pore water pressure in the ground. To ensure safe and comfortable travel by rail, an alarm system has been developed to protect railroad services. By monitoring groundwater compositions, the weathering degree of the ground where landslides take place on railroad tracks was suggested to predict the next occurrence of landslide.

1 INTRODUCTION

Landslides will occur in principle when soil blocks on slopes, wings, or cliffs become unbalanced as its weight of soil blocks change to make them stable by collapsing or moving. Heavy rainfalls and earthquake shocks are common factors to bring about landslides. The rainwater will be absorbed into the ground at rainfalls as the pore water. The ground becomes heavier due to the additional load from the pore water resulting in increases in the soil block weight to make the slope lose the balance. There are two sorts of slopes, however, although they are in the same geological condition when subjected to them rainfalls with the same volume in the same period. One of the two types collapses and the other does not. What makes a difference between the two types is how these slopes are weathered, in other words, if these slopes are ready to collapse or if the ground is weathered enough to collapse or not. In case the ground is well weathered, the pore water pressure rises quickly at rainfalls to make soil blocks not sturdy enough to keep standing in and after rainfalls. Thus, for slopes to be ready to collapse, slopes shall be weathered enough to expeditiously absorb the groundwater and become unstable to slide down after rainfalls. To know if a specific slope will collapse or not, therefore, the weathering degree of the slope should be checked. If information thereon is available, it would be possible to estimate the likelihood of landslides to occur. Usually, the ground is drilled to collect soil core samples for inspection. Unfortunately, however, it is difficult to apply this method to continuously observe the state of the ground inside and recognise how it is weathered or the progress of weathering. Primarily, the ground is weathered day by day without any brakes.

Factors that weather soil include the groundwater, winds, sunlight, shocks, and changes in humidity and temperature. The groundwater is effective to quickly weather soil because it moves in the ground 24 hours a day to come into contact with soil particles in the ground. If well weathered, the area of soil particles per unit weight becomes larger to enlarge surface areas of soil particles, where chemical reactions can take place. On the surface, inorganic ions like sodium, calcium, and sulphate are exchanged between the groundwater and soil particles. Thus, the concentrations of these ions suggest how large the areas are on the soil particle surface. If the concentration rises, it would show that the surface area becomes larger due to the progress of weathering and vice versa. It suggests, therefore, that ion concentrations indicate the condition of the slope ground, in which the pores become larger resulting in allowing much more ions to stay exchangeable between the groundwater and soil particles.

In short, ground movement caused by landslides can be expected according to the chemical composition of the groundwater coming from the landslide area. As mentioned above, soil particles in the ground remain in contact with the groundwater all the time to continuously exchange inorganic ions with it. The concentrations of specific ions in the groundwater just depend upon the surface area of soil particles. The larger the surface, the higher the concentrations. Thus, when soil particles are displaced even on a small scale in the deep and virgin ground, the groundwater composition changes.

As expected, laboratory tests also showed changes in the groundwater composition before soil samples collected from a landslide area were sheared or distorted. Furthermore, during over-five-year observation in several actual areas, displacement and distortion taking place on the ground surface and structures thereon were found whenever the specific ion concentrations in the groundwater immediately increased.

The targeted ground conditions are monitored indirectly by checking chemical composition of the groundwater from landslides. By using chemical sensors, the next occurrence of landslide can be estimated in advance. An automatic and remote system has been developed to watch the behaviour of landslide with chemical sensors and notify workers of the hazard. To tell the truth, however, it is not easy to selectively determine specific ions in water samples in the field at regular intervals. To overcome this inconvenience, ion-selective electrodes were employed, which work without any care at least for half a year.

On the basis of this fact, to prevent customers and their property from being damaged by ground disasters, a system was developed to predict disastrous incidents. The information on the prediction was transmitted to workers in charge of train operation to stop trains when necessary. Engineers and dispatchers can decide if they should make arrangements for holding trains prior to the occurrence of a disaster as per the information sent from sensors through telecommunication systems. With the detection system, all the data processed in a data logger are transmitted to PCs and mobile phones held by the workers who need to control train operation through the public mobile phone service. Normally, the groundwater composition is checked every one to three days. But it can be determined any time at their convenience or on demand. In case there are higher risks of seeing disasters in the very near future, the composition is checked any time as they require. The workers in charge can be with the exact concentrations of the specific ions indicating the possibility of landslide. If necessary, they may also take a look at the time profile of the groundwater composition prepared for easier understanding about the picture of changes in the composition. This information provided prior to the occurrence of landslide disasters enables easy and early preparation for in-advance-protection of railroad operation and subsequently fast relief of damaged facilities. The train operation will successfully be managed to keep customers and their property away from unfortunate incidents. This technology can also be applicable to other facilities, including roads and residences beside landslides.

This paper describes how the groundwater is monitored remotely from workers to obtain real time information on changes in the groundwater composition, by which the next occurrence of landslide can be predicted, and also the arrangement for stopping trains in accordance with the information on the groundwater composition. This is because it is important to use a natural phenomenon to create scientific knowledge applicable to engineering technologies. Japan Railways has been perusing the safety of railroad transportation to make customers enjoy railroad travel with much less risk.

2 EXPERIMENTS AND OBSERVATION

2.1 *Chemical observation*

To monitor changes in the chemical composition of the groundwater in a batchwise operation, its samples were periodically collected from landslide areas (Sakai et al., 1996 and 2000). The concentrations of sodium, potassium, magnesium, calcium, chloride, and sulphate were determined by ion chromatography with conductivity detection. To inspect the change in the groundwater composition, it is recommended that the groundwater be collected right from sliding surfaces. Unfortunately, it is not quite easy to successfully obtain such groundwater, in general. If it does not originate exactly from a sliding surface, the composition would be modified with the groundwater from other places. At any rate, the groundwater available inside landslide areas will come passing or originating from adjacent sliding surfaces. Usually, samples can be obtainable from drains already inserted into the ground by carefully deciding which pipe introduces drainage from the spots near a sliding surface. Eventually, a 100 ml of groundwater was sampled in a polyethylene bottle every one to two weeks.

2.2 *Instrumental measurement*

For continuous measurement of chemical composition of the groundwater, an ion-selective electrode was employed. The tip of the electrode was soaked into the flow of the groundwater introduced into a well to store the drainage from drainpipes for a while. To keep the ion sensor work for a long time, care was taken to avoid stream stagnation and eliminate interferences with the ion-selective electrode.

In general, the behaviour of landslide is practically measured with such instruments as wire-line extensometers, clinometers, and regular extensometers. However, when the instrumental measurement of landslide movement is not allowed due to inconvenient conditions of the field to be investigated, it is forced to take a second best measure to watch displacement or distortion taking place on the ground surface and existing structures. In other words, the width of the crack growing on the structure surface can effortlessly be measured. On the occasions mentioned in this paper, proper measurement suitable to the respective sites was employed so that the ground displacement caused by landslides can successfully be observed. That is to say, the displacement in the deep ground was measured with a clinometer; the progress in crack width appearing on ground surfaces was directly with a measure; or the displacement of structures was also instrumentally with a clinometer.

2.3 *Telecommunications for transference of the data chemical observation*

The changes in the voltage originating from ion-selective electrodes were converted into electric signals in microcomputers to save with a data logger. The data recorded with the logger were sent to PCs at track work offices and mobile phones at track workers in charge through commercial telecommunication systems. The data were processed to have the workers easily grasp the current situation of the hot landslide with a visual display. To urge them to decide if trains should be stopped, emergency functions were also furnished to make them aware of the event to be expected on railroad tracks in the immediate future by flashing lights set on the PC or mobile phone as well as making vibrations and noise.

3 EXPERIMENTS AND OBSERVATION

The purpose of this paper is to show the technologies to alarm operators for stopping trains at natural disasters including landslides. The technique can also be applicable to other facilities, namely, public traffic roads, public buildings, and private residences adjacent to landslide areas. Thus, the details of basic experiments are skipped at this opportunity, in order to fully describe the applications of this technology to railroad operation for its safety. However, the soil tests using soil cores sampled from landslides and their results are briefly mentioned to state how this technique was invented.

3.1 *Soil core tests*

Actually, by using soil core samples collected from the same landslide area as that where groundwater composition was observed, tests were carried through to confirm the relationship between groundwater and landslide behaviours for disaster prediction. The soil samples were tested to observe changes in the conductivity of the groundwater passing the soil core sample while the core is stressed. The original soil samples were dug out from landslide slopes and kept. The samples were carefully cut off to prepare columnar specimens to set in a triaxial compression apparatus. The groundwater collected from the same landslide area was poured at the top of the specimen to flow from the top to the bottom. To measure the conductivity of the groundwater reaching the bottom of the specimen, a conductivity detector cell was placed right below it. The detector cell was specially developed to easily detect the conductivity of the groundwater moving even at a low flow rate and even in a tiny volume. The strain when the specimen was stressed was measured by a displacement transducer, which was attached to the surface of the specimen. In the tests, the higher the rate of loading, the larger the gap in the conductivity of the groundwater. Around the point where the conductivity becomes the maximum, the specimen is under a yield condition. Thus, ion concentrations in the groundwater would change in the transit process from the elastic domain to the plastic zone. In this case, the higher the rate of loading, the larger the surface area of soil particles in the specimen, which the groundwater comes to be in contact with. The chemical action on the surface of the soil particles must be improved even at high rates of loading. As already estimated, the lower the rate of loading becomes, the much more slowly the conductivity reaches the maximum. This dependence agrees with the fact that the time when the specimen is at the yield point is very close to the time when the groundwater conductivity is at the maximum. The relationship between axial strain and time and that between the conductivity of groundwater from the specimen and time indicate that changes in the chemical composition of the groundwater are caused by the extension of distances between soil particles at their collapse in specimens. Thus, the actual displacement caused by a landslide would be predicted by monitoring changes in the composition of the groundwater passing the area where the landslide is still active.

3.2 *Relationship between landslide movement and groundwater composition*

Below explained are the features of the landslide sites inspected and the results given in this work, which was carried through by periodically but simultaneously observing the groundwater composition as well as ground displacement. All the areas cited in this paper are located just beside the lines run by Japan Railways.

Each area introduced in this paper is located at different places in Japan. But, they are similar to one another in regard to their geological conditions. This is because landslides are commonly seen in strongly weathered mudstone layers in Japan. The layers at all

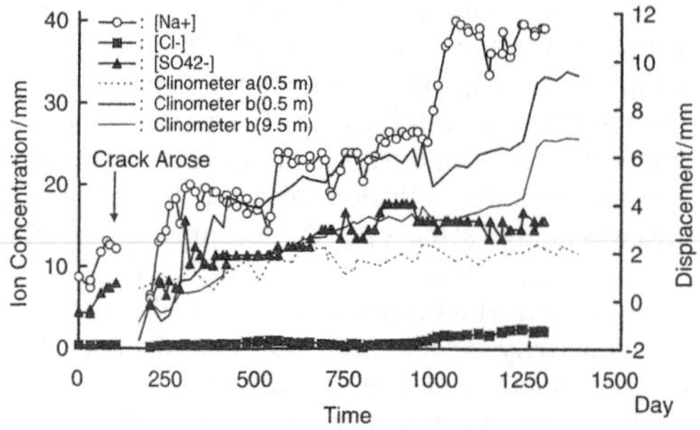

Figure 1. Relationship between changes in groundwater composition inside a landslide area and displacement appearing in clinometer holes caused by the landslide in the first instance.

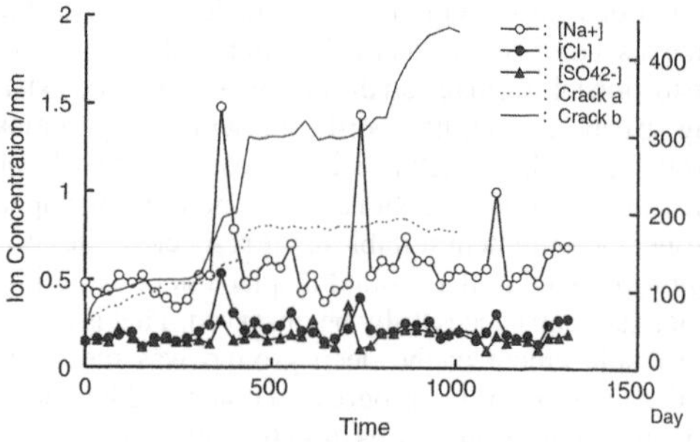

Figure 2. Relationship between changes in groundwater composition and landslide movement appearing as displacement of cracks growing up on the surface ground in the second instance.

the sites mostly consist of mudstone produced in the Tertiary period. Japanese cedars are usually implanted throughout slopes to foster a headwater forest. There are also some colonies of bamboo and bamboo grasses around. Moreover, a high level of water was observed all the time in drain wells constructed to squeeze groundwater from landslides where a large volume of groundwater was continuously supplied. In most cases, railway tracks are constructed right on the mudstone layer by cutting and opening the low-pitched slope developed by landslide or just on the landslide area with the toe cut off. The landslide ground has repeatedly been shifting on a small scale once every five to 20 years. The movement is very slow and intermittent but sturdy in the direction to the railroad track side from the top of mountains. This action occasionally makes the railroad bed swollen just at a point where the landslide crosses the railroad track resulting in track irregularities.

Clinometers were installed to instrumentally watch the behaviour of the landslide, which were set in the ground or on the top of the earth retaining walls built right beside the railroad track to restrain the landslide movement. Cracks found on slopes were also a good indicator to show the progress of landslide. The width was checked by measuring the space between two rectangular wooden plates placed in series, which were prepared by cutting the plate laid over the crack to allow the two portions move freely. One side of each plate was firmly fixed on the ground sandwiching the crack. The landslide behaviour was indirectly expressed by the change in the crack width. It is an easy way at a low cost to follow the behaviour of landslide.

To monitor the changes in the groundwater composition, sodium, potassium, magnesium, calcium, chloride, and sulphate in all groundwater samples collected in the landslide sites were carefully determined by ion-chromatography with conductivity detection. Specific ions sensitively and noticeably point out the behaviour of landslide, but chloride indicates the shifts in groundwater route only. The groundwater was obtained from clinometer holes as well as from the end of drains on earth retaining walls furnished to reinforce cliffs beside railroad tracks and those inserted into the ground to introduce the groundwater out of the landslide.

The relationship between the groundwater composition and landslide behaviour caught up by instrumental monitoring of displacements in the manners mentioned above are illustrated in Figs. 1 to 2. It was commonly found that increases in the sodium concentration appeared prior to the occurrence of ground displacement. The sulphate concentration in the groundwater did not increase remarkably while the sodium concentration therein varied in the same behaviour as that of sulphate. The variation in the sulphate concentration is basically less than that in the sodium concentration. This is because the background concentration of sulphate in the groundwater is fundamentally much lower than that of sodium. In any event, it was recognized that these concentrations increase before ground displacement occurs.

As a result, it was verified that the next occurrence of landslide was semi-quantitatively expected by observing changes in the ion concentration in the groundwater during a unit time length. To easily understand the changes in the sulphate concentration, the rate of change therein was expressed in Fig 3. When the value of the rate is unity, the sulphate concentration in the groundwater remains constant for two weeks. When the value is larger or smaller, the concentration increased or decreased during that period, respectively. Namely, the rate points out the change and its degree in the sulphate concentration. When the sulphate concentration increased over those days, the rate of change in the sulphate concentration exceeded 1.5. Furthermore, extremely remarkable increases in the sulphate concentration were found occasionally during the work period as indicated in Fig. 3. As one of the characteristics turning out in the investigated

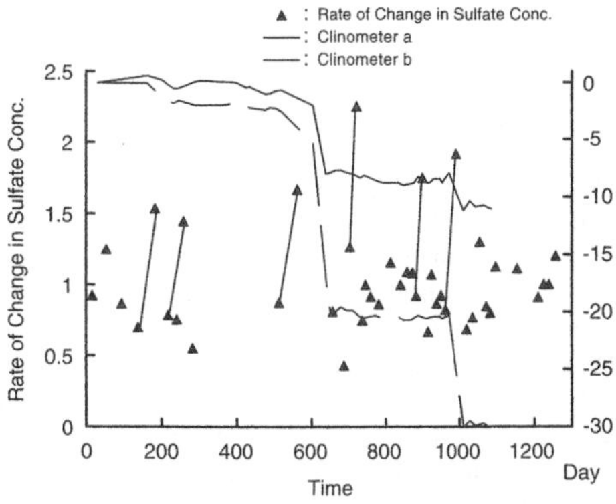

Figure 3. Relationship between the rate of change in sulphate concentration in groundwater and progress in displacement of earth retaining walls that was caused by landslide in the third instance.

landslide, these phenomena were definitely observed always in advance whenever a landslide attacked.

4 APPLICATIONS TO RAILROAD OPERATION

The results achieved in this work agree with the capability to predict the subsequent occurrence of landslide by periodically monitoring the groundwater composition with chemical sensors more easily, which is more efficient earlier than that by measuring the displacement and distortion in a traditional manner with wire-line extensometers, clinometers, and regular extensometers. However, a problem with the prediction by chemical technique is how the groundwater composition is continuously monitored remotely from work offices without any care for a certain length of time. Normally, inorganic ions such as alkaline metals, alkaline earth metals, halides, and oxo-anions are determined in flame spectrochemical analysis or by ion-chromatography in laboratories. However, these techniques do not work outside for a long time without being taken care of by operators. Thus, other methods are desired to observe the groundwater composition in the field.

Ion-selective electrodes are sensitive to specific ions in water samples. The electrode works without any care at least for half a year even in case it is left in flowing groundwater. An electrode of this type was installed, therefore, at a landslide site to find changes in the groundwater composition. Some inorganic ion concentrations in the groundwater as well as its temperature and conductivity were measured by an electrode, thermometers, and conductive detection cells, respectively, when required. Figure 4 schematically illustrates how the data were saved in a data logger to transmit to PCs at track work offices and mobile phones at workers responsible for train operation. The system including the electrode, data processors, and data loggers were powered by solar batteries as shown in Fig. 5. Thus, it is not needed to find a place where the commercial power source is available to the system. The groundwater composition can be measured anytime everyday. The frequency of once a day is enough. When it is required to predict landslide attacking in

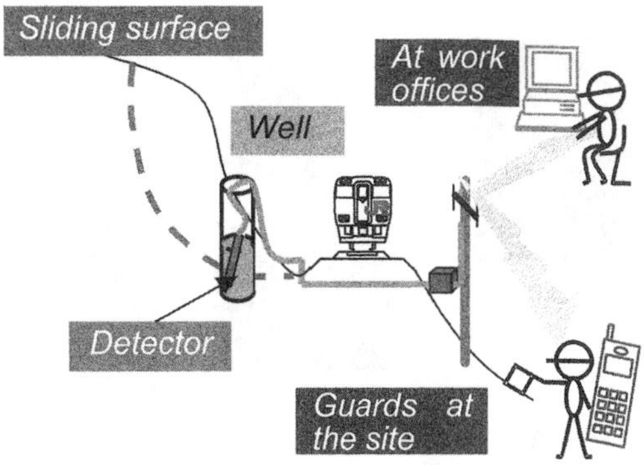

Figure 4. The telecommunication system for successfully sending information on the groundwater composition on demand to workers in charge of and responsible for train operation. The ion-selective electrodes soaked into the groundwater in the well periodically determine the concentrations of ions in the groundwater. The signals from the electrodes are processed to save the data in a data logger, which are transmitted to the relevant workers through public telecommunication services.

Figure 5. Ion-selective electrodes in a well, where the groundwater composition should be monitored. The groundwater was introduced through a pipe to squeeze itself from the landslide site and poured in a pail. The electrodes were left in it. The signal from the electrode was transmitted through a cable to a data logger placed outside the well for processing.

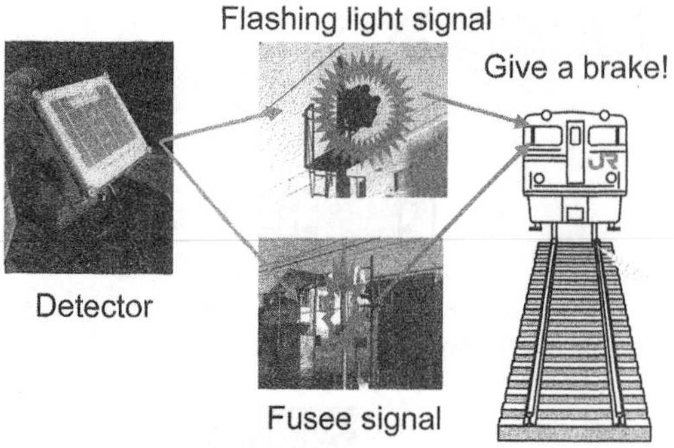

Figure 6. The facilities to immediately stop trains in the case of emergency. Flashing obstruct warning indicators of two types are implanted on railroad tracks. These indicators give a strong light flashing with LEDs, which are still noticeable even during daytime to make crews aware of obstacles on tracks.

the immediate future or check the groundwater composition in case a drastic change is observed, it can be checked every five minutes and send the data to the workers every one to 24 hours per day. The frequency of inspection depends on demand. Workers send commands to the system to set the sequential to monitor the composition and deliver the date to the recipients or the workers. Twenty channels are available to telecommunicate with a recipient at a time, including PCs and mobile phones. The displays equipped on the PC and mobile phones show the changes in the conductivity and turbidity of the groundwater beside its ion concentrations. For maintenance, the inner temperature of the electrode and the current solar battery voltage are also provided in the display.

For warning to stop trains, in case the increase in concentrations of specific ions in the groundwater exceeds three times the fluctuation of their concentrations for a fixed past period, an urgent notice to the recipients are fired by giving noise, vibration, and flashing lights at the PCs and mobile phones. If needed, the information on the excess over the limit of the fluctuation in the ion concentrations can be introduced even to safety systems to stop trains. To let dispatchers know what is going on in landslides beside railroad tracks in charge, the notice can also be delivered thereto. Furthermore, signalling systems may be combined with the groundwater monitoring system to make direct arrangements by flashing obstruct warning indicators implanted on tracks for protection of trains from being included in landslide disasters. See Fig. 6. The ion concentration changes at least two weeks prior to the event including ground distortion and displacement subjected to landslide. Thus, it is not necessarily needed to do so usually, but just in case, these functions are available for higher safety levels.

This technology was developed to successfully protect customers from inconvenience or danger from landslides. What is the most important and indispensable is to know in advance if ground disasters do or do not come in the immediate future. This is because it would mostly be impossible to hold ground disasters like landslides with a massive ground movement. It is required for us, therefore, to make most of information available to work for the enhancement of safety for human activities.

REFERENCES

Sakai, H, Murata, O, and Tarumi, T; (1996), A variety of information obtainable from specific chemical contents of groundwater in landslide area, *Proceedings of the 7th International Symposium on Landslide*, Vol. 2, p. 867–870, June 17–June 21, 1996.

Sakai, H, and Tarumi, H; (2000), Estimation of the next happening of a landslide by observing the change in the groundwater composition, *Proceedings of the 8th International Symposium on Landslide*, Vol. 3, p. 1289–1294, June 26–June 30, 2002.

Landslides and Climate Change – McInnes, Jakeways, Fairbank & Mathie (eds)
© 2007 Taylor & Francis Group, London, ISBN 978-0-415-44318-0

Networking for the future – addressing climate change effects on slope instability

T.A. Dijkstra & N. Dixon
Department of Civil and Building Engineering, Loughborough University, United Kingdom

ABSTRACT: CLIFF*S* (Climate Impact Forecasting For Slopes) was set-up as a network to define the research needs and form targeted approaches to solve problems associated with the potential effects of climate change on slope stability across the UK. Much of the information obtained for stability analysis of slopes is imprecise and there exists a very patchy understanding of the geotechnical environment. Current slope modelling approaches reflect these uncertainties – a situation exacerbated by a poorly understood hydrological basis of slope failure. Significant progress is being made (UKCIP08) in better understanding the uncertainties related to emissions scenarios, natural climate variability and climate modelling and in representing these uncertainties in probabilistic terms. The introduction of probabilistic climate scenarios will provide more information, but will also raise new challenges for impacts analysis and decision-makers. All involved in slope stability issues (i.e. academics, designers, contractors, stakeholders) need to achieve a better understanding of the limitations and opportunities, and make better use of improvements in climate modelling so that progress can be made towards defining best practice. The aim is to limit costs involved in managing unstable slopes in a changing environment. The paper discusses the above issues and identifies key research needs.

1 INTRODUCTION

Slope instability of both natural and constructed slopes has a significant impact on the built environment and infrastructure in the UK. The topography, geology, present and past climatic conditions and human modification of the landscape result in slope processes being active over a significant area of the UK.

Urban environments (both current and future developments) are at risk from the reactivation of movements along shallow relict slip surfaces. Currently some 90.000 homes (representing assets of more than 20 billion pounds) are on sites classified by the British Geological Survey (BGS) landslide hazard potential map as having a 'significant potential' for slope failure (Gibson and Hobbs 2006). Thus, many tens of thousands of people live with continuing slope instability or the threat of instability. This includes many population centres on actively eroding coasts (e.g. Ventnor, Lyme Regis, Holderness) and on inland slopes (e.g. London, Edinburgh, South Wales Coalfield). Planned urban extensions, such as that of the Thames Gateway, will involve construction on slopes already exhibiting marginally stability, potentially giving rise to significant increases in slope failure related construction and maintenance problems.

Thousands of kilometers of transport links and utilities are located in areas susceptible to failure of natural slopes. For example, recent investigations have established that almost 10% (more than 300 km) of the Scottish trunk road network is at potential risk from mass movements (Winter *et al.* 2005). In addition, construction often involves the formation of cut-and-fill slopes that can also become unstable. In England about one-third of the £60 billion highways asset value is potentially at risk from movement of unstable slopes (Glendinning *et al.* 2006)

1.1 *Linking climate change to slope instability*

Although a large array of parameters influence the stability of slopes, for both shallow landslides and flows the hydrogeology (pore pressure regime) is generally a major governing factor determining the deformation behaviour. Pore pressure variations provide a dynamic environment determined by a number of climate dependent variables such as rainfall, evaporation and temperature. In addition, climate variability and land management influence a further set of parameters such as vegetation cover, land use and drainage that, in turn, also affect the stability of slopes.

Directly or indirectly, climate change will influence the stability of our slopes. Despite a vast literature on the subject (see e.g. de Vita *et al.* 1998), further research is still required before a modelling capacity can be established to forecast the degree to which slopes will respond to this change.

1.2 *The need for networking*

The nature of the problem is such that it affects many different stakeholders and end-users (ranging from, for example, the Highways Agency to road users). The problem is also being approached from many different angles and with many different objectives in mind. It thus forms a very broad multi-disciplinary field in which geographers, mathematicians, statisticians, physicists, engineers, ecologists, hydrologists, etc. try to work out their own particular problem angles and seek to forge links to provide a broader solution than would be achieved individually. This is not always easy as specialists speak different (scientific) languages and do not always share the same philosophical approach to problem solving.

With this in mind the network CLIFF*S* (climate impact forecasting for slopes) was funded by the UK Engineering and Physical Sciences Research Council (EPSRC) in 2005 to bring together academics, research and development agencies, stakeholders, consultants and climate specialists. The main aim of bringing these people together is to stimulate an integrated research response to address the intricately linked problem of forecasting, monitoring, design, management and remediation of climate change induced variations in slope instability. The size of the task and the complexity and multi-disciplinary nature of the problem requires active participation of a wide group to assess the magnitude of the resulting impact on UK society and to identify appropriate management and remediation strategies To achieve a better insight into the links between climate change and slope stability in the UK, firstly there is a need to determine the information requirements and, secondly, a need to focus research efforts on targeted assessments of long-term scenarios. Although detailed processes or individual site conditions are being addressed, general process-response issues are still not well understood or researched – a problem exacerbated by poor communication in this multi-disciplinary field.

CLIFF*S* is managed from a base at Loughborough University and is supported by a large core group (see Figure 1). It currently has more than 150 members, mainly from the UK. It operates by organizing multi-disciplinary themed workshops and by providing a web-based information exchange facility. Workshop themes have included issues of risk and uncertainty, and aspects of the responses of natural and constructed slopes to changes in climate. Details of these workshops can be accessed at the network's website on cliffs.lboro.ac.uk.

British Geological Survey
British Geotechnical Association
British Waterways
Cementation Foundations Skanska Ltd
Geotechnical Consulting Group
Halcrow Group Ltd
Highways Agency
Imperial College London
Isle of Wight Council
Kingston University
Loughborough University
Mott MacDonald
Nottingham Trent University
Queen's University Belfast
UK Climate Impacts Programme
University of Bristol
University of Birmingham
University of Newcastle

Figure 1. CLIFFS core group.

2 SLOPE INSTABILITY

The design and management of infrastructural and urban assets affected by instability in natural and constructed slopes is carried out on the basis of specified standards and guidelines that force an assumption of static environmental conditions. Even detailed assessments may only address issues of slope stability on the basis of past groundwater records or observations made during a site investigation (often over a limited period of time). These stability assessments are only valid as long as these conditions are relevant, i.e. a steady state is applicable. However, the long design life of these assets (generally some 60 to 120 years) cover a period during which the climate is forecast to change significantly, therefore making it questionable whether the current steady state analyses are relevant at best, or misleading at worst (Derbyshire *et al.* 2001).

2.1 *Landscape response*

It is required to more closely look at the dynamic (meta-stable) responses of landscapes to changing external driving factors. The sequences and relative intensities of preparatory and triggering events have been found to be important driving factors for the determination of slope instability. However, triggering of slope deformation may generate form responses that significantly alter future threshold transgressions. Whether, in certain cases, a meta-stable dynamic system needs to be applied can be assessed by comparing the modelled responses to climate change using variable levels of sensitivity in this kind of system, with outcomes predicted on the basis of steady state assumptions of trigger and form response. The importance of including these more complex responses has

Figure 2. 1999 DEM of the Mam Tor landslide complex in Derbyshire, UK – a well-known example of infrastructure affected by mass movement (Walstra et al. 2006; Dixon and Brook, 2007).

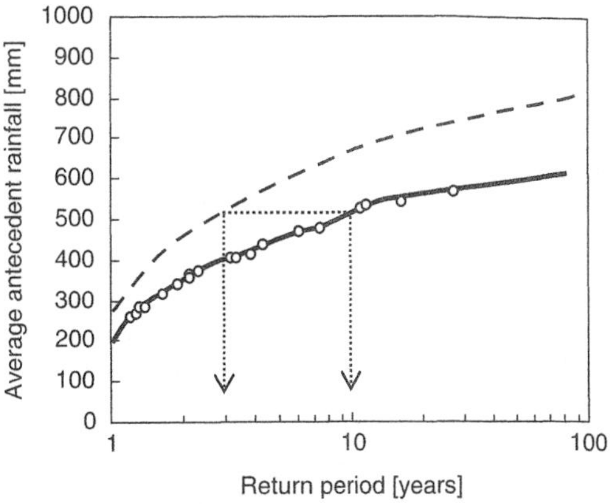

Figure 3. Forecasted change in the incidence of cliff recession due to smaller return periods of critical rainfall thresholds. The thick curve represents data from the period between 1895 and 2005 and the white dots indicate historical landslide occurrences. The broken line represents the changes in return period for average antecedent rainfall for 2080 derived from the UKCIP02 forecasts. A threshold amount of rainfall historically occurring once every 10 years reduces to once every 2 years by 2080, or, in terms of probability of slope deformation this is likely to increase from 0.1 (current) to 0.5 (2080 forecast; after Moore 2006).

already been illustrated by studies such as those of Bromhead *et al.* (1998), Collison *et al.* (2000), Buma (2000) and Dehn *et al.* (2000). Their assessments of slope stability in response to climate change appear to show that it is likely that the frequency of slope failure will reduce. Other analyses, such as Dixon and Brook (2007; Figure 2), take a different set of assumptions, and achieve different forecasts of the incidence of slope instability ranging from an increase in slope movement frequency, when changes in rainfall are taken as a dominant factor, to a decreased frequency when temperature changes are considered to more important. These studies are useful in broadening an understanding of how landscapes (both natural and constructed) may be affected by climate change. However, from the available literature it may be concluded that it is important to include, among others, additional elements such as climate-vegetation feedbacks, shrinkage cracks, material availability, and temperature in order to provide a more realistic forecast. Also, the potential effects on pore pressure variations have not, at the present time, been adequately modelled and assessed. It is likely that, when considering a (more realistic) dynamic approach, future stability models may produce very different outcomes for the long term forecasting of slope behaviour.

Climate change may also lead to accelerated coastal cliff instability and recession rates through increases in frequency and magnitude of toe erosion and cliff recession driven by sea level rise and slope failures due to elevated groundwater positions that are the result of higher effective winter rainfall. It is not easy to establish predictable links because of the highly episodic nature of these events. Work carried out on the Isle of Wight cliffs provides an insight into periodic slope movement in relation to rainfall (Figure 3). If forecasted change in rainfall characteristics for 2080 is superimposed on the current trend it is clear that return periods for movement are likely to reduce significantly. Cliff recession monitoring in the same area also suggests that recession rates are increasing (accelerating) along the entire cliff-line by about 1.3 to 3.3 cm/year indicating that the cliffs may be responding to rising relative sea-levels. These observations may be valid for relatively simple cliff systems. For complex systems significant further investigations are required (Moore 2006).

2.2 *Model choice*

Modelling approaches vary from very simple purely empirical models to physically based models (PBMs) capable of providing analyses of pore pressure responses to changing input parameters. Although empirical models may be attractive in that they generate outputs in a relatively simple fashion, they lack flexibility to address changes in processes as suggested in the narratives associated with the projected consequences of climate change. On the other hand, fully developed PBMs should allow the investigation of the effect of changes in climate parameters through the coupled modelling of hydrology and geotechnical responses of slopes. These models provide a capability to take into account the influence of transient hydrology from observed or generated time series and rainfall distributions. Historical or inferred environmental change can be translated through the hydrological response into changes in slope deformation (e.g. Dixon and Brook 2007). Using climate thresholds for rainfall-induced landslides, the models also provide a framework for the definition and updating of design criteria, the validation of thresholds against regional landslide inventories and the extrapolation of landslide hazard categories to regions with a lack of data.

The usefulness of PBMs remains limited by uncertainties associated with both parameters (upper and lower bounds) and processes. Data scarcity is also a problem, particularly concerning the hydrological trigger conditions and pore pressure changes. Further research is also needed with regards to model error and error propagation (see e.g. van Beek 2006). It is important that over-parameterization should be avoided so that simple models can be created that can be validated against historical data and that are sufficiently flexible to cope with forecasting on the basis of projected trends in key input parameters. As a consequence, it has been shown (e.g. Buma 2000) that it is likely that the intermediate complexities of semi-empirical models will deliver the best compromise.

2.3 *Input parameter problems*

Much of the information needed for slope stability analyses is imprecise, mainly because of problems of visual characterisation, very scarce (and often imprecise) measurements, geotechnical parameters reported as intervals rather than probabilities and the difficulties in obtaining expert judgements (experts can rarely unambiguously specify a unique probability distribution that represents their beliefs). The introduction of probabilistic climate scenarios (UKCIP08) will provide more information about uncertainties, but will also raise new challenges for impacts analysis and decision-makers (Hall 2006).

3 CLIMATE CHANGE FORECASTING

There are major differences in projected outcomes of climate change models and one of the main problems at present is that we do not know what the relative likelihood of each mission scenario is. Still, due to significant lag times in global climate processes it appears that models forecasting annual mean temperature rise result in trends that are very similar for the next 40 years. All emission scenarios appear to forecast change within in a narrow band – only after 2040 does a difference in model outcome appear that is dependent upon emission scenario (Figure 4). However, natural variability of climate on a more detailed scale is very difficult to model. Although significant spatial variation is forecasted to occur in the UK, the general trend is that the summers are likely to be drier and the winters wetter.

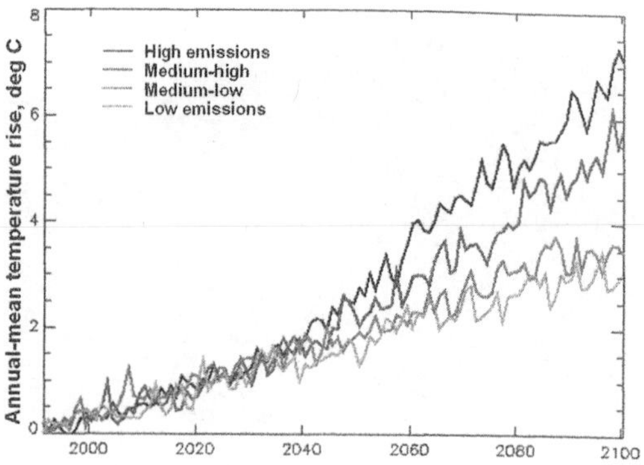

Figure 4. Mean annual temperature rise – significant deviations due to different emissions scenarios are only forecasted to have an effect after 2040 (from Jenkins 2006).

3.1 *Uncertainty and probabilities*

There are three main areas that introduce uncertainty in the forecasting of climate change. It is uncertain as to which emissions scenario will develop over time. Climate itself has a natural variability and there is uncertainty introduced in the way it is modelled (Jenkins and Lowe, 2003).

Running climate change models several times and then comparing the results for individual time slices shows that variations between runs are significant. The outcomes of model runs for winter precipitation (expressed as a deviation from a long term average) reflect the observed variations in winter rainfall, but forecasting annual rainfall for any particular year (beyond about 5-years from now) is not possible.

Natural variability is also reflected by models used for the forecasting of climate change. Running the same model several times gives results that are broadly similar in their patterns, but important regional variations occur, that reflect this type of uncertainty. In addition to this level of uncertainty, consideration should be given to the differences between the many models that try to forecast the same thing. One of the major problems with these analyses is that the relative likelihood of each of these outcomes remains unknown; both for the intra-model and the multi-model comparisons (Jenkins 2006).

3.2 *The near-future of climate model outputs*

What climate modellers are now trying to achieve is a move away from levels of uncertainty with the aim of determining probabilities associated with the model outcomes – a format that is potentially more useful to those working with derived forecast models, such as those associated with slope stability.

Currently, one of the options available for probabilistic modelling is to turn the UKCIP02 information

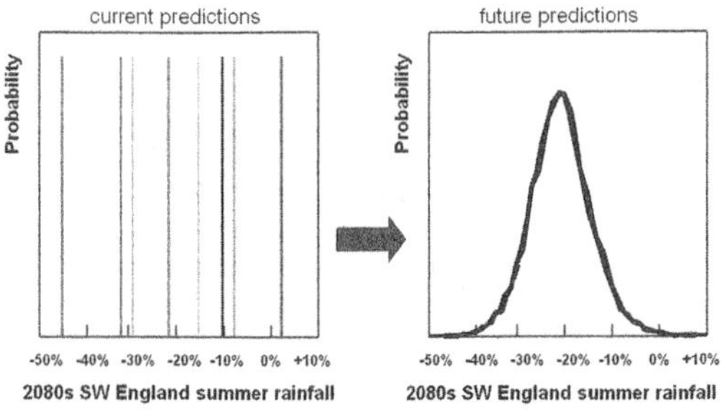

Figure 5. Moving from uncertainty to probabilities (Jenkins 2006).

into a set of probabilities. However, these probabilities are then based on just one model scenario, and therefore only represent a small portion of all available information.

The next step, currently underway, is to take multi-member ensembles of different models and to look at their individual probability distributions. These can then be combined into one compound probability distribution (see Figure 5).

It is clear that deterministic predictions are no longer defensible and that probabilistic predictions are the key to handling uncertainty. To achieve this, UKCIP08 scenarios are being developed where uncertainties regarding emission scenarios, natural variability and model approaches will be turned into relevant probabilities. The analyses will still be based on a 25 km resolution but efforts are also underway to achieve more effective downscaling which would be helpful to slope stability modelling.

Work is also progressing on combining Hadley ensembles with single predictions from other models. On the basis of daily diagnostics, probabilistic changes in extremes can be produced, as well as monthly means. Detailed daily time series can be produced using methods such as those advanced by BETWIXT and EARWIG (e.g. Watts *et al.* 2004).

4 CONCLUSIONS

4.1 *Generating useable outputs*

Good, solid, 'old-fashioned' models have worked well and many standards for design and development in the built environment are based on well-established working practice. It is not unlikely that new parameters (probabilities) and improved models capable of working in an uncertain, but dynamic, environment will lead to changes in costs of design, operation and maintenance. In order to convince managers that they need to manage their existing assets and design new ones using better, updated models the proposals for improvement need to be underpinned by the best science available.

4.2 *Key research needs*

Most of the discussion during the workshops organised by the CLIFF*S* network focuses on the attempts to quantify the effects of climate change on pore pressure responses and resultant rates of movement. Despite some good research in this direction there are still some fundamental questions to be raised. For example, how will climate change (in particular effective rainfall) affect slope stability? How well do we understand pore pressures in complex geological environments – what happens when slides move? How do pore pressure variations lead to changes in landslide movement? To what extent do changes in basic assumptions lead to changes in outcomes – e.g. is summer drying and resultant formation of shrinkage cracks enabling more efficient wetting up of potential slip surfaces during winter storms?

There is an urgent need for a fundamental understanding of these relationships prior to making predictions of future recession/slope stability. We need more field investigation and monitoring. For example, it is necessary to test pore pressures/water levels between models and real conditions. To test this we would need fundamental historical datasets that would include relationships between sea level rise and cliff recession, rainfall and pore pressures/past slope failures.

4.3 *Recent developments*

On many different fronts the detailed understanding of slope responses to climate change is being tackled. One such example is the EPSRC-funded BIONICS project, providing a facility for engineering and biological research for the purpose of determining the effects of climate change on the slopes of infrastructure embankments. This involves the construction and monitoring of an embankment representative of UK infrastructure subjected to different climates and a range of vegetation cover (see www.ncl.ac.uk/bionics). Over the coming years it will provide important insights into the response of embankments to changes in climate.

4.4 *Networking*

In terms of risk assessment approaches it is clear that it is essential to include climate change as another input in this methodology. The new probabilistic outputs generated by UKCIP08 would be very helpful in this context. It is important to realise that probabilities (whether generated by climate or slope modellers) are not static – they need constant updating. Systems are dynamic, and the question is how do we move along with it?

This paper provides only a small selection of the challenges faced by the diverse communities involved in managing slopes in the UK. For a comprehensive assessment of climate change dependent slope instability it is essential to operate in a multi-disciplinary field that involves specialists such as risk managers, statisticians, social scientists, geographers/geomorphologists, civil engineers, climate modellers, biologists, soil scientists, etc. It is clear that in this environment cooperation, information exchange and communication are of paramount importance to strike the right balance addressing research questions and solving problems. **CLIFFS** aims to offer an inclusive framework to progress understanding of how climate change will affect slopes in the UK. If you are interested in receiving the support of the network, please send an email to cliffs@lboro.ac.uk.

ACKNOWLEDGEMENTS

The authors would like to thank all members of the network for their inputs. In particular, we are grateful to our colleagues who have contributed to the CLIFFS workshops.

REFERENCES

Bromhead E.N., Hopper, A.C. & Ibsen M.-L. 1998. Landslides in the Lower Greensand escarpment in south Kent. *Bulletin of Engineering Geology and the Environment*, 57(2), 131–144.

Buma J. 2000. Finding the most suitable slope stability model for the assessment of the impact of climate change on a landslide in SE France. *Earth Surface Processes and Landforms*, 25, 565–583.

Collison A., Wade S., Griffiths J. & Dehn M. 2000. Modelling the impact of predicted climate change on landslide frequency and magnitude in SE England. *Engineering Geology*, 55, 205–218.

Dehn M., Burger G., Buma J. & Gasparetto P. 2000. Impact of climate change on slope stability using expanded downscaling. *Engineering Geology*, 55, 193–204.

Derbyshire E., Dijkstra T.A. & van Beek L.P.H. 2001. *Forecasting medium to long-term slope stability in a context of progressive environmental change*. In Sassa, K (Ed.) Procs. UNESCO/IGCP Tokyo Symposium on landslide risk mitigation and protection of cultural and natural heritage.

De Vita P., Reichenbach P., Bathurst J. C., Borga M., Crosta G., Crozier M., Glade T., Guzzetti F., Hansen A., & Wasowski J. 1998. Rainfall-triggered landslides: a reference list. *Environmental Geology*, 35(2–3), 219–233.

Dixon N. & Brook E. 2007 (in press). Landslide Response to Predicted Climate Change: Case Study of Mam Tor, UK. Landslides, Volume 4.

Gibson A. & Hobbs P. 2006. *BGS Landslide Data*. Presentation to the second **CLIFFS** workshop, Loughborough University, July 2006, available at cliffs.lboro.ac.uk.

Glendinning S., Mohammed R., Hughes H. & Davies, O. 2006. Biological and engineering impacts of climate on slopes (BIONICS): The first 18 months. In: *Engineering geology for tomorrow's cities*. Procs. 10th IAEG International Congress, United Kingdom, paper 348, theme 2.

Hall J. 2006. *Uncertainties in slope stability and impacts of climate change*. Presentation to the first **CLIFFS** workshop, Loughborough University, February 2006, available at cliffs.lboro.ac.uk.

Jenkins 2006. *New scenarios of UK Climate Change (UKCIPnext)*. Presentation to the first **CLIFFS** workshop, Loughborough University, February 2006, available at cliffs.lboro.ac.uk.

Jenkins G. & Lowe J. 2003. Handling uncertainties in the UKCIP02 scenarios of climate change. *Hadley Centre technical note 44*.

Moore R. 2006. *Climate Change impacts on Cliff Behaviour?* Presentation to the first **CLIFFS** workshop, Loughborough University, February 2006, available at cliffs.lboro.ac.uk.

van Beek L.P.H., Malet J.-P. & van Asch Th.W.J. 2006. *Modelling landslide hydrology*. Presentation to the first **CLIFFS** workshop, Loughborough University, February 2006, available at cliffs.lboro.ac.uk.

Walstra J., Chandler J.H., Dixon N. & Dijkstra T.A. 2006. Time for change – quantifying landslide evolution using historical aerial photographs and modern photogrammetric methods. In: *Geo-imagery bridging continents*. Procs. IAPRS, XXXV(B4), Commission IV, WG IV/6.

Watts M., Goodess C.M. & Jones P.D. 2004. *The CRU daily weather generator*. BETWIXT Technical Briefing Note 1, Version 2, February 2004.

Winter M.G., Macgregor F. & Shackman L. (Eds) 2005. *Scottish road network landslides study*. The Scottish Executive.

Landslides and Climate Change – McInnes, Jakeways, Fairbank & Mathie (eds)
© 2007 Taylor & Francis Group, London, ISBN 978-0-415-44318-0

Laser survey and mechanical modelling of chalky sea cliff collapse in Normandy, France

T.J.B. Dewez, J. Rohmer & L. Closset
Natural Risks and Land Management, BRGM – French Geological Survey, Orléans, France

ABSTRACT: Repeated terrestrial laser surveys (TLS) were conducted on the chalk cliffs of Mesnil Val, Normandy, to quantify erosion and calibrate models of rock collapse hazard. TLS is capable of measuring cliff topography with point density better than one point per 5 cm with precision of ca. 3 cm. In the period December 2005 to March 2006, 418 m^3 of chalk disappeared from the cliff. Comparison of TLS surveys provided a frequency-magnitude event distribution spanning six orders of volume magnitude (10^{-4} to 10^2 m^3) for 2202 individual patches of departed chalk. Erosion events ranged from chalk flakes only 5 to 15-cm-thick to 150-m^3-single-piece block. A simple geomechanical model was applied to predict large block collapse induced by gravity. Although, its physics need be refined, its predictive power makes it a reasonable first order hazard estimator.

1 INTRODUCTION

It is widely acknowledged that chalk cliffs along the English Channel are actively eroding (e.g. IFEN, 2006). So far, long-term cliff-retreat rates have been quantified by mapping the receding position of cliff tops from historical aerial-photographs or co-registered historical topographic maps (e.g. Costa 1998). Studies like Costa's give an integrated cliff retreat rate over a period of time of several decades to a century. Rosser et al. 2005 highlighted however, that average retreat rates do not relate unambiguously to phenomena visible on the ground on a day to day basis. The purpose of this paper is twofold: first, present preliminary results of repeated terrestrial laser scanner surveys of a chalk cliff, and second propose a first order geomechanical predictor model to highlight block collapse hazard. These two objectives fit into a broader project at BRGM that aims to provide grounds for mapping collapse susceptibility hazard along steep rock faces.

The stretch of coast along which the study takes place is located at Mesnil Val, Haute-Normandie, northern France. The Mesnil Val survey site is about 750 m-long and rises from 25-m to 75-m-high. The cliff face is aligned to the NE–SW and faces the English Channel to the NW. The cliff is made of relatively homogeneous Upper Cretaceous chalk belonging to the Half Lower Lewes Nodular Chalk formation at the base (Turonian, ca. 91 Ma ago), overlain by the Upper Lewes Nodular Chalk up and by Seaford Chalk formation (Coniacian, ca. 88 Ma ago) around the 35–40 m elevation (see inset in Figure 1). Stratification is sub-horizontal, with about 5 m of palaeo-topography at the interface between Turonian and Coniacian.

The cliff face contains two strong horizontal mechanical barriers. The first one is the hard-ground delimiting the transition between Turonian and Coniacian. It is underlined by a nearly continuous level of flint stones and lies at an elevation of ca. 10 m (Figure 1). The second hard-ground horizon is more conspicuous and is located at the transition between the Upper Lewes Nodular Chalk and the Seaford Chalk at an elevation of ca. 35 m (Protect Report, 2003). The cliff face is also traversed by vertical mechanical discontinuities, i.e. fractures striking parallel to the cliff face (NE–SW). These were inferred to come from relief stress (Protect Report, 2003). In places another set of steep oblique fractures cut the cliff perpendicularly and delimit large dihedra. These are exploited by collapse events that affect the entire cliff height (e.g. 23 June 2002, Protect Report, 2003) but were not seen yet during this study.

Average cliff retreat rates were quantified by Costa (1998) and amount to 0.24 m/year during the period 1947–1995. This article presents differences of chalk cliff topography, termed "erosion", as measured with a terrestrial laser scanner during the winter of 2005–2006.

2 METHODOLOGY

The cliff face was surveyed in December 2005 and in March 2006 with a terrestrial laser scanner (TLS). A time-of-flight TLS is a laser rangefinder equipped

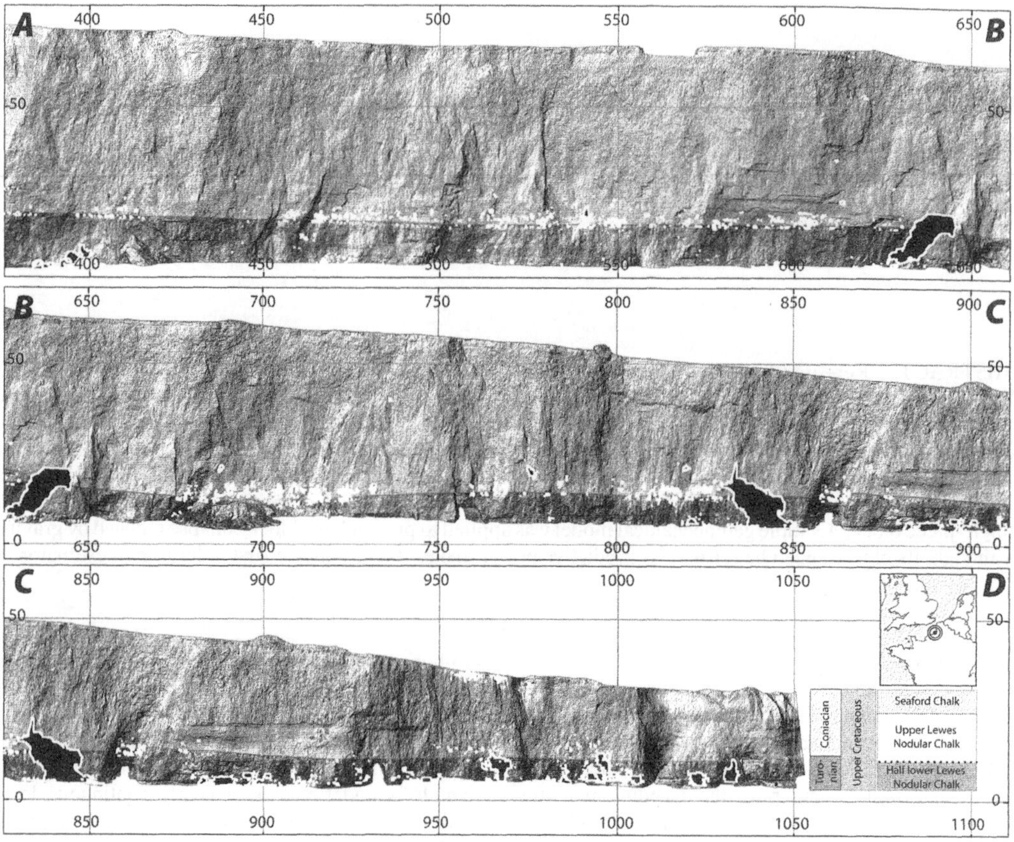

Figure 1. Shaded relief map of the Mesnil Val Upper Cretaceous chalk cliff, Haute Normandie, Northern France as measured with a terrestrial laser scanner. In black surrounded by white: material eroded between December 2005 and March 2006. Grey shading delimits Turonian (at the base) from Coniacian above. Note that the panels follow from top to bottom (A to D), that is from NE to SW. Local coordinates are in meters.

with a rotating system for scanning a surface with high-accuracy and high density. The Riegl LMS Z420i hired from the survey company ATM3D is capable of recording 360° topographic panoramas along 80°-wide strips, within a range of 1 km. It has a beam divergence of 0.25 mrad (i.e. a spot of 25 mm at 100 m or 2.5 cm at 1 km) and minimum angular increment as small as 0.004° in both vertical and horizontal planes. In this case study, overlapping panoramas were measured with scanning angular intervals of 0.05° at each station so that each point cloud counts about 3 million usable points.

2.1 *Terrestrial laser scanner survey*

Survey campaigns comprise 8 overlapping laser scanner stations installed at regular intervals on the intertidal chalk platform at low-tide. They were setup 100–200 m seaward of the cliff face. The 8 point clouds measured during a campaign were merged together in order to fill in hidden faces and produce a continuous 3D representation of the cliff face. The same stations were occupied at both campaigns to optimize survey comparisons.

Geographic referencing is a key issue in high-precision topographic comparisons. The comparison can only be as precise as the least precise referencing technique employed. Here, three devices were used to reference point clouds: D-GPS to position the permanent reference frame inside the National French Grid, EDM-total station to link temporary reflective markers to the permanent reference frame, and TLS to locate the point clouds respective to the temporary reflective markers. D-GPS precision will only be relevant if the survey pins materialising the permanent reference frame were lost. Otherwise, three principal factors limit comparison precisions: 1) intrinsic TLS point precision; 2) precision of TLS positioning inside the campaign reference frame (TLS dependant); 3) precision of control point positioning inside the permanent reference frame (EDM-dependant).

First, laser point precision depends on several factors: the precision of a laser point is a function of the distance of the object reflecting the laser beam, its reflectivity, its roughness, the incidence of the laser beam. On average, according to the laser's manufacturer specifications, we estimate TLS point precision to be about 0.017 m (2 sigma threshold of signal distribution, see Figure 3). Second, precision of TLS locations is a function of reflective target location precision inside the survey reference frame and target distribution around the TLS. The TLS processing software RieglScan solves TLS locations from the position of reflective targets by intersection. Relative positioning of scanner stations is generally achieved within 1–3 cm. Location of the scanner inside the permanent reference frame depends also on the precision with which the reflective control points have been measured with the EDM-total station. Conventional surveying precision of reflective targets is usually better than 3 cm but at times hampered by gusty winds which made the tripod unsteady and precluded accurate sighting of the targets. This problem turned out to be significant in the comparison between December 2005 and March 2006 where systematic offset of the cliff DSM grids were systematically offset. This bias was corrected by fitting a trend surface to the DSM difference surface and removing the systematic offset. Overall, we estimate the campaign precision to be of the order of 3 cm.

2.2 *Processing of laser point clouds*

Both survey campaigns produced continuous point clouds of about 11 million xyz data points over the length of 750 m of cliff face. From these points, solid models of the cliff were built for both epoch and subtracted to estimate the amount of material that departed the cliff face. The solid models were approximated to 2.5D grid with a TIN (triangular irregular network) linear interpolator. Grid cell dimension of 0.05 m is a compromise between point cloud density, manageability of grid sizes and estimated precision of the survey. According to point cloud neighbourhood statistics, three quarters of the points are closer together than 0.045 m, yet the variability in places is larger than 0.07 m from point to point (Figure 2).

3 RESULTS

3.1 *Comparison of cliff face topography*

The comparison was performed by subtracting the Digital Surface Model grid of March 2006 from that of December 2005. Negative differences therefore indicate erosion while positive differences reveal

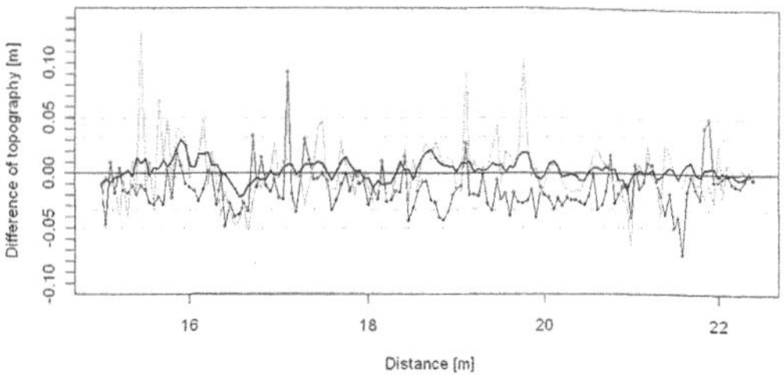

Figure 2. Difference of DSM grids of December 2005 and March 2006 as measured with a terrestrial laser scanner. Thin black dotted line: raw differences [March – December], note that this signal has a bias of about −0.015 m. Thick grey line: de-trended difference where a third order trend surface was subtracted to remove the systematic bias seen on the green curve. Thick black line: filtered differences with a 3×3 moving average filter to remove data spikes. Note that the average difference is now equal to 0 m and that oscillations are generally contained within ±0.017 m (at 2 sigma). This transect was chosen in a location where no erosion could be recognized.

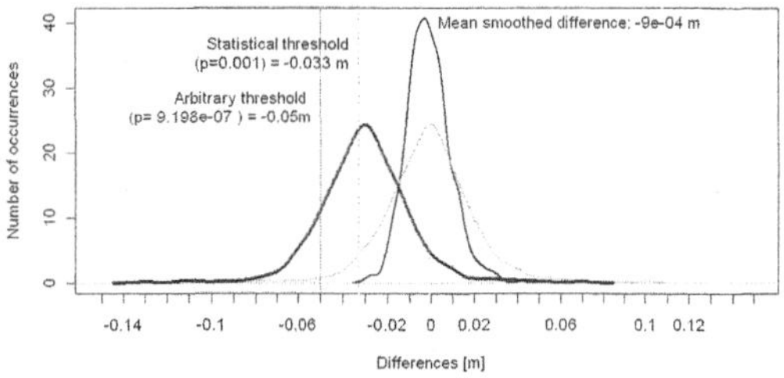

Figure 3. Distribution of raw and corrected differences as plotted on Figure 2 profiles. Dotted black line: Raw differences. Grey line: Detrended differences. Thick black: filtered differences with a 3 × 3 mean kernel.

accretion. Significant accretion only occurred at the foot of the cliff while significant erosion occurred throughout the cliff face as expected (Figure 1).

The best way to appreciate topographic differences is probably through representative cross-profiles (e.g. Figure 2, Figure 4 and Figure 5). In cross-section, topographic changes exhibit three types of signals: a long wave length signal (ca. 100–300 m, in green on Figure 2), a short wavelength signal (Figure 2, Figure 4 and Figure 5, ca. 0.15 to 10 m) and a very short wavelength signal (ca. 0.05–0.10 m).

The long wavelength signal has maximum amplitude of 0.06 m. This betrays the imprecise registration of TLS stations during both surveys as evoked above. This poor registration appears as a systematic bias between campaigns (it corresponds to the mean signal ca. 0.015 m on the profile plotted on Figure 2). This systematic bias was corrected by sampling the difference grid in 3000 random positions distributed where collapse events were not observed, fitting a third

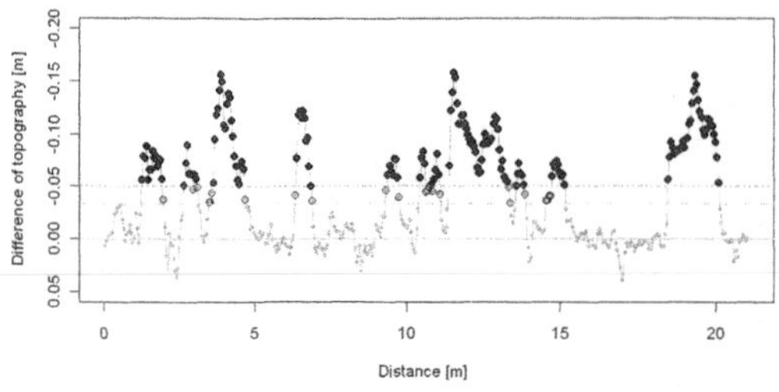

Figure 4. Cross-profile through a cliff section where chalk flakes detached from the cliff. The data corresponds to the filtered and de-trended difference grid seen on (thick black line). This type of erosion is probably due to frost and thaw cycles in a saturated chalk layer that shattered the outer surface of the chalk. Black dots: data points where topographic differences between March 2006 and December 2005 are larger than −0.05 m.

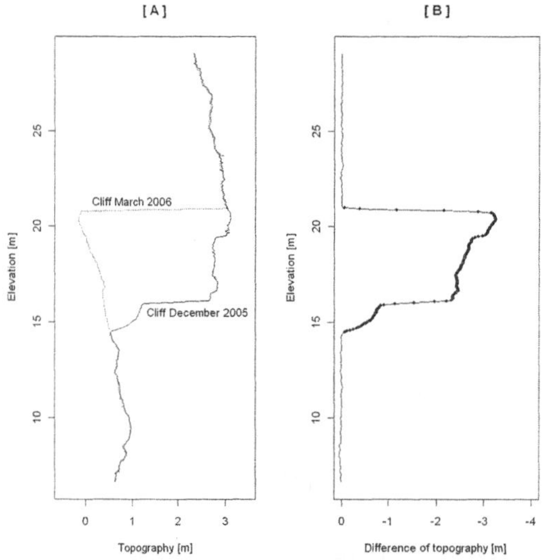

Figure 5. Cross-profile through a large collapse event involving chalk blocks up to 3.2 m-deep. A: profile before and after block collapse. B: profile of topographic difference. Note that noise is hardly visible around the 0 m difference abscissa mark whereas the block collapse signal (highlighted with black dots) is much larger. This is a section through the largest event recorded in March 2006.

order trend surface through these samples, and subtracting this trend surface from the DSM difference grid (see Figure 2 and Figure 3, green and blue curves, to appreciate the effect of the de-trending).

The very short wave length signal has amplitude of the order of +14 cm and −8 cm (Figure 2 and Figure 4). It comes from both slightly imprecise relative positioning of TLS stations during a single survey campaign, from laser point imprecision and from data decimation during the interpolation. This noise was nearly completely removed using a 3×3 moving-average filter. After smoothing, the distribution of the resulting signal became normally distributed about zero (Figure 3). This Gaussian distribution property offers the possibility either to set a statistical erosion detection thresholds based on the distribution of the signal, or pick an intuitive but arbitrary threshold. In this case, a statistical threshold of ±3.3 cm enables us to qualify erosion with a probability of misclassification of one in a thousand (p-value of 0.033 m is 0.001); an arbitrary threshold of 0.05 m has a probability of misclassification of one in nearly two million. Both these thresholds demonstrate the remarkable performance of TLS for monitoring rockfall events.

Finally, the short wave length signal, which is that of interest to us, highlights the places where material fell off the cliff (Figure 2, Figure 4 and Figure 5). Its amplitude, in this study, goes from 0.033 m to about 4.5 m. Note that the upper limit of this signal is only limited by the size of the largest collapse events identified in the three months period. It could have been bigger should another collapse have occurred.

Given the levels of precision, TLS is capable of confidently detecting particle erosion of about 3.3 cm-thick and 225 cm^2.

3.2 *Map of collapse events*

The topographic difference map, after de-trending and smoothing, provides a means to delineate areas where erosion was larger than 0.05 m (threshold retained initially for the analysis even though 0.033 m turned out to be sufficiently sensitive). During the winter 2005–2006, 2202 eroded patches have been flagged. This corresponds to an eroded surface of 562 m^2 for a total of 34322 m^2 of surveyed cliff face (i.e. 1.63% of total surface area). The total volume of eroded material amounts to 418 m^3 in 3 month. If this eroded volume were distributed on the entire cliff face, this would extrapolate to an average annual retreat rate of approximately 0.05 m/yr, which is five times smaller than the 0.24 m/yr average retreat rate computed by Costa (1998). This suggests that expected eroded chalk volumes ought to reach about 2060 m^3 per 3 months period or 8240 m^3 per annum for TLS sampling to be representative of longer term erosion rates. Thus, the sampling period of December 2005 to March 2006 has had 5 times less collapse events than is anticipated from the long term erosion rate.

The distribution of individual events (Figure 6 A) tells that small collapse events are much more frequent than large ones. In fact, three quarters of collapse events are smaller than $2.7e^{-3}$ m^3 (i.e. rock lumps of about 2.7 litres or ca. 5 kg); 19 events involve volumes larger than 1 m^3 (0.8% of events) and only two events are larger than 100 m^3 (i.e. 0.09% of events). Given the observed frequency/magnitude distribution, one may

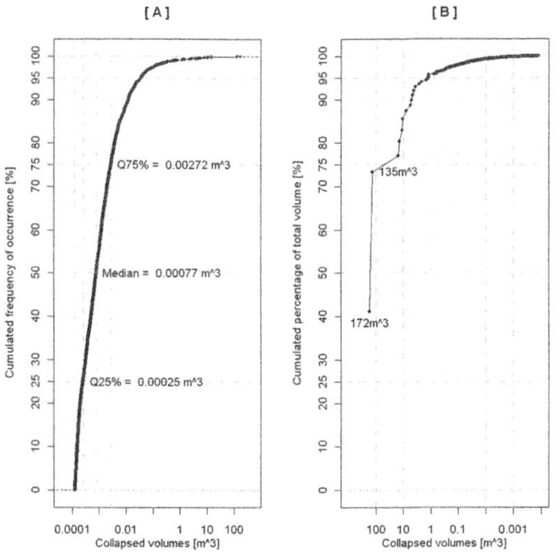

Figure 6. Frequency/magnitude distribution functions for the collapsed events that occurred between December 2005 and March 2006. A: Frequency of occurrence of an event for a given volume. B: Contribution of each event to the total volume of eroded chalk. Note the reversed abscissa scales.

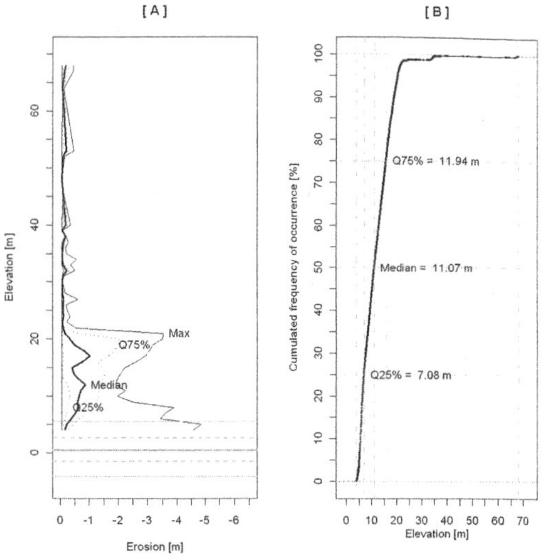

Figure 7. Distribution of eroded chalk thickness versus elevation along the cliff face. [A] Plot of erosion with elevation. [B] Plot of relative frequency of eroded pixels with respect to elevation. All cliff pixels with differences larger than −0.05 m have been considered. This graph (A) shows that thick collapse events (larger than 1.2 m, red line) concerned only the first 21 m, while 97% of recorded events (B) are contained below 21 m of elevation. Tidal distribution (horizontal grey lines on Graph A) reports elevation of the sea such as modelled by SHOM for Le Tréport harbour (3.5 km N of MesnilVal) for the period between both scanner surveys (06 December 2005 to 24 March 2006), from top to bottom: maximum, Q75%, median, Q25% and minimum of modeled tidal water height.

infer that there is a probability of less than one in a thousand that events larger than 100 m^3 occur in a three months period (Figure 6 A). The relative importance of these large collapse events however, account for about 75% of the total volume of chalk lost from the cliff face (Figure 6B). Further, ninety-five percent of the total volume concerns events larger than 1 m^3(Figure 6B).

It is worth noticing the position of thin erosion events (e.g. Figure 4). Many of them occur in the layer at the transition between Turonian and Coniacian. This layer has a very high water contents and these thin events probably relate to frost-induced flaking of the chalk.

Elevation of collapse events along the cliff face is also informative. Ninety-seven percent of eroded grid pixels stand below 21 m of elevation. Figure 7 represents the distribution of eroded material versus elevation. The different lines correspond to the distribution of the eroded pixels considered by 1 m elevation slices. At an elevation of 20 m, three quarters (75%) of the points saw erosion smaller or equal to 2 m, while no erosion was larger than 1.2 m above 21 m.

The influence of the sea is not straightforward. Between the TLS campaigns, three quarters of the tides reached only elevation of 2.62 m, which is below where the chalk is exposed. This water level distribution however is based on modelled tides at Le Tréport harbour (SHOM, 2006), 3.5 km to the North of Mesnil Val, and ignores currents, winds or storms effects. Further investigations will attempt to link environmental parameters to rock falls. To this end, it is worth noting that laser scanner point clouds also provide a relatively dense array points on the intertidal platform, which will be invaluable for modelling wave energy dispersal due to platform roughness.

At this stage, it is too early to infer the effect of climate change. Future work will attempt to separate the impact of the sea from that of meteorological influence by extending the comparison to sections of cliffs where man-made protection prevent waves to reach the foot of the cliff.

4 GEOMECHANICAL MODELLING

Comparison results prompt this question: given the collapse events observed for the winter period 2005–2006, is one capable of predicting future chalk collapse. Below we propose a preliminary geomechanical model that focuses on predicting large block collapse due to gravity.

The distribution of eroded chalk thickness is bimodal (Figure 8). The most numerous eroded pixels concern small events, but a secondary mode appears above 1.2 m of erosion (Figure 8). Therefore, our

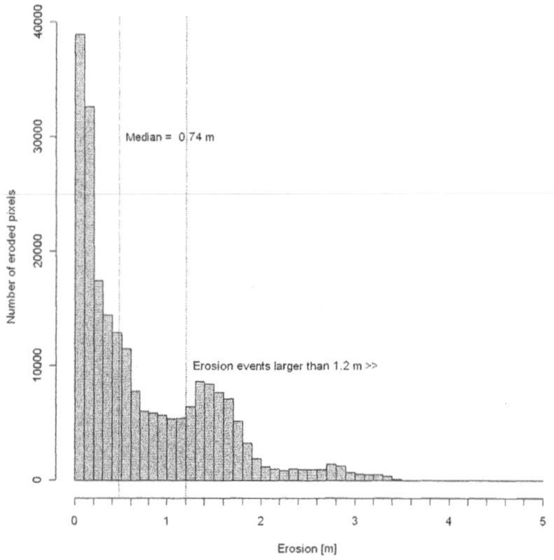

Figure 8. Histogram of eroded chalk thickness on Mesnil Val cliff between December 2005 and March 2006. Half of the eroded points concerned thicknesses of less than 0.25 m.

preliminary geomechanical model is trying to explain block failure driven by gravity for events where erosion reached at least 1.2 m. TLS survey comparisons indicated that these events account for 63% of total volume of chalk losses.

4.1 *Assessing rock block stability*

Topographic comparison data have shown that failure started from the maximal overhang (Figure 9) and that the failure surface along the upper section of the collapsing block is planar. These indicate that failure probably occurred when chalk weight exceeded its matrix tensile strength.

On this basis, the geomechanical model is constructed around four hypotheses: (a) the forces involved in the failure process are planar; (b) vertical slip failure occurs on the vertical sides of the corresponding block; (c) chalk is homogeneous and abides by Mohr-Coulomb criteria; and (d) gravity is the main destabilizing factor.

The model accounts for three forces exerted on a rock block (Table 1, and Figure 9):

In Table 1, a rock column has a width W (here simplified to 1 pixel width), a height H (number of pixels above the most incised point in the profile called maximal overhang) and a depth D (difference between a given pixel and the maximum overhang). These three dimensions are converted to meters through the size of grid pixels in meters. The force expressions account for the following chalk properties: c (chalk matrix cohesion), R_T (chalk matrix tensile strength), ρ (chalk density). g is the gravity acceleration.

Table 1. Forces exerted on a block of chalk.

Slip resistance force exerted along the vertical side of the rock column	$\tau =$ c.W.H
Tensile resistance force exerted along the upper side of the rock column	$F_T = R_T$.W.D
Gravity force exerted on the entire volume rock column	$F_G = \rho$.g.W.D.H

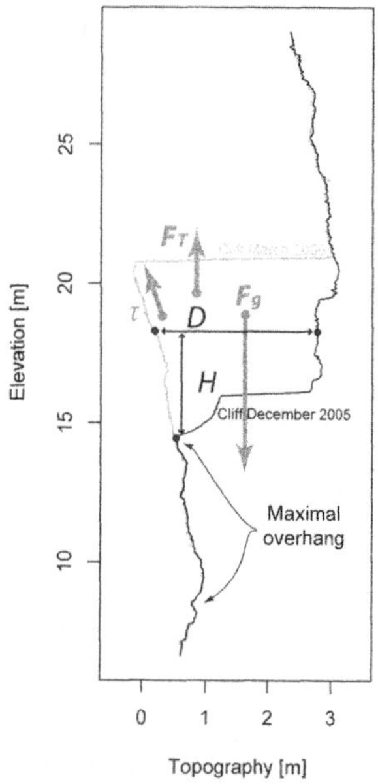

Figure 9. Geomechanical model for block failure assessment. D: Depth of block; H: Height of block; Ft: tensile force; τ: slip resistance; Fg: gravity force (see Table 1).

The stability of any chalk block is computed by evaluating the ratio S (called stability factor) between the gravity force exerted on a chalk column and the slip and tensile forces resisting along the sides of the same column.

$$S = \frac{F_G}{\tau + F_T} = \frac{g\rho H}{R_T a + cH} \quad (1)$$

4.2 *Mapping the stability factor at the scale of the cliff*

The stability analysis was carried out on a simplified 1 × 1 m cliff DSM grid of December 2005. Resampling of the TLS 0.05 × 0.05 m grid was performed by nearest neighbour decimation to preserve originally observed values. Computation of the stability factor was limited to below 21 m elevation, because this is

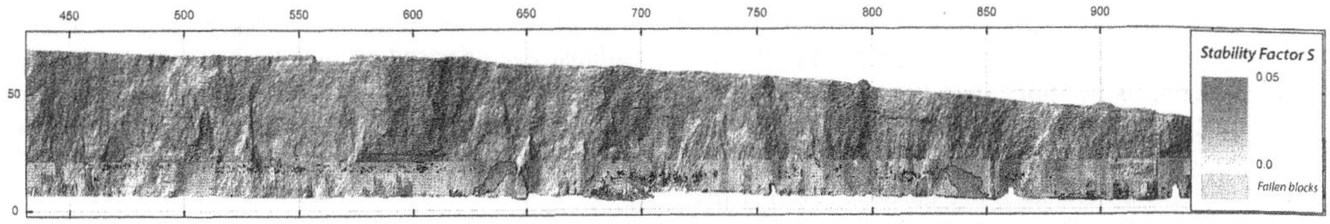

Figure 10. Map of the stability factor computed below 21 m of elevation. Colour gaps correspond to areas where there is no overhang. Areas where blocks were effectively eroded from the cliff face are in semi-transparent grey.

the domain where large collapse events were contained (Figure 7).

The rock properties chosen for the computation come from a previous mechanical study performed at Mesnil Val (Schoumacker, 2002): $c = 1$ Mpa ;$R_T = 3.5$ bars and $\rho = 1900\,\mathrm{kg/m^3}$.

Force balance was systematically computed along 1-meter-wide (1 pixel) vertical profiles. Rupture is presumed to occur when the cumulated weight of overhanging pixels exceed the combined resistance to slip of that rock slice and the tensile strength along the upper surface of the considered pixel. In other words, collapse should occur for S = 1.

4.3 *Stability factor results*

Surprisingly, the stability factor never reached a value of 1 (Figure 10). Rather, the maximum S value staid below 0.044 (gravity only reaches 4.4% of the assumed chalk resistance). Yet chalk collapsed from the December 2005 cliff face, as quantified precisely with the TLS comparison survey. Thus if the assumed rock resistance was overpowered, gravity alone cannot explain it; or alternatively, rock resistance is a lot smaller than that assumed.

This discrepancy is not too surprising since the physics of the geomechanical model is extremely simplistic. There are two main avenues to improve the model: (1) constraining precisely variations in geomechanical properties of the chalk and (2) monitoring the impact of the sea on the foot of the cliff. This would however require large amounts of laboratory and field efforts. So, let us examine the predictive nature of the stability factor.

4.4 *Towards a predictive block failure model*

The computed stability factor varies along the cliff face according to the amount of observed overhang, but the largest collapse events detected by TLS did not necessarily occur where the overhang was largest. And neither did it describe situations where physics could explain collapse phenomena. In other words, the stability factor cannot be used deterministically to predict collapse. So, what about considering it in a probabilistic fashion? For that, one needs to check whether the probability that chalk collapsed where S was large is higher than that where chalk did not collapse.

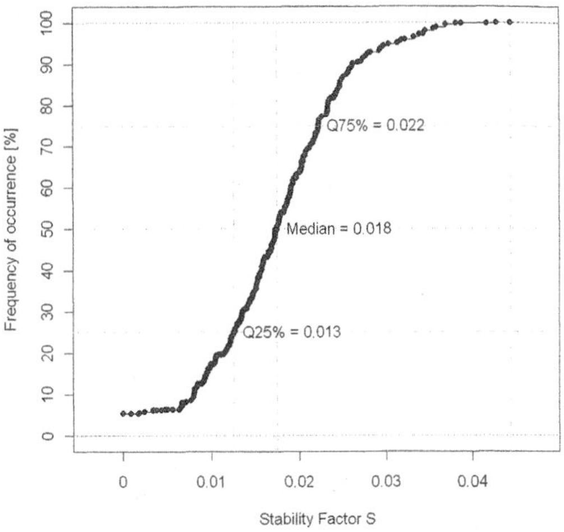

Figure 11. Cumulated distribution of stability factor S for collapsed pixels where eroded material was thicker than 1.2 m.

Figure 11 shows that S reached a value of 0.018 or higher for half the pixels where the chalk eroded is larger than 1.2 m. To define the predictive value of the stability factor, one may consider the so-called *likelihood ratio* LR (Lindley, 1961). LR is a ratio of two probabilities. The first probability counts how many times chalk collapsed (thicker than 1.2 m) when S exceeded a 0.018. The second probability counts how many times chalk did not collapse (thicker than 1.2 m) when S exceeded 0.018. LR gives the ratio between true "alarms" versus false "alarms". Computing LR is easily done with simple GIS grid map algebra. Here is the mathematical expression:

$$LR = \frac{\dfrac{N(S \geq 0.018 | D = 1)}{N(D = 1)}}{\dfrac{N(S \geq 0.018 | D = 0)}{N(D = 0)}} = \frac{0.5}{\left(\dfrac{894127}{4063629}\right)} \approx 2.272 \quad (2)$$

N is the number of pixels satisfying the conditions between brackets. D describes the collapse event, D equals 1 when there is a collapse event, D equals 0 for no collapse.

The result tells that when S goes above 0.018, whatever the actual physical reason, the probability of collapse increases by more than two. Aspinall et al. (2003) qualified this range probability as "just worth a mention" (Aspinall et al., 2003, Table 1, p.275). This result is somewhat encouraging because, even if the stability factor describes imperfectly the physics of the phenomenon, it can still be used as a first order approach to predict chalk collapse events.

If the physical model was refined, one would expect an even better chance of predicting when chalk is eroded off the cliff. Improvements will occur in two ways: first, TLS survey will continue at Mesnil Val, hence providing a more extensive data set of collapses, and second a more complete set of geomechanical surveys and tests will be carried out on chalk strata in the future, improving the knowledge of geomechanical behaviour of chalk in Mesnil Val.

5 CONCLUSION

This study examines a quantitative comparison of terrestrial laser survey (TLS) of a chalk cliff face of the French coast of the English Channel (Mesnil Val, Haute Normandie). TLS surveys are capable of describing the topography of a cliff face 750-meters-long with a precision of 3 cm for points posted every 5 cm. Comparison of such a detailed topographic surface between December 2005 and March 2006 revealed two types of erosion events: large block collapses and flakes peeling off the cliff. The first ones are rare, involving volumes of several cubic meters (up to hundreds) and account for 95% of the total volume of chalk lost in the 3-months-period. The second, flaking, are extremely frequent (99% of the recorded events) but contributed only for 5% of the total chalk volume lost. It is important to note that erosion observed during this three months period is five times smaller that anticipated from 50-years-averaged cliff top retreat rates.

A geomechanical model attempted to predict large collapse events based on the ratio between chalk resistance forces and gravity forces. Although the physics of the model could, and will, be refined, its result is an encouraging first order predictor of chalk collapse.

ACKNOWLEDGEMENTS

We would like to thank ATM3D for carrying out laser data acquisition and point-clouds pre-processing. This work was funded under BRGM's Directorate of Research project PDR06ARN53.

REFERENCES

Aspinall, W.P., Woo, G., Voight, B., Baxter, P.J., 2003, "Evidence-based volcanology: application to eruption crises", Journal of volcanology and geothermal research 128, pp 273–285.

Costa, S., 1998, Dynamique littorale et risque naturel : l'impact des aménagements, des variations du niveau marin et des modifications climatiques entre la baie de Seine et la baie de Somme, PhD Thesis, Univ. Paris 1.

IFEN, 2006, Un quart du littoral Français recule du fait de l'érosion, Le 4 pages de l'IFEN, Institut Français de l'Environnement, September 2006, vol 117, 1–4.

Jeffreys, H., 1961, "Theory of Probability, Oxford University Press", Oxford, 459 pp.

Lindley, D.V., 1971, "Making Decisions", Wiley, London, 207 pp.

PROTECT, 2003, PRediction Of The Erosion of Cliffed Terrains, Unpublished Periodic report EU contract EVK3-CT-2000-00029, Reporting period April 1 2002 – March 31 2003.

Schoumacker, L., 2002, "Essais physiques et mécaniques sur la craie de Val-Mesnil", LAEGO – ENSG, in French.

SHOM, 2006, Calcul de la Marée, http://www.shom.fr/ann_marees /cgi-bin/predit_ext/choixp? opt=12&zone=11&port=LE_TREPORT&date=&heure=&portsel=list, last visited 26 October 2006.

Landslides and Climate Change – McInnes, Jakeways, Fairbank & Mathie (eds)
© 2007 Taylor & Francis Group, London, ISBN 978-0-415-44318-0

Development of a landslide hazard assessment model for cut-slopes along the Tamparuli – Sandakan road in Sabah, Malaysia

S. Jamaludin, A.H. Shamsuddin, R. Majid & A. Mohamad
Slope Engineering Branch, PWD Malaysia Headquarters, Kuala Lumpur, Malaysia

ABSTRACT: Landslide hazard assessment using discriminate analysis was carried out to estimate the instability of slopes in areas underlain by metasediment formation was presented in this paper. 1341 cut slopes along Tamparuli – Sandakan road in the state of Sabah, Malaysia was selected to be used in the development of the slope assessment model in this study. Out of 29 slope parameters analysed using discriminate analysis, 12 parameters were found significant to be used in predicting landslides on cut-slopes along this road. Overall correct classification of both failed and not failed slopes produced by the new assessment model was 77.8%, at par with accuracy produced by other previous researcher's works.

1 INTRODUCTION

Landslides are one of the most destructive natural disasters in Malaysia besides floods. The socio-economic losses due to landslides are significant and set to continue growing as population growth sees development expanding to encompass further potentially unstable hillslope areas. From 1993 to 2004, there are numbers of major landslides reported in Malaysia, involving fill, cut and natural slopes with a total lost of more than 100 lives. The major controls on the landslides occurrence in highland areas of Malaysia are geometry, hydrology and geology of the slopes (Othman & Lloyd, 2001) and the triggering causal factor is mainly due to intense and prolonged rainfall (Aik, 2001). The most common types of landslides in Malaysia are shallow slides where the slide surface is usually less than 4 m deep and occurred during or immediately after intense rainfall (Othman & Lloyd, 2001). These slides commonly occurred in residual soils mantles of grade V and grade VI using the commonly used classification systems of Little (1969).

A significant proportion of landslide losses in Malaysia affect transportation facilities, especially along the Federal Route 22, from Tamparuli – Sandakan Road. These social and economic losses can be reduced by means of effective planning and management which involved hazard assessment, mitigating measures and warning systems (Schuster, 1995 and Dai et al, 2002). It is an important task to accurately estimate the hazard of landslides so that the relevant government officials can make sound decisions based on these values.

The landslide hazard can be estimated using various assessment methods. It is varied from subjective to objective and from qualitative to quantitative. Since decades ago numbers of papers proposing many different methods for an assessment to estimate landslide hazard have been published. A thorough examination of the methodological aspects, together with examples and exhaustive reference lists, can be found. According to Varnes (1984), van Westen (1993), Carrara et al. (1995), Soeters and van Westen (1996), Guzzetti et al. (1999) and Dai et al. (2002) there are four (4) methods of landslide hazard assessment, namely through a landslide inventory, the heuristic approach, the statistical approach, and the deterministic approach. Irigaray et al. (1996) discussed six methods of assessment namely percentage of rupture zones, intervals of 'critical slope angle', matrix, indexing, value of information and multiple regression. Rosenbaum et al. (1997), Ali Jawaid (2000), Tangestani (2003) and Ercanoglu & Gokceoglu (2004) described an attempt to use of 'fuzzy set theory' analysis for evaluating landslide hazard. Fractal dimension, a mathematical theory that describes the quality of complex shapes of images in nature, is also suitable to measure landslides complex topography as reported by Kubota (1996) and Yi et al. (2000).

In this study, the statistical approach using discriminate analysis was used to develop a predictive model to estimate the landslide hazards along the Federal Route 22, from Tamparuli – Sandakan Road.

2 AREA DESCRIPTION AND DATA COLLECTION

The study was carried out along the Federal Route 22, from Tamparuli – Sandakan Road (Figure 1). The study

Figure 1. The study site is along 297.55 km of the Tamparuli – Sandakan road in the state of Sabah.

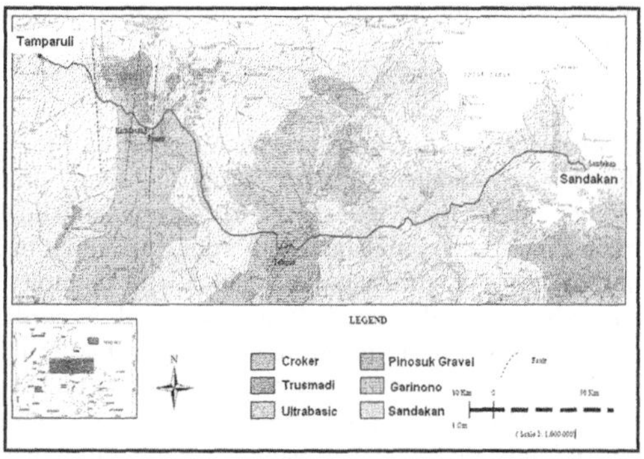

Figure 2. General Geology along the Tamparuli – Sandakan Road, Sabah.

started at km 0.0 from the Tamparuli round about and ended at km 297.55 of the Sandakan Road. The road traverses through the geologically young Crocker and Trusmadi Ranges, rising from an elevation of approximately 200 m to 1700 m.

Within the study area, there is much visual evidence supporting the fact that changes in land use (from natural to agricultural) on landslide prone areas immediately adjacent to the road is proceeding at an increasing rate. Changes to the climate pattern are much more difficult to confirm. However, with the impact of the last El Nino (in 1997) events changes to the commonly accepted seasonal patterns are being appreciated and worldwide there is acceptance of global climate change. Sign of bushfire occurrence within the study area is maybe due to this phenomenon.

In general the geological formations along the Tamparuli – Sandakan Road are complex (Figure 2) and can be summarised as follows; (i) Crocker Formation, (ii) Trusmadi Formation, (iii) Ultrabasic Formation, (iv) Garinono Formation, (v) Sandakan Formation and (vi) Pinosuk Gravels.

According to Stauffer (1967), who has examined outcrops of the Crocker Formation along the road, cutting from Kota Belud to Papar and Tenom, has recognized several distinct types as follows:

a. Flysh-type interbedded sequence
b. Laminate-type sequence
c. Red and green mudstone
d. Mass-flow sandstone
e. Slumped zone.

The Trusmadi Formation is predominantly argillaceous rock of middle to lower Eocene age. The formation comprises strongly folded and faulted alternating bands of siltstone, mudstones and shales in which the generally thinly bedded siltstones are subordinate to the more massive mudstones.

The ultrabasic rocks mainly consist of peridotites that show variable degrees of serpentinization. Vein and other more irregular lensoidal forms of pegmatite, gabbro, pyroxenite, chert and anorthositic varieties are also to be found within these ultrabasic rocks.

The Garinono Formation extends from the Sandakan Peninsula westwards to the northwestern part of Sandakan. This Formation consists of mainly slump breccia which sometimes contains pebbles of volcanic rocks, sequences of interbedded mudstone, minor sandstone and calcarenite. The Garinono Formation is of Upper Miocene age.

The Sandakan Formation area is underlain mainly by thick-bedded sandstone, interbedded sandstone, mudstone and thin carbonaceous seams of the Sandakan Formation. The Sandakan Formation is of lower Miocene to late Miocene age and was formed in synclines.

The Pinosuk Gravels were deposited during the late Pleistocene, approximately 37,000 year BP or older. It generally consists of two units: Lower and Upper Units representing two phases of deposition (Jacobson, 1970). The Lower Unit, consisting of shaped edged sandstone and ultrabasic rock, was deposited by glaciations whereas the Upper Unit made of rounded granodiorite was by ancient mudflow due to thawing of the glacial and ice cap at Mt. Kinabalu. Both tertiary rock formations are highly folded, faulted and fractured.

A key requirement of the study is the capture of data for all cut-slope features along the entire route. Such data must be of a scope and resolution suitable to allow assessment of hazard and of a quality to allow confidence in the generated results.

From a practical perspective, the length of the road and the size of many individual slopes mean that considerable care was required to ensure that an inventory of each feature was a practical possibility within both the budget and time schedule requirements. A further

Figure 3. Failure on Ultrabasic formation cut-slope.

Figure 4. Failure on Crocker formation cut-slope.

benefit associated with simplification of field data collection was that it aids accuracy and reduces the need for specialised personnel and improves repeatability (both by individuals and between individuals).

To overcome the problem, airborne laser mapping, LiDAR (Light Detection And Ranging) was used. LiDAR is a fast and reliable method of obtaining 3-dimensional data from a fixed wing airborne platform for the creation of a Digital Terrain Model (DTM) with a relatively high degree of accuracy. In addition, a laser DTM can be produced in a much shorter time frame than a similar product using conventional photogrammetric techniques.

Based on the end product of LiDAR works (DTM), the number, location and sizes of all slopes along the highway was pre-identified, without going to the sites. On the basis of this pre-identified slopes and its associated features have been allocated to the field teams on an informed prioritized basis. This was a critical element of ensuring practicable efficient field data collection.

Using this method, detail inventories of slope parameters for 1341 cut slopes along the Tamparuli – Sandakan Road was captured and then separated into two groups; 1090 failed slopes and 251 not yet failed slopes. From the results of field data collection, appeared that most of landslides (or failed slopes) occurred in the Crocker and Ultrabasic formation. Two example of landslide found are as shown in Figure 3 for landslide on Ultrabasic formation and Figure 4 for landslide on Crocker formation.

3 DATA ANALYSIS

29 slope parameters (such as slope angle, slope height, percentage slope uncovered, etc.) of 1341 cut slopes within the study area was analysed using discriminate analysis. A discriminate analysis is used to determine the best parameters or the most significant parameters discriminate between two or more naturally occurring groups. In this study, the discriminate analysis used to investigate which the best parameters discriminate between 'Failed' and 'Not Failed' slopes. The Statistical Package for the Social Science (SPSS) computer software was used as a tool in analyzing the data. The analysis result is a discriminate function or discriminate score equation as in Equation 1 below;

$$D_s = B_0 + B_1X_1 + B_2X_2 + B_3X_3 + \ldots + B_iX_i + \ldots + B_nX_n \quad (1)$$

where X_i are the values of the parameters and B_i the calculated coefficients. Before any further analysis can be performed, the success of the formula in separating the two groups must be tested (Suzen, 2002). For this purpose three tests can be used.

Firstly the variability between the two groups and within the groups, and the total variability of the data, is calculated. The ratio of the variability between the two groups and the variability within the groups is called the eigenvalues. It should be maximized for a good discriminate function.

Secondly the ratio of the variability between the two groups and the total variability is called 'Wilk's λ' (Wilk's Lambda). A small value indicates strong variation between groups and less variation within groups. A Wilk's λ of 1 indicates that there is equally great variation within groups as between groups (i.e. that the function does not discriminate).

Thirdly the χ^2 (Chi-square) test to determine if the two groups are significantly different was carried out.

Using significant slopes parameters, a discriminate function and its regression coefficient, scores of discriminate function for a series of 1341 slopes features in the models development can be computed. The next stage is to determine the value of scores that separate the two groups of slopes (failed and not failed).

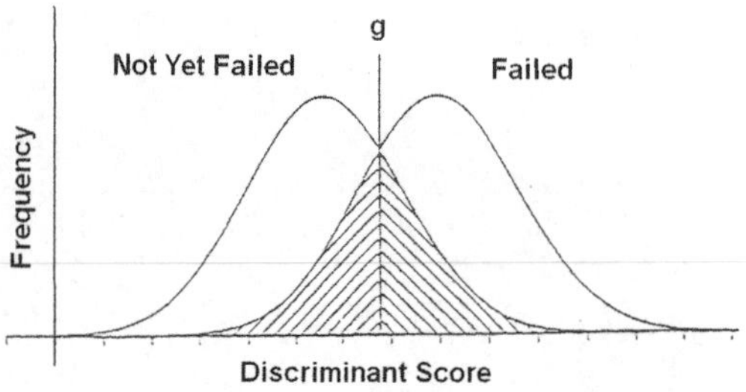

Figure 5. Two normal density curves with a common pooled standard deviation.

There are two normal density curves of scores that can be developed; i.e. for failed and not failed slopes. These two normal density curves are of different means and by assuming its standard deviations are the same (as shown in Figure 5), the process of assigning the separating values of both models becomes much simpler.

The point of intersection g can be used as the boundary that separates the two groups such that, in this case, a slope is,

Not Failed if $D < g$, otherwise Failed.

g can be defined in such a way that,

$$g = \frac{(Ds + Df)}{2} \tag{2}$$

where,
Ds = average score for not failed group
Df = average score for failed group.

Based on the computed discriminate scores for a series of 1341 slopes features and the g value separated the Failed and Not Yet Failed slopes, overall correct classification (or accuracy) of the model in predicting the actual condition within the model development data can be determine.

4 RESULTS AND DISCUSSION

The result of the discriminate analysis using SPSS software showed that there are twelve significant parameters that can discriminates the Failed and Not Yet Failed slopes, namely; slope height, slope angle, slope shape, slope plan profile, cutting topography relationship, presence of structure, main cover type, slope cover, percentage rock exposure, presence of corestone boulders, rock condition profile and ground saturation. Table 1 below shows the significant slope parameters and its discriminate function coefficients (canonical discriminate function coefficients produced by the SPSS).

Table 1. 12 significant parameter and its discriminate function coefficients for cuts slopes.

Slope parameter	Discriminant function coefficients
Slope height	0.027
Slope angle	0.02
Slope shape	0.163
Slope plan profile	0.354
Cutting topography relationship	0.278
Presence of structure	0.202
Main cover type	−0.172
Slope cover	0.472
Percentage rock exposure	0.017
Presence of corestone boulders	−1.266
Rock condition profile	0.249
Ground saturation	0.281
Constant	−4.293

Table 2. Results of three tests carried out as suggested by Suzen (2002).

Eigenvalues	0.312
Wilk's Lambda	0.762
Chi-square	214.664

The twelve-parameter equation for prediction model produced from the analysis is as follows;

$$D = 0.027(\text{height}) + 0.02(\text{angle}) + 0.163(\text{shape}) + 0.354(\text{plan profile}) + 0.278(\text{cutting topography}) + 0.202(\text{structure}) - 0.172(\text{main cover type}) + 0.472(\text{cover}) + 0.017(\%\text{ rock exposure}) - 1.266(\text{corestone boulders}) + 0.249(\text{rock condition profile}) + 0.281(\text{ground saturation}) - 4.293 \tag{3}$$

where D is discriminate function representing 'instability score' of the assessed slopes and for the calculation of D, the slope parameters in the bracket should be replaced by value or classes of slope variables.

During the analysis, all three tests as suggested by Suzen (2002) were automatically carried out by the SPSS software, and the results were as shown in Table 2 below.

Discriminant function of both the Failed and Not Yet Failed slopes then could be computed using this equation (Equation 3). The boundary of discriminate function separating these two groups (Failed and Not Yet Failed) was calculated using average of these two groups mean, which could be determined statistically, and the results was as shown in Table 3 below.

So the g value separating two groups is 0.426. Using this g value, the boundary condition separating Failed and Not Yet Failed slopes is as follow;

Not Yet Failed $D < 0.426$, otherwise Failed.

Table 3. Not Yet Failed and Failed groups mean.

Group	Group mean
Not Yet Failed	1.128
Failed	−.276
Average mean	0.426

Table 4. Accuracy of the developed model in predicting actual condition in the model development data.

(1) Number of assessed slopes	1341
(2) Numbers of actual landslide or failed slope	1090
(3) Number of failed slopes correctly classified	873
(4) Number of not yet failed slopes	251
(5) Number of not yet failed slopes correctly classified	170
(6) Overall correctly classified	1043
(7) % of overall correctly classified	77.8

Table 5. Accuracy of the models from previous works by other researches on landslide assessment.

No.	Country	Accuracy (%)	References
1	Italy	72.7 and 80.7	Carrara et al. (1995)
2	Italy	72.0	Guzzetti et al. (1999)
3	Bolivia	78 to 89	Péloquin & Gwyn (2000)

Accuracy of the developed model was determined based on overall correct classification of the model in predicting the actual condition (Failed and Not Yet Failed) within the model development data.

The new prediction model developed shows a good capability in predicting landslides along the study road. Comparing to the accuracy of the previous works by other researches using discriminate analysis as shown in Table 5, 77.8% overall correctly classified in the model development data produced in this study was at par with those works.

In this study the discriminate function (D) then will be converted into probability. The equations that have been used to transform the data from the individual discriminate function to probabilities of group membership (i.e. Failed or Not Yet Failed) have been derived through curve fitting, and it was simplified as shown in Table 6 below.

Following the calculation of probability score, the data is then categorized in qualitative terms for the purpose of interpretation and action. The hazard rating categories designed for this purpose are very low, low, medium, high and very high (Table 7).

The series of landslide hazard rating category for cut slopes along the Tamparuli – Sandakan Road was then presented in form of Hazard Maps. Figure 6 shows example of Hazard Maps for a section of the Tamparuli – Sandakan Road in the study.

Table 6. Conversion of D into probability, P.

Value of D	Calculation of probability, P
$D < -2$	$P = 0.05$
$-2 < D < 0.5$	$P = 0.0037D^3 + 0.0891D^2 + 0.3195D - 0.3531$
$0.5 < D < 4$	$P = 0.0105D^3 - 0.1275D^2 + 0.5152D + 0.2952$
$D > 4$	$P = 1$

Table 7. Probability score and category used in the study

Probability score	Hazard category
0.0–0.2	Very low
0.2–0.4	Low
0.4–0.6	Medium
0.6–0.8	High
0.8–1.0	Very high

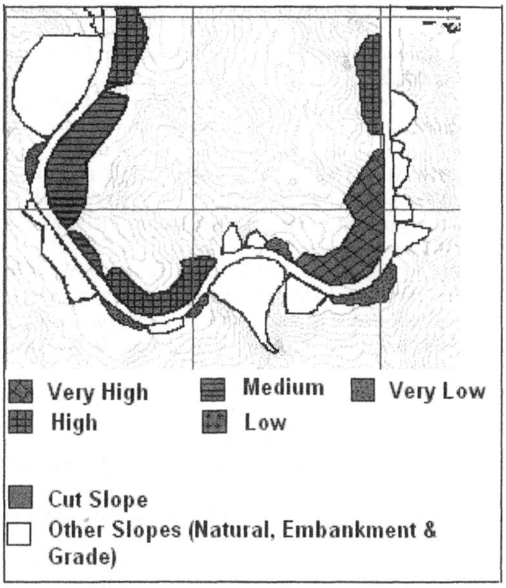

Figure 6. Typical landslide hazard map for a section produced from the study.

5 CONCLUSION

Through this study, twelve significant parameters influencing the landslides occurrence along the

Tamparuli – Sandakan road was identified. The twelve-parameter equations model developed using the discriminate analyses on twenty nine slope parameters appeared to show a good capability in predicting landslides. An accuracy produced was at par with other previous researcher's works.

Based on principle drawn by Varnes (1984) that landslide in the future will be most likely to occur under geomorphic, geologic and topographic conditions that have produced past and present landslide, this twelve-parameter equation can be extended to other slopes with similar geomorphic, geologic, and topographic conditions to produced Hazard Maps.

REFERENCES

Aik, N.C. 2001. Practical design aspects for slopes in mountainous terrains. *Proc. of National Slope Seminar*, Cameron Highland, Malaysia, pp 29.

Ali Jawaid, S.M. 2000. Risk assessment of landslide using Fuzzy Theory. *Proc. of 7th International Symposium on Landslides*, Cardiff. Thomas Telford. 31–36.

Carrara, A., Cardinali, M., Guzzetti, F. & Reichenbach, P. 1995. GIS-Based techniques for mapping landslide hazard. In *Geographical Information System in Assessing Natural Hazards*. Academic Publication, Dordrecht, Netherlands.

Dai, F.C., Lee, C.F. & Ngai, Y.Y. 2002. Landslide risk assessment and management; an overview. *Journal of Engineering Geology 64*. Elsevier, London. 65–87.

Ercanoglu, M. & Gokceoglu, C. 2004. Use of fuzzy relations to produce landslide susceptibility map of a landslide prone area (West Black Sea Region, Turkey). *Journal of Engineering Geology 75*. Elsevier, London. 229–250.

Guzzetti, F., Carrara, A., Cardinali, M. & Reichenbach, P. 1999. Landslide hazard evaluation: a review of current techniques and their application in a multi-scale study, Central Italy. *Journal of Geomorphology 31*. Elsevier, London. 181–216.

Irigary, C., Chacon, J. & Fernandez, T. 1996. Comparative analysis of methods for landslide susceptibility mapping. *Proc. 8th International Conference and Field Trip on Landslide*, Spain, Balkema, Rotterdam, The Netherlands. 373–383.

Jacobson, G. 1970. Gunong Kinabalu Area, Sabah, Malaysia. *Geological Survey Malaysia.*

Kubota, T. 1996. Study on the fractal dimensions and geological condition of landslides. *Proc. 8th International Conference and Field Trip on Landslide*, Spain, Balkema, Rotterdam, The Netherlands. 385–392.

Little, A.L., 1969. The engineering classification of residual tropical soils. *Proc. Specility Session on the Engineering Properties of Lateritic Soil, Vol. 1*, 7th Int. Conf. Soil Mechanics & Foundation Engineering, Mexico City. 1–10.

Othman, M.A. & Lloyd, D.M. 2001. Slope instability problems of roads in mountainous terrain: a geotechnical perspective. *Proc. of National Slope Seminar, Cameron* Highland, Malaysia, pp 11.

Peloquin, S. & Gwyn, Q.H.J. 2000. Using remote sensing, GIS and artificial intelligence to evaluate landslide susceptibility levels: Application in the Bolivian Andes. *Proc. of 4th International Conference on Integrating GIS and Environmental Modeling (GIS/EM4)*: Problems, Prospects and Research Needs. Banff, Alberta, Canada. pp 12.

Rosenbaum, M.S., Senneset, K. & Popescu, M.E. 1997. Assessing the likelihood of landslide-related hazards on a regional scale. In *Engineering Geology and the Environment*, Marinos, Koukis, Tsiambaos & Stoumaras (eds). Balkema. 1009–1014.

Schuster, S.R.L. 1995. Reducing landslide risk in urban areas – experience in the United State. In: *Urban disaster mitigation – The role of engineering and technology.* Cheng and Sheu (Eds.). Elsevier, pp 217–230.

Soeters, R. and van Westen, C.J. 1996. Slope instability recognition analysis and zonation. In Turner, K.T. and Schuster, R.L. (eds); *Landslides: Investigation and Mitigation*, Transportation Research Board National Research Council, Special Report no: 247, Washington D.C. 129–177.

Stauffer, P.H. 1967. Studies in Croker Formation, Sabah: *Borneo Reg. Malaysia Geol. Survey*. Bull 8.

Suzen, M.L. 2002. *Data Driven Landslide Hazard Assessment Using Geographical Information Systems and Remote Sensing.* Unpublished PhD Thesis, Middle East Technical University, Turkey.

Tangestani, M.H. 2003. Landslide susceptibility using the fuzzy gamma operation in a GIS, Kakan catchments area, Iran. *Proc. of Map India Conference 2003*. 7p.

van Westen, C.J. 1993. Application of Geographic Information Systems to Landslide Hazard Zonation. *ITC Publication 15*. International Institute for Aerospace Survey and Earth Sciences (ITC), Enschede, Netherlands, pp 245.

Varnes, D.J. 1984. Landslide Hazard Zonation: a review of principles and practice. *UNESCO, Natural Hazards*, No:3, pp 61.

Yi, S., Li. R., Pu, X. & Fu, S. 2000. The fractal characteristics of the temporal and spatial distribution of the Zameila Mountain landslide activities and its fracture structure in Tibet of China. *Proc. of 7th International Symposium on Landslides*, Cardiff. Thomas Telford. 1605–1608.

Landslides and Climate Change – McInnes, Jakeways, Fairbank & Mathie (eds)
© 2007 Taylor & Francis Group, London, ISBN 978-0-415-44318-0

Landslide and chalk fall management on the Folkestone to Dover railway line, Kent

G. Birch
Network Rail, London, UK

C.D. Warren
Halcrow Group Ltd, London, UK

ABSTRACT: The railway line between Folkestone and Dover (Kent) crosses an area of marginally stable coastal landslide, known as Folkestone Warren before continuing on to Dover through two major tunnels known as Abbotscliff and Shakespeare Tunnel constructed through precipitous chalk cliffs lying immediately adjacent the sea. Since the line was opened in 1844, the track has been severely damaged and broken by landslip and major ground movements on several occasions, most notably in 1915 when the line was closed for 4 years. The operation of the route has also been disrupted by periodic collapse of the High Cliffs that form the rear scarp of the Warren. These problems have resulted in considerable investment in site investigation, ground monitoring and the installation of a variety of remedial measures to stabilise the Warren, including drainage, toe-weighting and coastal protection. This paper summarises the main causes and mechanisms of historical events observed within the Warren and adjacent High Cliffs, and considers the potential for future events and sustainability of the railway, given the possible impact of future climatic change.

1 INTRODUCTION

1.1 *Geological and historical setting*

The Folkestone Warren is the name given to the area of coastal landslide between Folkestone and Abbotscliff, which occupies an area 2.7 km long and between 50 m and 350 m wide (Fig. 1). The back of the Warren is defined by the 'High Cliff', which is over 100 m high and composed of Lower and Middle Chalk. In plan the Warren is cuspate in shape and concave towards the sea. The hummocky lower ground within the Warren, known as the Warren Undercliff, comprises entirely landslipped material over which passes the railway. The geological sequence exposed comprises some 200 m of strata which dips gently at about 1 degree towards the north-east, commencing with the ferruginous sands of the Folkestone Beds (Lower Greensand), passing up through the Gault Clay and then into the Lower and Middle Chalk (Osman, 1917). These deposits are overlain unconformably by thin patchy deposits of clay with flints and red-brown sands with ironstone.

The selection of this dramatic coastal route for the railway between Folkestone and Dover was not without some controversy as the engineers were aware of the geological instability of the Folkestone Warren. However, this route was adopted in preference to a more costly and time-consuming inland alternative in order satisfy shareholders' aspirations of winning the race against a competing railway company to Dover, and thus secure the lucrative continental traffic.

Figure 1. View east along Warren past Warren Halt and Horsehead Point to the west portal of Abbotscliff Tunnel.

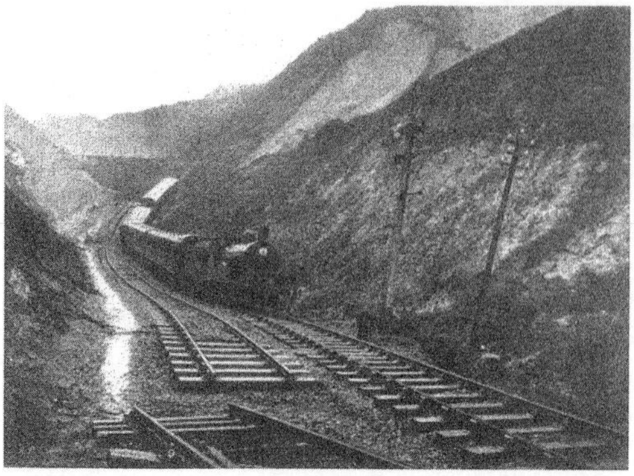

Figure 2. 1915 slip showing train crossing the western backscar of the 1915 movement located immediately east of Martello Tunnel.

The line was opened in 1844 amongst great pomp and circumstance with a declaration by the Inspector General of Railways that having inspected the construction of the railway with great attention, he took "…. great pleasure in assuring your Lordships, not only that the railway itself is in a perfectly safe and efficient state, but that no part of the works are exposed to the smallest danger, either from the eruptions of the sea, or from the fall of the cliff: though it was natural for the public to have their doubts, in the first instance, as to the success of so arduous an undertaking".

However, the Warren continued to suffer a further seven major landslips two of which closed the line, 1877 and again in 1915, when the line was closed for four years (Fig. 2).

1.2 *Investigations and monitoring*

Notable amongst those who strove to understand the nature and significance of the ground movements was Karl Terzaghi, an eminent father of the science of soil mechanics, who journeyed in early 1939 from his home town of New Orleans, Louisiana, to advise on how to handle the land instability both here at the Warren and at a landslip near Sevenoaks. His report to the Southern Railway Company in 1939 (Terzhagi, 1939) likened the situation to tackling a complex military battle. His analogy was to consider the Warren landslip as a large army and the railway engineers as a small but mobile opposing force and to consider holding the line against a surprise attack.

"This can be accomplished in two different ways. Either we excavate trenches along the entire line, spread our forces fairly uniformly over our fortified line, and wait for what is going to happen next or else we adapt our policy to the characteristics of the opponent. That means, instead of fortifying our line, we send out scouts to watch the movements of the enemy and we keep our own resources mobile and ready for concentrated action. The first procedure may have the advantage of creating in the civilian behind the line, a feeling of security; because the civilian sees that something rather spectacular is being done. At the same time, the procedure is obviously wasteful and dangerous, In addition, because we cannot judge whether or not our line will be strong enough to stand up against a wedge-action of unknown intensity. Hence there is no doubt that the second method deserves the preference, provided the scouts are trustworthy and thoroughly familiar with the local topography."

This philosophy has stood the test of time and the continued efforts to maintain the safety of the line are driven by the programme of monitoring and vigilant observation. The extensive work undertaken to date, including additional toe-weighting and drainage measures put in place to improve stability, have done much to reduce the propensity for major movements. However, ground movement is ongoing with fresh cracks appearing in several places in the Warren, most notably in the vicinity of Horsehead Point.

Following Terzaghi's advice a number of actions were pursued in order to advance the understanding of ground movements which included:

- test borings to survey the top of the Gault Clay and observe variations in groundwater, in particular, to ascertain the effects of subsequent drainage.
- laboratory investigations to determine the physical properties of the Gault Clay and any variations in properties over the depressed area.
- installation of accurate devises within drainage headings to measure their change in length in response to the ongoing ground movements.
- installation of offset target survey points along the railway line.

These investigations led to the implementation of a variety of measures designed to mitigate the land slipping processes, including:

- drainage including the construction of some twenty drainage headings from the foreshore.
- sea defence works including the construction of 3 kms of sea walls.
- beach replenishment and the introduction of shingle to counter erosion and the loss of toe-weighting material.
- stability enhancement and local re-profiling by moving spoil around within The Warren.

Four periods of study and ground investigation have been carried out since the Second World War, each following periods of renewed activity, namely during the period 1948–55 (Toms, 1946; Muir Wood, 1955; Hutchinson, 1969), between 1969–70 (Hutchinson et al. 1980, Muir Wood, 1994), between 1980–1983 (Trenter & Warren 1996; Warren & Palmer 2000) and more recently in 2001.

Figure 3. February 1988 chalk fall at Lydden Spout.

There is no doubt that following these further investigations and advances in technology we now have a better understanding of the Warren slips in terms of their three-dimensional character, geological make-up and general groundwater conditions so as explain the overall stability and movements occurring (Birch & Warren 2006). As a result we are now better informed as where to place any additional remedial measures to achieve optimum effect whether this be the installation of additional rock groynes, the placement of toe-weight material, piled solutions or more effective drainage arrangements (i.e. additional headings with drainage probe arrays, use of pumped wells or vacuum dewatering etc.). However, it should be pointed out that, even if such measures were installed, ongoing monitoring of the Warren and the adjacent Abbotscliff and Shakespeare Cliff, through which the railway runs in tunnels, and which are affected by chalk falls and sea erosion, will always be required. The coastline is ever evolving as evidenced in the past by the construction of Folkestone harbour wall which had a major impact on subsequent landslide activity in the Warren by cutting off the supply of shingle to beaches in the Warren (Hutchinson et al. 1980). An analogous effect was felt following a chalk fall that occurred in 1960 below Gallery 8 of Abbotscliff Tunnel, which effectively blocked the movement of sand further east leading to subsequent removal of a sand bar and Lydden Spout Pool. Subsequent erosion of the cliffs resulting in a fall of chalk in February 1988 (Fig. 3). Figure 4 showing the aerial view of the area in 1952 as compared to the view in 1980 subsequent to the 1960 chalk fall. A prominent set feature can be seen on the 1952 photograph behind the unstable chalk cliff prior to failure.

2 FORMS AND MECHANISM OF FAILURE

2.1 *Cliff falls*

Between the Warren landslip and Dover, the railway passes along the coast on seawalls built for that

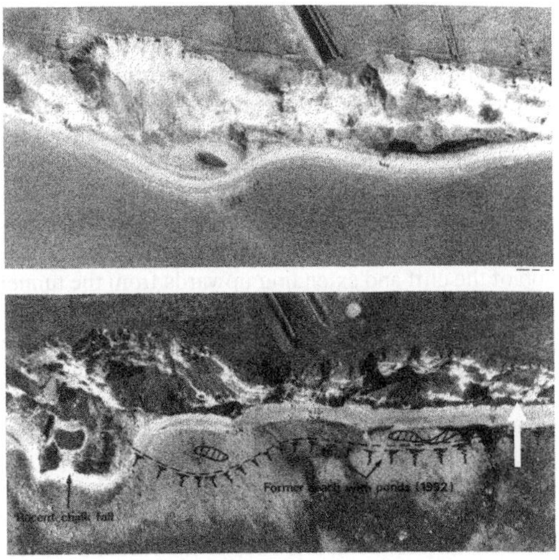

Figure 4. Aerial photographs of Lydden Spout area taken in 1952 (top) and in 1980 (bottom) showing chalk cliffs, condition of foreshore (beach deposits), and 1960 chalk fall. White arrow shows area of fall that occurred in February 1988.

purpose and in tunnel beneath the two promontories of Abbotscliff and Shakespeare Cliff. A third, less stable, promontory situated between the two tunnels, Round Down, was removed by explosives with the assistance of Her Majesty's Sappers & Miners in order to accelerate the completion of the line and avoid the cost of tunnelling.

The displaced chalk formed a platform at the foot of the cliff, which was used as the site for the first Channel Tunnel attempt in 1881, an aborted colliery in 1886 to 1905, the second Channel Tunnel attempt in 1974 and subsequently the final and successful scheme in 1987.

Observations from aerial photography and field reconnaissance have identified two dominant cliff fall mechanisms and four subordinate processes which can occur independently from or in association with the dominant mechanisms:

- major falls independent of undercutting.
- major falls triggered by undercutting.
- toppling failures.
- wedge failures.
- debris slides.
- local spalling.

The first group, major falls independent of undercutting, represent the most significant in terms of quantity of material involved and potential impact on the railway. The falls occur rapidly and often with sufficient momentum to continue seaward as a flow slide for anything from one to six times the cliff height. Aerial photos reveal a series of over-printed horse-shoe shaped bunds along the foreshore which are attributed to many thousands of years of cliff falls.

Recent falls that have occurred in the area include the February 1988 fall at Lydden Spout described above and a further major fall that occurred on the 26th January 1988 at the west end of Abbotscliff Tunnel which involved some 150,000 m^3 of chalk with a run-out distance of about 180 m from the base of the cliff (Birch & Warren 1996; Birch & Griffiths 1996). Prior to failure the top of the cliff had been showing signs of seawards movement, a tunnel shaft located at the top of the cliff and extending upwards from the tunnel crown being sheared by 800 mm at a depth of 37 m. The chalk in this area was also extensively jointed and the cliff was being exposed to wave erosion during periods of inclement weather despite the presence of a low concrete apron at the base of the cliff. The possibility of the cliff failing in this area was known following the investigations in 1982 and inclinometer boreholes were established both at the top and at the base of the cliff. The inclinometer at the top of the cliff and the top of the shaft eventually was lost during the chalk fall. Additional more recent minor chalk falls also occurred at the back of the Warren in January 2001 following the period of extensive rainfall that occurred up to this date. Whilst little can be done to prevent these large falls, the safety of the tunnels and the railway and householders on the top of the High Cliff needs always to be considered, regular inspection and monitoring being a necessity. Where the railway line is exposed to chalk falls, it is protected by a "chalk fall fence" to which a tension-sensitive wire is attached, turning the rail signal to danger in the event of debris hitting the fence.

2.2 *Warren landslips*

The Chalk and Gault landslipped strata in the Warren appear to reflect ancient multiple movements which commenced at the west end and moved progressively eastwards along the Warren towards Abbotscliff Tunnel. (Trenter & Warren 1996). The movements at the west end were of such a scale that the whole of the cliff and underlying strata appears to have moved bodily seawards with the cliff top then collapsing into the void so created (Warren & Palmer 2000). This would explain why the more rotational form of failure close to the sea (Slip 2 of Trenter and Warren 1996) contains a considerable thickness of Gault above the basal slip surface whereas the main non-circular multiple slip extending to the back of the Warren (Slip 1 of Trenter and Warren 1996) has a much reduced Gault thickness above the main basal slip surface. This situation is unusual when compared with the geological strata forming slips of type 1 and 2 further to the east.

East of Warren Halt, a "hinging mechanism" of movement is invoked, influenced by the easterly dip of the sulphur band, above which the basal slip surface usually occurs, and the passive resistance and confinement provided at the west end by the adjacent slip mass; this produced a series of backtilted blocks of relatively intact chalk arranged in "en eschelon" fashion with the backscar trending towards the High Cliff. Four such slips have been recognised in the Warren, east of Warren Halt. The nature of the movements was such that the greatest lateral displacement, rotation and downthrow occur at the east end furthest from the hinge point. The excessive nature of the rotation at the east extremity can be observed immediately offshore of Horsehead Point where the upthrust boundary between the Gault and overlying Glauconitic Marl (also known previously as the Chloritic Marl) dips steeply landward at 70 degrees (Fig. 5); further landward of this location shallower dips of 45 degrees are recorded in the nodular Melbourn Rock (Middle Chalk) forming the Horse's Head itself.

Figure 5. Glauconitic Marl/Gault boundary dipping landwards at 70 degrees immediately offshore of Horsehead Point. Gault to right of arrow.

Further East of Horsehead Point the landslip blocks appear to have slipped "en masse" and intact and have generally suffered less rotation. Little or no ongoing movement is apparent in this area possibly indicating the increased stability as a consequence of the increased passive resistance due to greater depth to the basal slip surface above the sulphur band (Trenter & Warren 1996). This is further enhanced by the increased thickness of the Gault Clay in this area such that dips on the sulphur band increase from 1 degree to 2 degrees east of Horsehead Point. No slip surfaces

Table 1. Summary of seawards movements for the periods 1962–1982 and 1983–2005.

Heading	Slip*	Movement mm 1962–1982	Movement mm 1983–2005
H1	WW	107	–
H2 new	WW	81	276
H2 old	WW	–	248
H4	WW	110	356
H5	WH	139	458
H6	WH	–	575
H6B	WH	195	569
H6A	HP	67	300
H7	HP	200	750
HP	HP	604	1692
H8	HP	1093	2085
East H8	WE	15	160
H12	WE	–	205

*WW Warren West; WH Warren Halt; HP Horsehead Point WE Warren East.

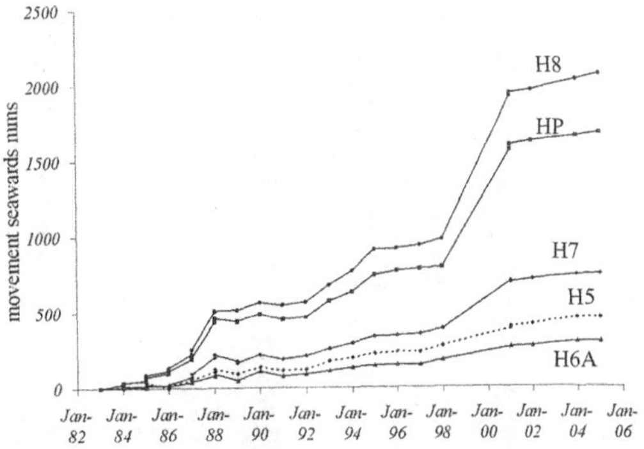

Figure 6. Movement within the Warren from 1983 to 2005 at Warren Halt (heading H5) and within the Horsehead Point Slip at headings H6A & H7, Horsehead Point (HP) and heading H8.

were observed on the foreshore suggesting that the movement is accompanied by upthrusting of strata on the foreshore (Muir Wood 1955) rather than emergence of actual slip surfaces.

Surveys of points established along the railway and on the seawall show movements since 1983 (refer Table 1) agree well with the form of slips, hinging mechanism and possible controls of pore water pressures acting on the slip surface on the overall stability previously identified by Trenter & Warren (1996) and Warren & Palmer (2000).

Measurements for points on the seawall within the Horsehead Point Slip from end of 1983 up to March 2005 show increasing seawards movements as we move west from heading 6A to near heading H7 beyond Horsehead Point (HP) towards heading H8 where the hinge backscar of the slip exits below the seawall onto the foreshore (Fig. 6). An acceleration in the movement was noted between the period November 1998, when the line became the responsibility of Railtrack to January 2001 when the surveying was resumed by Network Rail following increased movement in the area after a period of sustained rainfall at the end of 2000 (Birch & Warren 2006). Over this period of time, the point near heading H8 moved a total of 970 mm, signs of the excessive nature of the movement being observed on the toe-loading platform in this area (Birch & Warren 2006). Together with the readings reported by Trenter & Warren (1996), this would suggest a total movement since 1962 of over 3 m in the area of heading H8. In comparison, movements recorded for the Warren Halt Slip further west show a total movement since 1983 of 458 mm near heading H5 making the total movement of about 600 mm since 1962.

With respect to the nature of the basal slip zone immediately overlying the sulphur band within the Warren slips, two types can be recognised:

- Type 1: fissured sheared and slickensided material; typical where the basal slip surface is overlain by a large thickness of Gault (usually Slip 1 slides of Trenter & Warren 1996). The Gault is similar in character to the adjacent slipped Gault bed but possesses distinct lustrous and slickensided surfaces along which movement takes place e.g. shear zone in borehole 82/9 (Fig. 7)
- Type 2: fully or partly disturbed, softened and remoulded material; typical where large Gault thicknesses have been removed by landslipping and the Chalk almost rests on the Folkestone Beds (usually Slip 2 slides of Trenter & Warren 1996). Most of the original Gault Clay fabric is lost and the material has been reduced to a firm clay having a notably higher moisture content e.g. shear zone in borehole 82/5 (Fig. 7).

Further details on the form of the slip surfaces in the 1982 boreholes are given by Warren & Palmer, 2000. It is suggested that the nature of the material forming the basal slip will impact the nature of movements within any given slip mass. For example one might expect ongoing creep movements to occur where the basal slip zone is formed of fully softened and remoulded material as compared to generally slipped yet intact Gault Clay containing discrete slickensided surfaces, where stick-slip movements are much more likely. Further research into this aspect together and also on the actual pore pressures acting on the shear surfaces needs to be undertaken before the true nature of movements in the Warren and the stability within any given three-dimensional slipped mass can be adequately explained (Warren & Palmer 2000).

It should also be noted that slickensided surfaces have been observed in boreholes drilled at the High

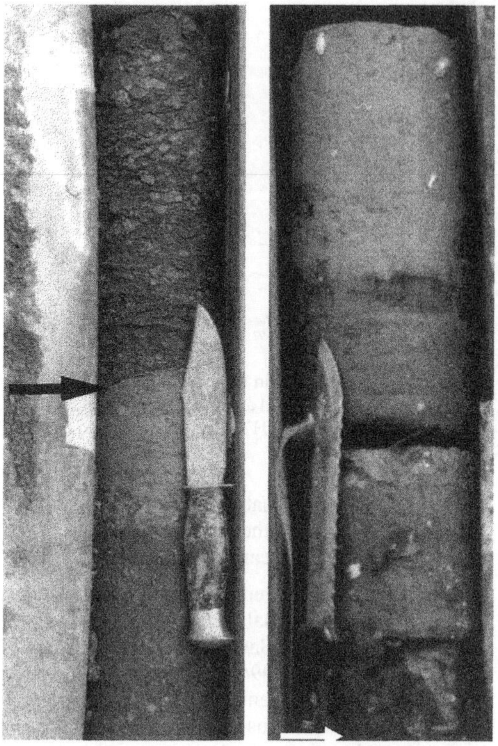

Figure 7. Type I slip zone in borehole 82/9 (left hand side: arrow shows basal failure surface) and Type 2 slip zone in borehole 82/5 (right hand side: comprising only 0.6 m of completely remoulded Gault Clay sandwiched between the Chalk and the sulphur band.; arrow shows basal slip surface.

Figure 8. Fossil hamites attenuatus and pyritic nacreous form of preservation which is characteristic of Gault Bed II.

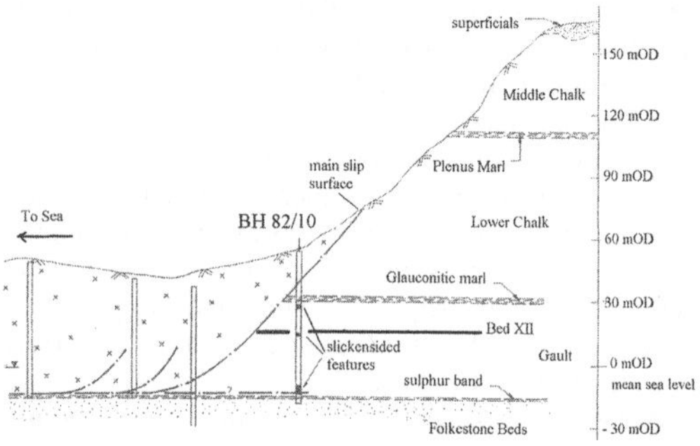

Figure 9. Slickensided features in borehole 82/10 below the High Cliff (modified from Hutchinson 1993).

Cliff through "undisturbed" in situ Gault. These features were observed both in the 1948 and 1982 series of boreholes, even those located outside the Warren landslip area below the west portal of Abbotscliff Tunnel. The Gault is made of 13 distinct beds and the use of characteristic micro-lithology and macrofossil zones enable the slickensided features to be correlated to the Gault Bed (Fig, 8, see also Warren & Palmer 2000). The delineation of such Gault beds not only assist in determining the magnitude of displacement within the Warren landslides but can also indicate whether any substantial displacement has occurred across the slickensided features observed in the Gault beneath the High Cliff. The studies indicated that the slickensided surfaces were usually encountered immediately adjacent the more competent glauconitic beds within the Chalk and Gault succession i.e. below the Glauconitic Marl, above and below Gault Bed XII and above Gault Bed I.

Minor displacements across the slickensided features were identified both in borehole 48/19 and 48/15 below the High Cliff at the west end of the Warren. In the former hole, the features were found at 15.2 m OD immediately above Bed XII and 8.8 m below the 1915 slip, a 0.45–0.6 m throw being reported. In borehole 48/15, the features were found at −17 m OD immediately above the sulphur band and some 40 m below the 1915 slip with a 1.2 m throw being reported.

Similar slickensided features have been reported in the 1982 boreholes, see Figure 9 for the features observed in borehole 82/10 at the west end of the Warren. A further detail regarding the slickensided features is provided by Hutchinson (1993).

Given that the majority of the features develop adjacent the more competent glauconitic beds in the Gault, this would suggest development along preferential planes of weakness in the Gault due to concentration of stress at these boundaries, Such stress concentration could either be caused in the past by tectonic forces e.g. intra-formational shears due to flexural bedding

Figure 10. Channel Tunnel construction and temporary cutting through Upper Gault Clay at Holywell Coombe. Movement recorded above Bed XII close to base of excavation in borehole located at arrow position on upslope cutting.

plane slip, or more recently due to unloading of the slope.

Additional evidence for a recent origin was provided during excavation of a 15 m cutting through in situ Gault Clay Beds XIII and XII at Holywell Coombe further north as part of the Channel Tunnel construction. During the course of the excavation, minor movements totalling 5–10 mm above the more competent glauconitic Bed XII lying close to the base of the excavation, was detected in a borehole located some distance upslope (Fig. 10).

3 MANAGEMENT STRATEGIES

3.1 *Monitoring*

For many decades the coast defences at between the Warren and Dover were maintained and repaired by a dedicated team based at Warren Halt. However, this luxury was lost in the late 1990's and repairs were carried out as required by specific contracts. This continued up to the winter of 1996 when part of the high wall at Warren east end collapsed. Emergency works, comprising reconstruction and rock revetment, were then undertaken to repair this section of wall. This collapse raised a number of issues concerning the long-term approach to maintaining the existing sea wall and coastal stabilisation measures and consequently a ten year strategy was developed. The strategy required a fresh "holistic" overview of the landslide processes, existing monitoring regime and maintenance strategy. This led to a rationalisation of the routine monitoring arrangements at the Warren:

- selected piezometers fitted with data loggers and monitored remotely by modem.
- surveying of ground movement along the railway formation and the sea wall.

Figure 11. Electrolevels installed along the railway at Warren Halt where the line currently experiences the most movement.

- condition surveys of the high seawall and apron.
- inspection of headings for movements, water flows etc..
- annual 'expert eye' walkover.
- monitoring of fissure systems behind the crest of the High Cliff.
- hazard assessment to focus efforts on zones of high risk in terms of safety to the railway line.
- installation of electrolevels on the tracks.
- extension to the "chalk fall" detection fence.
- long-term monitoring of cliff face alongside Abbotscliff and Shakespeare Tunnels to indicate de-stressing effects and possible future chalks.

The inspections and condition surveys are used to determine the maintenance priorities for the ensuing years, whilst the monitoring of the piezometers and ground movements is aimed at providing early warning of potential instability.

The electrolevels placed on the railway tracks (Fig. 11) provide "real time" monitoring of track twist and settlement. Threshold levels have been set on the electrolevels (Red, Amber and Green), which allows swift remedial action to be taken in the event of significant ground movements.

An automatic meteorological station has also been installed at Warren Halt which can be interrogated by a modem to determine the occurrence of extreme rainfall events that could trigger movements.

The density of undergrowth within the Warren below the High Cliff precludes comprehensive geomorphological mapping to a consistent level of detail throughout the Warren. However, there are a variety of stereo aerial photographs available dating back to the earliest RAF sorties in 1947, which have been used to determine the ground morphology from "snap shots" in time. A further means by which the morphology is revealed, albeit in a transient way, is the occasional clearance of selected areas of scrub to maintain the habitat for a rare species of moth, the fiery clearwing (Bembecia chrysidiformis), which is found only at Folkestone Warren. It is the latter clearance behind Horsehead Point that revealed the existence of graben features bounded by opposite facing normal faults, or "antithetic" faults, to accommodate the principally lateral nature of the ground movement.

Archive research has revealed photographs from earlier periods of ground movements which provide insight into the nature of the movements we see today. A notable observation is that the "fresh" ground movements observed over the winter of 2000/2001 are a re-activation of the same failure surfaces associated with the last major phase of ground movements in 1915. Indeed, the process rates had accelerated to the point where, if it had not been for the drainage and weighting works put in place following the 1915 events, we could well have seen similar catastrophic failures.

In terms of process rate there has been a good correlation observed between the Met Office calculations of Soil Moisture Deficit (SMD) and landslide activity. It is not so much the rainfall on the Warren itself which drives the activity, but the longer duration build up of groundwater within the Folkestone – Dover Chalk Block and this build up is mirrored by periods of low SMD.

Outside of the Warren are a number of processes which become active in response to prolonged periods of rainfall and associated low SMD. These are movements of portions of the Warren High Cliff and failures of the adjacent cliffs to the east of The Warren. Such features can be detected many decades in advance of collapse by settlements adjacent to the cliff crest described as "sets" (Hutchinson, 1969). The falls can involve many hundreds of tonnes of chalk which, under certain circumstances, can develop into flow slides with a run out distance in excess of the cliff height. Minor cracks have recently appeared behind the High Cliff along the Old Dover Road adjacent Capel Court and further east of the cliff from which 1915 Great Fall occurred. Additional investigation and regular monitoring of these cracks will be required in future to identify their origin and confirm whether they represent a early warning of a possible future large chalk fall from the High Cliff.

Whilst conventional topographic methods dominate the base surveys, the system is limited effectively to two sub-parallel survey lines, one along the railway track bed and the other along the seawall. These linear survey lines are supplemented locally by points aside from these two survey traces but ideally one would be looking to regularly scan the entire Warren Undercliff to complete the full 3D picture of ground movements. The principal constraint on this is vegetation density and its sensitivity to disturbance by intruding surveyors equipped with GPS. The relative high vantage points from around the crest of the Warren High Cliff provide scope for LiDAR scanning provided sufficient targets are permitted within the SSSI.

Folkestone Warren has been selected by the European Space Agency (ESA) along with a handful of international sites to trial the application of satellites for ground deformation mapping at the "engineering" scale. The technique, known as Interferometric Synthetic Aperture Radar (InSAR), uses radar pulses emitted from satellites which orbit the earth at an altitude of approximately 800 kms. Any difference in reflectance between successive images can represent ground deformation, which is displayed as interference fringes to which values can be assigned. Whilst accuracies of a few millimetres can be achieved, this is dependant on the following success criteria:

- suitability of satellite paths.
- assumptions on earth flow directions.
- good "pixel" coherence.
- good reflection characteristics.
- availability of a Digital Elevation Model (DEM) for the map base.
- availability of ortho-rectified aerial photographs to "drape" over the DEM.

The use of satellite interferometry can help with the geomechanical understanding of ground movements but practical considerations, notably the image processing time, preclude this as a system for real-time detection of ground movements.

3.2 *Capital works*

The combination of long term monitoring and future strategy leads to the execution of remediation works commensurate with commercial aspirations for the route's viability. Current and future capital works include:

- rock revetments.
- seawall and concrete apron maintenance.
- enhanced drainage by probe hole arrays.
- enhanced drainage by directional drilling.
- drainage interceptor galleries.

The recent phases of rock revetment work have been the single most positive contribution to maintaining the coastal defences which, in turn, protect the railway (Birch & Warren 2006).The revetment design provides the following key benefits:

- physical support to the sea wall.
- breaks up the destructive wave energy.
- traps sediment which reduces erosion.
- adds toe weighting against rotational slip surfaces.
- blends in well with the local environment (stone colour etc.).
- meets environmental approvals.

Figure 12. Photograph showing ongoing movement at Horsehead Point and damage to toe-loading platform following the wet winter of 2000/2001.

Given that we now have a good three-dimensional understanding of the slips arrangement and geology, it would be useful if further stability analysis was carried out so as to provide a better insight into the existing stability of each area and the effect any additional remedial measures (drainage and/or toe weighting) might have. Studies into sediment transport in the bay would also be useful. The results from both would then allow rock groynes to be sited so as to achieve the maximum benefit, for example in relation to the active slips at Warren Halt and Horsehead Point (Fig. 12).

The groynes would hopefully enable enough sand and shingle to become trapped on the west side of the groyne during the process of littoral drift and consequently add toe-weight material where it was most needed. East of Horsehead Point, protection of the seawall and close monitoring of the slips should suffice. West of Warren Halt and the concrete apron works, piled solutions along the seawall could be considered given that the basal slip surface lies at or just below sea-level in this area. The piles would extend through the slip material down into the Folkestone Beds. In relation to drainage, additional drainage would be beneficial at Warren Halt close to the railway and also at the west end of the Warren landward of the railway. Well pumping systems may also have to be considered so as to lower the groundwater tables in the landslipped blocks of chalk more effectively especially where such blocks are confined by impermeable slipped Gault further seawards such as at the Warren west end.

4 FUTURE CHALLENGES

4.1 *Climate change*

In recent years a wealth of coastal process data has been gathered by the Strategic Regional Coastal Monitoring Programme to which Network Rail have access on a data exchange basis through associate membership of the South East Coastal Group. The principal data streams are:

- LiDAR surveys of the coastline.
- ortho-rectified digital aerial photography.
- automated Beach Monitoring Surveys (ABMS) and beach profiling.
- Shoreline And Near-shore Data System (SANDS).
- offshore wave and tide monitoring.

These data streams represent a considerable improvement on the previous techniques used for monitoring beach sediment migration and long-term trends in beach level especially relevant to the positive effects of toe loading and negative effects of scour in front of the apron and seawalls. Additional benefits are in setting a long-term survey of coastal change and impact of particular storm events.

The consequences of climate change on the operation of the railway are manifest as changes in coastal process rates and changes in precipitation trends. Whilst routine maintenance and capital works schemes can be used to mitigate the adverse impact on man-made coastal defence structures, nothing can be done in the short to near term to those sections of the route which pass in tunnels behind the undefended coastline.

In terms of forecasting future change on the undefended sections of coast, the impact on the coastal processes has been considered for different scenarios and over three "Epochs" from the present day up to 100 years time. Considering the scenario "Baseline Case" for "undefended coastline" with "no intervention" the three epochs are as follows:

- First epoch; present to 2025: undefended cliff frontages will continue to erode at the same rate.
- Second epoch; 2025 to 2055: accelerated sea level rise, increased storminess, and accelerated toe erosion.
- Third epoch; 2055 to 2105: beaches will narrow, where cliff retreat is slower than the advance of the sea, leading to accelerated cliff erosion.

Thus, the general prognosis is for accelerated toe erosion in response to rising sea levels and increased storminess. This may increase the propensity for major cliff falls, effectively reducing the return period.

In terms of precipitation, the general conclusions are that the winters are getting wetter with a larger proportion of the precipitation falling in the heaviest downpours whilst the annual average precipitation may increase by between 0 and 15% by 2080 with the greatest change being in the South East.

The scenarios have been used, in conjunction with past rainfall trends and historical coastal retreat rates, to determine the future strategy for the line so that commitment can be made to the letting of franchises to train operators. Increase rainfall and rising sea levels will also adversely affect the stability of the Warren landslides due to increased groundwater levels in both the Chalk and Folkestone Beds. The latter are hydraulically connected to the sea, as indicated by the tidal response in certain piezometers installed in the sands, and hence rising sea levels could result in increased pore pressures acting on the main basal slip surface which directly overlie the sands (especially in areas where the cover of slipped Gault Clay above the slip surface is thick). In addition, the rise in sea level could also affect the existing drainage arrangements in the Warren, given that all headings extend landward into the landslide mass from the existing seawall lie at about 6.5 m OD, high tide being 4.5 m OD. Therefore drainage out of these of these headings could become increasingly difficult and pumped systems may have to be considered for more effective lowering of pore pressures on the slip base.

Thus, the policy if Network Rail is to maintain a "watching brief" on climate science and its influence on national bodies in climate research. Monitoring is a key element of asset management and the impacts of climate change are considered so that asset resilience can be improved when renewals become due.

5 CONCLUSIONS

The paper demonstrates the importance of looking at a wide range of sources in the investigation and understanding of landslide processes, in particular, it is important to ensure a balance between the geotechnical investigations and the geomorphological observations.

The application of modern technology has allowed the rationalisation of an unwieldy monitoring regime and the use of developing technology shows promise in enhancing the overall understanding of the ground movement mechanisms. In particular, the use of SMD as a forecasting tool and of satellite interferometry for ground deformation mapping, have been key developments in the monitoring and management of the landslide complex.

Given the current speculations on climate change, it is becoming increasingly important to have a monitoring strategy which, on the one hand builds on the established long-term history and, on the other hand optimises the best of the developing technology such that we can continue to operate a safe, punctual and dependable railway service.

ACKNOWLEDGEMENTS

The authors are indebted to Network Rail for permission to reproduce data and photographs and to work contributed by consultants Halcrow and W.S. Atkins and Sir Alan Muir-Wood, Professor J. N. Hutchinson and Professor E. N. Bromhead.

REFERENCES

Birch, G.P. & Griffiths, J.S. 1996. Engineering geomorphology. In C.S. Harris et al (eds), *Engineering Geology of the Channel Tunnel*: 64–75. London: Thomas Telford.

Birch, G.P. & Warren, C.D. 1996. The cliffs behind the Channel Tunnel workings. In C.S. Harris et al (eds), *Engineering Geology of the Channel Tunnel*. 75–87. London: Thomas Telford.

Birch, G.P. & Warren, C.D. 2006. Technical developments in the monitoring of the Folkestone Warren landslide complex. *International Association of Engineering Geologists, Conference*, Nottingham.

Hutchinson, J.N. 1969. A reconsideration of the coastal landslides at Folkestone Warren, Kent. *Geotechnique*, 19, 6–38.

Hutchinson, J.N., Bromhead, E.N. & Lupini, J.F. 1980 Additional observations on Folkestone Warren landslides. *Quarterly Journal Engineering Geology*, 13, 1–31.

Hutchinson, J.N. 1993. Deep-seated mass movements on slopes *Fourth Seminar on Deep-seated Gravitational Slope Deformation*. Florence.

Muir Wood, A.M. 1994. Geology and Geometry. Period Return to Folkestone Warren. *Proceedings 13th International Conference Soil Mechanics and Foundation Engineering*, New Delhi.

Osman, C.W. 1971. The landslips of Folkestone Warren and thickness of the Lower Chalk and Gault near Dover. *Proceedings Geological Association*, 28, 59–84.

Terzaghi, K. 1939. Concerning the Slides of Sevenoaks and Folkestone Warren. Memorandum to Southern Railway dated 21 June 1939.

Toms, A.H. 1946. Folkestone Warren landslips: research carried out in 1939 by the Southern Railway. *Proceedings Institution Civil Engineers*. Railway Paper, 19, 3–25.

Trenter, N.A. & Warren, C.D. 1996. Further Investigations at the Folkestone Warren Landslide. *Geotechnique*, 46, 589–620.

Warren, C.D., & Palmer, M.J. 2000. Observations on the nature of landslipped strata, Folkestone Warren, United Kingdom. *Landslides in research, theory and practice, Proc. 8th Intern. Symp on Landslides, 3, Cardiff*, 1551–1556.

Landslides and Climate Change – McInnes, Jakeways, Fairbank & Mathie (eds)
© 2007 Taylor & Francis Group, London, ISBN 978-0-415-44318-0

The Tessina landslide and the Civil Defence Plan

M.-G. Angeli & S. Silvano
IRPI CNR Perugia, Italy

P. Gasparetto
IQT Consulting S.r.l., Rovigo, Italy

L. Pedol
Municipality of Chies d'Alpago (BL), Italy

F. Pontoni
GEOEQUIPE Tolentino (MC), Italy

ABSTRACT: Italian law dictates that all municipalities should have a Civil Defence Plan concerning major risk, and that the Mayor is the local authority for Civil Defence. In April 1992 a dangerous mass movement, the Tessina landslide, caused a high-risk situation for two villages (Funes and Lamosano). On that occasion, adequate measures to safeguard the people exposed to the risk had to be considered, besides the need to monitor and check the evolution of the movement. The Ministry for Civil Defence assigned funds for the installation of monitoring and warning systems and during critical events various automatic alarm levels can be set (i.e. Pre-alarm, Normal Alarm, and Serious Alarm). These different alarm levels are directly integrated into the Civil Defence Plan, which includes evacuation of the population from the residential areas in the case of possible danger, by ensuring collaboration from several other boards and institutions. The Plan identifies the procedures to be activated in order to attain an optimised preparedness in case of a re-activation of the landslide. This Plan involves the national and local government offices, and volunteers, according to the outline reported in this paper. Such a plan may face any possible evolution of this phenomenon even in relation to its negative increase caused by current climate change.

1 EMERGENCY MANAGEMENT IN ITALY

In the 20th century numerous hydrological-geological disarray events caused the loss of at least 5381 lives in Italy and left over 100,000 people homeless or evacuated (Guzzetti, 2000). In the past 50 years, damage to assets and properties was calculated to be around 500–1000 million euros per year (Catenacci, 1986). Following these disastrous events, a growing interest has been recorded both at a national and international level in search for new strategies aimed at mitigating geological and hydrological risks.

In these years in Italy specific laws were promulgated for the management of hydrological-geological risks. Among them there is Law no. 183/89 (Norms for the reorganisation of soil conservation policies) and Law no. 225/92 (Establishment of the National Civil Defence Service for the forecast and prevention of great risks and rescue and emergency management).

Law no. 225/92 was introduced following the disastrous landslides which happened in Sarno and Soverato claiming the lives of several people; a specific law (DM 180/1998 converted into L. 267/1998) was promulgated, with the purpose of identifying and defining all over the national territory the areas which are particularly subject to hydrological-geological risk. These areas are divided into four classes (from R4, very high risk with the possibility of casualties or serious damage to buildings and infrastructures, to R1, low risk), on the basis of some fundamental guidelines for a common working methodology.

These disastrous landslides have given new impetus to prevention and mitigation initiatives and, in this context, a fundamental tool for the management of emergencies is the Civil Defence Plan, which involves the coordinated activities and procedures of Civil Defence to face every expected disastrous event in a defined area.

Among the general criteria of the Plan, it is defined that:

- the design phase must be separated from the planning one;

- the design is defined as the knowledge of the risk situations which can happen on a national scale and as the activities toward the reduction of the same risk situations (expectation and prevention). The design phase must take into account all the critical phenomena present in the nation and must foresee the definition of their possible solution with a specific care (attention) to time and to the availability of funds and material;
- the planning consists of the whole of operative procedures to carry out in case that a foreseen event occurs in a specific scenario; this is the basis of an emergency plan.

In more detail, the design activity forms part of the event foresight phase, that is the knowledge of the risks which are present in the specific area, and of the prevention phase consisting of the mitigation of the defined risk situations. Also, the design represents the reference point for the definition of the priority and the temporal trend of the civil protection activity in relation with the hazard of the disarray event and of the vulnerability of the area as well as of financial fund availability.

Depending on the foresight hazard severity, the national rules define three different levels for the management of the activities, as they are defined in the Plan.

At a national level (Civil Protection Dept.) the design concerns scenarios which are related to risks which, depending on their nature and area range, require the help of central state departments.

The design aims at defining first aid assistance to the people who have been struck by events, which on the basis of their strength and area range, must be faced with special means and authority. It also aims at coordinating the support of the National Service different Boards.

National emergency plans are divided into different kinds of risks and they refer to specific zones of the Italian territory which were defined with the help of the scientific structures and in any case they are enclosed in the national design.

At a regional/county level the design concerns scenarios which are related to risks which, depending on their nature and smaller area range, require the help of Regions or County. Law 225/92 does not envisage any emergency planning activities.

At a lowest level is divided into two area contexts:

At a district level the State Representative has to define the plan to face the emergencies on the whole concerning area. He also has to take care that the assistance activities are carried out on the basis of the district risk scenarios.

The municipality level envisages that the Town Hall participates in the organisation and achievement of civil protection activities, with particular care to the collection and updating of data concerning the population and of the maps. It also cooperates in defining the district foresight and prevention plan on the basis of specific guidelines. According to the law, the Mayor can endow his local community with civil protection services so as to guarantee public safety. He also has to inform the population about the risks both before and after the event and has to manage the emergency together with the other State Representatives if the event cannot be faced in an ordinary way.

Other organisations, which are responsible for specific features of the territory (for instance mountain areas), can act as a reference point as regards the emergency plan concerning their specific competence, also supporting the Mayor's activities.

It is clear that the Municipal Risk Plan is an essential aspect within the Emergency Plan; it should be activated in every Municipality, mainly in relation to the possible risk conditions identified by the Plans for the Territory Hydrological and Geological Arrangement and the scenarios envisaged for each foreseen hydrological-geological event.

The management of the Civil Defence Plan is under the responsibility of the Mayor, who represents the local authority for Civil Defence. In fact, beside the emergency plan, the Mayor should take into account all those activities which would permit the best possible arrangement of all the procedures envisaged by the plan.

2 THE TESSINA LANDSLIDE

In April 1992 a dangerous landslide, called Tessina landslide, caused situations of high risk for two villages: Funes and Lamosano (Figs. 1–3).

In that occasion, also adequate measures to safeguard the people exposed to the risk had to be considered, besides the need to monitor and check the movement's evolution (Angeli et alii, 1994; Mantovani et alii, 2000; Pasuto, Silvano 2004).

The Tessina landslide, which was first triggered in October 1960, is a complex movement with a source area affected by rotational and translational slides in the upper sector; downhill the slide turns into a mud flow through a steep channel. The landslide developed in the Tessina valley between altitudes of 1220 m and 625 m a.s.l., with a total longitudinal extension of nearly 3 km and a maximum width of about 500 m. The mud flow skimmed over the village of Funes and stretched downhill as far as the village of Lamosano.

The landslide involves the Lower Eocene Flysch Formation which consists of alternated marly-argillaceous and calcarenite layers with a thickness of about 1000 to 1200 m. This formation makes up the impermeable bedrock of the entire sliding area and crops out at the foot of Mt. Teverone, which is mainly made up of Fadalto Limestones (Upper Cretaceous).

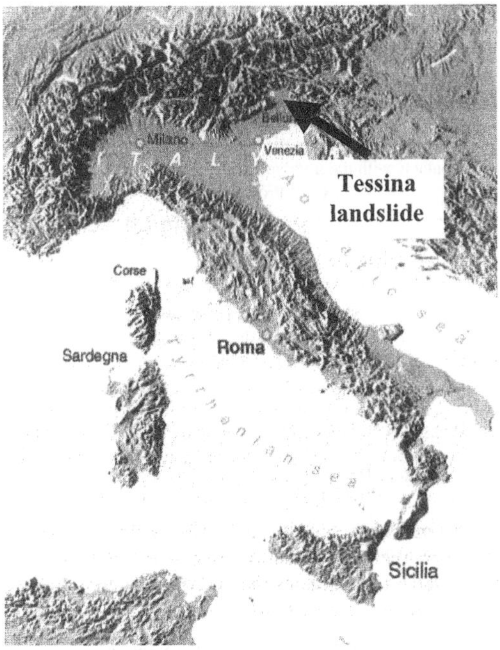

Figure 1. Location of Tessina landslide.

Figure 2. The Tessina landslide.

During the 1960s several reactivations, involving about 5 million m^3 of material, occurred causing the filling of the Tessina valley with thickness of material ranging from 30 to 50 m. These movements seriously endangered the village of Funes, which was situated on a steep ridge originally quite high above the river bed, but now nearly at the same level as the mud flow.

The collapsed sector of the April 1992 event occupies a 40,000 m^2 wide area, on the left hand-side of the Tessina stream, with an approximate volume of 1 million m^3. The movement corresponds to a rotational slide with a 20 to 30 m deep failure surface, affecting also the flysch bedrock. Initially it caused the

Figure 3. Funes village (The mudslide flows from top left to bottom right).

formation of a 15 m high scarp and a 100 m displacement downstream with consequent disarrangement of all the unstable mass and destruction of the drainage systems set up some years earlier.

The movements in this area continued with a certain intensity up to June, causing the mobilisation of another 30,000 m^2 with a total volume displaced of about 2 million m^3.

The material from this area, which is intensely fractured and dismembered, was channelised along the riverbed where, owing to its continuous remoulding and increase of water content it became more and more fluidified, thus giving rise to small earth flows converging into the main flow body.

After these events the inhabitants of Funes and Lamosano were evacuated.

The landslide evolution is still in progress and has led to a constant widening of the source area and increasing of the displaced volume which is now about 7 million m^3.

3 THE MUNICIPAL RISK PLAN

The Italian law states that all municipalities should have a Civil Protection Plan against major risks; and it also states that the Mayor is the main authority in charge of civil protection with all the duties envisaged by his position.

The plans follow the "Augustus" method, a Civil Protection Department manual, which gathers in a single operative document the guidelines to be considered.

The "Augustus" method clearly defines a simplified way of work to identify and activate the procedures to coordinate the civil protection answer. It also provides a tool for the emergency planning, which is flexible in relation with the risks to face.

In this way, the concept of available human and material resources is strongly pointed out. For this reason, responsible and supply functions must be present in the emergency plan so as to make the plan work actively, also through updating and periodic trainings.

The Civil protection Plan of the Chies d'Alpago Municipality was firstly defined in the 90's only for the Tessina landslide. Later the Plan was updated following the guidelines of the "Augustus" method.

When an emergency occurs within the municipal area, the Mayor takes control and co-ordinates rescue and assistance of affected populations implementing the necessary interventions and immediately informing the State Representative and the President of the Regional Council. He also has to take into account all those activities which will permit him to arrange in the best possible way all the procedures foreseen by the plan.

In this activity and during episodes of emergency management the Mayor is assisted by a number of work groups especially indicated in the plan.

The main Institutions involved in this Civil Defence Municipal Plan are (Pasuto, Silvano 2004): Municipal Operational Centre, which supports the Mayor's decisions and co-ordinates the emergency activities.

The involved bodies are:

- Technical-scientific co-ordination and planning
- Health, social assistance and veterinary care
- Voluntary activities
- Materials and means of transport and assistance
- Essential services and educational activities
- Census of damage to people and property
- Local operative structures
- Telecommunications
- Assistance to the population
- Relations with Surveillance Corps
- Relations with Police Corps;

Surveillance Commission, which organises surveillance, shifts and alerts the interested bodies in case of increase of flow velocity of the landslide accumulation zone.

The involved bodies are:

- Safety Commission
- Municipality of Chies d'Alpago
- Regional Department for Civil Defence
- Belluno Government Office
- Veneto Region Council
- Province of Belluno
- Mountain Municipalities' Consortium
- Fire Brigade Chief's Office
- CNR-IRPI of Padova
- Military Police (Carabinieri)
- Volunteers;

Safety Commission, which decides on the various activities to be organised and, in need, it starts evacuation operations and forewarns the Mayor, it

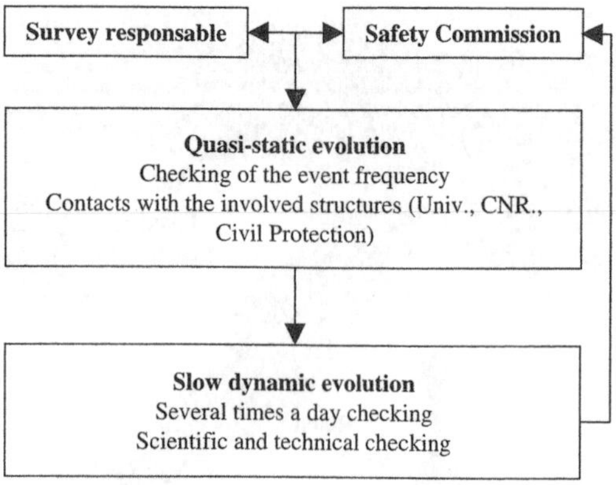

Figure 4. Procedures of the Municipal Plan of Civil Defence of Chies d'Alpago, Belluno, concerning quasi-static evolution and slow dynamic evolution risk levels.

co-ordinates the activities of voluntary associations, it prearranges an evacuation guide for the population and periodically updates the evacuation plan.

The bodies involved are:

- Municipal Councillor for Civil Defence
- Delegate of the Fire Brigade
- Technicians from the CNR-IRPI of Padova
- Delegate from the Region's Department for Civil Defence
- Delegate from the Region's Civil Engineers Board.

Owing to the particular evolution features of the Tessina landslide, the Plan consists of three different levels:

Quasi-static evolution: in which the person in charge of surveillance keeps in touch with the Safety Commission and organises activities checking up on the efficiency of the monitoring system. It also informs all the interested bodies (Fig. 4).

Slow dynamic evolution: the person in charge of surveillance increases the check up frequency and informs the Safety Commission on the evolution of the phenomenon. It also has the power to activate alarm procedures in case of movement extension (Fig. 4).

Fast dynamic evolution: all the procedures in the Municipal Risk Plan are activated (Fig. 5).

It is clear that a good Civil Protection Plan must be flexible and simple, that is, it must define:

- co-ordination and address for all the phases of the plan;
- simple procedures;
- definition of every responsibility in the intervention model;
- flexibility concerning the support function activities.

The Mayor must take into account both the Emergency Plan, which must be set up in case of necessity,

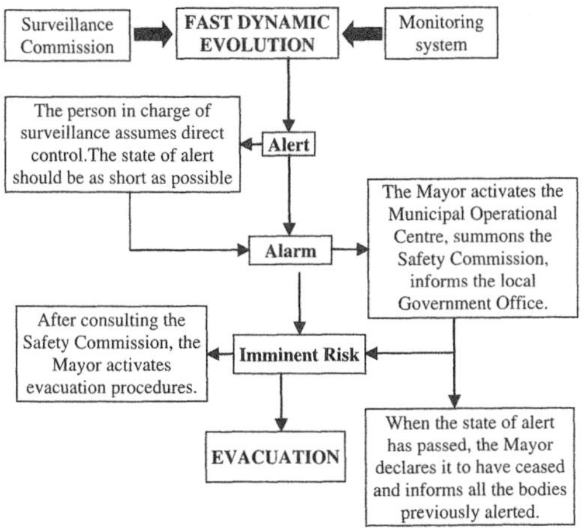

Figure 5. Procedures of the Municipal Plan of Civil Defence of Chies d'Alpago, Belluno (after Pasuto and Silvano 2004).

and all those activities which allow the Mayor to optimize the plan procedures.

Therefore, the Mayor has to foresee, define and verify all those situations which must be taken into account and must be solved before the start of the Emergency Plan.

That is, all the foreseen activities must be considered and must find a way of realization.

Besides this initial phase, the Plan has to include the updating of the population data (number, characteristic, place of residence), of the kinds of buildings, of the ready-to-use machinery, etc. which must be always available in the most accurate way.

As regards the Municipalità of Chies d'Alpago, in relation to the Tessina landslide, a list of people responsible for each of the following framework is ready:

- Technical-scientific co-ordination and planning
- Health, social assistance and veterinary care
- Voluntary activities
- Materials and means of transport and assistance
- Essential services and educational activities
- Census of damage to people and property
- Local operative structures
- Telecommunications
- Assistance to the population
- Relations with Surveillance Corps
- Relations with Police Corps
- Clearing centre

4 CONCLUSION

A Civil Defence Plan must be considered an activity which aims at organising the population assistance programme in the case of a disastrous event in advance. It is supposed to define a plan which has to indicate different kinds of events, it has to foresee their effects and as a consequence it has to include all the most effective operative procedures.

Such a flexible plan may also face any possible negative evolution caused by climate changes.

Nevertheless in 1988, Luis Theodore, Joseph P. Reynolds e Francis B. Taylornnel, on the basis of the analysis of many emergency plans, found out that most of the plans were not based on actual co-ordination and ready-to-use machinery but, during the emergencies, they created conflict among the involved authorities. What is worse is that they also found out that the population did not know what to do and, even worse, the people who were responsible for the actuation of the plans did not know about them.

They also defined a checking list to estimate the efficiency level of their own peculiar Plan. Among the most important points to be considered:

- Does the Plan cover all the emergency situations which can actually happen or, because of other reasons, does it consider only a limited range of situations?
- Have there been serious and unexpected drills or have they been mainly carried out for mass-media purpose?
- Does the plan indicate who will be responsible for official communication or, during the emergency; will any involved person feel authorized to give his own opinion?
- Is the plan based on existing means and structures or, is it based on the ones which are not yet ready?
- Does the plan clearly indicate who is in charge and who and what he is in charge of and who replaces him if he is absent?
- Has the plan been signed by all the people who are responsible for the emergency or will they feel free from any responsibility during an actual emergency?
- When was the plan last updated?

REFERENCES

Catenacci, V., 1986. Il dissesto geologico e geoambientale in Italia dal dopoguerra al 1990. *Memorie Descrittive della Carta Geologica d'Italia, Servizio Geologico Nazionale,* 47–301.

Angeli, M.-G., Gasparetto, P., Menotti, R.M., Pasuto, A., Silvano, S., 1994. A system of monitoring and warning in a complex landslide in northeastern Italy. *Landslide News*, 8: 12–15.

Gazzetta Ufficiale della Repubblica Italiana, 1999. Atto di indirizzo e coordinamento per l'individuazione dei relativi agli adempimenti di cui all'art. 1, commi 1 e 2 del decreto legge 11 giugno 1998, n. 180. Serie Generale, anno 140, n. 3, 5 gennaio 1999, 8–34.

Gazzetta Ufficiale della Repubblica Italiana, 1998. Misure urgenti per la prevenzione del rischio idrogeologico a favore delle zone colpite da disastri franosi nella Regione Campania. Serie Generale, anno 139, n. 208, settembre 1998, 53–74.
Guzzetti, F., 2000. Landslide fatalities and evolution of landslide risk in Italy. *Engineering Geology* 58: 89–107.
Mantovani F., Pasuto A., Silvano S., Zannoni A., 2000. Collecting data to define future hazard scenarios of the Tessina Landslide. *JAG* 2,1: 33–40.
Pasuto A., Silvano S., 2004. The management of geological-hydrological risk in Italy. In: Armand Colin (Ed), *Natural risk and natural development in Europe, Paris 22–25 October 2002,* 121–130.

Landslides and Climate Change – McInnes, Jakeways, Fairbank & Mathie (eds)
© 2007 Taylor & Francis Group, London, ISBN 978-0-415-44318-0

Landslides associated with the Soufrière Hills Volcano, Montserrat, West Indies

L.J. Donnelly
Halcrow Group Ltd., Deanway Technology Centre, Handforth, Cheshire, UK

ABSTRACT: The Soufrière Hills volcano is on the Caribbean island of Montserrat in the West Indies. It began to erupt in July 1995, after about 400 years of remaining dormant. Several different types of landslides have been observed to cause damage to land, structures, roads, utilities and structures. Landslides occur on both the active, expanding, extruding lava dome as well as other parts of the island beyond the current areas of activity. In extreme cases falls, slides and topples from the lava dome develop into pyroclastic flows, or surges, which move along the incised valleys which drain the volcano. Climate, and in particular rainfall, is a major factor controlling the stability of slopes and the generation of landslides. The erosion of recently deposited pyroclastic debris, during rainstorms and hurricanes, has resulted in the subsequent generation of lahars (an Indonesian term to describe a volcanic mudflow). The lahars deposits have filled river valleys, choking drainage catchment areas and burying infrastructure, including principal roads, bridges, bridges and villages. Throughout the island, beyond volcanic activity, landslides have been observed. These include both coastal and inland landslides, which are generated by a combination of earthquakes (tectonic and volcano-magmatic) and climatic events (such as rainstorm and hurricanes). One of the principal factors which influence the generation of landslides on Montserrat is rainfall some of which may be associated with tropical storms and hurricanes. Short and longer term changes in weather may have an adverse affect on the generation of landslides and the stability of both natural and engineered slopes. The objectives of this paper are to document and draw attention to the causes and principal types of landslides on Montserrat, by highlight cases studies, discuss possible mechanisms, and drawing attention to potential the long-term influences of climate change on landslides generation.

1 INTRODUCTION

Montserrat is a small island located in 'the Caribbean', in the northern part of the Lesser Antilles, in the British West Indies. It is situated within a volcanic arc formed along the junction where the Atlantic plate boundary is subducted beneath the Caribbean Plate. Montserrat is approximately 17 km long (north-south) and 10 km wide (east-west) and covers and area of about 102 km^2. The landscape is dominated by volcanic centres forming areas of high relief and steep topography, which have associated volcanic fall out materials and flow debris in different stages of erosion. The Soufrière Hills Volcano, located in the southern part of the island began to erupt in 1995 after being dormant for over 400 years (Figure 1). Further information on the geology of Montserrat and the 'recent' volcanic activity can be found in Macgregor (1938) Martin-Kaye (1959), Rae (1974, 1968), Perret (1939), Anon (1988), Aspinall *et al.* (1998), Shepherd *et al.* (1977), Druitt & Kokelaar (2002), MVO (2004a, 2004b, 1997), Sparks *et al.* (1998) Donnelly *et al.* (2006).

2 LANDSLIDE TYPES

Landslide (or landslip; the two words are synonymous) is the relatively rapid movement of a mass of rock, earth (soil) or debris (a mixture of rock and earth) down a slope, under the influence of gravity. For the purposes of this paper landslides may be classified in several ways depending upon, mode of failure, initial rupture surface, dominant form of displacement, behaviour of the rock and/or soil once movement has commenced and the subsequent deformation of the material. The type of 'materials' displaced during landsliding is recognised by the further division based on the following:

- Soil: fine engineering soil (silt, clay & organic material).
- Rock: *insitu* bedrock.
- Debris: coarse engineering soil, of sand size or greater and/or an admixture of gravel or boulders, (rock and soil).

The types of landslides which have been observed on Montserrat may be described as; volcanic

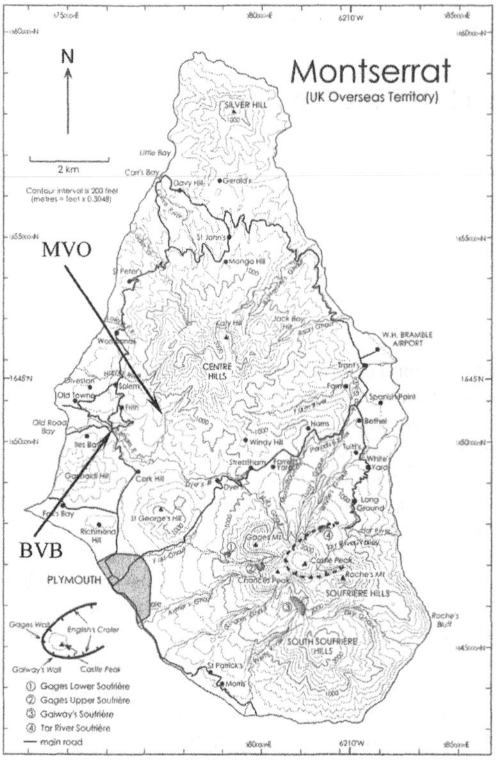

Figure 1. Topographic map of Montserrat showing the location of the Montserrat Volcano Observatory (MVO) and the Belham Valley Bridge (BVB) (source: Montserrat Volcano Observatory & British Geological Survey, after Donnelly *et al.* 2006).

landslides, falls, topples, slides and flows. Other types of ground movements associated with landslides include creep, cambering and subsidence, and each of these are briefly reviewed below.

2.1 *Volcanic landslides*

Volcanic landslides occur when a sizeable part of a volcanic edifice becomes unstable and slides catastrophically. These are rare but this type of landslide is believed to have occurred 4,000 years ago which generated English's Crater.

2.2 *Falls*

Falls involve the detachment of a soil or rock mass, from a slope, with little or no shearing; the material descends largely through the air by falling, bouncing or rolling. These occur on the expanding active dome, along road-side cuttings, the coastline and in the Silver Hills (Figure 2) (Carn *et al.* 2003, MVO 1974).

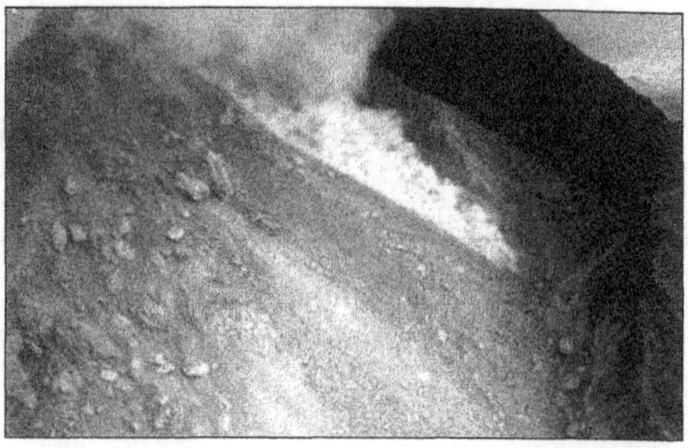

Figure 2. Rockfalls sliding down the andesite lava dome, filling and almost breaching English Crater (after Donnelly 1997).

2.3 *Topples*

Topples involve the movements out of the slope, of a mass of rock or soil, about a pivot point at the base of the affected mass. These also may be generated by the forward rotation, out of the slope, of a mass of rock or soil, about a pivot point at the base of the affected mass. Toppling failures have been observed on the extrusive lava dome, whereby 'spines' of andesite have been seen to fail by this mechanism, before subsequently sliding or 'bouncing' along the flanks of the dome.

2.4 *Slides*

Slides involve the down-slope movement (forward rotation) of a soil or rock mass along a failure ('slip' or 'shear' are synonymous) surfaces; the surfaces may be curved (rotational slides) or planar (translational slides). These have been observed on the expanding lava dome, along the coastline and inland where thick lateritic soils and weathered profiles have developed in bedrock. They also developed within fill placed on steep slopes during building and construction. Debris slides are the main type of landslide on Montserrat and are common on the flanks of ghauts (a local term for steeply incised drainage channels). Palaeo (prehistoric) relic landslide scarps and slipped masses exist in areas of high relief such as the Silver Hills and Centre Hills. Slickensided slip scarps and clay infilled smears were also exposed on the flanks of the Lower Belham valley during geotechnical ground investigations; these being indicative of the sliding of the valley slopes in the geological past (Figure 3).

2.5 *Flows*

Flows involve movement similar to that of a viscous fluid in which inter-granular movements predominate over failure surface movements. These are restricted to the areas where pyroclastic flows, or surges (also known as block and ash flows) and lahars

Figure 3. Debris slides generated by the rainfall induced remobilisation of 'recent' pyroclastic flow surge deposits, Gages Estate.

Figure 4. Failure of part of the extrusive, andesite lava dome caused by over steepening followed by gravitational collapse, exacerbated by rainfall, resulting in the generation of a series of pyroclastic flows, which first reached the Atlantic Ocean, in May 1996, producing a delta consisting of pyroclastic debris. Note how the surge is confined by the topography, within the Tar River valley (after Donnelly 1996a).

are generated. Flows are influenced by climate (and in particular rainfall) and the topography due to the presence of drainage channels which radiate from the upper slopes of the Soufrière Hills volcano (Figure 4).

2.6 *Creep*

Creep is the slow, continuous, down slope movement of soil and debris, characterised by the bending of trees on the flanks of the volcano and tilting of walls or other man-made structures. This occurs on the lower and middle slopes of active and dormant volcanic centres where ash has accumulated, or on the flanks of ghauts, which contain thick soil profiles. Soil creep is usually observed as a secondary process in areas where landslides are active and is not an extensive, destructive process.

2.7 *Cambering*

Cambering is a process that acts on rocks found in valleys where the lower and middle slopes consist of volcanic clays (or semi-consolidated ash) and the valley crest is capped by a harder, brittle rock such as andesite. Over thousands of years the clays gradually squeeze outwards from the valley sides towards the valley bottom (known as valley bulging). This lessens the support for the stronger rock above, which then sags towards the valley. As the harder rock cannot bend it breaks to generate fissures (known as 'gulls') that are aligned parallel to the valley sides. These are not common on Montserrat but examples do occur along ghauts, in the lower Belham valley and along the coastline. This process is not extensive and there have been no reports of damage to structures, but is may influence rock falls and the ravelling of cobbles and boulders from slopes.

2.8 *Subsidence*

Subsidence induced slope failures have been observed on the flanks of the volcano. Rock falls developed on the steep sides of the subsidence troughs. These depressions have now been filled and covered by 'recent' deposits of ash and pyroclastic flow deposits. The origin and evolution of these subsidence depressions are not yet fully understood, they may be related to the depletion of magma, collapse and void migration.

3 CAUSES

The main causes of landslides on Montserrat are the instability of the lava dome caused by oversteepening, earthquakes, rainfall, weathering, geomorphological processes (erosion) and the activities of man such as road building, construction, mineral extraction and land use changes.

3.1 *Instability of the lava dome*

Rock falls, rock avalanches, debris slides and debris flows on the lava dome have been generated by the extrusion, growth and over-steepening of the lava dome (facilitated by the escape of rapidly expanding volcanic gases), followed by gravitational collapse and the generation of fast moving, hot, pyroclastic surges, (accompanied by ash clouds and explosions). These rock falls, or 'incandescent rock avalanches' may be induced by, or exacerbated by rainstorms and increased rates of magma intrusion.

3.2 *Seismcity & earthquakes*

Seismic activity caused by regional tectonic events, or volcano-seismicity associated with the movement of magma, may cause landslides and rock falls on the lava dome or on recently constructed, poorly consolidated cuttings and embankments. Beyond the area of current volcanic activity, artificially induced shocks and ground vibrations however, (caused by blasting or civil engineering construction) may potentially result in vertical and horizontal stresses within slopes that cause landslides to develop.

3.3 *Climate (rainfall and changes in groundwater flows)*

Prolonged heavy rainstorms and hurricanes are one of the principal causes of landslides on natural and engineered slopes. As happened when Hurricane Hugo hit Montserrat in 1989. Long-term climatic change, may, ultimately, have a direct impact on the frequency and magnitude of landslide events. Projected increases in rainfall for example, could increase the number of new landslides that develop annually, and may cause the reactivation of recently failed slopes. Slopes, which are wetted following a period of prolonged drying, may also result in failure caused by the relaxation of inter-granular capillary forces in interstitial moisture. Any change therefore, in the surface and groundwater regime may induce landslides. Meteorological events, such as prolonged rainfall, tropical storms, floods, hurricanes, drought and changes to the water content of slopes may ultimately influence landslides.

3.4 *Weathering*

Weathering of the bedrock (physical, chemical, & biological) and the formation of thick lateritic soils (the regolith), changes the chemistry, geometry, morphology, geotechnical properties and engineering behaviour of slopes (affecting in particular cohesion, friction, effective stresses across discontinuities and load). This weakens the internal discontinuities and soil fabric and contributes to soil forming processes. Weathering and changes to the strength of the insitu rock mass will promote landslides. This will be exacerbated by rainfall and the drainage of slopes.

3.5 *Geomorphological processes*

Geomorphological processes by streams, sea, and winds change the geometry, height, aspect and angle of slopes. The undercutting of slopes for example by coastal and river erosion, removes the support from the toe of the slopes and therefore reduces the resistance to movement, this is particularly evident along Iles Bay. Failure may also be influenced by expansion, swelling, fissuring and softening of weak volcanic clays, ash and mudstones. The deposition of sediments changes the loading of slopes, water content, water pressure; and all of these processes influence landslides.

3.6 *Quarrying & surface mineral abstraction*

Rock aggregates, sand and gravel resources are necessary for the rehabilitation of Montserrat and in particular for the construction of homes, public buildings, other structures, for the maintenance of the island road network and the construction of new roads. Aggregate is currently obtained from the scavenging of andesite boulder and borrow pits have been excavated Geralds Hill, Silver Hills and Little Bay areas. Sand and gravel on the other hand is extracted from lahar sediments in the lower Belham valley. Both sand and gravel extraction and the quarrying of intrusive rocks have led to the generation of rock falls. All mining operations need to consider slope stability including the failure of mine waste tips and the initiation of the first-time failure of excavated quarry slopes. Mineral abstractions change the strength, water concentration, water pressure, removes physical support; all of which can influence slope stability.

3.7 *Building and construction*

Human activities; such as civil engineering, building houses and roads, tipping, digging, water abstraction or waste disposal, may change the course of surface water courses and influence groundwater flows. Tipping and waste disposal may generate leachates and promote increased run-off. Loading of the top of slopes by tipping, or the construction of buildings, increases the load of the top of the slope, therefore contributing to driving forces which promotes instability. The construction of roads can undercut slopes and cause the reactivation of historical landslides where the road cut crosses the toe of older slips (Figure 5). The leakage of water retaining structures, such as reservoirs, ponds, swimming pools, pipelines and sewers may also initiate failure. The construction of artificial rock and soil slopes, if not properly designed, may also result in their failure.

3.8 *Land use changes*

Changes in land use, including the removal of vegetation (destroyed by volcanic gas, ash deposition,

Figure 5. A localised earth-fall, rotational slip, on a granular soil slope under construction, at Gerald's Airport.

deforestation, burning or clearance by man) and the subsequent loss of root strength, can result in less water being extracted from slopes. The reduced interception of groundwater will increase surface run-off, contributing to erosion and landslide generation.

4 CONSEQUENCES

Landslides have caused some damage to land and structures on Montserrat, including homes, industrial premises, utilities and communication cables. To date this has not been extensive and widespread (apart from where pyroclastic flows and lahars have been generated) but does occur in some places. There does however, exist the potential for landslides to have a more severe affect on roads, land, property, infrastructure and people. As the island becomes more redeveloped (following its destruction by volcanic activity) and if the density of population and structures increase over the next generation, then the frequency of landslide cases may also increase particularly if steep slopes are to be developed, or if more cases are reported.

The immediate costs of a significant size destructive landslide could include the mobilisation of rescue and relief work, evacuation, provision of temporary replacement buildings. Associated costs would include the losses of available land, industrial infrastructure and operational businesses. The costs of investigation to determine stabilisation and repairs, and the implementation of these measures, would be significant. Preventative costs may involve research into the mechanisms of landslides, geomorphological studies, landslide hazard vulnerability mapping, formulation of planning policies related to development on providing technical advise on land prone to landslide hazards, coastal and inland protection and stabilisation schemes, design and construction of preventative measures including drainage, reprofiling of slopes and costs for future monitoring. A greater understanding of the long term climate change and the affects on landslides is also desirable and is probably currently under researched.

5 CLIMATE AND CLIMATE CHANGE

Landslides are intimately related to weather, climatic hazards and climatic change; either as a cause, as its potential consequence, or both. Heavy rainfall, flooding, storms, winds and waves, are related to extreme weather conditions, and are most likely to take place in Montserrat during the hurricane season (Smith Warner International 2003). Prolonged exposure to these climatic hazards could increase weathering and erosion rates, coastal erosion, increase surface and groundwater flows, alter slope gradients and reduce the strength of rock and soil on slopes and exposures. The long term results could be the reactivation of older, degraded landslides and the first time initiation of slope failure. Other types of geohazards could potentially be initiated by a landslide event and these include for example; tsunami (especially if the landslides enter the sea), induced seismcity (tremors), and secondary flooding caused by blockage of stream valleys. Long term climatic changes which increase the frequency and magnitude of rainfall in Montserrat, will, ultimately, increase lahar generation following the remobilisation of thick, unconsolidated, recently erupted volcanic debris on the upper and middle slopes of the volcano.

6 CASE HISORTY 1: THE MONTSERRAT VOLCANO OBSERVATORY (MVO) LANDSLIDE

The Montserrat Volcano Observatory (MVO) was established in 1995, soon after the occurrence of phreatic eruptions in English Crater. The observatory has been located in several specially adopted buildings, but since 2003, the MVO has occupied a purpose built observatory, situated approximately 6 km northwest of the volcano, at Flemings.

The volcano observatory is located at a moderate altitude on the middle slopes, of the southern flank of the Centre Hills. This area has not been directly affected to-date by the 'recent' volcanic activity, apart from occasional ash falls. Semi-humid scrub vegetation develops on slopes and in the low-lying, rain shadow of the northern part of the Centre Hills. Thick tropical vegetation dominates the well-watered parts and summit of the Centre Hills. Much of the geology this are is concealed by a mantle of vegetation and thick tropical soils. Tropical weathered laterite soils and smectite occur on Montserrat. These develop on flat to moderately steep slopes and intensely weathered bedrock. The soils influence surface and groundwater drainage, spring flows and seepages. These soils potentially introduce severe design considerations for new buildings and influence the stability of both natural and engineered slopes.

The MVO has been built on a plateau, immediately beyond the crest of a deeply incised ghaut which drains the Centre Hills. This valley is oriented approximately northeast to southwest and it eventually drains into Spring Ghaut, one of the main tributaries draining into the Belham valley.

6.1 *Overview of the MVO landslide*

By March 2005 a distinct landslide had developed at the MVO (Donnelly 2005b). The backscarp was adjacent to the main observatory structure but just a couple metres from the helicopter landing deck (also known as the 'helipad' or 'helideck').

The landslide is a debris slide, with component of rotational slip incorporated in the upper part of the slipped mass. It has developed on the flank of a steeply incised valley which has become over-steepened due to erosion. During the construction of the observatory's helicopter landing pad, waste material (consisting of granular loose rock and soil) was reported (anecdotally) to have been tipped over the crest of the valley side (Figure 6).

Figure 6. The MVO landslide in Spring 2005 (source, Montserrat Volcano Observatory & British Geological Survey in Donnelly 2005b).

The date when the MVO landslide first failed is not known. Waste material and debris was apparently tipped within the ghaut during construction of the MVO in 2001–2002, when the vegetation cover was also removed (for abseil training by local authorities). Anecdotal information, provided by MVO geologists and others, suggests that the landslide might have first occurred in August to October 2004 (Donnelly 2005b).

6.2 *Description and classification of the MVO landslide*

The main backscarp of the landslide is high-angled (70° to vertical), distinct, approximately 1.2 m high, 14 m long and curvilinear. This was complimented by smaller sub-scarps, 0.01 m to 0.02 m high and one forward tilted block, dipping at 20°. Some of the sub-scarps were dilated by 0.1 m forming small ground fissures and bridged by soil cover and vegetation. The MVO helicopter landing pad is situated immediately beyond the crest of the landslide, on an excavated platform. The distance to the edge of the helipad to the backscarp of the landslide varies from between 1.8 m to 3.2 m. The central part of the platform, beyond the landslide crest, and the helipad contains a subsidence depression (settlement), 80 mm deep approximately 3 to 4 m wide. Settlement of the material beneath the concrete pad created a 20 mm void at the edge of the pad (Figure 7 & Figure 8).

Figure 7. The MVO landslide showing the position of the backscarp in relation to the helicopter landing pad. Note the subsidence beyond the backscarp of the landslide crest and the unstable, forwarded tilted block in the centre of the landslide.

The run-out of the slipped mass was inaccessible without roped access and could not be measured, but was estimated by visual inspection to be about 75 to 100 m long. The gradient of the surface of the slipped mass is at least about 40°. It is distinct and the main body of the landslide is well defined. The failure occurred on densely vegetated valley slopes and

Figure 8. The location of the backscarp of the MVO landslide in relation to the helicopter landing pad and the main structure of the volcano observatory.

Figure 9. The main body of the MVO landslide consist of a slipped mass containing poorly sorted, loose, granular debris with cobbles, boulders and tipped waste material. Note the ravelling and small scale toppling failure on the backscarp and the dilated, sub-scarps, fissure and forwarded tilted block on the crest of the MVO landslide.

has destroyed completely the vegetation cover within the zone of failure, exposing weathered bedrock and regolith. The upper slopes, which provide the source rock for the debris slide, consist of a boulder-strewn slipped mass, with large andesite boulders up to 1 m wide. Scattered between boulders was a high proportion of domestic waste materials including wood, plastic, rubber and other types of debris which could not be identified. These appear to have been tipped over the slope and the host materials do not appear to have been compacted to an engineered specification (Figure 9).

The middle and lower slopes consist of a slipped mass consisting of loose, highly porous sand and gravel sized with abundant cobbles and boulders. Some boulders had toppled down the slipped mass and accumulated at the toe of the landslide, filling the stream channel. Parts of the slipped mass were draped with vegetation, contained surface scars, erosion chutes and rock piles which have accumulated on ledges or on the antislope (upslope) side of large boulders and other obstructions.

Poorly defined individual flow lobes occur at the toe of the slope, but these are likely to be temporary features, and may become destroyed by weathering, erosion and secondary mass wasting (slumping, sliding and slipping). Secondary, small scale displacements were common on the backscarp and slipped mass. This includes the frequent and semi-continuous falling and flaking of material (ravelling), including grains, cobbles and boulders. Finer debris, such as sand sized fragments, show evidence of down slope movement by surface water runoff (wash erosion). Furthermore, debris disintegration and flexural debris toppling occurs on the backscarp. These ground movements represent the progressively slow, but continual gravitational deformation and sliding of landslide due to the removal of the lateral support.

It is likely that the landslide will continue to fail and may be reactivated by rainstorms, or moderate to strong earthquakes. Secondary ground movement on the backscarp are likely to continue, interrupted by more dramatic and larger scale failures, probably during rainstorms, by surface run-off of water from the helipad onto the landslide, or if further tipping of waste takes place.

7 CASE HISTORY 2: THE BELHAM VALLEY LAHARS

The term lahar is of Indonesian origin used to describe flows involving mixtures of water and debris, or sediment-laden flow, in and around volcanoes (Verstappen 1992, Hungr *et al.* 1987 Macedonio & Pareschi 1992, Caruso & Pareschi 1993). This is a generic term, referring to an event, rather than a deposit. In the Belham valley, numerous lahars have been generated by the removal (erosion) of pyroclastic debris during rainstorms. These events are also referred to as mudflows (or sometimes hyperconcentrated flows) (Barclay *et al.* 2004).

The generation of lahars in the Belham valley appears to be principally controlled by sediment supply (from volcanic eruptions, air fall deposits and pyroclastic flows) and water availability (from rainfall), although gradient, channel confinement and slope morphology are also significant (Hodgson & Manville 1999, Lavigne & Thouret 2002, Lavigne & Suwa 2004).

A combination of increased sediment supply, caused by the addition of new eruptive materials, increased speed of run-off due to the removal of vegetation have contributed to the generation of lahars. From January 1999 to July 2004 there have been at least 40 days when tropical storms and hurricanes have

Figure 10. (Left) Lahars breaching the Belham valley bridge (source, Montserrat Volcano Observatory & British Geological Survey).

Figure 11. The former position of the lower Belham valley bridge (shown dotted) now covered by approximately 4 metres of lahar deposits.

remobilised the huge volumes of volcanic debris and lahars have been observed in the Belham valley, or detected instrumentally using the short-period seismic record (Barclay *et al.* 2004). Additional lahars were observed to have taken place between 2004 and 2006 (Donnelly 2005a). In general, the lahars coincide with increased rainfall usually between the months of May to November, or during tropical storms and hurricanes (Figure 10)

Of the lahars, in the period 1999 to 2004, the vast majority, about 63%, were associated with rainfall of greater than 20 mm in a 3 hours period and 24% occurred after greater than 10 mm in 3 hours (Barclay *et al.* 2003, Bouquet *et al.* 2003). Lahar events in 2005 and 2006 were also observed to have been triggered by rainfall. The volcanic and volcaniclastic deposits around the periphery of English Crater were remobilised and redeposited in the middle and lower flanks of the volcano. This caused the river valleys to be infilled by substantial quantities of lahar (mudflow) sediments.

Figure 12. The safety barrier of the Belham valley bridge, exposed in a trial pit, and buried 3.8 metres below current ground level. The typical lahar deposits can be seen to consist of granular, saturated, highly variable, well-rounded, spherical, cobbles and boulders (mainly andesite and dacite) and sub-rounded, elongate cobbles and boulders with preferential alignment and cross laminations.

In many places the former flanks of the Belham valley became breached. Numerous houses, the island's golf course, a jetty and a principal bridge were buried beneath huge volumes of sediments (Figure 11).

A geotechnical and geohazards ground investigation was conducted in April 2005, to identify a crossing point for a new bridge across the Belham valley. This investigation confirmed the bridge in the lower Belham valley was buried under 3.8 m of lahar sediments. It was further estimated that this section of the valley has been infilled by an estimated 10 to 15 m of lahar deposits (Donnelly 2005a, Donnelly *et al.* 2006) (Figure 12).

The lahar sediments within the valley consisted primarily of highly permeable, loose, saturated sand, cobble and boulders with occasional clay and silt zones. These were found to have with low bearing capacity, making these unfavourable for the citing of bridge foundations.

The underlying bedrock consists of moderately fresh, moderately strong, pyroclastic rocks (tuffs), which break-down on excavation to weathered sand with cobbles and boulders. These present more reasonable founding conditions. However, the high probability of encountering boulders (and volcanogenic groundwater) in the lahar deposits presented engineering challenges for foundation design and construction (Figure 13).

8 CONCLUSION

Montserrat is susceptible to naturally occurring landslides which can be primarily attributed to volcanic

Figure 13. Huge volumes of coarse lahar sediments filling the lower Belham valley, engulfing houses.

activity, earthquakes, steep slopes, high relief, thick soils and groundwater availability. Of these, groundwater, from rainfall is the principal cause of the majority of landslides.

The most dramatic rock falls and landslides are associated with the eruption of Soufrière Hills volcano. In extreme cases these develop into highly destructive, fast moving, hot pyroclastic flows, or surges. In addition to volcanic and tectonic processes, climate and meteorological conditions have a significant influence on the generation of landslides. Rainfall is one of the principal triggering events affecting the frequency and magnitude of both natural and man-induced landslides, on and away from the areas of current volcanic activity. The potential for landslides may be expected to increase during the hurricane season from June to November each year, or at other times of the year when sustained, prolonged, rainfall occurs with high intensity. Increased rainfall caused by 'long-term' climatic change and increased development of the islands infrastructure on moderate to steep slopes may, ultimately, potentially increase the frequency of landslides on Montserrat and this may need to be considered to ensure long term sustainability of planned infrastructure. Intense rainfall, sometimes associated with tropical weather systems, storms and hurricanes, can cause widespread debris slides as happened when Hurricane Hugo hit Montserrat in 1989. Debris slides primarily occur in natural drainage channels or on moderate to steep slopes, covered with a mantle of thick tropical soils. The Silver Hills and Centre Hills are particularly susceptible to debris slides, although a mitigating factor is the presence of dense vegetation. The vegetation may have a positive influence of the stability of slopes. Man-induced landslides may be triggered by the digging of inappropriately designed road cuts, by the cutting into slopes for building houses, during civil-engineering construction and by the uncontrolled, unregulated disposal or tipping of waste onto slopes, as demonstrated at the Montserrat Volcano Observatory. Here a debris slide occurred on a steep slope adjacent to the helicopter landing deck and this may be partially attributed to the small scale tipping of domestic and building waste onto slopes already vulnerable to landslides. Man induced landslides, such as those document in this paper, are likely to influence by both short and long term changes in climate. It is expected that increased rainfall, caused by climate change, may exacerbate both natural and man-induced landslide in Montserrat, on the active volcano and throughout the remainder of the island.

ACKNOWLEDGEMENTS

The author would like to acknowledge the Montserrat Volcano Observatory (Dr Vicky Hards); Halcrow Group Limited (Mr Howard Siddle); the British Geological Survey (Prof Martin Culshaw, Lee Jones, Bill McCourt); Mabey & Johnson Ltd (Carole Hewitt), the Government of Montserrat, Public Works Department (Conrad Dilkes).

REFERENCES

ANON, 1938. West Indian landslide kills hundreds. Engineering News-Record 121:685.

ANON, 1988. Geophysical Research Letters 1998, Memoir 21 of the Geological Society of London 2005, Journal of Petrology 2003.

ASPINALL, W. P., LYNCH, L. L., ROBERTSON, R. E. A., ROWLEY, K. C., SPARKS, R. S. J., VOIGHT, B., YOUNG, S. R. 1998. The Soufrière Hills Eruption, Montserrat, British West Indies: Introduction to Special Section, Part 2, Geophys. Res. Lett., 25, No 19., 3651.

BARCLAY, J., ALEXANDER, J., MATTHEWS, A. H., JOHNSTONE, J., CHIVERS, C., JOLLY, A., NORTON, G. 2004. Rain-fall induced activity in the Belham valley. Report for the Scientific Advisory Committee to the Government of Montserrat.

BOUQUET, T., BARCLAY, J., ALEXANDER, J. 2003. Recent changes in the lower Belham valley, Montserrat: investigating rainfall-induced lahars. British Sedimentological Research Group Annual General Meeting, Leeds, 20–22 December.

CARN, S. A., WATTS, R. B., THOMPSON, G., NORTON, G. 2003. Anatomy of a lava dome collapse: the 20 March 2003 event at Soufrière Hills Volcano, Montserrat, J. Volcano. Geotherm. Res., 120, 1–20.

CARUSO, R., PARESCHI, M. T. 1993. Estimation of lahar and lahar-run out flow hydrograph on natural beds. Environmental Geology, 22, 141–152.

DAVIES, J., PEART, R. J. 2003. A review of the groundwater resources of Central and Northern Montserrat. British Geological Survey, CR/03/257C.

DONNELLY, L. J. 1996a. The 12 May 1996 Tar River Pyroclastic Flows, Soufrière Hills Volcano, Montserrat, West Indies: Helicopter Observations. British Geological Survey, WN/96/26.

DONNELLY, L. J. 1996b. A Rapid Mineral Exploration Reconnaissance Survey for a Rock Aggregate Resource, Montserrat, West Indies: A Pre-feasibility Study in Conjunction with the British Task Force. British Geological Survey, WN/96/27.

DONNELLY, L. J. 1997. Rock falls and Pyroclastic Flows on Andesitic Domes: Observations on the Soufrière Hills Volcano, Montserrat, West Indies. British Geological Survey, WN/97/8R.

DONNELLY, L. J. 2005a. Belham valley bridge, Montserrat, West Indies, Geotechnical and Geohazards Investigation. Report of investigation undertaken by Halcrow Group Ltd.

DONNELLY, L. J. 2005b. Montserrat Volcano Observatory Landslide. Report of investigation undertaken by Halcrow Group Ltd.

DONNELLY, L. J., JONES, L., PALMER, M. & DILKES, C. 2006. Engineering geological and geotechnical aspects of the Soufrière Hills volcanic eruption, Montserrat. IAEG 2006, international conference, The Engineering Geology for Tomorrows Cities, Nottingham, 6–10 September 2006.

DRUITT, T. H., KOKELAAR, B. P. (eds). 2002. The Eruption of Soufrière Volcano, Montserrat, from 1995 to 1999. Geological Society of London, Memoirs, 21, 645.

HALCROW GROUP LTD 2003. Geotechnical Site Investigation, Gerald's Airport, Montserrat, West Indies. The Government of Montserrat in conjunction with UK Government and Department for International Development and the European Development Fund.

HODGSON, K., MANVILLE, V. R. 1999. Sedimentology and flow behaviour of a rain-triggered lahar, Mangatoetoenui stream, Ruapehu volcano, New Zealand. Geol Soc. Am. Bull., 111, 743–754.

HUNGR, O., VANDINE, D. F., LISTER, D. R. 1987. Debris flow defences in British Columbia. Geol Soc. Amer. Reviews in Engineering Geology, 7, 201–222.

LAVIGNE, F., SUWA, H. 2004. Contrasts between debris flows, hyperconcentrated flows and stream flows at a channel on Mount Semeru, East Java, Indonesia. Geomorphology, 62, 41–58.

LAVIGNE, F., THOURET, J-C. 2002. Sediment transportation and deposition by rain-triggered lahars at Merapi Volcano, Central java, Indonesia. Geomorphology 49, 45–69.

MACEDONIO, G., PARESCHI, M. T. 1992. Numerical simulation of some lahars from Mount St Helens. Journal of Volcanological and Geothermal Research, 54, 65–80.

MACGREGOR, A. G. 1938. The Royal Society expedition to Montserrat, B.W.I.: The volcanic history and petrology of Montserrat, with observations on Mt. Pele in Martinique. Phil. Trans. R. Soc. London., B, 229, 1–90.

MARTIN-KAYE, P. H. A. 1959. Reports on the geology of the Leeward and British Virgin Islands. Voice Publishing CO. Ltd., Castries, St. Lucia.

MVO 2004a. Assessment of the hazards and risks associated with the Soufrière Hills Volcano, Montserrat. Second report of the Scientific Advisory Committee on Montserrat Volcanic Activity, 1–4 March 2004: Part 1, Main Report.

MVO 2004b. Assessment of the hazards and risks associated with the Soufrière Hills Volcano, Montserrat. Second report of the Scientific Advisory Committee on Montserrat Volcanic Activity, 1–4 March 2004: Part 1, Technical Report.

MVO 1997. Deformation of the Galway's Wall and related volcanic activity, Special Report 2, November 1996 to March 1997.

PERRET, F. A. 1939. The volcano-seismic crisis at Montserrat, 1933–1937. Carnegie Inst., Washington, 512.

REA, W. J. 1968 Geology of the southern part of Montserrat, West Indies. Proc Geol. Soc. London, 1649, 115–116.

REA, W. J. 1974. The volcanic geology and petrology of Montserrat, West Indies. Journ. Geol. Soc. London, 130, 341–366.

SHEPARD, J. B., TANNER, J. G., MCQUEEN, C. M., & LYNCH, L. L. 1997. Final report: Seismic Hazards in Latin America and the Caribbean-seismic hazards maps of the Caribbean.

SMITH WARNER INTERNATIONAL 2003. Integrated Vulnerability Assessment of Montserrat, submitted to the Government of Montserrat, June 2003.

SPARKS, R. S. J., YOUNG, S. R., BARCLAY, J. ET AL. 1998. Magma production and growth of the lava dome of the Soufrière Hills Volcano, Montserrat, West Indies: November 1995 to December 1997. Geophysical Research Letters, 25, 3421–3424.

VERSTAPPEN, H.Th. 1992. Volcanic hazards in Colombia and Indonesia: lahars and related phenomena. In: Geohazards, McCall, G. J. H. et al., (eds). Chapman Hall, 33–42.

Session 5
Isle of Wight case studies

An insight into instability investigations, management and remedial measures, on the Isle of Wight. In light of the predicted climate change scenarios and increasing instability, the provision of new information in high-risk areas such as the Isle of Wight can assist planners and managers in preparing for the future.

Castlehave, Isle of Wight, UK
Courtesy Wight Light Gallery, Ventnor, Isle of Wight, UK.

Landslides and Climate Change – McInnes, Jakeways, Fairbank & Mathie (eds)
© 2007 Taylor & Francis Group, London, ISBN 978-0-415-44318-0

Ventnor Undercliff: Development of landslide scenarios and quantitative risk assessment

M. Lee
Geohazard Risk Consultant, York, UK

R. Moore
Halcrow Group Ltd., Birmingham, UK

ABSTRACT: Quantitative risk assessment has been undertaken to demonstrate the economic viability of different landslide management strategies for the Ventnor Undercliff, Isle of Wight, UK. An expert panel was convened to make judgements about the likelihood of different landslide scenarios and their impacts. The QRA was directed towards estimating the risk both under current conditions and in the future, taking account of the predicted changes in landslide behaviour due to climate change.

1 INTRODUCTION

The Isle of Wight Undercliff is an extensive coastal landslide complex with a permanent population of around 6500 located in the small towns of Ventnor, St Lawrence and Niton (for the most recent studies see Moore et al. 2007, Carey et al. 2007, Palmer et al. 2007). Contemporary ground movements within the town have been slight (e.g. Marsden, 2007). However, because movement occurs in an urban area, the cumulative damage to roads, buildings and services has been substantial. Over the last 100 years about 50 houses and hotels have had to be demolished because of ground movement. The average annual loss in the Ventnor area has been estimated by the local authority as exceeding £3 M.

The Isle of Wight Council has taken a major role in addressing the landslide problems, with the assistance of the Undercliff Landslide Management Committee (e.g. McInnes & Jakeways, 2002). An important element of the landslide management strategy has been a programme of coast protection improvements, using the geomorphological understanding of landslide behaviour to develop the business case for grant-aid from the Government (e.g. Clark et al. 1994). It is estimated that over £20 million will have been spent on improving the coastal defences along the Undercliff by 2006.

However, climate change and relative sea-level rise over the next century is expected to result in an increase in the landslide risk. This concern has led to a re-evaluation of the potential benefits of major coastal landslide stabilisation works in the Ventnor area. Quantitative risk assessment (QRA) was undertaken to determine whether this is likely to be a cost effective approach to managing the landslide risks.

Risk is generally expressed as the product of the likelihood of a hazard and its consequences (Royal Society 1992). Thus, for the Undercliff landslides the risk model was:

$$\text{Risk} = \text{Prob. (Landslide event)} \times \text{Consequences}$$

The risk assessment was directed towards the quantification of consequences (losses) associated with a number of different landslide reactivation scenarios. Estimates were presented of:

1. The current annual risk i.e. an estimate of the landslide risk under current conditions expressed as an annual value.
2. The present value of the annual risk spread over a particular period (e.g. 100 years). This was used to compare the benefits associated with different management strategies to a "do nothing" strategy.

This paper describes the general approach taken in developing the QRA, focussing on direct property losses (although infrastructure losses, emergency service costs, tourism losses and traffic disruption also formed part of the overall assessment). An expert panel was convened to make judgements about the likelihood of different landslide scenarios and their impacts. The QRA was directed towards estimating the risk both under current conditions and in the future, taking account of the predicted changes in landslide behaviour due to climate change. Further background

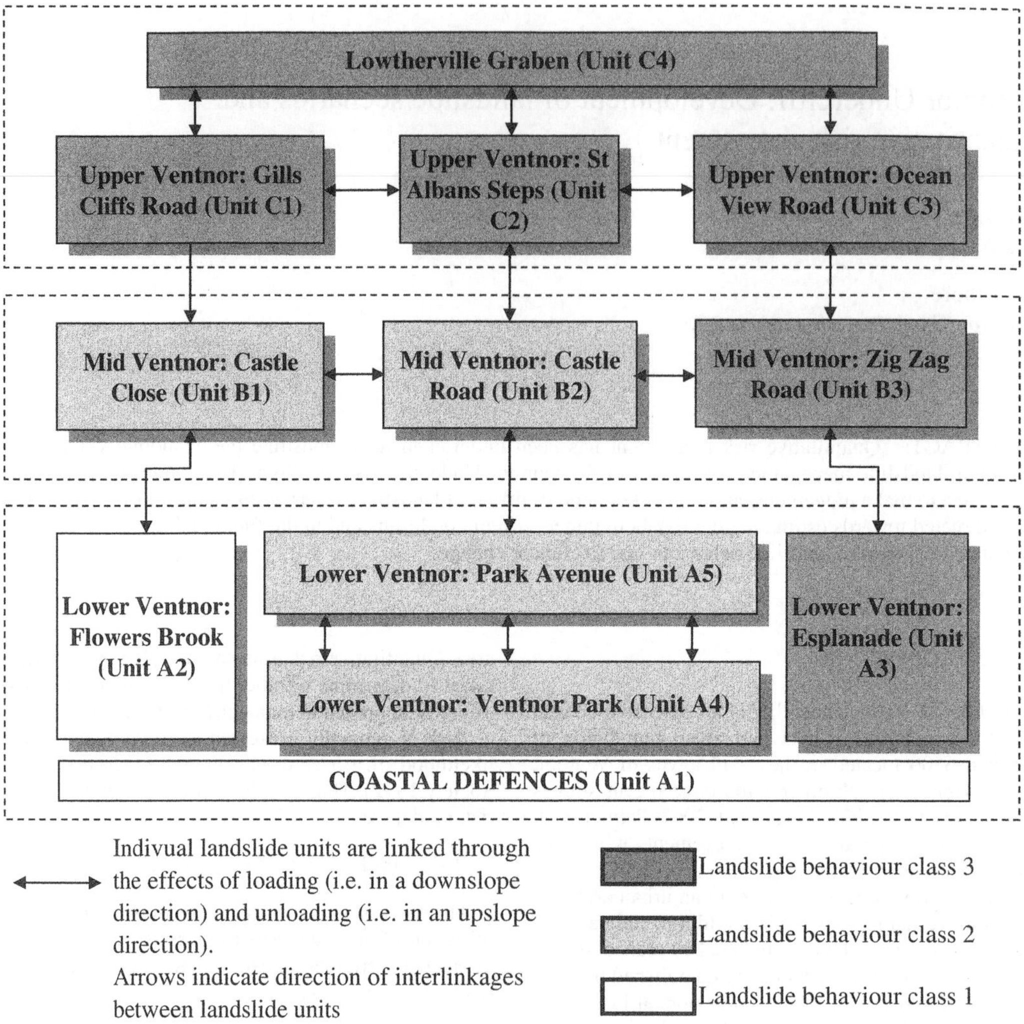

Figure 1. Schematic landslide hazard model of part of the Ventnor Undercliff.

on the key components – landslide hazard models, probability models, consequence models and the use of expert judgment – can be found in Lee & Jones (2004).

2 LANDSLIDE HAZARD MODEL

Whilst the geology has controlled the nature and pattern of landsliding within the Ventnor Undercliff (e.g. Lee & Moore 1991, Hutchinson & Bromhead 2002), the potential for reactivation of the system is influenced by the complex interaction with adjacent landslide units and their relative sensitivity to factors such as high groundwater levels and marine erosion. The Undercliff is sensitive to removal of support currently provided by the lower landslide units (Figure 1). Coastal defences along the Western Cliffs, in the form of a rock armour revetment, currently provides a degree of protection at the toe of the Undercliff landslide complex from active marine erosion, and further over-steepening and destabilisation.

Failure of these coastal defences will result in renewed erosion of the lower units (Units A2, A3, A4 and A5; see Figure 1). This could promote seaward movement of the lower landslide units, which in turn could result in unloading and destabilisation of the landslide units in mid-Ventnor (Units B1 to B3). This would ultimately lead to accelerated ground movement and retrogressive failure of the landslide units in upper Ventnor (Units C1 to C4).

A progressive reactivation model was developed for the Undercliff, in support of the QRA process.

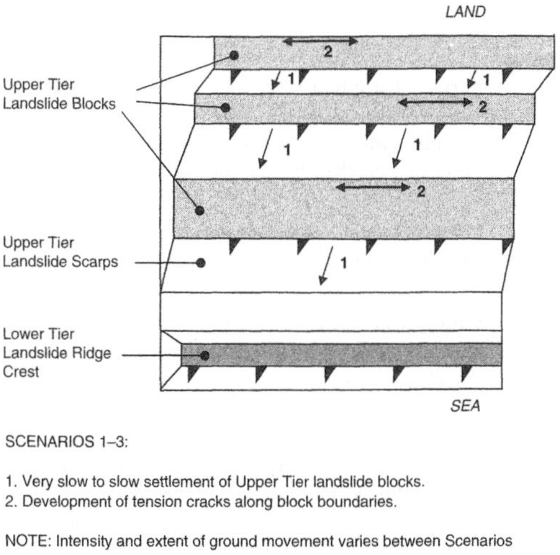
LAND
Upper Tier
Landslide Blocks
Upper Tier
Landslide Scarps
Lower Tier
Landslide Ridge
Crest
SEA
SCENARIOS 1–3:
1. Very slow to slow settlement of Upper Tier landslide blocks.
2. Development of tension cracks along block boundaries.
NOTE: Intensity and extent of ground movement varies between Scenarios

Figure 2. Schematic model: Scenarios 1–3.

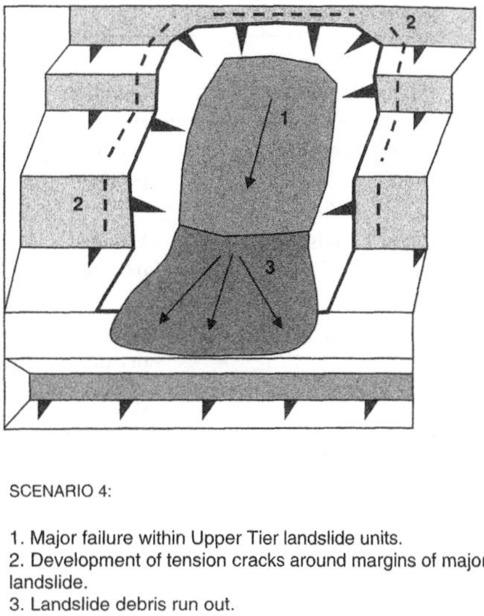
SCENARIO 4:
1. Major failure within Upper Tier landslide units.
2. Development of tension cracks around margins of major landslide.
3. Landslide debris run out.

Figure 3. Schematic model: Scenario 4.

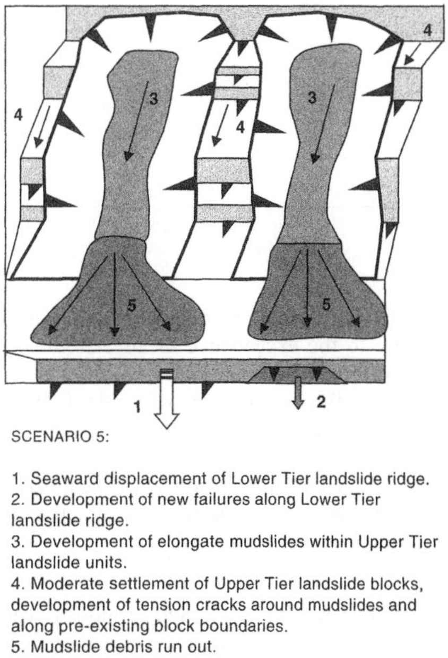
SCENARIO 5:
1. Seaward displacement of Lower Tier landslide ridge.
2. Development of new failures along Lower Tier landslide ridge.
3. Development of elongate mudslides within Upper Tier landslide units.
4. Moderate settlement of Upper Tier landslide blocks, development of tension cracks around mudslides and along pre-existing block boundaries.
5. Mudslide debris run out.

Figure 4. Schematic model: Scenario 5.

The model involves the following potential behaviour types:

1. Almost continuous extremely slow creep;
2. Episodes of significant slow to very slow ground movement;
3. Major landslide events, involving the development of new landslide units within the system;
4. Re-establishment of active instability i.e. the breakdown of the existing pattern of landslide units and the generation of a new Undercliff form (e.g. Hutchinson & Bromhead 2002).

Reactivation is driven by the chance occurrence of initiating events and subsequent system responses (e.g. Lee & Jones 2004). For example, there is a close association between the recorded pattern of landsliding in the Undercliff and prolonged periods of higher rainfall i.e. the occurrence of 'wet years' and the resulting high groundwater levels. Periods of accelerated ground movement in Upper Ventnor have been associated with winter rainfall conditions that can be expected about one year in 100 (Lee et al. 1998; Moore et al. 2007a).

Reactivation can also occur in response to the progressive effects of repeated toe unloading after a single initiating event (e.g. seawall failure or a wet year sequence). The net result is a spread upslope of active instability (retrogressive failure) as higher level slides are successively unloaded by the movements of the lower level slides which had previously provided passive support. The development of reactivation sequences is conditional on the unloading/removal of support to units due to the movement of the adjacent, downslope units. The response to unloading/removal of support to a unit may be delayed or lagged. It is expected that reactivation at a particular level will be conditional on the occurrence of suitable triggering events (e.g. wet year sequences and high groundwater levels).

Five scenarios or plausible sequences of events were developed for evaluation in the QRA (Figures 2–4):

Scenario 1 – almost continuous creep;
Scenario 2 – an episode of significant ground significant winter rainfall threshold level;

Table 1. Reference landslide events.

Scenario	Affected area	Duration	Speed/cumulative displacement	Surface disruption
1	Class 3 landslide units	Winter period	Very slow; up to 1.6 cm/year. <0.01 m	Localised minor creep, including development of tension cracks (up to 1 mm wide).
2	Class 3 landslide units	Winter period	Very slow; up to 1.6 cm/year. <0.1 m	Localised creep, including development of tension cracks (upto 10 mm wide).
3	Class 2 and 3 landslide units	Winter period	Slow; up to 1.6 m/year. <1 m	Widespread creep including development of tension cracks (up to 50 mm wide), with evidence of localised surface displacement (<1 m displacement).
4	Upper and mid Ventnor; landslide dimensions 250 m wide and 300 m long, with 250 m of run out.	<10 days	Moderate; up to 13 m/month. blocks10 m	Major deep-seated landslide event, involving development of a series of displaced landslide separated by steep scarp slopes and bounded by a pronounced lateral shear scarp. Widespread ground disruption within the slide area, with up to 10 m surface lateral/vertical displacements and tension cracks (up to 0.5 m wide).
5	Ventnor Undercliff	<10 days	Moderate; up to 13 m/month. >10 m	Extensive major landslide activity systems and creating significant changes to the Lower and Upper Tier geomorphology. Activity includes seaward displacement and major failure of the Lower Tier ridge, formation of a new landslide geometry and failure mechanism within the Upper Tier. Widespread ground disruption, with over 10 m surface lateral/vertical displacements and tension cracks (up to 1 m wide).

Scenario 3 – an episode of major ground movement in response to the exceedence of an extreme winter rainfall threshold level;
Scenario 4 – the occurrence of a major landslide event within the upper tier landslide units (C1–C4 and B1–B3);
Scenario 5 – a landslide reactivation sequence that leads, ultimately, to the establishment of active landsliding throughout the Undercliff.

Uncertainty over the scale and ground movement characteristics of these scenarios was accommodated by the definition of reference events (Table 1). These pre-defined events were used to provide benchmark conditions for estimating scenario probability and the development of consequence models.

The impact of these scenarios is expected to vary from place to place, according to the geomorphological setting (Lee & Moore 1991). The contemporary ground behaviour provides an indication of those parts of the Undercliff that are most likely to be affected by Scenarios 1 to 3. Areas that have experienced the effects of ground movement over the past 200 years will probably continue to be the most susceptible to these styles of movement in the future (Moore et al. 1995). Individual landslide units were assigned to a landslide behaviour class (Table 2) to indicate the extent to which they have been affected by movements in the past. In developing the hazard model, it was assumed that:

Scenarios 1 and 2; the impact will be limited to those units that have experienced regular damage in the past (i.e. Class 3).
Scenario 3 involves units affected by rare and regular damage (i.e. Classes 2 and 3).
Scenario 4 was assumed to affect both upper and mid-Ventnor (units C1–C4 and B1–B3), with the run-out causing damage in unit A5.
Scenario 5 involves the re-establishment of active landsliding throughout the Ventnor Undercliff.

It is noted that there are no precedents for scenarios 4 and 5 at Ventnor, although landslide events comparable to what could be expected have occurred at The Landslip at Bonchurch (Palmer et al. 2007b) and at Blackgang (Moore et al. 1998) in recent years.

Table 2. Landslide behaviour classes.

Class	Historical behaviour	Units
3	*Regular damage*; properties situated in these areas have been affected by differential shear, distortion and tilting. The cumulative effect of this movement has resulted in serious and severe damage to property.	A3, B3, C1–4
2	*Rare damage*; most properties situated in these areas have been largely unaffected by ground movement. However, in places, the cumulative effect of this movement has resulted in moderate and slight damage to property.	A4, A5, B1–2
1	*Negligible or no damage*; properties situated in these areas have been largely unaffected by ground movement.	A2

3 PROBABILITY MODELS

Judgments on the scenario probabilities were made by an expert panel that sat in open forum to identify and resolve the key issues related to the reactivation problem (see Lee & Jones 2004). The panel took into account the historical evidence, together with the site investigation and stability analysis results.

Probability models were developed to support the process for the different scenarios:

1. Threshold winter rainfall model; landslide activity is triggered by the winter rainfall exceeding a particular threshold level (Lee et al. 1998). This model was used for Scenarios 1-3 and 4a. A simple conditional probability model was used (see Table 3):

   Prob. (Scenario) = Prob. (Threshold Rainfall) × Prob. (Movement given Threshold Rainfall)

2. Conditional sequence model (Scenarios 4b); the occurrence of landslide scenarios involves an interrelated sequence of events driven by an initiating or triggering event (e.g. coastal defence failure, rapid sea-level rise). These sequences were modelled using event trees (Figure 5).
3. Environmental change model; recent studies (see Moore et al. 2007) have indicated that Scenario 5 is dependent on major environmental changes (e.g. the onset of periglacial conditions or major seabed changes in the vicinity of the lower tier landslide toes). Modelling of long term climate changes suggests that periods of periglacial conditions could be expected at least twice in the next 100 K years (e.g. Goodess et al. 1999). This predicted future frequency (1 in 50,000 years) was used as a best estimate of Scenario 5, and assumes that if climatic conditions deteriorate then the scenario will occur.

Table 3. Threshold movement model; estimated annual probabilities of Scenarios 1-4b.

Scenario	Annual prob. (threshold rainfall)	Prob. (movement given threshold rainfall)	Conditional prob. (scenario)
1	0.95	1	0.95
2	0.1	1	0.1
3	0.02	0.5	0.01
4a	0.002	0.5	0.001
4b	See event tree		0.000002

The 'best-estimate' judgments are presented in Table 4, which includes 'upper bound' and 'lower bound' estimates. These estimates give a broad indication of the expected scenario probability for a 'do nothing' option (i.e. no active landslide management) under the current environmental conditions.

4 CONSEQUENCE MODELS

The impact of a landslide scenario is determined by the ground behaviour (e.g. the intensity of the movements), the assets at risk, the exposure of the assets and their vulnerability to damage (Lee & Jones 2004). Exposure and vulnerability factors were used to calculate the possible damages compared with a total loss event where the assets at risk would be completely lost. Thus, for a landslide event of a particular probability and intensity:

Risk = Prob. (Event) × Consequence (Total Loss × Exposure × Vulnerability)

Individual landslide units provide the spatial framework for the QRA. Therefore, it is necessary to carry out this exercise for each landslide unit:

Risk (Unit C1) = Prob. (Event) × Consequence (Unit C1)

Risk (all units) = Prob. (Event) × $\sum$Consequences (all units)

Within Ventnor, property damage tends to occur at the boundaries between major landslide blocks where narrow bands of severe hazard can be recognised. Whilst one property may be severely damaged by differential movement, a nearby property may be largely unaffected. The degree of property damage has also varied between the different landslide units, reflecting their varying susceptibility to movement.

It was assumed that this variable *exposure* of property, both within and between landslide units, will be

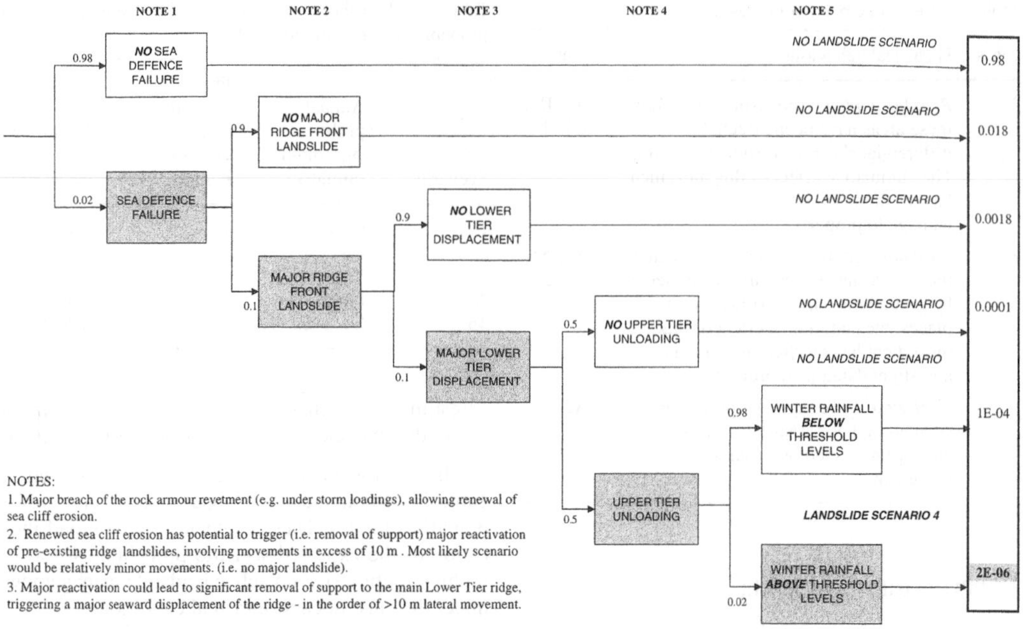

NOTES:

1. Major breach of the rock armour revetment (e.g. under storm loadings), allowing renewal of sea cliff erosion.
2. Renewed sea cliff erosion has potential to trigger (i.e. removal of support) major reactivation of pre-existing ridge landslides, involving movements in excess of 10 m. Most likely scenario would be relatively minor movements. (i.e. no major landslide).
3. Major reactivation could lead to significant removal of support to the main Lower Tier ridge, triggering a major seaward displacement of the ridge - in the order of >10 m lateral movement.
4. Lateral movement of the Lower Tier ridge could cause unloading of the Upper Tier landslide units. However, the significance depends on the Upper Tier sub-surface geometry and the support provided by the Park Avenue sediment infill.
5. The occurrence of a major landslide in the Upper Tier is reliant on high groundwater levels at the time of unloading. It is assumed that this could be the equivalent of a 1 in 50 year effective winter rainfall total. If groundwater levels are below this threshold, then Upper Tier movements will probably be confined to block settlement, creep and development of tension cracks.

Figure 5. Scenario 4b event tree: sea defence failure (major breach).

Table 4. Estimated annual probabilities of landslide scenarios.

Scenario	Best-estimate	Upper bound estimate	Lower bound estimate
1	0.95	0.99	0.9
2	0.1	0.2	0.04
3	0.01	0.02	0.005
4	0.001	0.002	0.0002
5	0.00002	0.0001	0.00001

reflected in different degrees of damage, ranging from negligible damage to severe damage and write-off (Table 5). The following method was used to accommodate the uncertainty with the location and severity of damage sites:

a) Average property values; the total number of properties within each unit was used to calculate an average property value (total property value divided by total number of properties);
b) Exposure factor; it was assumed that for each scenario that there will be an even distribution of damage across each landslide unit. A range of exposure factors were developed to account for the likely impact of each scenario on units from the three

Table 5. Landslide damage classes.

Damage class	Description
Negligible (N)	Hairline cracks to roads, pavements and structures with no appreciable lipping or separation.
Slight (S)	Occasional cracks. Distortion, separation or relative settlement apparent. Small fragments of debris may occasionally fall onto roads and structures causing only light damage. Repair not urgent.
Moderate (M)	Widespread cracks. Settlement may cause slight tilt to walls and fractures to structural members and service pipes.
Serious (Sr)	Extensive cracking. Settlement may cause rotation or slewing of ground. Gross distortion to roads and structures. Repairs will require partial or complete rebuilding and may not be feasible. Severe movements leading to the abandonment of the site or area.
Severe (Sv)	Extensive cracking. Settlement may cause rotation or slewing of ground. Gross distortion to roads and structures. Repairs will require partial or complete rebuilding and may not be feasible. Severe movements leading to the abandonment of the site or area.

Table 6. Exposure factors for landslide behaviour classes 1–3: proportion of properties affected by damage of different severity under each landslide scenario.

Landslide unit behaviour class	Scenario	Unaffected	Negligible to slight damage	Moderate damage	Serious damage	Severe damage	Write-off
3	1	0.75	0.20	0.025	0.02	0.004	0.001
	2	0.75	0.1	0.08	0.05	0.01	0.01
	3	0.25	0.25	0.15	0.15	0.1	0.1
	4	0.01	0.01	0.08	0.15	0.25	0.5
	5	0	0	0	0	0	1.0
2	1	1.0	0	0	0	0	0
	2	1.0	0	0	0	0	0
	3	0.75	0.1	0.08	0.05	0.01	0.01
	4	0.01	0.01	0.08	0.15	0.25	0.5
	5	0	0	0	0	0	1.0
1	1	1.0	0	0	0	0	0
	2	1.0	0	0	0	0	0
	3	1.0	0	0	0	0	0
	4	1.0	0	0	0	0	0
	5	0	0	0	0	0	1.0

landslide behaviour classes (see Table 2). Table 6 presents the proportion of properties affected by damage of different severity under each scenario. In Scenario 2, for example, it is assumed that 8% of all the properties within 'Class 3' landslide units would be affected by moderate damage, whereas in Scenario 3, it has been assumed that 10% of the properties within 'Class 3' landslide units would be damaged so badly that they would have to be written-off. These proportions were established through expert panel discussions, based on historical evidence.

c) Vulnerability factor; the economic implications of the variation in impact between landslide units and scenarios are reflected in different levels of loss. Vulnerability is defined as the level of potential damage, or degree of loss, of a particular asset (expressed on a scale of 0 to 1) subjected to a landslide of a given severity. Estimates of vulnerability were made by an expert panel based on the inferred relationship between the severity of damage and the proportion of the property value that would have to be spent in undertaking repairs (Table 7). For example, moderate damage is expected to result in losses equivalent to 10% of the property value, whereas write-off would result in 100% losses.

Thus, the consequences of a particular Scenario '*s*' in landslide unit '*u*' are calculated for each damage class shown in Table 6, as illustrated below for damage class '*d*':

Consequences (s, u, d) = Average Property Value × No. of Properties × Proportion Damage (Class d) × Vulnerability (Damage Class d)

Table 7. Vulnerability of property: property damage as a proportion of the total property value for each damage class.

Unaffected	Negligible-slight	Moderate	Serious	Severe	Write off
0%	1%	10%	25%	50%	100%

For example, if there are 10 properties with an average value of £100 k in landslide unit A3 (class 3) then the severe damage impact of Scenario 3 would be:

Consequences = (100 × 10 × 0.1) × 50% = £50 k

The overall consequences of Scenario '*s*' in landslide unit '*u*' are the sum of the consequences of all damage classes, for both property and services:

Consequences (s, u) = ∑Consequences (all Damage classes)

Consequence models were also developed for highway and infrastructure losses, along with the emergency service costs, tourism impacts and traffic delay likely to be associated with the landslide scenarios.

The overall consequences of scenario '*s*' are the sum of the consequences for all landslide units affected by that particular event:

Consequences (s) = ∑Consequences (all Landslide units)

This exercise was repeated for all the scenarios to generate a total risk value for the Undercliff.

5 IMPLICATIONS OF CLIMATE CHANGE

The predicted impacts of climate changes over the next century are expected to increase risks from landsliding. The UKCIP02 climate change scenarios (Hulme et al 2002) give estimates of future potential change to mean precipitation and temperature for southern England (Table 8). The predicted changes in climate, especially winter rainfall (25–30% wetter under the High Emissions scenario by the 2080s), will lead to changes in the probability of landsliding in many areas.

The application of these change scenarios to the Ventnor climate data gives an indication of the likely magnitude of change to mean effective rainfall (e.g. Moore et al. 2007a). The estimates indicate a 5 to 6% increase in mean monthly effective rainfall under the Low scenario and a 12 to 25% increase for the High scenario. The greatest estimated change occurs in the December to February period.

The current probability of mean monthly effective rainfall of 100 mm is 0.1 (or 1:10 years). Under the UKCIP02 Low scenario the probability is largely the same but under the High scenario the probability is increased to 0.2 (1:5 years). The potential change for more extreme conditions is greater with, for example, the probability of mean monthly effective rainfall of 150 mm increasing from the current 0.005 (1:200 years) to 0.02 (1:45 years) for the UKCIP02 High scenario. The estimated increase in mean monthly effective rainfall is expected to result in an increase in the frequency or probability of landslide events, assuming the distribution of events will be similar to the historical record.

These potential increases in rainfall event frequency equally apply to the established relationship between effective rainfall and ground movement susceptibility (e.g. Moore et al. 2007a). It follows, therefore, that the estimated scenario annual probabilities will increase over time. Table 9 presents a comparison between the 'current' estimates and the estimates that might apply in 2080, based on the UKCIP02 High scenario.

In the QRA process, the effect of climate change was modelled in two ways:

1. By repeating the 'do nothing' risk assessment, using the future best estimates for Year 99 (i.e. 100 years in the future).
2. By modelling the incremental effects of gradual climate change over the next 50 years, for the 'do nothing' option. This was achieved by calculating the risk for ten decades: Years 0–9, 10–19, 20–29, 30–39, 40–49, 50–59, 60–69, 70–79, 80–89, 90–99). In each of these cases the scenario probability was taken to be the mid-point value i.e. the estimated probability in year 4 was used for the first decade, year 14 for the second, and so on.

6 ECONOMIC EVALUATION

Allocation of public resources for landslide management is a political decision, influenced by the need to find an acceptable balance between investments in a wide range of competing public services. Three tests are usually applied to decision-making about the allocation of public expenditure:

- the scarcity of resource require that, as a nation, investments are made that give the highest returns;
- decisions to invest public funds must be accountable and justifiable; and

Table 9. Estimates of annual probability of scenarios for the 'do nothing' option.

Scenario	Current best-estimate	Future best-estimate
1	0.95	0.99
2	0.1	0.273
3	0.01	0.027
4	0.001	0.002
5	0.00002	5.5E−05

Table 8. UKCIP02 climate change predictions for the southern England coast for the 2080s (from Hulme et al 2002).

	Low emissions scenario		High emissions scenario	
Period	Temperature change °C	Precipitation change %	Temperature change °C	Precipitation change %
Winter	1.5 to 2	15 to 20	3 to 3.5	>30
Spring	1.5 to 2	0 to 10	3 to 3.5	0 to 10
Summer	3 to 3.5	−40 to −30	>4.5	> −50
Autumn	2.5 to 3	−10 to 0	>4.5	−10 to 0
Annual	2 to 2.5	−10 to 0	4 to 4.5	−10 to 0

- decisions must be based on a rational comparison between the available options.

Economic evaluation provides a mechanism for comparing the benefits of landslide management with the costs incurred, to determine:

- whether the benefits exceed the costs; and
- the strategy that is expected to deliver the greatest economic return i.e. the most efficient use of resources.

The decision to invest in landslide management should depend on a thorough appraisal of the benefits of risk reduction over the expected lifetime of the scheme or project. The benefits of management are the difference between the losses that would be incurred without a scheme or project and the losses that would be incurred when the scheme fails and landslide activity is renewed. An alternative way of looking at this is that the scheme reduces the probability of movement and reduces the risk over its expected lifetime and not just in a single year.

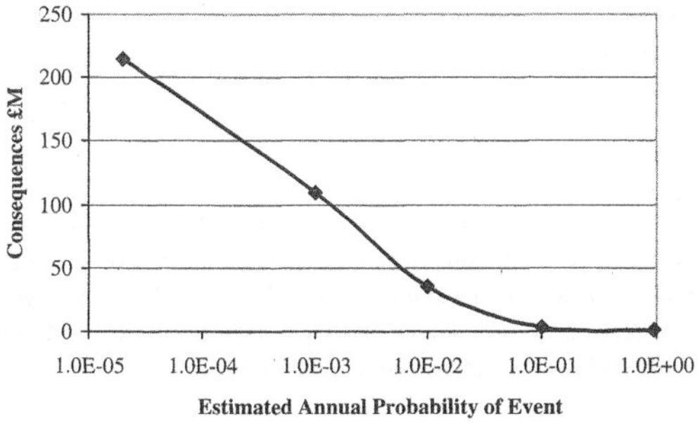

Figure 6. Landslide damage curve: do nothing option.

Thus, a stabilisation scheme with a design life of 100 years will reduce the risk over a period of 100 years. Thus the benefits of this management option are:

Landslide Management Benefits = Without Project Losses (Year 0–99) – With Project Losses (Year 0–99)

Economic theory indicates that the value of risk reduction in future years is worth less than that achieved at present. It is, therefore, necessary to express all future risks in terms of their Present Value (PV), by discounting (see MAFF 2000, Lee & Jones 2004):

With Project Risk (Years 0–99) $= \sum$ *(Prob. Event × Losses × Discount Factor (Year 0–99))*

Figure 6 presents a landslide damage curve that relates total losses to the scenario probability, for the "do nothing" option. As any of the scenarios could occur in a given year, depending on the winter rainfall, it was necessary to calculate the average annual damage that takes account of each possible combination of event probability and total loss (Table 10).

The present value (PV) risk over the next 100 years, for the 'do nothing' option, can be calculated from: *Risk (Years 0–99)* $= \sum$ *(Average annual risk × Discount Factor (Year 0–99))*

= Average annual risk × 29.86 = 4.64 × 29.86 = £77M

The economic risks are expected to increase by about 0.25 % per year, as a result of climate change (Table 11).

The benefit*s* of a landslide management strategy are the reduction in risk, expressed in monetary terms, compared with a 'do nothing' case. The *costs* include all the costs incurred during the investigation, planning and design, construction and operation of the strategy.

Table 10. Do nothing option: economic risk: summary of landslide damage for different scenarios.

	Scenario and estimated annual probability of event					
Do nothing option	Scenario 1 0.950	Scenario 2 0.100	Scenario 3 0.010	Scenario 4 0.001	Scenario 5 0.00002	Infinity 0.000
Damage category	Damage					
Property	1260.45	3622.64	34228.27	104284.67	196215.48	100000
Indirect losses (e.g. tourism)	0	10	98.78	790.27	977.96	10000
Traffic disruption	0	0.00	36.77	1341.95	1341.95	2500
Emergency services	0	0.00	800.00	2720	16600.00	5000
Other	0	0	0	0	0	0
Loss of Life	0	0.002	0.002	0.08	0.02	0.0002
Total damage £M						
Area (damage % frequency)	1.26	3.63	36.16	109.14	215.14	117.50
Total £M	**4.64**					

Table 11. Economic risk: effects of climate change.

Variations of the best-estimate case	Economic risk £M	
	Annual risk	PV risk (Year 0–99)
Current situation	4.64	138
Incremental effects of climate change	c.5.6	168
Future situation (Year 99 as present day)	8.7	260

Table 12. Estimated annual probability of scenarios associated with different landslide management strategies.

Scenario	Do nothing (Best estimate)	Do something (Landslide management)	With project (Landslide stabilisation)
1	0.95	0.5	0.02
2	0.1	0.02	0.002
3	0.01	0.01	0.0002
4	0.001	0.001	0.00008
5	0.00002	0.00002	0.00002

Both benefits and costs should be considered over the strategy lifetime and, hence, need to be brought back to their Present Value by discounting.

Two strategy options were evaluated:

- a 'do something' case which involves limited intervention to attempt to reduce rather than control the problems, preventing water leakage on unstable slopes, and the provision of early warning systems. This is the current management practice; and
- a 'with project' landslide stabilisation works (coast protection and deep drainage) that provide a significant reduction in risk.

Both the 'do something' and the landslide stabilisation option are expected to reduce the risk by reducing the annual probability of the landslide scenarios (rather than reducing the consequences; Table 12).

When compared with the 'do nothing' option, the economic risks have been reduced by the ongoing commitment to active landslide management in the Undercliff (i.e. the 'do something' option; Table 13). The landslide risks could be further reduced by improving the stability of the Undercliff through landslide stabilisation measures (i.e. the 'with project' option).

Preliminary assessments indicate that a landslide stabilisation scheme is economically viable, with a Benefit: Cost Ratio (BCR) in excess of 3. It will, of course, be necessary to refine these conclusions through detailed investigations, studies and designs.

Table 13. Summary of economic risk for different management options.

	Annual risk	PV risk (Year 0–99)
Do Nothing Option	4.64	138
Do Something	2.18	65
With Project	0.1	3

Table 14. Summary of benefit: cost analysis for different management options.

	Costs and benefits £M		
	No project: Do nothing*	Option 2: Do something**	Option 3: With project
PV costs PVc	–	10**	27.00
PV damage PVd	138.00	65.00	3.00
Total PV benefits PVb		73.00	135.00
Net Present Value NPV		63.00	108.00
Average benefit/cost ratio		7.3	5.0
Incremental benefit/cost ratio			3.65

Notes: * the do nothing damages are based on an assessment of the current situation and do not take account of climate change. **Costs of do something are PV Costs associated with an estimated annual expenditure of £0.5 M.

The Benefit: Cost ratio (BCR) is a widely used measure of economic cost-effectiveness, and summary results for the 'do something' and the 'with project' landslide stabilisation options are presented in Table 14.

Other useful measures include the Net Present Value ($NPV = Risk\ Reduction - Costs$) and the Incremental Benefit: Cost ratio which represents the change in costs and benefits between options.

The 'With Project' Option appears to be economically viable, having a BCR of 5.0 and an incremental BCR (compared with Option 2) of 3.65. At this stage, there are clearly considerable uncertainties regarding both the cost of a landslide stabilisation scheme and its potential effectiveness.

7 UNCERTAINTY

Incomplete knowledge of the landslide scenarios and their consequences introduces uncertainty into the risk assessment process. Amongst the most important sources of uncertainty are associated with the availability of data, especially knowledge of:

- past phases of landslide activity their timing and the association with causal factors such as wet year sequences;
- the interaction between landslides and the assets at risk, especially the failure criteria for buildings under landslide loads;
- the effectiveness of stabilisation scheme options, in terms of reduction in the scenario probability.

Uncertainty is an inevitable component of the risk assessment process. However, the presence of uncertainty should not obscure the fact that it is possible to generate a preliminary indication of the general levels of landslide risk based on the information currently available.

These uncertainties prevent *accurate* prediction of the landslide risk. However, the current level of precision and understanding is believed to sufficient to enable an adequately informed decision to be made about whether the risks at individual sites are unacceptable or tolerable, and whether stabilisation works are likely to prove cost effective. Sensitivity testing of the assumptions could be carried out to develop upper and lower estimates to highlight the range of expected risk levels for different project options.

ACKNOWLEDGEMENTS

The authors are grateful for the assistance of colleagues at Halcrow and the Centre for the Coastal Environment, Isle of Wight Council, for their contributions to the work reported in this paper. The project was funded by the Isle of Wight Council and the Department for Transport.

REFERENCES

Carey, J., Moore, R., Petley, D.N. & Siddle, H.J. 2007. Prefailure behaviour of slope materials and their significance in the progressive failure of landslides. This Conference.

Clark, A.R., Lee, E.M. & Moore, R. 1994. The development of a ground behaviour model for the assessment of landslide hazard in the Isle of Wight Undercliff and its role in supporting major development projects. Proceedings, 7th Congress of IAEG, Lisbon. Vol VI; 4901–4913. Balkema.

Goodess, C.M., Watkins, S.J., Burgess, P.E. & Palutikof, J.P. 1999. Assessing the Long-term Future Climate of the British Isles in Relation to the Deep Underground Disposal of Radioactive Waste, Nirex Report N/010, United Kingdom Nirex Limited, Harwell, 230pp. (Available from: http://www.nirex.co.uk/educate/educate.htm).

Hulme, M., Jenkins, G.J., Lu, X., Turnpenny, J.R., Mitchell, T.D., Jones, R.G., Lowe, J., Murphy, J.M., Hassell, D., Boorman, P., McDonald, R. & Hill, S. 2002. Climate Change Scenarios for the United Kingdom: The UKCIP02 Scientific Report. Tyndall Centre for Climate Change Research, School of Environmental Sciences, University of East Anglia, Norwich, UK.

Hutchinson, J.N. & Bromhead, E.N. 2002. Isle of Wight landslides. In R.G. McInnes and J. Jakeways (eds.) Instability: Planning and Management, Thomas Telford, 3–72.

Lee, E.M. & Moore, R. 1991. Coastal Landslip Potential Assessment: Isle of Wight. Report to the Department of the Environment.

Lee, E. M. & Jones, D.K.C. 2004. Landslide Risk Assessment. Thomas Telford.

Lee, E.M., Moore, R. & McInnes, R.G. 1998. Assessment of the probability of landslide reactivation: Isle of Wight Undercliff, UK. In D. Moore and O. Hungr (eds.) Engineering Geology: The View from the Pacific Rim, 1315–1321.

McInnes, R.G. & Jakeways, J. 2002. Managing ground instability in the Ventnor Undercliff, Isle of Wight, UK. In R.G. McInnes and J. Jakeways (eds.) Instability – Planning and Management, Thomas Telford, 739–746.

Ministry of Agriculture, Fisheries and Food (MAFF) 2000. Approaches to Risk. FCDPAG4 Flood and Coastal Defence Project Appraisal Guidance. MAFF Publications.

Moore, R., Lee E.M. & Clark A.R. 1995. The Undercliff of the Isle of Wight: a review of ground behaviour. ISBN 1 873295 70 7 Cross Publishing.

Moore R., Clark A.R. & Lee E.M. 1998. Coastal cliff behaviour & management: Blackgang, Isle of Wight. In: Maund JG & Eddlestone M (eds) Geohazards in Engineering Geology, Geological Society, London, Engineering Geology Special Publications, 15, 49–59, 1998.

Moore, R., Carey, J., McInnes R.G. & Houghton, J. 2007a. Climate change, so what? Implications for ground movement and landslide event frequency in the Ventnor Undercliff, Isle of Wight. This Conference.

Moore, R., Carey J. & Turner, M. 2007b. The Ventnor Undercliff: a new ground model and implications for climate change induced landslide behaviour and risk. This Conference.

Palmer, M.J., Carey, J. & Turner, M.D. 2007a. Lithostratigraphy of the Ventnor Undercliff and determination of critical horizons through borehole geophysics. This Conference.

Palmer, M.J., Moore, R. & McInnes, R.G. 2007b. Reactivation of an ancient landslip, Bonchurch: event history, causes, mechanisms, climate change and landslip potential. This Conference.

Royal Society 1992. Risk: Analysis, perception and management. Report of a Royal Society Study Group. The Royal Society, London.

Landslides and Climate Change – McInnes, Jakeways, Fairbank & Mathie (eds)
 ISBN 978-0-415-44318-0

Climate change, so what? Implications for ground movement and landslide event frequency in the Ventnor Undercliff, Isle of Wight

R. Moore & J.M. Carey
Halcrow Group Ltd, Birmingham, UK

R.G. McInnes & J.E.M. Houghton
Isle of Wight Council, Ventnor, UK

ABSTRACT: The town of Ventnor is located on the south coast of the Isle of Wight in an area of ancient coastal landslides known as the Undercliff. The ancient landslide complex is marginally stable and has been subject to ground movement and occasional landslide events in some locations causing significant damage to property and infrastructure. Since 1992, the local authority has made a significant investment in real-time monitoring and analysis of hydrogeological data at key sites, which has considerably advanced understanding about the relationships between climate and ground behaviour response. These data reveal close relationships between antecedent rainfall conditions, groundwater levels and ground movement rates, confirming that prolonged periods of heavy rainfall and associated elevated groundwater levels are a fundamental trigger of ground movement and landslide events. Latest climate change predictions point to significant increases in winter rainfall frequency, intensity and amount, drier summers and rising sea level which are likely to prove particularly challenging in terms of managing ground instability in the Undercliff. The implications of climate change predictions for the Undercliff are both spatial and temporal; firstly, there are concerns that hitherto marginally stable areas of the Undercliff may become unstable due to reactivation of ground movement and the occurrence of new landslides, secondly, in areas previously affected by ground movement or landslides, the frequency and rate of ground movement and landsliding is expected to increase. The paper presents historical and contemporary data to demonstrate the relationships between rainfall and ground movement and uses these to project the likely impacts of climate change predictions on the future stability of the Undercliff. The paper concludes that climate change poses a very real and significant challenge to the future management of the Undercliff and other similarly marginally stable ancient landslides in southern Britain.

1 INTRODUCTION

1.1 *The Undercliff*

The Undercliff is an extensive area of ancient landslides that extend some 12 km from Luccombe to Blackgang on the south coast of the Isle of Wight (Fig 1). In cross-section the landslides extend up to 1 km landward from the shoreline and form a steep terraced landscape with impressive views of the English Channel. Apart from its stunning location, the special climate of the Undercliff is another factor which led to development of the area in the mid to late 1800s. The main residential areas include Bonchurch, Ventnor, St Lawrence and Niton which have a permanent population of more than 6,000. Yet, the Undercliff is one of the most unstable geological settings in the UK, and faces substantial challenges from coastal slope instability, particularly in the context of future climate change predictions.

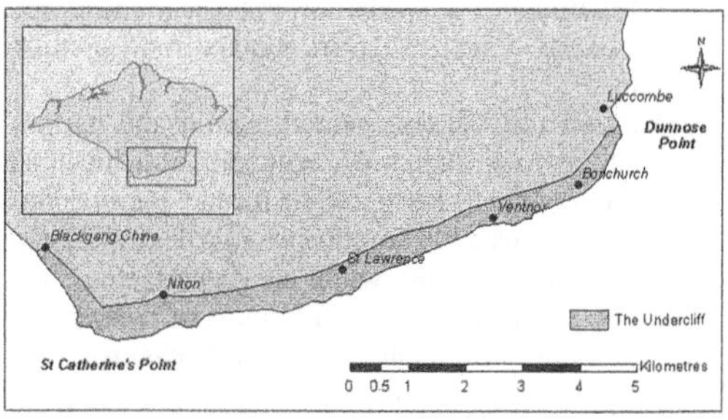

Figure 1. The Isle of Wight Undercliff.

1.2 *Causes of landslides and ground movement*

Landslides occur when the force of gravity exceeds the strength of the soils and the rocks that form the slopes. Slope failure occurs to restore the balance between the destabilising forces (stresses) and the resisting forces

Table 1. Causes of landslides.

Internal causes	External causes
Materials:	**Removal of slope support:**
• Soils subject to strength loss on contact with water or as a result of stress relief (strain softening)	• Undercutting by water (waves and Stream incision)
• Cohesive soils which are subject to strength loss or gain due to weathering	• Washing out of soil (groundwater)
• Soils with discontinuities characterised by low shear strength such as bedding plames, faults, joints etc.	• Man-made cuts and excavation
Weathering:	**Increased loading:**
• Physical and chemical weathering of soils causing loss of strength (cohesion and friction)	• Natural accumulations of water, snow, talus
• Slope ripening and soil development	• Man-made pressures (e.g. fill, tips, buildings)
Pore-water pressure:	**Transient Effects:**
• High pore-water pressures causing a reduction in effective shear strength. Such effects are most severe during wet periods or intense rainstorms.	• Earthquakes and tremors
	• Shocks and vibrations

(shear strength) along a zone of weakness or shear surface. Therefore, a landslide may be regarded a dynamic process that changes a slope from an unstable to a more stable state.

The causes of landslides are well documented by others (i.e. Jones and Lee 1994; Moore et al. 1995). They are generally separated into two types:

1. Pre-conditioning factors which work to make the slope increasingly susceptible to failure without actually initiating it, and
2. Triggering factors which initiate movement.

All landslides are caused by a combination of preparatory and triggering factors. The interrelationship of these factors controls the location and timing of events. When considering the actual causes of landsliding this relative simplicity gives way to complexity as there is a great diversity of causal factors. In broad terms, however, they may be divided into internal causes that lead to a reduction in shear strength and external causes which lead to an increase in shear stress (Tab. 1).

The Undercliff shows marked spatial and temporal sensitivity to ground movement and landslides that reflect a legacy of past geological and geomorphological processes and formative events. These are the pre-conditioning factors which are described in sections 2–4 below along with the transient events that trigger ground movement and landslide events.

1.3 *Climate change*

The UK Climate Impacts Programme (UKCIP) has published downscaled (50km grid) predictions of climate change for the UK to 2080 (UKCIP02; Hulme et al. 2002) based on established global emission scenarios (IPCC 2000). UKCIP02 provides:

- a summary of the changes that are already occurring in global and UK climate
- four alternative climate change scenarios for the UK, including information about changes in average climate, in some selected daily weather extremes, and in average and extreme sea levels around the coast, and
- the main uncertainties influencing confidence in the scenarios.

In summarising the historical changes in UK Climate, UKCIP02 states that "winters across the UK have been getting wetter, with a larger proportion of precipitation falling in the heaviest downpours, while summers have been getting slightly drier. The average rate of sea-level rise during the last century around the UK coastline, after adjustment for natural land movements, has been approximately 1 mm per year. Although the last decade has seen an increase in gale frequency in the UK, this increase is not unprecedented in the historic (sic) record".

The historical evidence presented demonstrates that the proportion of winter to summer rainfall has increased over the last 30 years based on records dating back 240 years. For example, the year 2000 had the wettest autumn in England and Wales when nearly twice the seasonal average of 257 mm was recorded (Hulme et al. 2002). There has also been an increase in rainfall intensity which has been attributed to increasing winter rainfall totals across the UK over the last 40 years (Fig 2).

UKCIP02 present four alternative future climates for the UK, known as the Low Emissions, Medium-Low Emissions, Medium-High Emissions and High Emissions scenarios. These are credible descriptions of how the climate may change in the future. No probabilities are provided for these scenarios and UKCIP do not suggest that any is more likely than another. While they represent a range of future climates, the UKCIP02 scenarios do not capture the entire range of future possibilities. Nevertheless, they are regarded

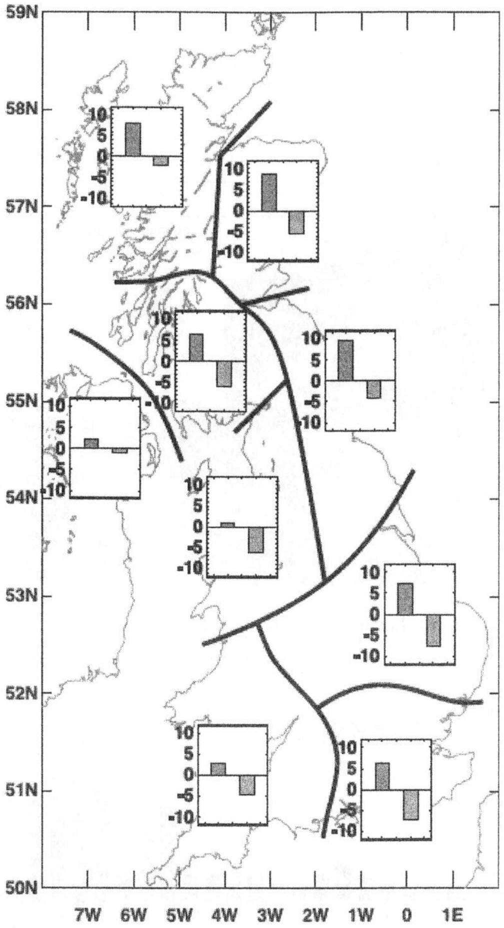

Figure 2. 1961–2000 trend in the fraction of total precipitation contributed by most intense rainfall events for winter (left hand bar) and summer months (right hand bar) (Hulme et al. 2002).

as representing the most likely range of future climates for the UK based on current knowledge of global emissions and their likely impact on future climate in northern Europe.

The key climate variables influencing coastal behaviour are:

1. sea level
2. nearshore waves, and
3. effective rainfall.

The future extreme sea level scenario combines changes to mean sea level, tidal regime, local land movements and storm surges. The latest historical data from tide gauges available through the Permanent Service for Mean Sea Level[1] show a relative rise in sea level (including isostatic land movement) between 1.3 and 2 mm/yr for the Undercliff.

The medium-high 2080 scenario rise in extreme sea level for the Undercliff is given in Table 2. The potential increase of 0.84 m in extreme sea level is of concern to existing coastal defences along the Undercliff, as current levels expected once in 100 years may occur on an annual basis by 2080.

A wave hindcasting analysis was undertaken for the region using wind data output from the Met Office Hadley Centre RCM (Hosking & Moore 2002). The results demonstrate a potential increase in extreme wave heights from the southwest but not the southeast. The wave data were transformed inshore which demonstrated there was potential for changes in sediment drift patterns which could have significant implications for future shoreline management.

Effective rainfall (precipitation minus evapotran spiration) provides a measure of the amount of rainfall directly contributing to groundwater levels within slopes. Historical records from Ventnor indicate that effective rainfall has increased by around 25% over the past 170 years (Fig 3), which has been linked to an increase in the frequency of landslides and ground movement in the Undercliff. The UKCIP02 scenarios predict a similar increase in winter rainfall to 2080.

In summary, the key UKCIP02 climate change projections to 2080 for the Undercliff include:

- temperature increase of 2° to 3.5°C
- winter rainfall increase of 20 to 30%
- summer rainfall decrease of 50%
- summer soil moisture decrease of 40%
- sea level increase of 26 cm to 84 cm, and
- extreme sea level 10 to 20 times more frequent.

So what are the implications of these predicted climate changes on the future stability of the Undercliff? The remainder of this paper presents the evidence linking past climate and rainfall records with the occurrence of landslides and ground movement; these relationships are then applied to evaluate the likely impact of future climate change scenarios on ground movement and landslide frequency in the Undercliff.

2 SPATIAL SENSITIVITY OF HISTORICAL LANDSLIDES IN THE UNDERCLIFF

Historically, the most active areas in the Undercliff have been located at Blackgang and the Landslip, at the western and eastern limits, respectively. The section of Undercliff between Niton and St Lawrence has, in recent years, experienced several major landslide events involving reactivation of the pre-existing ancient landslides (Fig 4). Elsewhere, the Undercliff has historically been less susceptible to ground movement and landslide reactivation events (Moore et al. 1995; McInnes 2000).

[1] www.nbi.ac.uk/psml/datainfo/rlr.trends

Table 2. Predicted extreme sea levels by 2080.

Global sea level rise	Local (UK) additional sea level rise	Isostatic land movement	Tidal Regime	Surge changes (1 in 50 year)	Total change for 1 in 50 year extreme sea level
0.41	0.04 m	0.08 m	0.01 m	0.3 m	0.84 m

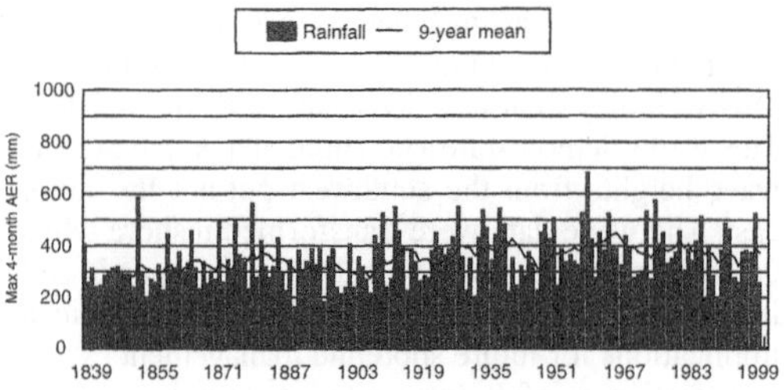

Figure 3. Effective rainfall trend at Ventnor.

Archaeological evidence suggests that during prehistoric times there were major periods of active landsliding within the area of Ventnor (Lee & Moore 1991). Discoveries of skeletal remains buried by landslide debris or crushed by rocks are numerous (e.g. Norman 1887 & Whitehead 1911).

Reports of ground movement and landslide events are recorded as early as 1781 at Binnel and Steephill when huge fragments of rock and earth fell from cliffs (Worsley 1781). Detailed reviews of these records are available in Lee & Moore (1991). The frequency distribution of historical landslides in southern England, including the Undercliff (Fig 5) shows a marked increase in the number of events during the 1900s. This trend probably reflects the spread of development on potentially unstable coastal slopes and a rise in the number of reported impacts of ground movement on development rather than landslide events. This is particularly the case for Ventnor, where historical records over the past 200 years indicate that the town has been continuously affected by ground movement (Moore et al. 1995).

Hutchinson & Bromhead (2002) identified several factors that control the spatial pattern of landsliding in the Undercliff, including:

- the presence of a broad concave down-fold (syncline) in the rocks which controls the height of the Gault relative to sea level. The two most active areas at Blackgang and the Landslip occur where the Gault is at its highest elevation at the shoreline
- the presence of extensive 'debris aprons' forming the lower slopes of the Undercliff. The debris aprons comprise an assortment of Chalk and Upper Greensand debris and were most likely deposited during the early Holocene, between 13–10,000 years ago, after the last glaciation; and
- the varying relative relief along the Undercliff.

Figure 4. Landslide reactivation in the Undercliff.

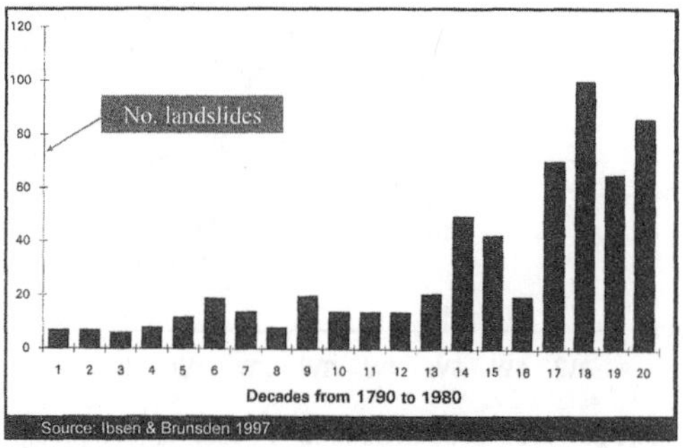

Figure 5. Coastal landslide event frequency in southern England.

Geomorphology and ground behaviour maps are now available for the entire length of the Undercliff (McInnes 2000). These maps define discrete landslide units based on surface morphology, geology and landslide mechanisms. The ground behaviour maps integrate past records of landslide and ground movement events and describe the present-day stability/susceptibility of each landslide unit to change, as follows:

- inactive ancient landslides
- intermittently active ancient landslides
- locally degrading ancient landslides
- degrading landslides
- active landslides.

The main preparatory factors and controls on the spatial distribution of landslides and ground movement include the progressive removal of passive support from the toe of the Undercliff by coastal erosion. Since the mid to late 1800s, the influence of construction activities, such as cutting and filling of slopes, and use of soakaway drains, has also influenced the spatial sensitivity of the Undercliff to ground movement in developed areas.

Coastal erosion has progressively reduced the overall stability of the Undercliff in response to sea level rise during the Holocene and is undoubtedly a major cause of contemporary ground movement and landslide activity in the Undercliff (Lee et al. 1998). The distribution of debris aprons is a significant factor in providing 'natural' passive support to the ancient landslides above. In places, debris aprons are absent where they have been eroded away by the action of the sea, removing passive support and making the ancient landslides above more vulnerable to reactivation.

The sea cliffs formed at the toe of the Undercliff are formed mostly of landslide debris comprising an assortment of chalk, greensand and locally, Gault clay. The landslide debris offers some resistance to the action of waves and rates of sea-cliff retreat are typically 0.3 m per year across the frontage. *In situ* soft sandstones (Sandrock) form the high sea cliffs at the extremities of the Undercliff at Blackgang and Luccombe, and rates of retreat can be much higher between 1–3 m year.

3 TEMPORAL LANDSLIDE FREQUENCY IN THE UNDERCLIFF

The temporal frequency of landslide events and ground movement is controlled by transient high groundwater levels and porewater pressures arising from prolonged and intense rainfall. Since development of the Undercliff in the mid to late 1800s, water leakage from mains supply and waste disposal networks has, in some locations, inadvertently surcharged natural groundwater levels.

Rainfall records are available for the Undercliff since 1839. A composite data set has been assembled from several weather stations which have operated in the Undercliff as follows:

- St Catherine's Point (1920–1988)
- Royal National Hospital (1839–1950)
- Ventnor Cemetery (1932–1988), and
- Ventnor Park (1926–1983; 1988–06)

To calculate effective rainfall, estimates of evapotranspiration were made using the methods of Thornthwaite for historical records and Penman-Montieth for direct measurement of solar radiation at the automatic weather station installed at Ventnor Park in 1992.

The rainfall records show a notable seasonal variation in mean monthly rainfall (Fig 6) comprising a drier period between February and August, and a

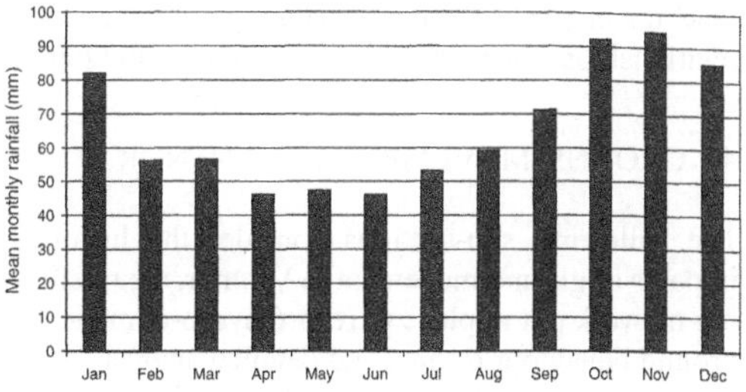

Figure 6. Mean monthly rainfall at Ventnor (1839–2005).

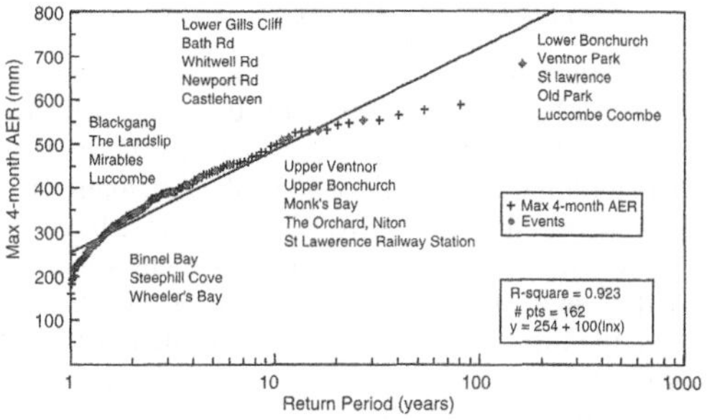

Figure 7. Relationship between 4-AER and reported landslide events.

wetter period between September and January, with maximum mean monthly rainfall typically occurring in November.

The maximum 4-month antecedent effective rainfall (4-month AER) total in any given year is statistically the most significant climate parameter related to incidents of ground movement and landsliding in the Undercliff (Fig 7).

Analysis of this parameter illustrates an apparent trend of increasing 4-month AER, particularly since 1910 (Fig 3). The 9-year moving mean identifies four periods or cycles of significantly higher effective rainfall (>350 mm per 4-month period) between 1874–1890, 1910–1920, 1925–1940 and 1955–1975.

The relationship between ground movement events and rainfall in the Undercliff has established the basis of a simple forecasting tool involving a comparison of the 4-month AER leading up to the initiation of landslide and ground movement in the Undercliff (Lee and Moore 1991; Moore et al. 1995). This relationship was refined to account for the spatial distribution of marginally stable and unstable areas in the Undercliff and to provide a probabilistic framework for quantitative risk assessment for Ventnor (Lee et al. 1998).

With the investment of real time monitoring of climate and ground behaviour response in Ventnor, and at other key sites in the Undercliff, it is now possible to examine in detail the relationships between rainfall, groundwater levels and ground movement rates and how, under current climate change predictions, these relationships can be expected to change in the future.

4 GROUND MOVEMENT IN VENTNOR

The following sub-sections consider the historical impacts of ground movement in Ventnor, the monitoring network put in place to record hydro-climatic and ground behaviour events, and detailed analysis of the data to establish relationships of the ground behaviour response to rainfall and ground water.

4.1 *Impacts of ground movement in Ventnor*

The impacts of coastal instability and ground movement on urban development within the Undercliff have been well documented (Lee & Moore 1991). Historical records indicate that throughout the development of Ventnor, over the past 200 years or so, the area has been affected by widespread landslide movement which has resulted in a range of problems. The movement has taken the form of very slow and intermittent creep of the ground, the cumulative effects of which, over many years, has caused significant damage to buildings and infrastructure. There have also been occasional more rapid landslide movements which have resulted in serious damage. Over the last 100 years about 50 houses and hotels have had to be demolished in Ventnor because of ground movement.

The most notable impacts of ground movement can be observed at the Lowtherville graben in Upper Ventnor (Fig 8). Anecdotal evidence suggests that the graben has developed since the early to mid-1900s. Ground movement in the area was first reported by Edmunds & Bisson in 1954. Today the graben extends some 500 m in length, 20 m across, and has subsided vertically by about 2–4 m. The main access road to Newport and other services cross the graben and are continually subject to damage due to ground movement. Since the 1980s, over 10 properties have had to be demolished in the area affected. The development of the graben is cause for concern as it may represent the early signs of major reactivation of the Undercliff in what is now a densely urbanised area (Hutchinson & Bromhead 2002; Moore et al. 2007).

The most significant episode of landslide activity at Ventnor in historical times occurred in the winter of 1960, following 713 mm of effective rainfall over the 4-month period between Aug-Nov, which has a return period greater than 1:170 years. This remains the most extreme rainfall event since records began. The landslide response caused widespread damage to property and infrastructure throughout the town and some properties had to be evacuated (Fig 9).

A systematic damage survey of property, walls and roads was carried out in 1990 as part of the Department of the Environment's landslip potential study at Ventnor (Lee & Moore 1991). The survey recorded damage according to five categories from negligible to severe. Results showed that approximately 23% of the area was affected by negligible to slight damage, 61% by moderate damage and 16% by serious or severe damage due to ground movement. Some sites were particularly noted for the impacts of ground movement, including:

- Newport and Steephill Down Road (location of the Lowtherville graben) in upper Ventnor
- Ocean View Road and Lower Gill's Cliff Road in mid-Ventnor, and
- Belgrave Road, Bath Road and the Esplanade in lower Ventnor.

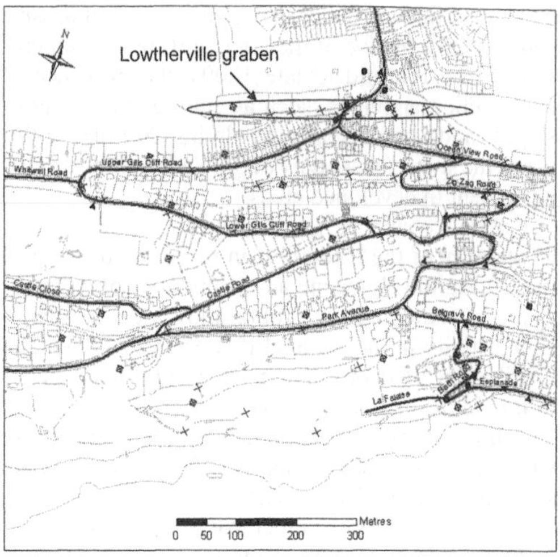

Figure 8.

Figure 9. Severe landslide damage, Bath Rd, 1960.

Table 3. Ground movement rates at Ventnor.

Location	Period	Rates of movement (mm/yr)
Newport Road	1982–1983*	19.7 to 39
Newport Road	1988**	28
Newport Road	1995–2005***	6.4 (*crack extension*) 33.2 (*settlement*)
Lowtherville graben	1982–1983*	26.9 to 29.9
Lower Gill's Cliff Road	1982–1983*	8 to 16.5
Ocean View Road	1982–1983*	6
Bath Road	1982–1983*	0
Bath Road	1988**	0 to 2
Bath Road	1995–2005***	−0.06 (*crack closure*) 22 (*settlement*)
Esplanade	1949–1988	20 (*heave*)

Note: *after Chandler(1984); ** after Woodruff (1989); ***Council records.

4.2 *Monitoring network at Ventnor*

A central element of the Council's landslide management strategy at Ventnor and other critical sites in the Undercliff, involves the maintenance of an extensive network of survey markers and automatic or manually-read monitoring devices. The network of survey markers and other installed devices is shown in Fig 8. In 1992, the following equipment was installed:

- an automatic weather station at Ventnor Park
- automatic crackmeters and settlement cells at the Lowtherville graben and Bath Road, and
- automatic vibrating-wire piezometers in two existing boreholes at the nearby Winter Gardens

The network was expanded in 2002 and 2005 with two boreholes fitted with manually-read inclinometers to monitor deep-seated displacement of the landslide; five other boreholes are fitted with multi-level piezometers to monitor groundwater levels. Inclinometer readings from Upper Ventnor (BH2) show 30 mm of movement at 94–98 m below ground level between September 2002 to August 2004 (Moore et al. 2007). No detectable deep-seated movement has been measured in BH4 which is located on the Western Cliffs, Ventnor Park. However, it is noted the inclinometer was only recently installed in June 2005 and may not have fully equilibrated bedded in with the surrounding ground.

In June 2003, a network of 50 permanent ground markers (pgms) was installed (Fig 8) and accurately surveyed using Global Positioning System (GPS). Repeat surveys were carried out in 2005 and 2006 to measure any movement of the markers; the results do not reveal any significant movement of the pgms over this short period.

The recent ground movement data have been collated with historical monitoring records (Table 3).

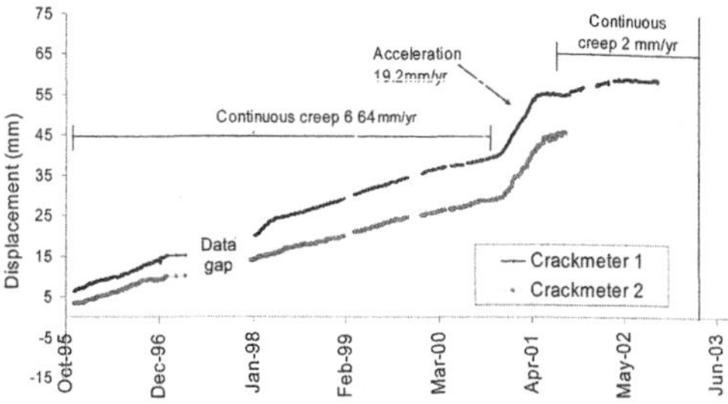

Figure 10. Crack displacement at Lowtherville.

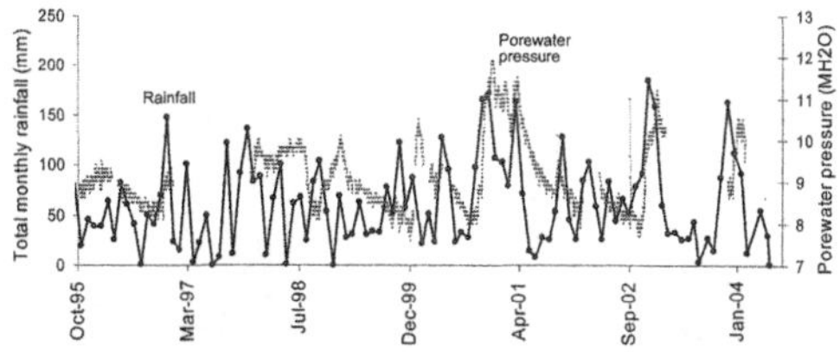

Figure 11. Rainfall and porewater pressure response.

Most sites show rates of horizontal movement of less than 6 mm/yr. Where there are overlaps with the direct and indirect measurements, the short-term movement rates appear consistent with long-term trends. The one exception to this is the Lowtherville graben where short-term movement rates up to 40 mm/year are greater than the long-term trend of 6 mm/year.

The automatic crackmeters and settlement cells have been recording data every three hours since 1992. Analysis of data for the Lowtherville graben shows a relatively continuous rate (6 mm/yr) of horizontal displacement, with a notable period of accelerated movement from November 2000 to February 2001 (Fig 10). Vertical displacement rates (33 mm/yr) are 5 times greater than crack extension, a characteristic feature of graben development.

4.3 *Ground behaviour response to rainfall*

The influence of antecedent effective rainfall conditions as a major control on landslide and ground movement activity was demonstrated in section 3. Prolonged periods of heavy rainfall act to raise groundwater levels and increase pore water pressures acting on shear surfaces which lead to a decrease in effective strength and potential failure.

Daily rainfall has been automatically recorded at Ventnor Park since 1992. A continuous record of daily groundwater levels recorded at the Winter Gardens is available for the same period. Comparison of the two datasets reveals a close relationship between rainfall and groundwater level (Fig 11); there is a notable 1 month lagged response between peak monthly rainfall and groundwater level.

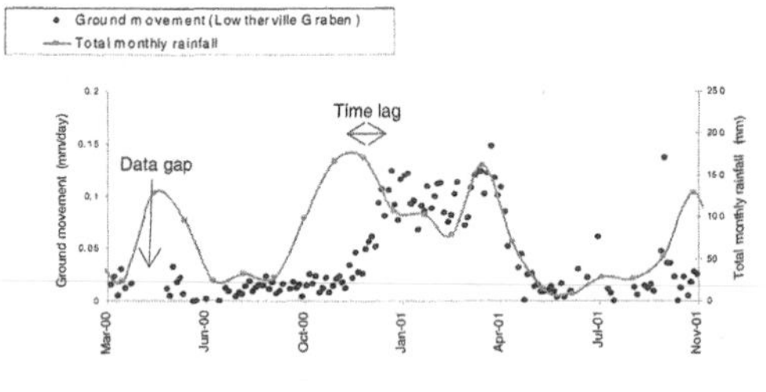

Figure 12. Ground movement response to rainfall.

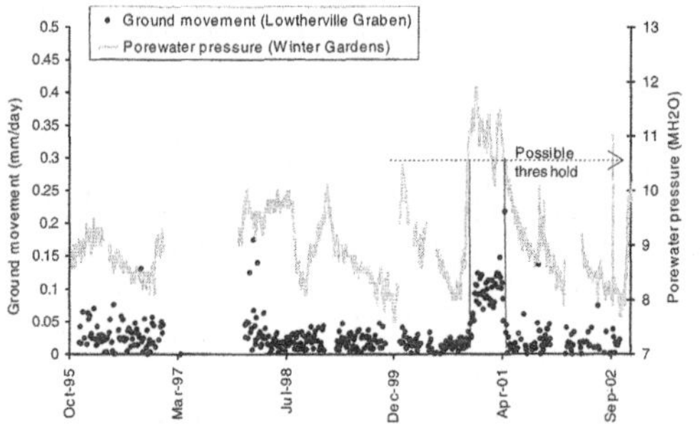

Figure 13. Ground movement response to porewater pressure.

A lagged response is also apparent between rainfall and ground movement (Fig 12). The data show the relationship between monthly rainfall and daily ground movement associated with a 1:17 return period 527 mm 4-month AER event between October 2000 and January 2001. A 3-month lag between peak monthly rainfall and ground movement acceleration is apparent at the Lowtherville graben, which occurred in December to March, resulting in damage to Newport Road and other assets in the area.

Fig 13 shows the ground movement response to porewater pressure by comparing crackmeter 1 at the Lowtherville graben with ground-water levels recorded at the Winter Gardens. Although the two sites are separated by a distance of about 1km, a good correlation is apparent indicating there may be close spatial continuity of the hydro-geological regime in the Undercliff. The data show that ground movement velocity is generally below 0.1 mm/ day. During the autumn of 2000, groundwater levels rose to their highest level so far recorded since 1992, and ground movement rates increased above 0.1 mm/day. The ground behaviour response is reversible with ground movement velocity reducing with groundwater level. Further, it is noticeable that the acceleration and deceleration of ground movement occurs during peak groundwater levels, indicating that ground movement is likely to be controlled by a specific porewater pressure threshold. These results confirm there are strong relationships between antecedent rainfall, groundwater levels, porewater pressure and ground movement rates. They provide real-time evidence that supports the established relationship between historical landslide event frequency and the 4-month AER.

5 SO WHAT ARE THE IMPLICATIONS OF CLIMATE CHANGE IN THE UNDERCLIFF?

The implications of climate change predictions for the Undercliff are both spatial and temporal; firstly, there are concerns that hitherto marginally stable areas of the Undercliff may become unstable due to reactivation of ground movement and the occurrence of new landslides, secondly, in areas previously affected by ground movement or landslides, the frequency and rate of ground movement and landsliding is expected to increase. The main consequence of predicted climate change on the stability of the Undercliff is likely to be an increased risk of damage to assets due to ground movement, particularly in built up areas, such as Ventnor.

The UKCIP02 predictions point to significant increases in winter rainfall frequency, intensity and amount, drier summers, sea level rise and more frequent storms. These will affect the two main causes of landslides in the Undercliff:

1. cliff erosion and removal of passive support, and
2. antecedent effective rainfall and groundwater levels.

5.1 *Cliff erosion and removal of passive support*

It is difficult to predict what effect a 0.84 m rise in sea level will have on the rate of erosion for unprotected cliffs; the cliff response is complex and depends on a number of factors, not least the cliff geology, nature of the beach and shore platform, and hydrodynamics. There are few studies that have considered the effects of past sea level rise on historical rates of recession. Historical data from the Suffolk and Holderness coast suggests that 20–30 mm of erosion has occurred for every 1 mm rise in sea level since the early 1950s (Lee pers. comm. 2006). Clearly, these data require further analysis and comparison with other sites to check on their broad applicability. Nevertheless, taking 30 mm as an upper bound erosion rate increase for every 1 mm rise in sea level, 0.84 m sea level rise would add 25 m to the total projected recession to 2080 using historical erosion rates. If this were applicable to the Undercliff, the current recession rate of 0.3 m/year could have more than doubled by 2080 due to the effects of 0.84 m sea level rise and more frequent storms.

Whilst the actual effects of sea level rise on future erosion rates are uncertain, the spatial sensitivity of the Undercliff to toe erosion, landslides and ground movement is better understood. The ground behaviour maps referred to in section 2, identify areas of inactive

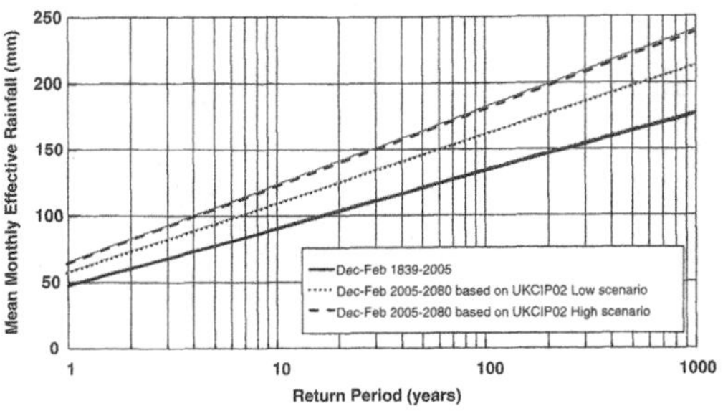

Figure 14. Predicted increase in mean monthly rainfall by 2080.

ancient landslides, active and degrading landslides, and locally or intermittently active landslides for the entire Undercliff; these maps can be used to locate where future landslide reactivation events are most likely.

With the exception of the inactive ancient landslides, which cover a large part of the Undercliff, an increased frequency and spread of landslide reactivation events can be expected in areas previously affected by local or intermittent ground instability; the majority of these areas are subject to active shoreline retreat, and the effects of future sea level rise and higher rates of erosion can only exacerbate the landslide response.

The majority of the inactive ancient landslides are either protected by extensive natural debris aprons or coastal protection structures which will counter the effects of future sea level rise and accelerated rates of erosion to a greater or lesser degree. Maintaining the current standard of protection in years to come will be a major issue for coastal defence management.

5.2 *Antecedent effective rainfall and groundwater*

Whilst the summer soil moisture balance is expected to fall by up to 40% this change will be manifest at the ground surface in the form of desiccation cracks and is unlikely to have much influence on groundwater levels fed by deep aquifers, such as the Undercliff (Moore et al. 2007).

The predicted increase in winter rainfall by up to 30% is of greater concern as this is likely to lead to an increase in the likelihood of extreme antecedent effective rainfall conditions. Added to this, the UKCIP02 predict an increase in the intensity of rainstorms which ultimately trigger landslide events.

A measure of the potential impact of an increase in winter monthly rainfall is provided in Fig 14. The graph shows the historical frequency of Dec-Feb mean monthly effective rainfall for the Undercliff and applies the UKCIP02 Dec-Feb percentage increases for the low and high scenarios. The results indicate that a current 1:200 return period event could become a 1:60 or 1:30 return period event in 2080 for the low and high scenarios, respectively. This implies that extreme monthly rainfall will become 3 to 6 times more likely in the future.

A similar effect is observed when applying the UKCIP02 winter rainfall increases to 2080 using the established 4-month AER relationship with histori cal landslide events (Fig 15). This relationship shows that a current 1:200 return period event could become a 1:10 year event by 2080. This implies that an equivalent rainfall event such as that experienced in the winter of 1960, which has a current return period greater than 1:170 years, could have an expected frequency of about 1:10 years by 2080. It is, therefore, very important to provide the necessary additional risk management measures, such as drainage, as part of the ongoing Undercliff Landslide Management Strategy.

6 CONCLUSION

The Undercliff shows marked spatial and temporal sensitivity to ground movement and landslides that reflect a legacy of past geological and geomorphological processes and formative events. Historical and contemporary data are presented to demonstrate the strong relationships that exist between antecedent rainfall, groundwater and ground movement response. These relationships are used to project the likely impacts of climate change predictions on the future stability of the Undercliff. In summary these include:

- a significant increase in the frequency of extreme 4-AER and accelerated ground movement rates in marginally unstable areas
- landslide reactivation events in previously unaffected areas due to the progressive effects of coastal erosion and transient high groundwater levels
- reduced lag times between rainfall, groundwater and ground movement response due to increased rainstorm intensity.

There is little doubt that climate change poses a very real and significant challenge to the future management of the Undercliff and other similarly marginally stable ancient landslides in southern Britain. The value of monitoring of hydro-climate and ground movement response is clearly demonstrated in this paper. The relationships and understanding derived from analysis of these data provide an opportunity to design robust early warning and response strategies in the event of major ground movement in more developed areas.

ACKNOWLEDGEMENTS

The authors would like to thank colleagues at Halcrow and the Isle of Wight Coastal Centre for the Environment who have assisted in this project; we also acknowledge the valuable contributions of former

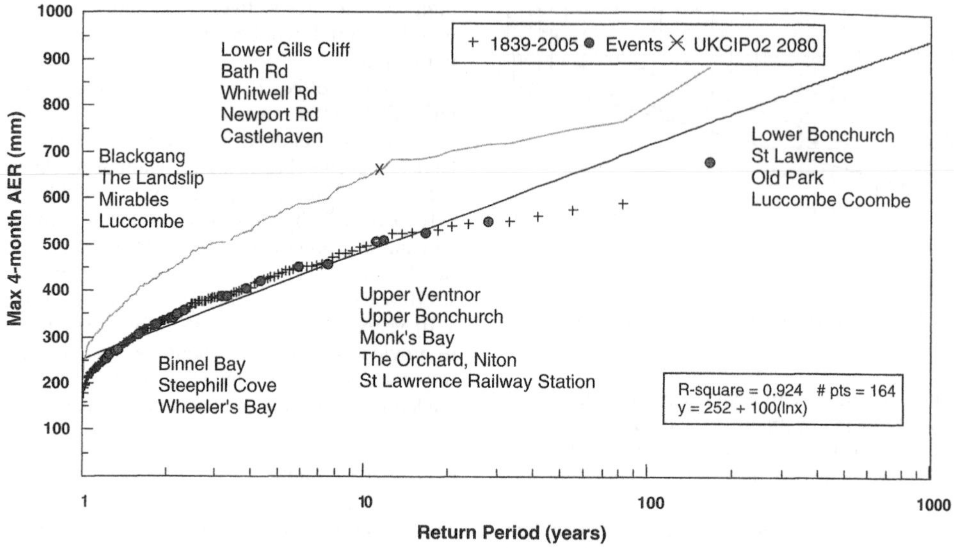

Figure 15. AER curve for 2080 based on UKCIP02 Med-High winter rainfall predictions.

colleagues and consultants at Geomorphological Services Limited and Rendel Geotechnics during the development of the landslip potential approach and implementation of the Undercliff Landslide Management Strategy in the early 1990s. Finally, we are grateful to the Isle of Wight Council who have invested significantly in the continued maintenance, monitoring and analysis of hydro-climatic and slope monitoring instruments at Ventnor and other key locations in the Undercliff. Without this investment, the advancement in the understanding of Undercliff ground behaviour would not be possible.

REFERENCES

Chandler M.P. 1984. The coastal landslides forming the Undercliff of the Isle of Wight. PhD thesis, University of London.

Edmunds F.H. & Bisson G. 1954. Geological report on road subsidence at Whitwell Road, Ventnor. Unpublished report.

Hosking A. & Moore R. 2002. Preparing for the impacts of climate change on the central south coast of England. International Conference on Instability, Planning and Management. Ventnor, Isle of Wight, May 2002.

Hulme M., Jenkins G.J., Lu X., Turnpenny J.R., Mitchell T.D., Jones R.G., Lowe J., Murphy J.M., Hassell D., Boorman P., McDonald R. & Hill S. 2002. Climate Change Scenarios for the United Kingdom: the UKCIP02 Scientific Report. Tyndall Centre for Climate Change Research, School of Environmental Sciences, University of East Anglia, Norwich, UK.

Hutchinson J.N. & Bromhead E.N. 2002. Keynote Paper: Isle of Wight landslides, In McInnes R.G. & Jakeways J. (eds.) Instability Planning and Management: seeking sustainable solutions to ground movement problems. Proc. Int. Conf, 3–70, Ventnor. Thomas Telford.

Ibsen M.L. & Brunsden D. 1997. The nature, use and problems of historical archives for the temporal occurrence of landslides, with specific reference to the south coast of Britain, Ventnor, Isle of Wight. *Geomorphology*, 15, 241–258.

Jones D.K.C. & Lee E.M. 1994. Landsliding in Great Britain: a review for the Department of the Environment. HMSO.

Lee E.M. & Moore R. 1991. Coastal Landslip Potential Assessment, Isle of Wight Undercliff, Ventnor. Technical Report prepared by Geomorphological Services Ltd for the Department of the Environment, research contract PECD 7/1/272.

Lee E.M. & Moore R. 2007. Ventnor Undercliff: development of landslide scenarios and quantitative risk assessment. This conference.

Lee E.M., Moore R. & McInnes R.G. 1998. Assessment of the probability of landslide reactivation: Isle of Wight Undercliff, UK.

McInnes R.G. 2000. Managing ground instability in urban areas: a guide to best practice. IW Centre for the Coastal Environment.

Moore R., Turner M.D., Palmer M. & Carey J.M. 2007. The Ventnor Undercliff: a new ground model and implications for climate induced landslide behaviour and risk. This conference.

Moore R., Lee E.M. & Clark A.R. 1995. The Undercliff of the Isle of Wight: a review of ground behaviour. Cross Publishing.

Norman M.W. 1887. A popular guide to the geology of the Isle of Wight. Ventnor.

Whitehead J.L. 1911. The Undercliff of the Isle of Wight past and present. London.

Woodruff M. 1989. Monitoring of ground movement in Ventnor. Report to South Wight Borough Council.

Worsley R. 1781. The history of the Isle of Wight. London.

Landslides and Climate Change – McInnes, Jakeways, Fairbank & Mathie (eds)
© 2007 Taylor & Francis Group, London, ISBN 978-0-415-44318-0

Monitoring coastal slope instability within the western Undercliff landslide, Isle of Wight, UK

A. Gillarduzzi, A.R. Clark & D.S. Fort
High-Point Rendel, London, UK

J.E.M. Houghton
Isle of Wight Council, Isle of Wight, UK

ABSTRACT: The Undercliff landslide system located on the southeast coast of the Isle of Wight is one of the largest developed landslides in western Europe approximately 12 km in length and extending up to 0.6 km inland from the coast. Coastal erosion, high groundwater levels, rainfall and susceptible geology all contribute to the reactivation of this landslide with many properties and infrastructure being at risk from ground movement. This paper describes the approach and procedures adopted at a number of vulnerable sites along the Undercliff to monitor and process inclinometer, weather and groundwater data to enable a current assessment of landslide behaviour within the present climate conditions and to act as part of a warning system before proposed stabilization works are implemented.

1 LOCATION

The study site is located parallel to the southeast coast of the Isle of Wight and it is crossed by a section of the A3055 Undercliff Drive of approximately 2 km in length between the villages of Niton and west of St Lawrence. The site is centred on coordinates 50°34′53″N and 1°16′07″W (Figs. 1–2).

The A3055 Undercliff Drive is an important road transport link that connects the small towns of Niton to the west to St Lawrence, Ventnor and Bonchurch to the east and serves a number of villages and properties along the route. The A3055 is also a major route for tourist traffic to the west of the Island providing access to areas of outstanding natural beauty.

2 STUDY SITE

The study site covers four main separate sites along the Undercliff Drive that have historically been significantly affected by instability and ongoing

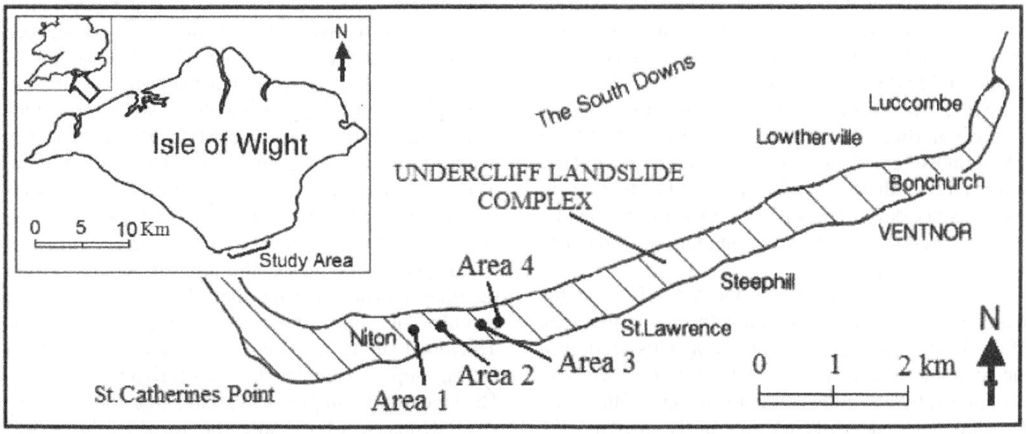

Figure 1. Location map.

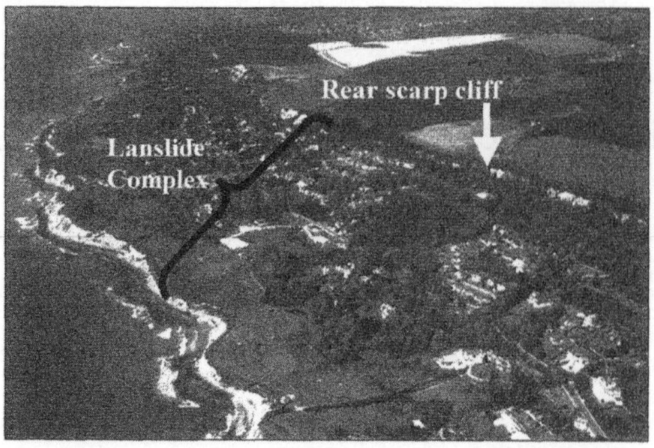

Figure 2. The western Undercliff.

Figure 3. Catastrophic failure of the A 3055 Undercliff Drive, in Area 1, during the winter of 2001/2002.

movement and are referred to as Area 1 (Beauchamp), Area 2 (Mirables), Area 3 (Undercliff Glen) and Area 4 (Woodlands).

Area 1: Beauchamp is located approximately 1 km to the east of Niton. This section is characterised by significant ongoing compound-rotational landslide movement and the road, which has a history of distress, failed catastrophically during the winter of 2001/2002 (Fig. 3). A new realigned section of the A3055 was reconstructed in the summer of 2002 as emergency temporary works following this event.

Area 2: Mirables is located approximately 250 m to the west of Mirables, c.400 m east of Area 1. In March 2001 cracks were identified in the carriageway and in a wall on the southern side of the road and further cracking has since developed with a significant step within the carriageway caused by the ongoing movement of deep-seated sliding.

Area 3: Undercliff Glen is located 1.5 km to the west of St Lawrence and immediately to the west of the Undercliff Glen Caravan Park. This section of the road shows evidence of instability as a result of the reactivation of a landslide downslope of the road. The slopes at Undercliff Glen previously failed in 1926 resulting in the relocation of the road to its present alignment. Further large-scale slope failures occurred during the winter of 2000/2001 such that the landslide back scarp has reached to within 2 m of the road. In response to this emergency situation the westbound carriageway was closed and traffic management system put in place and a restriction limiting the width and weight of vehicles was imposed on the road. The slopes at this site are actively unstable.

Area 4: Woodlands is located 200 metres to the east of Undercliff Glen and is characterised by a significant depression of the road and associated cracking of the pavement. This section of the A3055 has been subject to on-going settlement over many years. The settlement has been attributed to shallow multiple-rotational and translational failures beneath and downslope of the road triggered by the extremely high groundwater levels at this site.

The general instability of the Undercliff and accelerated periodic reactivation of the landslide, within the limits of the study site, pose a risk to various properties and infrastructure, including the A3055 Undercliff Drive.

3 GEOLOGY AND GEOMORPHOLOGY

The geology of the southern coast of the Isle of Wight comprises Upper and Lower Cretaceous strata. The youngest deposits include Chalk, Upper Greensand (Chert Beds, Malm Rock and Passage Beds) and Gault Clay, which overlie the older deposits of the Lower Greensand Formation (Carstone and Sandrock) (Hutchinson & Bromhead 2002).

The study site lies within the Undercliff landslide system, which is one of the largest developed landslide systems in Western Europe approximately 12 km in length and extending up to 0.6 km inland from the coast.

The inland extent of the Undercliff is generally marked by a prominent south facing rear-scarp cliff. The top of the scarp rises to elevations ranging between about 105 mOD and 150 mOD. The Chalk is mostly absent above the rear-scarp in the site area resulting in an Upper Greensand plateau with massive cherty sandstone strata (Chert Beds and Malm Rock) generally exposed in the near-vertical cliffs. The upper section of the rear-scarp cliff is unaffected by deep landsliding, although shallow slides in weathered strata, soil erosion and creep and rockfall may occur occasionally.

The geomorphology of the site and borehole investigations indicate that the main Undercliff landslide complex lies immediately below the cliff and it is developed along a series of variably steep, south facing slopes separated by sub-horizontal benches. The Undercliff comprises a variable thickness of landslide deposits, which take the form of large and small landslides of various types plus benches, blocks, debris, mudslide deposits, talus, etc. This debris varies in characteristics, thickness and lithology depending on its source materials and the degree of disturbance it has undergone.

The landslide is mainly formed by Upper Greensand debris and disturbed Gault Clay (known locally as "Blue Slipper"), underlain by intact Gault Clay and sandstones (Lower Greensand). Of particular importance is the presence of thin clay layers within the Sandrock (Lower Greensand), which together with the Gault Clay largely influence the stability of the entire area (Hutchinson 1991). The landslide complex can be subdivided into two tiers:

Upper Tier consists of a broad zone of multiple rotational slides giving rise to linear benches separated by scarp slopes. This unit comprises back-tilted blocks of Upper Greensand and Chalk that were deposited over a long period of time following movement of the landslide features located downslope. This has led to the removal of lateral support for the cliff resulting in the development of open joints ("vents") mainly parallel to the cliff and subsidence (increasing seaward) of the intervening blocks.

The Lower Tier consists of a sequence of compound slides.

The two tiers are separated by a prominent steep largely un-vegetated east-west Gault Clay scarp that is an often prominent feature across the whole of the Undercliff. The toe of the shallow mudslides in the Upper Tier often coincides with the top of the Gault Clay scarp. Relic landslide debris at the foot of the inland scarp obscures the Passage Beds underlying the Gault Clay.

Recent instability movement of the Undercliff landslide has exposed the in situ Lower Greensand (Carstone and Sandrock) at the base of the Gault Clay scarp. The Carstone is locally argillaceous and variably cemented coarse sand and grit with quartzite pebbles, while the Sandrock sequence is traditionally subdivided into five lithological units predominantly sandstone and including thin low permeability strata.

The coastal slopes are bounded by vertical to sub-vertical sea-cliffs cut into in situ strata and landslide debris. The geological structure of the area includes the St Lawrence syncline (Hutchinson & Bromhead 2002) superimposed on the east-west axis of the main folding in the area between the west-central Wight and the Undercliff. The axis of the syncline plunges gently to the south in the vicinity of St Lawrence. On either side of this axis the elevation of strata increases, leading to significant variations in geological exposures and influencing the mechanisms of landsliding along the coast. The strata in the site area generally dip gently to the southeast at 1°–2°. The Upper Greensand strata are strongly jointed by two principal sets of near-vertical joints one of which is parallel to the rear scarp.

Hydrogeologically the Undercliff landslide is closely connected with the Southern Downs (Shaw & Packman 1988) to which the entire input of water is by precipitation. The watershed of the Southern Downs is located in close proximity to its southern edge (i.e. the Undercliff back-scarp) and consequently only a limited amount of near-surface water runs southward towards the Undercliff landslide complex. Sub-vertical fractures and the slightly southerly dip of the strata allow deeper infiltration towards the landslide with development of spring lines mainly the Upper Greensand Passage Beds, which are generally concealed by debris. The infiltration of groundwater through fracture networks, probably at least partially exploiting existing buried valleys, is considered the principle cause of the localised accelerated movement over discrete sectors of the Undercliff complex.

4 EVOLUTION OF INSTABILITY

The first major compound failures in Cretaceous clays and sandstones probably commenced offshore of the present Undercliff in the Late Quaternary (Hutchinson 1987). The present Undercliff is a complex of ancient deep-seated landslides that was formed as a result of landsliding which took place after the last glacial period c. 8500yrs BP (Tomalin 2000) following major changes in climate and sea level and consequent effect on marine erosion along the southern coast of the Isle of Wight. A second phase of instability, c. 5000–4000yrs BP, probably followed and continued over the centuries up to date with alternated periods of relative stability and phases of intense movement (Ibsen 2000). A detailed description of the evolution of the Undercliff is presented elsewhere (e.g. Hutchinson & Bromhead 2002).

Previous studies (e.g. Moore et al. 1995) indicate that some sites along the entire Undercliff are actively unstable or close to instability although considerable areas are relatively stable.

Ground movements along the entire Undercliff may be sub-divided into three categories:

- very slow intermittent movement where services and property are largely unaffected by movements;
- slow but significant ground movement, which has caused moderate and occasionally serious damage to property, infrastructures and services;
- recurrent, severe and often dramatic accelerated movement, generally related to climatic extremes, which can lead to problems for property, infrastructures and services. These sites are situated almost exclusively within the largely undeveloped study area.

5 MONITORING

5.1 *Historical records*

The activity of the Undercliff landslide, over the last 223 years is inferred from records maintained by local authorities (e.g. survey, road repair, etc.) and statutory undertakers, local newspaper, postcards, historical photographs, engravings, Ordnance Survey maps, oral records, direct observations and technical papers (e.g. Moore et al. 1995) and it is consequently relatively well known although generally limited to the areas of easy access (e.g. Upper Tier) and those more developed.

5.2 *Monitoring instrumentation*

Remediation of landslide damage to the A3055 has been a longstanding maintenance requirement for the Isle of Wight Council (IoWC). Various early studies of the Undercliff have been carried out during the years (e.g. Moore at al. 1995, Hutchinson et al. 1985, Chandler 1984). The detailed systematic study of the area, with the scope of identifying the causes of instability to develop a landslide management strategy and remedial design for the IoWC, commenced in the early-nineties. This included the review of available records, reports and documents followed by a programme of detailed investigations involving systematic geomorphological and geological mapping, review of historical air-photographs, assessment of ground movement rates, and surveys of damage caused by ground movement. This was followed by detailed topographic survey and various stages of ground investigations including updated geomorphological and water features mapping, pumping tests and the installation of groundwater and meteorological monitoring and inclinometers

The monitoring system installed along the study site (Tab. 1), commencing from April 2001, includes 65 separate items of individual instrumentation. In the summer of 2006, 49 instrumentations were active while the remainder had been severed, blocked or were no longer accessible.

The meteorological data for the area are gathered, every month, from two automatic weather stations. The older, operating from April 1992 is installed at Ventnor Park (i.e. central Ventnor), 4.9 km to the northeast of the centre of the site, while the second, commissioned in April 2004, is installed at Castlehaven as part of the "Castlehaven Coast Protection Scheme, Niton, IoW" (Clark et al. 2007) and is located only 1.3 km to the southwest of the site and is therefore considered to provide the most site specific meteorological data.

The main purpose of instrumenting and monitoring the site was: to obtain ongoing information on landslide behavior to establish the ground model and landslide mechanisms; to identify patterns of evolution of the landslide systems over several years to assist in the prediction of their behavior in the future and to provide information for the design of planned remedial works.

Table 1. Monitoring instrumentations installed along the Undercliff within the study area.

Type of Monitoring instrumentation	Total No.	Active No.	Inactive (2006) No.
Weather station♦	2	2	0
Inclinometer	15	10	5
Trial pit piezometer*	1	1	0
Well piezometer*	3	1	2
Well piezometer♦	4	4	0
Vibrating-wire piez.♦	14	14	0
Drive-in piezometer*	17	11	6
Standpipe piezometer* in borehole	9	3	6

*Manual dip monitoring ♦Installed with automatic datalogger.

The Undercliff landslide complex has a number of important environmental designations that affect and constrain the options available of stabilisation of the area. These designations include Area of Outstanding Natural Beauty (AONB), Heritage Coast, Site of Importance for Nature Conservation (SINC) and Site of Special Scientific Interest (SSSI) and it is adjacent to a Special Area of Conservation (SAC). The monitoring has also the purpose to create a benchmark conditions database to assess environmental changes at the site.

Although not an objective of its original design, the monitoring system, since 2005, is used as part of an early warning system to give prior indication of possible destructive landslide events that may damage the A3055 Undercliff Drive, properties, services and pose at risk the public. This specific use of the monitoring system has required the definition of a simple mechanism to validate the monitoring data through cross checking and the identification of site-specific trigger levels and response actions for the IoWC.

The instrumentation monitoring regime, with the exception of the weather stations, is kept flexible so it may be adjusted to meet changing circumstances. The monitoring data are gathered manually and automatically using single channel dataloggers to provide a quantitative database. Geotechnical walkover surveys are carried out on a regular basis to estimate the evolution of the site and confirm instrumental observations. During the winter period, when historically accelerated slope movement has taken place, the frequency of monitoring is increased to track the development of the slopes while, during the summer period, only minimum, intermittent monitoring is carried out. Although this regime is optimal for the purpose of

Table 2. Sequence used to identify specific factor thresholds triggering accelerated slope stability within the study site.

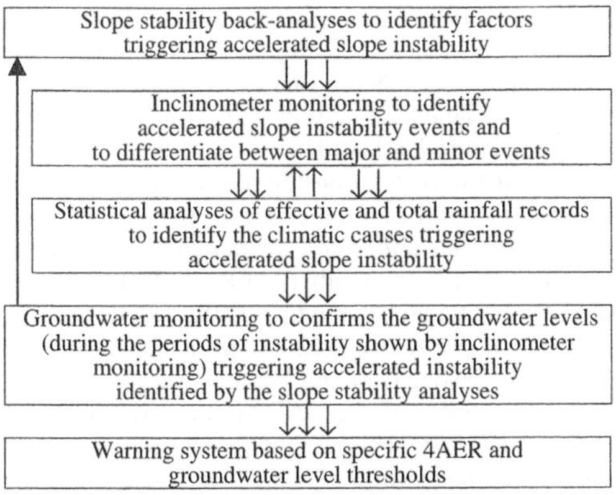
Slope stability back-analyses to identify factors triggering accelerated slope instability
Inclinometer monitoring to identify accelerated slope instability events and to differentiate between major and minor events
Statistical analyses of effective and total rainfall records to identify the climatic causes triggering accelerated slope instability
Groundwater monitoring to confirms the groundwater levels (during the periods of instability shown by inclinometer monitoring) triggering accelerated instability identified by the slope stability analyses
Warning system based on specific 4AER and groundwater level thresholds

identifying the winter risk of instability and to optimize the resource available to IoWC, this frequency is not always ideal to identify longer-term patterns of evolution of the Undercliff landslide complex. Ideally this would require regular monitoring over the entire year.

5.3 *Monitoring data interpretation*

The monitoring data interpretation has been carried out on groups of instrumentation (i.e. inclinometer, groundwater and meteorological monitoring) based on the use of Excel spreadsheets, specialist software and graphical presentations.

The following procedure (Tab. 2) is used for the interpretation and validation of monitoring data and slope stability analyses and for the identification of specific factor thresholds triggering accelerated movement along the Undercliff.

6 SLOPE STABILITY ANALYSES

The stability of each area along the study site has been back analysed to identify the cause of accelerated slope instability.

A number of ground model sections along the direction of maximum movement were generated from a dedicated topographical survey. The stratigraphy of the landslide has been interpolated from borehole logs and geomorphological mapping. However, as ground investigation information becomes less detailed south of the A3055 Undercliff Drive due to difficulties in access due to difficult terrain within the landslide, assumptions about the stratigraphy in these areas have been made based on a number of observations including geomorphological mapping and the nature and extent of the Gault Clay scarp.

Various scenarios have been considered in the back analyses including the assumption that both the Upper Tier rotational slides and the Lower Tier translational slides are at limit equilibrium under average winter groundwater levels, with residual shear strength mobilised along the shear surfaces. The analyses, under these limiting conditions, confirm that accelerated slope movement only occurs as a result of a trigger event that can include high groundwater levels, removal of the toe of the landslide by erosion of the sea cliffs, loading of the head of the landslide through failure of the rear scarp or human activity.

Extensive ground investigations, the monitoring described below and the development of an accurate ground model have confirmed the main landslide failure mechanisms, at this stage of the landslide evolution, as being reactivation of the landslide due to high groundwater levels and recession of the sea cliffs due to marine erosion. The periodic rise in groundwater level is predominantly related to rainfall but it is also influenced by uncontrolled discharge of surface water drained from roads and probably also by domestic sources.

7 INCLINOMETER MONITORING

During the various ground investigations carried out within the study site since 2001, 15 borehole biaxial inclinometers, ranging in depth from 35.1 m to 52.3 m (total 456 m), were installed at key locations with the purpose of monitoring subsurface ground movement with time at various levels within the landslide. The location of the inclinometers was influenced by the severe access problems caused by the landslide terrain and private property limits and they are therefore not always at the best position to identify maximum accelerated slope movements. Since installation, five inclinometers, installed in particularly active areas, were severed and blocked and were abandoned; some were re-drilled. The inclinometers are considered to provide the best and most accurate factual information to detect changes in rate of ground movement along the slip surfaces.

The interpretation of the inclinometer data is supported by the use of the following graphs:

- Absolute Position Graph showing the absolute position of the inclinometer tubes with time and depth;
- Cumulative Displacement Graph (CDG) showing the amount, levels and direction of slope movement along the inclinometer tube length allowing the identification of the development of new slip surfaces;
- Time Plot Graph (TPG): showing the inclinometer displacement versus time, at specified depths

intervals, in proximity to identified slip surfaces identified in the CDG and over the entire inclinometer length. The displacement has been calculated over a 3 month period (3TPG) because of the limited, and sometimes sporadic, database not allowing a higher level of accuracy.

Inclinometer records and consequently the 3TPGs for the study area are available only for the most recent, recognized accelerated slope instability events (Fig. 4). The inclinometers recorded accelerated movement at various depths on three main occasions.

A large magnitude instability event occurred during winter 2002–2003 with peak movement recorded in January 2003. This was preceded by a progressive increase in displacement along the identified slip surfaces and creep, which commenced at least 3 months earlier and it was followed by movement for 6 to 9 months. This instability was manifest almost contemporarily in all areas and depths and coincided with exceptional wet weather, which led to widespread landslide activity along the entire Undercliff, causing distress along sections of the A3055 although no catastrophic failure of the road occurred.

The 3TPG of BH3 (Area 1), which commences from July 2001, also shows what is probably the "tail" of a previous large magnitude slope instability event occurred in winter 2000/2001 when unprecedented rainfall led to significant land movement and renewed distress at several locations along the A3055. This event, in early 2001, caused catastrophic failure at two different sites. To the west, in Area 1, the road was breached by landsliding at Beauchamp House resulting in its closure and the house was condemned.

To the east, at the location of Undercliff Glen Caravan Park (Area 3), the back scarp of a major landslide reactivation moved closer to the A3055 westbound carriageway and the road showed significant cracking.

A low magnitude instability event occurred in late winter 2003–2004 and generally resulted in substantially less intense movement compared to winter 2002–2003. In this case the inclinometers indicate that the movement often started at a different time along the study site. The movement recorded in most inclinometers was initially at shallow depth (e.g. inclinometer I1, I2, I4, I5) and then progressively, within various months, developed at greater depth.

The 3TPG of BH3 (Area 1) shows an event of similar magnitude that occurred in winter 2001–2002.

Unfortunately there are no other inclinometer data for the same period to be compared with and this instrumentation was sheared and abandoned in spring 2003 not allowing a comparison with the winter 2003–2004 instability event.

Very low magnitude instability occurred in late winter 2004–2005 and generally resulted in movement over a short period of time compared to the 2003–2004 event.

8 METEOROLOGICAL MONITORING

Previous studies (e.g. Hutchinson & Bromhead 2002, Moore et al. 1995, Lee and More 1991, DOE 1991) and anecdotal evidence indicate that heavy rainfall over relative long periods are one of the main triggers of accelerated landslide activity and ground movement along the entire Undercliff.

To confirm the relationship between ground movement and rainfall, the antecedent effective and total rainfall for the site were calculated over 5 month (5AER, 5ATR), 3 month (3AER, 3ATR), 2 month (2AER, 2ATR) and 1 month (1AER, 1ATR) periods, from April 1992 to August 2006, using meteorological data from Castlehaven and Ventnor weather stations. The comparison of the 4AERs (Fig. 5), inclinometer data (3TPGs), (Fig. 4) and direct observations of accelerated instability allows defining the following three magnitude instability events and relative 4AER thresholds.

- Major instability event: consistent accelerated landslide events over the entire study area but with substantially different magnitude and intensity from place to place. The movement commences when the 4AER exceeds 375 mm, accelerates when the 4AER is over 470 mm and peaks with the maximum 4AER. The time lag between the 4AER exceeding the 470 mm threshold and accelerated movement is minimum (i.e. generally <3 months);
- Minor instability event: minor accelerate movement events over the entire study relatively consistent in magnitude from place to place. The movement commences when the 4AER exceeds 375 mm and the time lag between 4AER exceeding the threshold and accelerated movement is between 3 to 9 months, with shallow movement (c. 0–15 mbgl) and creeping-type movement occurring within 3 months followed by movement at depth (c. >15 mbgl) over a longer period of time.
- Recurrent winter instability: minor accelerated instability triggered by seasonal winter rainfall with the 4AER threshold less than 375 mm and including creeping-type movement and rapid, very small magnitude, instability accelerations from time to time.

The 4AER and 4ATR thresholds for Minor and Major landslide events calibrated are summarized in Table 3.

The 4AER provides (Fig. 5) the best fit with the identified accelerated landslides events although the 3AER still allows identifying the major and minor events.

The 2AER graph (Fig. 5) indicates that major events are triggered by effective rainfall over short period of a couple of months and consequently these events are likely to be highly sensitive to climatic variability and

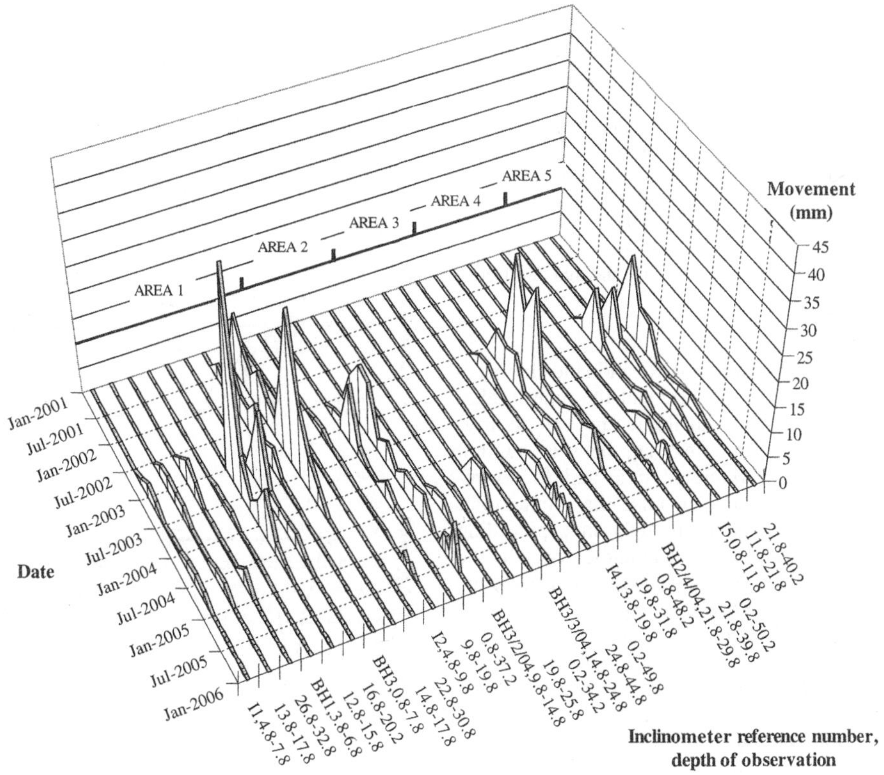

Figure 4. Time plot graph of inclinometers records for the study site. The displacement has been calculated over a 3 month period (3TPG) because of the limited, and sometimes sporadic, database not allowing a higher level of accuracy.

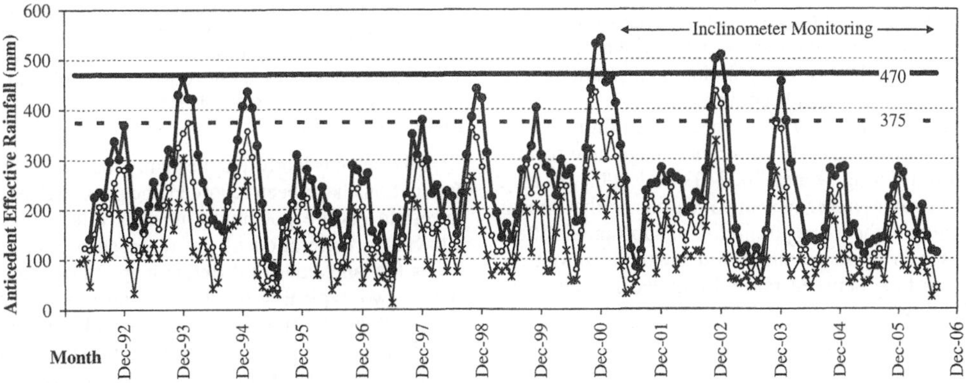

Figure 5. Antecedent Effective Rainfall (AER) series and Four Month Antecedent Effective Rainfall (4AER) thresholds causing landslide reactivation. Continuous horizontal line = 4AER major landslide reactivation threshold; Horizontal dotted line = 4AER minor landslide reactivation threshold; Other lines 4AER (bold line), 3AER (empty circle) and 2AER (X symbol).

to weather extremes. This is also confirmed by the 1ATR.

It should be commented that both the AER and ATR correlations with accelerated landslide reactivation may be complicated by other influences. The Undercliff landslide system is very complex and includes various unstable areas behaving differently and the instability may not be directly related to rainfall alone but triggered in combination or solely as result of marine erosion or other factors.

Table 3. Climatic series thresholds causing landslide reactivation in various scenarios.

	Landslide accelerated reactivation thresholds:	
Statistical analysis	Minor event mm*	Major event mm*
4AER	375	470
4ATR	395	470

* Approximate mm of rain for the ATR and mm of effective rain for the AER.

The correlation between AERs and landslide reactivation does not take into account the time it takes for effective rainfall to infiltrate the soil to trigger the instability. The time lag between rainfall and infiltration is variable in time and for example more water is needed to raise the groundwater table following periods of drought in a preceding period. The lag is also dependent and dictated by the site-specific geology, geomorphology, permeability, vegetation coverage, etc. For example minor instability events at Mirables (Area 2) are expected when the 4AER exceeds c.260 mm, and therefore below the average 4AER for the study area. Accelerated landslide reactivation is considered more likely to occur after long period of relative stability in which the shear stress is allowed to build up. The duration of accelerated movement is related to the length of the 4AER exceeding the specific threshold with heavy rain generally required over a long period to maintain accelerated slope instability. The quantity of effective rainfall causing slope movement is largely dictated by the point of evolution of the instability (e.g. first landslide, re-mobilisation of instability) and location along the Undercliff landslide complex.

Therefore although it is expected that future accelerated slope instability of similar magnitude to that occurred during winter 2002–03 and 2003–04 may lead to inclinometer movement similar to that shown in Figure 4, there is a possibility of substantial differences in terms of magnitude and rate of the movement and lag time between rainfall and accelerated slope instability.

9 GROUNDWATER MONITORING

Various groundwater monitoring instrumentation has been installed within boreholes and trenches to measure groundwater and piezometric levels throughout the landslide mass (Tab. 1). The database for the manually monitored instrumentation is sometimes fragmentary and consequently the extrapolated water level trends can only provide indicative information of the real groundwater behaviour.

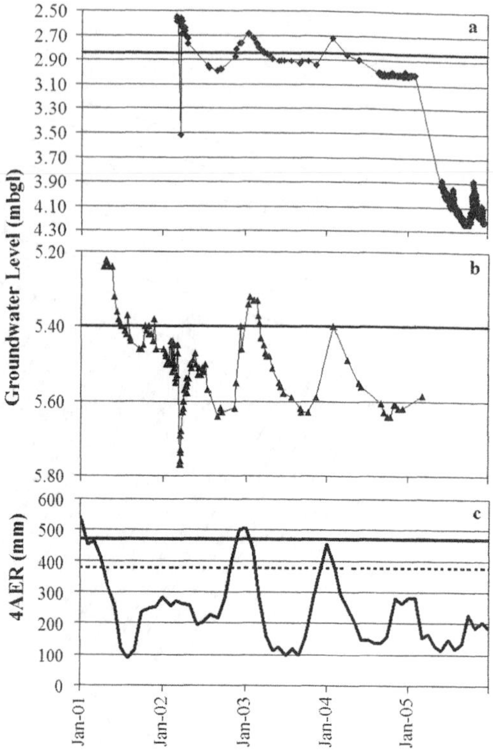

Figure 6. Groundwater monitoring records for two different piezometers installed in different positions along the landslide in Area 3 (a & b) and the 4AER for the Undercliff area (c) including the instability threshold shown in Figure 5. The recorded seasonal groundwater fluctuations occur consistently for the two monitoring point despite marked differences in the magnitude of the seasonal fluctuation and in the groundwater levels. Continuous horizontal line (a & b) = groundwater level threshold causing accelerated landslide reactivation.

Although it is likely that the main groundwater level fluctuations have been recorded, the peak highest and lowest levels are potentially missing effectively not allowing a calibration of the recorded groundwater levels for landslides of different magnitude. Figure 6 shows two typical ground water level monitoring graphs along the study site together with thresholds, based on inclinometer monitoring and slope stability analyses, corresponding with the occurrence of accelerated slope movement over the entire study area.

10 DISCUSSION

The Undercliff is a complex feature with various landsliding styles and activities related and depending on the geology, geomorphology, hydrogeology, site evolution, climate, vegetation, sea erosion, rockfalls, etc., including anthropogenic factors (e.g. tree clearance, reforestation, road drainage, construction of properties, regrading, etc.). Although site-specific observations and an instrumental monitoring database are available for various sections of the western Undercliff, mainly along the Upper Tier in proximity to the A3055 Undercliff Drive, the behavior of large areas is less well known.

The deterministic approach described above uses climate, instability and groundwater monitoring data to identify the mechanism triggering accelerated instability within the study site. The monitoring data have also been used to validate the slope stability analyses for the site.

Landslide movement records derived from factual inclinometer monitoring data (Fig. 4) have been correlated with rainfall data (Fig. 5) from nearby weather stations. The available information clearly indicates that high antecedent effective and total rainfall over 4 month periods, leads to high groundwater levels in the Undercliff slopes. This triggers accelerated slope instability with a lag time between rainfall and commencement of accelerated movement dependent on the magnitude of the 4AER and 4ATR.

Monitoring over several years has established that much of the study area undergoes some degree of ground movement each year, generally of the order of a few millimeters or centimeters, and most of this movement occurs during winters when the 4AER exceed a threshold of c.375 mm or a 4ATR of c.395 mm and further accelerates when the 4ATR and 4ATR are over c.470 mm (Tab. 3). The sensitivity of the Undercliff stability to the natural variability of the amount of effective and total rainfall is quite high and a difference of 95 mm of 4AER and 75 mm of 4ATR leads to substantially different magnitude instability events.

A comparison of the meteorological (i.e. 4AER and 4ATR) and deformation (i.e. inclinometer) monitoring records supplemented by visual inspections of the slopes have allowed the identification of three classes of instability with different magnitude. However there are limited data of inclinometer monitoring linked to weather and consequently it is not possible to identify a correlation between each class of instability and expected movement. This uncertainty is largely due to the pattern of past instability events along the western Undercliff which is the result and combination of particular and unique weather and environmental conditions, site specific geology, hydrogeology and geomorphology and additional forcing conditions that are difficult to evaluate (e.g. distribution of destabilising forces, loads, etc) and existing conditions (e.g. existing slip surfaces, active scarps, bare ground, open cracks, etc). The behaviour of renewed slope instability is also influenced by the location, size and effect of past events, which are often unknown.

The complexity of the study area, the broad pattern of landslide movements (i.e. different type, extent, etc.) indicate also that a substantial monitoring system, as the one presently installed, may not provide a fully comprehensive guide to accurately predict the occurrence of landslides for all parts of the western Undercliff. Therefore the assessment of potential future movement has largely to rely on a statistical average approach and expert judgment but informed by the monitoring data.

The methodologies described in this paper have been formalized in a series of procedures for the Isle of Wight Council to process, report and interpret the results of monitoring instrumentation against a simple warning and alarm system for possible future slope movement until stabilization works of the area are carried out. This system, which is intended to be operated for only few years, is based on previously identified approach, which relies on the extrapolation of past trends and identification of specific 4AER and a groundwater level threshold determining pre-alarm and alarm situations for different areas within the study site. The exceedence of the various thresholds leads to specific actions for the Isle of Wight Council including increased frequency of monitoring, a more complex cross checking of the data and advanced interpretation, inspections of the site, restriction of access, closure of road, information to the public and ultimately evacuation of the area.

Studies for the UK (Hulme et al. 2002) strongly suggest that in the future the climate will be less predictable and characterized by higher frequency of climatic extremes. A greater seasonality is expected on average resulting in warmer and drier summers and wetter winters. The western Undercliff could be potentially more vulnerable to the increased variability of winter rainfall leading to increase landslide activity characterized by major events. Conversely the predicted increases in winter rainfall could be mitigated by increases in evapotranspiration consequently leaving the frequency of accelerated landslides, especially minor instability, substantially unchanged as modeled elsewhere in southeast England (Collison et al. 2000).

There is strong evidence that climate change has already taken place and may accelerate in future (IPCC 2001). The variability of the directly observed and extrapolated monitoring (i.e. effective rainfall) data for the study site is generally high and consequently any effects of actual climate change hides the effects caused by natural variability of the data.

The effect of direct factors (e.g. storms, high rainfall, soil in dry condition, opening of dry cracks, changes in vegetation, lengthening of the growing season for plants, etc.) and indirect factors (e.g. further development of the area due to improved summer weather, influence by anthropogenic factors, land use patterns, rate of soil weathering, etc) have substantial effect on the stability of a complex landslide such as the Undercliff. The present prediction of possible future instability in relation to climate variability is partially subjective and the introduction of further variables, directly or indirectly caused or related to a rapidly evolving climate, leads to a wide range of possible scenarios and in due course may require revision to the AERs, ATRs and groundwater level thresholds.

Therefore, although at the moment the methodologies described above based on climatic parameters to determine the risk of landslide reactivation are a proven landslide forecast instrument along the Undercliff they carry various uncertainties for a satisfactory use in the longer term.

The maintenance of an accurate continuous instrumental database is considered fundamental to update, improve and revise the forecast method and to validate its use in the future.

11 CONCLUSION

A number of sites along the western Undercliff have been monitored over a series of years to enable a current assessment of landslide behaviour. A coherent relation between slope stability back-analysis, inclinometer, groundwater and weather data have been used to define a warning and alarm system for possible future slope movement until stabilization works of the area are carried out. The high variability and complexity of the study area pose various limitations in the prediction of future instability events. The climatic changes foreseen in the coming years are expected to affect and alter the factors triggering and influencing the instability that at present are fixed and are destined to become additional variables in the future. Therefore climate change is expected to reduce the predictability and to modify the effect of accelerated landslide activity along the western Undercliff. The continuous instrumental monitoring of the area is considered of primary importance to identify variations in the landslide behaviour with the purpose to update, improve and validate the landslide forecast methods and to revise and adapt it to the new environmental conditions.

ACKNOWLEDGEMENTS

The authors would like to gratefully acknowledge the co-operation and assistance of the Isle of Wight Council.

REFERENCES

Chandler, M.P. 1984. The coastal landslides forming the Undercliff of the Isle of Wight. Ph. D. Thesis (unpublished), Imperial College, University of London.

Clark, A.R., Fort, D.S., Holliday J.K., Gillarduzzi, A., Bomont, S. 2007. Allowing for climate change; an innovative solution to landslide stabilisation in an environmentally sensitive area on the Isle of Wight. International Conference: Landslides and climate change – challenges and solutions, 21–24 May 2007, Ventnor, Isle of Wight, UK, Rotterdam: Balkema.

Collison, A., Wade, S., Griffiths J., Dehn, M. 2000. Modeling the impact of predicted climate change on landslide frequency and magnitude in SE England. *Engineering Geology*, Volume 55, Number 3, February 2000, 205–218. Amsterdam: Elsevier Science.

Department of Environment, 1991. Coastal landslip potential assessment of the Isle of Wight Underclliff, Ventnor, by Geomorphological Services Ltd.

Hulme, M., Jenkins, G.J., Lu, X., Turnpenny, J.R., Mitchell, T.D., Jones, R.G., Lowe, J., Murphy, J.M., Hassell, D., Boorman, P., McDonnald, R., and Hill, S. 2002. Climate change scenarios for the United Kingdom: the UKCIP02 scientific report, Tyndall Centre for Climate Change Research, School of Environmental Sciences, University of East Anglia, Norwich, UK.

Hutchinson, J.N. 1991. Theme lecture: The landslides forming the South Wight Undercliff. International conference on slope stability engineering developments and applications, R.J. Chandler Ed157–168, London: Thomas Telford.

Hutchinson, J.N. 1987. Some coastal landslides of the southern Isle of Wight. In K.E. Barber (ed). *Wessex and the Isle of Wight. Field guide.* 123–135. Cambridge: Quaternary Research Association.

Hutchinson, J.N., Chandler, M.P., Bromhead, E.N. 1985. A review of current research on coastal landslides forming the Undercliff of the Isle of Wight, with some practical applications. Proc. Of the consequence and problems associated with the coastline, 17–18 April 1985, 1–16, Isle of Wight, Newport.

Hutchinson, J.N., & Bromhead, E.N. 2002. Isle of Wight landslides. In R.G. McInnes & J. Jakeways (ed.), *Instability planning and management,* 3–70. London: Thomas Telford.

IPCC, 2002. Climate change 2001: Summary for policy makers. The Intergovernmental Panel on Climate Change third assessment report. *Available at http:www.ipcc.ch/index.htm*

Ibsen, M-L. 2000. The impacts of climate change. In: R.G. McInnes & D. Tomallin (ed). *European Commission LIFE project, LIFE-97 ENV/UK/000510. Isle of Wight Centre for Coastal Environment UK and partners, Vol. 1. Chapter 6.*

Lee, E.M. & Moore, R. 1991. Coastal landslip potential assessment: Isle of Wight Undercliff, Ventnor. Technical report published by the Department of the Environment, UK. DoE Publications.

Moore, R., Lee, E.M., Clark, A.R. 1995. The Undercliff of the Isle of Wight, a review of ground behaviour. Rendel Geotechnics. Cross Publishing.

Shaw, P.J. & Packman, M.J. 1988. Chalk-Upper Greensand groundwater investigation. Report on Hydrogeological Investigation of the Southern Downs aquifer. Southern Water Authority, Isle of Wight Division.

Tomallin, D. 2000. Ventnor Undercliff–palaeo-environmental overview, Isle of Wight UK. In: R.G. McInnes & D. Tomallin (ed). *European Commission LIFE project, LIFE-97 ENV/UK/000510. Isle of Wight Centre for Coastal Environment UK and partners, Vol. 2, Palaeo-environmental area P4.*

Landslides and Climate Change – McInnes, Jakeways, Fairbank & Mathie (eds)
© 2007 Taylor & Francis Group, London, ISBN 978-0-415-44318-0

Reactivation of an ancient landslip, Bonchurch, Isle of Wight: Event history, mechanisms, causes, climate change and landslip potential

M.J. Palmer
Halcrow Group Ltd, Swindon, Wiltshire, UK

R. Moore
Halcrow Group Ltd, Birmingham, UK

R.G. McInnes OBE
Isle of Wight Council, Ventnor, Isle of Wight, UK

ABSTRACT: The Landslip is located between Luccombe and Bonchurch on the south coast of the Isle of Wight. The area includes a wide variety of interests and amenities including residential, access, ecology and wildlife, recreation and geology. Earliest reported movements took place in 1810, although the landslip is considerably older. A major reactivation of part of the Landslip took place in 1995 which was followed by a second large-scale slope failure in 2001. The 2001 event resulted in the landward retrogression of the rear-scarp and extensive displacement and widening of the area affected in 1995. The retrogressive nature of the recent event gave cause for concern regarding the ground instability risks to property, local access and the A3055 strategic coastal road located upslope of the rear-scarp. Acting on these concerns, the Isle of Wight Council commissioned a site investigation in 2001 comprising geological and geomorphological mapping, a GPS ground movement survey, commissioning of a ground investigation comprising several deep boreholes and installation of automatic slope monitoring. These investigations have advanced understanding of the ground conditions, mechanisms, and causes of failure within the Landslip. The paper presents the results of the site investigation and identifies the future challenges of managing potential landslide reactivation (retrogressive) events in response to the effects of climate change.

1 INTRODUCTION

The Undercliff on the south coast of the Isle of Wight is an area of extensive ancient landslides that extends some 12 km from Luccombe to Blackgang (Fig 1). In cross-section the landslides extend up to 1 km landward from the shoreline and form a steep terraced landscape with impressive views of the English Channel. Apart from its stunning location, the special climate of the Undercliff is another factor which led to development of the area in the mid to late 1800s. The main residential areas include Bonchurch, Ventnor, St Lawrence and Niton which have a permanent population of more than 6,000. Today, the Undercliff is a major tourist destination. Yet, the Undercliff is one of the most unstable geological settings in the UK, and faces substantial challenges from coastal instability, particularly in the context of climate change.

The area known as 'The Landslip' is located at the eastern end of the Undercliff between Luccombe and Bonchurch (Fig 2). The Landslip occupies the full cross sectional area of the Undercliff, from head scarp to the sea, a distance of between 200 to 300 m, and encloses an area of around 24 ha. The Landslip is elevated above a 30–40 m high sea-cliff, from which it rises to 100 m above Ordnance Datum (AOD) beneath the prominent head scarp, known locally as the Devil's Chimney, which is a near-vertical cliff between 15–20 m high.

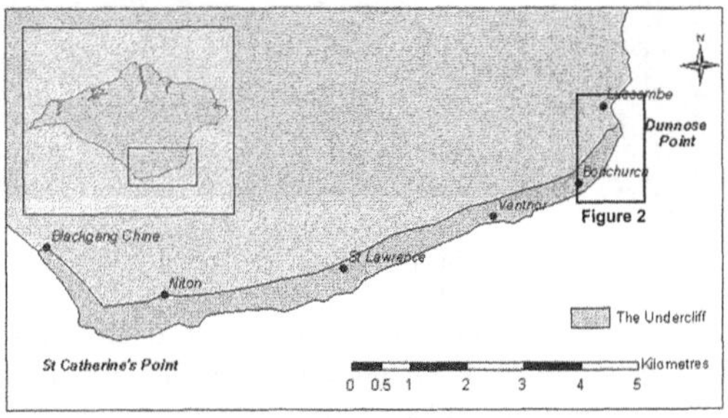

Figure 1. The Isle of Wight Undercliff.

Figure 2. The Landslip.

'The Landslip' by name would appear to have been recognised as inherently unstable for some time. It is first mentioned in the Doomsday Book of 1086. Today, the area is an important local amenity and is designated a SSSI and forms part of the Isle of Wight Heritage Coast. Access to the area is via coastal footpaths from Bonchurch, Luccombe and the Devil's Chimney (via the A3055 at Smugglers Haven). The Landslip has diverse amenity interests including geology, ecology and wildlife, and recreation. For the most part, the Landslip is covered by dense woodland.

Major landslide events were recorded in the Landslip in 1810 and 1818. In 1995, a major reactivation of the western part of the Landslip led to the destruction of the coastal footpaths and temporary closure of access to the area. The event raised concerns about public safety and maintenance of access to the coastal paths in the Landslip in the long-term.

A major landslide affected the same area in 2001. This event resulted in retrogressive failure of the head scarp (and recession of the Undercliff) towards a number of properties and the A3055 strategic coastal road raising concerns about the future sustainability of some properties and the A3055.

Landslide reactivation events prior to this had largely remained within the ancient Undercliff 'footprint' and had not resulted in retrogressive failure of the head scarp on such a scale. Therefore, the 2001 event marked a major change in behaviour of the Undercliff at the Landslip, resulting in retrogressive failure of the head scarp.

As a result of this concern the local authority commissioned a site investigation comprising geological and geomorphological mapping, a GPS ground movement survey, commissioning of a ground investigation comprising three deep boreholes and installation of automatic slope monitoring. The site investigation was needed to determine the ground conditions, construct a landslide model and determine landslide risk scenarios accounting for the impact of future climate change predictions.

The paper presents the results of the site investigation and identifies the future challenges of managing landslide reactivation (retrogressive) events at the Landslip in response to climate change.

2 HISTORY OF GROUND MOVEMENTS

The Landslip has a long history of major landslide events dating back to 1810. Prior to this there are no records of ground movement or landslide events although it is likely that ground movement and major events have occurred throughout time (i.e. the Holocene) given the presence of features of inherent instability in the area. It is worth noting that the inherent instability of the area is also likely to have been a reason why the area was not developed during Victorian times as was the case for Bonchurch and Ventnor nearby.

Major landslide events were recorded in the Landslip in 1810 and 1818 (Table 1). The 1810 event was described "for three days successively the earth heaved and sank" (Adams, 1856) and involved an area up to 12 ha. The event in 1818 was reported to have affected a much larger area (20 ha). The precise location of these events is not known.

Reports of ground movements and damage in the area over the period 1900–1950s were depicted on engravings, postcards and reported in the local press (Tab 1). The 1904 and 1910 events were captured on several postcards; they were labelled 'The Great Landslip' at Luccombe, but it is apparent from the image (Fig 3) that these events were probably related and located at Dunnose Point (Fig 2). The next major event at the Landslip took place in February 1995, which was followed by a further major landslide in March 2001. These latest events are located at the western limit of the Landslip and affect a different area to the events reported in the early 1800s.

2.1 *The February 1995 landslide event*

In early February 1995 a large landslide was reported to have occurred in the area known as the Landslip. Early reports of the event indicated that a significant area of the Landslip had been affected by the landslide. The event was reported to have resulted in the loss of cliff-top land due to failure of the rear-scarp, along with considerable settlement and disruption of the slopes and amenities within the Landslip. Large quantities of landslide debris were reported to have been deposited on the foreshore, beneath the high sea cliffs, as a consequence of the ground movements above.

Table 1. Historical events at the Landslip.

Date	Description	Reference
1810	12 ha of The Landslip "for three days successively the earth heaved and sank"	Adams, 1856
1818	20 ha of The Landslip Major deep seated failure	Barber, 1834 Wilkins & Brion, 1859
Feb 1904	Active mudslide above sea-cliff	Postcard
Feb 1910	The 'Great Landslip' captured on a postcard at the time – shows a major deep-seated failure of the landslip at Dunnose Point	Postcard
Jul 1916	"Some paths in the Landslip had fallen away"	Anon, 1916
Jan 1926	"Heavy shifting of land over the sea-cliff	Anon, 1926
Dec 1952	Movements damaged access path	Anon, 1952

Figure 3. Postcard of the Great Landslip 1910.

Although the reports of the event suggested a fairly sudden slope failure, there were no reports of any casualties; bearing in mind the area affected is crossed by several coastal footpaths. The landslide caused total destruction of the footpaths and other amenities in the area affected.

The extent of the landslide was confined to the lower slopes of the Landslip. At its widest, above the sea-cliff, the landslide was around 400 m wide and extended 260 m upslope.

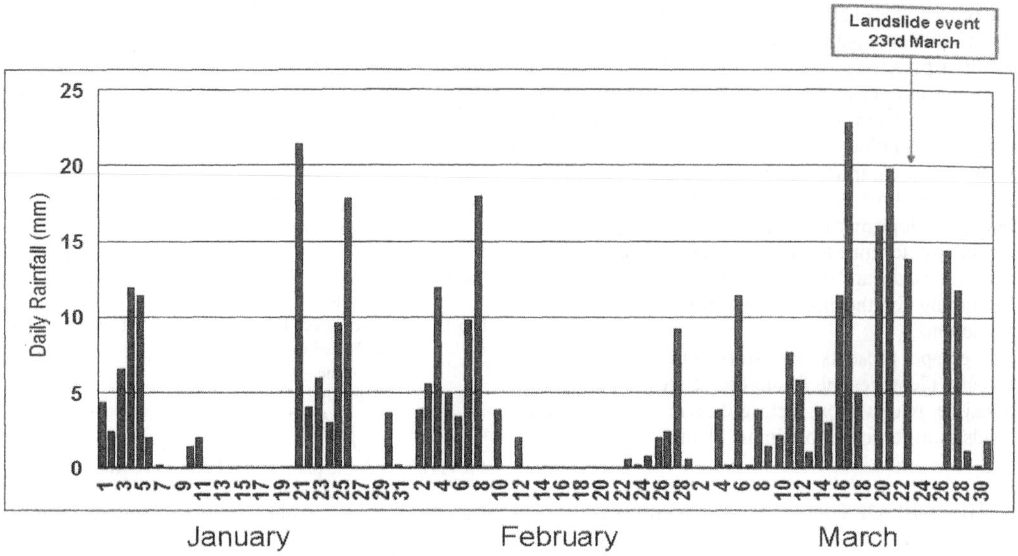

Figure 4. Rainfall and landslide event sequence in 2001.

The area affected was about 4 ha which is much smaller than the events reported in 1810 and 1818. The boundaries of the failure coincided with several morphological boundaries depicted on the pre-failure published geomorphological map for the area (Moore *et al.* 1995). The head scarp defining the landward limit of the Landslip and Undercliff at this point was only partly affected by the 1995 event, where local rear-scarp failure over a distance of 20 m resulted in 10 m of recession and loss of private land.

Post-failure evidence of backtilted blocks and displaced footpaths indicated that the landslide involved deep-seated rotational failure and displacement of pre-existing landslide debris. The amount of horizontal displacement was estimated to be about 50 m and it was also apparent there was considerable settlement and reduction in elevation of the landslide mass. The displacement of the landslide had two main consequences:

1. It removed support from a small section of head scarp resulting in localised cliff-top recession; and
2. It displaced a considerable amount of landslide debris over the sea-cliff forming lobes of debris on the foreshore, which were rapidly eroded and washed away by the sea.

The causes of the event were attributed to long-term recession of the sea-cliff, removing passive support to the slopes above, and antecedent rainfall conditions, which ultimately triggered the event. Monthly rainfall totals in January and February 1995 were twice the average, and a particularly wet period prior to the event comprising 136 mm of rainfall in 16 days, was attributed to triggering the landslide early on Friday 3 February.

2.2 *The march 2001 landslide event*

The landslide event reported on the 23rd March 2001 occurred in the same area affected by the 1995 event. The 2001 event resulted in significant upslope retrogression of the rear scarp, resulting in partial loss of the field downslope of East Miramar, and extensive displacement and widening of the area affected by the 1995 landslide event (Fig 4).

The landslide occurred soon after a particularly wet winter period from October 2000 to January 2001. During this 4-month period, a total of 530 mm of antecedent effective rainfall (rainfall less evapotranspiration) was recorded, which ranks as the 10th highest 4-month winter effective rainfall total since records began in 1839. The equivalent return period for this winter rainfall event is 1:16 years. This was a more extreme event than the one that triggered the 1995 failure, which had a return period of 1:5 years.

Figure 4 presents daily rainfall recorded at Ventnor Park, approximately 4 km to the west of the Landslip. The figure shows that the timing of the landslide event followed a continuous rainfall period that began on the 22 February, reaching a peak intensity of 10–25 mm per day between the 16–23 March.

Further background and discussion of the results of the 2001 landslide investigation follows.

3 INFORMATION SOURCES

The 1995 event occurred shortly after publication of the 'Undercliff of the Isle of Wight: a review of ground behaviour' (Moore *et al.* 1995). This report included detailed maps of the geomorphology, ground behaviour and planning guidance for the Landslip for the first time.

Other publications and reference sources of relevance to the area include the reports and maps published by Geomorphological Services Ltd (Lee and Moore 1991) in connection with the Department of Environment R&D project 'Coastal Landslip Potential Assessment: Isle of Wight Undercliff, Ventnor'. There are few other sources which consider the geomorphology and geotechnics of the Landslip with the exception of Chandler (1984) and Mathews (1997). Hutchinson *et al.* (1981) and Posford Duvivier (1991) provide geotechnical assessments for the coastal landslides at Bonchurch and Monks Bay, about 1 km distant to the southwest of the Landslip. Consultancy reports on the 1995 and 2001 events were commissioned by the Isle of Wight Council from High Point Rendel and Halcrow Group Ltd, respectively.

4 GROUND INVESTIGATION

The site investigation commissioned by the Isle of Wight Council after the 2001 event comprised detailed geomorphological mapping, installation of a GPS control network and permanent ground movement markers, rotary core drilling of three boreholes, laboratory testing of core samples, and installation of automatic slope monitoring equipment. Existing borehole records for an adjacent site at East Miramar were also located and made available to the study.

The site investigation was carried out in several phases as follows:

1. Review of existing data;
2. Site survey and geomorphological mapping;
3. Ground investigation in December 2001 to provide key data on subsurface geology and soils;
4. Installation of an automatic slope monitoring system; and
5. Interpretation of the results to define the mechanisms of failure and future landslip potential.

5 GEOLOGY

The generalised stratigraphy of the Undercliff is shown in (Fig 5) and comprises Lower and Upper Cretaceous strata. These consist of the Gault which is underlain by weak sandstones of the Lower Greensand and overlain by the massive cherty sandstones of the Upper Greensand and Chalk. The Gault, known locally as the 'Blue Slipper', provides critical weak layers upon which the Landslip has developed. The strata dip to the southwest by c2° (White 1921) which has a significant control on the hydrogeology and stability of the coastal slopes.

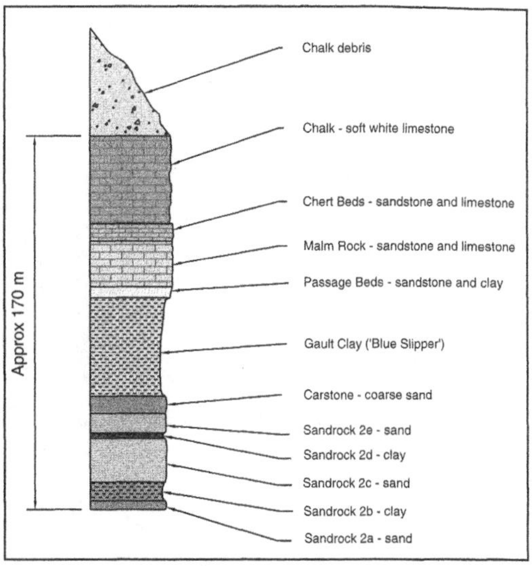

Figure 5. Generalised stratigraphy.

Further discussion about the ground conditions and landslide mechanisms is presented in sections 7 and 8, below.

6 GEOMORPHOLOGY

The geomorphology of the Undercliff at the Landslip comprises four distinct landforms (Fig 6):

1. Chalk Downs
2. Upper Greensand bench
3. Undercliff Landslide features
4. Sea cliffs.

The Chalk Downs form the higher ground above 120 m AOD immediately landward of the Undercliff landslide complex. The Downs reach an elevation of around 240 m AOD at Bonchurch Down.

The Upper Greensand (UG) bench is located at the base of the Chalk downs between 115–129 m AOD. The bench was most likely formed as a consequence of the removal of the overlying Chalk through solifluction processes at the end of the last glaciation, around 13,000 years ago. The Undercliff landslides have removed support to the UG bench resulting in stress relief and the opening of joints (locally known as vents) and settlement of the UG strata. The Devil's Chimney provides a good example where historical settlement and opening of a major joint within the UG has been utilised for construction of access steps down the precipitous rear-scarp of the Landslip.

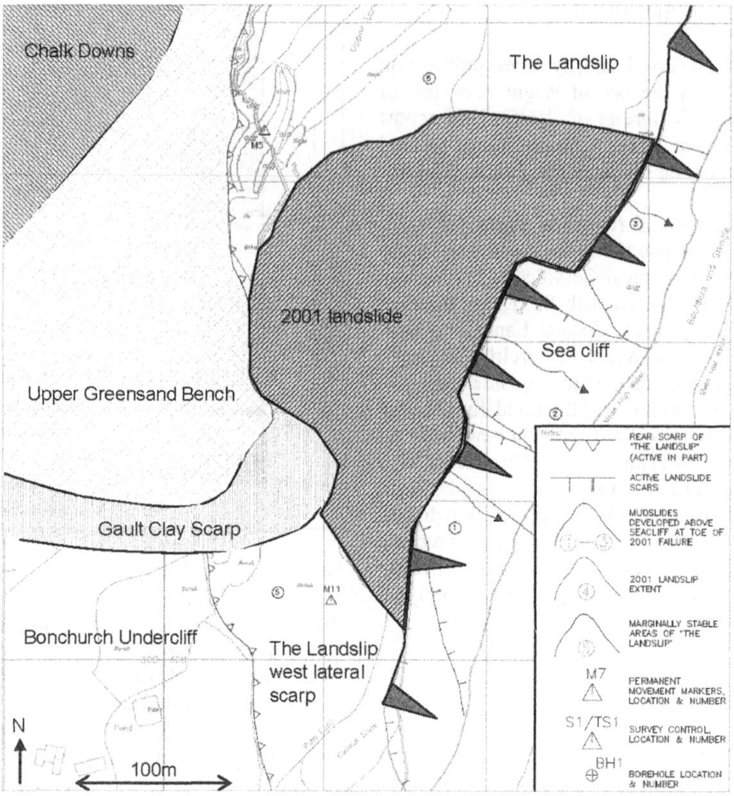

Figure 6. Summary geomorphology map.

The UG bench is a significant feature as it marks a narrowing in the Undercliff landslide complex (Fig 6); a prominent spur is apparent between the Landslip to the northeast and the Undercliff at Bonchurch to the southwest. The spur is unloaded on both sides and its leading edge which has resulted in settlement of the bench and damage to property and services in the area affected.

The landslide features are characterised by rotational failure of the UG strata above the Gault. The landslides may occur at multiple levels within the Gault sequence, although a lithological boundary between the lower 'silty' and upper 'plastic' Gault, about 15 m from the base of the sequence, has been observed elsewhere in the Undercliff to support shear surfaces (Bromhead *et al.* 1991; Moore *et al.* 1995). The landslides forming the 'Landslip' are perched above high sea cliffs and consequently there is no passive support to the unstable slopes above.

Figure 6 is a summary geomorphological map of the landslide area from a site survey carried out in November 2001. The map reveals several key features of the recent landslide events:

- extent of the 2001 landslide event;
- three well-defined elongate and seasonally active mudslides conveying landslide debris over the near-vertical sea cliff onto the beach;
- areas of 'marginally stable' land that represent the extent of the Landslip or pre-existing landslide complex surrounding the recent active landslide area; and
- a tension crack in the grounds of East Miramar some 65 m behind the existing rear scarp.

The significance of these features is discussed further in sections 7 and 8.

The sea cliffs are up to 40 m high and are formed of *in situ* Sandrock, and where present, the Carstone. The cliffs are undercut by the action of waves which has led to significant cliff retreat in recent years, estimated between 1–2 m per year. The soft sands offer little resistance to wave erosion and subaerial weathering resulting in frequent cliff falls and slides.

The conveyance of landslide debris over the cliffs from the landslides above further exacerbates the erosion and retreat of the sea cliffs. Debris derived from the upper landslides and cliff erosion is deposited at the base of the sea cliff in accumulation lobes, which are rapidly removed by the action of waves.

7 GROUND CONDITIONS

Three deep boreholes were drilled in November 2001, the locations of which are shown in (Fig 6). A summary of the geological succession is presented in (Fig 5). A cross-section through the landslip based upon a surveyed section line is shown in (Fig 8).

In boreholes 2 and 3, located behind the rear scarp of the Landslip, the Malm Rock and Passage Beds are of reduced thickness compared with borehole records at Ventnor (Palmer *et al.* 2007). In borehole 1, which is also located above the rear scarp, the reduced thickness of the Malm Rock and Passage Beds appears to be the result of previous landslide movement; significantly, no Chert Beds were observed in landslide debris or the crest of the rear- scarp to an elevation of 118 m AOD.

Borehole 1 encountered colluvium from ground level to 11.7 m depth. The colluvium is formed of Chalk Marl debris and most likely derives from solifluction of the adjacent steep Chalk Downs in the early Holocene. The colluvium infills a depression, possibly created by settlement of a back tilted block, as displaced Malm Rock is exposed in the face of the current rear scarp. The reduced thickness of the Malm Rock and Passage Beds, and the maximum thickness of Gault (44 m) encountered in the borehole, coupled with the geomorphology map evidence indicates the basal shear surface of the 2001 landslide was most likely located in the upper part of the Gault.

The geomorphology mapping identified a tension crack at ground surface approximately 65 m behind the existing rear scarp. This tension crack can be traced and connects to a scarp in which the Chert Beds are exposed behind Miramar Cottages to the west and the Devil's Chimney to the east. The feature is evidence that prior settlement of the Upper Greensand bench has taken place and that extensional movement appears to have taken place recently. Previous boreholes located near Miramar Cottages, although not conclusive, also suggest that this area is underlain by rotated landslide blocks overlain by colluvium.

The laboratory testing programme was primarily limited to soil characterisation and index tests, and a profile was established for the full thickness of the Gault to assist interpretation at the site, and elsewhere in the Undercliff.

The Gault can be divided into a number of divisions based on plasticity index (Palmer *et al.* 2007). (Fig 7) depicts the liquid and plastic limit test results on the samples from Bonchurch. There is a gradual increase

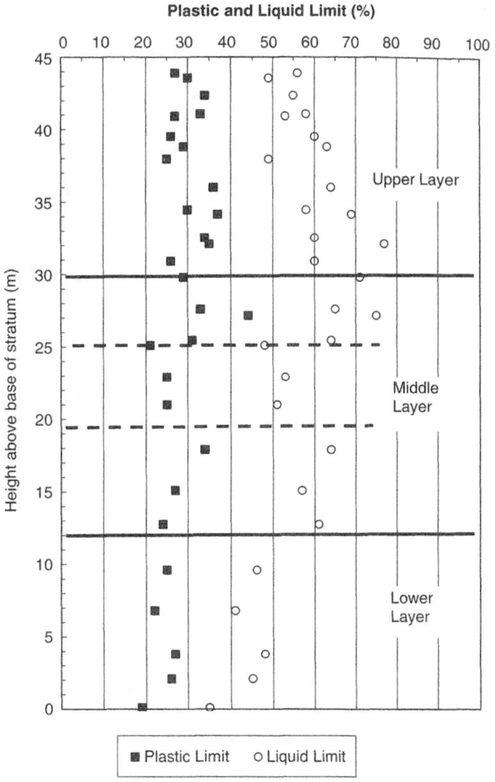

Figure 7. Plastic and liquid limits of Gault.

in plasticity between 12–30 m above the base of the Gault which defines the middle layer. Plasticity is less in the generally siltier upper layer between 30–44 m from the base.

A relatively continuous steep scarp slope, up to 20 m high, known as the Gault Clay scarp (Moore *et al.* 1995), can be traced though much of the Undercliff, west of Ventnor. The scarp slope is often the site of active mudslides and small rotational slides and is believed to mark the surface exposure of the lower, silty layers of the Gault Formation (Bromhead *et al.* 1991; Moore *et al.* 1995).

On the western lateral shear of the 2001 landslide, a fresh exposure of Gault was observed near the junction with overlying UG landslide debris (Fig 6). The Gault exhibited flexural and large displacement shears, with iron staining along open fissures and fresh weathering and spalling of debris. This exposure is of particular interest as it is the only exposure known of the 'Gault Clay scarp' with evident large displacement shears.

The Gault Clay scarp at Bonchurch occurs at a higher elevation than in other parts of the Undercliff. The location, relative to the litho-stratigraphy recorded in the boreholes indicates that it is positioned near to

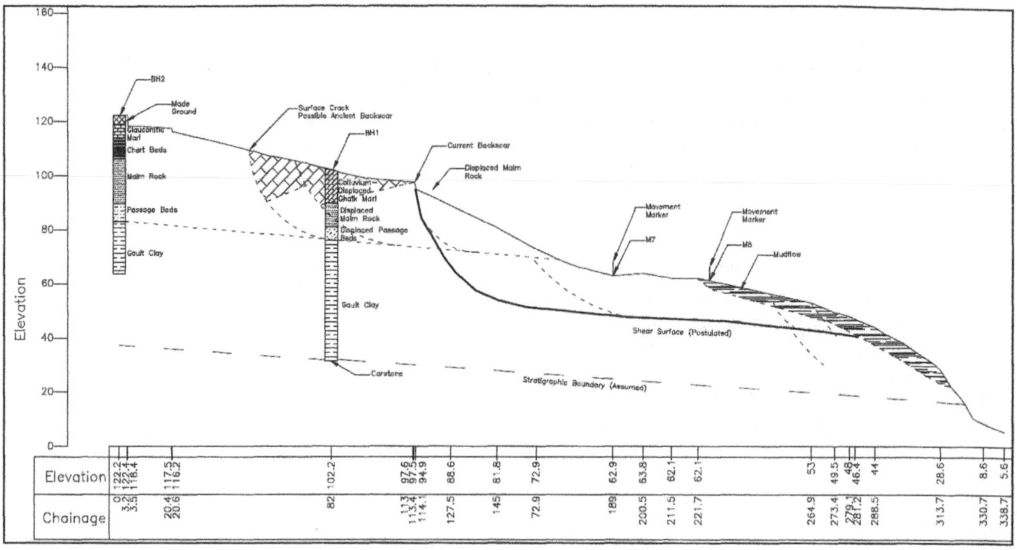

Figure 8. Landslide model.

the base of the upper (silty) layer i.e. approximately 30 m above the base of the Gault (Fig 7). This differs from the interpretation in the western part of the Undercliff (Bromhead *et al.* 1991) where the critical failure horizon occurs at the middle and lower Gault interface, approximately 12 m from the base of the Gault.

Active mudslides developed above and upon the sea cliffs obscure the contact between the lower Gault, Carstone and upper Sandrock beds. Projection of the top of the Carstone outcrop to the sea cliff would indicate contact at the cliff face at approximately 17 m AOD. Oblique photography of this section suggests that the outcrop may be slightly higher in the cliff face. A higher outcrop elevation could mean that the dip of the beds reduces seawards or, more likely, that the basal shear surface at this location occurs at a lower level in the Gault sequence.

8 LANDSLIDE MECHANISMS

The ground model (Fig 8) indicates that the basal shear surface of the upper landslide occurs at the top (or near the top) of the Gault. There is an apparent seaward dip of the strata of some 4 degrees to the southwest which is greater than the generalised dip of strata reported by White (1921). The dip of strata and shear surface are clearly a major contributor to the instability of the landmass.

The site investigation has confirmed that the Landslip involves deep-seated rotational failure of the upper Gault and overlying massive Upper Greensand rocks and colluvium. The rotational failures are degraded by mudslides which convey landslide debris over the *in situ* sea cliffs below. The rotational failures and mudslides probably form basal shear surfaces at multiple levels with the Gault, reducing in elevation towards the sea cliff.

These observations are analogous to the landslide mechanisms reported at Blackgang (Moore *et al.* 1998) following a major landslide of the upper parts of the Undercliff in 1997/98.

9 SLOPE MONITORING AND WARNING

Figure 6 shows the location of a permanent ground movement marker network that was installed during the site investigation. These markers are intended to be re-surveyed to measure ground movement rates within the main slide complex which is subject to seasonal displacements and degradation.

In addition to the ground movement markers, automatic tiltmeters and specially designed wire extensometers have been installed upslope of the landslide rear-scarp and at the Devil's Chimney to provide continuous real-time monitoring and early warning of ground movement. The tiltmeters are designed to monitor rotational movements of the UG bench and rear-scarp which are anticipated if extensional movements and rotational failure of the upper Landslip were to retrogress upslope. The wire extensometers monitor horizontal displacement across tension cracks and the Devil's Chimney. Data are recorded automatically and transmitted to the Council via the telecoms network. Since their installation no significant ground movement has been recorded.

The automatic monitoring network is programmed to raise a warning to the Council and Police when ground movement exceeds a pre-defined threshold in several linked devices.

10 LANDSLIDE POTENTIAL

The 2001 landslide event at the Landslip resulted in local failure of the rear scarp and partial loss of the field downslope of East Miramar. The event marked a step change in behaviour of the Undercliff involving retrogressive failure of Upper Greensand bench which had largely been unaffected by major deep-seated landsliding. The rear scarp of the Undercliff now forms a steep slope which has raised concern about the potential for further retrogressive failures upslope that will result in further losses of private land, property and the A3055.

The results of the investigation have established the probable mechanisms and causes of failure. The principal causes of failure have been attributed to toe erosion of the soft rocks forming the sea cliffs, and extreme rainfall and associated increases in groundwater levels. The prognosis for the future of the Landslip is for ground movement and degradation of landslides to continue seasonally, and for an increase in the frequency of further landslide events given current predictions for sea level rise and higher winter rainfall due to climate change (Moore et al 2007). Based on these predictions the following can be expected to influence the timing and extent of ground movement and events at the Landslip over the next 100 years:

1. Toe erosion at the base of the sea cliffs, resulting in at least 1–2 m recession per year and possibly more given predicted increases in sea level and frequency of storm surges
2. Drier summers, which will cause greater desiccation of the clay soils, opening tension cracks and allowing rapid percolation of surface water
3. Wetter winters, with a 20–30% increase in winter rainfall, resulting in a higher frequency of extreme rainfall events with potential to trigger landslides.

Seasonal degradation of the landslide has taken place since first initiation in 1995, causing repeated damage to footpaths and facilities located in the western part of the Landslip. The ongoing development of mudslides on the lower coastal slope give clear indication that further significant land movements and recession events are highly likely.

Minor collapses of the rear scarp were noted to be continuing during the investigation and significant tension cracks were apparent in the field. The height of the rear scarp has increased due to the recent failures and is unsupported. From borehole 1, the materials exposed in the rear scarp are weak and susceptible to erosion. Therefore it is likely that further failure and recession of the rear scarp will continue in the short-term.

The observation of an open tension crack in the ground surface approximately 65 m behind the existing rear scarp indicates that the UG bench has undergone pre-failure in the past and that extensional movement appears to have been reactivated by the latest events. Automatic wire extensometers have been located across the tension crack to monitor and warn of any significant displacement across the crack.

It is important that monitoring and observation of the tension crack continues as the potential for failure of the UG bench and spur between the Landslip and Bonchurch Undercliff (Fig 6) cannot be ruled out. Here ground movements could accelerate rapidly and involve a significant area and volume of material; run-out of landslide debris onto the lower Undercliff at Bonchurch could have significant impact on property and other assets in this area. Without further detailed analysis of the stability risk scenarios it is not possible to be more precise about the sensitivity and likelihood of future recession events and their consequences.

11 MANAGEMENT OPTIONS

The site investigation has advanced understanding of the mechanisms and causes of recent failures with the Landslip. The ground model combined with the geomorphological framework provides a reliable basis for determining future landslide hazard and risk scenarios. The scope of the present work has precluded carrying out analytical slope stability based risk analysis of the site which accounts for present-day stability conditions and future risk drivers, such as climate change.

Such analytical work is essential to understand the sensitivity of the ground model to the effects of coastal erosion and rainfall, particularly given predicted increases in sea level and winter effective rainfall due to climate change. Projection of future rates of change, landslide frequency and extent, and retrogressive potential scenarios should be made to identify the assets at risk and projected timing of potential losses.

Management options available to landowners and the Council include landslide stabilization, monitoring, planning and building control and raising awareness of the hazards and risk.

ACKNOWLEDGEMENTS

The authors acknowledge the assistance of colleagues at Halcrow and the Centre for the Coastal Environment, Ventnor. The site investigation reported was funded by the Isle of Wight Council.

REFERENCES

Adams W.H.D. 1856. *The history, topography and antiquities of the Isle of Wight.*

Anon 1916. Newspaper article, *Isle of Wight Mercury*, 14 July.

Anon 1926. Newspaper article, *Isle of Wight Mercury*, 29 Jan.

Anon 1952. Newspaper article, *Isle of Wight Mercury*, 5 Dec.

Barber T. 1834. *Picturesque illustration of the Isle of Wight.*

Bromhead E.N., Chandler M.P. & Hutchinson J.N. 1991. The recent history and geotechnics of landslides at Gore Cliff, Isle of Wight. pp189–196, *Slope Stability Engineering*, Thomas Telford.

Chandler M.P. 1984. The coastal landslides forming the Undercliff of the Isle of Wight. *PhD Thesis*, University of London.

Hutchinson J.N. 1981. Report on the coastal landslides at Bonchurch, Isle of Wight. Technical report to Lewis & Duvivier.

Lee E.M. & Moore R. 1991. Coastal Landslip Potential Assessment, Isle of Wight Undercliff, Ventnor. *Technical Report by Geomorphological Services Ltd for the Department of the Environment, research contract PECD 7/1/272.*

Matthews M.C. 1997. Geological and geotechnical report on the south east Isle of Wight: geotechnical study of the eastern extremity of the Undercliff (the Landslip). Unpublished BSc report.

Moore R., Lee E.M. & Clark A.R. 1995. The Undercliff of the Isle of Wight, A review of ground behaviour. Cross Publishing.

Moore R., Clark A.R. & Lee E.M. 1998. Coastal cliff behaviour and management: Blackgang, Isle of Wight. In: Maund J.G. & Eddleston M. (eds) Geohazards in engineering geology. Geological Society of London Engineering Geology Special Publications 15;49–59.

Moore R., Carey J.M., McInnes R.G. & Houghton J. 2007. Climate change, so what? Implications for ground movement and landslide event frequency in the Ventnor Undercliff, Isle of Wight. *This Conference*.

Palmer M.J., Carey J. & Turner, M.D. 2007. Lithostratigraphy of the Ventnor Undercliff and determination of critical horizons through borehole geophysics. *This Conference.*

Posford Duvivier (1991) Geotechnical appraisal of ground investigation for Monk's Bay coast protection, Isle of Wight.

White H.J.O. (1921) A short account of the geology of the Isle of Wight. Memoir of the Geological Survey of England and Wales. HMSO.

Wilkins E.P. and Brion J. (1859) A concise exposition of the geology, antiquities and topography of the Isle of Wight.

Landslides and Climate Change – McInnes, Jakeways, Fairbank & Mathie (eds)
 ISBN 978-0-415-44318-0

The Ventnor Undercliff: Landslide model, mechanisms and causes, and the implications of climate change induced ground behaviour and risk

R. Moore, M.D. Turner, M.J. Palmer & J.M. Carey
Halcrow Group Limited, Birmingham, UK

ABSTRACT: The paper presents a calibrated ground model for the Ventnor Undercliff for the first time based on the results of two ground investigations carried out in 2002 and 2005. The ground model is supported by surface and sub-surface monitoring data which identify the most probable failure mechanisms and causes of ground movement in the Ventnor Undercliff. The results of the investigation have reduced uncertainty about the landslide mechanisms, causes and ground behaviour at Ventnor. This allows the implications of climate change induced ground behaviour and risk to be assessed with greater confidence; the results indicate that the risks are increasing and that accelerated ground movement and potential landslide events are more likely in the future. The implications of climate change induced ground behaviour and risk poses very significant challenges for the future. The local authority has been co-ordinating and implementing a landslide management strategy involving planning and building controls, engineering works, monitoring and dissemination of information. What is clear from the results of the recent investigations and stability analysis is that coastal landslide stabilisation will be needed to prevent a significant decline in stability of the Ventnor Undercliff; this could be achieved by preventing coastal erosion and deep drainage to reduce groundwater pressure beneath the landslide.

1 INTRODUCTION

The Ventnor Undercliff is located on the south coast of the Isle of Wight and forms part of an extensive area of ancient coastal landslides between Luccombe and Blackgang (Fig 1). In cross-section the landslides extend up to 1 km landward from the shoreline and form a steep terraced landscape with impressive views of the English Channel.

Apart from its stunning location, the special climate of the Undercliff is another factor which led to development of the area in the mid to late 1800s. The main residential areas include Bonchurch, Ventnor, St Lawrence and Niton which have a permanent population of more than 6,000. The Undercliff is one of the most unstable geological settings in the UK, and faces substantial challenges from coastal instability, particularly in the context of climate change.

Several studies have specifically addressed the nature and causes of instability at Ventnor (Chandler 1984; Lee & Moore 1991; Hutchinson & Bromhead 2002). These are of relevance to the current investigation as ideas on the mechanisms, causes and historical ground movement event records within the Undercliff provide an important starting point in trying to unravel the complexity of the landslide units at Ventnor.

Parts of the deep-seated landslide complex are believed to have formed in pre-glacial times; a main contention of the paper is that retrogressive failure of the Ventnor Undercliff has taken place during the Holocene (i.e. 13,000 years BP to present). The latest stage of the landslide development has resulted in the opening and subsidence of a feature known locally as the Lowtherville graben which defines the present-day landward extent of the Undercliff at Ventnor; anecdotal evidence suggests this feature has developed since the early 1900s.

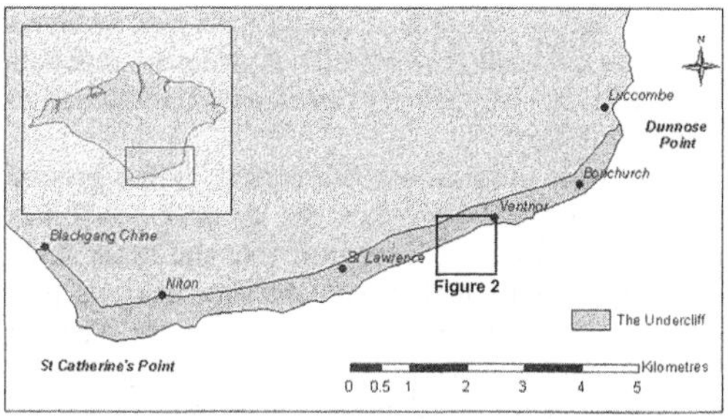

Figure 1. The Isle of Wight Undercliff.

The occurrence of widespread ground movement in the Ventnor Undercliff has resulted in a range of problems to the community. Over the last 100 years, some 50 houses and hotels have had to be demolished due to the effects of ground movement and many more

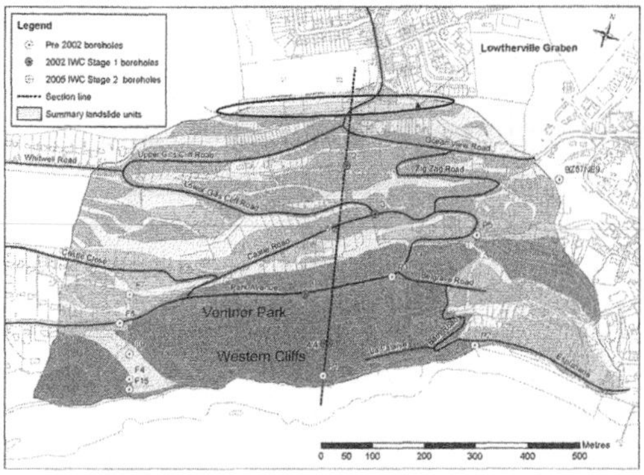

Figure 2. Ventnor Undercliff study area.

have sustained significant damage. The annual costs of landslide damage and management measures are estimated to exceed £3 million per year (McInnes 2000).

Previous studies of the Ventnor Undercliff (op. cit.) recognised the need for detailed sub-surface investigations to verify the mechanisms and causes of landsliding. In 2002 and 2005, the local authority commissioned two ground investigations to determine the mechanisms and causes of the Lowtherville graben and Ventnor Undercliff landslides (Fig 2). The ground investigations were needed to prepare an interpretative report on the ground conditions and landslide mechanisms, and to develop a quantitative risk analysis (QRA) to evaluate cost benefit of different management options, including coastal slope stabilisation (Halcrow 2006).

This paper summarises the results of the ground investigations and describes the ground conditions encountered. Review of the borehole and other slope monitoring data coupled with stability analysis provides a calibrated landslide model for the Ventnor Undercliff for the first time. Other relevant papers on the Ventnor Undercliff presented in this conference include Palmer et al. (2007) on the litho-stratigraphy of the landslide geology; Carey et al. (2007) on the pre-failure landslide behaviour; Moore et al. (2007) on the implications of future climate change on ground movement and landslide event frequency in the Ventnor Undercliff; and Lee and Moore (2007) on the development of landslide scenarios and quantitative risk assessment.

2 GEOLOGY AND HYDROGEOLOGY

The Undercliff is situated on the southern limb of the Southern Downs of the Isle of Wight and comprises a sequence of interbedded sedimentary rocks which dip seaward by about 1.5–2°. The sedimentary rocks were

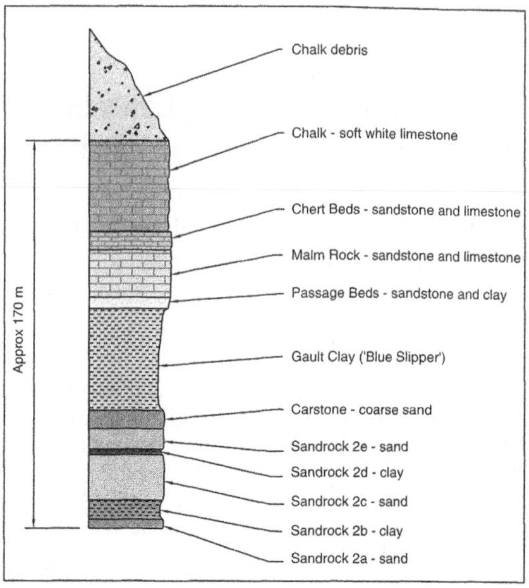

Figure 3. Summarised geology of the Undercliff at Ventnor.

laid down during the Cretaceous period, approximately 80 to 120 million years ago.

The ground investigations at Ventnor have provided a rare opportunity to obtain geological information to a considerable depth (up to 150 m) in the upper parts of the Undercliff landslide complex. Existing boreholes in the region (see Fig 2) are limited to site specific investigations and do not provide sufficient information to adequately describe the ground conditions and landslide mechanisms along an appropriate stability section line.

The 2002 & 2005 investigations comprised five deep rotary and open-cored boreholes, engineering and geophysical logging of materials, laboratory testing of samples and installation of borehole inclinometers and piezometers. The staged investigations were of high cost owing to the depth of boreholes investigated; this was compounded by severe access constraints to preferred borehole and monitoring sites in a densely developed part of the town.

A detailed description of the litho-stratigraphy encountered at Ventnor is provided by Palmer et al. (2007). The stratigraphy was interpreted from the engineering and geophysical logs which were calibrated between the boreholes. A detailed knowledge of the litho-stratigraphy is fundamental for the interpretation of borehole and geophysical gamma records when formulating the ground model. The use of geophysics was particularly valuable where the strata have been disrupted by the landslide.

The general sequence of undisturbed strata at Ventnor is shown in Fig 3. The Lower Chalk overlies the Upper Greensand which comprises the Chert Beds,

Malm Rock and Passage Beds. Up to 45 m of Gault Formation is present between the overlying strata and Sandrock Formation and Ferruginous Sand Formations beneath. The Sandrock Formation is divided into a number of sand and clay rich units; importantly, the clay units have been identified as potential failure horizons whilst the sand layers support confined aquifers which have a significant influence on the stability of the Undercliff. The *in situ* strata are mantled by variable thicknesses of Chalk and Upper Greensand debris (derived from solifluction of the adjacent Downs during the Quaternary) and form notable thick deposits in the area of Ventnor Park and the Western Cliffs.

A stratigraphic description was provided by Lee & Moore (1991). This classification divided the *in situ* strata into six units, numbered 1 to 6. The recent investigations have confirmed this classification and provide further detail and spatial comparison within some of the units (Palmer et al. 2007).

The general sequence of strata was disrupted during the formation of the Undercliff landslides. Despite this, strata thicknesses appear to be reasonably consistent, provided account is taken of the landslide block geometry and where strata have been lost due to shearing and displacement. Typical thicknesses recorded at Ventnor are described by Palmer et al. 2007 and compared with deep borehole records from Bonchurch in the eastern part of the Undercliff and other published data. The stratigraphic classification provides a valuable aid in determining where strata may have been thinned or lost due to landsliding. Ground calibration is achieved by comparing anticipated strata thicknesses with those recovered from individual borehole or geophysical logs.

In the upper parts of the Ventnor Undercliff, at BH2 (Fig 2 and Fig 6), a basal shear surface has been confirmed by inclinometer records at the base of the Gault, about 100 m below ground surface. At BH5, located in the mid section of the Undercliff, there is evidence of a disturbed zone at the base of the Gault, although this was not evident in BH3, nearby. Whether the basal shear surface is at the base of the Gault or deeper (i.e. the Sandrock 2d unit) at this location is not yet proven. Presence of slickensides and evidence from geophysical logs and index testing suggest that the topmost section of the Gault in BH3 and BH5 is absent (Palmer et al. 2007). The loss of the upper sections of the Gault at this location may be due to complexities of the subsurface geometry and general disruption of the landslide blocks or due to shallow landsliding during the Quaternary.

The entire section above the Sandrock 2d unit was absent at BH1 (Park Avenue; Fig 2 and Fig 6) and those boreholes seaward of this location (Palmer et al. 2007); instead, a thin bed comprising a melange of Gault, Passage Beds and chalk was recovered above the Sandrock 2d interface. Assuming that deep failure of the Undercliff extends from the Lowtherville graben to the base of the Gault in the upper and mid-slope sections, stepping down to Sandrock 2d in the lower-section, provides evidence that the Ventnor Undercliff is a two-tier landslide, as suggested by Lee & Moore (1991).

BH1 revealed a notable thickness of apparent chalk debris; Chalk material was recorded to 55 m depth and at least the upper 27 m was described as chalk debris. It is not possible to determine whether the material between 27 and 55 m depth was intact but displaced Lower Chalk, or chalk debris.

Palaeosols were identified within the chalk debris at 2.9 m and 20.6 m below ground surface providing opportunity to date the soils. At the shallower depth, poorly preserved pollen was linked to the Late Bronze Age or a later period. At 20.6 m, a well preserved pollen assemblage was dated between 9500 to 9000 years BP (i.e. Boreal period, Flandrian chronozone Ib). However, small numbers of spruce pollen were found which were not native during the Boreal. The pollen may be derived from earlier sediments, which would indicate a much earlier age of Middle Pleistocene. These dates lend support to the view that the lower parts of the Ventnor Undercliff landslides are considerably older than its upper parts (Hutchinson & Bromhead 2002).

Boreholes (BH4 and 4I) located at Ventnor Park (Fig 2 and Fig 6) confirm the presence of chalk debris overlying Lower Chalk and Upper Greensand. The litho-stratigraphy of the Upper Greensand and Lower Chalk units recorded at this location suggests that the sequence represents an intact but displaced block rather than landslide debris. Beneath the Upper Greensand, approximately 15 m of disturbed Gault was found above the basal shear surface. The Gault sequence comprised lithorelicts in a remoulded matrix and is, therefore, notably different from the Gault recovered in boreholes BH2, BH3 and BH5 where the fabric was largely intact apart from occasional slickensides.

The sand-rich horizons of the Sandrock (notably units 2a and 2e), the Upper Greensand, Lower Chalk and chalk debris are more permeable (and water bearing) than the clay-rich units of the Gault and Sandrock. The hydrogeology and connectivity of these units is complicated; there are four main groundwater tables (GWT):

- GWT1 is the uppermost water table above the Gault which is hydrostatic
- GWT2 is active beneath the Gault and above Sandrock 2d which is an aquiclude
- GWT3 occurs in Sandrock 2c which is confined by Sandrock units 2b and 2d which are both aquicludes, and
- GWT4 occurs in Sandrock 2a which is confined below unit 2b.

GWT1 is recharged from rainfall percolating through the porous caprocks of the Southern Downs.

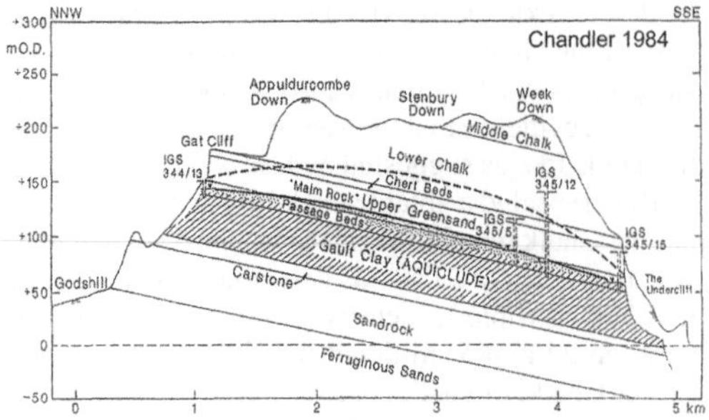

Figure 4. Hydrogeological model of the Southern Downs.

GWT1 is separated from the lower aquifers by the relatively impermeable Gault; at BH3, GWT1 is about 50 m above Ordnance Datum (AOD) whilst at BH5, GWT2 is about 30 m AOD; the water table and pressure gradient that applies to the Gault has been taken to vary linearly between GWT1 to GWT2.

GWT2 & 3 are confined aquifers recharged from inland where the Lower Greensand units outcrop (Fig 4) to an elevation of about 100 m AOD. Given the regional dip of strata to the SSE, drainage of the confined aquifers is towards the Undercliff and there is potential for artesian pressures to develop. The pressure in GWT3 will differ from GWT2 as different hydraulic gradients are involved. Also, it is noted that GWT3 'discharges' 20 m below sea level which is likely to prevent effective drainage of this unit.

The distribution of the different GWTs in the Undercliff probably influences the location of the major slip surfaces within the Gault and control the stability of the landslide blocks forming the Ventnor Undercliff. Warren and Palmer (2000) describe the complex three-dimensional nature of the landslips in the Gault at Folkestone Warren, where the aquifer systems in the Chalk and Folkestone Beds (beneath the Gault) result in varying porewater pressures along the slip surface. This is further complicated by observations of depressed porewater pressures noted by Hutchinson et al. (1980) on slip surfaces within the Gault at Folkestone Warren.

3 GROUND BEHAVIOUR RECORDS

Ventnor has had a long history of ground movement events which have caused significant damage to property and other infrastructure in the town (Lee and Moore 1991; Moore et al. 2007). The movement has taken the form of very slow and intermittent creep of the ground, the cumulative effects of which, over many years, has caused significant damage to buildings, services and other infrastructure. There have also been occasional accelerated ground movements, such as in the winter of 1960, when serious damage was experienced at several key locations in Ventnor.

The most notable impacts of ground movement can be observed at the Lowtherville graben in Upper Ventnor (Fig 2). Anecdotal evidence suggests that the graben has developed since the early to mid-1900s. Ground movement in the area was first reported by Edmunds & Bisson (1954). Today the graben extends some 500 m in length, 20 m across, and has subsided vertically by about 2–4 m. The main access road to Newport and other services cross the graben and are continually subject to damage due to ground movement. Since the 1980s, over 10 properties have had to be demolished in the area affected.

The development of the graben has been cited as a major cause for concern as it represents the early signs of reactivation of the Undercliff landslide complex in what is now a densely developed area (Hutchinson & Bromhead 2002; Moore et al. 2007). The impact of future predicted climate change is likely to exacerbate the problems at the graben and other key sites, increasing the risks to the town (Lee & Moore 2007).

A key element of the local authority's Landslide Management Strategy involves monitoring of a network of surface and sub-surface slope monitoring equipment to improve understanding of the relationships between rainfall, groundwater and ground movement. The monitoring system at Ventnor includes:

- borehole inclinometers in BH2 and BH4
- crackmeters and settlement cells at the Lowtherville graben and Bath Road
- piezometers and slip rods in BHs 1 to 5, and boreholes at the nearby Winter Gardens, and
- permanent ground markers which are regularly surveyed by GPS.

A detailed review of the ground movement characteristics and the relationship to groundwater and antecedent rainfall conditions is provided by Moore et al. (2007). A summary is provided here to assist interpretation of the Ventnor Undercliff landslide model.

Table 1 summarises typical rates of historical ground movement at Ventnor from 1932 to present; data for the graben show horizontal movement is typically about 6mm/yr compared to vertical movement which is greater at about 40 mm/yr. Accelerated ground movement occurs during extreme winter rainfall conditions which resulted in more than 3 times the yearly average in the most recent extreme rainfall period in the winter of 2000 (Moore et al. 2007).

Installation and survey of some 50 permanent ground markers across the site during the period 2003 to present reveals a widespread pattern of movement. Whilst the actual displacement rates are measurable in millimetres over this short period the pattern is similar to the measured rates of ground movement calculated

Table 1. Measured ground movement rates at Ventnor.

Location	Period	Rates of movement (mm/yr)
Newport Road	1982–1983*	19.7 to 39
Newport Road	1988**	28
Newport Road	1995–2005***	6.4 (crack extension) 33.2 (settlement)
Lowtherville graben	1982–1983*	26.9 to 29.9
Lower Gill's Cliff Rd	1982–1983*	8 to 16.5
Ocean View Road	1982–1983*	6
Bath Road	1982–1983**	0
Bath Road	1988**	0 to 2
Bath Road	1995–2005***	−0.06 (crack closure) 22 (settlement)
Esplanade	1949–1988	20 (heave)

Note: *after Chandler(1984); **after Woodruff (1989), **Council records.

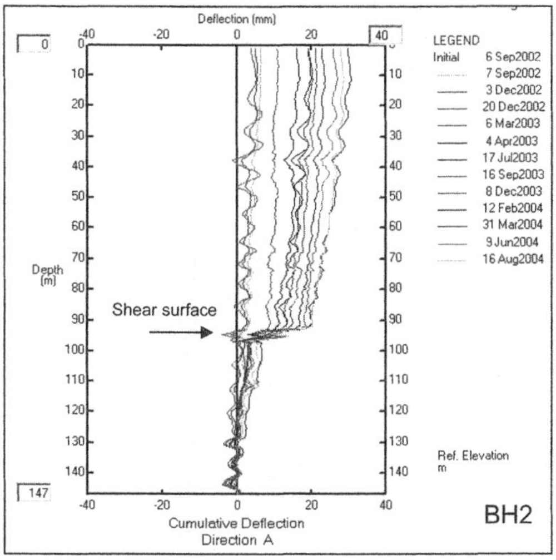

Figure 5. Inclinometer readout revealing basal shear surface.

from re-survey of Ordnance Survey benchmarks over the past 100 yrs or so (Lee & Moore 1991).

Of the two inclinometers at the site, BH2 has recorded movement at a depth of 94 to 98 m below ground surface, at the base of Gault (Fig 5). Between Sep 2002 and Aug 2004, about 30 mm movement had been recorded at the basal shear surface. These rates are consistent with those recorded at the graben where the basal shear surface cuts to ground level. So far, no deep-seated movement has been observed in BH4I; however, it is noted that the inclinometer was only recently installed in 2005 and that the instrument may not have fully equilibrated with the surrounding soil to detect any noticeable movement. No detectable moment has been identified with the slip rods in boreholes 1 to 5 but again this could simply reflect the limited monitoring period since their installation in 2003 and 2005.

4 UNDERCLIFF LANDSLIDE MODELS

Interpretation of the mechanisms causes and ground behaviour in the Undercliff depends on developing a consistent explanation which accounts for all of the features (geological, geomorphological, geotechnical, ground behaviour) in a logical and scientific way. Models of the Ventnor Undercliff landslides have been developed by a number of authors in the past which are conveniently summarised by Hutchinson & Bromhead (2002). The most favoured models include multiple rotational failures in the Gault and compound (non-rotational) failure in the Sandrock. These failure mechanisms have different shear surface depths.

A number of authors have made connections between the failures in the Undercliff with Folkestone Warren (e.g. Hutchinson 1965; Hutchinson et al. 1981 and Chandler 1984). At Folkestone Warren, progressive failure of the Gault is a major control on landslide development, with lateral expansion due to long term unloading resulting in shear surface propagation at the base of the Gault. The opening of joints in the caprock above the Gault occurs as a result of lateral expansion and tensile stress that form fissures and/or vents. The joints provide detachment surfaces for large coherent blocks which undergo rotational failure on a curvilinear shear surface linking the base of Gault with the ground surface rupture.

The Sandrock failure models were developed from the results of ground investigations of the Undercliff at St Catherine's Point (Chandler 1984; Bromhead et al. 1991). At this site, the landslide is composed of two linear ridges upslope of a broad debris apron. Site investigation revealed that whilst these ridges appeared to have been developed by rotational failure, movement had actually occurred by non-rotational sliding along a basal shear surface developed in Sandrock 2b, 50 m beneath the base of the Gault.

Lee & Moore (1991) and Moore et al. (1995) presented landslide models for the Ventnor Undercliff based on the geomorphological evidence and limited borehole investigations in proximity of the site. The model was generally considered to be two-tier, involving:

1. multiple rotational failure of the Gault and overlying strata in the upper section landward of Park Avenue, and

2. compound failure on clay layers within the Sandrock in the lower section seaward of Park Avenue.

Without the benefit of site-specific subsurface ground investigation, preliminary stability analysis was carried out by Lee & Moore (1991). The analysis considered four potential levels for the basal shear surface based on various assumptions with the ground model, groundwater levels and properties of soils. The results demonstrated that a two-tier landslide model was more likely although there was considerable uncertainty with the most probable depth of basal shearing in the Gault and Sandrock which could only be resolved with subsurface data. Following the investigations in 2002 and 2005 this is now possible.

5 LANDSLIDE MODEL FOR VENTNOR

Fig 2 shows the stability section line that was used to locate boreholes for the purpose of calibrating a ground model, identifying the principal failure mechanisms (landslide model) and for undertaking stability analysis. The section is generally representative of the Ventnor Undercliff landslide system and is aligned to the maximum dip of strata. The section line is not necessarily orthogonal to the coastline or the surface geomorphology, however, slight changes in the orientation of the section were found to have little effect on the calculated stability of deep shear surfaces.

The strata boundaries were plotted on the section from the borehole records and geophysical logs and cross referenced to the litho-stratigraphic framework (Palmer et al. 2007). All boreholes along the section line penetrated the base of the landslide complex into the Sandrock beneath. Correlation of the Sandrock units were tied back to landward of the Undercliff to reconstruct the *in situ* Upper Greensand and Lower Chalk strata above, using the stratigraphic framework.

Fig 6 presents a landslide model for the Ventnor Undercliff based on the findings of the recent ground investigations. The retrogressive two-tier landslide model comprises a distinct upper and lower landslide section, as follows:

1. The upper landslide section is the area between the Lowtherville graben and Park Avenue, Fig 2. The sequence is characterised by distinct rotational failure blocks seated at the base of the Gault that have undergone displacement to varying degree. The landslide blocks are capped by chalk debris, Lower Chalk, Malm Rock and Passage Beds over a substantial thickness of Gault, under which lies Carstone and Sandrock.

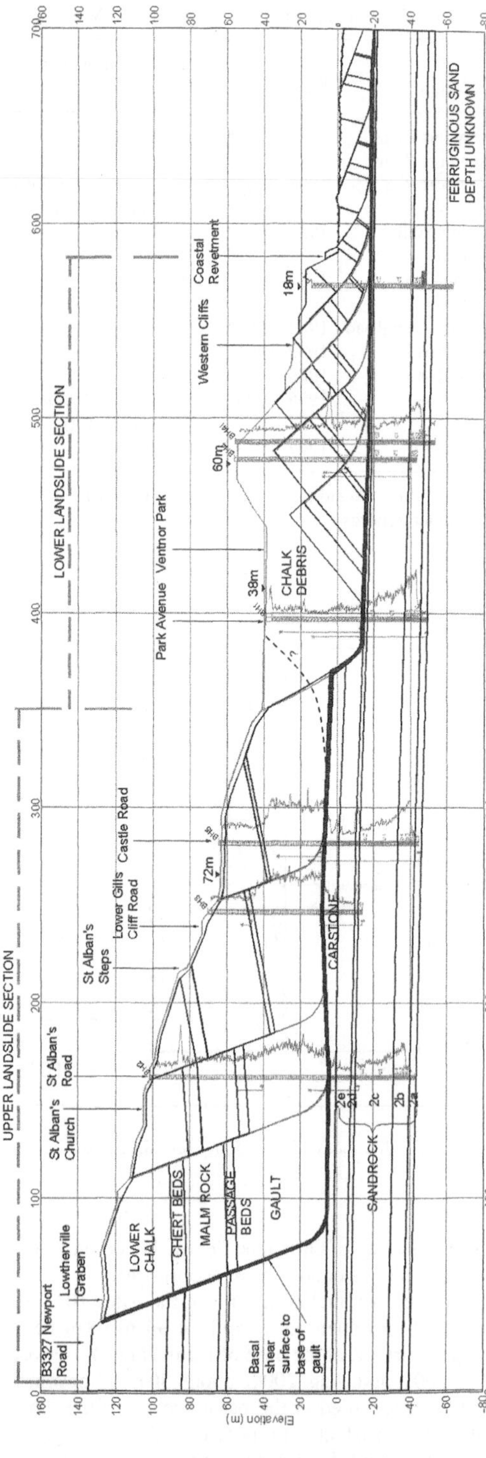

Figure 6. Ventnor Undercliff landslide model.

2. The lower landslide section comprises the area from Park Avenue to the sea. It is distinguished by a prominent ridge at Ventnor Park and fronted by the Western Cliffs (Fig 2). The ground along Park Avenue is formed of a thick sequence of chalk debris, whilst the seaward block is thought to comprise a remnant multiple rotational landslide block sequence draped by chalk debris. Notably, the lower landslide section is seated upon a thin layer of disturbed Gault above Sandrock 2d.

The landslide blocks may act singly or interactively. It is not yet possible to clarify this although future monitoring will provide the basis for confirming the model. The strata of the lower landslide section are highly disrupted and are fundamentally different to the upper landslide section; the absence of nearly all the Gault sequence is particularly notable. This interpretation adds support to the view that the lower landslide section forms the head of an ancient landslide system that once extended offshore prior to submergence by rising sea level during the Holocene.

The interpretation of the landslide stratigraphy is reasonably well defined in the upper section of the Ventnor Undercliff provided a multiple rotational model is invoked. The interpretation correlates key strata (or marker horizons) with the key surface morphology to identify probable boundaries between the deep-seated rotated blocks. The model shows greater displacement and backtilt of the landslide blocks downslope, a characteristic of retrogressive multiple-rotational landslides and clay extrusion.

The model correlates evidence of deep failure within the upper section that is manifest at the ground surface by distinct morphological features, such as scarp slopes, indicating that the Lowtherville graben is the uppermost and most recent failure of a series of deep shear surfaces in that area. The model assumes that the development of the upper landslide section has occurred subsequent to the formation of the lower landslide section.

The origin, nature and composition of the chalk debris and landslide blocks forming the lower section remain partly conjectural. The large thickness of chalk debris in BH1 was unexpected and the structure of the landslide block at BH4 shows chalk debris over largely intact chalk, with significant differences between BH4 and BH4I, even though these are only about 10 m apart. The nature of the contact between the chalk debris and slope above is also a matter of conjecture. This may be a shear surface but, may also be chalk washed against extruded and rotated Gault.

Only a restricted amount of information is known about the landslide stratigraphy out to sea. The work of Clark et al. (1994) shows this submarine slope is mantled by planed off remnants of old, deep-seated landslides to a depth of at least 25 m and possibly 40 m. This would suggest that at least the seabed between the current shoreline and the St Catherine's Deep, about 2 km offshore, is formed of pre-glacial slide remnants and debris. While the St Catherine's Deep is sufficiently far from the shore to have little effect on the Undercliff landslides, it would be desirable to confirm that there are no major features of seabed instability within influencing distance.

The main uncertainty with the landslide model is the configuration of the multiple rotational landslide blocks between the boreholes. The basal shear surface is reasonably well defined, although it was anticipated that monitoring of the inclinometer and slip indicators would have revealed the depths of ground movement, this has only been achieved at BH2. Consequently, stability modelling of several potential failure surfaces has been carried out (see below), based on geomorphology and suspected shear zones encountered within the boreholes. These cannot be considered definitive until proved by further monitoring of borehole deformations and surveys of surface displacements.

6 STABILITY ANALYSIS

Stability analysis of the landslide model (Fig 6) was conducted to verify the probable mechanism(s) of failure and the sensitivity of parameters through back analysis. The landslide model was also used to evaluate the potential effects of coastal erosion and climate change (groundwater levels) on present-day stability conditions (see section 8). Stability analysis can also be used to evaluate the benefits of various stabilisation measures on improving the stability of the Ventnor Undercliff. The following failure mechanisms were analysed:

1. failures seated in the Gault and extending from the Lowtherville graben to Park Avenue, with a shear surface cutting to ground surface upslope of the Ventnor Park block
2. a 'two-tiered' mechanism seated in the Gault in its upper section and stepping down to Sandrock 2d in the lower section
3. failures involving the chalk debris and the landslide blocks forming the Western Cliffs.

The shear strength parameters and bulk densities used in the analyses were derived from laboratory testing and published information. Residual shear strength parameters were assigned to those strata where failure surfaces were confirmed or could be present. The analysis used the residual strength parameters summarised in the table below.

Soil unit	ϕ'_r	Soil unit	ϕ'_r
Chalk	27.5	Carstone	32
Chert Beds	30	Sandrock – sand	30
Malm Rock	30	Sandrock 2d – clay	13
Passage Beds	22	Sandrock 2b – clay	10.6
Gault – upper layer	16.5	Gault debris	15
Gault – middle layer	9		
Gault – lower layer	15		

Notes: $c'_r = 0\,kN/m^2$ Unit weight = $18.9–22\,kN/m^3$.

There was a wide scatter in the residual shear strength tests carried out on samples of Gault from the boreholes and the choice of appropriate shear strength parameters for stability analysis was not straightforward (Palmer et al. 2007). Difficulties are compounded by the variation of shear strength with test method. In this analysis, the shear strengths adopted for the three layers of the Gault were moderately conservative and based on the available test results for each. Back analyses were carried out for likely failure modes and the resulting shear strengths compared with the range of values obtained by laboratory testing.

The analyses account for four groundwater regimes that potentially affect the analyses:

- GWT1 affecting the top of the Gault and all materials above
- GWT2 in Sandrock 2e, a confined aquifer between the Gault and Sandrock 2d
- a non-hydrostatic water table in the Gault governed by the pressure gradient between GWT1 & 2, and
- GWT3 a confined aquifer in Sandrock 2c.

The groundwater levels used in the analysis were representative of winter conditions when ground movement is most likely (Moore et al. 2007).

Factors of safety were computed for the selected non-circular shear surfaces in terms of effective stress using the Morgenstern and Price method and Slope/W. The Morgenstern and Price method of stability analysis provides reasonable correlation between the computed factors of safety (FoS) equal to one for a slope which is known to be marginally stable. The analyses proceeded by first calculating factors of safety for the 'existing conditions' using the models and input parameters described. These were then further explored by back analysis and sensitivity tests of the shear strength and water tables.

Results for landslide mechanisms 1 & 2 (op. cit.) demonstrate that the critical shear surface is the 'two-tiered' mechanism with a FoS equal to 1.13, whilst the FoS for a shear surface rising to Ventnor Park was equal to 1.22. The difference is small but nevertheless significant in that it indicates a lower probability for a major landslide event within the upper landslide section (Lee & Moore 2007).

Consideration was given to potential failure of Sandrock unit 2b, as this unit was identified as the principal basal shear surface for the St Catherine's model (Bromhead et al. 1991). The FoS for this surface was 1.23 and back-analysis yielded residual strength of $\phi'_r = 5.5°$ to achieve FoS=1, which is too low to be credible.

Results for landslide mechanism 3 at the Western Cliffs demonstrate FoS slightly above unity (1.05). There is active movement in this area, indicated by movement marker monitoring and by records of damage to paths, for which a factor of safety of unity is indicated. The sensitivity of the model to coastal erosion was considered. Assuming a historical rate of erosion of 0.2 m/year operating over 80 years, the effects of removing some 16 m of the toe of the cliffs was analysed. This was shown to decrease the factor of safety by 7% to 0.98 which highlights the need to maintain and improve the standard of protection of the recently constructed sea defences.

7 VENTNOR UNDERCLIFF EVOLUTION

An outline of the likely late Quaternary history of the Undercliff is provided by Hutchinson (1987) and Lee & Moore (1991). They suggest that the coastal landslides would have been mantled by thick deposits of soliflucted debris (mostly Chalk from the Southern Downs) during periglacial periods; the extensive debris aprons of the Undercliff will have formed at this time only to be largely removed by erosion due to rising sea levels during the interglacial periods. A key point to note is the cyclical loading of the pre-existing landslides will have had a significant destabilising effect particularly when the debris aprons had been substantially removed by erosion.

The last major phase of periglacial activity has been linked to the Late-glacial Interstadial, around 11300–12100 years BP. This period can be regarded the last 'formative' event during which substantial loading of the Undercliff will have taken place. The Flandrian Transgression followed the decay of the Late Devensian Ice Sheets; the transgression was characterised by rapid sea level rise from −120 m about 14000 years BP to −20 m by about 8000 years BP, since which the rate of sea level rise has been much slower (Hutchinson 2002). The rise in sea level and associated erosion resulted in the substantial removal of the landslide systems and debris aprons that formerly extended offshore and has shaped the Undercliff that exists today.

The actual date of the 'original' Undercliff landslides is unknown. They probably initiated south of the present shoreline, on the dip slope of the Upper and Lower Cretaceous strata which extends to the St Catherine's Deep some 2.8 km offshore in water depths of about −60 m AOD (Fig 7). Removal of support from

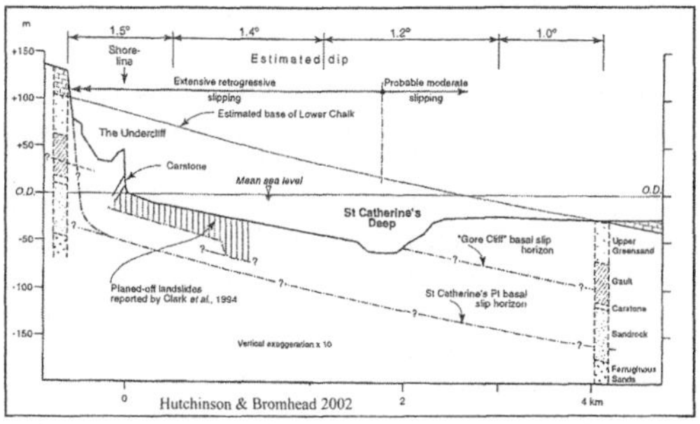

Figure 7. Relation of the Undercliff to St Catherine's Deep.

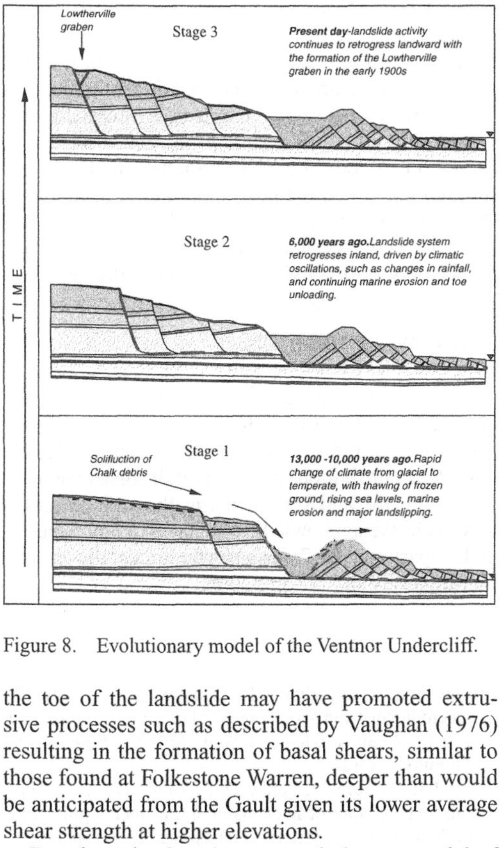

Figure 8. Evolutionary model of the Ventnor Undercliff.

the toe of the landslide may have promoted extrusive processes such as described by Vaughan (1976) resulting in the formation of basal shears, similar to those found at Folkestone Warren, deeper than would be anticipated from the Gault given its lower average shear strength at higher elevations.

Based on the forgoing, an evolutionary model of the Ventnor Undercliff landslide development is postulated (Fig 8). This model comprises an event sequence through the Holocene to the present day which begins at the time of the last 'formative' event about 13,000 years BP. The subsequent climatic changes and rapid rise in sea levels have caused major changes in the spatial and temporal sensitivity of the landslides and ground behaviour response.

The evolutionary model assumes that by around 13000 years BP, the landslide toe lay seaward of its current location (Fig 7) but would have migrated some distance north of its initiation point at St Catherine's Deep (Hutchinson 2002).

At that time, sea levels were so low that the English Channel would have been dry land. Permafrost conditions prevailed with rainfall and seasonally thawed ice shed as surface runoff causing major erosion and solifluction of the chalk downland. The soliflucted chalk debris formed thick deposits on the lower slopes, as described by Hutchinson (1987), and would have occurred at a time when the lower slopes of the Undercliff were effectively dry with low pore water pressures.

With increasing temperature the permafrost thawed to form groundwater which will have recharged the Lower Greensand. The effect of rising groundwater level and porewater pressures at the landslide toe will have caused the factor of safety to reduce below unity and failure recommenced.

The displacement of the lower landslide block at Ventnor Park would have required a large increase in porewater pressure. Development of significant porewater pressure seems unlikely if groundwater was able to drain from Sandrock 2e but could have developed if drainage was prevented. Drainage could have been impeded where Gault debris occurs above Sandrock 2d, where it would form an aquiclude. In that event, porewater pressures could have risen above 100 m OD causing failure; this semi-hydraulic failure would have left Sandrock 2d little disturbed, as observed.

In summary, the combination of rapid loading, elevated groundwater levels, and toe erosion driven by rapid sea level rise during the period 13000 to 10000 years BP is believed to have caused reactivation of the compound failure on Sandrock 2d displacing the compound block about 50 m seaward.

The displacement of the compound block would have opened up a depression in the current area of Ventnor Park, which was rapidly infilled with solifluction debris which continued for some time after. The formation of the depression is confirmed by a palaeosol encountered at 20.6 m depth in BH1 (Section 2); the palaeosol has been dated as pre-Boreal, about 10000 years BP, the period when rapid infilling took place.

A compound failure of this nature and the relatively modest 50 m distance it travelled would have removed support and exposed the *in situ* Gault of the upper landslide section. It is postulated that the development of a quasi-rotational retrogressive failure would have occurred at the base of the Gault (Stage 1, Fig 8) at this time. The initial failure could have been rapid, initiated by seepage erosion and liquefaction of the water-bearing Sandrock (GWT2), restrained only by the 'undrained' strength of the overlying Gault, with limited resistance offered by the major weak joints of the overlying caprocks. Subsequent failures may have initiated but were not entirely mobilised due to the recovery of porewater pressures in GWT2 and the pressure gradient with GWT1 within the Gault. The rate of development of the multiple rotational failures of the upper section will have been governed by the long-term 'drained' behaviour of the Gault, and the propagation of shear surfaces up dip at the base of the Gault.

Rapid deposition and consolidation of chalk debris towards the end of Stage 1 would have substantially infilled the depression opened up by the compound failure, providing passive support to the multiple rotational failure(s) of the upper section that had initiated at this stage.

Further retrogressive failures are postulated to have occurred during the Mid Holocene wet period as groundwater levels were temporarily elevated (Stage 2) and sea levels had substantially reached present-day levels. The multiple rotational landslides would have retrogressed upslope as a result of erosion and unloading (displacement) of the lower compound block, and development of high porewater pressures acting at the base of the Gault (GWT2). General extrusive processes such as described in Vaughan (1976) may also have contributed.

The retrogression of the multiple landslide complex is postulated to have continued up dip since the mid-Holocene to its current position (Stage 3), with the most recent multiple rotational block developing in approximately the last 100 years, as manifest with the development of the Lowtherville graben. The development of the Lowtherville graben is compelling, and based on this postulated evolutionary sequence, demonstrates a ground behavioural response that dates back to the Late-glacial, some 13000 years BP.

While the above provides a possible evolutionary landslide model, it is recognised that there are still aspects that are not perfectly understood. Nevertheless, the evolutionary model offers some context to the concern that the stability of the Ventnor Undercliff could decline rapidly due to the effects of future predicted climate change, potentially resulting in the initiation of rapid failure of the upper section or the entire Ventnor Undercliff (Moore et al. 2007; Lee & Moore 2007). Given the evolutionary model presented, the likelihood of rapid failure of the upper section seems unlikely provided the lower compound landslide block and infilled depression continue to provide passive support as they have done so for the past 10000 years. There is no precedent for such a rapid behavioural response of the upper landslide section even considering the formative events that occurred around 13000 years BP. Nevertheless, further work is ongoing to evaluate the pre-failure behaviour of the Gault (Carey et al. 2007).

8 IMPLICATIONS OF CLIMATE CHANGE

Further discussion of the potential impacts of climate change on future ground behaviour at Ventnor is presented by Moore et al. (2007). In summary, it is anticipated that increases in antecedent effective rainfall and ground water levels at the site will result in transient and progressive reduction in FoS through time; coupled with the potential loss of toe support through coastal erosion due to sea level rise, there is real concern that the marginal FoS of the two-tier landslide could be reduced to unity over the next 100 years. Any decline in the FoS is likely to result in more frequent and higher rates of ground movement with consequential damage to property and infrastructure at key sites, such as the Lowtherville graben. The ground behaviour response to climate change over the next 100 years has been estimated to increase the economic risks to the town by 0.25% per year (Lee & Moore 2007).

The only effective means to prevent a decline in FoS will be to consider stabilisation of the Undercliff through combined toe protection and deep drainage. Prevention of erosion and control of the porewater pressures acting on the basal shear surface will secure the future stability and sustainability of the town.

ACKNOWLEDGEMENTS

The authors would like to thank colleagues at Halcrow and the Isle of Wight Coastal Centre for their assistance with this project; we also acknowledge the valuable contributions of former colleagues and consultants at GSL and Rendels during the earlier stages of this work. We acknowledge the Isle of Wight Council and Department for Transport who funded the ground investigations in 2002 and 2005.

REFERENCES

Bromhead E.N., Chandler M.P. & Hutchinson J.N. 1991. The recent history and geotechnics of landslides at Gore Cliff, Isle of Wight. pp189–196 *Slope Stability Engineering*, Thomas Telford.

Carey J.M., Moore R., Petley D. & Siddle H.J. 2007. Pre-failure behaviour of slope materials and their significance in the progressive failure of landslides. *This conference.*

Chandler M.P. 1984. The coastal landslides forming the Undercliff of the Isle of Wight. Unpublished PhD thesis, Imperial College, University of London.

Clark A.R., Lee E.M. & Moore R. 1994. The development of a ground behaviour model for the assessment of landslide hazard in the Isle of Wight Undercliff and its role in supporting major development and infrastructure projects. Proc. 7th Int. Cong. Int. Ass. Engng. Geology, Lisbon, 6, 4901–4913.

Edmunds F.H. & Bisson G. 1954. Geological report on road subsidence at Whitwell Road, Ventnor. Unpublished report.

Halcrow 2006. Ventnor Undercliff, Isle of Wight Coastal Instability Risk: Interpretative Report and Quantitative Risk Analysis. Technical report to Isle of Wight Council.

Hutchinson J.N. & Bromhead E.N. 2002. Isle of Wight landslides, keynote paper, p3–70. In: McInnes R. and Jakeways J. (eds.) Instability, Planning and Management: seeking solutions to ground movement problems. *Proceedings of the International Conference organised by the Centre for the Coastal Environment, Isle of Wight Council, Ventnor, Isle of Wight. Thomas Telford.*

Hutchinson J.N., Bromhead E.N. & Lupini, J.F. 1980. Additional observations on Folkestone Warren landslides *Quarterly Journal Engineering Geology*, 13, 1–31.

Hutchinson J.N. 1965. A reconnaissance of the coastal landslides in the Isle of Wight. Note No. EN 35/65. Watford: Building Research Station.

Hutchinson J.N. 1987. Some coastal landslides of the southern Isle of Wight. In: Wessex and the Isle of Wight. Field Guide (ed. Barber K.E.) 123–135. Quaternary Research Association, Cambridge.

Hutchinson J.N., Chandler M.P. & Bromhead E.N. 1981. Cliff recession on the Isle of Wight SW coast. Proc. 10th Int. Conf. on soil mechanics and foundation engineering, Stockholm, 1, 429–434.

Lee E.M. & Moore R. 1991. Coastal Landslip Potential Assessment, Isle of Wight Undercliff, Ventnor. Technical Report prepared by Geomorphological Services Ltd for the Department of the Environment, research contract PECD 7/1/272.

Lee E.M. & Moore R. 2007. Ventnor Undercliff: development of landslide scenarios and quantitative risk assessment. *This conference.*

McInnes R.G. 2000. Managing ground instability in urban areas: a guide to best practice. IW Centre for the Coastal Environment.

Moore R., Carey J., McInnes, R.G. & Houghton J. 2007. Climate change, so what? Implications for ground movement and landslide event frequency in the Ventnor Undercliff, Isle of Wight. *This conference*. Balkema.

Moore R., Lee E.M. & Clark A.R. 1995. The Undercliff of the Isle of Wight: a review of ground behaviour. Cross Publishing.

Palmer M.J., Carey J.M., & Turner M.D. 2007. Lithostratigraphy of the Ventnor Undercliff and determination of critical horizons through borehole geophysics. *This conference*.

Vaughan P.R. 1976. The deformations of the Empingham valley slope. pp451 to 461 of Appendix to paper by Horswill, P and Horton, A entitled: cambering and valley bulging in the Gwash Valley, Empingham, Rutland, *Phil. Trans. Roy. Soc. A238.*

Warren C.D., & Palmer, M.J. 2000. Observations on the nature of landslipped strata, Folkestone Warren, United Kingdom. Landslides in research, theory and practice, Proc. 8th Intern. Symp on Landslides, 3, Cardiff, 1551–1556,

Landslides and Climate Change – McInnes, Jakeways, Fairbank & Mathie (eds)
© 2007 Taylor & Francis Group, London, ISBN 978-0-415-44318-0

Litho-stratigraphy of the Ventnor Undercliff and determination of critical horizons through borehole geophysics

M.J. Palmer
Halcrow Group Ltd, Swindon, Wiltshire, UK

J.M. Carey
Halcrow Group Ltd, Birmingham, UK

M.D. Turner
Halcrow Group Ltd, Cardiff, UK

ABSTRACT: Ventnor is an area of notable landsliding. Concerns over the stability of a number of areas have in the past been investigated where possible with rotary cored boreholes. Landslide risk assessments which take account of future climate change will require further ground investigations of the landslide complex. Within Ventnor this may require the drilling of boreholes in excess of 100 m, at significant cost. This paper reviews and updates the published information regarding the litho-stratigraphy of the geological succession in Ventnor and demonstrates how geophysical gamma-logging can be used to correlate the succession thus providing an economic and useful investigative technique in place of rotary coring. The characteristics and index properties of the Gault are also described from the available data, particularly in relation to the occurrence of critical shear surfaces in this stratum.

1 INTRODUCTION

Ventnor is an area of notable landsliding. Extensive geomorphological mapping has been undertaken for the area and subsurface investigations have been executed within the area known as the Undercliff (Figure 1).

A review of past work on the Undercliff landslides together with a summary of the broad styles of landsliding identified is provided by Hutchinson and Bromhead (2002).

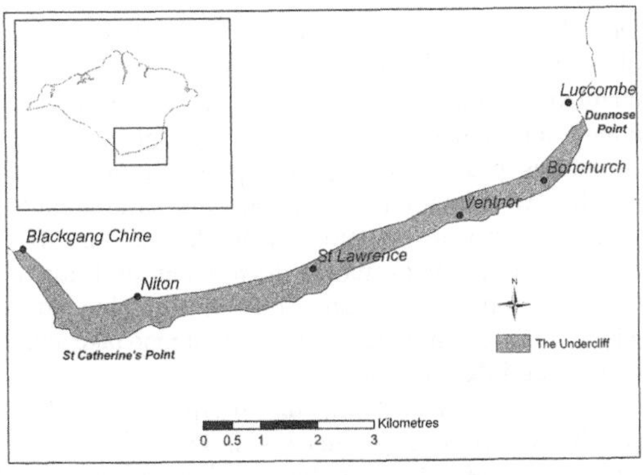

Figure 1. The Undercliff, Ventnor, Isle of Wight.

As climate change predictions and landslide risk assessments raise questions about the stability of areas of the Undercliff further ground investigations will almost certainly be required. Within Ventnor this may require the drilling of boreholes in excess of 100 m, at significant cost to the Isle of Wight Council and disruption to residents and businesses.

The establishment of accurate geological ground models for the different areas of the Undercliff plays an important role in risk assessment.

Therefore the formulation of a reference litho-stratigraphy is beneficial in assisting both the planning and interpretation of subsurface investigations.

The availability of a characteristic geophysical logging signature enables information to be captured without the need for coring. This can reduce the time and cost of investigations and also provide additional information where perhaps previously it would not be feasible to collect.

2 LITHO-STRATIGRAPHY

2.1 *Geological structure*

The geological structure of the Ventnor area comprises the southern limb of the *en echelon* asymmetrical

Table 1. The Primary lithostratigraphic units at Ventnor and there varying thicknesses.

STAGE	FORMATION	UNIT	THICKNESS (m) Lee & Moore (1991)	Ventnor	Bonchurch	LITHOLOGICAL DESCRIPTION
			-	27 (55m possible)	-	**Chalk Debris**
CENOMANIAN	Lower Chalk	*6b*	>5.5	15-31.4		**Chalk Marl Member** Grey marly chalk generally with no flints
		6a	2.1	3.1-3.4	4.4-5.0	**Glauconitic Marl** Calcareous glauconitic sand distinctive basal unit heavily bioturbated
UPPER ALBIAN	Upper Greensand	*5c*	8.2-8.8	6-8	7.8-8.7	**Chert Beds** prominent Chert and limestone bands in glauconitic sandstones
		5b	23.3	20.5-22	16.5	**Malm Rock** buff and grey coloured fine sand and sandstone
		5a	12.0	2.1-5.3	6.7-8	**Passage Beds** grey and buff coloured silts and clays
	Gault	*4b*	44.0	45.1; 36-42	44.2; 36.4	Very stiff dark grey (silty) clay
LOWER ALBIAN		*4a*		3.1-8.8	7.8	Very stiff brown (silty) clay
	Carstone	*3*	10.5	4.1-9.2	>0.1	Grey brown angular sand with pebbles and clayey sand interbeds
	Sandrock	*2e*	6.4	8.5-10.1		Light grey and buff sand
		2d	3.0	2.3-2.6		Dark grey or mottled sandy clay or sand
		2c	17.7	20.0-21.3; *(ii) 5.3-6.2*		Light grey fine to medium sand with glauconitic partings
UPPER APTIAN		*(ii/i)*		*(i) 14.5-615.4*		Bioturbated, dark grey glauconitic clayey silt/ sand with light grey sand partings
		2b (i-iv)	5.1	8.3-9.3		*see insert box*
		2a	10.8	3.6-4.6*		Light grey weakly cemented sand
LOWER APTIAN	Ferruginous Sand	*1c*	>3	>11.4		Dark grey green glauconitic sands

Insert box:

2b(iv)	3.3-3.8	Dark green grey clayey sand
2b(iii)	0.9-2.6	Mainly clay with some bioturbation
2b(ii)	1.5-3.0	Sandy clay overlying dark coloured sand (sand absent in BH1)
2b(i)	0-1.1	Brown grey clay (translational to 2a) unit only present in Bh1

Sandown Anticline with Cretaceous strata dipping at about 1.5 to 2° to the south-southwest. Superimposed on the generally southerly dip there is the St Lawrence syncline. However owing to extensive landsliding there is only limited reliable information on the relative positions of undisturbed strata within the Undercliff area.

2.2 *Generalised litho-stratigraphy*

A generalised stratigraphic description was provided by Lee and Moore (1991) as part of an assessment of coastal landslide potential for Ventnor. This classification divided the strata, usually present in Ventnor into six units, numbered 1 to 6. For consistency this classification has been retained but, based on careful examination of cores across the Undercliff area, it has been expanded and updated to include greater detail within some of the units (Table 1).

Despite the occurrence of extensive displacement within the Undercliff landslide complex, strata thicknesses in general, appear to be reasonably consistent, providing account is taken of where shear surfaces has caused changes in thickness. Typical thicknesses recorded across the landslip are described in Table 1. These values provide a very useful aid when determining where strata may have been lost due to landsliding. Clarification is obtained by comparing anticipated strata thicknesses with those recovered from individual borehole or geophysical logs.

2.3 *Downhole geophysical gamma logging*

Downhole geophysical gamma logs have been obtained from a number of deep boreholes. This has enabled a composite geophysical profile to be generated and cross referenced. The geophysical log signatures provide a graphic representation of the lithological variation throughout the stratigraphic sequence. They are operator independent, and if sufficiently distinct can overcome the problem of different interpretations of the same lithostratigraphic units which may occur when individual under-take core logging at different times.

Sufficient confidence in the interpretation of the geophysical profile permits boreholes to be drilled without the need for sampling, coring and an interpretation made solely from the downhole gamma log.

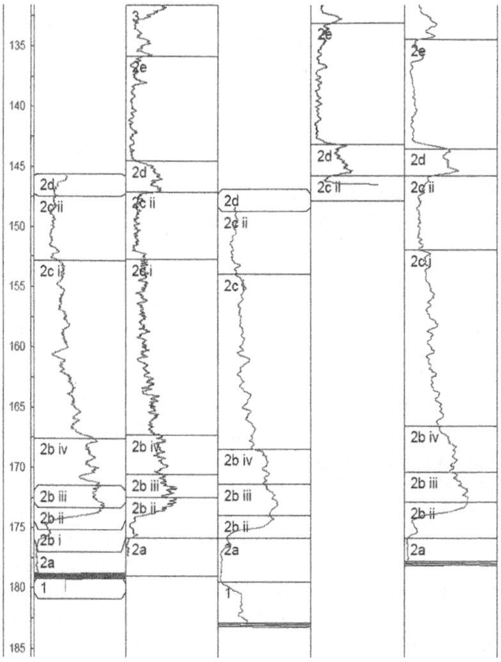

Figure 2. Geophysical gamma log of the Sandrock Formation. The logs have been normalized relative to the top of the Sandrock unit 2a. (The depth is given in meters).

However, correlation between individual horizons within the Gault is less distinct, although there is a notable distinction between displaced, but largely intact Gault and disturbed Gault (displaying a characteristic fabric of lithorelics in a remoulded matrix).

3 FERRUGINOUS SAND

The Ferruginous Sands (unit 1) are typically dark grey green fine to medium glauconitic sands when recovered in boreholes. This formation is not affected by landslipping in the Undercliff.

4 SANDROCK FORMATION

Figure 2 illustrates the gamma log signatures of the Sandrock Formation. The most significant horizons in the Sandrock Formation with regards landslide ground models are units 2d and 2b (iii). These units comprise clay beds that contain slickensided and polished surfaces representing flexural shearing and landslide displacements.

None of the boreholes examined provided a complete sequence through the Sandrock due to either landsliding or limited borehole penetration. However, nine lithostratigraphic units have been identified by considering a composite sequence (Table 1).

Table 2. Summary of index water contents in Sandrock 2b.

Parameter	Range	Median value
Plastic limit	18 to 31%	25%
Liquid limit	36 to 74%	60%
Plasticity index	18 to 44	36.5%

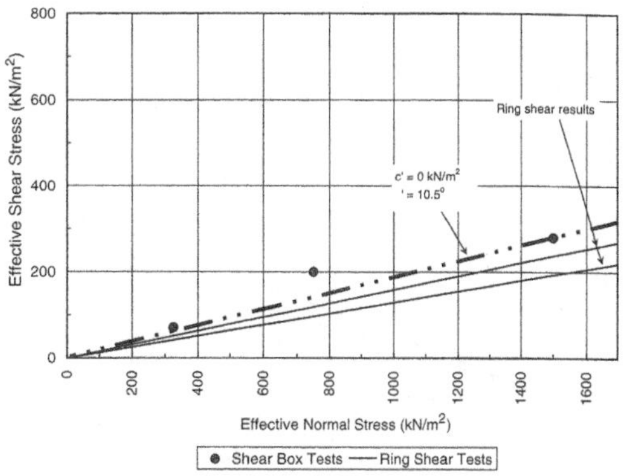

Figure 3. Residual shear strength test results of Sandrock 2b.

4.1 *Sandrock unit 2a*

The Sandrock unit 2a is typically a light grey weakly cemented sand with pockets and partings of lignite.

4.2 *Sandrock unit 2b*

Sandrock unit 2b (ii) typically comprised a sandy clay in the upper part overlying a dark coloured sand, which rests directly upon unit 2a. Sandrock unit 2b(i) was commonly absent. Locally the dark coloured sands of unit 2b(ii) are absent and the brown grey clay of unit 2b(i) is present.

Table 2 summarises the index properties from the Sandrock 2b unit.

One set of three shear box tests and two ring shear tests have been conducted to determine residual shear strength parameters. These results are plotted on Figure 3.

4.3 *Sandrock unit 2c*

Sandrock unit 2c typically comprises light grey-green weekly cemented fine to medium sand with glauconitic striations. In some boreholes closely spaced laminae of dark grey-green sandy clay are present. The base of unit 2c generally comprises dark grey bioturbated sand with lenses of light grey sand.

Table 3. Summary of index properties in Sandrock 2d clay units.

Parameter	Range	Median value
Plastic limit	18 to 26%	20%
Liquid limit	37 to 62%	52%
Plasticity index	18 to 42	37%

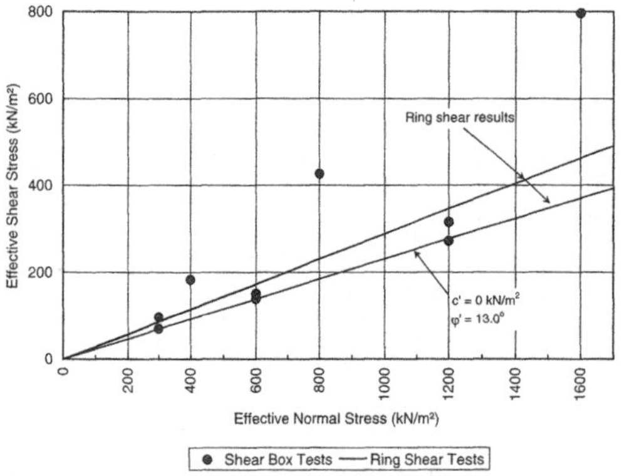

Figure 4. Residual shear strength test results of Sandrock 2d.

4.4 *Sandrock unit 2d*

This is a variably dark grey clay, dark grey mottled green sandy clay or sand occasionally with subrounded gravel of black chert and thin laminae of fine sand. This unit commonly contains slickensided surfaces within the clay, but these are easily masked if contaminated by the sand partings.

At Ventnor, horizontal to sub-horizontal polished surfaces with slickensides were noted in all boreholes. A summary of the index properties from the Sandrock 2d unit is provided in Table 3.

The results from 60 mm shear box tests to determine residual strength are shown on Figure 4. The results from two ring shear tests, providing ϕ_r' values of 13.1 and 16.2° are also included.

4.5 *Sandrock unit 2e*

The upper unit of the Sandrock Formation comprises light grey (or white) and buff sands with occasional clay laminae.

5 CARSTONE FORMATION

The Carstone, Unit 3, is typically a grey brown angular sand with occasional fine gravel, interbeded with sandy clay and clayey sand. The base of the unit is often marked by a thin pebbly band.

Thickness variations recorded in boreholes appear to be a condensed sequence as geophysical logs depict the same signature over a shorter length.

6 GAULT FORMATION

The maximum thickness of the Gault (Unit 4a and 4b) is typically 45 m in the Undercliff area.

Despite the importance of the Gault Formation in landsliding it remains poorly described in the Ventnor area due to the absence of complete exposures. Gale *et al*, (1996) presented the stratigraphy of the Gault at Redcliff east of Sandown. At that location the thickness of the Gault is only 28.5 m, although the underlying Carstone is some 22 m thick. The combined thickness is very comparable to that observed in Ventnor (Table 1).

Gale *et al*, (1996) identified that the Gault succession at Redcliff and Cowes (i.e. in the north and east of the island) contained discrete sand and clay rich units. The succession in the south and west of the island (including Ventnor) is notably sandier, and the gradual lithological transition to the Upper Greensand is characterized by a separate unit called the Passage Beds (unit 5a).

In the boreholes drilled at Ventnor, it is not easy to identify discrete lithological units or marker horizons in the Gault such as described by Gale *et al*, (1996) at Redcliff. As noted above, at Ventnor distinct lithological units are less apparent and are transitional.

A change from the typical dark grey colour present in the upper portion of the Gault unit (unit 4b in Table 1) can be observed, with the base of the Gault (unit 4a) becoming a 'grey brown' colour, this change may be sharp or transitional.

Typically in the upper portion the Gault appears less plastic and more silty and fissured.

6.1 *Geophysical gamma log*

Figure 5 shows the gamma geophysical log through the Gault Formation (but ignores sequences that clearly display remoulding and lithorelics as a consequence of landslide disturbance). The sequence is normalized relative to the base of the Gault Clay.

The occurrence of the brown colour change (unit 4a) at the base of the Gault sequence does not appear to be represented within the geophysical signature.

Approximately 30 m above the base of the Gault there is a characteristic reduction in the gamma log signature. This 'upper layer' can, with difficulty, also be identified from borehole core although the location of the boundary with the underlying 'middle layer' is generally transitional.

The correlation in the geophysical signature between boreholes over the lower 12 m of the sequence ('lower layer') is not particularly consistent.

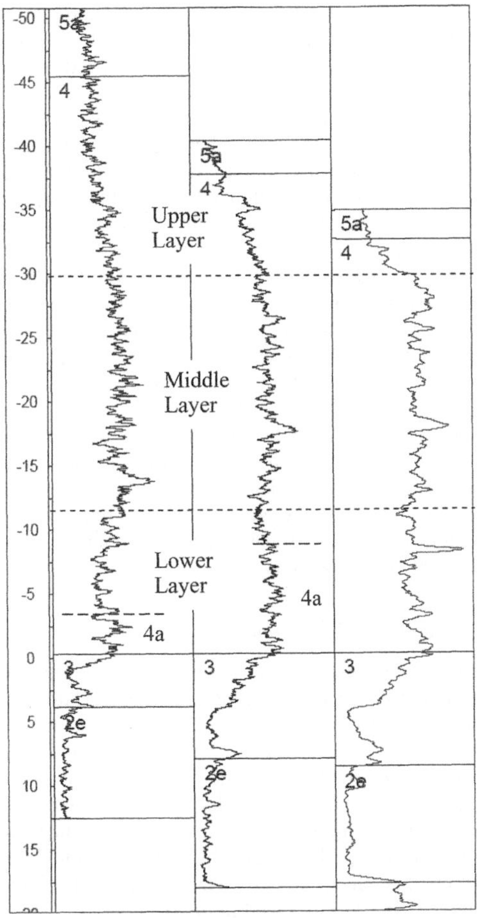

Figure 5. Geophysical gamma log of the Gault Formation at Ventnor. The logs have been normalized relative to the base of the Gault Formation (unit 4a) (depth scale is in meters).

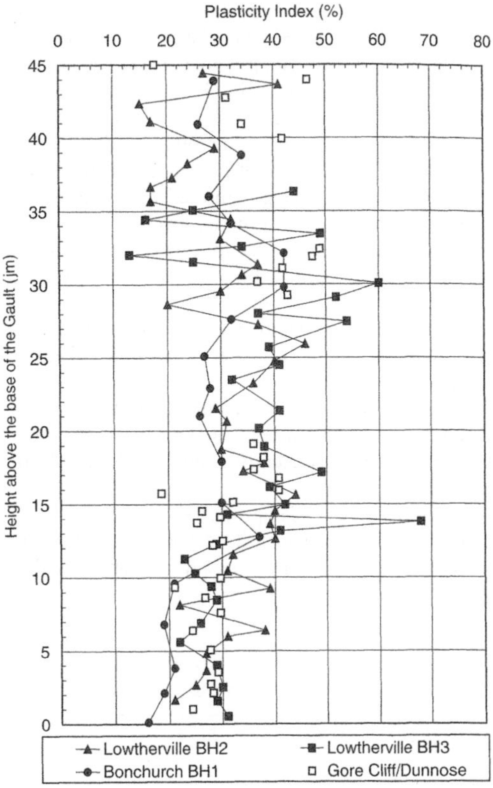

Figure 6. Plasticity index of the Gault Formation.

Gale *et al*, (1991) report that density geophysical logs appeared to correlate better with the lithological succession at Redcliff than gamma or resistivity logs and the use of this technique may enhance the geophysical correlations obtained from Ventnor.

The identification of units within the Gault Formation at Folkestone using lithological and palaeontological divisions to assist in determining the landslide ground models of the Folkestone Warren landslides has been described by Warren and Palmer, (2000). However, at Ventnor the lithological units are less distinct and there are fewer fossils, which makes this technique less definitive.

6.2 *Index properties*

Bromhead *et al*, (1991) measured index properties in the available exposures of the Gault at Gore Cliff at the west end of the Undercliff and similar measurements were made at Dunnose in the east by Chandler, (1984). By this means, Bromhead *et al*, (1991) divided the Gault in the Undercliff into an upper, more plastic stratum and a lower, less plastic stratum. A slip-prone horizon was invoked at the base of the upper stratum to coincide with the change in plasticity, approximately 15 to 18 m thick.

The sequences described by Bromhead *et al,* (1991) and Chandler, (1984) are on incomplete sequences through the Gault and the section at Dunnose only included the lowest portion of the sequence. Investigations at Ventnor (Moore *et al*, 2007) and Bonchurch (Palmer *et al*, 2007) have provided index properties through full thicknesses of the Gault. These results are shown in Figures 6 and 7, normalized from the base of the strata, together with the earlier results from Gore Cliff and Dunnose.

The reduction in plasticity observed at Gore Cliff is more transitional at Ventnor and Bonchurch where a notable reduction is not apparent until 12 m above the base of the Gault (see Figure 6).

Numerous slickensides are present at varying levels within the Gault sequence which may represent flexural slip or landslide shears. Hutchinson and

Bromhead, (2002) attributed the major slip horizon to be 15 m above the base of the Gault which coincided with the increase in plasticity index. However, at Ventnor the major slip horizon in Gault was proven with inclinometers to be at the base of the Gault (Moore *et al*, 2007). However, the reduction in Gault thickness and the apparent absence of part to all of the 'upper layer' from the gamma logs (Figure 5 also suggest that other slip surfaces exist at higher levels.

At Bonchurch landsliding appears to have occurred within the upper part of the sequence (Palmer *et al*, 2007).

In general terms, samples of higher natural moisture content tended to associate with samples of higher plasticity.

A number of other divisions in the lower part of the sequence can be discerned based on plasticity index (Figure 6) and more clearly from liquid and plastic limits (Figure 7).

Variations in the index properties of the Gault Formation are more easily identified when considering the trend of the results as a whole, rather than individual boreholes.

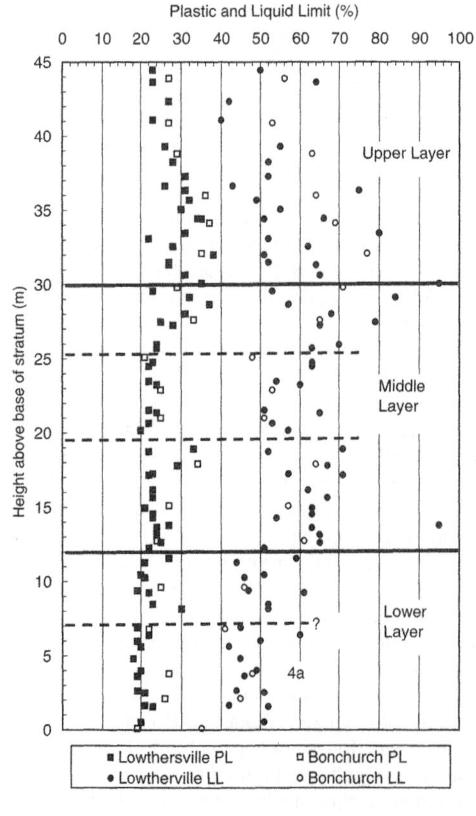

Figure 7. Plastic and liquid limits of the Gault together with suggested index test based subdivisions.

There is a notable reduction in index properties 30 and 45 m above the base of the Gault and this corresponds with the 'upper layer' identified by the geophysical gamma signature in Figure 5.

The sequence from 12 to 30 m above the base of the Gault Formation (termed here the 'middle layer') could be further subdivided based upon liquid limit properties. It is postulated that these subdivisions may also be apparent in density geophysical logs for future investigations.

Particle size distributions show the Gault to be a slightly sandy clayey silt, containing 15 to 35% of clay. On the basis of the clay content, the Gault at Ventnor appears coarser than other parts of the UK where clay contents range up to 65%, often with the silt and clay content being similar.

In accordance with descriptions defined by BS5930, (1999), the majority of the samples comprise clays of intermediate and high plasticity. Other samples fall within the ranges of silts of intermediate to high plasticity and clays of low and very high to extremely high plasticity.

6.3 *Residual shear strength*

The results of ring shear tests conducted in the Gault at Ventnor provide values ranging from 4.9° to 18.8° with a median of 10.4° (Moore *et al*, 2007). This compares to the range given in Hutchinson *et al*, 1991 of 7.5° to nearly 20° for the St Catherine's Point landslip.

The ring shear test results and equivalent values of liquid limit for the samples concerned are superimposed on the relationship between liquid limit and residual shear strength established by Mesri and Capeda-Diaz, (1986) (see Figure 8).

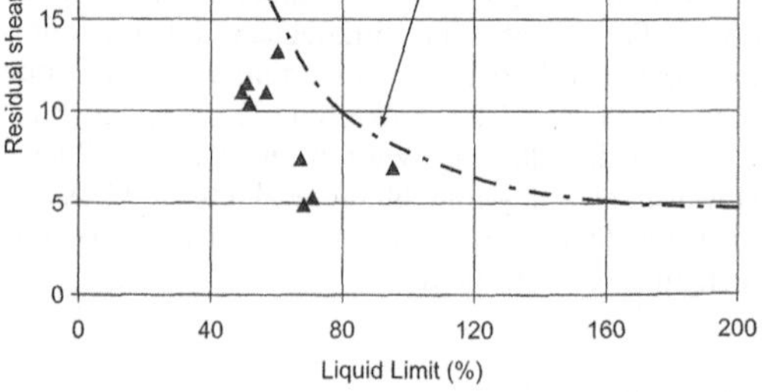

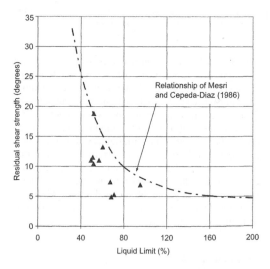

Figure 8. Relationship between liquid limit and residual shear strength at Ventnor.

It is noted that the ring shear test results mostly plot well below this relationship with liquid limit, as would the values given by Hutchinson *et al*, (1991).

Furthermore, particle size distribution test results do not indicate the high clay contents that would be compatible with such low ϕ_r' values (e.g. Skempton 1985). This may relate to the depths, and hence high stresses involved, which are large in comparison with those normally relevant to most slope stability problems.

There is a volume of evidence that the residual strength determined in the ring shear apparatus gives lower values than determined in shear box and those which operate in landslides as calculated by back-analysis. Skempton (1985) observed that back analysis of slopes in which failure had occurred yielded residual strengths that were within the limits of variation of shear box tests, and these were some 1 to 2° higher than those from ring shear tests. In tests on London Clay and Upper Lias Clay, Chandler and Skempton (1974) noted larger differences, of the order of 4 to 6°. Back analysed values of ϕ_r' for Gault at the Folkestone Warren landslide were compared by Hutchinson *et al*, (1980) to values from ring shear tests. The value from back analysis of the slope using high water tables was 9.4°, whereas ϕ_r' from ring shear lay between 6.2° and 11.2° with some scatter. Ring shear results were on average a little lower than back-analysed values. Varley *et al*, (1996) provides a further example from the Castle Hill landslide at Folkestone where results of ring shear tests were noted as being typically 1 or 2 degrees lower than those from shear box tests. Consequently it is suggested that for the purpose of stability analysis that 2 degrees should be added to the values determined from ring shear tests in the Gault Formation.

A summary of the adjusted ring shear test results suggested for slope stability analysis related to the height above the base of the Gault Formation are given in Table 4.

7 UPPER GREENSAND

The Upper Greensand (Units 5a, 5b and 5c) typically comprise glauconitic sands, sandstones and siltstones with a maximum total thickness of 34 to 44 m i.e. similar in proportion to the Gault Formation. It is subdivided into Passage Beds, Malm Rock and the Chert Beds, the characteristic geophysical gamma logs of which are provided in Figure 9.

Table 4. Residual shear strengths determined from adjusted ring shear test results at Ventnor.

Layer	Approximated height above base of Gault Formation (m)	Residual shear strength
Upper	30 to 45	16.5°
Middle	12 to 30	9°
Lower	0 to 12	15°

7.1 *Passage Beds*

The Gault Formation, where not affected by landslipping, passes in an unbroken sequence into the Passage Beds. The Passage Beds, as the name suggests, represents a transition from the overlying buff sandstone of the Malm Rock to the Gault. The base of the Passage Beds is normally identifiable by the occurrence buff or grey brown silts above the grey clays of the Gault Formation. An increased occurrence of buff coloured silts marks the transition into the overlying Malm Rock.

7.2 *Malm Rock*

The middle section of the Upper Greensand sequence is termed the Malm Rock and comprises buff and grey coloured fine sands, sandstones and siltstones which are locally glauconitic with occasional irregular calcareous bands. Records within the Undercliff record a

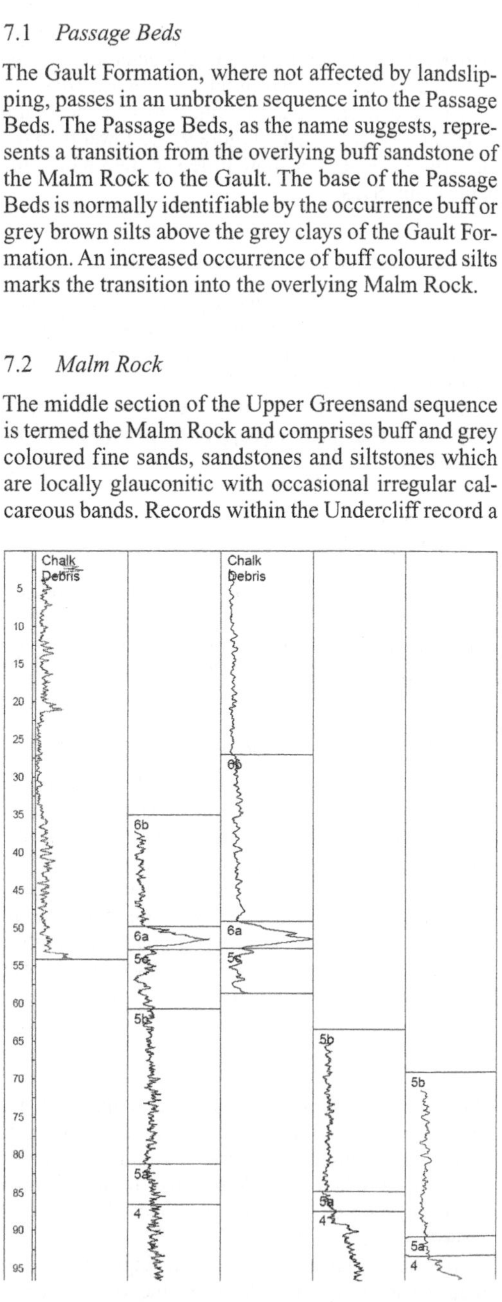

Figure 9. Geophysical gamma log of the Upper Greensand and Lower Chalk and chalk debris at Ventnor (depth scale is in metres).

prominent, 1.5 m thick, massive sandstone at the top of this unit known as "freestone" (from the absence of joints). However, at Ventnor this bed is not readily identifiable and it would appear that the freestone unit is not as marked as earlier records suggest throughout the Undercliff.

The lower 5 to 10 m of the Malm Rock and the underlying Passage Beds are characterised by the presence of the fossil worm tubes *Rotularia concava*.

The geophysical gamma signatures of the Malm Rock and Passage Beds (Figure 9) are lower than the underlying Gault, but show little variation within the units themselves.

7.3 *Chert Beds*

The Chert Beds as the name suggests contain numerous black chert and limestone bands in glauconitic sandstones and siltstones and form prominent exposures when weathered.

The Chert Beds and the overlying Glauconitic Marl create a distinctive geophysical gamma signature (Figure 9).

8 LOWER CHALK

Two members of the Lower Chalk (Unit 6a and 6b) are identifiable at the top of succession typically intercepted by boreholes in Ventnor Undercliff area, namely the Glaucontic Marl and the Chalk Marl Members.

8.1 *Glaucontic Marl member*

The Glauconitic Marl comprises highly glauconitic and bioturbated calcareous sands. The base of this unit is heavily bioturbated and the junction with the underlying Chert Beds represents a bored surface.

A very characteristic 'spike' is observed in the geophysical signature. When coupled with the signature of the Chert Beds this has proved valuable in identifying the presence of a displaced sequence in the lower slopes of the Ventnor landslide.

8.2 *Chalk Marl member*

This Member typically comprises grey marly chalk, generally with no flints. Towards the base it becomes glauconitic although the true stratigraphic boundary with the Glauconitic Marls may be higher than has been adopted here for practical identification purposes.

At times it may be difficult to distinguish between hand specimens of Chalk Marl and the Malm Rock of the Upper Greensand and care is need in this matter, particularly when the materials are displaced or disturbed. The use of the geophysical logs and placing the strata in context with Glauconitic Marl and Chert Beds can aid in this interpretation.

9 CHALK DEBRIS

The occurrence of chalk debris is common in the Undercliff either forming a surface mantle over the underlying landslide strata or infilling graben-like structures. Prominent among these is the thick sequence of chalk debris in the Park Avenue where at least 27 m thickness is recorded (Moore *et al*, 2007).

In borehole core it is very difficult to distinguish between chalk debris and displaced Chalk Marl. The geophysical gamma logs suggest that chalk debris has a lower signature than the Chalk Marl (Figure 9).

At Ventnor, borehole core through a thick sequence of chalk material provided some indications that the whole sequence may not be comprised of chalk debris but may in part be semi-intact but displaced Chalk. A later borehole in conjunction with geophysical logs enabled a sequence of Chalk Marl, Glauconitic Marl and Chert Bed to be identified beneath the chalk debris and demonstrated how knowledge of the litho-stratigraphy and geophysical profile proved invaluable in the understanding of this sequence.

10 SUMMARY

A clear understanding in the litho-stratigraphic sequence of the strata at Ventnor provides an invaluable tool for interpretation of the landslide ground models throughout the Undercliff. Logging of borehole cores should take account of often subtle variations in lithologies. Frequently, 'engineering' logging without the inclusion of 'geology' results in poorly defined stratigraphy, which in turn leads to in inaccurate ground models.

The use of geophysical gamma logging and comparison to a characteristic signature assists in the interpretation of the ground model. In the Sandrock Formation the gamma signature can allow considerable cost savings by reducing the amount of coring required.

In the Gault Formation characteristic profiles of gamma signatures and index water contents provide evidence for subdivision of the Gault, which will lead to an improved understanding of the ground models for future landslide studies in the Ventnor Undercliff.

ACKNOWLEDGEMENTS

The authors wish to thank the Isle of Wight Council for permission to publish this paper.

REFERENCES

Bromhead, E.N. Chandler, M.P. & Hutchinson, J.N. 1991. The recent history and Geotechnics of landslides at Gore Cliff, Isle of Wight. In Chandler, R.J. (ed.), *Intern. conf. on Slope Stability Engineering – Developments and Applications,* 189–196, London: Thomas Telford.

BS5930. 1999. Code of practice for site investigation. *British Standard*, BSI.

Chandler, M.P. 1984. The coastal landslides forming the Undercliff of the Isle of Wight. *PhD Thesis*, University of London.

Chandler, R.J. & Skempton, A.W. 1974 The design of permanent cutting slopes in stiff fissured clays. Geotechnique 24, No 4, pp457–466.

Gale, A.S. Huggett, J.M. & Gill, M. 1996. The stratigraphy and petrography of the Gault Clay Formation (Albian, Cretaceous) at Redcliff, Isle of Wight. *Proc. Geologists' Association* 107 p287–298.

Hutchinson, J.N. & Bromhead, E.N. 2002. Keynote Paper: Isle of Wight landslides In McInnes, R.G. & Jakeways, J (eds.), *Instability Planning and Management – Seeking sustainable solutions to ground movement problems. Proc. of the intern. conf,* 3–70, Ventnor. Thomas Telford.

Hutchinson, J.N., Bromhead, E.N. & Lupini, J.F. 1980. Additional Observations on Folkestone Warren Landslides. *Quarterly Journal Engineering Geology*, 13, 1–31.

Hutchinson, J.N., Brunsden, D. & Lee, E.M. 1991. The geomorphology of the landslide complex at Ventnor, Isle of Wight. *in Slope Stability Engineering,* Thomas Telford Ltd 213–218.

Lee, E.M. & Moore, R. 1991. Coastal Landslip Potential Assessment, Isle of Wight Undercliff, Ventnor. *Technical Report prepared by Geomorphological Services Ltd for the Department of the Environment, research contract PECD 7/1/272.*

Mesri & Capeda-Diaz. 1986. Residual strength of clays and shales. *Geotechnique* 10, 269–274

Palmer, M.J., Moore, R. & McInnes, R. 2007. Reactivation of an ancient landslip, Bonchurch: event history, causes, mechanisms, climate change and landslip potential. *This Conference.*

Skempton, A.W. 1985. Residual strength of clays in landslides, folded strata and the laboratory. Geotechnique 35,1, 3–18.

Moore, R, Turner, M.D., Palmer, M.J. & Carey, J.M, 2007. The Ventnor Undercliff: a new ground model and implications for climate change induced landslide behaviour and risk. *This Conference.*

Varley, P.M., Warren, C.D. & Avgerhinos, P. 1996. Castle Hill west landslip. *Chapter 20 from Engineering Geology of the Channel Tunnel.* Thomas Telford Ltd.

Warren, C.D., & Palmer, M.J. 2000. Observations on the nature of landslipped strata, Folkestone Warren, United Kingdom. *Landslides in research, theory and practice, Proc. 8th Intern. Symp on Landslides, 3, Cardiff,* 1551–1556.

Session 6
Responding to climate change impacts at the coast

The predicted impacts of climate change present significant challenges to future coastal and geohazard management and will pose increased levels of risk to many communities and assets. It is essential to carry out research into climate change impacts now, so that appropriate proactive decision-making can be implemented through the planning process in the coastal zone to control or reduce risk.

Shanklin Cliffs, Isle of Wight, UK
Courtesy Wight Light Gallery, Ventnor, Isle of Wight, UK.

Landslides and Climate Change – McInnes, Jakeways, Fairbank & Mathie (eds)
© 2007 Taylor & Francis Group, London, ISBN 978-0-415-44318-0

Preparing for climate change impacts at the coast: Identifying patterns of risk and prioritising a response

H.E. Fairbank & J. Jakeways
Centre for the Coastal Environment, Isle of Wight Council, Ventnor, Isle of Wight, UK

ABSTRACT: The likely impacts of climate change at the coast are potentially devastating to both the built and natural environments. Mounting evidence suggests that natural hazard events such as landsliding, erosion and flooding may occur more frequently and as increasingly intense episodes, due to the effects of climate change. It is necessary to identify and prepare for the likely impacts of climate change at the coast, and to implement risk management strategies, in order to reduce the escalating risks and growing financial burden on coastal communities. To assist in the achievement of this aim, "RESPONSE" was a 3-year project (2003–2006) led by the Isle of Wight Council's Centre for the Coastal Environment, UK, with nine partners in the UK, France, Italy and Poland. The Project received financial support from the EU LIFE-Environment Programme. The Project developed advice and guidance for practitioners and decision-makers to assess and reduce the likely impacts of climate change along a defined stretch of coastline, to understand and communicate the likely pattern of future change including areas of high risk. This can contribute to the development of sustainable and practical management strategies to mitigate the impacts of climate change and assist prioritisation of resources.

1 INTRODUCTION

Coastal settlements are particularly vulnerable to the impacts of climate change. In addition to meteorological factors, the coastal zone is susceptible to changes in oceanic parameters, especially sea level rise and wave climates. Despite this, development in the coastal zone has continued to expand and land and property at the coast remains valuable. As coastal populations have increased so has the need to defend them from natural hazards, often using hard engineering solutions. Structures such as sea walls and groyne fields have not only helped to sustain these lands from erosion and flooding, but have also provided the impetus for more coastal land to be developed (Eurosion, 2004). In view of the anticipated impacts of climate change (including rising sea level and the growing incidence of current and anticipated future risks and storms of increasing intensity) and the continuing pressure for expansion of development into areas of risk, governments need to prepare and maintain coastal hazard and risk management plans, and communicate the growing incidence of current and anticipated future risks to decision-makers, to inform sustainable policy decisions.

1.1 *The European RESPONSE project*

The RESPONSE project (Responding to the risks from climate change) was a three-year project (2003–2006) supported by the LIFE financial instrument of the European Community, launched in Brussels in December 2006. Nine partner organisations in the United Kingdom, Italy, France and Poland participated in the project, led by the Isle of Wight Council's Centre for the Coastal Environment, UK. The Partners are experts in managing the impacts of natural hazards and risks for coastal communities around the European coastline. The Project Partners were:

1. Isle of Wight Centre for the Coastal Environment, Isle of Wight Council, UK;
2. Consiglio Nazionale delle Ricerche, IRPI Institute, Perugia, Italy;
3. Bureau de Recherches Geologiques et Minieres (BRGM), France;
4. Regione Marche, Italy;
5. Provincia di Macerata, Italy;
6. Provincia di Pesaro e Urbino, Italy;
7. Scarborough Borough Council, UK;
8. Maritime Office Gdynia, Poland;
9. Standing Conference on Problems Associated with the Coastline, Regional Coastal Group for central southern England (SCOPAC), UK

The aim of the RESPONSE project was to assist organisations managing the coastline in assessing and prioritising the risks arising from climate change impacts on natural hazards at the coast, to inform the planning process. The project results and outputs

Figure 1. Selsey, West Sussex, UK. The area is protected from flooding by a groyne-stabilised beach and a seawall.

illustrate a regional-scale qualitative assessment of coastal risk in the context of climate change, fully transferable to any region or sediment cell, together with many examples of good practice.

2 CLIMATE CHANGE AT THE COAST

Climate change is a global issue that is of increasing political concern. Over the next 100 years climate change and sea level rise will result in an increase in the occurrence of damaging events. However, it is uncertain how the responsible authorities will manage the increased risks. The RESPONSE project outputs provide advice and guidance to local and regional authorities on managing natural hazards at the coast.

In many European countries there are no statutory powers to protect coastal communities from the impacts of coastal erosion and ground instability. For financial, environmental and technical reasons, the natural processes of erosion, flooding and instability cannot be prevented in all areas. In locations where it is unacceptable or unachievable to provide protection against natural hazards, alternative management strategies must be considered to mitigate the risk. Increasingly, communities are being forced to accept a certain level of risk, and to adapt to the consequences of living in a hazardous environment. Politically, it can be very difficult to inform communities that risk prevention is not possible. It is likely that as a result of climate change, in future years engineers, planners and politicians will face difficult decisions. The most authoritative reports on the science of climate change are those produced by the Intergovernmental Panel on Climate Change (IPCC), the latest in 2007, which bring together the leading scientists from around the world. The IPCC was established in 1988 by the World Meteorological Organisation (WMO) and the United Nations Environment Programme (UNEP), to assess the scientific and technical literature on climate change, the potential impacts of changes in climate, and options for adaptation to and mitigation of climate change (Watson *et al.* 1997). Climate change models predict that the effects of climate change will include a general increase in temperature, more winter precipitation and less summer precipitation. The IPCC (Watson *et al.* 1997) identify coastal zones as "sensitive regions" due to the potential effects of sea level rise and changes in storm activity.

There remains a great deal of uncertainty as to how our coastal zones will be affected by climate change impacts. Sea level rise is well evidenced and such a continued effect can be prepared for given adequate planning. However, the impact of less predictable factors, such as potential changes in storm frequency and intensity, is difficult to plan for. As a result, it is these unknown parameters that are of greatest consequence to coastal zones. Despite these uncertainties, a sensible analysis of the risks does not allow us to sit back and wait, when preventative action can provide the most cost effective approach to reduce risks.

3 THE RISING COSTS OF COASTAL NATURAL HAZARDS

Despite uncertainty over the precise impacts of climate change at the coast, there is a recognition that the costs associated with the management of natural hazards are likely to increase.

The economic costs of emergency action, remediation and prevention in reference to coastal hazards such as erosion, flooding and landsliding represent a significant burden to the communities affected, often local or regional authorities with limited resources.

A major finding of the Eurosion report (Eurosion, 2004) was that the cost of reducing the coastal erosion risks is mainly supported by national or regional budgets, hardly ever by the local community and almost never by the owners of assets at risk or by the party responsible for coastal erosion. This is emphasised by the fact that coastal erosion risk assessment has not been incorporated in the decision-making processes at the local level in many areas and communication of risk information to the public can also remain poor. This issue highlights the importance of the RESPONSE project in its aim to develop sustainable strategies for managing coastal natural hazards (including erosion and landsliding) that can inform land-use development and planning by ensuring decisions are compatible with specific local coastal conditions and also future challenges.

The construction of coastal protection measures requires the allocation of a significant amount of public expenditure, costs that are increasing with the recognition that the impacts of natural hazards are set to increase both as a result of climate change and the continued pressure of development in marginal and vulnerable areas.

In addition to the effects of sea level rise on coastal erosion and flooding, many areas may also experience increased coastal landsliding, which can sometimes result in dramatic and rapid changes with widespread direct and indirect impacts. Coastal communities may be vulnerable to increased hazards as a result of reactivation of presently relic cliffs. This is potentially a major risk where relic landslides occur, e.g. around Lyme Regis, Dorset, and within the Undercliff landslide complex and on the north coast of the Isle of Wight. Evidence of landslide reactivation on the Isle of Wight was observed in the exceptionally wet winter of 2000/01, which resulted in intensification of reactivations at a number of locations and was an indication of the conditions that might be expected to occur more frequently in the future. The A3055 road was completely severed over a 100 m section by a major landslide within the Isle of Wight Undercliff during spring 2001. The landslide appears to have occurred due to prolonged high groundwater levels following heavy rainfall in autumn 2000. The landslide represents a reactivation and major seaward and downslope movement of an ancient landslide that already existed within the Undercliff. It produced a new scarp some 10 m to 20 m in height that severed this important coastal route.

The risk from landsliding in coastal locations is set to increase in the face of climate change, not least because developments, through choice or necessity, are still being situated in vulnerable areas.

Natural hazards are having an ever-burgeoning impact on the insurance market. Recent decades have seen a large increase in losses associated with natural hazards not only due to the fourfold increase in the global population (frequently concentrated in coastal zones) but also because the number of great natural catastrophes has increased threefold (Munich Re, 2003). The insurance industry is also assessing future costs, for example the Association of British Insurers has highlighted that increased costs are unavoidable: "Global damages from a 0.5 metre rise in sea-level have been estimated as $24–42 bn per year. Adaptation – in the form of coastal defences – could bring these costs down to $8–10 bn per year." (ABI, 2005).

Although the loss/damage to property is a significant economic cost of natural hazards, the environmental loss of the undermining of coastal dunes/natural sea defences is of greater overall cost as it has the potential to impact on several thousands of square kilometres and millions of people who are protected by these kinds of defences. Traditionally hard engineering techniques have been implemented as a means of protecting the community. The Eurosion report (2004) found that over the past hundred years the limited knowledge of coastal sediment transport processes at the local authority level has resulted in inappropriate measures of coastal erosion mitigation. This is problematic as we are increasing the costs, economic, environmental and social, by bad practice and in effect transferring not solving the problem. Only recently has significant progress begun to be made towards incorporating softer engineering solutions such as beach nourishment and managed realignment, where possible, to reduce the impact on the natural environment, and encourage natural feedback and protection.

A recent report by the European Environment Agency states that in terms of a management strategy, the main objective should be to shift from coastal defence and beach management to sediment management. Modern methods of "soft" coastal engineering that reinforce natural buffers against the rising tides (such as dunes and salt marshes) and the protection of key sources of sediment, will help maintain coastal sediment balance and the stability of coastal systems (European Environment Agency, 2006).

The geology and topography of the European coastline presents an enormous variety of coastal conditions, natural hazards and problems resulting from historic development in unsuitable or marginally stable locations. The interaction between coastal erosion, ground stability and development results in a need for sustainable planning and legal frameworks to ensure that appropriate development takes place. It is necessary to adopt sound approaches to planning and development of coastal land taking account of the physical and human environment and in view of the anticipated impacts of climate change.

4 THE RESPONSE PROJECT MAPPING

The effects of the predicted changes in climate outlined above are causing considerable concern to those responsible for coastal risk management. Without

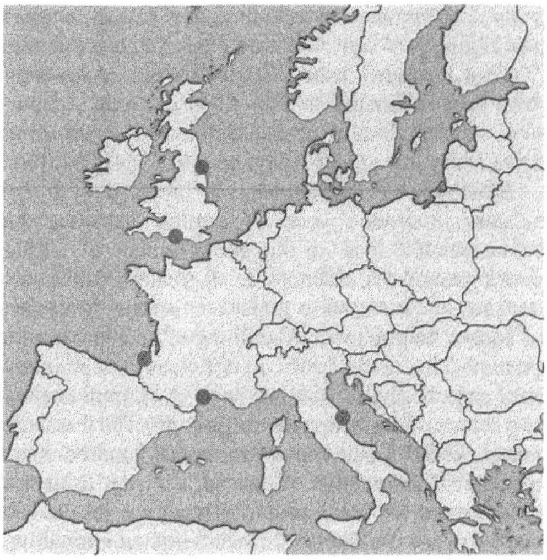

Figure 2. Map showing the location of the five RESPONSE study areas. Map © European Community 2006.

more effective and integrated coastal planning and implementation, consequences for the coastal zone could be severe.

4.1 *Identifying patterns of risk*

The RESPONSE project illustrates a regional-scale approach to coastal risk management in the context of climate change, through the mapping of coastal evolution and risk and provision of planning guidance based on as assessment of increasing risks due to climate change.

The regional-scale geomorphological approach for mapping coastal evolution and risk was demonstrated in five European study areas:

- Central-southern coast, England (on the Channel);
- Regione Marche coast, Italy (on the Adriatic);
- Aquitaine coast, France (on the Atlantic);
- Languedoc Roussillon coast, France (on the Mediterranean);
- North Yorkshire coast, England (on the North Sea).

These five study areas were selected to represent the range of morphological types and conditions found along the coast of Europe, and each area also contains a diversity of landforms such as landslides, barrier beaches, hard cliffs and estuaries. Each Study Area was approximately 100–400 km in length and was chosen based on the Sediment Cell boundaries or an effective administrative boundary.

The methodology for developing coastal evolution and risk maps is based upon a geomorphological appraisal of a regional length of coastline. The approach promotes a process by which the coastline can be classified into sections, determined by its geomorphological characteristics, the vulnerability of coastal settlements and its sensitivity to forcing factors.

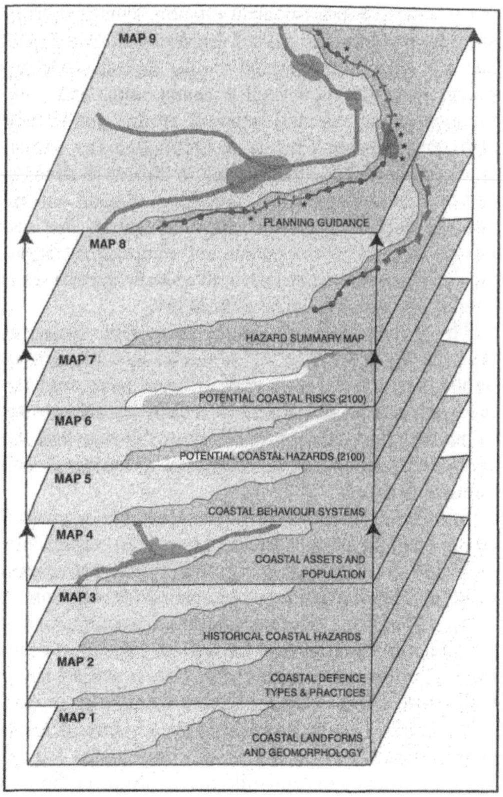

Figure 3. Sequence of RESPONSE coastal evolution and risk maps (GIS or paper-based assessment).

Knowledge of the sensitivity of a coastline to coastal and climate change enables engineers, planners and decision-makers to anticipate impacts that could emerge over future decades and prioritise resources and management efforts that need to be undertaken to minimise the risks or to mitigate possible consequences.

The RESPONSE mapping methodology provides an assessment of risk based on the development of risk maps. Figure 3 illustrates the sequence of nine maps. The series of maps comprise three phases of work:

- Information gathering; Maps 1–4;
- Interpretation: Maps 5–6;
- Outputs for decision-makers: Maps 7–9.

Figure 4 provides an example of Map 7 for the Regione Marche study area in Italy.

For the central-southern coast of England study area, two non-technical summary maps have been produced for decision-makers, which include additional

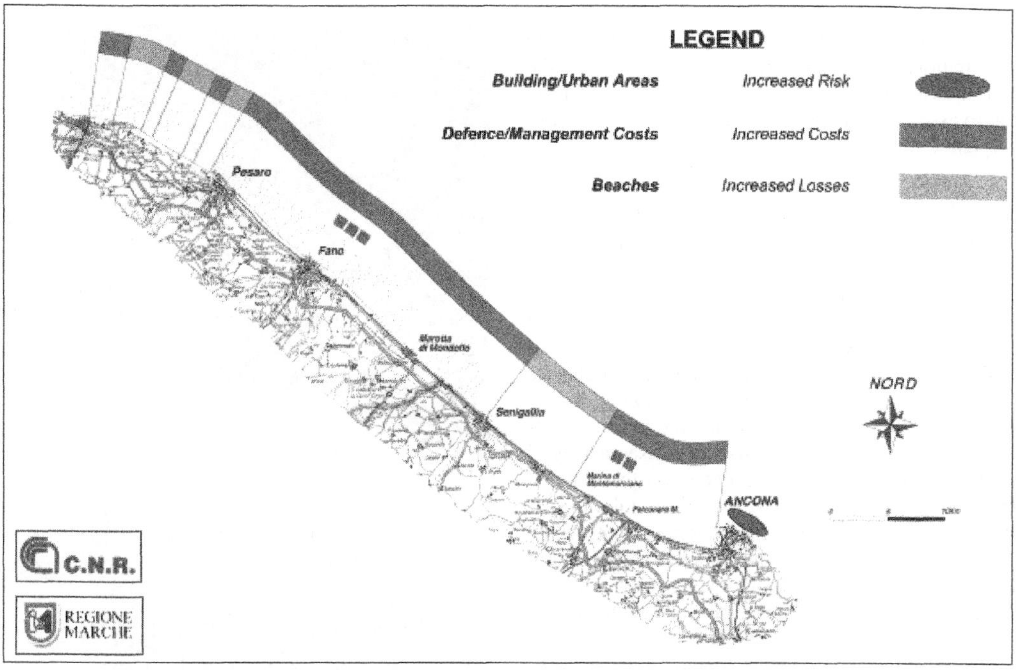

Figure 4. Extract of Map 7: Potential coastal risks, Regione Marche, Italy coastal study area.

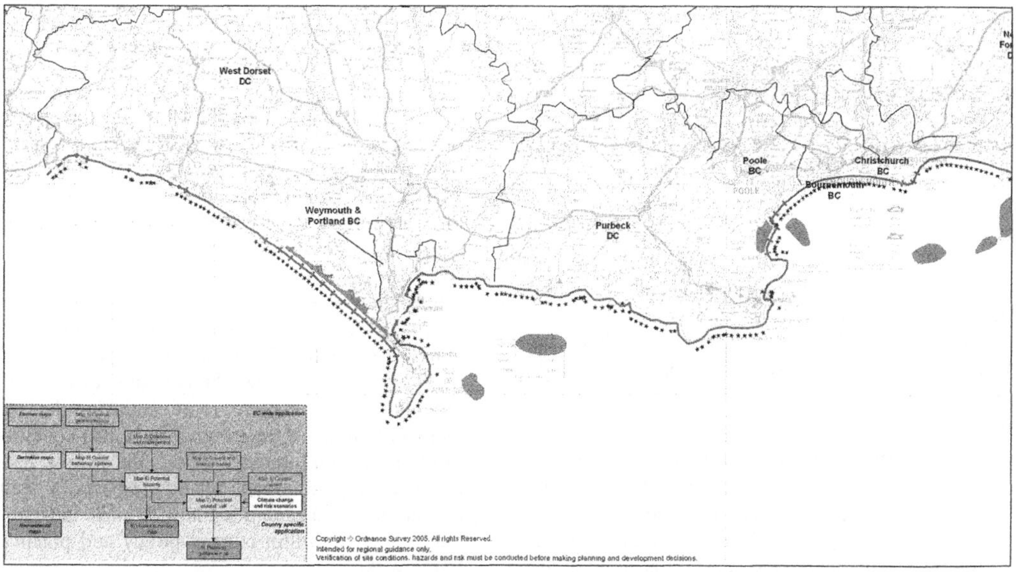

Figure 5. Extract of Map 9: Planning Guidance, Central southern England coastal study area.

information and guidance for planners (Figure 5). The RESPONSE maps provide an effective mechanism for collecting and presenting broad-brush information for strategic planning, providing guidance on where potential problems may arise and where attention should be focused.

The aim of the RESPONSE maps is to provide an overview of the pattern of hazard activity along a stretch of coastline. It is important to note that the maps present a relative risk assessment technique at a regional scale and are not prepared with the aim of assessing hazard and risk at a localised level e.g. to assess individual sites for development control purposes.

4.2 *Prioritising a response*

The RESPONSE maps allow the potential impacts of climate change and the future of the coastline to be communicated clearly to decision-makers and planners. The maps are based on an understanding that the coastline has always evolved and will continue to do so. It is essential to understand the impacts of climate change now so appropriate decision-making can be implemented through the planning and political processes. Understanding the costs and consequences of inaction allows cost-effective decisions to be made and justified, contributing to long-term solutions. It will be impossible to develop strategies that negate climate change because the future level of risk cannot be defined to that level of accuracy. It is therefore only possible to attempt to decrease risk and vulnerability natural hazards to an acceptable level.

Further information on the methodology for producing coastal evolution and risk maps can be found in paper "Response: applied earth science mapping for evaluation of climate change impacts on coastal hazards and risk across the EU" by P. Fish *et al* in this volume. The two French Study Areas are also discussed in paper "Application and implementation of the RESPONSE methodology to evaluate the impact of climate change on the Aquitaine and Languedoc Roussillon coastlines" by C. Vinchon *et al* in this volume.

5 RESPONSE PROJECT OUTPUTS

The RESPONSE project had produced an Information Pack, entitled "Responding to the risks from climate change at the coast". The pack contains advice for managing risk in coastal areas of Europe and is available in English, Italian and French. The information pack contains:

- A Training Pack: Mapping coastal risks in a changing climate: The 48-page Training Pack demonstrates how to produce maps that show the future pattern of coastal evolution and risks across a region (or sediment cell), which take account of the impacts of climate change. These maps can provide an understanding of the pattern and scale of coastal change and assist in targeting resources effectively. The Training Pack also contains evidence on the importance of taking cost-effective action now to prevent worsening impacts in a changing climate. Illustrated in Figure 6;

Figure 6. RESPONSE Training Pack (in English).

- A Good Practice Guide: The Good Practice guide (80 pages) contains advice on sustainable coastal risk management and examples from around Europe and the world of how the growing impact of natural hazards on coastal communities can be addressed and reduced. Illustrated in Figure 7;
- CD-Rom of resources: The CD-Rom contains full information on each stage of the mapping process (summarised in the Training Pack). It also contains Case Studies from the RESPONSE Project Study Areas in the UK, France and Italy, which provide practical experience of producing the risk maps in a range of environments. It also contains full briefings on (amongst others) the costs of natural hazards and climate change impacts (a "call for action"), and examining how to make the mapping effective within current policy and practice frameworks;
- DVD film: A 15-minute film introducing the viewer to the subject of coastal risks in a changing climate, and presenting the potential of the RESPONSE Project publications for use in coastal zones around Europe. It is designed for both non-technical and specialist audiences, and was filmed in the study areas in the UK, France and Italy.

Figure 7. RESPONSE Good Practice Guide (in English).

6 CONCLUSIONS

A certain amount of future climate change is now inevitable in response to the effects of anthropogenic influences in recent decades.

Sustainable coastal risk management measures should involve understanding and working with natural coastal processes wherever possible and promoting adaptation to coastal and climate change. Without more effective and integrated coastal planning and implementation, consequences for the coastal zone could be severe.

Coastal erosion, landslide and flood risks must be taken fully into account to aid sustainable land-use planning and management. Authorities are urged to prepare and maintain up-to-date planning policy guidance documents for coastal hazards and risks.

The most effective risk management strategies are those that have been implemented with strong support from local politicians, stakeholders and the community. This support is aided by the preparation and publication of well-illustrated informative guidance aimed at the educated layman.

The RESPONSE project outputs provide a good example of effective risk management in the context of climate change (which can be transferred to any area, indeed they were designed with this aim in mind) and of using maps to communicate risk to technical and non-technical decision-makers, to assist the incorporation of long-term (100 year) sustainability considerations into planning and policy frameworks.

The mapping of the five regional study areas highlighted some interesting lessons:

1. The mapping was proved to be a most effective methodology for transmitting information;
2. As the mapping developed, high risk areas were highlighted, and although these areas of concern were sometimes known, the impacts of climate change had not necessarily been accounted for;
3. Social, economic and political changes over the next 100 years could be as significant as climate change in determining future coastal risk and how coastal risk is managed, so the threat of climate change should focus attention upon the needs for forward planning and preparedness;
4. Waiting for overwhelming evidence of physical impacts of climate change will miss the opportunity to reduce and avoid future risks in the most cost-effective and achievable way;
5. The investigations highlighted disequilibrium in terms of sediment supply along the coastline in some instances, a process which could worsen as climate change progresses. This highlighted the essential role of the 'sediment cell' approach, considering both sediment inputs into the system as well as interventions, transfers and losses.

The benefits of the RESPONSE project lie in the development of a fundamentally transferable methodology which will enable, in qualitative terms, local authorities and stakeholder groups across the EU to assess the impacts of climate change on their coastal zone and develop a response which is targeted and cost-effective. Moving towards quantitative benefits, the RESPONSE approach is based upon the principle of preventative action – that taking appropriate precautionary action now is less expensive than taking no action, and paying much higher economic costs later on in emergency action, remediation and prevention.

The transferability and reproduction potential of the RESPONSE Project is ensured by the development of its methods and results in a range of EU coastal environments and a range of regulatory and administrative frameworks through Partnership working, to reflect the diversity of situations around the coastline of the European Union. The regional and local approach of the Project also enhances its transferability and reproduction potential by targeting its results and training materials towards the end-user groups directly responsible for coastal management, climate change impacts and land-use development and planning, who are usually local and regional authorities.

ACKNOWLEDGEMENTS

The authors are grateful to the RESPONSE project partners in the UK, Italy, France and Poland for their expertise and experience; together with a team of technical experts from consultants Halcrow and Dr. Mark Lee, who contributed to the REPONSE project between 2003 and 2006.

REFERENCES

ABI. 2005. Financial Risks of Climate Change, Summary report, June 2005. Association of British Insurers, with Climate Risk Management and Metroeconomica, UK.

European Environment Agency. 2006. The changing faces of Europe's coastal areas, Luxemburg.

Eurosion. 2004. Living with coastal erosion in Europe: Sediment and Space for Sustainability, PART I: Major findings and Policy Recommendations of the EUrosion project. Luxemburg, European Commission.

Halcrow. 2001. Preparing for the impacts of climate change. Standing Conference on Problems Associated with the Coastline (SCOPAC).

IPCC. 2001. Climate Change 2001: Impacts, Adaptation and Vulnerability, Cambridge University Press, Cambridge.

McInnes, R.G., Jakeways, J. & Fairbank, H. 2006. EU LIFE *'Response'* project Final Report for European Commission, Ventnor, Isle of Wight.

McInnes, R.G. 2006. Responding to the risks from climate change in coastal zones, a good practice guide. Ventnor: Isle of Wight, UK.

Fairbank, H. & Jakeways, J. Mapping coastal risks in a changing climate. A training pack, Ventnor, Isle of Wight, UK.

Munich Re. 2003. Natural catastrophes and climate change – the insurance industry's fears and response options. Münchener Rückversicherungs-Gesellschaft, Germany.

Watson, R.T., Zinyowera, M.C. & Moss, R.H. 1997. IPCC Special Report on the Regional Impacts of Climate Change. An Assessment of Vulnerability, Intergovernmental Panel on Climate Change.

Landslides and Climate Change – McInnes, Jakeways, Fairbank & Mathie (eds)
© 2007 Taylor & Francis Group, London, ISBN 978-0-415-44318-0

Response: Applied earth science mapping for evaluation of climate change impacts on coastal hazards and risk across the EU

P.R. Fish & J.L. Moss
Halcrow Group Ltd, Birmingham, UK

J. Jakeways & H. Fairbank
Isle of Wight Council, Ventnor, UK

ABSTRACT: Response is an EU-funded demonstration project that aims to develop a methodology for assessing climate change impacts on coastal hazards and risk. The methodology described employs commonly available digital datasets in a Geographical Information System (GIS) to assess regional-scale levels of coastal risk through production of a series of maps. The methodology allows the impact of a range of climate change and coastal management scenarios to be assessed. This paper provides details of the Response methodology. Outputs of the methodology comprise factual data maps and thematic maps and also non-technical summary maps aimed at coastal risk management, such as planning guidance. Factual data maps include information on coastal geomorphology, defences, hazards and coastal assets. This information is used to produce the thematic maps, which comprise coastal behaviour systems, potential coastal hazards and potential coastal risks. The Response methodology has been developed for application to any European coastline. This paper discusses the strengths and limitations of the methodology that were encountered when it was applied to the south coast of England, and highlights possible future developments.

1 INTRODUCTION

The coastline of Europe has been subject to increasing development in recent centuries and provides a valuable resource, in terms of tourism, industry and natural environments. In more recent years, the European Commission has encouraged improvements in coastal management, and this has led to development of a range of strategies designed to support sustainable development through risk management.

A key driver in future coastal risk management policies and practices is undoubtedly the impact of climate change and rising sea-levels. As it is recognised that a policy for coastal defence is undesirable for the entire coastline, it is essential that future coastal behaviour is fully understood when considering coastal risk management. The EU has therefore funded a series of projects to support policy-makers (McInnes et al. 2000; European Commission, 2004).

RESPONSE comprises the latest EU-funded demonstration project that aims to develop a methodology for assessing the hazard and risk of climate change at the coastline.

Central to the Response approach is the development of a series of maps that provide factual information and guidance to coastal managers. Input data comprise a series of factual element maps which are then combined and interpreted to create a series of derivative maps that show coastal behaviour systems and potential coastal hazards and risks. Summary maps that highlight key coastal hazards and provide planning guidance can also be created, according to the specific requirements of local planning decision makers. The associations of these maps are highlighted in Figure 1.

This paper summarises the results of the project which comprise, technical guidance for application of the Response methodology and information on the

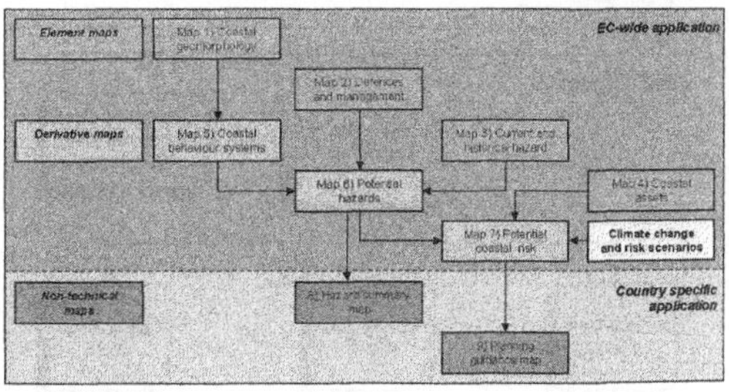

Figure 1. Association of component maps in the RESPONSE methodology.

scientific principles behind the methodology, including of coastal behaviour, hazard and risk. The application of the Response methodology is described using a series of descriptive, interpretative and non-technical maps covering the south coast of England as a worked example.

2 UNDERLYING PRINCIPLES

Effective Responses to the challenges presented by climate change will involve scientists, decision-makers and the public, working in partnership to undertake risk-based decision-making through the following processes (Figure 2):

1. Understand how the levels of risk currently imposed on coastal communities could change in the future (i.e. risk assessment)
2. Communicate this risk information to decision-makers in a form that is readily accessible to a non-scientific audience (i.e. risk communication)
3. Evaluate the significance of the changes in risk and establishing whether current risk management strategies and policies need to be modified or replaced (i.e. risk evaluation); and
4. Implement the risk management strategies and policies, through a combination of state and local government funding, private investment and public participation (i.e. risk management).

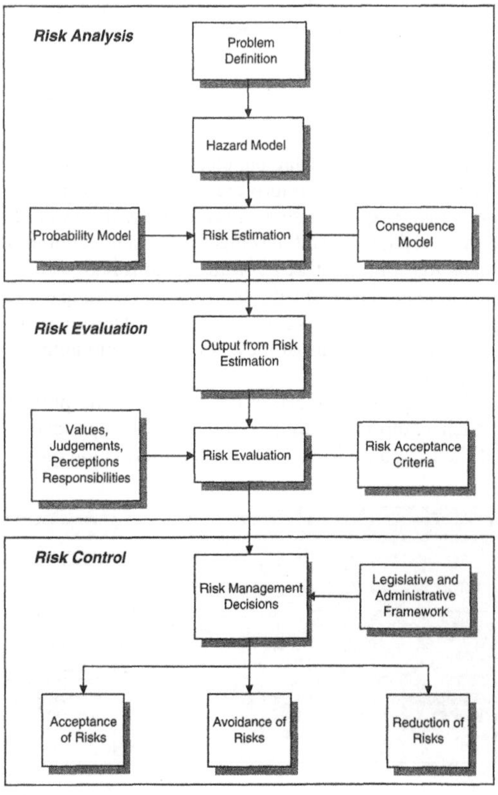

Figure 2. The risk assessment process.

The Response study has demonstrated that maps, and in particular those developed using a GIS database, are a very useful means of communicating risk information, especially to non-technical audiences. However, it must be borne in mind that this is essentially the end product of a risk assessment procedure that is dependent on reliable models of coastal behaviour, hazard, consequences and risk.

2.1 *Coastal behaviour*

The key message of the Response project is that understanding the shoreline behaviour is essential for predicting the Response to climate change and sea-level rise. Earth surface systems provide the framework for developing this understanding, most appropriately at the scale of coastal behaviour system and sediment transport cells (Lee & Clark, 2002). These are conceptual models that can be used to describe how sediment and energy transfers provide inter-linkages between the different landforms along a shoreline.

Coastal behaviour describes the morphological Response to variations in the balance between the forcing factors (e.g. wave energy) and the resistance of the materials. Over time, the morphology of the sub-systems represents a balance between energy inputs and materials. However, morphology also influences the available energy (e.g. beach height, shoreline orientation).

Models that may be used to understand future coastal behaviour can be:

- Qualitative generalisations about future behaviour based on historical trends or analogues
- Simple quantitative models of system Response (e.g. change in rate or event frequency), e.g. the Bruun Rule and probabilistic cliff recession models
- Complex numerical models, incorporating random forcing and non-linear behaviour, e.g. Monte Carlo simulation.

2.1.1 *Coastal response to sea-level change*

Relative sea-level rise will result in an increase in wave and tidal energy arriving at the shoreline. The landward end of the shore profile (e.g. the beach face or salt marsh cliff) would be exposed to a higher energy environment, leading to erosion. The net result could be a landward migration of the shore profile, so that the profile position and form is maintained relative to sea-level.

There is currently limited information the relationship between rate of sea-level rise and rate of shoreline

Response (i.e. erosion). Recently published data from a variety of soft rock cliffs in southern England suggests that around 2 cm of erosion has occurred for every 1 mm of relative sea-level rise (Lee, 2005). Despite these results coastal erosion it not necessarily the corollary of relative sea-level rise and the reverse is possible where coastal sediment supply is different. For example, Hutchinson & Gostelow (1976), demonstrated coastal advance, through salt marsh development and cliff abandonment, during the mid-Holocene (around 6,000 years ago), at a time when sea-levels were rising at around 4 mm/yr, but when coastal sediment supply was probably significantly greater.

2.2 *Hazard models*

In the context of Response, hazards are defined as the potential for future coastal processes to adversely affect humans and the things that people value. Among the most important coastal hazards are: flooding of low-lying coastal areas, e.g. from storm surges or tsunami; erosion of coastal cliffs, including coastal landslides; and wind blown sand from dune systems.

Hazards are the agents that can cause loss and contribute to the generation of risk. Therefore, changes in the frequency of these hazards results in a change in the likelihood of the hazard (i.e. 'hazardousness'), and not a change in the scale or likelihood of adverse consequences.

In the Response methodology, simple hazard models have been developed to answer questions such as: what could happen (the nature and scale of future events); where could it happen (the spatial framework); and why such events might happen (the circumstances associated with particular events).

Hazard assessment should not be restricted to determining the magnitude and frequency characteristics of the main damaging events (e.g. a major landslide) but should, ideally, seek to identify possible sequences of hazards that could develop from an initial event.

2.3 *Consequence models*

Adverse consequences include the direct impact on people (i.e. loss of life or injury), direct and indirect economic losses and intangible losses. All assets occurring in an area that could be adversely affected by a hazard are known as 'elements at risk'. These elements are extremely diverse in nature and can be divided into the following major groupings:

- Populations; detriment is usually expressed in terms of loss of life or injury, although the longer-term consequences of ill-health can also be considered;
- Structures, services and infrastructure; the value of these physical assets can usually be determined from estate agents or local authority housing tax bands and from the owners/operators for services and infrastructure. Damage can be total or partial (i.e. repairable);
- Property; including the contents of houses, businesses and retailers, machinery, vehicles, domesticated animals and personal property.
- Activities; whether for financial gain or pleasure. The main components are business, commerce, retailing, entertainment, transportation, agriculture, manufacturing and industry, minerals and recreation.
- Environment; including flora, fauna, environmental quality and amenity.

2.4 *Risk estimation*

Typical approaches to risk estimation are:

- Qualitative risk estimations where both likelihood and adverse consequences are expressed in qualitative terms.
- Semi-quantitative risk estimations are combinations of qualitative and quantitative measures of likelihood and consequence. More usually it is probabilities of frequency that are known, or assumed, while levels of consequence remain uncertain.
- Quantitative risk estimations, where values of detriment are combined with probabilities of occurrence.

Different legislative requirements will dictate that different types of risk assessment need to be undertaken; some qualitative (e.g. in France and Italy), some quantitative (e.g. UK, DEFRA PAG 4). A variety of scales will be relevant, from high level assessments to detailed assessments.

Qualitative methods are of value where the available resources or data dictate that more formalised quantitative assessment would be inappropriate or even impractical. For example, in the Response methodology, a qualitative measure of risk has been obtained by combining a measure of the likelihood of a hazard occurring with the severity of consequences to provide a ranking of risk levels (Figure 3). Although rankings are value judgements, experienced specialists are able to make realistic assessments of the likelihood of events and consequences, based on an appreciation of a particular environment, together with knowledge of the particular site.

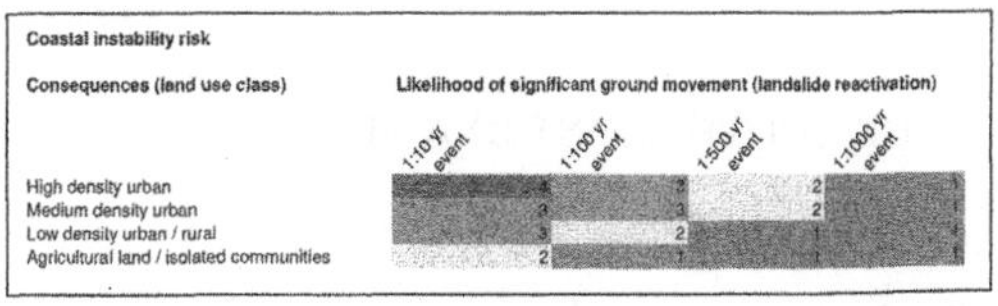

Figure 3. Ranking of risk levels for coastal instability (4 = high risk, 1 = low risk).

2.5 *Risk communication*

Many parts of the coast are subject to risks associated with erosion, landsliding or flooding. However, in many countries decision makers have often taken insufficient account of these risks. This is largely due to a lack of awareness of the physical environment and the limited availability of suitable technical information to support decision making.

One of the main problems facing coastal management is that few planners have an earth science background and few earth scientists have a planning background, and hence there is often a communication gap between the two groups. However, both groups often share a common skill: familiarity with maps. The shared background in geographical skills provides an opportunity for presenting technical information in a format that is readily appreciated by planners. However, such maps need to address planning concerns and not seek to present technical detail

Maps for coastal planners need to be:

- Concise and clear summaries of key earth science information as it relates to key planning issues;
- Highlight potential problems so that users are aware of the factors which may restrict development opportunities in an area;
- Indicate the types of planning response that may be appropriate to take account of particular physical conditions and hazards.

Coastal hazard and risk assessments should be focussed towards preparing a combination of thematic maps at a general scale, which become increasingly focused on key planning issues as they are developed from the basic factual information. The process of map preparation follows the assessment procedures outlined earlier:

- Element maps; depicting factual information on specific earth science related topics;
- Derivative maps; drawing on the basic data to define characteristics of particular interest;
- Non-technical summary maps compiled from the element and derivative maps which highlight the general characteristics of an area in terms of, for example, the risks to development.

The remainder of this paper will illustrate the Response process using the south coast of the UK as an example.

3 UK SOUTH COAST EXAMPLE

3.1 *Framework*

The Response approach comprises the development of a series of maps that provide factual information and guidance to coastal managers (Figure 1). Maps one to four are 'element maps' that show coastal geomorphology and hydrodynamic processes (Map 1), coastal protection measures (Map 2), historical coastal hazards (Map 3), and coastal assets of the natural and built environment (Map 4). These maps have then been combined to create derivative maps showing Coastal Behaviour Units (Map 5), and coastal hazards (Map 6). Map 7 shows future coastal risks, and can be adapted to highlight the impact of various scenarios, encompassing changes in coastal management or changes in climate or sea-level. Map 7 shows the possible coastal risks by the year 2100 under two scenarios; that of 'business as usual', with no change in coastal management and limited climate change, and a 'worst case', where all coastal defences are lost and climate change is significant enough to increase the historical rate of coastal Response (i.e. erosion). In both cases the risk of erosion, flooding and reactivation of coastal landslide complexes is assessed using a relative risk matrix that combines the economic value of the coastline being assessed and the likelihood of occurrence of each hazard. Production of these maps follows a standard approach which can be applied throughout EU Member States.

In addition to these seven generic maps, additional non-technical summary maps can be produced to highlight various applications of the assembled data, and are designed for coastal managers or planners. These additional maps provide planning guidance, appropriate to English planning regulations, and a non-technical summary of hazards. Because planning laws will vary across the EU, it is likely that these maps will be country-specific, the approach highlights how the data and methodology can be applied, or expanded to provide additional guidance.

A Geographical Information System (GIS) provides an excellent tool for the assembly and analysis of the spatial data required for the Response methodology. ArcView GIS has been used in this example.

However, because of the variable nature and quality of the input data, which may range from accurately calculated and detailed floodplain outlines to imprecisely-located historical hazard events, caution is needed when presenting the data in GIS. Therefore, it is essential that the scale of presented maps is appropriate to the accuracy of input data.

3.2 *Map 1 coastal geomorphology*

The first of the factor maps describes a 'system model' for the study coastline. Key elements are the forcing factors (i.e. hydrodynamics) and system state (i.e. characteristics and form of the coastline). This type of model provides a framework for understanding and representing the relationships between different system components, and forms the basis for classification of Coastal Behaviour Units (Figure 4).

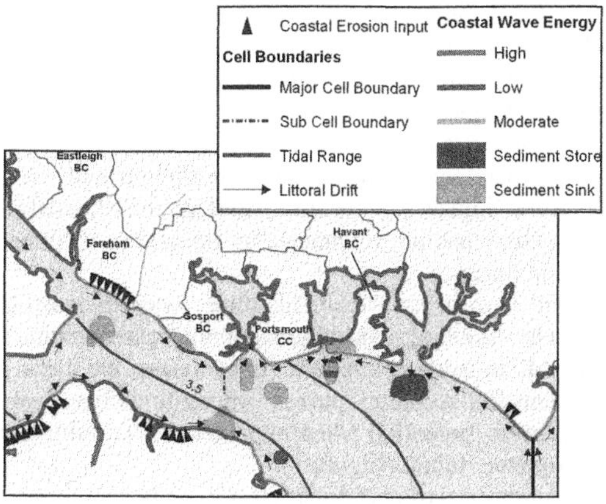

Figure 4. Coastal geomorphology.

Variations in forcing factors in the system model can generate changes along the coast, and hence the impact of climate change can be investigated. Because the magnitude and frequency of energy inputs to the coastal system tend to be random and the precise nature of future climate change is unclear, uncertainty will always exist in the behaviour of the system.

These maps provide a representation of factual data that has been derived from a variety of sources, including SCOPAC (Halcrow 2001) and FutureCoast (Halcrow, 2002). The map provides information on the spatial distribution of landforms and the nature of forcing factors. Key information on the bathymetry of the coastline is included, as is information on the 'structure' of the coastal sediment system, including sediment cells and sediment stores and sinks.

3.3 *Map 2. coastal defence management types and practices*

Much of the coastline of Europe is not able to operate 'naturally' because of the impact of coastal defence structures. The distribution and nature of defences needs to be understood as they have an effect on the coastal morphology and change the impact of forcing factors. Defences have value in terms of the protection afforded to assets such as towns or agricultural land, and also in terms of the economic cost of their construction and maintenance. This is important when assessing the potential coastal risks if the defences fail. Coastal defences may degrade in the future leading to an increase in the probability of failure through time and an associated increase in the risk to the asset protected. Management policies may change in the future, with the potential for defences to be removed or realigned to allow coastlines to begin to operate naturally.

The data on Map 2 has been derived from information previously published (Halcrow, 2001).

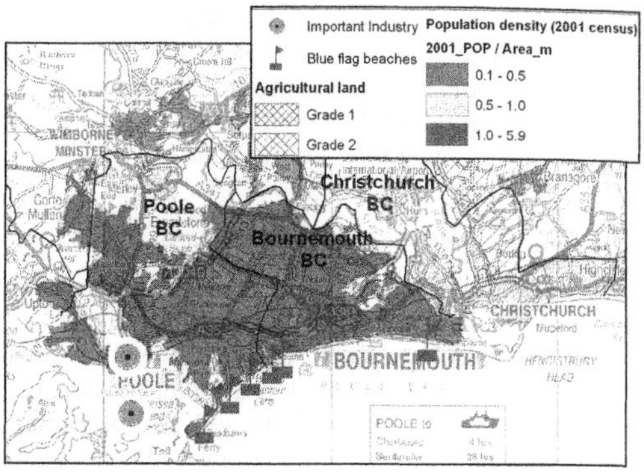

Figure 5. Coastal assets.

3.4 *Map 3. Historical coastal hazards*

Records of past and present coastal hazards highlight hotspots of activity and provide a good indicator of the pattern and nature of future hazards. Historical records of the location and nature of past and present hazards are presented on Map 3. The map includes three types of data; precisely delimited flood envelopes showing current areas of land affected by tidal and fluvial flooding sourced from the UK Government's Environment Agency; information on the approximate locations of past coastal retreat or past landslide events from historical records sourced from local council records; and information on the approximate extent of active landsliding from a variety of sources.

Flood plains are derived from accurate, ground-truthed GIS outputs of numerical modelled datasets. The location of current and historical events is indicated by points on the map, which can be cross-referenced to a database which provides full information. These data were not collected with GIS use in mind, and therefore locations of historical events are represented by the centre point of a feature rather than its spatial extent.

3.5 *Map 4. Coastal assets*

Risk results from the combination of hazards with their consequence, which may include loss of life, property or habitat. These features are presented on Map 4, which, for clarity, is subdivided into maps which show the human assets and natural assets.

Information on human assets includes the mapped population centres, displayed in terms of population density, blue flag beaches and areas of important industry (Figure 5). These data are derived from UK Government GIS data and population statistics derived from the 1990 census.

The map of natural assets comprises data on Sites of Special Scientific Interest, Special Protection Areas and wetlands designated as Ramsar sites,

Areas of Outstanding Natural Beauty, National Nature Reserves, Special Areas of Conservation and candidate Special Areas of Conservation. These data are all derived from fully attributed GIS layers sourced from the UK Government (Natural England).

3.6 *Map 5. Coastal behaviour systems*

Based upon the landforms and their interactions (Map 1) Coastal Behaviour Systems (CBS) for the study coastline have been defined. Previous work (SCOPAC) has been adapted to account for episodic (i.e. landsliding) and progressive soft cliff behaviour, which are considered important over the next 100 years. This assessment identified six CBS types (Table 1).

The landforms present along different coastlines should be reviewed to consider how well they comply with the SCOPAC CBSs and their constituent landforms.

Should additional CBS be required they should be determined by the coastal forms present and their generic sensitivity to climate change factors. Possible refinements include differentiation of coastal lowlands from barriers, and separate treatment of spits, inlets and tidal lagoons.

3.6.1 *Map 6. Future coastal hazards (2100)*

Global climate change projections for the next 100 years have been published by the Intergovernmental Panel for Climate Change (IPCC, 2001). The UK benefits from higher resolution projections produced by the United Kingdom Climate Impacts Programme (UKCIP, 2002, 2005).

Table 1. Composition of Coastal Behaviour Systems.

Element	Shore-face	Shoreline	Back-shore
Hard cliff	Steep	Fringing boulder beach and/or shore platform	Hard Cliff
Soft cliff progressive erosion	Gentle	Fringing sand, shingle or mixed beach	Soft Cliff
Soft cliff episodic erosion	Gentle	Fringing sand, shingle or mixed beach	Soft Cliff / landslide
Lowlands & barriers	Gentle	Fringing sand, shingle or mixed beach, free-standing shingle barrier, fronting sand or shingle beaches	Lowland
Spits, inlets & tidal deltas	Gentle	Inlets, tidal deltas, free-standing shingle beach	Lowland
Estuaries & tidal rivers	Gentle	Tidal flat, saltmarsh, tidal river	Lowland

UKCIP indicates sea level rise of up to 80 cm by the 2080s (UKCIP, 2005, Figure 6) and that under the medium high scenario mean annual temperatures may rise by 4° and winter rainfall may increase by as much as 30% above today's levels (UKCIP, 2002).

As discussed above, changes in climate will have a direct impact on the rates and nature of coastal processes, leading to changes in the nature of future coastal hazard.

The hazard map identifies three types of hazard, namely: flooding, coastal erosion and reactivation of coastal landslide complexes. In each case, the hazard is identified as being 'current' where no defences are present or 'potential' where the hazard is conditional on defence failure (Figure 7).

Future hazard classifications are made using maps of coastal behaviour systems, coastal defences and past and present hazard events (Figure 8). Coastal erosion hazard is classified according to the expected average recession rate over the next 100 years from either progressive or episodic failure. The coastal erosion hazard classes are <0.5 m/yr, 0.5 to 1 m/yr and >1 m/yr, and reflect the possible areas of coastal hinterland that could be lost over the next 100 years. For the study coastline, it has been possible to assign each cliff unit a recession rate derived from the Futurecoast database (Halcrow 2001). If similar data is unavailable

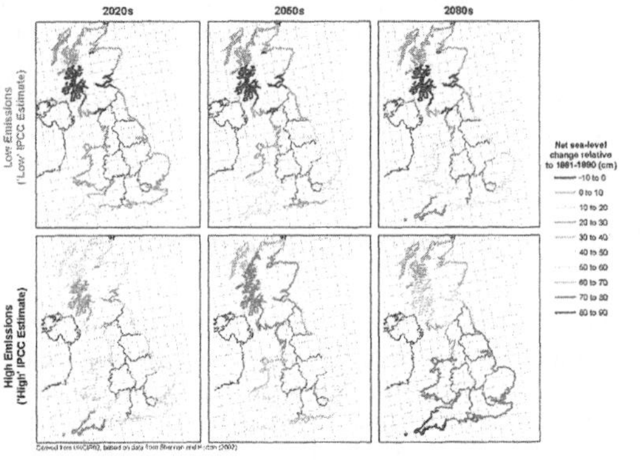

Figure 6. Net sea level change for the UK (after UKCIP, 2005).

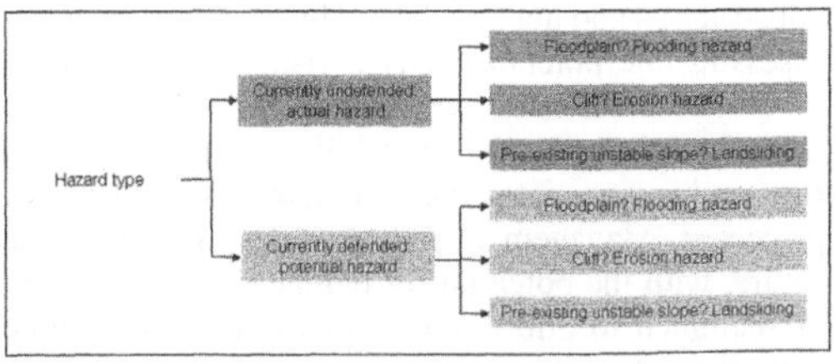

Figure 7. Hazard classification, based on CBS and defences.

in other EU countries, qualitative assessments of cliff behaviour can be made using expert judgement.

Coastal landslide complexes, such as the Ventnor Undercliff and Lyme Regis, which have episodic behaviour over many thousands of years, are separately identified. These features are identified on the map of current and historical hazards.

Areas at risk from flooding are identified by the intersection of mapped flood plains with the coastline, which have been identified on Map 3.

In areas that are not classed as a floodplain, cliff or landslide, no hazards are considered to be present.

3.7 *Map 7. Potential future coastal risk (2100)*

3.7.1 *Risk matrix*

There is no baseline quantitative measure of risk against which changes can be compared (such as present value of properties), therefore an approach, using a matrix of relative risk ratings has been developed (Figure 9).

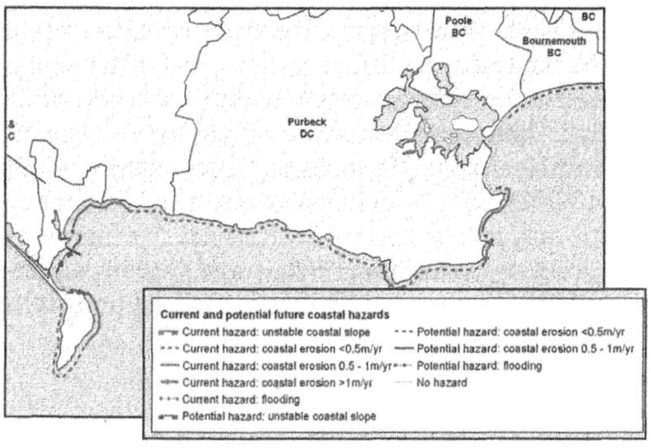

Figure 8. Future coastal hazards.

Relative risk classes and definition:

4	Very high risk
3	High risk
2	Medium risk
1	Low risk

Erosion risk

Consequences (land use class)	Degree of hazard (progressive and episodic erosion): >1m erosion/yr	0.5 - 1 m erosion/yr	<0.5m erosion/yr	No erosion
High density urban	4	3	3	2
Medium density urban	3	3	2	1
Low density urban / rural	3	2	2	1
Agricultural land / isolated communities	2	2	1	1

Flood risk

Consequences (land use class)	Likelihood of flooding: 1:10 yr flood	1:100 to 1:200 yr flood	ca.1:1000 yr flood (5 to 10m OD)
High density urban	4	3	2
Medium density urban	3	2	2
Low density urban / rural	3	2	1
Agricultural land / isolated communities	2	1	1

Coastal instability risk

Consequences (land use class)	Likelihood of significant ground movement (landslide reactivation): 1:10 yr event	1:100 yr event	1:500 yr event	1:1000 yr event
High density urban	4	3	2	1
Medium density urban	3	3	2	1
Low density urban / rural	3	2	1	1
Agricultural land / isolated communities	2	1	1	1

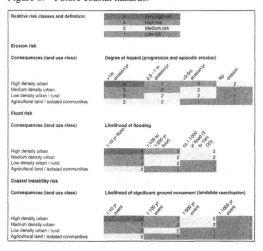

Figure 9. Matrices for the assessment of risk from erosion, flooding and coastal instability.

This approach assesses risk by comparing the economic consequences of flooding, coastal erosion or reactivation of coastal landslides with the likelihood of different magnitude events.

Economic consequences are measured using the population density for each section of the coastline. This approach is based on DEFRA's PAG3 (no date) documentation and identifies four classes of population density as follows:

- High density urban (>1 person/metre^2)
- Medium density urban (0.5 to 1 person/metre^2)
- Low density urban/rural (0.1 to 0.5 people/ metre^2)
- Agricultural land with isolated communities (<0.1 person/metre^2)

The breakpoints for definition of these four groups have been selected to approximately define the four quartiles of the data set. They are appropriate for use in this relatively densely populated study coastline, but can be adjusted to suit other coastline regions of Europe.

Likelihoods of hazard events are measured differently, according to the hazard type, as follows:

- Coastal erosion is quantified by expected average recession rate over the next 100 years (i.e. accounting for both episodic and progressive losses). Classes are >1 m/year; 0.5 to 1 m/year; <0.5 m/year and no erosion (due to coastal defences);
- Flood risk is quantified by the expected return interval of flood events. Unprotected floodplains are expected to flood every 10 years; areas defended against flooding are expected to be protected against the 1:100 year event; and other areas adjacent to floodplains between 5 and 10 m OD are expected to be flooded every 1000 years; and
- Coastal instability risk is measured by the return interval of ground movement events, as follows: 1:10; 1:100; 1:500 and 1:1000 year events.

Using these two criteria it is possible to assign each cell of a matrix a relative risk rating. This study uses a four-point classification from 1 (Low Risk) to 4 (Very High Risk), but additional classes can be added if required in different study areas.

Each cell is assigned a risk rating based on expert judgement. Errors and bias can be minimised if the task is undertaken by a number experts. The risk matrices produced for erosion, flooding and coastal instability are shown in Figure 9.

3.7.2 *Scenarios and assessment of risk*

The methodology developed allows for any number of scenarios to be tested, but for simplicity, the impact of two scenarios on the coastline is assessed here:

- 'business as usual', with limited coastal Response to climate change and no change in coastal management practices over the next 100 years;

– 'worst case', with climate change having a direct and negative impact on coastal processes and dramatic changes in coastal management practices, with all defences being lost over the next 100 years.

In the business as usual scenario, the population density and degree of hazard for each segment of the coastline (i.e. the current and future hazards shown on Map 6) is visually assessed and a risk score is added in the GIS.

For erosion risk, the degree of erosion hazard is derived from Map 6, with all potentially eroding coasts currently protected by defences, being classed as 'no erosion'. The highest risk score can only be achieved at locations of greatest population density and greatest erosion hazard, and conversely, the lowest risk scores are achieved where erosion rate is low or absent in locations of medium to very low population density (Figure 9).

Flood risk is determined by the presence or absence of flood defences and the population density of floodplain areas. The highest risk score is applied to urban areas situated on undefended floodplains. Lower risk scores are applied if defences are present, and if population density is lower (Figure 9).

Coastal instability risk is assessed using the likelihood of significant ground movement as the measure of hazard. High risk scores are therefore applied to locations where population density is highest and significant ground movement is expected as a 1:10 year event, while lower scores are applied where the likelihood of reactivation or population density is lower (Figure 9).

The risk of the different hazards cannot be directly compared because risks are assessed relatively. Therefore the mapping shows the relative risk scores for erosion, flooding and coastal instability separately identified with the prefix E, F or C. This means that very high risks of flooding (F4) and erosion (E4) are not the same and cannot be considered to be comparable.

To assess the impact on coastal risk of the worst case scenario, the same series of relative risk matrices can be used. To account for the impact of climate change at each coastal segment the degree of hazard (i.e. average erosion rate, likelihood of flooding or likelihood of significant ground movement) is increased. Using the matrix, this is achieved by moving one cell towards the right along the same Consequence (land use class) row (Figure 9). An example of the worst-case scenario is shown in Figure 10.

Depending on the structure of the matrix, this could lead to an increase in risk although such a result is not always the case. For example, erosion hazard of <0.5 m/year in a low density urban/rural location under a business as usual scenario gives a risk rating of E2. However, in the worst case scenario, by increasing erosion hazard to 0.5 to 1 m/year (i.e. moving one cell to the right) results in the same risk rating of E2.

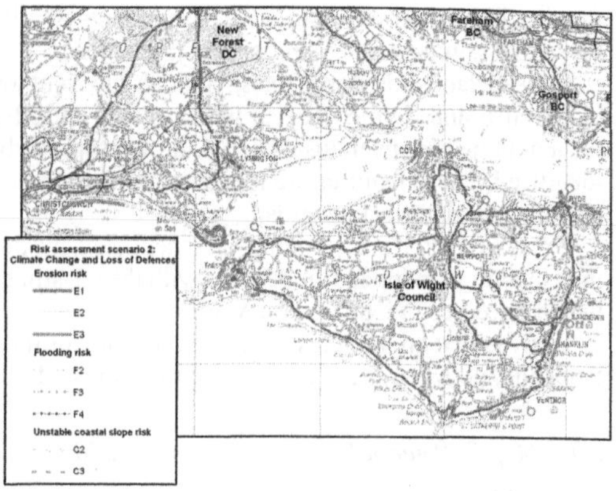

Figure 10. Future coastal risks under a worst case scenario.

The impact of removal of coastal defences on erosion rate and coastal instability is assessed by selecting a most likely recession rate/reactivation return period for the unprotected cliffs/landslides, and then moving one cell to the right to account for the impact of climate change. Data on unprotected cliff/landslide behaviour is available in the Futurecoast cliffs database (Halcrow, 2002), but expert judgement can also be applied where such information is unavailable. The impact of defence removal on flood risk is achieved by assuming all floodplain areas will be inundated on a 1:10 year basis.

4 CONCLUSION

Effective coastal zone management requires decision-making that takes full account of past, present and future conditions. This can be achieved by guiding future developments towards more suitable locations, where levels of risk are deemed acceptable. Coastal managers and planners clearly need to be well-informed if they are to make appropriate decisions regarding future coastal development and management of existing assets at risk from coastal hazards. Decision-making based on full understanding of coastal behaviour trends and likely impacts of future climate change is clearly desirable.

The Response methodology provides a valuable map-based GIS tool to assist in regional-scale decision-making. The methodology outlined above is adaptable to other EU coastlines, where there may be different coastal behaviour systems, coastal hazards, consequences and projections of future climate, and can be extended to provide non-technical guidance to a range of end-users.

The provision of end-user guidance maps constructed from the relevant earth science data provides an essential and auditable basis for decision-making. It is intended that Response will greatly assist responsible management of the EU coastline that accounts for the impact of climate change.

ACKNOWLEDGEMENTS

The authors are grateful for the assistance of Dr Mark Lee and colleagues in the Isle of Wight Council and Halcrow during the 3-year project, and in the preparation of this paper. The Response project benefited from financial support from the European Commission's LIFE Environment Programme.

REFERENCES

DEFRA. No date. Flood and Coastal Defence Project Appraisal Guidance Notes 3 and 4. London: HMSO.

Halcrow. 2001. *Preparing for the impacts of climate change*. Standing Conference on Problems Associated with the Coastline (SCOPAC).

Halcrow. 2002. *FutureCoast*. CD-ROM for DEFRA.

Hutchinson, J.N. & Gostelow, T.P. 1976. The development of an abandoned cliff in London Clay at Hadleigh, Essex, *Phil. Trans. Roy. Soc. Lond.* A283: p557–604.

European Commission. 2004. *EUrosion – Living with coastal erosion in Europe*. Luxembourg, European Commission.

IPCC. 2001. *Climate Change 2001. Synthesis report*. Cambridge: Cambridge University Press.

Lee, E.M. 2005. Coastal cliff recession risk: a simple judgement-based model. *Q. J. of Eng. Geol. and Hydrol.* 38: p89–104.

Lee, E.M. & Clark, A.R. 2002. *Investigation and management of soft rock cliffs*. London: Thomas Telford.

McInnes, R.G. 2006. *Responding to the risks from climate change in coastal zones, a good practice guide*. Ventnor: Isle of Council.

McInnes, R.G, Tomalin, D. & Jakeways, J. 2000. *Coastal change, climate and instability*. EU Environment LIFE project. Ventnor: Isle of Council.

UKCIP. 2002. *Climate Change Scenarios for the United Kingdom: The UKCIP02 Scientific Report*. Norwich: Tyndall Centre for Climate Change Research, University of East Anglia.

UKCIP. 2005. *Updates to regional net sea-level change estimates for the UK*. Oxford: UKCIP.

Landslides and Climate Change – McInnes, Jakeways, Fairbank & Mathie (eds)
© 2007 Taylor & Francis Group, London, ISBN 978-0-415-44318-0

Application and implementation of the RESPONSE methodology to evaluate the impact of climate change on the Aquitaine and Languedoc Roussillon coastlines

C. Vinchon, S. Aubié, Y. Balouin, L. Closset, D. Idier, M. Garcin & C. Mallet
BRGM, France

ABSTRACT: Application of the RESPONSE methodology, which was originally devised for the south-central coast of England, to two French regions required a slight adaptation of the methodology to fit the regional specificities and data availability and to meet the expectations of the end user. Results from both regions highlighted that coastal risks are very likely to increase due to the impacts of climate change and to the intensity of development along this fragile coastline. It was also noted that some mitigation may be possible due to the potential for sediment storage. The need for spatially integrated management of the coastline and its defences will be crucial in the coming century.

1 INTRODUCTION

Within the framework of the LIFE Environment demonstration project RESPONSE, whose objectives and methodology are detailed in this volume (Fairbank, 2007 and Fish, 2007), Aquitaine and Languedoc Roussillon coastlines (France) were studied in terms of the likely impacts of climate change.

The initial methodology as proposed by Cooper *et al*, 2002 and Hosking *et al*, 2002 for SCOPAC (the regional coastal group of central-southern England) and modified for the RESPONSE Project (Fish, 2007), was adapted to the specificities of the two French case study regions. The adaptations are briefly described, the results for each region are presented and an inter-regional comparison is made, underlining the concomitant role of both geomorphology and development.

2 RESPONSE METHODOLOGY AND ITS IMPLEMENTATION

2.1 *RESPONSE methodology framework*

RESPONSE (Fairbank, 2007, this volume) aimed to provide a regional scale mapping of coastal hazard and risk evolution in view of climate change, in order to produce hazard and risk evolution maps for decision-makers so that the impacts of climate change may be taken into account in further development planning.

The main stages of the proposed methodology (Fairbank, 2007, Fish, 2007, this volume) are as follows:

- Collecting and mapping coastal geomorphology, coastal dynamics, existing defences, known hazards data
- Defining with this information "coastal behaviour systems (CBS)", described from sea to land, which represent a coherent response to coastal hazards
- Adapting regional climate change scenario(s), based on IPCC(2001) hypotheses and models, including sea level rise, rainfall and temperature, to local conditions when available
- Evaluating the degree of change in coastal hazards (erosion, marine flooding, landslides), that could occur due to those scenarios of climate change
- Evaluating the degree of change in coastal risks, when assets are involved in hazard change
- Identifying areas where this change in risk might be most critical ("hotspots")

2.2 *Implementation of the methodology in the French pilot regions*

The RESPONSE mapping was applied as proposed in the methodology described in Fairbank, 2007 and Fish, 2007 (this volume).

However, adaptation of the RESPONSE methodology was necessary due to the geomorphological specificities of the French studied regions, and also in relation to the available data to fulfil the proposed methodology. Interaction with end-users of the project guided the decision of mapping scale (1/100 000), and also which climatic scenario would be considered.

More specifically, it was necessary to adapt the geomorphological descriptions, climate change scenarios, risk definition and the method used to identify "hotspots".

2.2.1 *Specificities of the geomorphology*

The coastline of both regions was described following the geomorphological features such as proposed in Fish, 2007, by collating data from the EUrosion project, IFREMER, ONF and SHOM databases for geomorphology, coastal processes and existing defences, diverse sources for historical hazard events (P.Durand,1999, Oliveros et al 2004), SMNLR and OCA data bases. The collation of this data was validated by a panel of local experts (BRGM, SMNLR, EIDméditerranée), and was also validated in the field when necessary.

Several geomorphological features were added to the original methodology to describe the Aquitaine dune system along the sandy coast. Shoreface features such as subtidal sandbanks and bedrock beaches are present in both regions. A differentiation was also made between hilly hinterlands (Pays Basque) and cliffs. Figure 1 provides the complete list of features identified in the French pilot regions.

2.2.2 *Climate change scenarios*

The available regional climatic change models (MeteoFrance, Tyndell, ACACIA) provide coherent values with the IPCC models in term of temperature and rainfall. MéteoFrance values were therefore used for these parameters. Table 1 lists the values that were considered.

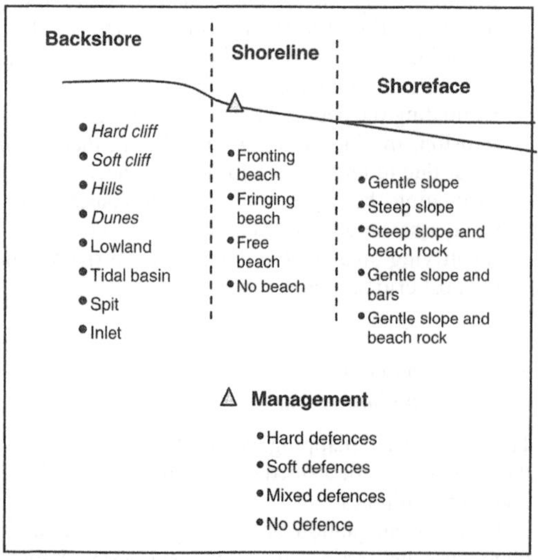

Figure 1. Geomorphologic features used to describe coastal behaviour systems of the coastlines of Aquitaine and Languedoc Roussillon.

Though temperature changes will not have a direct impact on coastal hazards, they may lead to changes in assets at risk, by increasing coastal populations and tourism.

Changes in rainfall may influence water table fluctuations which act as a major triggering factor in cliff collapses and landslides.

No regional model was available for sea level rise data. The IPCC 2001 models (figure 2) were used to provide estimates of sea level rise, to which was added for each region, the highest regional tide level, and the highest known storm surge. The pessimistic hypothesis of 88 cm sea level rise in 2100 was chosen, upon request of the end-users and in line with the project aim to stress critical spots.

In Aquitaine, summing potential sea level rise, highest tide and highest known storm surge provided a potential flooding level of 5.8 m above IGN datum.

Table 1. MéteoFrance values for changes in temperature and rainfall in case of doubling C02 concentration.

		Aquitaine	Languedoc Roussillon
Temperature change (°C)	Winter	+1 to +2	+2 to +4
	Spring	+1 to +3	+2 to +3
	Summer	+2 to +3	+2 to +3
	Fall	+2 to +3	+ 2 to +4
Rainfall change (mm/d)	Winter	+1 to +2	+0.2 to +0.6
	Spring	−0.2 to +0.6	−0.6 to 0
	Summer	−0.2 to +0.2	−0.2 to +0.2
	Fall	−0.2 to −0.6	−0.6 to +0.6

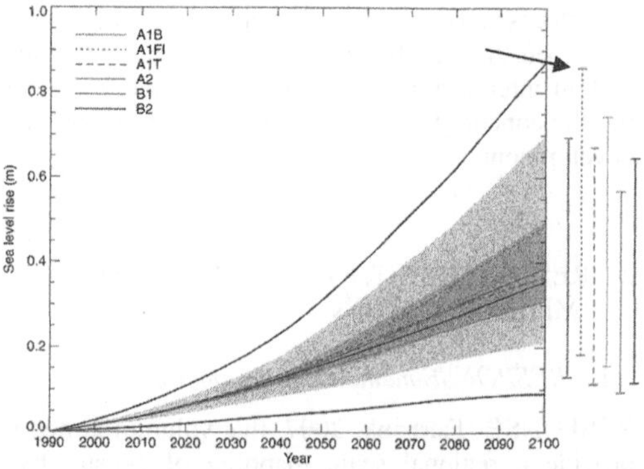

Figure 2. Mean global rise in sea level between 1990 and 2100 for various socio-economic scenarios computed using the seven coupled air/ocean models (IPCC, 2001). Uncertainties in the results of each scenario are indicated to the right of the graph.

In Languedoc Roussillon, it could reach 2.64 m above IGN datum.

No reliable models exist at this stage to predict changes in storm activity within the coming century. In the evaluation of hazard changes, it was estimated that storms will increase in frequency and intensity, which would lead to an increase in wave energy on the coastline.

2.2.3 *Evaluating and scoring the hazard change*

On the basis of the established Coastal Behaviour Systems, and the climate change scenarios (described above), the evolution of coastal hazards was cross evaluated by a panel of experts from BRGM. Hazard was categorised using a grading system – drastic or significant decrease, no change or significant or drastic increase – for the probability of:

- Change in marine flooding hazard,
- Beach erosion, integration the adaptation capacity given by a available sedimentary stock (dunes or sandbars)
- Cliff erosion by ground movement

System changes were considered possible in the case of beach or sandbar disappearance.

It was considered in the grading system that no guarantee of defence upgrading could be guaranteed, and that they had to be considered as non efficient in the future. Also, the behaviour of offshore sandbars is not known precisely, so it was considered that they might disappear.

2.2.4 *Mapping changes in risk*

Asset data was extracted and collated from Corine Land Cover data base (1990) to map urban, tourist, industrial and agricultural assets. Environmental assets were derived from DIREN GIS but proved to be non discriminatory as all the shoreline was identified as deserving of preservation.

Data on the value of assets at risk due to climate change was difficult to obtain in the French pilot regions. To represent this data, assets within a 300 m buffer along the coastline were mapped (average value for coastal erosion, McInnes 2005) where erosion hazard or potential marine flooding is likely increase drastically.

2.2.5 *Definition of "hotspots"*

One aim of RESPONSE was to highlight, for decision-makers, specific stretches of coastline where it will be necessary to focus resources and undertake further studies to mitigate the impacts of climate change.

In the French pilot regions the present day so-called "hotspots" in terms of present-day erosion and/or marine flooding risk were first identified.

This information was then incorporated into the hazard change map, allowing an estimation of where present day hotspots may encounter an increased hazard, and where new stretches of coast may become critical. The final step was to allocate priorities to hotspots where the hazard increase will result in assets experiencing increased levels of risk.

3 RESULTS OF RESPONSE MAPPING

3.1 *Aquitaine*

3.1.1 *Description of the coastline*

The Aquitaine coastline is composed of a 200 km long sandy coast from the Gironde to the Adour estuaries, and of soft rocks cliffs, from Adour to the Spanish boarder.

The Arcachon Basin was not included in this study as it is controlled by a semi-closed tidal bay dynamic and its behaviour in regard to climate change requires more detailed study.

The sandy coast is dominated by large and high (up to 20 m) dune systems, built since the Holocene, bordered by ridge-and-runnels beaches, sloping gently seaward, and resulting in subtidal sandbars (figure 3), according to a dominant longshore sediment drift, from north to south from the Pointe du Verdon to Cap Breton and from south to north from Anglet to Cap Breton. This dune system is locally cut by small coastal river outlets, with low lying flood plains in the hinterland.

The Basque cliffed-coast is composed of cretaceous and tertiary sediments (chalk, clays, sandstone), where differential weathering and alpine geometry produce points, coves and bays.

Settlements are concentrated on the Basque coast (urban and tourism) and on most of the rivers outlets of the sandy coast (urban, tourism, industry, following a development plan set out in the 1960s (MIACA, 1967)

Along most of the sandy coast, present day erosion processes are known, apart from the accreting seaward side of the Pointe du Cap Ferret.

Hard defence works, built to prevent erosion, are localised at points of historical breaching, such as the Pointe du Verdon to Soulac where breakwater systems were built and have been improved since the beginning of the 20th century, and on most coastal outlets where

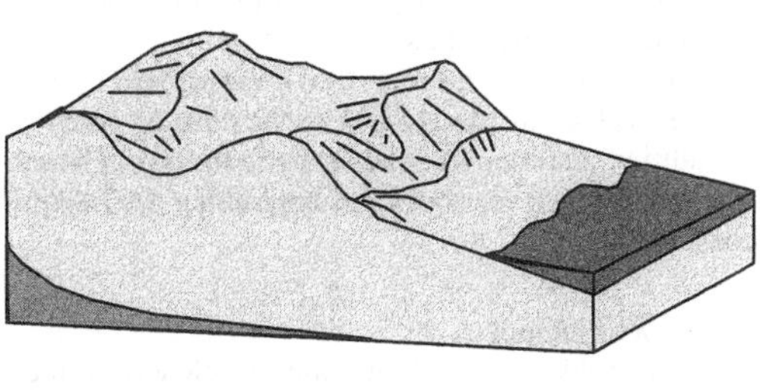

Figure 3. Aquitaine dune CBS model.

sets of groynes and dykes were built to protect existing assets and prevent outlet drift. Recent experimentation of soft defences (sediment hydraulic by pass, beach nourishment, and geotextiles) are still being evaluated.

Numerous ground movements occur along the cliffed-lined coast of Pays Basque, due to differential weathering and hydrogeological processes (Mallet et al 2005) and induced cliff engineering (base abutments, paling, seawalls, shotcrete etc.). Up to now, coves and bay beaches have been sheltered from erosion.

The marine flooding risk has up to now been localised in low-lying zones, linked to coastal river plains of Aquitaine coast.

Present day hotspots are known by stakeholders as subject to erosion risk. Primarily this concerns the coastal towns along the sandy coast, where the existing transversal defences often aggravate the processes by trapping longshore drift. There are numerous examples of this issue along the heavily developed Basque coast.

3.1.2 *Hazard change evaluation*

On no point along the Aquitaine coastline was it considered that coastal hazards are likely to decrease.

The dunes systems along the sandy coast are considered as likely to undergo a significant increase in erosion and the disappearance of subtidal sandbars; however the volume of sediment storage represented by dunes and sandbars gives these geomorphological systems a fairly good capacity for adaptation and resilience towards the initial morphology, except if development has prevented the availability of sediment supply (Amélie/mer, Soulac). Erosion processes might increase more drastically along urbanised low-lying outlets, if it is considered that existing defences will not be upgraded.

Increasing water table fluctuations and storm events are also likely to significantly increase the erosion process, already important on the cliffs of the Basque coast, if existing defences are not upgraded. Also, bays and coves have little sediment supply apart from weathering product of the cliffs; sandy or sand and pebble beaches are narrow and lying on bedrock. The impact of sea level rise and more frequent and stronger storms are likely to induce beach disappearance.

Due to sea level rise and likely increased frequency of storm surges, marine flooding will increase significantly at existing hotspots in Aquitaine, and these hotspots may expand in size.

Most of the present day critical hotspots will remain critical with an increasing risk in the coming century. In addition, attention must be drawn to the risk of beach disappearance in the coves and bays along the Basque coast.

3.1.3 *Risk change*

Asset distribution in Aquitaine is closely linked to the MIACA strategy established in the 1970s.

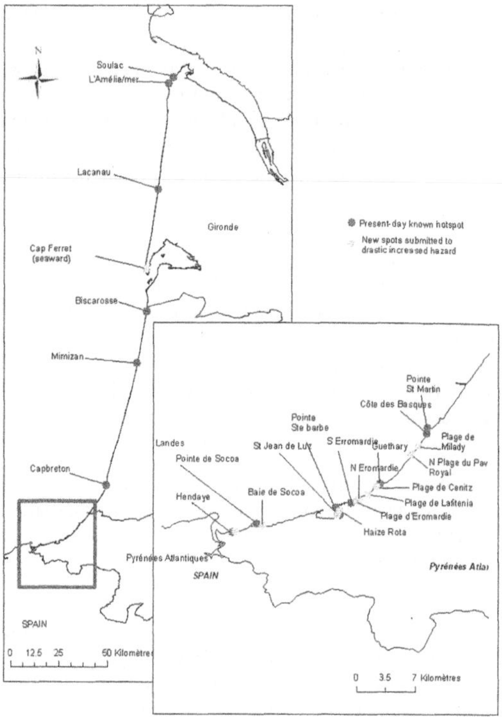

Figure 4. Present-day and future hotspots in Aquitaine. New "hotspots" due to increased hazard are localised in Pays Basque; most of them implies assets at risks.

Development along the sandy coast is concentrated around tourist towns, which settled along coastal rivers outlets (such as Vieux Boucaux, Cap Breton etc) identified as vulnerable to a drastic hazard change (both erosion and flooding hazard), or on the very edge of the dune system (Soulac, L'Amélie/mer etc) which is already experiencing significant erosion and where development limits the potential for sediment storage. Leading to a concertina effect of both parameters, most of the coastal towns of the sandy coast are likely to have assets at risk.

The development of the Basque coast is continuous and historical (mid 19th century), and concentrated up to 500 inhabitants/km2. It has long been a point of focus for landslide management and will continue to be in the coming century. The sheltered bays and coves of this coast have a limited potential for sediment storage in the case of erosion.

Due to the past development strategy of Aquitaine coast (MIACA 1967), and the concertina effect of geomorphology and existing assets, most of the present day "hotspots" will continue to be a point of concern in the coming century, considering the management strategy of no upgrading or inappropriate upgrading of defences. The currently sheltered bays and coves of

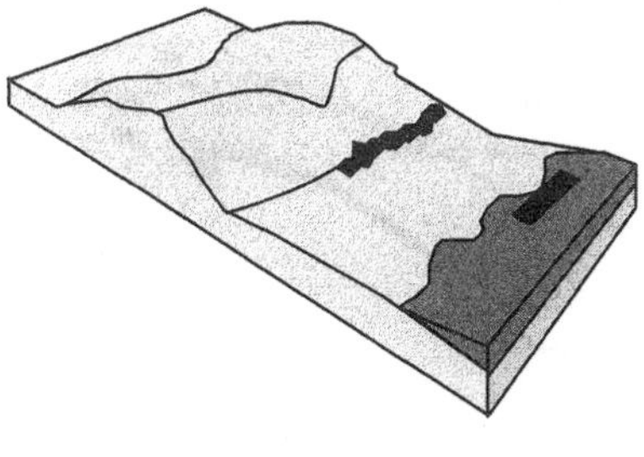

Figure 5. Languedoc Roussillon Lido CBS model.

Pays Basques may become new hotspots, with the possible disappearance of the fringing beaches (figure 4).

3.2 *Languedoc Roussillon*

3.2.1 *Description of the coastline*

The Languedoc Roussillon coastline consists primarily of a quaternary sandy coast extending from Camargue in the west to Argeles, and a hardrock cliffed-lined coast from Argeles to the Spanish boarder.

Apart from a few rocky sections (Sète, Agde) marking the limit of sedimentary cells, the sandy coast consists predominantly of "lidos", described as narrow and low dune systems limited landward by low-lying marshes or salt lagoons, and seaward by a fringing beach, and a gentle to steep sloping shoreface with sandbars (figure 5). The dune system is interrupted by narrow and initially divagating outlets (rivers or lagoons) called "graus".

The southern rocky coast consists of hard Cambrian rocks, sloping steeply into the sea, and sheltering a few small coves and bays, with sand and pebble beaches.

The sandy coast was highly developed during the development strategy of the Mission RACINE in the 1970's and 80's.

On the contrary, little development has been undertaken along the cliff-lined coast, except for small resorts such as Collioure, in sheltered bays.

The presence of assets and the existence of erosion and marine flooding have made it necessary to construct heavy defence works along the sandy coast (groynes and breakwaters), which might have in several places locally aggravated the problems. Future projects must take into account the concept of integrated management and seek a balance between "hard" and "soft" defence, for example, planning partial retreat of the assets on the lido of Sète. Along the cliffed coast there is less of a problem due to coastal erosion or marine flooding and no defence works have been constructed (except for harbour structures).

Erosion hotspots have been identified along the sandy coast by a regional study undertaken by SMNLR towards a regional strategy against erosion, and marine flooding (figure 6).

3.2.2 *Evaluation of hazard change*

As is the case along the Aquitaine coast, at no point are coastal hazards likely to decrease.

Along most of the sandy coast, the sedimentary store was considered low, or not available due to urbanisation, and the morphology of the lido prone to erosion or even breaching. If the existing defences are not upgraded or improved, this coastline is considered as likely to undergo a drastic change in erosion hazard, except when dunes are higher and sedimentary stores are available, or when sandbanks provide sand. Gradation of probable increases in the marine flooding hazard was evaluated on the basis of the 2.64 m above IGN datum potential sea level rise during storms.

On the cliff-lined coast, beaches may be subject to an erosion hazard increase, having little to no sediment storage to enable a response to sea level rise and storm beach erosion. Limited land will be subject to an increase in marine flood hazard.

When considering the length of coastline likely to be subject to an increased hazard in Languedoc Roussillon, the concept of critical zones has to be enlarged to stretches of coast more than to hotspots. Those stretches of coast will need to be managed in an integrated approach (figure 6).

3.2.3 *Risk change*

Asset distribution along the Languedoc Roussillon coast is spread all along the sandy coast, as a result of the development strategy of the 1970–80's. Urban and tourism and infrastructure assets are often close to the sea, whereas most of the industrial and agricultural assets are in the hinterland, in the surroundings of the lagoons.

There is limited development along the southern rocky coast, apart from a few tourist towns.

The increase of erosion and marine flooding along the sandy coastline may affect many urban and tourism assets, as well as communication structures. If a hazard increase effect stretches of coast, within the frame of a sedimentary cell, existing assets are more localised. However, when considering protection of the assets, it is necessary to follow an integrated approach, up to the sedimentary cell scale.

Figure 6 illustrates present day hotspots and their expected evolution in the coming century, if the above hypotheses (pessimistic scenario for climate change, and no upgrading or improvement of the defences) are considered.

4 CONCLUSION

4.1 *Comparison of the results in both pilot regions*

The RESPONSE approach in both pilot regions demonstrates that coastal risk is a reality today and is likely to increase in the coming century.

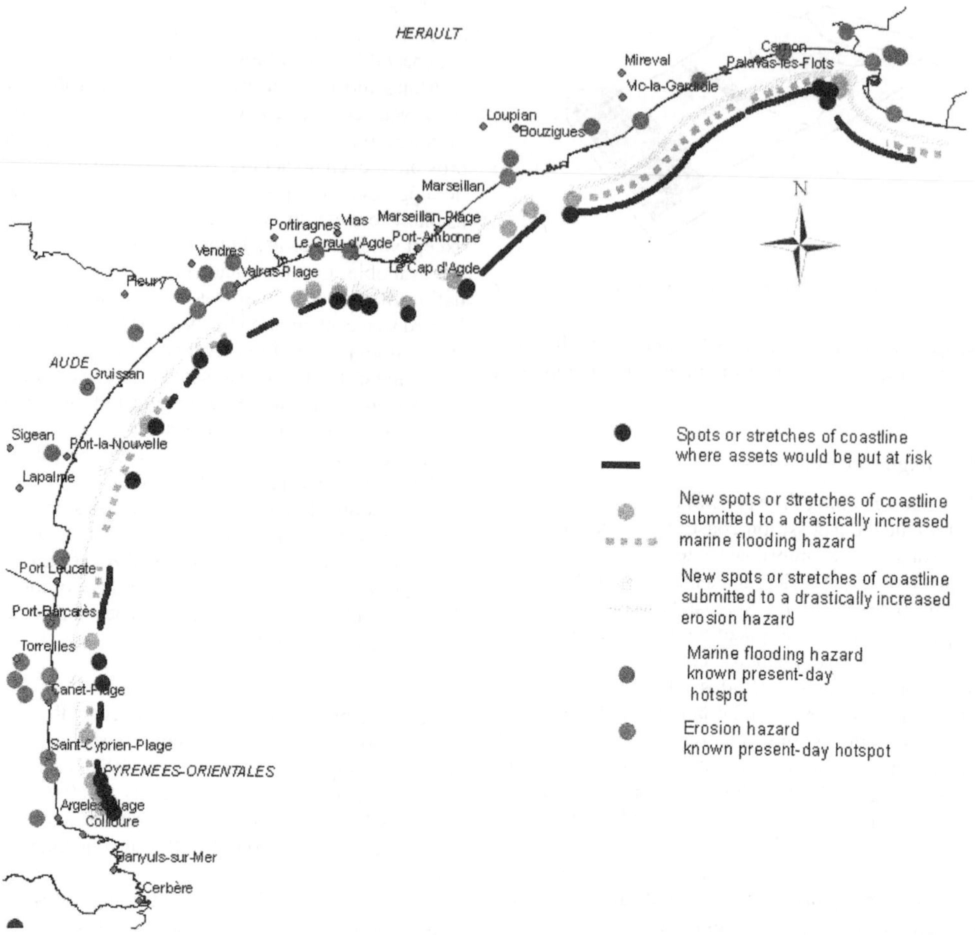

Figure 6. Evolution of the "hotspot" concept in Languedoc Roussillon. Hazard increase will concern larger stretches of coastline, and put at risk mostly tourism and urban assets.

The concertina effect between low-lying coastlines and development appears to be the major element leading present-day and potential future coastal risks (erosion and marine flooding). Different development strategies in both regions have induced a different risk distribution. Also, linked to this development and geomorphology, the existence and availability of sediment is likely to allow a relatively good resilience of the Aquitaine sandy coast in the coming century, whereas the Languedoc Roussillon coast is less resilient.

Ground movement hazards affecting the cliff-lined coasts are closely linked to the geo-mechanical and hydrogeological properties of the cliffs and a potentially increased energy of the sea. Climate change should result in changes in rainfall distributions, which might modify these properties and increase the risk of ground movement.

The soft rock cliffs of Pays Basque are already known to be at risk, whereas fewer hazards have been identified in the Roussillon hard rock cliffs. Climate change impacts will emphasise that difference. Again, asset distribution appears concomitant to a higher hazard. In addition to known hotspots, presently sheltered coves and bays might undergo a disappearance of their beaches, due to a lack of sediment supply.

4.2 *Lessons learned*

To adapt to the specificity of the French pilot regions (geomorphology, available data and models and expectation of the end users), the RESPONSE methodology was applied on the basis of the more pessimistic scenario in terms of climate change and management of existing defences, and by considering the availability of sediments to evaluate the resilience capacity of the coastline. The risk was evaluated in a qualitative approach.

In reference to other potential RESPONSE pilot regions, it demonstrates the transferability of the method and its capacity to be adapted to the regional specificities and regional scale information, to determine the likely impacts of climate change on coastal risks.

In both French regions, this work has provided a regional overview of the likely impact of climate change on coastal hazards and risks and the need to take that impact into account in further development planning, and establish a preliminary hierarchy in the zones of concern.

It also highlights, particularly in Languedoc Roussillon, that coastal risk management needs to be managed following an integrated approach, at least at the scale of the sedimentary cell, even if assets at risk are restricted to smaller stretches of coast.

It demonstrates the impact that intense development during the last century has had on coastal risk along a fragile coastline. This vulnerability must be taken into account in further planning as well as in managing the defence strategy of existing assets.

However, the limits of this approach must consider in the regional scale results:

- interaction between coastal behaviour systems have not been considered
- the sediment availability is estimated on a qualitative basis
- climate change scenarios are based on hypothesis and models that will need further validation
- the risk change evaluation lacks data on the value of assets, their vulnerability and their future evolution.

The results must now be reviewed by regional stakeholders. To aid the decision-makers at a more local scale, further quantification and modelling is required.

ACKNOWLEDGMENTS

The RESPONSE project was part of the LIFE Environment demonstration program and received European funding.

It was carried out from September 2003 to September 2006, under the leadership of Isle of Wight Centre for Coastal Environment, with the partnership of research organisations in UK (Halcrow ltd), France (BRGM) and Italy (CNR) and end-users in the UK (SCOPAC, Scarborough Borough Council), France (maritime services), Italy (Regione Marche, Provincia de Macerata, Provincia di Pesaro y Urbino) and Poland (maritime office of Gydnia)

This partnership allowed an effective synergy and refinement of the methodology throughout the project.

The application of RESPONSE methodology in Aquitaine and Languedoc Roussillon was made possible by the help of the Maritime service of Languedoc Roussillon, EIDMéditerranée, the Maritime service of Gironde, the Departmental direction of Equipement of Landes in collecting data, and sharing of their coastal environment knowledge. The authors wish to thank them.

The authors wish to thank Nicole.Lenôtre Carlos Oliveros and Rodrigo Pedreros, in BRGM for their expert advices, participation to cross-evalutions, and reviewing.

REFERENCES

Baede A.P.M., Ahlonsou E., Ding. Y., Schimel D. (2001) – 1. The Climate system: an overview, p. 85–106.Third assessment report of the IPCC(2001).

Church J.A., Gregory J.M. *et al.* (2001) – 11. Changes in sea level, p. 639–684. Third assessment report of the IPCC(2001).

Cooper N.J., Jay H. (2002) – Prediction of large-scale coastal tendency: development and application of a qualitative behaviour based methodology. *Journal of Coastal Research*, Special Issue 36, p. 173–181.

Cubash U., Meehl G.A. *et al.* (2001) – 9. Projection of future climate change, p. 525–582. Third assessment report of the IPCC(2001).

Durand P. (1999) – L'évolution des plages de l'ouest du golfe du Lion au XXe siècle, PhD thesis, 1999, univ. Lumière Lyon 2.

EUROSION European programme (2004) – Vivre avec l'érosion côtière en Europe. Espaces et sédiments pour un développement durable. Conclusions de l'étude EUROSION. European Community Office of official publications.

Fairbank H.E., Jakeways J. 2007, Preparing for climate impacts at the coast: identifying patterns of risks and prioritizing a response (This volume).

Fish P, Moss J.L, Jakeways J, Fairbank H. (2007,) Response: applied earth science mapping for evaluation of climate change impacts on coastal hazards and risks accros the Eu. (This volume).

Folland C.K., Karl T.R *et al.* (2001) – 2. Observed climate variability and change, p. 99–181. Third assessment report of the IPCC(2001).

Hosking A., McInnes R. (2002) – Preparing for the impact of climate change on the Central South Coast of England: A framework for future risk management. *Journal of Coastal Research*, Special Issue 36, p. 381–389.

Hulme, M. and Carter T.R. (2000) – The changing climate of Europe" in Assessment of potential effects and adaptations for climate change in Europe: the Europe ACACIA project (M. L. Parry, ed.), 320 pp, The Jackson Environment Institute, University of East Anglia, Norwich, U.K.

Mallet C. *et al.* (2005) – Synthèse des études réalisées sur les instabilités de la côte basque entre 2001 et 2005. Rapport BRGM/RP-54012-FR, 32 p, 25 fig.

Mearns L.O., Hulme M. *et al.* (2001) – 13. Climate scenario development, p. 741–761. Third assessment report of the IPCC(2001).

Oliveros C., Lambert A. (2004) – Etude des phénomènes de submersion marine sur le littoral de la commune des Saintes Maries de la Mer. BRGM/RP-52902-FR, 116 p., 37 fig., 8 tabl., 1 pl. hors texte.

Parry M.P. (ed.) (2000) – Assessment of potential effects and adaptations for climate change in Europe: the Europe ACACIA project (M.L. Parry, ed.), 320 p., The Jackson Environment Institute, University of East Anglia, Norwich, U.K.

SMNLR (2003) – Orientations stratégiques pour la gestion de l'érosion en Languedoc-Roussillon.

Watson R.T. *et al.* (2001) – Changement climatique 2001. Rapport de synthèse, Résumé à l'intention des décideurs. Version française. Third assessment report of the IPCC (2001).

INTERNET REFERENCES

http://www.eurosion.org/ (geomorphology, defense works) http://www.ifremer.fr/anglais/ (coastal processes) http://www.shom.fr/ (bathymetry) http://www.onf.fr/ (defense works, dunes topography,hazard events) http://littoral.aquitaine.fr/ (hazard events,…) http:// www.ifen.fr/donIndic/Donnees/corine/produits.htm (assets distribution) http://siglittoral.3ct.com/) IPLI (Inventaire Permanent du Littoral (assets distribution) http://www.cru.uea.ac.uk/~timm/climate/ateam/TYN_CY_3_0.html (climate change european models) http://www.meteofrance.fr (climate change in France model)

Landslides and Climate Change – McInnes, Jakeways, Fairbank & Mathie (eds)

Erosive processes and related hazards in the coastal zone

M.-G. Angeli
I.R.P.I. CNR, Perugia, Italy

S. Castelli
Te.Ma. snc Faenza (RA), Italy

A. Galvani
Facoltà di Scienze Economiche, Università di Bologna, Bologna, Italy

P. Gasparetto
IQT Consulting, Rovigo, Italy

F. Marabini
ISMAR, Marine Geology Institute, C.N.R., Bologna, Italy

A. Mertzanis
Technological Educational Institution of Lamia, Greece

F. Pontoni
Geoequipe STA, Tolentino (MC), Italy

ABSTRACT: The aim of this work, considering the evolution of the Adriatic Coastal area during the last 50 years, is to show how it is possible to face and contrast the erosive phenomena at the beginning of the event. A late intervention is, of course, more expensive and cannot guarantee the final result. The selected area permits a consideration of different physiographic coastal units: the Venice lagoon; the Po River delta; the sandy coast with the Padan Plain behind; and the cliff from Cattolica to Ancona.

1 INTRODUCTION

Since the beginning of the XXth century, the world's coastal zones have been affected by a widespread shoreline regression that reached the critical stage after 1950. This situation is in contrast with the general trend of the coastal zone accretion during the XIXth century.

The intensity of human activity and the scale of settlements in coastal zones around the world, accounts for the rapid instability of the coastal environment.

Increased economic development, without regard for its future impacts, tends to worsen the already precarious situation even more.

The Italian coastal zone along the Adriatic Sea is a good example of the above mentioned state of the environmental destabilisation.

2 THE EVOLUTION OF THE ADRIATIC COASTAL ZONE

The upper Adriatic coast of Italy, from the Venice lagoon to the Ancona promontory provides an example of the above mentioned state of environment destabilisation.

Travelling from North to South, the thin Venetian coastal zone, with the lagoon behind is located; the Po river delta follows, where development is continuous toward the coast. From here to Cattolica a continuous sandy coast, with the Po plain behind.

From Cattolica to the promontory of Ancona, the coast is a narrow, sandy and gravely strip, which in some places lies at the foot of a cliff belonging to the Apennine Mountains, often very close to the sea.

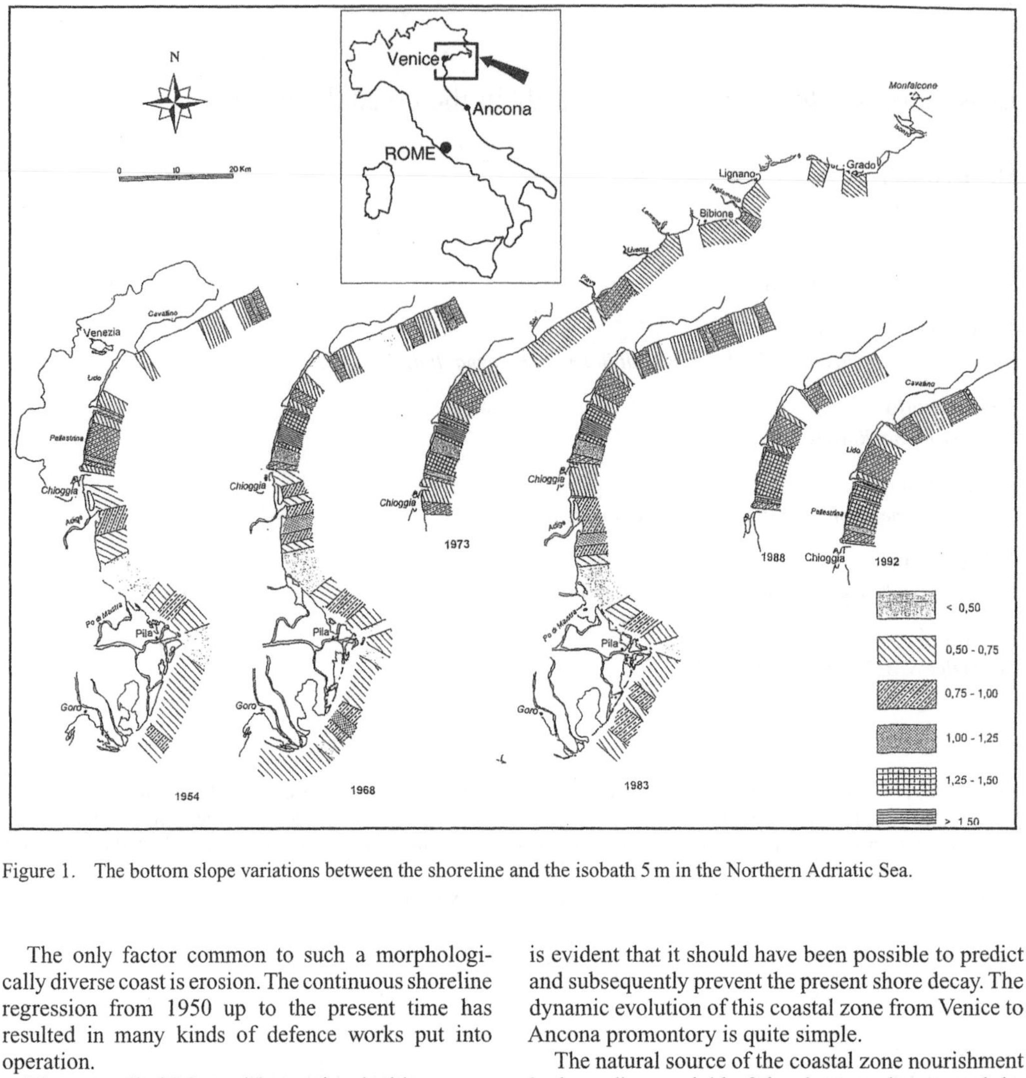

Figure 1. The bottom slope variations between the shoreline and the isobath 5 m in the Northern Adriatic Sea.

The only factor common to such a morphologically diverse coast is erosion. The continuous shoreline regression from 1950 up to the present time has resulted in many kinds of defence works put into operation.

The sea walls ("Murazzi") associated with groynes to protect the Venice lagoon; dykes to defend the lowland behind the shoreline and longard tubes in the Po River delta; breakwaters from the Po delta to Ancona promontory lie sporadically along coastal stretches mainly protected by groynes, or star-shaped concrete elements established on piles or by underwater barriers, constructed of synthetic sacks filled with sand and laid down in a cell, similar to a system where the cell is artificially replenished with sand.

All these protective works, constructed at different times and impelled by necessity, protect the coast without guaranteeing its future stability. Moreover, since they were built in the course of erosive process, their cost was astronomical.

If the evolutional trend of the whole coastal area from Venice to Ancona promontory is considered, it is evident that it should have been possible to predict and subsequently prevent the present shore decay. The dynamic evolution of this coastal zone from Venice to Ancona promontory is quite simple.

The natural source of the coastal zone nourishment is the sediment yield of the rivers to the sea and the material derived from the landslides in the cliffs. The long-shore current moves from south to north.

This natural system has completely changed during the last 50 years due to human activity and anthropomorphic climate change (alternation of cold/wet and warm/dry periods).

The final result is an insufficient sediment yield for the nourishment of the coastal zone and a prevailing of the destructive action of storm waves during the negative climatic conditions.

If the evolutional trend of the whole Adriatic coastal stretch is considered, it is possible to demonstrate some significant geomorphological parameters derived from the numerous previous studies made during the last 50 years. The increased bottom slope between the shoreline, and the 5 m isobaths (the wave

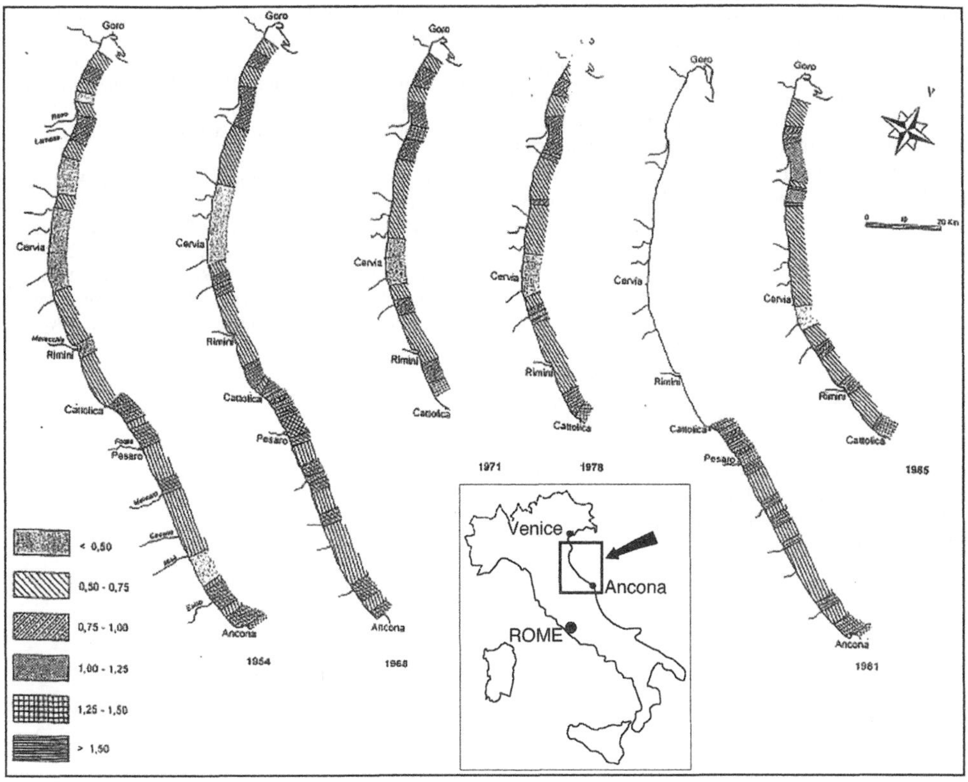

Figure 2. The bottom slope variations between the shoreline and the isobath 5 m in the Central Adriatic Sea.

breaker zone in the Adriatic Sea) is of particular interest to predict the shoreline regression before its happening (fig. 1, 2).

The comparison among the considered variation of the bottom slope in the Adriatic Sea shows that if the bottom slope is <0.50% the coast is stable; the interval 0.50–0.75% shows and unstable situation; the interval 0,75–1% shows a very dangerous situation for shoreline stability and with a bottom slope > 1% the storm waves attack and destroy the backshore with irreversible effect.

These parameters are valid, of course, only for the considered coastal area, depending on the local geomorphological situations, but the methodology can be applied elsewhere in the world's coastal zones.

It is quite evident, using this methodology; it is possible to undertake preventive action, following the evolutive prediction, to tackle erosive phenomena at its beginning.

3 CONCLUSIONS

It is quite evident, from the studies made, that it is possible to undertake preventive action to avoid erosive phenomena which, today, are very dangerous and very difficult to solve.

Not only in Italy, but in many other countries, coastal protection structures are built only after the setting in motion of the erosive process, and in many cases when it is irreversible.

A late intervention is, of course, more expensive and more difficult.

The comparison among changes of bathymetry and consequently shoreline regression shows that it would have been possible to predict the present situation many years ago and undertake appropriate protective action that could have improved the chances of maintaining the environmental equilibrium.

It is incomprehensible why this obvious principle is never used in the coastal environmental defence.

Based on past experience, it is recommended to establish a continuous and systematic survey of those environmental parameters capable of showing evolutional variation in coastal zones. Management in this way will avoid being taken by surprise when erosive processes develop.

It would be opportune that coastal town officials should take on this responsibility.

Briefly, it should be remembered that for a basic survey for possible future interventions, it is sufficient to measure very simple parameters such as those described previously, which should serve not to quantify the interventions but to indicate when the situation is developing into dangerous levels for the equilibrium of a coastal environment. The cost of such a service is relatively low, but is would save in the future, guaranteeing at the same time greater possibilities of success in protecting our coasts.

REFERENCES

Bartholin T.S., 1984. Dendrochronology in Sweden, in Morner N.A. and Karlen W. (eds.). *Climatic changes on a yearly to millennial basis. D. Reidel P.C.*, Dordrecht, pp. 261–262.

Brückner E., 1980. Klimaschwankungen seit 1700 nebest Bemerkungen über die Klimaschwankungen der Diluvialzeit, *"Geographische abbandlungen"*, B. IV. H. 2, Wien, pp. 153–184.

Cancelli A., Marabini F., Pellegrini M., Tonnetti G., 1984. Incidenza delle frane sull'evoluzione della costa adriatica da Pesaro a Vasto. *Mem. Soc. Geol. It. 27*, pp. 555–568.

Carbognin L., Marabini F., 1989. Evolutional trend of the Po river delta (Adiratic Sea, Italy). The 28th International Geological Congress, Washington D.C., U.S.A., July 9–19, Vol. I, proceeding, pp. 238–239.

Marabini F., 1985. Evolutional trend of the Adriatic Coast (Italy) – *Coastal zone 85, IV Symposium on coastal and ocean management, 30/7-2/8/85,* Baltimore U.S.A, Proceedings, Post conference Volume, pp. 428–439.

Marabíni F., Veggiani A., 199I. Evolutional trend of the coastal zone and influence of the climatic fluctuations. *Atti del C.O.S.U. II,* Long Beach, U.S.A., 2–4 April 1991.

Landslides and Climate Change – McInnes, Jakeways, Fairbank & Mathie (eds)
© 2007 Taylor & Francis Group, London, ISBN 978-0-415-44318-0

Lyme Regis phase II coast protection & slope stabilisation scheme, Dorset, UK – the influence of climate change on design

D.S. Fort, P. L. Martin & A.R. Clark
High-Point Rendel, London, UK

G.M. Davis
formerly West Dorset District Council, UK

ABSTRACT: Lyme Regis is situated on one of the most unstable and actively eroding stretches of coastline in the UK. Over the centuries, various coast defence structures have been constructed to protect the town against attack from the sea and to provide increased stability. These structures have been under constant threat due to storm attack, foreshore lowering and landslide activity. Large parts of the developed town have been constructed on pre-existing coastal landslip systems, many of which are still undergoing movement, which over time causes damage to properties and infrastructure due to the cumulative effects of the ground movement. Occasionally rapid landslide events take place which may be a threat to public safety and destroy property and infrastructure assets in a relatively short period. An appreciation of climate change is fundamental to dealing with coastal management issues and for the development of acceptable engineering solutions, which meet the requirements of sustainability in terms of engineering, environment and economics. This paper describes the historical climate setting of Lyme Regis, considers future trends, and describes some of the implications of climate change in understanding the risks to the town and how these have been addressed in the design and construction of the new coast protection and slope stabilisation works.

1 INTRODUCTION

Lyme Regis on the West Dorset coast, a gateway town to the Dorset Jurassic Coast World Heritage Site, is situated on one of the most unstable and actively eroding stretches of coastline in the UK (Figs 1 & 2). Over the centuries, various coast defence structures, including sea walls and the famous 13thC Cobb, have been constructed to protect the town against attack from the sea and provide increased stability for the town. These structures are under constant threat due to storm attack, foreshore lowering and landslide activity.

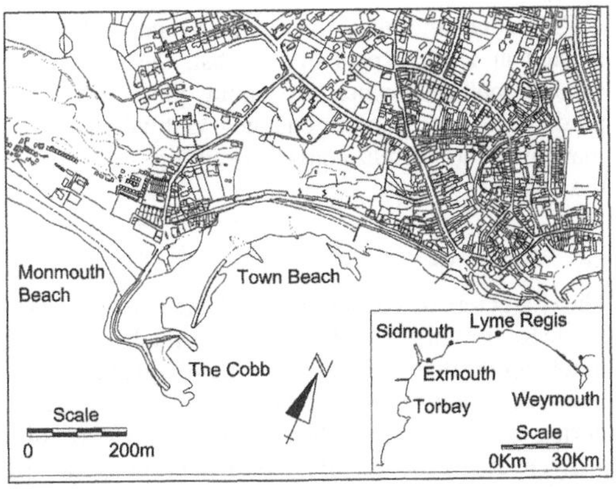

Figure 1. Location plan.

Figure 2. Lyme Regis from the south-west.

Figure 3. Example of effects of landsliding.

As well as being located on an historically actively eroding stretch of coastline, large parts of the town have been constructed on pre-existing coastal landslip systems at several different levels within the town, extending up to 1 km inland from the coast. Landslide activity promoted by periods of cold or wet climatic conditions in the past has left a legacy of relict landslide systems which are susceptible to reactivation at present and in the future. Many of these landslip systems are still undergoing slow movement, which over time causes damage to properties and infrastructure due to the cumulative effects of the ground movement (Fig 3). From time to time, rapid landslide events take place which may destroy assets including houses, roads and sea walls in a relatively short period and pose a threat to public safety. Not surprisingly, ongoing instability and the constant risk of damage to the town are one of the principal concerns in managing this part of the coast.

The town is underlain by Lower Jurassic clays and mudstones, which are particularly susceptible to landsliding (Table 1), (Fort et al 2000, Sellwood et al 2000). The position of the principal landslide shear surfaces is controlled by beds of strong limestone at different levels within the weaker clays and mudstones. The Cretaceous Upper Greensand, which underlies the higher parts of the town, is also prone to landslide activity.

The principal triggers of today's and future landslide reactivation on the coastal slopes are marine erosion and sea level rise at the toe together with high groundwater levels resulting from rainfall. A third cause of instability has been due to inappropriate construction and development within the town.

Current trends of climate change in Southern England, including increasing rainfall, are predicted to continue and this has serious implications for the future of the town. The only realistic management options for protecting the town against instability in the long term involve engineering schemes on a large scale, the design criteria for which must allow for possible future climate changes.

Over the centuries, and particularly since the mid 18th century, attempts have been made to control marine erosion and landslipping through the construction of sea walls, foreshore structures, retaining walls and drainage systems. The result is a patchwork of structures of various dates, types and condition, many of which are close to the end of their useful life.

Such are the threats to the future of the town, the local authority, West Dorset District Council, supported by their consultants, has implemented a long-term programme of engineering works in order to improve stability and to reduce the risk of landslide disasters occurring in the future.

2 PAST CLIMATE CHANGE

2.1 *Introduction*

Climate is inherently variable. Meteorological instrument records for the last two centuries indicate distinct changes in temperature and precipitation. For example, there was a relatively cool, dry period in Britain in the 1940's, whilst four out of the five warmest years on record in central England occurred in the 1990's (SCOPAC 2001).

Determining climate changes from before the advent of reliable instrument records is more difficult. However, there are many lines of evidence (for example ice core data, tree ring sequences, and historical records) which indicate that the climate was constantly fluctuating. Prior to around 12,000 years BP, during the last (Quaternary) Ice Age, there were huge and rapid changes in climate, and Northern Britain was periodically covered by ice sheets. More recently, in Roman times, the weather in Britain is thought to have been a little cooler than it is today. A relatively stable period of warm, settled weather between about 800 and 1300AD, know as the Mediaeval Warm Period, gave way to a cooler, wetter and more unsettled climate in the period 1300 to 1850, known as the Little Ice Age (Fagan, 2000).

An understanding of the climatic conditions which existed at Lyme Regis in the past, the changes in weather patterns which are taking place today and predicted future climate changes, is fundamental to managing the current coastal instability problems in the town.

2.2 *The ice age*

The characteristics of some of the landslide deposits at Lyme Regis indicate that they were formed in a climate much colder than at present, in the periglacial conditions which existed periodically in Southern England during the Pleistocene. The products of periglaciation at Lyme Regis include areas of ground weakened by frost action to over 8 m below ground level, sheets of solifluction debris produced by freeze-thaw action and

Table 1. Summary of strata and stratigraphy at Lyme Regis.

Period	Formation with Typical description *(Limestone marker strata shown in italics)*	Approximate thickness (m)
QUATERNARY/RECENT	**Superficial and landslide deposits** derived from Cretaceous and Lower Lias, includes alluvial deposits, beach deposits, talus, Landslide Deposits etc. For the study Landslide Deposits classified as either: Mudslide Deposits, Landslide Blocks, Disturbed Lias, Debris.	Variable
LOWER CRETACEOUS	**Upper Greensand** – Dense yellow brown fine silty sand with chert beds in upper 10 m	18
Unconformity	**Gault** – Dense greyish green silty sand	1
JURASSIC (LOWER LIAS)	**Black Ven Marls** – Stiff to hard, dark grey clay and very weak to moderately strong laminated mudstone, with occasional beds of limestone and thin layers of 'beef' (fibrous calcite) *– Stellaris Nodule Bed (25–29 m above base of BVM)* *– Upper Cement Bed (16–20 m above base of BVM)* *– Lower Cement Bed (10–13 m above base of BVM)*	35
	Shales with Beef – Stiff to hard grey clay and very weak to moderately weak, dark grey laminated mudstone, with thin layers of 'beef' and occasional beds of limestone or stronger calcareous mudstone. *– Birchi Bed (Top of Shales with Beef)* *– Mid Shales with Beef (12–14 m above Grey Ledge)* *– Table ledge (5–7 m above Grey Ledge)* *– Fish Bed (1–2 m above Grey Ledge)*	25
	Blue Lias – Interbedded sequence of mudstones, limestones and siltstones. Limestones – thin/medium bedded, light grey and argillaceous. Generally moderately to extremely strong. Mudstones – grey/dark grey thinly laminated to very thinly bedded, locally calcareous and weak to moderately weak. Also occurs as stiff/very stiff fissured clays. Siltstone – very weak to strong, grey, calcareous. *– Grey ledge (Top of Blue Lias)* *– Glass Bottle (0.5–1 m below top of Blue Lias)* *– Top Quick Ledge (1.5–2 m below top of Blue Lias)* *– Venty Bed (2.5 m below top of Blue Lias)* *– Best Bed (5 m below top of Blue Lias)*	19

landslides and other forms of mass movement deposits formed as a result of softening of clay in thawing ice-rich ground (Hutchinson & Hight 1987). Essentially, the effects of frozen ground and associated high groundwater pressures during the Ice Age have left a legacy of ancient mass movement deposits, disturbed ground and shear surfaces which are prone to reactivation due to the triggering mechanisms that cause present day landsliding.

2.3 *Postglacial warming*

As the climate improved during the postglacial period, the landslide systems eventually reached a stage when there was a reduction in activity and the slopes became more stable and vegetated. However, the global warming that took place also led to a rise in sea level. As the sea level rose, marine erosion created new sea cliffs in the landslide slopes and led to the progressive reactivation of the relict landslide systems, which continues to the present day. In other words, postglacial warming is the fundamental driver of the current coastal instability problems.

2.4 *The little ice age*

Large landslide systems are present on the coastal slope behind the main frontage at Lyme Regis and extend about 250 m inland from the current high water mark (Fig 4). Radiocarbon dating of vegetation buried beneath the landslide deposits in this area gives calibrated dates in the range 1400AD to 1700AD. This falls within the period known as the Little Ice Age, when the climate was cooler and wetter than it is today. This has serious implications for the future, as it suggests that a small change in climate, such as that which is taking place at present, could reactivate widespread, destructive landsliding.

Figure 4. Aerial view of the slope behind the main frontage at Lyme Regis. Most of the development in this photograph is located on landslide. The wooded areas mark the backscars of landslides which were active during the Little Ice Age.

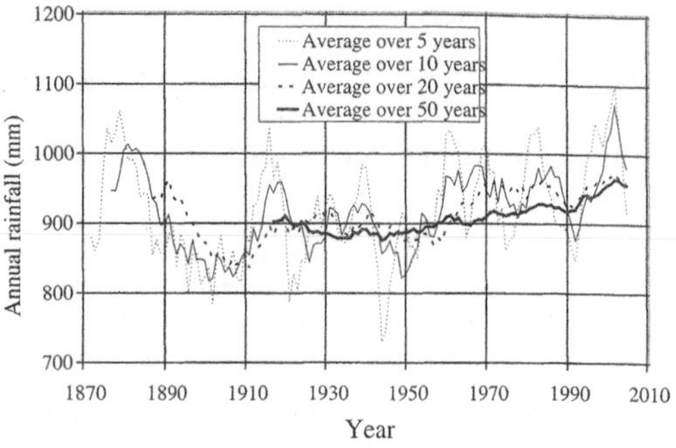

Figure 5. Average annual rainfall at Lyme Regis 1873–2005, for various antecedent periods. Note the dry spell in the 1940's and the trend of increasing rainfall since the 1950's.

3 CURRENT AND FUTURE CLIMATE CHANGE

There is little doubt that Britain is currently undergoing a period of significant climate change. In addition to the warming trend, the weather is also becoming wetter. At Lyme Regis every one of the consecutive years 1993 to 2002 was wetter than the long-term average, and this has coincided with serious landslide activity, particularly in the years 2000 to 2002 (Fig 5). Studies such as those carried out at the Hadley Centre (SCOPAC 2001) predict that in Southern England:

- It will continue to get warmer.
- Rainfall will continue to increase in winter, but may decrease in the summer, with an overall increase in both total and effective rainfall.
- Sea level will continue to rise.
- There is the possibility of an increase in the number of extreme rainfall and storm events.

Essentially, these predictions would appear to be continuations of current trends but with expected increased rates of change.

4 UNDERSTANDING THE COASTAL INSTABILITY PROBLEMS

4.1 *Instability studies*

In the development of coastal instability management options at Lyme Regis one of the principal initial objectives was to understand the nature of the instability problem itself. To achieve this a series of multidisciplinary studies have been carried out involving a considerable investment in acquiring and interpreting data on geology, ground stability, coastal erosion and other issues that may have an influence on the implementation of stabilisation schemes (Clark et al 2000, Cole et al 2002). Elements of these studies have included:

- Geological and geomorphological mapping.
- Ground investigation.
- The monitoring of slope movements and groundwater levels.
- Interpretation of the instability affecting the town in terms of the evolution of the coast, both in the long-term and short-term.
- Prediction of future landslide behaviour.

The studies have given some insights into how climate change has influenced the evolution of coastal landsliding in the past, and how it may continue to do so in the future. One of the most important aspects of the weather in terms of landslide activity is rainfall. The landslides at Lyme Regis are sensitive to changes in groundwater levels, such that high groundwater levels due to heavy rainfall and infiltration often trigger landslide activity. Temperature is also relevant as it has an influence on potential evapotranspiration and hence the amount of rainfall which can actually recharge the groundwater, with lower temperatures allowing more of the rainfall to infiltrate into the ground.

4.2 *Monitoring*

Extensive monitoring of ground movement and groundwater has been undertaken at Lyme Regis since

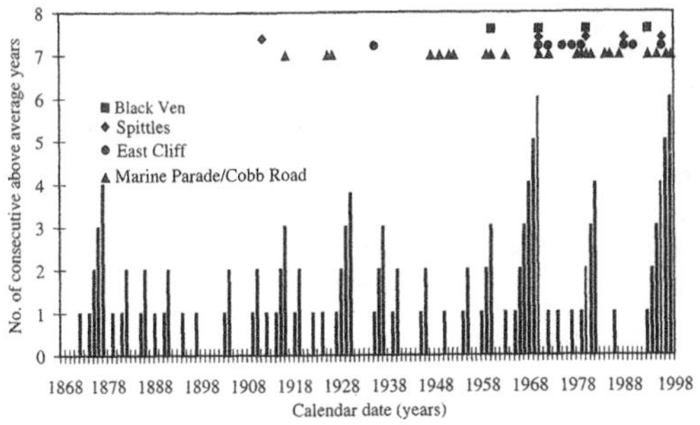

Figure 6. The number of years with above average effective rainfall for 1868–1998 and the incidence of landslides.

the mid 1990's as part of the design studies (Fort et al 2000, Davis et al 2002), whilst daily rainfall records are available for the area since 1868. Historically, the known pattern of landslide occurrence at Lyme Regis and adjacent areas closely reflects the frequency of years with above average effective rainfall (Fig 6), (High-Point Rendel 2000, Brunsden 2002).

The adopted monitoring strategy has been set up to provide quantitative data on the location, rate, magnitude and distribution of ground movements for the various landslide systems and groundwater data across the site. In addition to providing an understanding of the complex multilayered groundwater regime and landslide models, another objective of the geotechnical monitoring was to establish the interrelationship of ground movement with piezometric levels and rainfall, with the ultimate aim being the forecasting of landslide events based on antecedent rainfall and groundwater levels.

Over 240 individual monitoring points have been in operation since 1997 comprising; ground surface movement monitoring (GPS/total station monitoring of ground markers), sub surface movement monitoring (borehole inclinometers and slip indicators), movement monitoring of structures, groundwater monitoring (standpipe piezometers), rainfall and drain flows. The collection and interpretation of this data has been used in the management of the instability of the town including the development of stabilisation measures described below.

4.3 *Landslide recession scenarios*

Following the determination of landslide mechanisms and their triggering events such as rainfall, increased groundwater levels, coastal erosion etc at Lyme Regis, there becomes a need to be able to assess the annual likelihood of a landslide event (i.e. landslide reactivation) occurring, and how this likelihood may change through time including the affect of climate change. To undertake this a structured use of expert judgement and subjective probability assessment using an event tree method was used to assess risks at Lyme Regis, (Lee et al 2000). This approach requires the identification of a range of landslide reactivation scenarios, each involving an interrelated sequence of events driven by an 'initiating event' (e.g. high groundwaterlevels or seawall failure), and 'propagating conditions' (e.g. progressive loss of toe support). In this approach the annual probability of wet years/high groundwater levels was estimated from historical trends of annual effective rainfall. At Lyme Regis there have been 8 "wet year" sequences in 130 years, indicating an annual probability of about 1 in 16 (0.06), with duration of 3–6 years. The frequency of these sequences appears to have increased over the last 3 decades, suggesting a current annual probability of around 1 in 10 (0.1). Since the late 1950s these wet year sequences have broadly coincided with landslide activity years on the coastal slopes at Lyme Regis and adjacent areas.

5 THE PHASE II ENGINEERING SCHEME

In the 'do-nothing' coastal management scenario, the future of Lyme Regis would appear bleak. There is no reason why the historical prevalence of landsliding should not continue, and the predicted continuation of the current climate changes, together with the deterioration of the existing coast protection systems, suggests that the situation could become substantially worse in the relatively near future. The level of threat is such that the only realistic management option to secure the future of the town in the medium to long term is the implementation of coast protection and slope stabilisation works on a large scale.

Such works include several phases of foreshore protection and slope stabilisation measures in order to deal with the relatively complex landslide systems. The main initiative to improve the stability of the town is the "Phase II Lyme Regis Environmental Improvement Scheme", which commenced in March 2005 with completion due in Spring 2007 with an estimated final total cost of around £25 million. This scheme (Fig 7) includes:

- Beach replenishment – comprising 114,000 tonnes of imported sand and shingle beach material.
- A new 250 m long sea wall incorporating an extended access road/promenade.
- New foreshore structures to maintain the beach comprising two new masonry faced jetties of 57 m and 76 m in length, and two realigned and extended rock armour breakwaters.

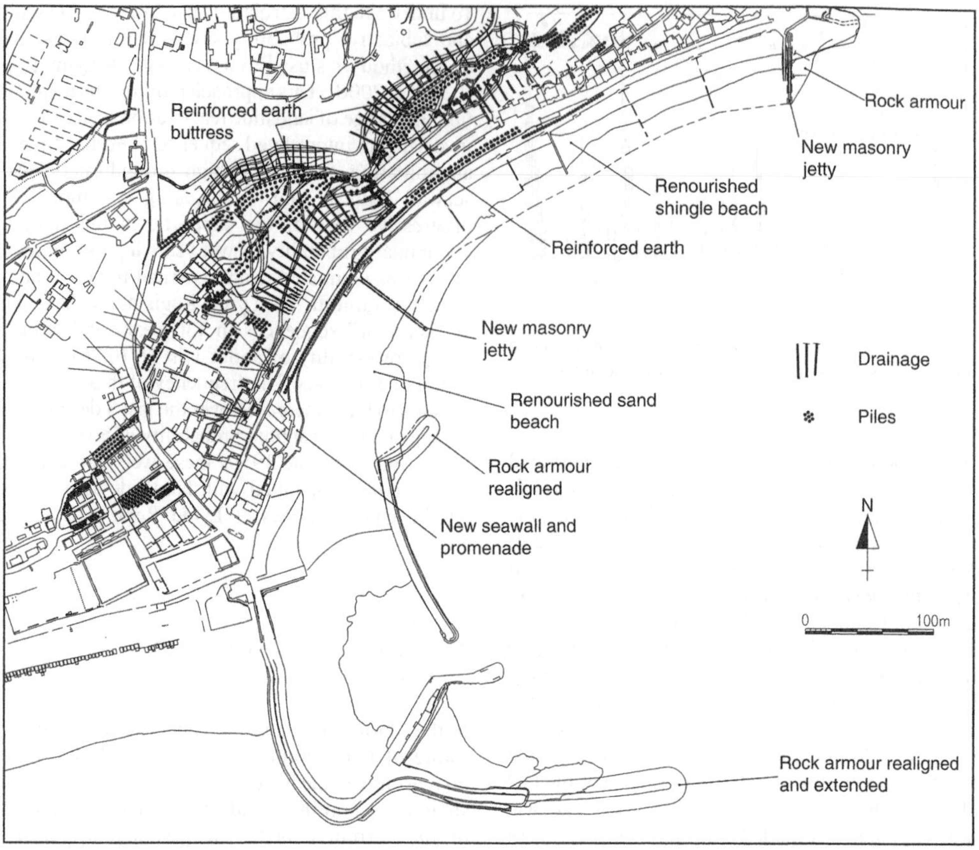

Figure 7. Phase II coast protection & slope stabilisation works.

- Slope strengthening – including the installation of 1,300 bored piles to act as dowels, and soil nailing.
- Extensive drainage improvements of different types; including trench and counterfort drains, cut-off drains, sub-horizontal drilled drains, improved highway drainage.
- Local earthworks, slope regrading, reinforced soil, soil buttresses and retaining structures.
- Hard and soft landscape works.

The design of the works needed to be relatively robust in order to allow for the predicted changes in climate and sea level rise which may take place over the design life of the scheme. As large and complex engineering schemes of this type generally take several years to implement, there was a risk that destructive landsliding could occur before the works are complete. In order to reduce this risk, a series of smaller-scale advanced stabilisation works, involving piling and drainage, were carried out in 2004 in the most critical areas.

6 SLOPE STABILISATION WORKS – DESIGN APPROACH, ASSUMPTIONS AND CRITERIA

6.1 *Design approach*

The detailed design of the slope stabilisation works has been carried out with the aim of ensuring that damage to the coastal defences and associated assets does not occur due to landslide movement over a period in excess of 50 years. The design has endeavoured to take into account all foreseeable events that may adversely affect the stability of the landslide systems. These effects include those that may occur during the construction of the stabilisation works and those that may occur at any time thereafter during the design life of the stabilisation measures. Such events are listed below:

During Construction:

- Small cuts and fills associated with making access for the construction plant.
- Live loads from the construction plant.
- Excavation of trench drains and counterforts.

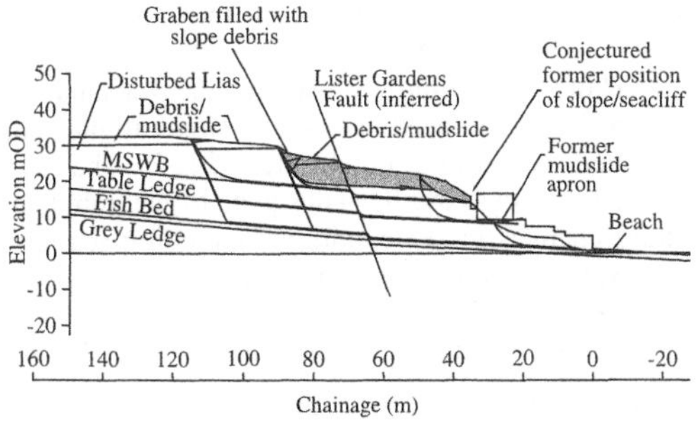

Figure 8. Example Ground Model showing strata controlled block sliding on geological marker horizons.

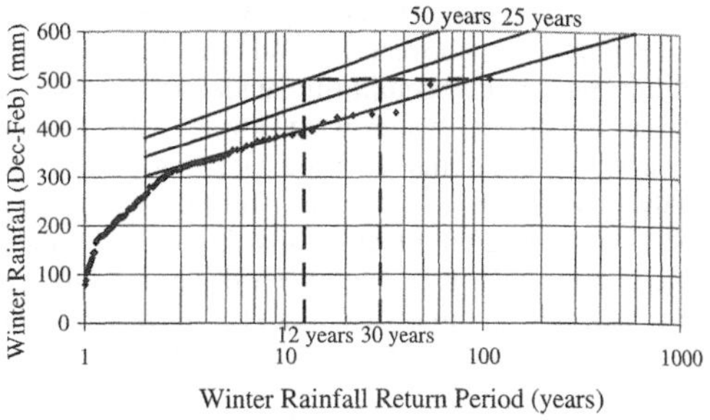

Figure 9. Winter rainfall against return period illustrating how the frequency of events might be influenced by climate change.

- Temporary stockpiling of arisings and imported materials.
- Construction of buttress support to the upper slopes.
- Reprofiling for landscaping purposes.

After Construction:

- Changes in groundwater levels/porewater pressures due to seasonal effects and the effect of climate changes e.g. increase in rainfall.
- Instability of the existing steep coastal slopes resulting in loading of the head of the landslides.
- Decline in the long term efficiency of drains installed as part of the stabilisation works.
- Live loads from vehicular traffic.

Where possible the designs are such as to minimise the need for long term maintenance and monitoring.

6.2 *Design assumptions*

The following assumptions relating to climate change were made in the development of the design:

The landslides in the form of block slides on or above the Mid Shales with Beef (MSWB) limestone bed (Fig 8) in the site area are particularly active and the landslide monitoring has indicated that the principal landslide movements occur in response to seasonal rises in water level. It is therefore envisaged that climate change during the 50-year design life of the scheme will have an adverse effect on landslide stability. Figure 9 presents a plot of winter rainfall against return period. The figure shows the rainfall totals that may be expected to be equalled or exceeded, on average, for particular recurrence intervals and reveals an almost linear relationship for return periods above 2. A crude estimate of the impact of global warming can be gained by increasing the winter water rainfall by 12.5% over the next 25 years and 25% over the next 50 years. Assuming that the pattern of rainfall events remain similar, it is possible that what is now a 1 in 100 year rainfall total (around 520 mm) could become the equivalent of a 1 in 30 year rainfall total in 25 years (annual probability = 0.03) and a 1 in 12 year rainfall total in 50 years (annual probability = 0.08). This suggests that there could be a significant increase in the frequency of wet years (a possible 8 times increase) and, hence, potential landslide triggering events over the next 50 years.

In light of the above it was decided that the design of the stabilisation measures should take into account the potential for groundwater level to rise. It is not considered possible to predict accurately the potential rise. However, it can be assumed with some confidence that water level will not rise higher than ground level, i.e. artesian conditions are unlikely to occur. The design has therefore addressed a worst credible case of groundwater at ground level with hydrostatic water pressures on the active shear surfaces, e.g. MSWB. With regard to potential failures on deeper limestone beds e.g. Table Ledge or Fish Bed, it was not considered credible to consider groundwater rising to ground level. This is because of the effect of underdrainage provided by the more permeable limestone strata and saturation of the relatively impermeable mudstones at higher level which will result in greater run off rather than greatly increased piezometric levels at the potential failure surfaces.

It is also assumed that routine maintenance of the drainage systems will be carried out to a reasonable standard as set out in a site-specific geotechnical slope management, monitoring and maintenance manual.

6.3 *Design criteria*

In all parts of the site the principal hazard to the coastal defences is ongoing block slide movement or activation/reactivation of block slide movement and shallow mudsliding on the frontal slopes (Fig 8). Monitoring has indicated that landslide movement generally occurs in response to seasonal rises in groundwater level.

It has been assumed in the design that both the active block slides on or above the MSWB and the mudslides are currently at limit equilibrium under pessimistically assessed "current" winter water levels, with residual shear strength mobilised on the shear surfaces. Back analyses were carried out to determine the operational shear strengths giving factors of safety close to unity. Stabilisation measures have been designed with the aim of achieving target "Factors of Improvement" on the assumed limit equilibrium value, namely:

- a Factor of Improvement of 1.25 assuming "current" winter water levels i.e. not including any benefit from drainage measures;
- a Factor of Improvement of 1.05 assuming the worst credible groundwater conditions that may occur during the design life of the scheme, allowing for drawdown where sustainable drainage has been installed.

In addition, a Factor of Improvement of at least 1.1 prior to the start of counterfort construction and buttress construction assuming pessimistically assessed "current" winter water levels has generally been applied.

The above criteria were applied to most of the site where the active block slide movements were on or above MSWB. For other parts of the site where it is considered necessary to provide measures to safeguard the coastal defences from potential block slide movement on Table Ledge or Fish Bed, or local rotational failure where there is little or no evidence for active landsliding under current conditions, a different approach has been adopted. At these locations the approach has been to determine the groundwater conditions that would bring the postulated block slides or rotational failures to a state of limit equilibrium using residual strength parameters, and to then determine the stabilisation forces required to achieve a factor of improvement. For these other areas the aimed for Factor of Improvement for potential failures on the limestone marker horizons is 1.05.

In addition to the Factor of Improvement criteria, consideration has been given in the design to the ground movements that may occur as the stabilisation measures develop their resistance to counter the landsliding in order to limit post construction deformations.

7 FORESHORE WORKS – DESIGN APPROACH, ASSUMPTIONS AND CRITERIA

A need was also identified to improve the coastal defences to arrest their deterioration and to reduce foreshore lowering and reactivation of the landslide systems. The selected foreshore works included beach replenishment, comprising 114,000 tonnes of imported sand and shingle beach material, the construction of a new 250 m long sea wall and foreshore beach holding structures to maintain the beach including two masonry faced jetties, and two realigned and extended rock armour breakwaters (Fig 7). The coast protection works have been designed to accommodate future changes in sea conditions including sea level rise, wave conditions and storm surges etc.

In developing the foreshore works extensive investigations and studies have been carried out including:

- Bathymetric and seabed surveys
- Foreshore investigation and mapping of geology and rock ledges
- Seawall structures condition surveys
- Numerical and physical (1:50 scale) hydraulic modelling by HRW – 2001/03
- Ecological studies – 2001/04
- Environmental Impact Assessments – 2001/03
- Landscape design of jetties & seawalls – 2003/04
- Identification of suitable sand, shingle and armourstone sources
- Public consultation

Sea level rise and an increase in the frequency of extreme high water levels as a result of combinations of high tides and storms need to be considered in the design of coast protection measures. As yet, British tidal gauge records show no clear evidence of acceleration in the rate of sea level rise. Allowances for SW England given by DEFRA for the design of coastal defences with an effective life beyond 2030 was 5mm per year. However the advice has recently been updated by a DEFRA supplementary note to operating authorities on climate change (DEFRA 2006). For SW England the revised net sea level rise for 1990–2025 is 3.5 mm/year, for 2025–2055 8 mm/year, for 2055–2085 11.5 mm/year and for 2085–2115 14.5 mm/year. This now reflects an exponential rise in sea level and replaces the previous straight line graph representation of 5mm/year. The same advice suggests indicative sensitivity ranges for extreme wave height of +5% for 1990–2025 and +10% for 2055–2115. On the south west coast (Devonport) it has been estimated that, with an effective sea-level rise of 0.29 m, the current 100 year water level (annual probability = 0.01) would, by 2050, be the equivalent of a 3 year return period water level (annual probability = 0.33) (SCOPAC 2001).

The design criteria for Lyme Regis used in the studies and modelling are listed here:

- wave and water level conditions to forecast beach response were tested for joint return periods of 1 in 10, 1 in 50 and 1 in 200 years,
- both extreme storm and swell wave condition were tested,
- offshore wave directions were modelled from the SW, S and SE,

Table 2. Some climatic episodes and their implications for coastal management at Lyme Regis.

Climatic episode	Effects	Implications
Cold periods during the Quaternary	Periglacial conditions in Southern England. Widespread landsliding and mass movement, disturbance and weakening of the ground.	Existence of large areas of landslide deposits prone to rapid reactivation. Design of stabilisation works needs to take account of existing shear surfaces and disturbed ground.
Postglacial warming	Sea level rise leading to formation of new sea cliffs and reactivation of relict landslides.	Sea level rise and the associated marine erosion is the fundamental driver of the current coastal instability problems.
Little Ice Age	Cool and wet weather promoted widespread development of coastal landslides.	The presence of pre-existing landslides prone to reactivation. Suggests that a relatively small change in climate in the future could lead to the re-establishment of widespread destructive landslide activity.
Present pattern of warming and increasing rainfall.	Continuing sea level rise and marine erosion. Higher ground water levels and increasing frequency of landslide events.	Urgent need for the implementation of coast protection and slope stabilisation works to protect Lyme Regis. Need for a landslide warning system and local civil emergencies plan.
Predicted further warming and increasing rainfall.	Further sea level rise, higher groundwater levels and increasing landslide activity.	Coast protection and slope stabilisation works need to be designed to withstand future deterioration in conditions. Landslide warnings may need to be issued more frequently.

- still water levels varied between 1.95 m and 3.52 m OD,
- offshore significant wave heights varied between 2.47 m and 5.98 m,
- an extreme event with a return period of 1:2000 was used to check armour stability,
- for the design life of 60 years used for the forshore works, a sea level rise of 300 mm (5 mm/year) was taken into account. (This compares to the revised DEFRA criteria of 425 mm for 2005–2065).

8 LANDSLIDE WARNING SYSTEM

The District Council operates a three tier landslide warning system, similar to the Environment Agency's flood warning system, to alert the public and emergency services if it is considered from visual observation or instrument readings that a landslide could be about to take place (Cole & Davis 2002). A local civil emergencies plan has been established in order to deal with the effects of a destructive landslide should one occur.

The landslide warning system was put into effect during the wet winters of 2000–2001 and 2002–2003. Fortunately, although large landslides occurred on either side of Lyme Regis during these times, no properties were lost within the town itself. However, if the present pattern of increasing rainfall continues the issuing of landslide warnings may become a more frequent occurrence in the future.

9 CONCLUSION

An appreciation of ongoing climate change is essential in gaining an understanding of the coastal instability problems at Lyme Regis and developing engineering solutions in order to secure the long term future of the town. A summary of some of the implications of climate change is given in Table 2.

ACKNOWLEDGEMENTS

The authors of this paper have received much assistance in understanding the implications of climate change on the adopted solutions from consultants and individual experts. The particular contributions of the following are also gratefully acknowledged: Professor Denys Brunsden, Professor Eddie Bromhead, Tony Bracegirdle, Dr Mark Lee, the team from High-Point Rendel especially Hugo Wood and Colin Reed, and West Dorset District Council, including Nick Baker for the preparation of the figures.

REFERENCES

Brunsden D. 2002. The Fifth Glossop lecture – Geomorphological roulette for engineers and planners: some insights into an old game. *Quarterly Journal of Engineering Geology and Hydrogeology*, 34.

Clark A.R., Fort D.S. and Davis G.M. 2000. The strategy, Management and Investigation of Coastal Landslides at Lyme Regis, Dorset. In Landslides in research, theory and practice, *Proc.8th Int. Symp. of Landslides*, Cardiff. Thomas Telford. London. 279–286.

Cole K, Davis G.M, Clark A.R. and Fort D.S. 2002. Managing coastal instability – a holistic approach. *Instability – Planning and Management*, Thomas Telford. London. 679–686.

Cole K and Davis G.M. 2002. Landslide warning and emergency planning systems in West Dorset, England. *Instability - Planning and Management*, Thomas Telford. London. 463–470.

DEFRA 2006. FCDPAG3 Economic Appraisal – Supplementary Note to Operating Authorities – Climate Change Impacts, October 2006.

Davis G.M., Fort D.S. and Tapply R.J. 2002. Handling the data from a large slope monitoring network at Lyme Regis,

Dorset, U.K. *Int. Conf. Instability, Planning and Managment*, Thomas Telford. London. 471–478.

Fagan B. 2000. *The Little Ice Age: how climate made history, 1300–1850*. New York – Basic Books.

Fort D.S., Clark A.R., Savage D.T. and Davis G.M. 2000. Instrumentation and monitoring of the coastal landslides at Lyme Regis, Dorset, U.K. In: *Landslides in research, theory and practice*, Proc.8th Int. Symp. of Landslides, Cardiff. pp Thomas Telford. London. 573–578.

High-Point Rendel. 2000. Landslide Recession Scenarios report, Ref R/H439/1/1 dated July 2000. HPR report prepared for WDDC.

Hutchinson J.N. and Hight D.W. 1987. Strongly folded structures associated with permafrost degradation and solifluction at Lyme Regis, Dorset. In: *Periglacial processes and landforms in Britain and Ireland*, JE Boardman (Ed) 245–256.

Lee E.M., Brunsden D., and Sellwood M. 2000. Quantitative Risk assessment of coastal landslide problems, Lyme Regis, UK. In: *Landslides in research, theory and practice*, Proc.8th Int. Symp. of Landslides, Cardiff.. Thomas Telford. London. 899–904.

Lee E.M. and Clark, A.R. 2002. *Investigation and management of soft rock cliffs*. Thomas Telford. London.

SCOPAC. 2001. Preparing for the impacts of climate change. Report prepared by Halcrow Group Limited, University of Portsmouth, University of Newcastle and UK Meteorological Office.

Sellwood M., Davis G.M., Brunsden D and Moore R. 2000. Ground models for the coastal Landslides at Lyme Regis, Dorset, U.K. In: *Landslides in research, theory and practice*, Proc.8th Int. Symp. of Landslides, Cardiff. Thomas Telford. London. 1361–1366.

Landslides and Climate Change – McInnes, Jakeways, Fairbank & Mathie (eds)
© 2007 Taylor & Francis Group, London, ISBN 978-0-415-44318-0

Downdrift erosion and the frequency of coastal landsliding

S. Brown & M.E. Barton
School of Civil Engineering and the Environment, University of Southampton, UK

ABSTRACT: Increased rates of landsliding and coastal recession ensue on the downdrift coast when a formerly ample supply of beach material is impeded by a barrier such as a groyne field, strongpoint or jetty. While the phenomenon is well known as the terminal groyne effect, the actual mechanisms associated with this can be complex where there is a long history of piecemeal coastal protection and the original littoral drift supply was variable. Rising sea levels will force selective protection for vital infrastructures with retreat allowed elsewhere, thus downdrift erosion will become of increasing concern. Research studies examining the forms and rates in which it occurs are being undertaken at Southampton University as a timely contribution for future coastal zone planning.

1 LANDSLIDING IN CLIFFS

With active toe erosion, a cliff is maintained in a constant state of potential instability. Recession of the cliff takes place in accordance with the rate of toe erosion and the hardness of the cliff material. A good beach and a continuing supply of beach material provides protection against most sea conditions but not against storms and exceptional tides. The combination of an inadequate beach and soft rocks provide conditions for frequent degradation by landsliding and rapid recession of the cliff line.

2 DOWNDRIFT EROSION

The presence of a barrier to littoral drift, such a groyne field or jetty protecting a harbour, depletes the beach on the downdrift side. Numerous examples of downdrift erosion due to the introduction of a barrier are recorded by Komar 1983, Carter 1988, French 2001 and others. A classic example of a barrier is seen in Fig. 1 showing the long groyne constructed at Hengistbury Head during the period 1937–39. The photograph dates from 1955 and shows an extensive sand bank accumulated on the updrift side and depletion on the downdrift (eastward) side. The resulting increased rate of erosion and cliff recession of the cliffs downdrift eventually stimulated the introduction of additional groynes and rock armouring to slow down the threat to the eastern end of the headland (Rose 1986; Parker and Thompson 1988). Despite its length and the clearly visible effect, the long groyne is not a total barrier: Cooper et al (2001) point out that sediment in suspension can move around it and under storm conditions overtopping occurs leading to sediment accumulation against the recently constructed rock groynes to the north east.

3 SET-BACK

The juxtaposition of shoreline protection and an adjacent unprotected shore will show up as a set-back in the line of the unprotected shore. Where increased erosion has resulted from the interruption of littoral drift and consequent beach depletion, then the rate of recession downdrift should show up as an increase in the recession rate. This can be seen in the central part of Christchurch Bay where the rock armoured groyne field at Barton-on-Sea starves beach material from the unprotected Becton coast to the east (Fig. 2). The rate of recession from 1869 (the date of the earliest large scale Ordnance Survey map of the area) is shown in Fig 3 (from Brown 2005). Marine erosion owing to natural beach depletion in the area increased during the 1930s (Stopher and Wise 1966) with the period from 1931 to 1967 showing an average cliff top recession rate of 1 m/year. The protection works at Barton-on-Sea went through a long series of modifications with rock armouring being introduced from 1968 onwards. Over the period from 1967 to 2001, recession curves show a marked increase to an average of 1.4 m/year.

4 MECHANISM OF CLIFF EROSION

Fig 4 shows a view looking east from the rock armoured extension of the terminal groyne at Barton-on-Sea towards the unprotected Becton shoreline.

Figure 1. Oblique aerial photograph of Hengistbury Head looking north towards Christchurch Harbour, taken in 1955, showing the beach accumulation to the west of the long groyne and increased erosion on the downdrift side. The photograph original was taken by Messrs. Aerofilms, now Messrs Simmons Aerofilms who kindly granted permission for its reproduction.

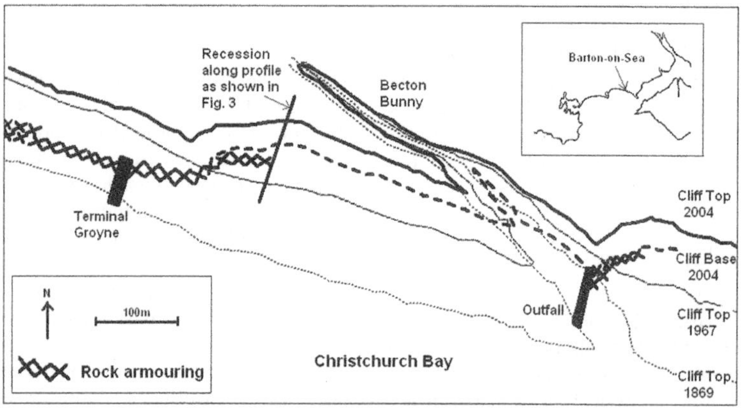

Figure 2. Plan of the unprotected shoreline downdrift of the terminal groyne at Barton-on-Sea. The current location of the cliff top and base are shown along with the cliff top position in 1869 and 1967. Between the terminal groyne and the rock armoured outfall, a small stream valley known as Becton Bunny cuts through the cliff line.

In the middle distance can be seen a accumulation of talus at the foot of the cliff which was derived from two moderate slumps of the Lower Headon Bed sands, silts and clays forming the cliffs in this area. The sea is beginning to recede from the high spring tide level, but even at its highest it did not reach the in-situ strata but certainly was attacking the talus accumulation. Observations indicate that the principle mechanism of cliff recession is this erosion of the talus produced by landsliding and only exceptionally in severe storm conditions are the outcrops of in-situ strata attacked by wave action.

5 SCOUR DUE TO RIP CURRENTS ADJACENT TO THE TERMINAL DEFENCE WORKS

Figure 4 shows the waves reaching the shore behind the rock armoured extension of the terminal strongpoint. Originally, this rock armouring was placed along the

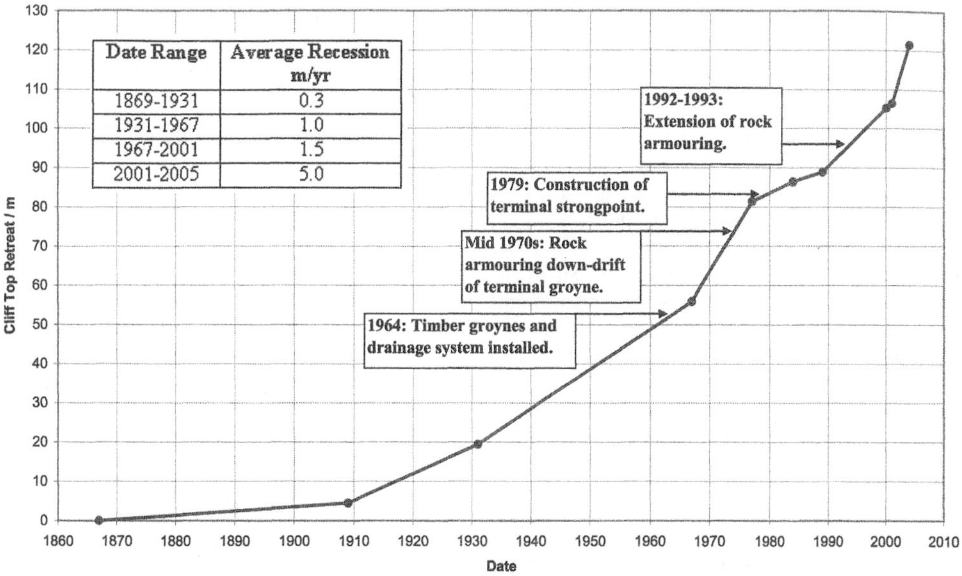

Figure 3. Recession of the cliff top from 1869 along the line normal to the Christchurch Bay shoreline shown in figure 2. Changes in recession rates can be related to the construction of the various modifications of the coastal defences.

Figure 4. View from the rock armoured extension of the terminal strongpoint at Barton-on-Sea looking eastwards across the undefended shoreline towards Becton Bunny and the rock armoured outfall. The scour channel developed behind the rock armoured extension (which was formerly adjacent to the cliff base) can be seen in the foreground. A large talus accumulation, produced by two cliff slumps, can be seen in the photograph centre. A steep beach berm protects the cliff base but the talus is being eroded by wave action.

berm at the back of the beach but subsequent scouring by rip currents has created a gulley down into the bedrock. Such features have been produced in physical models by Silvester and Hsu (1997) and have been notably observed with a similar structure by Granja and Carvalho (1995). The danger of such scouring is the likelihood of outflanking the protective works, a danger which tends to stimulate further extensions of the protection into the downdrift area but without affecting a cure (Granja and Carvalho 1995). The effect of this scour channel has on the recession at Becton can be seen in the upturn of the recession rate to 5 m/year from 2001 to 2004 as shown on Fig. 3.

6 BEACH VOLUMES DOWNDRIFT

The Becton coast seen in Fig. 4 represents a littoral cell between the rock armouring in the foreground and a rock armoured outfall at a distance of about 600 m to the east. Under normal circumstances little sediment is able to escape this cell and it might be expected that under the influence of the dominant longshore currents, the beach would form a parabolic shape, such as shown by Silvester and Hsu (1997), with the updrift end, starved of sediment, being reduced to a narrow berm fronted by a shore platform in the in-situ strata. This rarely happens and more usually, the beach is fairly symmetrical around this small cell. The explanation as to why this happens is as follows. Immediately downdrift of the terminal strongpoint, the beach ceases to be subject to process control of littoral drift but becomes subject to supply control. The normally dominant eastward longshore movement of material by wave action cannot be supported owing to the lack of beach material immediately east of the strongpoint. However, the more infrequent westward drift is less limited by problems of supply and hence this westward drift prevents total depletion of the beach at its western end. Beach surveys have been undertaken since 1989 but it is found that the offshore/on shore volumes are greatly in excess of longshore volume changes and mask small trends in possible overall beach depletion.

7 DOWNDRIFT EROSION AND ACCELERATED CLIFF LANDSLIDING

Interruptions to the supply of beach material by defence works around the East Anglian coast leading to increased erosion rates in the unprotected areas have been documented by Clayton (1989). Extensive landsliding occurs naturally at Overstrand but the increased rates of movement there are attributed to the reduction in littoral drift due to the groyne fields at Cromer (Hutchinson 1976). Similar observations concerning interruptions to littoral drift along the Holderness coast have been described by Mason and Hansom (1988). As well as small scale, but frequent, landsliding in soft cliffs, the reactivation of large landslide complexes have also been ascribed to human interference with littoral drift. Thus Hutchinson et al (1980) refer the famous 1915 reactivation of the whole of the Folkestone Warren landslide to the construction of Harbour works at Folkestone in the preceding years which interrupted the natural supply of beach material from south west to north east. It is impossible in many of these cases to say what acted as the final trigger for movement, whether it was excess pore pressures generated by heavy rain or the marine erosion of a talus slope which had afforded some stability by toe weighting and passive pressure support but all such effects become critical near the point of equilibrium.

8 DOWNDRIFT EROSION AND RISING SEA LEVELS

With rising sea level full coastal protection everywhere is not an option and hard choices concerning the areas to be protected against marine erosion will need to be made. It is thus evident that as part of long-term managed re-alignment, the terminal groyne effect will become entrained into the policy as some, but not all, coastal defences are strengthened. Existing crenulate bays will be enlarged downdrift of newly strengthened defences. New crenulate bays will develop as former defence works are abandoned in schemes of managed retreat. The developing crenulate bays with cliffed coastlines will show increased landslide activity but which, in some mitigation, will contribute beach material according to their lithological composition.

9 RESEARCH AND POLICY INITIATIVES

Studies of the rates and forms of crenulate bay development in cliffed coastlines will be essential for future coastal planning. Knowing what to expect and the rate at which cliff top facilities including buildings and infrastructures can be expected to disappear will be a key component of decision making. Selective protective works could be used to control the crenulate shape, ensure that the strengthened defences are not outflanked and aim to deflect maximum erosion into those parts which could supply the greatest volume of beach material. Curiously, study of downdrift erosion in cliffed coastlines has been neglected in the past but we are sure that it will become an aspect of Coastal Engineering that will be of increasing importance as coastal management priorities begin to change. The need for such studies will come faster than many may think as the world moves nearer the crucial tipping point of climate change.

REFERENCES

Brown, S. 2005 *Downdrift Erosion and Recession at the Becton Bunny Cliffs, Christchurch Bay*. Unpublished MRes Dissertation, University of Southampton.

Carter, R.W.G. 1988 *Coastal Environments*. Academic Press, London.

Clayton, K.M. 1989. Sediment input from the Norfolk Cliffs, eastern England – A century of coastal protection and its effect. *Journal of Coastal Research* 5 (3), 433–442.

Cooper, N.J., Hooke, J.M. & Bray, M.J. 2001 Predicting coastal evolution using a sediment budget approach: a case study from Southern England. *Ocean & Coastal Management.* 44, 711–728.
French, P.W. 2001 *Coastal Defences: Processes, Problems and Solutions.* Routledge, London.
Granja, H.M. & Carvalho, G.S.*D. 1995 Is the coastline protection of Portugal by hard engineering structures effective? Journal of Coastal Research.* 11, 1229–1241.
Hutchinson, J.N. 1976 Coastal landslides in cliffs of Pleistocene deposits between Cromer and Overstrand, Norfolk, England. In *Janbu, N. et al (eds.) Laurits Bjerrum Memorial Volume, Contributions to Soil Mechanics.* Norwegian Geotechnical Institute, Oslo. 155–182.
Hutchinson, J.N., Bromhead, E.N. & Lupini, J.F. 1980 Additional observations on the Folkestone Warren landslides. *Quarterly Journal of Engineering Geology* 13, 1–31.
Komar, P.D. 1983 Coastal erosion in response to the construction of jetties and breakwaters. In *Komar, P.D. (ed.) Handbook of Coastal Processes and Erosion.* CRC Press, Florida. 191–204.
Mason, S.J. & Hansom, J.B. 1988 Cliff erosion and its contribution to a sediment budget for part of the Holderness coast, England. *Shore and Beach* 56, 30–38.
Parker, D.J. & Thompson, P.M. 1988 An extended economic appraisal of coast protection works: a case study of Hengistbury Head, England. *Ocean and Shoreline Management* 11, 45–72.
Rose, M. 1986 Bournemouth counts cost of coastal protection *. Surveyor 166 (no. 4898), 12–15.*
Silvester, R. & Hsu, J.R.C. 1997 *Coastal Stabilisation.* Advanced series on Ocean Engineering, 14. World Scientific Publishing, Singapore.
Stopher, H.E. & Wise, E.B. 1966 Coast erosion problems in Christchurch Bay. *Journal of the Institute of Municipal Engineers* 93, 328–332.

Landslides and Climate Change – McInnes, Jakeways, Fairbank & Mathie (eds)
© 2007 Taylor & Francis Group, London, ISBN 978-0-415-44318-0

Managing the impact of climate change on vulnerable areas: A case study of the Western Isles, UK

L.A.R. Richards & P.J. Phipps
Mott MacDonald, Croydon, Surrey, UK

ABSTRACT: During January 2005 a storm with winds reaching 106 mph struck the west coast of the Western Isles (or Outer Hebrides) of Scotland. The storm caused large scale damage to the coastline, ultimately leading to loss of life. The coast of the Western Isles is dominated by relict features formed by changes in sea level over the last 10 thousand years. Barrier islands, shingle ridges and machair are all relict sedimentary features which will be altered by modern and future sea level rise. The communities of the Western Isles could be adversely affected by the potential loss of the barrier islands and shingle ridges because the coastline will become more vulnerable to coastal processes. Due to the low lying nature of the machair hinterland it may inevitably become more prone to coastal inundation with rising sea levels. The potential impacts of climate change and changing sea levels have come into focus in this area due to the severity of the January 2005 storm. A concern of those who manage coastal environments is the increasing frequency with which high magnitude storms will affect sensitive coastlines. This paper will present the characteristics of the storm, its effects and the impact on coastal features of the Western Isles within the context of a changing climate and sea level rise. The findings are based on: local evidence and expert judgment; historic information of the coastline and meteorological records; as well as predicted future patterns of climate change. The difficulties relating to coastal management of this area are acute, and it will be vital to apply sustainable management practices now, so that future challenges can be readily met. This paper will outline such potential management approaches for this and similar coastlines in a changing climate, as well as the potential effects of mismanagement of the area.

1 INTRODUCTION

1.1 *Rationale*

The storm of the 11th and 12th January 2005 resulted in wide spread damage of coastal areas across the Western Isles. In many areas roads were either eroded or blocked with debris, services were cut, agricultural and semi-urban land was flooded and ultimately lives were lost. In response to the damage caused, the Local Authority (Comhairle nan Eilean Siar) commissioned Mott MacDonald to undertake a review of the twelve most affected sites and where appropriate make applications for funding to the Scottish Executive under the Coast Protection Act 1949.

This paper considers in the context of a changing climate the coastal evolution of the Western Isles along with potential impacts of future climate change and the role of sustainable management in supporting the vulnerable west coast of the islands.

2 STUDY AREA SETTING

2.1 *Geographical location*

The Western Isles, or Outer Hebrides are a chain of islands off the north-west coast of Scotland. The main islands are Lewis, Harris North Uist, South Uist, Benbecula and Barra, while smaller islands include Eriskay, Vatersay, St Kilda etc.

The areas of coastline considered here are all on the west coasts of the islands of Benbecula, South Uist, Barra and Vatersay (see Figure 1).

2.2 *Geology and soils*

The geology of the Western Isles is predominantly Lewisian Gneiss, which represents some of the oldest rock in the UK and possibly in the world (NCC 1977), having been dated as forming before 2800 million years ago (Moorbath et al 1975). The rock has been

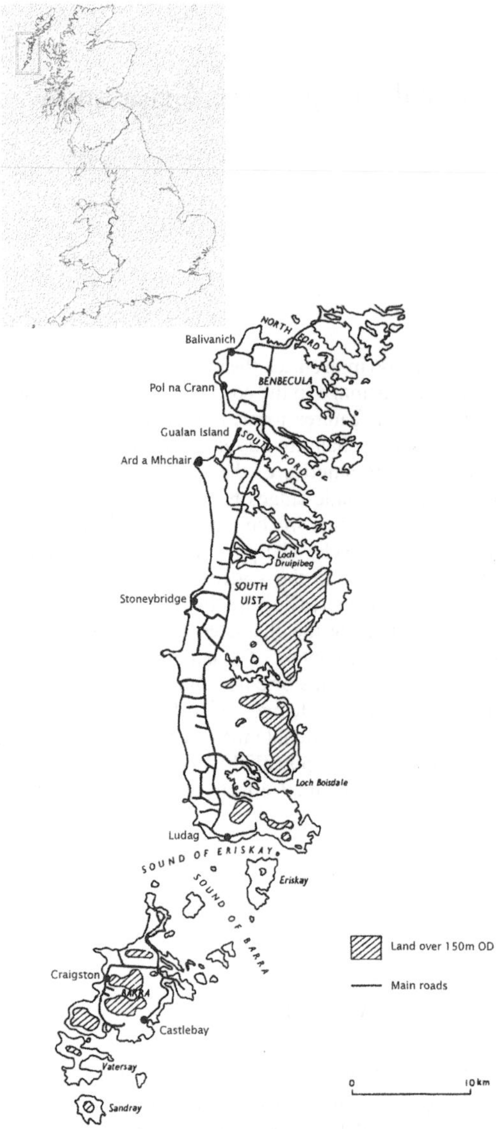

Figure 1. Map of the study area including site locations referred to in the text. The position of the Western Isles is shown relative to the UK.

subject to high pressures and temperatures through its history and as such is hard and resistant to weathering.

The gneiss which makes up the majority of the islands is described by the British Geological Survey (1992) as course grained, with a banded or streaky appearance. Within the gneiss there are rocks of a different character which are mainly of sedimentary origin. These features can occur as well defined belts of material extending for over 15 km, down to small discontinuous lenses a few metres in length. The Lewisian

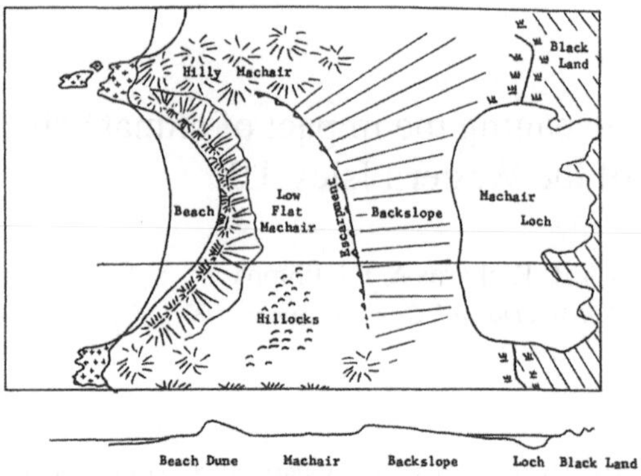

Figure 2. Typical plan view and cross section through machair land. From Ritchie 1971.

rocks have also been intruded by dyke swarms over their history so that a number of swarms of different ages are known to exist.

However, it is the erosion of superficial deposits over the top of the Lewisian Gneiss which is leading to coastal retreat in the Western Isles. As erosion of the mantling deposits proceeds it either leaves behind a series of irregular gneiss outcrops on the shore, or often a more formal gneissose wave cut platform.

2.2.1 *Machair*

Superficial sand deposits, called machair, mantle the Lewisian Gneiss and form much of the coast. Machair is one of the rarest habitats in Europe and is protected by UK and European Legislation. It is mainly found on the high energy low-lying western coasts of Scotland and North West Ireland.

Machair is low grassy land on calcareous, often shell rich sand which has a rich and diverse floral habitat and supports breeding and wintering waterfowl and wildfowl of international importance.

Machair is characterised by an assemblage of interconnected habitats progressing from mobile or semi-fixed dunes, through coastal plains and 'blacklands' (a mixture of calcareous sand and peat) to acidic moorlands of the interior (See Figure 2). Formed through accretion of wind blown shell sands, the machair encompasses a transition of habitats from base-rich to acid over a short distance, thereby creating a species-rich mosaic of grasslands and fens.

2.2.2 *Peat*

The coastline is dominated by machair, which tends to have peaty podzols to its landward side (Angus 1998), within the study area peat beds can sometimes be observed in outcrop on the beachs. Peat also plays a key role in the environment of the lakes and blacklands which form the hinterland (see Figure 2).

Table 1. Extreme water levels and local chart datum in metres above Ordnance Datum (HRW 2006).

Return period (years)	10	25	50	100	250	Chart datum
Kinlochbervie:	3.63	3.74	3.80	3.89	3.97	−2.50
Stornoway:	3.22	3.33	3.39	3.47	3.55	−2.71
Ullapool:	3.39	3.49	3.55	3.66	3.76	−2.75
Tobermory:	3.31	3.45	3.52	3.63	3.71	−2.39
At 57.48°N 5.87°E:	3.36	3.45	3.50	3.60	3.70	

2.3 *Marine conditions*

There are few tidal gauges or wave rider buoys in the Western Isles and thus tides and extreme water level heights were extrapolated during the review process from nearby ports such as Stornoway, Ullapool, Tobermory and Kinlochbervie (see Table 1). Balivanich is estimated to have a spring tidal range of 3.8 metres.

The wave climate of the Outer Hebrides is severely depth limited. The offshore platform to the west of the Outer Hebrides is a relict of the old coastal plains which have been flooded since Neolithic times (4000–2000BC). This eroded land provides a gently sloping platform far out to sea which forces incoming waves to shoal and lose energy before breaking on the coastline.

During a storm or particularly high tide event wave heights may increase to be greater than normally anticipated owing to the increased depth of nearshore water (and consequently the decreased shoaling effect). In the future, sea level rise and climate change will intensify the wave climate in Western Isles.

Sea level rise will result in an increased vulnerability to coastal erosion, and erosion rates should increase as the coastline will be subject to greater wave energy (UKCIP 2002). Increased storminess will result in a higher frequency and severity of storms. For example, a storm which would currently have a likelihood of 1 in 10 year event may, in the medium term have a probability of (say) a 1 in 5 year event for the same magnitude event. The increase in storminess will induce a more severe and variable wave climate and potentially increase the amount of rainfall.

In areas as prone to gales as the Western Isles the population have adapted to the prevalent conditions and, for the majority of the time, can function despite interruptions by stormy weather. If the level and frequency of storminess increases the population will become subject to increasingly severe storms more often, this will have a detrimental effect on how the population functions.

2.4 *Coastal recession rates*

Available historical resources were researched in order to produce best estimate coastal recession and change scenarios. Researched resources included Ordnance Survey maps, Admiralty charts and aerial photographs.

The coastline shown on the First Edition Ordnance Survey maps for the study area was surveyed in 1878. It is apparent that the coastline was not surveyed in the 1902 Edition but was re-surveyed for the National Grid survey in 1967. General re-survey work was undertaken in the early 2000s, but this did not include the coast. The lack of historical coastal change data was problematic in developing accurate historical term recession scenarios for the study area.

Aerial photographs were reviewed from 1946, 1956, 1959, 1963 and 1987. In general the coastline has been increasingly developed with housing. Original track ways have been upgraded to roads without being moved from their historic position by the coast. Overall, the coastline has been eroding, with associated loss of vegetation quite evident.

The general recession rate across the study area is an average of 0.2 metres per year. However, local variations occur and maximum recession rate of up to 0.4 m per year have been identified. Such observations can be supported by longer term changes. The amount of land lost from the west coast since Neolithic times (4000–2000BC) is estimated to be 1.25 km (Barber 2003). Thus erosion has been occurring on the west coast at an average rate of around 0.2 to 0.3 metres per year since the beginning of the Neolithic. It is recognised that different areas have eroded and accreted at different rates throughout time and that estimates such as these can only be utilised when the limitations to the information are understood.

3 SOCIO-ECONOMIC FACTORS

3.1 *Population*

The population of the study area is sparse. The towns, villages and hamlets on the islands are generally concentrated on the low flat land close to the western coastline as opposed to the more rugged hillsides on the east of the islands. The low flat inhabited land is still farmed by a traditional crofting system, its fertility a product of centuries of traditional farming practices. Within future climate change scenarios the population of the islands may potentially become more vulnerable to sea level rise and extreme events.

The most populated island in the study area (the islands of Benbecula, South Uist Barra and Vatersay) is South Uist, the largest Island (see Table 2). However, Benbecula has the largest town in the study area, Balivanich. Table 2 shows that the population of the islands has been declining through recent years. The level of population decline for the large Scottish island groups (Western Isles, Shetlands and Orkney) was higher in the 1991 to 2001 period than the Scottish

Table 2. Population of the Western Isles. GROS 2003.

Area	1991 Total	Change 1981 to 1991	2001 Total	Change 1991 to 2001
Lewis	20,159	−3%	18,489	−8%
Harris	2,222	−11%	1,984	−11%
North Uist	1,815	+1%	1,657	−9%
Benbecula	1,803	−4%	1,249	−31%
South Uist	2,285	−6%	1,951	−15%
Barra & Vatersay	1,316	−4%	1,172	−11%
Totals	29,600	–	26,502	−10%

Table 3. Population decline of the Western Isles (CNES 2006).

Year	Persons Present	Persons Resident
1901	46,172	
1911	46,732	
1921	44,177	
1931	38,986	
1939	38,529	
1951	35,591	
1961	32,609	
1971	29,891	
1981	31,884	30,713
1991	29,370	29,600
2001	–	26,502
2011*	–	22,446
2018**	–	21,725

*2011 figure is a population projection calculated by the GROS and is based on figures from 2000.
**2018 figure is a population projection calculated by the GROS and is based on figures from 2002.

average (CNES 2006). The decline in population of the Outer Hebrides over the last ten years (between 1995 and 2005) was the highest of any Local Authority in Scotland at −8.5% (CNES 2006).

The figures quoted in Table 3, are for all of the Western Isles. Historically the census would count the number of people present on the islands on a certain night. More recently the census has recorded the number of people that reside on the islands. It is important to note that the population has almost halved in the last 100 years.

3.2 *Landuse*

The majority of the land in South Uist, Benbecula, Barra and Vatersay is not actively utilised as it comprises high inhospitable hills, boggy land or lochs. The land which is actively used is either for housing or agriculture.

Owing to the sparse population and the dissipated nature of the population centres across the islands the roads may be single track with local passing places, but they are a vital link to maintain an integrated island community.

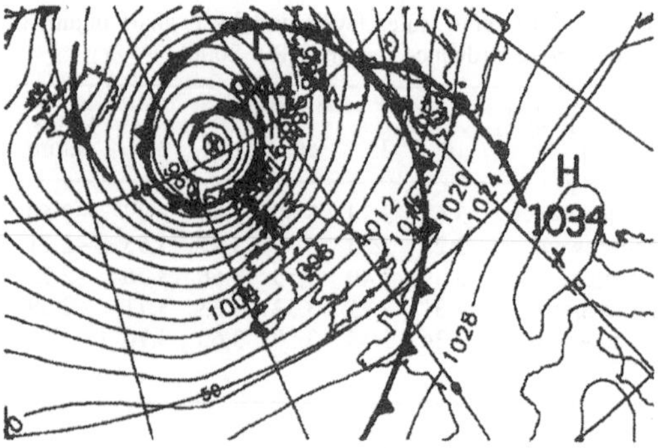

Figure 3. Meteorological chart showing the storm at 0000 hrs on 12th January 2005. (c) Crown Copyright 2006.

3.3 *Industry*

The biggest employer in the Western Isles is the public sector (admin, education, health etc) followed by retail and construction. Fishing and farming also employ a large number of the population, although most people who maintain crofts are actively involved in other industries to supplement their income. Each year tourism contributes more money to the local economy (CNES 2006).

4 EVENTS OF JANUARY 2005

On the evening of the 11th and 12th of January 2005 a large storm battered the west coast of Scotland causing severe damage (see Figure 3). Large scale erosion occurred, roads were closed, trees blown over and flooding was experienced.

On the Western Isles the storm was manifest by gusts of up to 92 knots (106.0 mph) (Angus and Rennie 2006), which coincided with a high spring tide and large tidal surge (HRW 2006). Although the 'depth' of a depression is not decisive in determining the damage a storm will cause it is worthwhile to note that the storm of January 2005 had lower air pressure at its centre than the storms which devastated southern England in 1987 and the storm which caused large scale damage in Norfolk and The Netherlands in January 1953 (Angus and Rennie 2006). The low air pressure at the centre of the storm acted on the Atlantic to create a large storm surge, which on landfall caused widespread damage. The peak calculated surges for the local standard ports (see Table 1) are around 1.5 m (HRW 2006).

The peak of the storm in terms of wind speed and waves coincided with the exceptionally high spring tide and surge on the evening of the 11th January.

The combined high sea level and wind/wave conditions resulted in a joint probability significantly higher than 1 in 200 years (HRW 2006) for that event.

Eye witness accounts of the storm from the local police force refer to roads being overwashed by water and becoming impassable. At the height of the storm one police patrol car (a Ford Focus) was surrounded by seawater up to its bonnet at Gramsdale (north west Benbecula, east of Balivanich). Windblown debris also made driving treacherous. For example, another police patrol car was nearly struck by half a caravan being blown across the road (Angus and Rennie 2006). In the village of Stoneybridge on the west coast of South Uist the level of the loch increased causing houses to be cut off from one another and from the main emergency and evacuation routes. About seven miles north of Stoneybridge at Ard a Mhchair a family trying to escape their flooded house by car were washed out to sea and drowned.

Comhairle nan Eilean Siar reported that among the population of Benbecula, South Uist, Barra and Vatersay there was no recollection in living memory of a storm causing as much damage as the January 2005 event. Large storms have been recorded in the late 1800s and early 1900's on the Western Isles. For example a storm in March 1921 left trawlers port-bound in Barra leading to massive food shortages for the island. At Stornoway gusts of up to 87 mph were recorded. In the press at the time it was reported to have been 'the worst storm for 40 years', claiming three lives.

Following the January 2005 storm several stretches of coastal road were blocked by debris, flooded or eroded. Over a year on the recovery from this is still ongoing.

5 SPECIFIC SITES

In early 2006 the coastal environment of twelve sites on the Western Isles were investigated. Six of these sites were bought forward for full project appraisal and applications for grant aid from the Scottish Executive under the Coast Protection Act 1949. Full details of these six sites are detailed below (see also Figure 1).

For many of the locations and schemes that were investigated it was considered that coastal protection or management is needed to prevent further coastal erosion from degrading the crest height on the coastline. On some parts of the coast the highest point in the hinterland tends to be the crest of the coastal ridge adjacent to the beach (Figure 2). As a result, when coastal erosion occurs the crest height is lost and the land becomes more prone to coastal inundation. In Balivanich on Benbecula this potential loss of the crest height would impact on the largest village in the study area by making it more prone to coastal flooding. At Pol na Crann, south of Balivanich, loss of the crest level would result in the more frequent inundation of the road and flooding of the hinterland.

The primary focus of the works proposed at Craigston and Ludag Road were to protect or consider re-routing major roads which are vital to the transport and emergency infrastructure of the islands. The roads have been built in their traditional position (as seen on historic maps) as a result they are increasingly prone to coastal erosion.

Gualan Island is a barrier island between Benbecula to the north and South Uist to the south. To the east of Gualan is the South Ford, which is spanned by the South Ford Causeway, part of the Western Isles spinal road link. Gualan Island is 2.6 km long and consists of three distinct sections. The northern and southern thirds of the island have a broad dune system behind a shingle upper beach, with a salt marsh on the eastern side. The central third consists of a shingle ridge, with occasional areas of dune behind. The central section was badly impacted during the January 2005 event, and has subsequently been bulldozed into a narrow and unnatural ridge shape.

Gualan Island dissipates the wave energy which enters South Ford resulting in a generally quiescent coastal climate on the shores of the ford. As a result the coast of South Ford is a relatively highly populated area with a number of villages, single houses and roads close to the water level. If Gualan is lost the full force of any Atlantic swell or waves may adversely affect the coast of South Ford.

The marine conditions in South Ford are complex. Gualan has a damping effect on the wave climate, as well as affecting the flow of water (tides, surges etc) in and out of the Ford. The causeway which carries the main spinal road through the islands is located on the eastern side of South Ford has only one culvert. This causeway is considered by the local population to be having detrimental effects on the South Ford, for example, preventing water escaping during storm or surge events. Currently such scenarios are conjectural and the Comhairle implementing further studies to understand the hydrodynamics of the South Ford area.

Stoneybridge is located on the west coast of South Uist. The area under consideration is where the local road swings to the west of Loch Altabrug, and is only approximately 30 m from the coastal edge. The beach is sandy, with a shingle ridge that has been formed into an artificial knife edge shape. During the January 2005 storm the ridge was flattened, with shingle and cobbles dispersed up to 100 m inland, at this time the shingle was blocking the road and thus was bulldozed back into position, with an exaggerated crest height. The area behind the ridge is machair, which slopes gently downwards towards Loch Altabrug.

6 CLIMATE CHANGE

6.1 *Historical change*

When trying to tackle a subject that is as emotive as climate change it is easy to forget that climate change has been occurring throughout Earth's history and that we are presently in a rare interglacial period. The current interglacial has been evident since the end of the last ice age 10 thousand yrs ago, with sea levels rising due to naturally driven change ever since.

The Western Isles have been subject to glacial rebound due to the isostacy of the Earth's mantle and crust. As a result of crustal rebound there are a number of raised beaches on the islands (Angus 1998). The crust below the Western Isles was covered by a relatively thin layer of ice and has now achieved isostatic equilibrium, with future rises in sea level potentially leading to submergence of these islands (Shennan and Horton 2002).

Large quantities of glacial sediments were deposited on the continental shelf to the west of the Western Isles following the retreat of the glaciers and the beginning of sea level rise (Ritchie 1971). Post glacial rises in sea level mobilised the glacial deposits, mixed them with shell debris and swept them onshore to provide the sand and boulders which characterise the west coast of the islands (Dawson et al 2004). Characteristic features formed by this process include barrier islands, shingle ridges and machair (Angus 1998).

6.1.1 *Machair formation and maintenance*

The machair which covers large areas of the study area is itself a product of climate change. The machair is the sand washed on shore by the process described above, in the present day the sands are vegetated and form part of a mosaic of constituent habitats.

The diversity of the machair is dependant on human action, namely the application of seaweed and the narrow strip rotation system of agriculture that is traditionally employed. Seaweed inhibits wind blown erosion in winter and adds organic matter to the naturally highly mineral soil. Cattle are put onto the land (except in spring) to trample the seaweed into the soil and their dung provides habitat for invertebrates. The most imminent threat to the diversity of the machair system is from the risk of human cultivation being suspended. Sustaining the socioeconomic viability of the local communities is consequently a fundamental objective for the environmental protection of the machair.

6.1.2 *Sedimentary features*

Barrier islands such as Gualan are formed through the movement of sediment onshore, with the supply of sediment ceased the barrier island is at risk of being breached and eroded away. Shingle barriers, such as those seen on the coastline near Stoneybridge, are prone to 'rolling back' and dissipation of the material which forms the ridge now that the supply of sediment has been cut off and the system has become relict. The barrier islands and shingle ridges along the west coast of the Outer Hebrides are a natural response to previous sea level changes during the Quaternary.

There is evidence that, due to the transitory nature of the sands that make up the machair and associated coastal landscapes of the Outer Hebrides, there have been large scale blowouts of dunes and erosion of the machair throughout history. For example McDonald 1811 states that "The depredations of the sea and storms are alarming...many places have lost a quarter of a mile in breadth...more than 6000 Scotch acres by sand drift" (as referred to in Ritchie 1971). Today the machair is quasi-stable with a few localised blowouts occurring in the larger dunes; this is a relatively stable state, compared to observations from the 18th and 19th centuries (Gilbertson et al 1999). These erosive events cannot be 'blamed' on tourism or anthropogenic climate change and it should be accepted that it is the inherent nature of a coastline to erode and accrete. However, with the spectre of climate change having a greater impact in the future, and current farming practices including the intensive rearing of cattle and sheep on the machair grassland dunes it is vital that the coast is managed appropriately to avoid rapid, devastating erosion.

The climatic systems which have been influencing this coastline until the recent past have been largely characterised by natural climate change. However, the climate of the present and foreseeable future will be dictated by anthropogenic climate change

6.2 *Future changes*

Anthropogenic climate change refers to the change in the Earth's climate which has been being driven by the release of CO_2, methane and other gasses through human action. This change in the Earth's climate is likely to lead to greater extremes of weather and a climate vastly different to what we have experienced ourselves. Emissions of greenhouse gasses have been growing rapidly since the industrial revolution. In 1750 the concentration of CO_2 in the atmosphere was 280 parts per million (ppm) in 2000 the concentration of CO_2 in the atmosphere now stands at 368 ppm this represents an increase of a third and is the highest level of CO_2 in at least 420,000 years, and likely the past 20 million years. (IPCC 2001b) The 1990's were likely to have been the warmest decade in the last millennium (IPCC 2001 a). The changes observed in this period, such as general warming, increasingly severe droughts, hurricanes and typhoons are to continue for at least several decades due to the greenhouse gasses already emitted (Van Aalst 2006) and because emissions of greenhouse gasses are unlikely to completely cease in the immediate future. Current climate predictions point to increasingly warm average temperatures world-wide and an intensification of the hydrological cycle but also, more worryingly, increases in the number and severity of extreme events. Extreme events include droughts, heatwaves, hurricanes and severe storms.

One effect of global warming on Western Isles is likely to be that storms similar to the January 2005 event become more commonplace. The January 2005 storm (or any other single event such as Hurricane Katrina) should not, in itself, be seen as an indication or proof of climate change occurring. The study of the storm can only be related to climate change in that events of this severity are likely to become more common in the future. For example, the number of deep atmospheric depressions passing across the UK is predicted to increase by 40% by 2080's. However, the likely severity, tracks and effects of future storms are particularly uncertain (Angus and Rennie 2006).

An increase in sea level will also have a detrimental affect on the coastline of the Western Isles in its own right. Higher sea level will lead to an increase in the sea level during extreme events, an increase in the severity of the wave climate and coastal inundation. Once the sea level is higher the protection afforded by the continental shelf to the west of the islands will be compromised. If the future is going to be characterised by increased sea level and an increasingly severe wave climate the erosion, overtopping and overwashing of coastal dunes and machair will be increasingly likely. Such scenarios would result in sediment removal, reduction in height of the coastal edge and a potential change in the machair habitat due to ingress by saline water.

When coupled with increased sea level and an area as vulnerable as the Western Isles the importance of sustainable coastal management becomes key.

7 SUSTAINABLE MANAGEMENT

The effects of climate change will most likely make living on the Western Isles less and less sustainable. Faming practices could potentially be interrupted, roads may be lost, communities could become more fragmented and the already decreasing population would be at greater relative risk from increasingly severe events similar to the January 2005 storm.

7.1 *Short to medium term management*

The naturally occurring coastal structures that have dissipated storm and wave energy – such as the Stoneybridge ridge and Gualan Island are currently at risk due to rapid roll back, breaching and loss. A question facing those who try to manage coasts with naturally occurring barriers, such as Gualan, or Chesil Beach is what lengths should managers persist with extending the longevity of transitory coastal features? This may only be an issue where a local population or other asset is present. In the case of the Stoneybridge shingle bank the local assets are a road and low grade land which is not used for agriculture. The social significance of the road to the local populous is high and it is believed that the ridge acted to prevent wave overtopping, reducing the flood level of the loch. It is proposed to use large boulders to bolster the lea of the shingle ridge and retard further role-back, thus preventing loss of the road service. The approach is low-impact which will have little long term effect on the coastal dynamics of the area. The scheme will act to minimise the impacts of any storms without giving a false sense of security to the local population.

It appears that Gualan Island may continue to thin and loose vegetation as it rolls back. The loss or breach of Gualan would probably have a detrimental effect on the coastline of South Ford. The proposed management option for Gualan is to carry out sand re-nourishment, dune thatching and planting and to encourage the local community to maintain the island.

The schemes for Gualan and Stoneybridge are short term low impact schemes aiming to increase the 'health' of the naturally occurring coastal structures which already protect the shoreline. They are not long term solutions to the problems that these islands will encounter as their climate changes.

7.2 *Medium to long term management*

In the long term it may be that the way of life on these islands slowly becomes unsustainable. In this respect, a parallel can be made with the islands of St Kilda to the west of the Western Isles. Although the study area is not near this stage of depopulation and isolation, St Kilda was evacuated in 1930 after a number of crop failures, outbreaks of influenza and emigration of the young (Fleming 2000). Although the evacuation of St Kilda was more due to economic difficulties than climatic obstacles it is possible that climatic forces may make the current lifestyle of the Western Isles unsustainable in the future.

In the medium term, it may be possible to carry out adaptive management of the various issues that arise. A large number of the roads and thus escape and emergency access routes to the villages run close to the coastline. Many of these roads were eroded or closed by debris following the January 2005 storm. It would be considered appropriate for medium term planning to encourage the building of roads away from the coastal edge and to re-route them inland or on higher ground so that the spinal road runs through the centre of the islands away from potential coastal flooding and overwashing with a number of 'access roads' running perpendicular to the coast from the trunk road into various villages. The relative benefit of this method is not easily conveyed through cost benefit analysis, but it would be a strategic approach to create a more sustainable road network which would also aid in emergency response to extreme events such as storms and coastal flooding.

The population could be moved inland of their current position and thereby reduce their vulnerability to coastal erosion and from inundation during storm events. However, this may not be appropriate because, like so many coastal areas at risk through climate change such as The Netherlands, Guyana etc, the most fertile land is close to the coast.

The fusion of coastal risk assessment and planning regulation enforcement to prevent building and development in vulnerable areas of the coast has been used on the Undercliff of the Isle of Wight UK. This approach would be a useful tool on the Western Isles, where it could be used to prevent building on land below the (say) 5 meter contour line.

8 CONCLUSIONS

The coastline of the Western Isles is more prone to gales and similar storms than most of the coast of Europe. To a certain extent the population has adapted to this fact and are able to manage in their current environment effectively. However, the ease with which the population deal with storm events in the future may be undermined by sea level rise and increased storminess. An illustration of this is that the January 2005 storm had a detrimental effect on infrastructure and housing of the islands, ultimately leading to loss of life.

The key to preventing further irreparable damage to the very existence of the islands as a functioning system is adaptive and sustainable management of the coast, a goal held by vulnerable coastlines globally. The price of mismanagement of the coastline will certainly be spiralling management and defence costs as unsustainable schemes are implemented and maintained. With this in mind, the use of 'no-regrets' low impact solutions often becomes the preferred management method. The defence of the majority of the coast of South Ford may be possible – simply by maintaining Gualan Island through dune thatching and planting. This scheme (for example) should not give a false sense of security to the population of South Ford. In other areas the machair is degrading through aeolian and human processes, making it more vulnerable to coastal erosion and inundation.

The coastline of this study area is a direct product of Quaternary sea level change. Future, accelerated climate change will have a considerable effect on the increasingly vulnerable, already depleting population. Now is a crucial time for implementing coastal and landuse management if the communities of the Western Isles are to be maintained in to the future.

REFERENCES

Angus, S. 1998. The Outer Hebrides: The Shaping of the Islands. Harris The White Horse Press.

Angus, S. and Rennie, A. 2006. The Natural Heritage impact of the storm of 11th January 2005 in the Uists and Barra, Outer Hebrides. Scottish Natural Heritage. (Second Draft).

Barber J. 2003 Bronze Age Farms and Iron Age Farm Mounds of The Outer Hebrides, Scottish Archaeological Internet Report 3, www.siar.org.uk

CNES 2006. Comhairle nan Eilean Siar. http://www.cne-siar.gov.uk

Dawson, S. Smith, D.E. Jordan, J. Dawson, A.G. 2004. Late Holocene Coastal Sand Movements in the Outer Hebrides, NW Scotland. Marine Geology 210. pg 281–306

British Geological Survey 1992. The Geology of the Outer Hebrides. A Memoir for 1;100 000 (solid edition) geological sheets, Lewis Harris Uists and Barra

Fleming, A. 2000. St Kilda: Family Community and the Wider World. Journal of Anthropological Archaeology 19, pg 348–368

Gilbertson, D. Schwenninger, J. Kemp, R. Rhodes, E. 1999. Sand-drift and Soil Formation along an Exposed North Atlantic Coastline: 14,000 Years of Diverse Geomorphological, Climatic and Human Impacts. Journal of Archaeological Science 24, pg 439–469.

GROS 2003. Scotland's Census 2001: Statistics for Inhabited Islands. Occasional Paper Number 10. General Register Office for Scotland, Edinburgh.

HRW (HR Wallingford) 2006. Western Isles Coast Protection Study. Wave and Tidal Conditions on 11th 12th January 2005. HR Wallingford Report TN CBM5582/01. March 2006.

IPCC (Intergovernmental Panel on Climate Change) 2001 a. Summary for Policy Makers. Climate Change 2001: Synthesis Report, Contribution of Working Groups I, II And III to the Third Assessment Report of the Intergovernmental Panel on Climate Change. Cambridge University Press, Cambridge.

IPCC 2001 b. Climate Change 2001: The Scientific Basis, Contribution of Working Group I to the Third Assessment Report of the Intergovernmental Panel on Climate Change. Cambridge University Press, Cambridge.

Nature Conservancy Council (NCC), Geology and Physiography Section 1977. The Outer Hebrides – Localities of Geological and Geomorphological Importance. Newbury

Moorbath, S. Powell, J.L. and Taylor, P.N. 1975. Isotopic evidence for the age and origin of the grey gneiss complex of the Southern Outer Hebrides Scotland. Journal of the Geological Society of London, 131 pg213–222.

Shennan, I. and Horton, B 2002. Holocene land-and sea-level changes in Great Britain. Journal of Quaternary Science 17, 5–6 pg 511–526

Ritchie, W. 1971. The Beaches of Barra and the Uists. A survey of the beach, dune and machair areas of Barra, South Uist, Benbecula, North Uist and Berneray. Department of Geography, University of Aberdeen Press.

UKCIP – United Kingdom Climate Impacts Program, Predicted Future Climate Change Scenarios from 2002. http://www.ukcip.org.uk/scenarios/guidance/sci_report.asp

Van Aalst, M. K. 2006. The Impacts of Climate Change on the Risk of Natural Disasters. Disasters 30(1) pg 5–18.

Landslides and Climate Change – McInnes, Jakeways, Fairbank & Mathie (eds)
© 2007 Taylor & Francis Group, London, ISBN 978-0-415-44318-0

Allowing for climate change; an innovative solution to landslide stabilisation in an environmentally sensitive area on the Isle of Wight

A.R. Clark, D.S. Fort, J.K. Holliday & A. Gillarduzzi
High-Point Rendel, London, UK

S. Bomont
TP Geo, France

ABSTRACT: A major landslide at Castlehaven on the Isle of Wight occurred as a reactivation of pre-existing landslides as a consequence of both coastal erosion and high groundwater levels caused by the very wet and stormy winter of 1994/95. The Castlehaven landslide lies within the Undercliff landslide complex and a strong relationship between landslide movement, groundwater level and the 4 month antecedent effective rainfall (4AER) has been developed for the area. In the site area movement occurs when the 4AER exceeds approximately 380 mm. Castlehaven also lies within an environmentally sensitive area with both national and international designations. The selection of remedial measures was influenced by the need for environmental protection but also needed to be robust to cope with predicted future changes in climate including increased storminess and winter total and effective rainfall, rising groundwater levels and sea level changes. A scheme including siphon wells, and electro-pneumatic wells, both used in UK for the first time, was designed and constructed to reduce winter groundwater levels to the equivalent summer levels at which movement is substantially reduced thus improving landslide stability and protecting valued environmental assets.

1 INTRODUCTION

Castlehaven, situated on the south coast of the Isle of Wight, is the site of the recently completed construction of a coast protection and landslide stabilisation scheme which won the 2005 Fleming Award of the British Geotechnical Association. This paper will describe the details of the scheme and how the impacts of predicted climate change have influenced the design and successful implementation of the works.

The site is located at the southern most point of the Island at Reeth Bay, south of the village of Niton (Fig. 1) and lies within the area known as the "Undercliff". This is a complex landslide system which fringes the whole south coast of the Island and is some 12 km in length and up to 0.6 km in width and contains the town of Ventnor and other areas of development including St. Lawrence, Bonchurch and southern Niton. The area is considered to be the largest urbanised landslide in western Europe.

Landsliding within the Undercliff is an ongoing post glacial phenomenon and has caused significant distress to the development and infrastructure throughout the area over recent centuries. The historic record of landsliding in the area has been well documented (Ibsen 2000 a-b, Hutchinson 1991, Ibsen & Brunsden 1996) with landslides of different sizes being a frequent event caused by the combination of marine erosion at the toe of the sea cliffs combined with recurrent periods of high rainfall, high ground water levels and susceptible geology. Several "great" landslide events occurred in the Undercliff in the very wet and stormy winter of 1994/95 including at Castlehaven where the reactivation of the pre-existing postglacial landslides has affected an area up to 400 m inland from

Figure 1. Aerial photograph of site.

the coast. If left unchecked this landslide had the potential to cause direct and consequential losses estimated in £20 million at 2004 prices.

To address the problem of landsliding and to reduce the risk to public safety, property and infrastructure a coast protection and landslide stabilisation scheme was designed. However the choice of the design and the implementation of the scheme was further complicated by the fact that the area is highly designated and protected for the environmental quality of the area and consequently the final scheme not only had to address the technical issues of preventing erosion and improving the stability of the landslides, including provision for predicted climate change, but had to incorporate design principles and methodologies that did not compromise the environmental quality of the area. The details of the environmental issues and the process undertaken to develop a scheme which met these criteria and the lengthy process required to achieve approval through the consultation and planning process has been the subject of a separate paper (Clark et al. 2002).

2 GEOLOGICAL BACKGROUND TO SCHEME

The regional solid geology of the area dips towards the south at approximately 2 degrees however locally the dip is towards the southeast due to the influence of the shallow St. Lawrence syncline (Hutchinson & Bromhead 2002). The strata consist of an interbedded sequence of high and low permeability Cretaceous rocks including the Upper Greensand which forms the inland cliff of the landslide backscarp some 500 m inland from the coastal cliffs. This is underlain by the Gault Clay, which is notorious locally for being susceptible to instability and is known as the "Blue Slipper". This in turn is underlain by the Carstone and the Sandrock of the Lower Greensand. The coastal cliffs at Reeth Bay are formed by intact near vertical Sandrock which has been subjected to marine erosion and on the basis of historic map analysis and local history the cliff line has retreated into the pre-existing landslide complex by up to 40 m since records of 1862.

This undermining of the toe of the landslides and steepening of the slope, combined with high rainfall and groundwater levels are the triggers for the reactivation of landslide movement within the Castlehaven area and the Undercliff in general.

3 GROUND MODEL

In order to address the options for remediation of the instability a ground model was developed on the basis of geomorphological mapping and borehole investigations in 1996 and 2002 to assess the ground

Table 1. Scope of Ground Investigations.

Activity	Year: 1996	2002
	No.	No.
Geomorphological Mapping	yes	yes
Boreholes		
Rotary Core (100 mm)	5	7
Rotary Core (300 mm)		2
Rotary Open Hole (100 mm)		4
Trial Pits		
Shallow	11	7
Deep	5	
In-site Testing		
Natural Gamma	6	
Neutron – Neutron	6	
Gamma – Gamma	6	
Caliper	6	
Absolute/Differential Temperature	6	
Absolute/Differential Conductivity	6	
Flow	6	
Permeability Tests		
Pumped Well Step & Constant		2
Rate Discharge		
Borehole Variable Falling Head		8
Trial Pit Variable Rising Head		2
Instrumentation		
Standpipe Piezometer	4	
Vibrating Wire Piezometer		3
Inclinometers	2	2

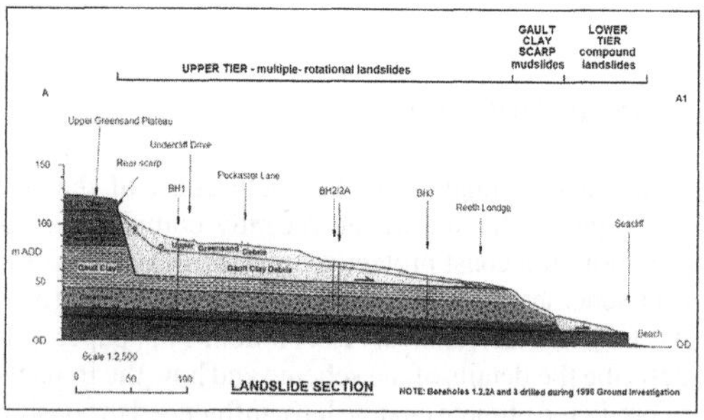

Figure 2. Geological section.

conditions, groundwater behaviour and aquifer characteristics of the area. The scope of these investigations is summarised in Table 1.

The landslide debris in the upper tier varied in thickness across the site increasing inland and consisted of an upper zone of Upper Greensand debris up to 10.5 m in thickness overlying a zone of Gault Clay debris up to 25 m in thickness. The overall geological section of the site area based on geomorphological mapping and the borehole interpretation is shown in Figure 2.

The landslides can be divided into an upper tier or inland area of multi-rotational failures separated from a seaward or lower tier of compound landslides by exposure of a Gault clay scarp. The upper tier landslides were determined from the degree of disturbance of the Gault Clay within the exploratory boreholes which separated the slipped "Gault Clay Debris" from the intact Gault and from inclinometer data. This model confirmed the preliminary landslide model developed from the geomorphological mapping undertaken soon after the major landslide of 1994/95 (Moore et al. 1995).

The landslide was also characterised by the widespread occurrence of springs, seepages and sinks, discontinuous surface and groundwater flow from north to south with a hydraulic gradient of around 12% towards the sea and by low to moderate permeabilities of the landslide materials. Local zones of high flow were also suspected along a fossil valley axis cutting through the escarpment and behind back tilted blocks.

4 SCHEME DEVELOPMENT AND DESIGN

4.1 *Overview*

In view of the overall thickness of the landslide debris within the model coupled with the severe environmental constraints imposed on construction in the area by the environmental consultees and the planning conditions it was evident that improvement in the stability of the upper tier landslide could not be achieved by major construction and the installation of physical strengthening of the ground with techniques such as piles, deep counterforts, regarding, etc.

Accordingly a design concept was developed which consisted of two elements, firstly the prevention of the erosion by the sea and landward retreat of the Lower Greensand cliff by construction of a rock armour toe and apron and secondly improving the stability of the upper tier landslide by drainage techniques. The design concept requirement was a fifty year design life as determined by Department for Environment, Food and Rural Affairs (DEFRA), the government funding department.

4.2 *Allowing for climate change*

Since the initial study phase it was obvious that predicted climate change (e.g. IPCC 2002, Hulme et al. 2002, Hulme & Jenkins 1998, Bray et al. 1997) had the potential to alter coastal processes and slope stability at Castlehaven and consequently the design of the scheme had to be robust and to allow for this scenario. The approach adopted at this site included the identification of all the factors influencing the stability of the area followed by a determination of the extent to which climate change could influence them. The key climate variables considered to influence the stability of the site and in turn to be influenced by climate change are:

- Sea level;
- Nearshore waves;
- Effective rainfall;
- Infiltration and additional leakage from pipes, soakaways and septic tanks and increased run off from roads and houses.

The study highlighted the complexity of the site and the complex inter-relation of the key triggering factors. In particular the effect of climate change on the groundwater and hydrogeology of the slopes was anticipated to be complex. Therefore the design of the foreshore and of the slope stabilisation works was based on two different approaches.

- The foreshore works were designed on the basis of the most recent predictions (at the time) in sea level rise for the future and wave modelling.
- The design of the slope stabilization had to be flexible ad adaptable to a relatively wide range of scenarios.

In terms of the foreshore works a sea level rise of 6 mm/year (Environment Agency 1999) was allowed with an overall sea level rise of 0.3 m in 50 years.

The provision for increased storminess and near shore waves was allowed and the rock armour stability was based on physical modelling and 1 in 100 year return period wave conditions. The slope stability design philosophy included the ability to cope with increased rainfall and groundwater levels raised to existing ground levels (i.e. total saturation).

The close relationship between rainfall, groundwater levels and landslide movement within the Undercliff, particularly to the 4-month antecedent effective rainfall (4AER) level has been recognised for some time. However the threshold levels at which significant movement is initiated varies at different locations within the Undercliff due to varying sensitivity to movement based on local ground conditions and physiographic features (Hutchinson 1991, Moore et al. 1995).

At Castlehaven since the major movements in the spring of 1994 and the winter of 1994/95 there have been subsequent significant movements during the winters of: 1997/98, 1999/2000, 2000/01 and 2002/03. No significant movement was observed during the summer periods or dry winters. There is a continuous rainfall record for the area since 1992 based on a Ventnor weather station supplemented in 2004 by an automatic station at Castlehaven set up as part of the works. Analysis of this data shows a good correlation between the onset of movement and a 4AER in excess of c. 380 mm (Fig. 3).

At the site groundwater and the hydrogeology of the landslide is influenced by rainfall and infiltration, recharge at the rear of the landslide from the main Upper Greensand aquifer perched on the underlying impermeable Gault and confined aquifers within the Sandrock. The area is also influenced by artificial

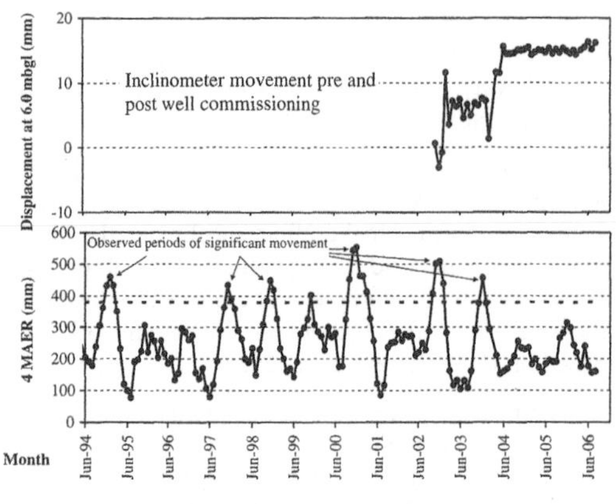

Figure 3. 4 Month AER and Ground Movement.

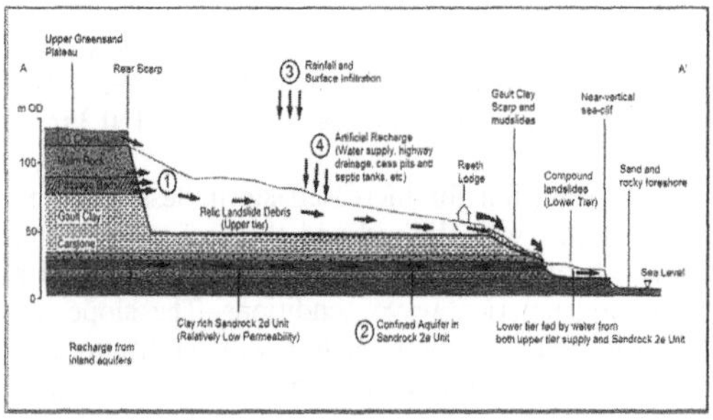

Figure 4. Sources of groundwater.

recharge from leaking sewers, septic tanks, gutters, road runoff, etc., (Fig. 4).

Groundwater levels have been monitored in standpipe piezometers since 1996 and in both standpipe and vibrating wire piezometers continuously since 2002. In December 2002 movement was again recorded by two inclinometers installed during the 2002 ground investigations and by surface movement and datum point monitoring when the 4AER exceeded c.380 mm which was equated to a triggering groundwater level at the site.

The considerable limitations placed on the scheme for environmental reasons, including the elimination of ground improvement construction options meant that the only feasible option was to lower groundwater. Thus the main criteria for the improvement in the stability of the landslide was to lower and maintain groundwater locally within the upper tier of the landslide to a level comparable to that experienced during previous summer levels when no significant movement occurs. A further constraint also precluded lowering of the groundwater below equivalent summer levels because of the potential of damage or stress caused to the local flora. This approach maintains partial activity of the spring line that occurs at the Gault Clay scarp and causes localized slope movement which is important for unstable slope habitat creation.

Thus the remedial measures available could not completely stabilise the landslide and the design approach was to achieve a target improvement on the current factor of safety based on a required minimum groundwater drawdown to achieve a minimum "Factor of Improvement" in excess of 1.10. Accordingly a groundwater control system was selected and designed that would eliminate the peak groundwater levels during periods of prolonged wet weather that trigger landslide movement. Furthermore the system had to be flexible and robust to be able to cope with the anticipated increase in "climatic events" that could occur as a result of a changing climate and the consequential seasonal increase in groundwater levels.

The ground model indicates landslide debris thicknesses varying up to a maximum in excess of 35 m at the rear of the landslide. The groundwater levels similarly varied throughout the site with maximum recorded winter levels of approximately 6 m and 1 m below ground level at the rear and the front of the upper tier respectively. Slope stability analyses indicated that the groundwater drawdown required to achieve the required factor of improvement was between 11.4 m and 7.7 m at the rear and front of the landslide respectively. The analyses were conducted on a number of cross sections for typical slope and basal shear geometries using laboratory determined residual shear values, worst case ground water conditions with an assumed existing factor of safety marginally above unity.

To achieve the required level of drawdown a series of lines of deep pumped wells was considered the most appropriate approach and a number of pump options were considered including electrical submersible, electrical pneumatic, electro-pneumatic and gravity fed siphon wells. Based on the thickness of the landslide debris that had to be penetrated by the wells, the level of drawdown required and a cost-benefit analysis of the whole life costing of each option, a combination of deep electro-pneumatic and gravity fed siphon wells was adopted. The details of the siphon drain concept are described in Bomont et al. 2005 and the main principles of the system are described below.

4.3 *Siphon drain principle*

Small diameter siphon drains are placed in vertical drilled drainage wells. These drains are generally spaced at between 3 m to 6 m centres and are sufficiently deep to provide the required drawdown. The wells are dewatered using the siphon principle and are able to drawdown under the influence of gravity up to approximately 8 mbgl. Siphon tubes are introduced into a permanently water filled reservoir at the base of the well with an outlet down stream at an outlet manhole, situated down slope (Fig. 5).

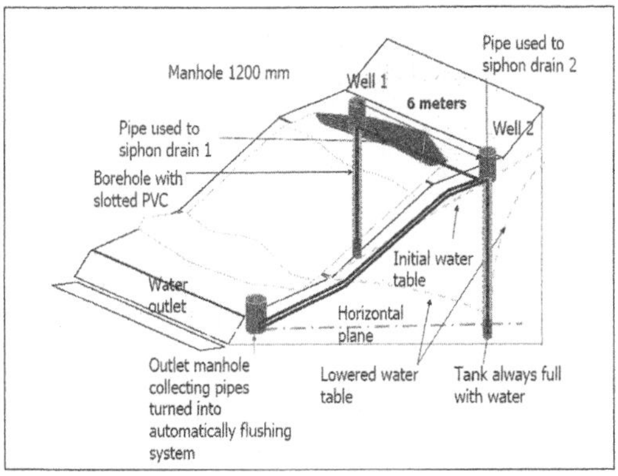

Figure 5. Section of siphon drain principle.

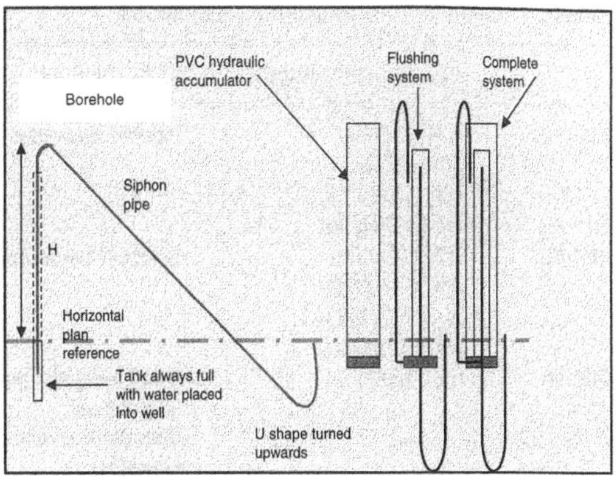

Figure 6. Siphon flushing system.

If the water level rises in the well, the siphon will flow and abstract water out of the well. The flow will continue until the water level in the well falls to the level of the outlet, provided that the flow rate in the siphon is sufficient to keep the siphon primed. As the water rises towards the top of the siphon the pressure falls, and may approach a prefect vacuum. In the upstream section the low pressure causes small bubbles to appear. Theses bubbles tend to coalesce into larger ones further downstream. Two forces act on the bubbles, firstly buoyancy and secondly hydraulic force due to the flow in the pipe. If buoyancy becomes the major force, the bubbles will collect at the summit of the pipe and combine into a single large bubble, which in time would break the siphon flow. This is avoided by using a system that automatically flushes out bubbles by turbulent flow.

The flushing system consists of a number of PVC pipes at the downstream end of the siphon pipe which acts as a hydraulic accumulator. When there is no flow in the siphon pipe the water level will be nearly the same as at the upstream end in the drainage well. When the water in the PVC accumulator reaches a certain level, the stored water is quickly emptied by a simple flushing system requiring no energy. The sudden lowering of the level of water in the accumulator causes flow in the siphon pipe. This rapid flow flushes out any air bubbles within the siphon (Fig. 6). It is important that the pipe work in the accumulator is sized to achieve a sufficient flow rate and duration to remove the air bubbles from the siphon tube. The flow continues until the siphon is clear of air bubbles when the flushing system stops the flow. The water level will then rise in the well and flushing system up to the predetermined level in the accumulator. The flow is then primed and the flushing cycle starts again.

One of the reasons for the selection of the system, which had never been used in the UK before although at various sites in Europe (Gress 1996), was it's simplicity of operation and cost since the siphon does not require any source of power to operate. In addition the siphon system is extremely flexible and is able to respond to changing conditions. The system as designed was capable of pumping at least twice the required rate to allow for increased inflow due to higher predicted rainfall intensities. Furthermore in the event of major climatic and significant ground water changes, the system if required could be retrofitted with additional tubing to increase the discharge output of the system.

The diameter of the siphon tubes can be varied from 10 mm and 20 mm achieving flows of around 2.9 m^3/day up to 24 m^3/day per well respectively and sufficient space was allowed for such a retrofit of multiple tubes within the design of the ducting of the drain system.

4.4 *Electro-pneumatic drain principle*

The electro-pneumatic drain has been developed to stabilise landslides by intercepting groundwater at greater depths or lowering groundwater to lower levels than that capable using siphon drain techniques. The electro-pneumatic drain design principle is similar to that for the siphon drains with a network of manholes and ducting for electrical cabling and pipework for water discharge. The wells are equipped with slotted PVC well casing of 110 mm internal diameter and centralisers and fine gravel filter to ensure its central location and filtering of incoming water.

The system is driven by an air compressor and the air reservoir tank that can be located up to 3 km away from the control panel and electrical supply.

When the water fills the well, it enters the pump by a non-return ball valve and fills it. The low level sensor is reached, the water level continues to rise, and when it finally reaches the high level sensor, an electrical connection is made by water conductivity. These two sensors are linked to a relay, which opens a solenoid and compressed air passes through the inlet air tube

Table 2. Summary of permeability calculations.

	Test no/depth (mbgl)	Calc. method	Calc.permeability (m/s)	Av. permeability (m/s)
BH 102	1/4.9 (UGD)	BS593 Hvorslev	1.29×10^{-6}	1.42×10^{-6} (UGD)
	2/8 (UGD)		1.55×10^{-6}	
	3/11.7 (GCD)		3.95×10^{-6}	2.02×10^{-6} (GCD)
	4/18.5 (GCD)		9.65×10^{-8}	
BH P2	1/5.4 (UGD)	BS5930 Hvorslev	2.95×10^{-6}	2.60×10^{-6} (UGD)
	2/8.5 (UGD)		2.30×10^{-7}	
	3/12.5 (GCD)		1.20×10^{-7}	1.25×10^{-7} (GCD)
	4/18 (GUD)		1.30×10^{-7}	
BH 101	Full depth	Back analysis: pump test data & Dupuit- For- cheimer equation	10 m spacing 20 m spacing	3.79×10^{-6} 4.43×10^{-6}
TP3	Rising head Pit at 4.5 m (gravely clay)	BRE 365 Method CIRIA 113		1.58×10^{-6} 1.13×10^{-6}
TP5	Rising head Pit at 3.4 m (sandy clay)	BRE 365 Method CIRIA 113		5.67×10^{-6} 4.12×10^{-7}

to the electro-pneumatic drain, filling it with air and pushing water out of the through the outlet tube. A non return ball valve prevents any water coming back into the pumping chamber. An electrical cable, which is connected to the water level detector, is linked through the duct to a control panel. The control panel contains a relay and solenoid that controls the operation of the compressed air-supply pumps. The solenoid switch controls the compressed air supply allowing it to pass from the air compressor and its air tank to the compressed air inlet tube to the pump.

5 SCHEME DETAILS

5.1 *Slope works*

An extensive series of instrumented in situ permeability and pumping tests were carried out to define the aquifer characteristics and as anticipated within a complicated heterogeneous landslide mass the hydrogeological conditions were anisotropic and varied throughout the site. In addition field trials to confirm the practical operation and performance of the siphon well system were carried out. The results confirmed that the required abstraction rates and groundwater lowering could be achieved. Estimates of permeability were calculated from a number of in-situ tests and from back analysis of pump test data. Results of these tests are summarised in Table 2. They indicate a range of permeabilities for the Upper Greensand (UGD) and Gault Clay (GCD) landslide debris of between 1.29 to 2.95×10^{-6} m/s and 3.95×10^{-6} m/s to 9.65×10^{-8} m/s respectively. The back analysed value of 4.43×10^{-6} m/s compared very well with the value estimated from previous studies of GDC of 4.9×10^{-6} m/s. The design requirements were specified in a schedule of minimum drawdown or trigger levels to be achieved in each siphon and electro-pneumatic wells and these were included in the performance specification.

The designed drainage system was constructed between September 2003 and October 2004 and comprised 151 deep drainage wells installed to depths up to 25 mbgl. The wells are equipped with slotted uPVC well casing of 110 mm internal diameter and centralisers and fine gravel filter to ensure its central location and filtering of incoming water. A total of 116 siphon wells were supplemented with 35 electro-pneumatic wells. The wells were connected by a system of trench drains, ducts and manholes, predominantly sited within the public highway for ease of access for construction and subsequent maintenance. They were arranged in three overlapping lines across the landslide with the elecro-pneumatic wells (Line 1) sited at the rear of the landslide where the required drawdown was greater than the capability of a siphon system (Fig. 7).

Each siphon well (Lines 2 & 3) was installed from the base of a manhole between 1.5 m and 3.5 m deep which connected the trench drains and improved the theoretical drawdown to approximately 12 m. The electro-pneumatic pumps were required to draw down to a minimum of 15 mbgl.

The total drainage installation quantities included a total installed well depth of 3180 m, 4800 m of twin wall plastic pipe, 1700 m of drainage trench, 206 manholes, 7500 t and 18,000 t of gravel trench backfill and spoil disposal respectively. Of a total construction cost for the project of £4.5 million, the wells, siphons and underground compressor chamber were £1.1 million and the drainage and infrastructure £1.4 million.

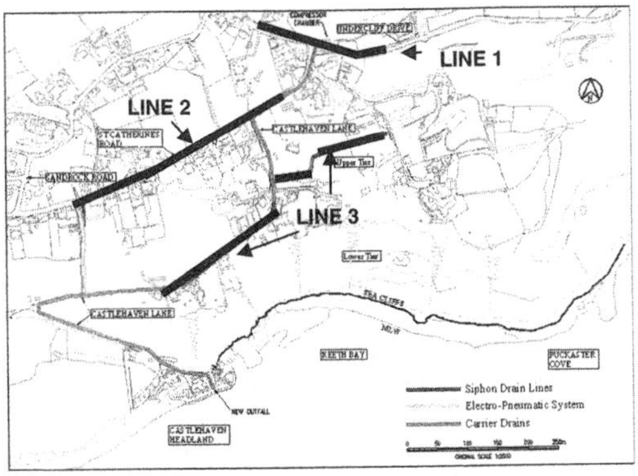

Figure 7. Layout of wells and drains.

5.2 *Foreshore works*

The foreshore coast protection works which cost £2.0 million consisted of a rock armour buttress revetment with a 3.5 m wide buried toe placed in front of the erodible 8 m high Sandrock cliff which formed the foreshore and the toe of the landslide. The rock armour, which consisted of 45,000 t of Carboniferous limestone, was imported by barge from Calais in northern France.

The primary armour ranged in size from 3 to 6 t placed on 5 to 500 kg sized core stone. The revetment had a 1 in 1.5 front slope with a 3.5 m crest at an upper level of 5 m above ordnance datum.

6 SCHEME PERFORMANCE AND POST CONSTRUCTION MONITORING

It was recognised that the performance of the drainage system would require regular monitoring and maintenance during its design life to ensure that the design objectives and assumptions were met and the system dealt effectively in the long term with the consequences of climate change and continued to comply with the performance specification.

A comprehensive system of instrumentation was installed as part of the construction works and comprised:

- Pressure transducers and dataloggers installed in observation wells between rows of siphon drain and electro-pneumatic pumping wells to measure water levels (9 no.);
- Vibrating wire piezometers installed in selected siphon and electro pneumatic wells and linked to multi-level dataloggers to monitor water level drawdown in the wells (42 no.);
- Vibrating wire piezometers installed and sealed at discrete levels within boreholes and linked to individual dataloggers to measure pore water pressures within the landslide (6 no);

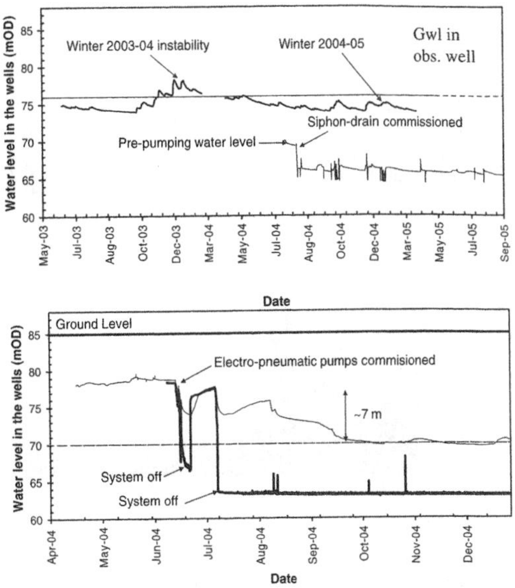

Figure 8. Groundwater response to pumping.

- Borehole inclinometers installed at various locations within the landslide to monitor lateral subsurface ground movements (4 no.);
- Survey of a network of ground markers by a global positioning system;
- An open channel flow logger to measure the discharged water flows within drainage pipe works (2 no.);
- An automatic weather station to monitor local meteorological input parameters during the contract and Defects Correction Periods.

The pumping system was commissioned in July 2004 and immediately there was a response within the pumping wells and in the observation wells situated between the lines of pumping wells. Typical results of the pumping and the drawdown of groundwater levels are shown on Figure 8.

The total discharge to the sea outfall has been monitored as part of the Environment Agency's conditions for discharge consent to the sea outfall.

The rate of discharge to the outfall with time for all pumps and drains, except the western limit of Line 2, is presented in Figure 9 and compared with the daily rainfall over the same period.

It is evident from these records that there is a rapid response to rainfall and that the monitored flows of up to 350 m^3/day do not exceed predicted long term base flow rates of approximately 441 m^3/day or maximum allowable rates of 500 m^3/day prescribed by the Environment Agency.

During the construction period the data from the instrumentation was reported initially two weekly

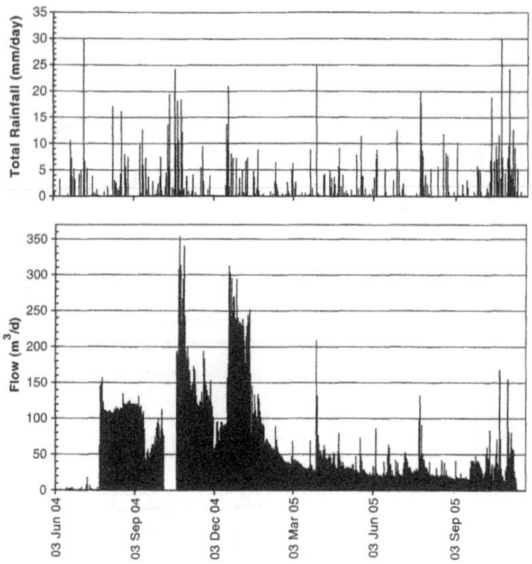

Figure 9. Rainfall and well system output.

reducing to monthly. Up to 1 year after construction the contractor and the specialist supplier (TP Geo) carried out the two monthly performance review which compared the water levels in each well with the minimum drawdown levels (Trigger Levels) given in the performance specification. Following the maintenance period the scheme responsibility transferred to the Coastal Authority (Isle of Wight Council – IOWC) and on site training was given to IOWC staff on all aspects of the scheme. In addition a detailed procedure for scheme performance, monitoring and maintenance was prepared. The performance review procedure included two water levels the "Alert" and "Action" level at 1 m and 2 m above the Trigger level respectively in each pumping well. If monitoring identified Action or Alert levels were operating in either single or groups of wells, specified checking procedures, data interrogation and as necessary maintenance responses were identified. Of particular importance was monitoring of the possibility of siltation, iron or calcium carbonate encrustation or bio-fouling which could impede the long term performance of the wells. Flushing of the system during the regular maintenance which is scheduled every 6 months, should deal with any siltation problems and additional purging with low concentration solutions of mild biodegradable organic acid are able to reduce bio-fouling.

With regular monitoring and maintenance it is anticipated that the siphon and electro-pneumatic well system will effectively improve the stability of the landslide throughout its design life whilst at the same time minimising the impact on sensitive environmental assets yet being able to respond to any changes in the climate that increase seasonal rainfall, groundwater levels and wave climate.

7 CONCLUSION

The major landslide movement at Castlehaven was caused by a combination of coastal erosion and high precipitation and groundwater levels. A close relationship between a 4 month antecedent effective rainfall of c. 380 mm, groundwater levels exceeding trigger levels and the onset of movement was established for the site area. Because of the high level of environmental protection at the site and an embargo on conventional ground strengthening techniques a design that reduced groundwater levels to equivalent summer levels, when movement did not occur, was designed and constructed using a system of siphon and electro-pneumatic wells. The design was robust and included for the effects of climate change. It allowed for ground water at ground surface and was capable of delivering twice the design flow rate in the event of predicted higher rainfall intensities. The siphon system is effective in use and environmentally sustainable; it is gravity fed and needs no power to operate. The system could also be retrofitted with additional capacity to further respond to climate change. The design approach aimed at improving the stability of the area to protect property and infrastructure but at the same time the environmental assets of the area were safeguarded by retaining appropriate levels of water within the slopes to maintain the important habitats.

ACKNOWLEGEMENTS

The authors would like to gratefully acknowledge the co-operation and assistance throughout the project of the Client, the Isle of Wight Council, the contractor Van Oord UK Ltd including T J Brent Ltd and TP Geo of France; Tony Bracegirdle geotechnical advisor to High-Point Rendel and the many colleagues at High-Point Rendel particularly Colin Reed, designer of the foreshore works and the construction supervision team of David Hattersley, Dan Squires and Peter Grice.

REFERENCES

Bomont, S., Fort, D.S. & Holliday, J.K (2005). Two applications for deep drainage using siphon and electro pneumatic drains. Slope works for Castlehaven Coast Protection Scheme, Isle of Wight (UK) and slope stabilisation for the Railways Agency, France. In, Proceedings of the International Conference on Landslide Risk Management. 18th Annual Vancouver Geotechnical Society Symposium.

Clark, A.R., Storm C.V, Fort, D.S & McInnes, R.G. 2002. The planning and development of a coast protection scheme in an environmentally sensitive area at Castlehaven, Isle of

Wight. Procurement International Conference on Instability, Planning and Management. pub. London:Thomas Telford. pp. 509–518.
Environment Agency, 1999. Overtopping of Seawalls, Design and Assessment Manual, Technical Report W178.
Gress, J.C.1996. Dewatering a landslip through siphoning drain – Ten years experiences. Proc.7th International Symposium on Landslides, Trondheim.Pub.
Hulme, M., Jenkins, G.J., Lu, X., Turnpenny, J.R., Mitchell, T.D., Jones, R.G., Lowe, J., Murphy, J.M., Hassell, D., Boorman, P., McDonnald, R., & Hill, S. 2002. Climate change scenarios for the United Kingdom: the UKCIP02 scientific report, Tyndall Centre for Climate Change Research, School of Environmental Sciences, University of East Anglia, Norwich, UK.
Hulme, M. & Jenkins, G.J., 1998. Climate changes scenarios for the UK: Scientific report. UKCIP Technical Report No. 1. Climate Research Unit, Norwich.
Hutchinson, J.N. 1991. Theme lecture: The landslides forming the South Wight Undercliff. International conference on slope stability engineering developments and applications, R.J. Chandler Ed., 157–168, London: Thomas Telford.
Hutchinson, J.N., & Bromhead, E.N., 2002. Isle of Wight landslides. In R.G. McInnes & J. Jakeways (ed.), Instability planning and management, 3–70. London: Thomas Telford.
Ibsen, M. & Brunsden D. 1996. The nature, use and problems of historical archives for the temporal occurrence of landslides, with specific reference to the south coast of Britain, Ventnor, Isle of Wight. Geomorphology 15. pp 241–258.
Ibsen, M-L. 2000a. The impacts of climate change and instability – some results from an EU LIFE project. In R.G. McInnes & J. Jakeways (ed.). Instability planning and management, 609–616. London: Thomas Telford.
Ibsen, M-L. 2000b. The impacts of climate change. In: R.G. McInnes & D. Tomallin (ed). European Commission LIFE project, LIFE-97 ENV/UK/000510. Vol. 1. Chapter 6.IPCC, 2002. Climate change 2001: Summary for policy makers. The Intergovernmental Panel on Climate Change third assessment report.
IPCC, 2002. Climate change 2001: Summary for policy makers. The Intergovernmental Panel on Climate Change third assessment report. Available at http:www.ipcc.ch/index.htm.
Moore, R., Lee, E.M., & Clark, A.R. 1995. The Under cliff of the Isle of Wight, a review of ground behaviour. Rendel Geotechnics. Pub. Cross.

Session 7
Safer societies and sustainable communities

Through effective risk management and informed, forward-looking planning decisions, we are endeavouring to improve the safety and sustainability of our communities. The insurance industry is also playing an important role in hazard management, but to the industry natural hazards are an ever- burgeoning issue. These issues must be considered in the context of the impacts of climate change.

Alum Bay, Isle of Wight, UK
Courtesy Wight Light Gallery, Ventnor, Isle of Wight, UK.

Landslides and Climate Change – McInnes, Jakeways, Fairbank & Mathie (eds)
© 2007 Taylor & Francis Group, London, ISBN 978-0-415-44318-0

Economic assessment of natural risks due to climate change. The case of an Italian mountain region

M. Brambilla & P. Giacomelli
Università degli Studi di Milano, Italy

ABSTRACT: This paper introduces the approach used to analyse the consequences of climate change on the Adda River basin (Lombardy, northern Italy); the area offers three main reasons of interest: it is one of the biggest in Italy; it is located in the richest region in the country and, thanks to its geomorphologic heterogeneity, could be affected by a wide range of natural hazards. The northern part, Valtellina, is a mountain area characterised by several hazards. The aim is to quantitatively assess the consequences of climate change on the socioeconomic system. The quantitative cause – effect approach is applied; climate change is the cause, and the effects are the outcomes on the social system. Such effects are described as "direct effects", directly tied up with "physical damages", and "indirect effects", due to the interruption of economic activities. Particular attention will be paid to extreme events; first of all, landslides.

1 INTRODUCTION

The assessment of environmental risks can be considered an important challenge for scientific research. Many aspects related to this topic need to be studied in more detail: the attempts to anticipate the risks (prevention rather than remedy), and therefore to forecast them, to assess the relationships between causes and effects, to balance the benefits with the costs associated to the control of risks.

The socioeconomic assessment of damages due to natural hazards is getting even more important as progressive climate change is threatening the safety of people and assets.

Extreme events such as floods, tornadoes and hurricanes are becoming stronger and stronger, and their increasing power could seriously compromise not only the structures involved, but also socioeconomic assets.

In the 45 years following the second world war, natural disasters have cost the Italian government 33,000,000 million liras (about 25,000 million euros in 2006), and landslides account for nearly 37% of lost lives (Catenacci 1992).

As a matter of fact, the interest towards direct and indirect damages due to natural hazard is increasing in the public opinion and decision-makers are asking for socioeconomic scenarios to face this occurrence.

The RICLIC project (Regional Impact of Climate Change in Lombardy Water Resources: Modelling and Applications) has been funded by the Regional Agency for Environmental Protection, the University of Milano, Bicocca and the main no profit foundation for Environment to assess consequences of climate change. In the project, socioeconomic damages caused by climate change in the region are investigated. This paper aims to explain the approach that it is going to be used to seek solutions to this problem.

The paper will analyse the approach applied in the RICLIC project to assess economic damages. The 1987 landslide in Valtellina will be taken as an example. The quantification of related socioeconomic damages will explain landslide consequences from a different point of view.

2 SITE SETTING

The RICLIC project has been created with the aim to assess regional impact of climate change on Lombardy water resources. In detail, the study will analyse the most likely impacts on one of the most important river basins in the northern part of Italy: Adda River. Identifying three macro areas, different from a geomorphologic point of view, as from a socioeconomic one, could be useful in order to develop the socioeconomic assessment. The upper part, Valtellina, is a mountain area, a large glacial valley whose economy is mainly based on viticulture, zootechnics and tourism. The central part hosts Lake Como; mainly because of its geophysical characteristics, the presence of a very deep lake with mountains overhanging its shores, the main inhabited centres are settled along the coastline. There are towns with tourist note and some important industrial sites. The lower part begins from the town of

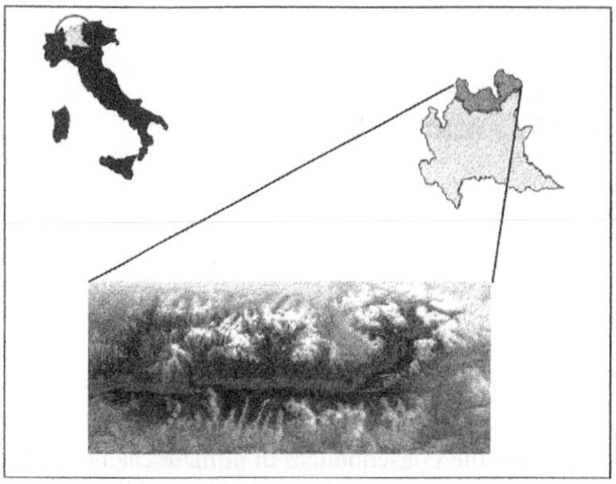

Figure 1. Valtellina location.

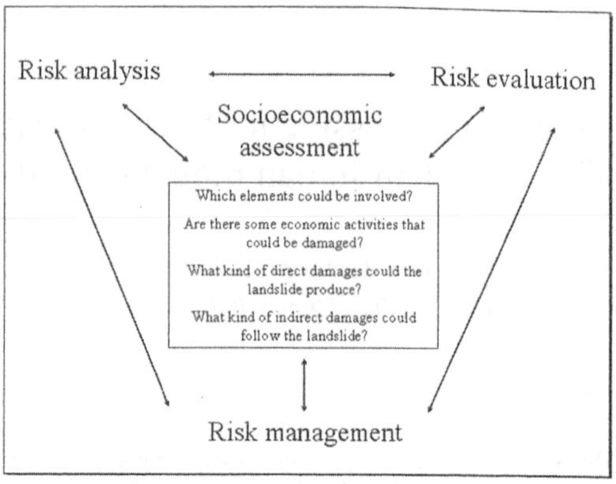

Figure 2. Integrated holistic concept of risk assessment, modified from Bell & Glade (2004).

Lecco, where the Adda River flows out from the lake, and ends in Castelnuovo Bocca d'Adda, where it enters the Po River. In this southern part, the Adda River flows through the Pianura Padana, the main industrial area of Italy. Figure 1 shows the study area.

Defining different scenarios is therefore useful in order to assess economic damages. In this paper only the possible consequences on the upper part, the most susceptible to landslide phenomena, will be discussed.

3 SOCIO-ECONOMIC ASSESSMENT

Socioeconomic assessment of environmental risk is a difficult, but very important matter in the management of territory.

The approach often used in risk analysis evaluates risk from hazard, vulnerability and value of elements involved (Varnes 1984). Nevertheless, this approach could not be easily used in a socioeconomic assessment, because it is not easy to determine the value of elements involved. Moreover, it could be difficult to agree with the identification of money value for human life, for ethical as well as practical reasons. Besides, the socioeconomic research has a different approach towards natural hazards that is centred on people and the different behaviour that people have towards them.

In the last years a great number of new approaches toward the estimation of economic damages due to natural hazard can be found in literature (Bell & Glade 2004, Bell at al. 2005, Australian Geomechanics Society 2000).

The economic approach discussed in this paper is based on the quantitative cause – effect correlation (Giacomelli 2005), which was developed and applied in the ALARM project (funded by EU, 2002–2004).

Such correlation is developed on three parallel levels: climatic, physical and socioeconomic levels.

After the determination of climate change scenarios, several physical effects of climate change on the investigated area will be identified (floods, landslides, drought).

The aim of the socioeconomic analysis is to determine the main economic drivers in the study area, the most important elements for the development of local economy and the weaknesses of the regional economic system.

In the socioeconomic cause – effect method, the phenomenon analysed is the cause, described by its physical characteristics, identified by its physical data, and effects are the structural damages to the elements directly involved by the phenomenon. After the assessment of physical effects, the socioeconomic analysis supports the geomorphologic analysis in the determination of economic effects. In Figure 2 it is possible to understand how socioeconomic assessment completes the holistic concept of risk assessment (Hollenstein, 1997).

First of all, the analysis starts with the study of meteorological data, carried out with the collection of a series of historical pluviometric data, maximum and minimum temperatures and their fluctuation in the last century. These meteo – climatic data will be useful for the analysis conducted by the other work packages involved. The impact of climate change on surface and underground water resources, the consequences of glaciers melting on the release of persistent organic pollutants (POP), the effects on agriculture and the hypothesis of rationalisation of water resources to optimise their management will be studied.

All the experts together will analyse, at least, physical consequences on the territory as a whole, combining their result to offer realistic scenarios on which estimate economic damages. One probable physical phenomenon will be considered for every scenario. The analysis starts with the description of the phenomenon itself; the study has to be made by technicians with an historical experience in geophysical analysis, and the first assessment must produce quantitative and spatial information that aim to understand the physical dimension of the phenomenon itself.

The second step is directed towards the assessment of physical effects; using GIS support, geophysical data are matched on socioeconomic data, which represent the key elements potentially involved. With geographic instruments it is possible to identify which elements (roads, houses, dams, and productive plants) are threatened in each step of the emergency.

According to this approach, landscape elements are not only considered as elements at risk, but first of all they are constituent of the socioeconomic structure and their role allows society to keep itself alive; the attention must be twofold for the elements on which is based the economic system: as a matter of fact, the interruption of their usual activities could have effects on the social equilibrium of the area.

This remark explains the importance of indirect damages in risk assessment. During the emergency, the socioeconomic damages could be not understood, but they can affect the regional economic system for long time.

4 THE IMPORTANCE OF WATER RESOURCE

What climate change could really cause on water resources is very difficult to predict, but studying the historical series it is possible to record that with the increasing in temperature, the managing of water resources could become even more difficult.

Interest in this study does not result only from damages that extreme phenomena could bring on the elements involved, but also from social implication connected to the management of this resource, especially during long periods of drought, when decision makers have the delicate problem to decide how to exploit water resources.

One of the most important social problems connected to water shortage is the use of water to produce hydroelectric power, instead of civil and productive uses. Every year, in summer, a water crisis takes place; all the local authorities in charge of the water management system are called together to decide how to solve the situation.

After the identification of physical damages, this analysis will be focused on the evaluation of economic damages due to climate change. This analysis is directed towards the assessment in two main scenarios: water shortage and water floods. Whereas management of floods involves the assessment of direct and indirect damages, the scenarios interested by water shortage are more important for the assessment of indirect damages on a regional scale; as a matter of fact, the management of water resources involved a great number of interests, from a political and an economic point of view, and the choice between different solutions could compromise the social equilibrium.

5 CLIMATE CHANGE

According to the results achieved by the team of meteorologist working in this project, in Lombardy region there are strong evidences of climate change. From the eighties, all the weather stations sited in Lombardy have registered abnormal conditions of temperature that proved a significant climate change at a regional scale. This alteration is revealed by an increase in heat waves frequency in summer and thermal zero elevation in winter, with snowfall at higher altitude. Moreover, the occurring climate change is the main responsible of the increasing extreme events, such as drought and floods. These phenomena must not be individually considered, because their occurrence involves other phenomena: first of all landslides.

6 CASE HISTORY

To explain the importance of socioeconomic analysis in the occurrence of landslide, the one that occurred in Valtellina valley on July 1987, from the mountain of Pizzo Coppetto is given as an example. Such example is given to understand how socio-economic analysis could help decision-makers in managing natural hazards and to comprehend how much the area is vulnerable to natural phenomena.

6.1 *Valtellina flood*

In summer of 1987 in Valtellina a series of natural calamity took place. After some days of heavy rain (on 17th July, fell 305 millimetres of rain down in 24 hours: a quarter of the total water that usually rains in this valley within the whole year), on 18th July 1987 in the town of Tartano, a huge mass of debris destroyed two residence buildings, killing eleven people. Afterwards, the Adda River overflowed and flooded the town of Ardenno, in the central part of Valtellina, interrupting connections with the eastern part of Sondrio Province; also the railway was swept away by the water. Many people were evacuated from their houses. Many other villages were flooded.

Bridges, farms, cultivated fields, graveyards and sheds were swept away: hundreds of animal carcasses floated on the huge lake that took up part of Valtellina. Also Sondrio, the chief town of Province, was in danger: Mallero torrent seemed to overflow. The same thing occurred in Morbegno, while the Adda River overflowed, flooding the valley floor in the industrial area between Talamona and Morbegno.

Meanwhile, the inhabitants of the upper part of Valtellina were evacuated. Moreover, the connections with Swiss were interrupted: the Piattamala custom was completely unfit for use.

On 28th of July a landslide broke off from the Cime di Pedasco, Pizzo Coppetto (3066 meters high mountain), swept away and completely destroyed Sant'Antonio Morignone (Valdisotto).

The town was previously evacuated and this caution saved the majority of people; despite it, the landslide swept a team that was working on reinforcement in Sant'Antonio and some inhabitants of the village of Aquilone were killed, wrongly judged out of danger.

Nobody could expect the pressure wave due to forty millions of cubic meters of debris flow, and the landslides power itself, that climbed the mountain on the other side of the valley for several hundred meters. The strength of landslides was so big that the debris climb up on the opposite mountainside like a huge wave, destroying the town settled at the bottom of the valley. The debris blocked the usual stream of the Adda river, forming an artificial lake that impend over the valley below. The threat of "Vajont effect" was felt: a new fall of material from the mountain could produce a new catastrophic wave of flood. Within the whole month of August, civil protection succeeded in bring the valley to normality, and in keep the lake under control, slowly emptying the valley from water and debris.

Valdisotto suffered the greatest amount of damages: 947 people were evacuated during the flood, 144 houses and a hundred of rural buildings were destroyed by the landslide, leaving 407 people homeless. Several productive activities were involved leaving 50 people out of work. Two months were needed to build temporary tracks; meanwhile, alternative routes were used.

Damaged areas are not restored yet; only some works of land management are already been carried out.

7 FUTURE SCENARIOS

As explained in the previous paragraphs, the climate change that is already being carried out is showing its effects with the intensification of extreme events. In this area, giving its physical and socioeconomic characteristic, the relevant intensification of extreme events due to climate change could bring damages even more important that those described for the landslide of Pizzo Coppetto.

Following the information of Sondrio Province, that rules Valtellina, is quite difficult to assess the economic damages that followed this event, even after twenty years. The government set aside 2,400,000 millions liras (about 2400 millions euros 2006) for the "rebuilding and development plan" but Lombardy region is still working on the land management programme.

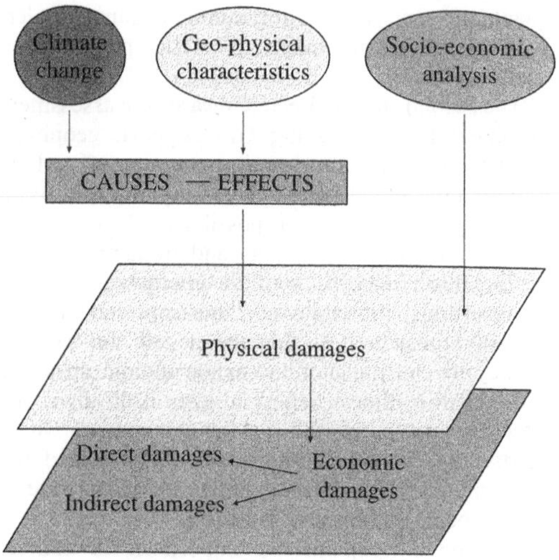

Figure 3. Methodology scheme.

Nobody is able to exactly quantify in economic terms the cost of Valtellina landslide. Certainly, this flow of money was not only seen as a compensation of damages, but was also a new source of earning, that brought a big amount of money to the areas involved.

8 METHODOLOGY

The method applied in the RICLIC project is based on the quantitative cause – effect correlation, where cause refers to the phenomenon itself, and effects concern all the economic damages, either direct or indirect. In Figure 3 is our methodological approach is briefly described.

The assessment of socioeconomic effects in the evaluation of consequences of natural hazards needs the contribution of physical experts who study climate and environmental characteristics. The most important characteristic of this project is the presence of different research teams, experienced in different aspects of global warming. To get to successful results, the information obtained has to be shared, and conclusions have to be drawn together. In Figure 3, the different informative levels, separately studied by the different groups, are in the circles; in the first step, the socioeconomic study is directed towards the gathering of data and the comprehension of historical series, drivers and weaknesses. The main target of the project is the characterisation of the most important elements, and their role in the developing of regional economy. In the second step the conclusions drawn by the different groups are shared and all the groups together determine the physical effects. The determination of physical effects has the aim to forecast the effects on the territory with changing climate conditions.

The approach used to study the economic assessment of damages and risk is the same previously used with the ALARM project: the cause – effect approach (Giacomelli 2005); while in the previous case the analysis was carried out only for a small area, a little mountain town, in this case the study is directed towards the comprehension of regional impacts, on an area with different socioeconomic characteristics. The comparison between the different working packages has produced a new informative level, the so called, "physical effects", whose results will be superimposed by those derived from the socioeconomic analysis. For this analysis it would be useful to use a GIS support, in order to better understand the connections and relationships between environmental and socioeconomic scenarios.

After the identification of physical effects, the socioeconomic analysis will help the study of the economic damages. Economic damages are divided into two categories: direct damages, defined as the cost of recovery or rebuilding of exposed elements, and indirect damages, coming from the interruption of economic activities of elements directly or indirectly involved. This distinction is important because, in case of involvement of structures that are important for the socioeconomic equilibrium, their temporary unavailability could compromise the usual activities, with economic indirect damages that may be even bigger than direct ones. It is also important to specify that indirect damages could weight upon a larger area and last for a longer period of time.

These characteristics determine the importance to analyse indirect damages.

9 TOURISM IN VALTELLINA

Tourism represents more than one third of the overall added value produced in Valtellina, (3836 million euros in 2004; source UNIONCAMERE). As a matter of fact, the service sector accounts for 69.1% of the total province income; this is an important result, if considered that service sector in Valtellina is tourism oriented.

The development of tourist activities in Valtellina experienced its greatest growth in the second part of twentieth century. The beginning of tourist industry was tied with tourist flows coming from the plain of Lombardy. With the development of communication lines and the extension of holiday periods, people gradually chose to change their destinations and to spend their longer holidays in mountain resorts far away. Weekend tourism and commuter tourism are still an important resource for Valtellina. Besides, a great number of houses are used as holiday houses, and it is not possible to gather statistical data about these tourist facilities.

As happened to the most important mountain resorts of Alpine bow, tourist activities are becoming different in the last decades; hence Valtellina started to offer to tourist demand a larger and larger number of sport activities, in summer and winter season, matching them with other attractiveness, first of all thermal tourism. Valtellina has hundreds years of experience in this field, that draws tourists every month of the year.

Another important characteristic of tourism in Valtellina is that this economic activity is not an economic system apart, but an economic integrated system with different activities; in fact the seasonal nature of tourism does not allow resident to make it their only job.

9.1 *Infrastructure*

Valtellina has a peculiar geomorphological structure: a long mountain chain in an east – west direction separates the large valley from the pre-alpine area. This conformation gives Valtellina its distinctive characteristics, which make it different from the other Alpine valleys, generally settled in a north – south direction.

For this reason, the State Highway SS 36 is the faster and most convenient road to reach Valtellina from Milano, the main centre in northern Italy; this highway reaches Lecco, on the eastern branch of Lake Como, and follows the eastern coast of lake up to Colico. After Colico, the road splits into two parts, one going in Valchiavenna towards the Spluga Pass, and the other entering Valtellina with the State Highway SS 38 towards Stelvio Pass.

Moreover, the only railway connection between Valtellina and the plain passes from Colico. This factor helps to understand the importance of this junction for Valtellina socioeconomic equilibrium.

10 THE VALTELLINA LANDSLIDE: EFFECT ON TOURISM

It is interesting to study statistics based on tourism in Valtellina in the last twenty five years. The whole area is characterised by a growing trend in the number of tourist facilities, as well as an increasing number of tourists. The study of tourist presences data allows an understanding of the impact that Valtellina landslide produced, not only on the structures involved, but also on the economy of the valley. The aggregate trend of tourist presence in Sondrio Province can be analysed in Figure 4.

Tourist economy has been characterised by a constant growth in time in the last twenty five years. Nevertheless, if the analysis is focused only on the two main tourist resorts situated in Alta Valtellina, Bormio e Valfurva, the result is different.

Alta Valtellina represents the most important tourist area, also thanks to Stelvio National Park and the biggest Alpine Himalayan glacier. Focusing the attention on the period elapsed between 1980 and 2005,

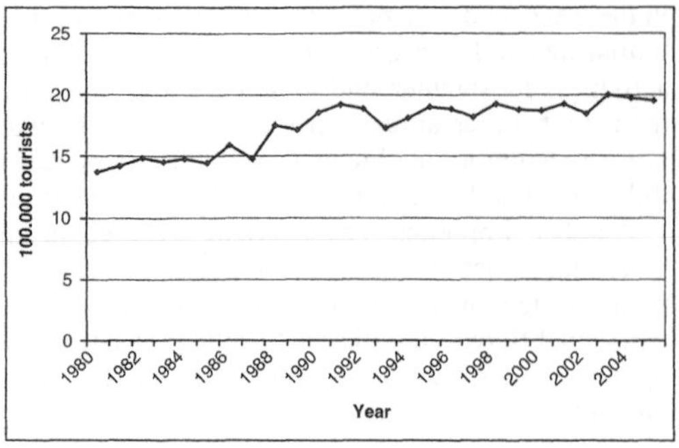

Figure 4. Tourist presence 1980–2005.

Table 1. Variation ratio in tourist presence.

	Bormio %	Valfurva %	Sondrio province %
1980			
1981	−7	13	4
1982	2	14	4
1983	−4	−14	−2
1984	3	3	2
1985	4	9	−2
1986	4	15	9
1987	−17	−25	−8
1988	22	23	16
1989	−9	14	−3
1990	13	9	8
1991	3	−9	3
1992	−7	−6	−2
1993	−16	−1	−9
1994	40	1	5
1995	6	3	4
1996	−1	−1	−1
1997	1	−7	−4
1998	2	7	5
1999	−1	−1	−3
2000	−6	−6	−0.3
2001	2	4	3
2002	−3	−15	−5
2003	4	9	8
2004	−5	−2	−1
2005	0.2	−19	−1

it is possible to see that effects due to the Valtellina landslide have effected the Alta Valtellina more than upon Valtellina as a whole. In 1987, both Bormio and Valfurva suffered the outcomes of Pizzo Coppetto landslide in terms of a decrease in tourist flow. As shown in Table 1, the greatest percentage fluctuation in the last twenty five yeas is registered both for Bormio and Valfurva in 1987, but there are not evidences of the same fluctuation for Sondrio Province as a whole. The phenomenon did not weigh upon the whole province in the same way. This is due to the interruption of State Highways SS 38, which connects this important tourist area with Colico.

Furthermore, in 1988 the growing trend restored the former tendency. This result highlights the importance of roads in the determination of indirect effects. The interruption of the most important route leading to the upper part of the valley caused a decrease in tourist flow in the Alta Valtellina.

As the data shows, the negative effect did not last for long. By 1988 the tourist flow had already recovered the growing trend, begun in the first half of the eighties.

A first analysis has assessed that such a decreasing in tourist flow generated to Alta Valtellina nearly 3.6 million euros (2006) of indirect damages.

It is also supposed that tourist flow decreased in the whole province after Valtellina landslide, because of the fear that people developed towards such a phenomenon.

11 CONCLUSION

The problems related to the assessment of damages due to climate change in a mountainous area have been analysed. Starting from the landslide that affected Valtellina in 1987, the effects registered on tourism have been shown.

The approach that is used in the RICLIC project has been presented. The contribution of each work package has outlined the scenarios of climate change, studied to assess the effects. The integration between the different work packages is basic to underline the weaknesses and the emergency of the physical system. Once different scenarios are outlined, the assessment of direct and indirect damages will produce the economic assessment of climate change. At present, only the initial steps of the work (data collection and analysis) are being performed; this paper aims to present the approach and to collect remarks and suggestions to improve the analysis and to get better results.

The greatest landslide that affected Valtellina is taken as an example to present an ex-post assessment of economic consequences, to understand the importance of a socioeconomic point of view in natural hazards assessment and management and to understand the dimension of economic damages. Moreover, this is an example of what could follow an intensification of extreme events in this area, even more often with climate change.

As the tourism activity produces more than one third of the whole added value, a phenomenon like the one that happened in Valtellina could seriously compromise the general economy of Sondrio Province, especially in the occurrence of climate change, which is bringing about an intensification of extreme events.

The results that will follow this analysis will contribute to hazard management and could help decision makers in the implementation of land management planning. Without a careful and effective management of territory, the increase and intensification of extreme effects could represent a significant threat to the economy and the equilibrium of this part of the valley.

12 PROJECT DESCRIPTION

This research is carried out in the framework of RICLIC project, funded by Università degli Studi di Milano Bicocca, Fondazione Lombardia per l'Ambiente (Lombardy Environment Foundation) and Regional Agency for Environmental Protection.

This paper has been discussed by both authors. Paolo Giacomelli wrote the paragraphs 2, 3 and 11 and Marta Brambilla wrote the paragraphs 1, 4, 5, 6, 7, 8, 9 and 10.

REFERENCES

Aleotti, P. & Chowdhury, R. 1999. Landslide hazard assessment: summary review and new perspectives. *Bulletin of Engineering Geology and the Environment* 58: 21–44.

Ashby, G. 2002. Development of a Risk Management Strategy for Part of State Highway 73 in the South Island of New Zealand. *Presented at the New Zealand Society of Risk Management Conference, Wellington, New Zealand, October 2–4.* 13p.

Bell, R. & Glade, T. 2004. Multi-hazard analysis in natural risk assessment. In Brebbia, C.A. (ed.), *International Conference on Computer Simulation in Risk Analysis and Hazard Mitigation, Rhodes, Greece.* WIT Press. 4: 197–206.

Bell, R. et al. 2005. Risks in defining acceptable Risk Levels. In: Oldrich, H. et al. (eds), *International Conference on Landslide Risk Management, Vancouver, Canada:* 7p. Balkema.

Blöchl, A. & Braun, B. 2005. Economic assessment of landslide risks in the Swabian Alb, Germany – research framework and first results of homeowners' and experts' surveys. *Natural Hazards and Earth System Sciences*, 5: 389–396.

Cardinali, M. et al. 1993. Progetto AVI – Relazione di sintesi. *Consiglio Nazionale delle Ricerche – Gruppo Nazionale per la Difesa dalle Catastrofi Idrogeologiche*: 39 p.

Cardinali, M. et al. 2002. A geomorphological approach to the estimation of landslide hazards and risks in Umbria, Central Italy. *Natural Hazards and Earth System Science*, 2: 57–72.

Catenacci, V. 1992. Geological and geoenvironmental failure from the post-war to 1990, Italy. *Memorie Descrittive della Carta Geologica d'Italia* 47: 301 p.

Chowdhury, R. & Flentje, P. 2003. Role of slope reliability analysis in landslide risk management. *Bulletin of Engineering Geology and the Environment* 62: 41–46.

Committee on Assessing the Costs of Natural Disasters, National Research Council (USA), 1999. *The Impacts of Natural Disasters. A framework for loss estimation.* National Academy Press.

De Amicis et al. 2003. La valutazione del rischio di frana. *Aestimum*, 42: 31–52.

Finlay, P.J. & Fell, R. 1997. Landslides: risk perception and acceptance. *Canadian Geotechnical Journal* 34: 169–188.

Finlay, P.J. et al. 1999. Landslide risk assessment: prediction of travel distance. *Canadian Geotechnical Journal* 36: 556–562.

Giacomelli, P. 2003. The economic evaluation of landslide's risk. *Discussion paper Prepared for IV ALARM Meeting*, Santander.

Giacomelli, P. 2005. *Economic evaluation of risk. The case of a mountain area.* Aracne ed.

Glade, T. 2001. Landslide hazard assessment and historical landslide data – an inseparable couple? *Advances in Natural and Technological Hazards Research.* eds. Kluwer Academic Publishers: Dordrecht: 153–168.

Glade, T. 2002. Ranging scales in spatial landslide hazard and risk analysis. In *Risk Analysis III*, CA Brebbia (ed.). WIT Press.

Glade, T. et al. 2005. *Landslide Hazard and risk.* Wiley.

Guzzetti, F. et al. 1994. The AVI Project: A bibliographical and archive inventory of landslides and floods in Italy. *Environmental Management* 18, Number 4: 623–633.

Guzzetti, F. et al. 1999. Landslide hazard evaluation: a review of current techniques and their application in a multi-scale study, Central Italy. *Geomorphology* 31: 181–216.

Guzzetti, F. 2000. Landslides fatalities and the evaluation of landslide risk in Italy. *Engineering Geology* 58: 89–107.

Hollenstein, K. 2005. Reconsidering the risk assessment concept: Standardizing the impact description as a building block for vulnerability assessment. *Natural Hazards and Earth System Sciences* 5: 301–307.

Lee, S. et al. 2001. Regional Susceptibility, Possibility and Risk Analyses of Landslide in Ulsan Metropolitan City, Korea. *Geoscience and Remote Sensing Symposium, 2001. IGARSS '01. IEEE 2001 International* 4: 1690–1692.

Quadrio Curzio, A. 1993. *Valtellina: profili di sviluppo.* Sondrio: Credito Valtellinse.

Varnes, D.J. 1984. Landslide hazard zonation: a review of principles and practice. *Natural hazards,* 3. Unesco Press, Paris, 63 p.

Landslides and Climate Change – McInnes, Jakeways, Fairbank & Mathie (eds)

Interpreting detailed site issues at a strategic level through local policy

W.E. Perera & C. Mills
Isle of Wight Council, UK

ABSTRACT: This project investigated the interpretation of detailed procedures for policy decision making policy at a strategic level. Local Development Frameworks were introduced through the Planning and Compulsory Purchase Act 2004, to provide scope for an approach to spatial planning that is visionary, wide-ranging, participative and deliverable. They provide local authorities with an opportunity to take a fresh look at their areas; and develop strategic approaches to spatial planning that deliver sustainable development and reflect the aspirations of local communities. Sustainability appraisal is the tool through which to assess the potential social, environmental and economic effects of the Core Strategy. Our experience shows that it is difficult to use the sustainability framework other than as a generic tool, because of the strategic nature of Core Strategies. It is likely that at the strategic level, similar difficulties will be encountered with other processes, such as Appropriate Assessment and Strategic Flood Risk Assessment.

1 INTRODUCTION

The Introduction of Local Development Frameworks has coincided with recognition within the planning profession of the need to address climate change.

Climate change impacts upon physical environments – it has a clear spatial context. But the link between physical impacts and the response of governments and local authorities is still developing through policy tools such as LDF's.

However there is no doubt that climate change will have physical impacts and given the geographical diversity of areas, impacts will vary between and within regions. We may be able to easily see or measure the impact it can and will have through sea level rise, landslides, changes in rivers or trees and crops; but just as important are the impacts that it will have on the places themselves.

There is documented evidence throughout history of humans adapting to physical environments, but in the 21st Century is our response more likely to be governed by choices made at a regional or local level and regulated through policy making and control?

2 DEVELOPING AND MANAGING ADAPTATION

A recent consultation by DEFRA (DEFRA 2005) sought views on the development of an adaptation policy framework. They found that throughout the UK, adaptation, although widespread, appeared to be ad hoc and reactionary, rather than proactive or anticipatory. This research concluded that adaptation needed to be embedded in a variety of agendas including economic, regeneration and social.

Even though this research focused entirely on adaptation to climate change the conclusions in terms of response would equally apply to specific issues related to climate change, such as landslides.

2.1 *The role of planning*

Given the results of the DEFRA findings it would appear that one of the major roles that planning policy and LDF's in particular could provide is that of being proactive and anticipatory, particularly given that the new framework is supposedly visionary and wide-ranging.

The central role for government's and local council's should be leadership – both in terms of setting agenda's and applying it to their own practices and the LDF is a critical tool for this role. The recent Government White Paper on strong and prosperous communities (DCLG 2006) stressed that leadership should be the single most important driver of change and improvement in local authorities.

In a recent article in Planning Magazine, (Morris, 2006) Planning was described as having a substantial legacy, with a clear role to encourage buildings and infrastructure to take account of climate change, particularly by limiting development in floodplains. The article discusses the Stern Review of the Economics of Climate Change and goes on to point out that the planning system will be a key tool for encouraging both private and public investment in locations that are vulnerable to climate risks.

The review makes the link between the role of planning and its influence over the market in terms of providing certainty. For example, policy on sustainable building practice will be needed prior to the market providing sustainable products.

The former Office of the Deputy Prime Minister (ODPM), now know as the Department for Communities and Local Government (DCLG) document "Sustainable Development Action Plan" states that a good starting point for creating sustainable communities is that they are well planned. Hence LDF's!

There are a variety of action plans, strategies and statements, all of which have a role to play in the development of sustainable communities.

Government sees the newly re-branded Sustainable Community Strategies as being the key to the local delivery of sustainable communities, delivered through local area agreements. But local area agreements are focused primarily on service provision, and there is little obvious link between this and addressing more strategic issues such as climate change.

Planning Policy Statement 12: Creating Local Development Frameworks (PPS12) sets the scene, ensuring that LDF's "must" have regard to other relevant policies and strategies at local and regional levels, particularly community strategies (ODPM 2004). It goes on to show how LDF's need to establish a clear chain of conformity with national planning policy and regional spatial strategies. This is supplemented by PPS1 – delivering sustainable development (ODPM 2005).

With regard to climate change in particular PPS1 states "development plan policies should take account of environmental issues such as mitigation of the effects of and adaptation to climate change through the reduction of greenhouse gas emissions and use of renewable energy" (ODPM 2005).

The role that regional agencies have through developing RSS combined with the roles of Local Authorities, with their associated Local Strategic Partnerships and delivery plans such as Local Area Agreements would therefore appear to be fundamental to addressing climate change or landslide issues.

Later this year (2006) the UK government is planning to consult on a PPS on Climate Change and Planning. It is likely that this will set out a national policy framework for plan making at a regional and local level. It is likely to focus on how Government expects participants in the planning process to work towards the reduction of carbon emissions in the location, siting and design of new development.

The draft South East Plan (RSS) has an accompanying climate change mitigation and adaptation implementation plan which sets out the actions required by different organisations and stakeholders to mitigate and adapt to the effects of climate change. In particular Policy CC2 of the RSS deals with adaptation, guiding strategic development to locations offering greater protection from impacts such as flooding, erosion.... However it goes on to state that adaptation and mitigation is still at an early stage in the region.

Directly linked to the LDF process, sustainability appraisal under the 2004 Act has been designed to incorporate the full requirements of EU Directive 2001/42/EC on the "assessment of the effects of certain plans and programmes on the environment", known as the Strategic Environmental Assessment (or SEA)Directive. It is perhaps this process that gives us the greatest opportunity to be proactive and anticipatory about physical process such as landslides.

Community involvement at different stages of the planning process is seen as having considerable benefits through improving understanding of the issues facing developers, local communities, stakeholders and the local planning authority (LPA), and helping to reduce conflict. It is aimed at making the whole planning system more transparent and to encourage people who would not normally have been involved in planning the communities of the future.

But community involvement is a very fashionable term. Together with terms such as participation, consultation and partnership it is used by many people without a real understanding of what it means, and it can easily mean different things to different people.

2.2 *Putting theory into practice*

From the variety of guidance that is available to us, there would appear to be strong links from central through to local levels to start to provide a consistent approach to climate change and its associated problems. But is this working in practice?

PPS1 should mean that everyone needs to look at climate change whilst considering regional and local level plans, but very few RSS are approved and the emerging PPS on climate change will be developed too late to have a direct impact on their content.

In practice, whilst the issues of climate change has been considered through the sustainability appraisal (SA) of the Island Plan Core Strategy, it has been difficult to use the sustainability framework other than as a generic tool because of the strategic nature of the document.

SA also works to a different time horizon – not only looking at the policies contained within the Core Strategy, but in addition, the longer-term effects of the plan through to 2030.

In addition, when dealing with the impacts of climate change we are inevitably dealing with uncertainties. What we do know are the consequences of certain things. For example the consequences of flooding are well known. Hence the outcomes of an increase in flood frequency and magnitude can be determined with considerable confidence, even if the probability of flooding events remains uncertain. Planning has historically dealt with things that are planned to happen, or actually happen with some degree of certainty; dealing with probabilities and uncertainty needs a complete change of approach and may be difficult to drive forward.

Coastal areas are particular exposed to the impacts of climate change through risk of erosion and land instability and sea level rise. Whilst the issues facing coastal areas may well be discussed in detail in Shoreline Management Plans (SMP's) there is little direct linkage between SMP's and LDF's at present. There needs to be greater linkage between the two. We need to develop more partnership working with services and agencies. This means the direct involvement of planners in the production of SMP's as well as linking the two processes. Strategies and documents need to be developed together, not remain the remit and priority of particular services or organisations.

A common theme of the environmental assessment processes that are coming to bear on the new system of plan-making is the early consideration of key issues in the development of options and policies for such plans. Draft DCLG guidance suggests that as with Sustainability Appraisal (incorporating the requirements of EU Directive 2001/42/EC on the assessment of the effects of certain plans and programmes on the environment through a Strategic Environmental Assessment) Appropriate Assessment, or perhaps as it should properly be referred to as 'Habitats Regulations Assessment', should develop its scope at the same time as SA, and then be used as one of the tools that will help inform the selection of options. The experience of applying this retrospectively to the Island's Core Strategy has not been an easy one. The problems, timing and retro-fitting aside (which limit the actions you can take to ensure a plan or policy has no adverse effect) are related to scale. The Core Strategy sits uncomfortably between regional, and area or site specific plan-making, in terms of assessment of impact. It is debatable when interpreting PPS12, how specific a core strategy should be in the allocation of development, particularly housing numbers. If such a plan indicates levels of growth, timescales and broad areas without being explicit, then it is difficult to carry out impact prediction. This places a greater emphasis on lower tier plans delivering both development and environmental assessment.

The relationship between the core strategy and lower tier plans when considering Strategic Flood Risk Assessment (SFRA) is a similar story. It may be sufficient for a core strategy to identify the need for an SFRA prior to site allocation, but PPS25 is clear on the requirement for SFRA, both to inform the SA of local development documents and to "inform the sequential approach to flood risk in the development allocation...process" (ODPM 2005). SFRAs should be undertaken for DPDs where allocations are explicit, to inform how suitable development zones are, and lead to a trickle-down of information to the methodologies of Flood Risk Assessments of specific sites.

Whichever way you look at it, Area Action Plans and site allocation DPDs are going to be key in delivering the real results intended from the array of assessment tools that planning is now expected to apply.

However, there is another planning angle which should not be forgotten when considering the effects of such assessments, and that is that solutions may lie in other DPDs. For example Development Control policies may be used to stipulate a whole host of generic policies such as sustainable drainage, grey water recycling, green roofs and other sustainable build options, that lead to more acceptable development, in more sustainable locations, that are more robust and "future-proofed" for the predicted changes in our climate.

Key to this process is how to get the local community involved, and it is proving a difficult challenge. The new planning system is all about frontloading. Early engagement is stressed as a way of bringing out all the issues and options before any of the difficult decisions are made. But it is not easy to get people involved at the Core Strategy stage.

Communities are interested in detail and action, the visionary aspect of core strategies makes it difficult to get people interested in wanting to discuss broad issues and options. This is illustrated by the fact that the Island Plan Core Strategy only received 200 responses.

At the time of writing we have not progressed an Action Plan to the options and preferred options stage, and perhaps it is only though reaching this next stage that we will see how willing local communities are to take part in developing plans for their communities.

3 CONCLUSION

Whilst there would appear to be a well documented chain of conformity and process for consideration in the development of local strategies and policies, in practice the linkages and seamless processes are not yet embedded in working practices.

The potential barriers to any meaningful discussion at the core strategy level are:

- Uncertainty about effects and impacts
- Complexity and challenges
- Different planning horizons
- Lack of joined up policy within and between different levels of decision making.

It is clear that it is the role of a wide range of agencies to implement adaptation and mitigation, and there is an associated responsibility of all agencies concerned to ensure that the local mechanisms and commitment that are needed are in place.

Public involvement is vital as communities need to feel they have been properly engaged in discussion about action to start dealing with the mitigation of the effects of climate change.

REFERENCES

Defra, 2005, Adaptation Policy Framework – a summary of the responses to consultation

DCLG, 2006, Government White Paper – Strong and Prosperous Communities

ODPM, 2004, PPS12: Creating Local Development Frameworks

ODPM, 2005, PPS1: Delivering Sustainable Development

ODPM, 2005, PPS25 (consultation): Development and Flood Risk

Morris, H, 2006, Stern warns over climate change, Planning.

Landslides and Climate Change – McInnes, Jakeways, Fairbank & Mathie (eds)
© 2007 Taylor & Francis Group, London, ISBN 978-0-415-44318-0

Drawing a line with coastal communities

J. Riby
Head of Engineering and Harbour Services, Scarborough Borough Council

J.G.L. Guthrie
Coastal Engineer, Royal Haskoning

S. Rowe
NECAG SMP2 Project manager, Scarborough Borough Council

ABSTRACT: It used to be far simpler; Engineers used a 0.3 mm ink pen, Planners a felt nib and Architects a soft B pencil. Was there a better inherent understanding of what was being portrayed, or were expectations just less? Now defining the line takes on a different perspective and significance. In developing a Shoreline Management Plan, one of the key issues is the definition and management of risk, in terms both of flooding and coastal erosion. However, the essential difference arising from one's point of interest remains. The paper considers different aspects of this relating to different areas of Scarborough Borough Council's coastline and areas adjacent to this, taking examples of hard rock cliffs, defended and undefended areas vulnerable to major instability of soft coastal slopes and in developed areas where the defence system has to be seen as a width rather than a single line. From this the paper explores the problems of delineating a line of risk; a line that potentially raises concern or even blight, but also acts, possibly to reflect best science and policy guidance. It is argued that what is needed is a distinction as to what is being portrayed rather than merely a precision arising from detailed analysis; drawing a line based on community understanding, a line with perspective.

1 INTRODUCTION

The southern limit of the NECAG (North East Coastal Authorities Group) SMP2 area is defined by the massive steep chalk cliffs of Flamborough Head. Overlain by a thin till cap, these cliffs are probably eroding at less than the nominal 0.1 m/year described in the SMP. Given the scale of the cliffs and with falls creating in many areas a deposit of rock at the toe of the cliff, partially protecting the cliff from direct erosion, this rate may be little affected by sea level rise over the next century. The main interest and value affected by erosion are the populations of cliff-nesting birds and the open scrub and agricultural land at the cliff top. With the single exception of the Flamborough Lighthouse, man has wisely never encroached within an area of perceived risk. The precision of assessing erosion rates has never been more than of academic interest.

Further north, within the area of the SMP, the colliery waste tipped frontages of the Durham Heritage coastline is possibly more complex. Rapid erosion, of up to 25 m/yr initially, has been recorded, of the artificially placed material, since tipping ceased in the early 1990s. This has slowed with erosion of 8 m/yr within 2 years of ceasing tipping and now being recorded as possibly as little as 0.5 m to 1 m/yr. As the blanket of material (large amounts of which were removed as part of the impressive regeneration scheme Turning the Tide) has been removed, the old harder Magnesian Limestone cliffs and headlands have once again start to dominate development of the frontage. Having been provided with this artificial protection over much of the last century, erosion of these natural cliffs may, in 10 to 30 years, recommence; although at what rate remains highly uncertain. Even determining when exposure of the old coastline may seriously occur is uncertain and with little previous record (the last unprotected maps being in 1858) there is little, apart from comparison with similar frontages, upon which to base future rates.

However, such uncertainty is not critical. The creation, through planning, of a buffer zone, extending back between 200 m and 400 m from the cliff, has generally, with a few local exceptions, provided a breathing space of centuries before conflict between natural the evolution of the coast and risk to property and life become a significant issue.

In both cases, the drawing of an erosion line, if of limited accuracy, for the next 20, 50 or 100 years is far from controversial and is little more than a nominal indication that there will be continued pressure on the coast and a reminder of the continuing erosion.

In other areas, such a line takes on a different significance. In many areas the uncertainty and complexity of defining the line (or lines) is as great or greater than in the case of the Durham coastline and such complexity, in the context of this uncertainty, is as much in relation to what a line means, as it is to defining the physics of coastal slope retreat.

While there is a growing expectation for people to be better informed, to have clear definition of risk, the understanding of the information and risk means different things to the individuals and organisations with interest in how our coast is managed. The paper considers several different sections of the coast from within the area of Scarborough Borough Council's (SBC) and adjacent sections of coastline.

2 A STARTING POINT

The aim of the Shoreline Management Plan is twofold: to identify areas at risk and to discuss and define policy for coastal engineering to manage this risk. The plan is looking forward 100 years but considering policy development in epochs of the next 20 years, 50 years, through to 100 years (and more) ahead. At best we tend to be working on monitoring which has occurred over the last 10 to 15 years, this often being no more than comparison of single profile data of cliff crest retreat. In a very few areas more detailed monitoring of sections of the cliff face have quite recently been started, but this has been ongoing for no more than 4 years. Examination of historical maps, with all the associated inaccuracies, gives a further indication of change possibly over the last 150 years, but even this, when looking, for example, at the records of the Durham coast, can often only give an assessment of past performance in relation to conditions prevailing over that period of time. There are, in addition therefore, changes in geomorphology and processes and even changes in resilience, as different geology becomes exposed or is eroded.

We are also faced with the uncertainties of land mass tilting and climate change, or sea level rise and the further uncertainty of how this will impact on erosion, coastal slope stability and retreat. Some work has been undertaken in addressing this latter issue but, in terms of a unifyingly clarity we are still some way off.

The approach adopted in the NECAG SMP2 has been to work with local information where possible, making individual assessments from this of current erosion and instability for each frontage of the 170 km long area of the coast. Unless specific assessments can be or have been made of the response to sea level rise, a generalised factor has been applied to current erosion rates to allow for sea level rise. Various simple approaches to this were considered; considering possible energy factors resulting from increased wave energy, giving factors between 50% and 400%, depending on the existing foreshore levels, simple roll back assessments; maintaining the coastal profile in relation to still water level, to relationships based on comparison of existing rates of sea level rise and future predicted rates. While considered on a case by case basis and taking into account the critical aspects in relation to policy, the latter simple, if precautionary, ratio of sea level rise was taken as a starting point. In the absence of other local considerations, the basic relationship has been assumed, that the Future recession rate =

Historical recession rate × future sea level rise
Historical sea level rise

This gives factors of ×2 and ×2.5 for years 50 and 100 years, respectively, on current estimates of erosion for the NECAG coastline. While accepting this probably represents an upper bound rate, this has provided a consistent baseline condition from which to assess policy. The following case examples highlight were there has been a need to vary from this baseline assessment and the implications and perceptions created by defining any line or even zone.

3 FLAT CLIFFS – FILEY BAY

This stretch of coast comprises the wide sand beach of Filey Bay backed by till cliffs of varying heights. The northern limit is marked by Filey Brigg. The southern end of Filey Bay runs to the chalk cliffs and associated shore platform of Flamborough. The coast is undefended apart from a stretch of sea wall in front of Filey Town. The top of the beach varies in elevation from 3 m OD near Filey Town to 2 m OD near Reighton. The till sequence in the cliffs of Filey Bay (30–50 m high) has been divided into an Upper Till Series and a Lower Till Series split by up to 3 m of sand and gravel (Edwards, 1981). The Upper Till Series is clay-rich and the Lower Till Series has a fine sand/silt matrix. Cliffs cut into this till are subject to erosion rates up to 0.3 m/yr (average *c.* 0.25 m/yr) (Institute of Estuarine and Coastal Studies, 1991; Mouchel, 1997; Halcrow, 2002b). These values are a long-term trend masking any short-term episodic failures.

Sand within the bay is fed by erosion of the till cliffs in Filey Bay itself. However, not all the sediment is suitable for retention within the bay. Halcrow (2002b) estimated that the average annual sediment supply to Filey Bay is between 4,179 and 23,176 m^3, of which

1,354 to 5,597 m^3 is derived from the ongoing erosion of the cliffs. The remainder is contributed from larger scale episodic event. Beach elevation data has been collected since 2002 by SBC.

There are steep till cliffs over a 1.5 km stretch of coast south of Filey, with a generally steep scarped toe, indicative of the general slow erosion, and simple slumping, or landside to the cliff face. Within the section, and just south of the Primrose Valley ravine, there is a small collection of holiday chalets close to the crest of the slope. This is the start of a larger development of private housing and the major holiday and caravan park extending over 1 km inland and continuing some 1 km to the south along the coast. While the simple landslip cliffs extend 500 m to the south of Primrose Valley, the character of the cliffs then change in front of Flat Cliff. Here there has been more major complex failure of the coast creating a series of narrow terraces from the high cliff behind through to a much lower, but still eroding slope to the beach. The initial 500 m of this area has been developed since the 1900's as an estate of some 40 properties. The properties of Flat Cliff generally lie within, or more correctly upon this area of former major coastal cliff failure but the access road to the lies with the transition between this and the more simple land slip section of coast to the north. In profile the cliffs tend to be destabilised by erosion at their toe or increased water permeating from the land above. As such the cliffs may remain relatively stable for long periods of time, with sudden and, in some locations, quite major movement. During consultation the erosion rates at Flat Cliff were queried because over the last several years local observation had suggested very little change. However, in this area it has been indicated that further erosion, potentially in the order of 5 m would set off failure slides extending back to the rear, higher cliff line. Based on typical erosion rates this occurrence could be within a 10 to 20 year period, but because of the episodic erosion of the toe could be sooner or later.

However, sea level rise will tend to allow increased wave action to the toe of the cliffs and the bay will attempt to readjust by eroding back. Over the 100 year period of the SMP2 it is likely that more major failures could arise affecting considerable areas above the higher cliff, taking out a major section of the caravan park's support infrastructure. Services to both the residential community and to the holiday park will also be affected and with the loss of the sewage pump station within the Flat Cliffs area these will be affected sooner than much of the area provided with these services.

Within the SMP2 the erosion of the toe of the cliff, in this case based on readjustment of the profile in relation to anticipated sea level rise, is 5 m, 20 m and 40 m, by years 2025, 2055 and 2105 respectively. Translating this into areas at risk, however, suggests potential risk zones of 15 m, 50 m and over 100 m, over the

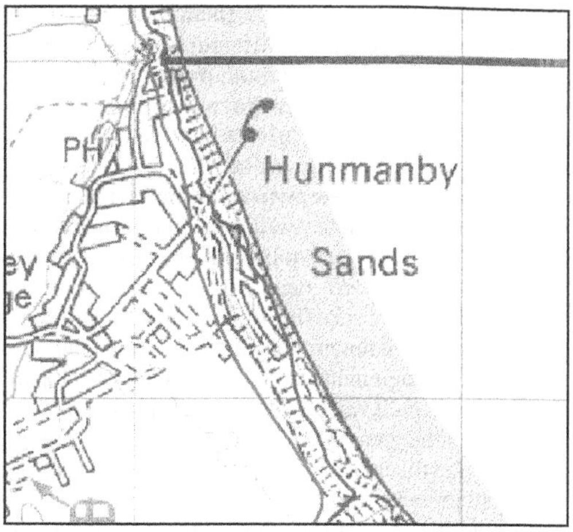

Figure 1. The area of Flat Cliffs.

same epochs. Such estimates can only be indicative, depending initially on episodic erosion; where an average rate is potentially misleading, and in the medium to long term on actual geotechnical influences on slope failure. The very position of the line changes necessarily from one of the existing toe, to one of defining where impact of erosion may occur.

Leaving aside the uncertainty of the physics, the critical issues; in terms of policy and management, are whether or not invention is justified in defence of Flat Cliffs and how both those people affected by that decision and the holiday park can adapt to the future threat.

There is scant justification for defence, and a draft policy for no active intervention (at the time of writing this paper). The immediate issue is one of the access, and the identification that this may be lost affects both the residents and the holiday park. This becomes more important than the precision or accuracy of a line defining timescales of when that loss may occur. Similarly, the slightly longer term warning that properties will be at risk and that essential features of the holiday park are considered to be at risk in the future, allow decisions to be made in advance as to how to respond. However, clearly, from the perspective of individual's expectations or the management by organisations, what the line really means differs.

In the case of shoreline management it is the overall trend and possible severity of that trend which indicates whether defence of the area is likely to be sustainable, either in terms of economics or in terms of longer term investment and impact; this latter consideration being as much one of where immediate defence decisions will lead, rather than any sudden impact that will arise. The banding of estimates, ranging from: no change and therefore no need for action, through minor pressure and, therefore, an acceptability of intervention subject to economic justification, to quite severe change with significant potential impact and commitment and, therefore, a policy not only of not intervening but also of prohibiting private intervention, are the basis for the interpretation of the line.

In the case of property owners, the line is interpreted quite specifically, that they will lose their property over the next 25 years, given a policy of no active intervention, and that they will need to plan and take action around this. In the case of the holiday park, and to some degree land use planners, the interpretation might be that future management and development of the park should be adapted, such that core facilities should be moved from the general area of risk and that planning needs to take this in account in assessing proposals for any development.

Only in the case of shoreline management and the long term organisational planning can the lines be interpreted with a degree of confidence, meeting the needs of the decision making required. In terms of private occupiers, interpretation of a line or even a probabilistic zone is potentially misleading. Although, it is also possibly unrealistic to assume that intensive further study would substantially improve upon the accuracy.

4 CAYTON BAY

Further to the north of Filey Bay is Cayton Bay. While in some respects similar to the situation of Flat Cliffs, the bay is more tightly controlled, geomorphologically, and has areas where there is ongoing active coastal slippage. Cayton Bay contains a wide sand beach bounded to the north and south by rock headlands. A series of faults run through the bay (British Geological Survey, 1998c) resulting in a range of lithologies being exposed in the unprotected cliffs, controlling its spatial and temporal development and the scale of land slipping. The northern part of the bay comprises Oxford Clay Formation overlain by 5–30 m of till; the central part of the bay is dominated by till cliffs (20–30 m high, divided into an upper sandier unit and a lower muddier unit) whereas the southern end is Ravenscar Group sandstones and mudstones. The northern part of the bay has been subject to numerous historical landslips, primarily developed in the till but with a basal shear in the underlying Oxford Clay Formation. The cliffs are generally unprotected throughout the bay and its crenulate shape indicates that the beach plan is tending towards an equilibrium form under existing conditions. Overall, an erosion rate of some 0.3–0.4 m/yr is likely for the cliffs in the bay.

There is a strong possibility that Osgodby Point, the main control feature to the north of the bay, will become detached from the coast as erosion continues

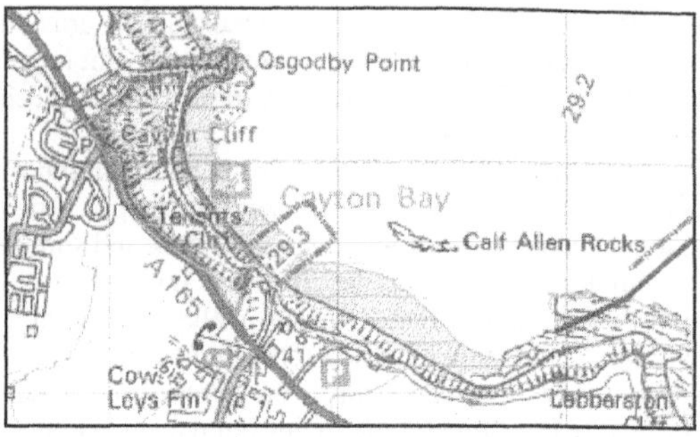

Figure 2. Cayton Bay.

and, while still acting as a sort of breakwater, the influence of the Point will be considerably reduced. Severe erosion along the Cayton Cliff toe might then be expected. The timing of such a development could well be within the period of the SMP and potentially within the next 50 years. Much would depend on the extent of the hard geology of the point, in terms of the continued protection it would provide. The Cayton Cliff is already considered to be a relatively unstable heavily saturated slope, vulnerable to both local and more major surface slippage. At present the toe of the slope is at an unsteady equilibrium, with slope slippage tending to move the toe forward and coastal processes tending to erode the toe back to the more stable bay alignment. This process continually steepens the coastal slope and increases the opportunity for a major landslip affecting the scarp at the top of the slope.

The Cayton Cliff, owned by the National Trust and an important recreational area, is also an area of SSSI, believed to have the richest invertebrate fauna of ground beetles and soldier flies associated with soft-rock cliffs in the whole of northern England. This resource depends on the continued action of natural processes on the soft cliffs, providing cracks and bare ground for the various different invertebrate requirements. Furthermore, a relatively recent housing development runs along the crest of the Cayton Cliff but is vulnerable to erosion and existing land slippage both within Cayton Bay and from the area to the north of Osgodby Point.

Some 500 m south of Cayton Cliffs, within a relatively more geotechnically stable section of the bay, there are modest coast protection works to the toe of the coastal slope, providing defence to a couple of properties and also providing the main access to the foreshore of the bay. Between here and the Cayton Cliffs, there is a more resistant deposition of rocks and boulders beneath Tenants' Cliff. Despite continued erosion, this strong point is likely to remain a key control feature within Cayton Bay. The terraced coastal slope behind Tenants' cliffs has been assessed as being quite stable in the short to medium term, although further erosion of the cliff face could eventually result in further major land movement and, although arguably at present not associated with coastal erosion, the back scarp, supporting the main A class coastal road, remains vulnerable to failure purely because of weathering.

There are significant uncertainties in assessing a totally realistic picture of how the coast will develop, with uncertainty linked to sea level rise, geomorphological development and geotechnical behaviour. However, even ignoring these and assuming these could reasonably be assessed, within the context of this paper, there remains a complexity of issues purely associated with the definition of what is being defined by a line of retreat.

From the perspective of the natural environment, and indeed the management of the area by the National Trust, the natural coastal slope of Cayton Cliff is inherently important and would remain so regardless of its stability or effective retreat. From this perspective the erosion of the toe is significant. From the perspective of householders at the crest of the Cayton Cliffs, it is the anticipated safety of themselves and their property which is the issue; this being determined in part by the threat of erosion but certainly more immediately by the underlying geotechnical instability of the slope and retreat at the crest. With respect to the highway authority, decisions have to be made in relation to the current support to the road but also, in the longer term, the risk of more major instability caused by coastal erosion. With respect to the owners of property currently protected in the centre of the bay and in relation to maintaining appropriate access to the bay, it is much more about the erosion of the foreshore, threatening the sustainability of defences, than the subsequent instability of land behind.

From the perspective, finally of coastal management, it is about all these. Any line will be open to misinterpretation and the distraction of a line possibly draws attention away from the discussion of the implications of management policy. The line, in this case, can only really be drawn in words.

5 COWBAR – STAITHES

This penultimate example highlights both the earlier issue of scale, rather than precision of erosion predictions, but also the value of continuing the process of consultation and re-evaluation throughout the development of a specific SMP process and beyond.

The North York Moors National Park coast generally comprises undefended high vertical cliffs composed of Lower Lias shales and sandstones backing a rock shore platform covered with a veneer of pebbles and cobbles. The cliffs rise towards the centre of the unit reducing in height at the western and eastern ends. The central high section of cliff is topped by Middle

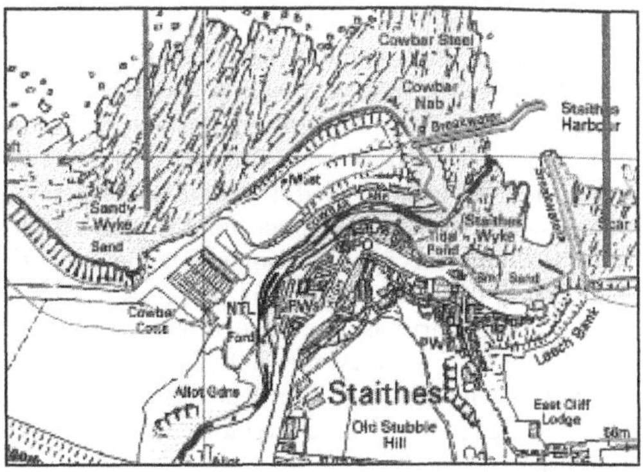

Figure 3. Cowbar Cliffs and Staithes Village.

Lias sandstones with ironstone bands. At Cowbar the cliff comprises several strata with a depth of a few metres of till capping. The overall exposed face being 40 m in height.

At the crest of the cliff are a two rows of cottages, in front of which runs the coastal road and only vehicular access to the northern section of Staithes Village. Recent appraisals of defence (High-Point Rendel 2000) conclude a potential rate of erosion of some 0.25 m/yr. Based on this rate, there was an assessed risk both to the road and to the properties behind. The principal element of risk was associated with the loss of the road and the significant impact on the village and harbour at Staithes. However in addition to this was the threat to a number of the cottages and the associated costs arising from any alternative re-routing of the road. A good justification was provided for defence of the crucial pinch point in front of Cowbar Cottages and a 50 year defence strategy was implemented, with slope stabilisation and the construction of a rock revetment to the toe of the cliff.

Taking this forward in the SMP2, allowing a baseline increase in retreat due to sea level rise, gave a potential worse case erosion of some 60 m in the area. In reassessing policy for the frontage, there was concern that there would not merely be increased pressure on the relatively limited length of exposed defence; being significantly in advance of the adjacent cliff line, but that there would also be the possible outflanking of the North Breakwater to Staithes Harbour. While supporting the findings of the 50 year strategy, the draft SMP2 highlighted the longer term threat to properties and concluded that, subject to further monitoring, the long term policy in the latter period of the SMP2 would most probably be for retreat of the existing line of defence, providing a more sustainable approach along a retired line to the main length of the headland; but resulting then in the loss of properties and the realignment of the road.

Concurrent with the development of the SMP2, monitoring of the cliff has been ongoing, undertaken by Durham University (Lim 2005). At the draft stage of the SMP2, the initial conclusions of this work became available. Over a four year period, erosion rates of 0.025 m/yr had been determined, further supported by a detailed analysis of historical mapping. Furthermore, the work has provided improved understanding of the detailed cliff mechanics for the area. While fully accepting the rigorous monitoring and analysis undertaken, the SMP necessarily proposes a slightly more precautionary baseline rate of erosion of 0.1 m/yr with a factor of ×2 over the 100 year period of the SMP. A further constant 5 m episodic risk distance based on the findings of the monitoring has also been allowed.

This has resulted in a proposed change in policy over the latter period of assessment. Significantly, this change comes more from the increased ability to maintain the existing defence, in relation to erosion of the adjacent cliff line, than from change to possible economic benefits of defence. In the discussion of this example, it is appreciated that the critical decisions being made derive not strictly from an absolute accuracy of retreat but rather in terms of the scale of pressure and risk in maintaining a defence.

6 SCARBOROUGH

Scarborough South Bay represents possibly the most complex area of all in defining a meaningful line of retreat. The frontage benefits from coast protection over all of its length. This defence includes at the northern end the Harbour, with its importance as a safe refuge for vessels but also its significance to tourism and the life of the town. The defence continues along the main promenade, with the central sea front area of Scarborough, through to the defences in front of the Spa. In terms of coastal processes, the whole area must be seen as a unit; the Harbour providing protection to other sections of the coast and allowing a longitudal accretion of the beach. Behind the defences and to the rear of the sea front development are varying sections of coastal slope, rising up to the town centre or to crescents of housing inherited from Victorian and Edwardian times. In some undeveloped areas the stability of the slopes is maintained quite directly by the existing coastal defence, in others the question of stability is lost in the plethora of housing and streets cladding the natural slope.

While the variation in the geotechnical behaviour of the different coastal slopes makes them quite distinct, the issue of their future stability is stitched together by the use of the coastal fringe and by the interactions between sections of the foreshore. There is little reliable historic information as to long term coastal change and current monitoring and investigations have been essential in gaining any assessment of future risk.

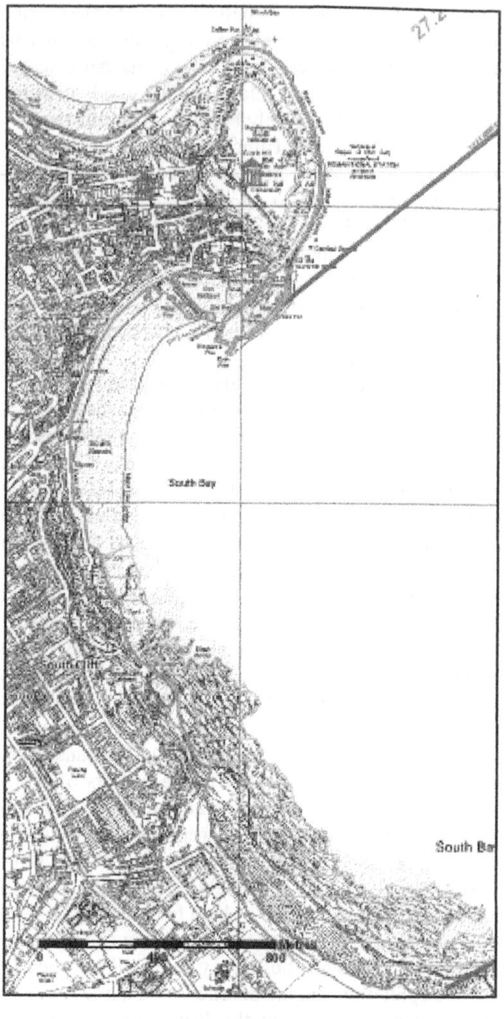

Figure 4. Scarborough South Bay.

At the shoreline, the main concerns at present are associated with ageing defences and flooding arising from overtopping and the potential increase in this due to sea level rise and increased storminess. Primarily associated with this, is the value of the sea front to the town, the recreational importance of the beach, the tourism significance of the sea front and the cultural value of the whole frontage. In providing defence and assessing areas of risk, the aspects of change in the crest of the beach, defining not merely a defence but also a recreational asset, the line of the actual man made defence against primary coastal erosion or proposals for integrating flood defence within the landscape of the promenade and, finally, the possibility of coastal slope and crest failure, make definition even of an initial line upon which to base a single distinct line of retreat virtually impossible. Clearly, in terms of assessing damages and impacts, considering the scope of impact is not only possible but necessary, but this has to be done at a scale of any specific examination. The opportunity for misinterpretation and subsequent decisions being made by organisations or individuals, with respect to a single line of risk is significant.

7 CONCLUSION

There is rightly an expectation that people should be better informed as to the risk of erosion and the implications of this in relation to both their interests and in respect of a transparency of how shoreline policy affecting them has been derived. There is furthermore a need for planning authorities to be able to assess risk to development and to current business needs, in granting permissions or in taking a forward view of their areas. There is also a need for central government and coast protection authorities for good information as to risk and the economic consequence of no active intervention, managed realignment or the investment in coastal defence, or at the more local level in assessing impact arising from management. There are fundamentally different questions to be asked relating to different aspects and different scales. However, it is accepted that these are expectations which have to be delivered based on improved knowledge and present estimations of understanding.

However, this paper has set out to demonstrate that this delivery is no simple matter and can be as misleading to one individual as it is essential to another. There is a continuing development of knowledge of how the coast is evolving and a still greater uncertainty as to how it will evolve. In many cases the answer is in understanding a complex physical interplay of different factors; from simple pressure on the coast in response to wave energy, to the developing behaviour of the shoreline in response to this and in response to sediment supply, climate change or geomorphological threshold change, through the resilience of coastal material and changing resilience of material subsequently exposed, to the uncertainty of geotechnical behaviour of coastal slopes. With monitoring, investigation and analysis this understanding has been reasonably addressed in some areas, sufficient to address some of the questions being asked and may with a degree of certain emerge for other areas.

Associated, however, with this practical uncertainty has to be an appreciation of the different issues and interests and the different scales of decisions being made with respect to this physical behaviour.

The need for definition of risk lies in the conflict between the natural or managed evolution of the coast and man's many interests and uses of the coastal zone. With this comes the individual perceptions of this risk and the individual definition of what then is important to know. While drawing lines representing specific aspects of risk and evolution are valuable; even essential, to specific analysis and need, there is a constant concern of misinterpretation of such information. No line can necessarily be complete without discussion, not just of its limits in terms of accuracy, but also its intent and the implications and criticality to management decisions which flow from its definition.

The discussion within the NECAG SMP2, driven by the approach and attitude of the project team led by Scarborough Borough Council, aims to provide this connection between meaning and definition. The real challenge is to persuade people to read the document and not just to look at the pictures.

REFERENCES

British Geological Survey. 1998c. Scarborough. Sheet 54. Solid and Drift Geology.

ICES 1991. Filey Bay Environmental Statement. University of Hull.

Mouchel. 1997. Shoreline Management Plan. Huntcliffe (Saltburn) to Flamborough Head – Sub-cell 1d.

Halcrow 2002. Filey Bay Coastal Defence Strategy Study. Report to Scarborough Borough Council, October 2002.

High-Point Rendel. 2000. Cowbar Coast Protection and Cliff Stabilisation Staregy. Report to Redcar and Cleveland Borough Council

Lim,M. Petley,D. Rosser,N. Allison,R. Long,A. & Pybus,D, 2005. Combined Digital Photogrammetry and Time-of-Flight Lasar Scanning for Monitoring Cliff Evolution. The Photogrammetric Record 20(110):109–129.

Landslides and Climate Change – McInnes, Jakeways, Fairbank & Mathie (eds)
© 2007 Taylor & Francis Group, London, ISBN 978-0-415-44318-0

Sleeping demons & terrified horses: Determining the onset of instability

M.E. Barton
School of Civil Engineering and the Environment, University of Southampton, UK

ABSTRACT: The paper draws attention to the problems of trying to predict the time when a slope, or quasi stable landslide, progresses into full scale instability. Rockfalls and large scale complex coastal landslides are highlighted as phenomena posing particular difficulty with a time-wise prediction. The consequences of long term sea level rise due to global warming for large scale coastal landslides mirrors a return to the conditions which will have existed during the later stages of the Flandrian transgression.

1 INTRODUCTION

We may distinguish between the hazard of landsliding and the prediction of a landslide. The former can indicate the degree of threat and is best shown by mapping hazard zones (Hutchinson & Chandler, 1991, Hutchinson, 1992). The prediction of a landslide means that in theory we ought to be able to say when it will occur. The distinction can be illustrated by reference to the determination of the factor of safety (Fs) from a stability calculation. The result will show us how near we are to the point of equilibrium and if all the parameters are reasonably correct, then if Fs <1, sliding must occur by the terms of the mechanics. However, time does not appear in the stability equation so the calculation does not provide a time-wise prediction. We expect slight errors in the parameters and a calculation showing Fs $= 1.0$ is going to be an approximation. We may see telltale signs of movement but when a sudden or catastrophic movement may occur is not derived from analyses of stability.

Using the analogy of the demon in the title, we can say that the premonitory signs of movement are stirrings of the demon. Geotechnics has not advanced sufficiently far, however, for us to say when, precisely, the demon may awaken from his slumber. Leroueil (2004) has provided a comprehensive review of the geotechnical phenomena that may occur in a slope before failure with a list of the common triggering or aggravating factors. The objective of this paper is not to rival that excellent review but to highlight just some of these factors and illustrate the problems that they pose for determining the onset of instability. Climate change and especially sea level rise are also highlighted since they provide a ready means by which landslide demons can be roused from their slumber.

2 TOPPLING AND ROCK FALLS

Weathering and creep of rocks under high stress leads to destructuring and eventual fatigue. Any additional forces derived from seepage, freezing or thermal cycling hastens fatigue and toppling will occur when sufficient component parts of the rock mass have weakened the overall strength. Nevertheless the processes are slow and influenced by seasonal factors such that measurements of deformation show irregular rates. Inclinometer, tiltmeter and strain gauge records typically show fluctuations around a long term trend. Where an increase in the rate of movement gives a clear indication that the tertiary creep stage is reached then it is possible to use various empirical methods of predicting the eventual collapse (Frederico et al. 2004). However, as pointed out by Zvelebil (1991) rates of movement can also decrease towards the time of sudden failure and nullify the use of such predictive methods.

Gore Cliff forms the back scarp to the Isle of Wight Undercliff landslides between Blackgang and Niton (Bromhead et al. 1991). It is the site of numerous spectacular rockfalls including the dramatic and famous fall in 1928 which destroyed a length of the A3055 coastal road. The scar of a rockfall which took place in 2001, critically loading the Undercliff, activating a mudslide and forming a new deep gulley is illustrated in Fig. 1. Gore Cliff is replete with numerous joint sets including stress relaxation joints parallel to the cliff face. Monitoring of a joint by Hutchinson et al. (1991) has shown seasonal movements. Periodic movements of the Undercliff landslides must relieve passive support at the base of the scarp and make rockfalls inevitable. Near the point of failure, the succession of minor falls, deformations and noises

Figure 1. Rockfall from Gore Cliff and part of the talus slope at the back of the St. Catherine's Point landslide with its associated mudslide gulley seen in the foreground. This took place on either late 29th or early 30th October 2001. The site is to the east of the historically recorded falls from Gore Cliff (Bromhead et al 1991) and located where the aspect of the cliff changes from south west facing to south east facing. The direction of the fall was into the Undercliffs fronting the south west facing side of Gore Cliff, an area already scoured by past mudslides, notably that set off by the famous 1928 rockfall. This new rockfall in turn set off a new mudslide, the track of which scoured a deep gulley at the edge of the car park forming the termination of the Sandrock Road (the former A3055 coastal road before the 1928 rockfall). Countless generations of students used to cross this area before 2001 in order to reach the exposure of the Gault Clay on the eastern side of the 1928 mudslide channel. Health and safety considerations made it inadvisable to venture across this new mudslide gulley, both from the treacherousness of the underfoot conditions and also from the very real risk of further rockfalls from the weakened talus slope seen at the top right.

recorded for the 1928 fall can be used to indicate that a large scale failure is imminent (Hutchinson et al. 1991) but predicting a time well in advance of such occurrences is not feasible.

3 PORE PRESSURE CHANGES

Fluctuations in pore pressure are the means by which a previously marginally stable landslide can be tipped beyond the point of stability. A network of piezometric points within the confines of a small landslide and linked to an automatic monitoring system will show the progressive onset of instability and provide advance warning of failure. A very large landslide extending through a range of strata and with a complex morphology will have a correspondingly complex hydrogeological regime (as in Bertini et al. 1991). In places high pore pressures may be inducing local instability but drainage elsewhere may provide enough resistance for the complete landslide to remain stable.

As an example of a large, complex landslide we can take the St. Catherine's Point landslide, well described by Hutchinson et al. (1991). This consists of a compound slide surface extending down to a clay seam within the Sandrock formation and an apron of colluvial debris at the toe as shown in Fig 2, from Hutchinson et al. (1991). The latter used the estimated upper and lower piezometric levels to determine critical values for the average residual shear strength (Ør') around the basal slip surface. An alternative way of presenting the stability conditions is to assume an average value of Ør' and to represent the pore pressure in terms of the overall average value of pore pressure ratio r_u which has the convenience of a linear relationship with the factor of safety Fs (Bishop & Morgenstern, 1960). Fig 3 shows Fs versus r_u, taking Ør' as the value derived from stability analysis with high pore pressures in the main part of the slide but the low piezometric level in the apron. For convenience, Janbu's method of stability is used and hence the values calculated will differ from the more accurate values of Hutchinson et al. (1991). On-going movements in the St. Catherine's Point landslide, involving both the Undercliff masses and the apron (Hutchinson et al. 2002), shows that it is hovering around the point of instability. Fig. 3 illustrates how slight pore pressure fluctuations could provide the tipping point to bring total failure.

Unloading (as by excavation or previous landsliding) of clay soils depresses pore pressures which will be followed by a slow equilibration to conditions set by the ambient seepage regime. Success in understanding and forecasting this response has been achieved using numerical methods allied to carefully measured soil parameters (Vaughan 1994, Potts et al. 1997).

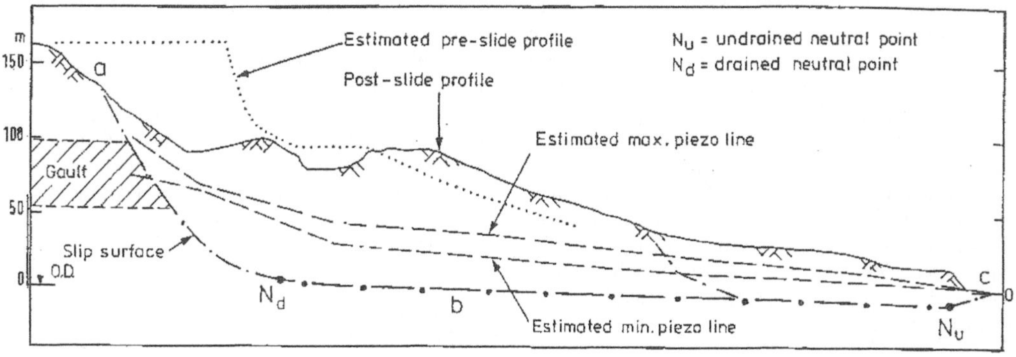

Figure 2. Catherine's Point landslide from Hutchinson, Bromhead & Chandler 1991. This landslide comprises the main slide containing the large slump blocks and a debris apron at the toe. The authors provide an upper and lower boundary for the estimated piezometric elevations and show the basal slip surface connecting with a small curved segment returning to beach level.

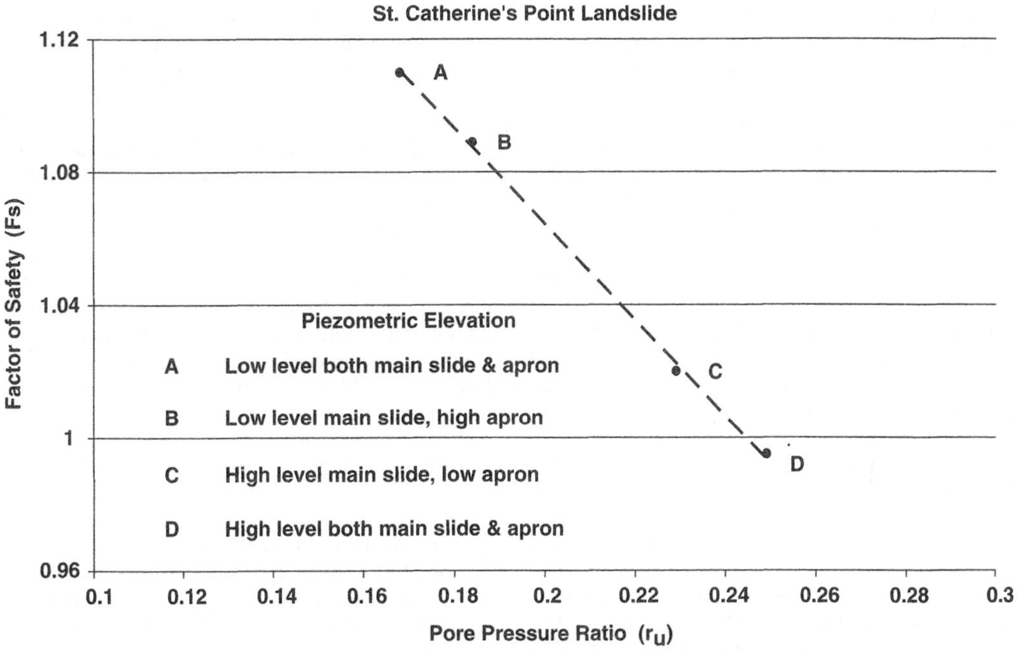

Figure 3. An interpretation of the relation between the factor of safety and the average value of the pore pressure ratio r_u for the main slip surface passing beneath both the main slide and the apron using Janbu's approximate method of slope analysis. The average value of residual shear strength φ_r' around the slip surface is assessed from the assumption that with the highest piezometric line for both main slide and apron, the slope will just fail.

Precise time prediction of a slope failure is not possible well in advance of the event since the response of the pore pressures is controlled by meteorological factors (Vaughan & Walbanke, 1973). With large and complex slides, the response will be complicated by fluctuations of deep groundwater levels. With slip surfaces passing through formations such as the Sandrock, as in the case of the St. Catherine's Point landslide, equilibration is unlikely to be a factor since the main part of the slip surface is a thin clay seam contained within highly permeable locked sands. Any depressed pore pressures within the clay seam should have been equilibrated well in the past.

A mechanism that causes a temporary pause in the progression of a cliff top slump to a large catastrophic slide is that of "re-engagement". Hutchinson (1969) illustrated this with respect to the chalk scarp at Folkestone Warren. It involves the accommodation of a large mass to a curved slip surface and results in a small step at the cliff edge and which may remain in place for long periods. Small "hung-up" cliff top slumps are common features of actively degrading cliffs (Barton et al. 1983). The re-engagement mechanism could also be associated with the Lowtherville Graben at Upper Ventnor. Hutchinson & Bromhead (2002) comment that this feature is clear evidence of the potential for a slide of considerable size. The Lowtherville Graben represents a sleeping demon with the potential to cause very serious disruption.

4 MONITORING, EARLY WARNING AND TERRIFIED HORSES

Various methods of monitoring slope behaviour can also be allied to early warning systems (Dunnicliff & Green 1988, Clark et al. 1996). These methods not only compensate for the inability to make precise time-wise predictions but also provide a means of acquiring fundamental knowledge about the processes taking place within potentially unstable slopes. Pore pressure changes and deformations can be monitored over long periods but when a landslide is imminent, it is possible that other signals may provide useful information.

Leroueil (2004) draws attention to unusual animal behaviour and gives a graphic account, translated from the Norwegian original, of horses which refused to cross an area which within a short time became the site of a serious quick clay landslide. Leroueil (2004) asks whether there may be signals which animals detect which so far have eluded our best instrumentation. This question is not easy to answer since by their nature, horses are animals whose response to danger is to flee and there could be many reasons for the horses to be worried other than directly by physical signals from an impending landslide. If there were direct signals beyond the range of human sensitivity they could involve intermittent vibration (van Genuchten 1988, Rouse et al. 1991), geoacoustic signals from rock fracturing (Cadman & Goodman 1967, McCauley 1976, Novosad et al. 1977) or gaseous emanations from opening joints and fissures (Selley 2006). Investigation of these signals by observing the behaviour of animals is unlikely to be productive and could lead onto the wilder shores of fanciful speculation as in Brown (2006) and Sheldrake (1999). A more sensible approach is to explore the full range of the electromagnetic spectrum as in McCann & Forster (1990) and Busby et al. (2002).

The inability to give precise time-wise predictions of slope failure is compensated in studies of landslide hazards by adopting a probability approach (Lee & Clark 2002). Such an approach lends itself readily to slope and landslide management by providing Engineers and Managers with a system to establish priorities for action. Such a system maybe the only practical way to proceed even though the determination of probability may involve elements of subjective judgement (Lee 2005). Such approaches represent the practicalities of managing slopes and landslides but it still leaves room for more fundamental investigations to determine the nature and distribution of the various processes that take place in a slope as it progresses towards the onset of instability.

5 CLIMATE CHANGE AND SEA LEVEL RISE

Changes to the rainfall pattern will also involve changes to hydrological regimes with possible serious consequences resulting from increased pore pressures (Hosking & Moore 2002). Some consideration of these consequences can be obtained from examination of the various methods available for relating meteorological data to landsliding and groundwater pressures as given by Lee et al. (1998) and Polemio & Petrucci (2000). The uncertainty of knowing exactly how regional climates may evolve in a warmer world creates an equal uncertainty concerning seepage regimes. Some regions will become wetter and others dryer but there is less uncertainty over the fact that eustatic sea level will inevitably rise as a consequence of global warming.

The Intergovernmental Panel for Climate Change (IPCC) predictions for eustatic sea level rise by 2100 give a range which depends on the emissions scenario used for modelling but provides a mean value of about 0.4 m (Houghton et al. 2001). The Third Assessment Report of the IPCC gives predictions beyond this but the uncertainty increases from having to guess the emissions scenario to use for the calculations. We can use a fairly high but still realistic emissions scenario to estimate eustatic sea level rise during the next millennium as shown in Fig. 4. Full details of how these curves were derived from the IPCC report are given in Barton (2006). The social and political consequences of these forecasts are extremely serious and indicate a world that will be a very different place from the one that currently exists.

Even if shorelines are protected from wave attack (as by emplacing large quantities of armourstone), we can see that the effect on the seepage regimes of large coastal landslides must inevitably upset their stability. Considering the St. Catherine's Point landslide as our example, an increase in the drainage exit levels within the apron will eventually raise the pore pressures to the highest estimated piezometric level. The result as indicated by Fig. 3 brings the factor of safety to less than 1.0. Even if the piezometric levels in the main slide remain low, a slide of the apron will remove the passive pressure support to the main slide and which then in turn will also become unstable. Similar considerations

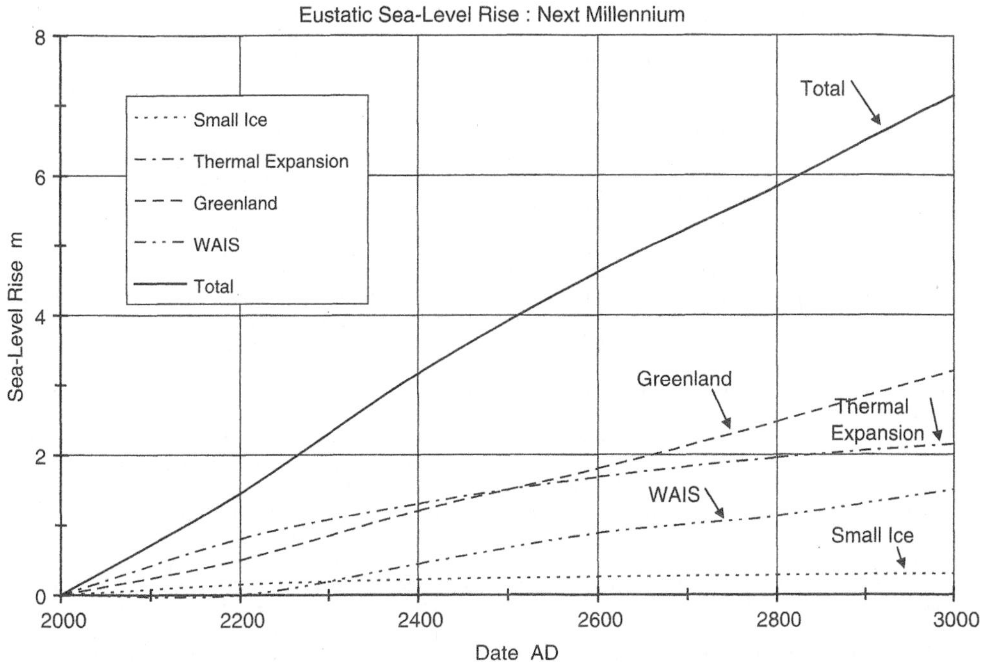

Figure 4. Predicted sea level rise during the next millennium based on data from the IPCC Report on Climate Change 2001 (Houghton et al 2001) and assuming a high but not unrealistic emissions scenario. The main components of sea level rise are shown with WAIS standing for the West Antarctic Ice Sheet. Initially the WAIS does not make a significant contribution until much of the surrounding ice shelves are depleted. It is assumed that the East Antarctic ice sheet stays in mass balance. While the mass balance of the ice sheets creates a degree of uncertainty, it should be noted that a large element of thermal expansion is already committed as a result of the increased warmth of the near surface ocean water penetrating to greater depths.

will apply to other large landslide complexes and thus no reliance can be placed on their future stability over the very long time scales covered by the IPCC Report.

6 CONCLUSIONS

- Our understanding of landslides in terms of their geology, geomorphology and geotechnics has dramatically advanced over the last century but except for certain conditions this understanding does not yield the means to predict precise times when total slope failure will occur.
- The natural processes leading to the onset of instability in large scale landslides are numerous in space and time and often obscure. It is their cumulative result which tips the balance to total failure.
- Instrumentation and monitoring not only provides advance warning of failure but provides the means by which to progress fundamental understanding of slope processes and their distribution in space and time.
- Long term sea level rise, as predicted by the IPCC, indicates that during the course of the coming centuries, many coastal landslide demons will be woken from their slumber.

ACKNOWLEDGEMENTS

Thanks are due to Dr. J.M. Lovegrove for the photograph shown in Figure 1. Thanks are also due to David Mears, Warden of the Knowles Farm National Trust site for supplying me with the date of the Gore Cliff rockfall shown in Figure 1.

REFERENCES

Barton, M.E. 2006 Do coastal cities have a sustainable long-term future? *Engineering Geology for Tomorrow's Cities: Proceedings 10th International Congress of the IAEG: (Culshaw, M.; Reeves, H.; Spink, T. & Jefferson, I..; eds).* Paper No. 263, CD ROM, Geological Society, London.

Barton, M.E.; Coles, B.J. & Tiller, G.R. 1983 A statistical study of the cliff top slumps in part of the Christchurch Bay coastal cliffs. *Earth Surface Processes and Landforms* 8, 409–422.

Bertini, T.; Cugusi, F.; D'Elia, B.; Lanzo, G. & Rossi-Doria, M. 1992 Slow movement investigations in clay slopes. *in Landslides: Proceedings 6th International*

Symposium on Landslides (Bell, D.H. ed.): Balkema, Rotterdam.1, 329–334.

Bishop, A.W. & Morgenstern, N.R. 1960 Stability coefficients for earth slopes. *Géotechnique* 10, 129–150.

Bromhead, E.N.; Chandler, M.P. & Hutchinson, J.N. 1991 The recent history and geotechnics at Gore Cliff, Isle of Wight. *In Slope Stability Engineering (Chandler, R.J. ed.)* Thomas Telford, London. 189–196

Brown, D.J. 2006 Etho-geological forecasting: unusual animal behaviour & earthquake prediction. http://animalsandearthquakes.com/etho-geo.htm. Accessed 6/11/2006.

Busby, J.; Gourry, J.C.; Senfaute, G.; Pederson, S. & Mortimore, R. 2002 Can we predict coastal cliff failure with remote, indirect measurements? *in Instability: Planning and Management (McInnes, R.G. & Jakeways, J. eds.).* Thomas Telford, London. 203–208.

Cadman, J.d. & Goodman, R.E. 1967 Landslide Noise *Science* 158, 1182–1184.

Clark, A.R., Moore, R. & Palmer, J.S. 1996 Slope monitoring and early warning systems: application to coastal landslides on the south and east coast of England, UK. *in Landslides, Proceeding 7th International Symposium on Landslides (Senneset, K. ed.)* Balkema, Rotterdam 3, 1531–1538.

Dunnicliff, J. & Green, G.E. 1988 *Geotechnical Instrumentation for Monitoring Field Performance.* John Wiley & Sons, New York.

Frederico, A.; Popescu, M.; Fidelibus, C. & Internò, G. 2004 On the prediction of the time of occurrence of a slope failure: a review. *Landslides: Evaluation and Stabilisation (Lacerda, E.; Fontoura & Sayäo, eds.)* Taylor & Francis Group, London. 2, 979–983.

Hosking, A.S.D. & Moore, R. 2002 Preparing for the impacts of climate change on the central south coast of England. *in Instability: Planning and Management (McInnes, R.G. & Jakeways, J. eds.).* Thomas Telford, London. 601–608.

Houghton, J.T.; Ding, Y.; Griggs, D.J.; Noguer, M.; van der Linden, P.J.; Dai, X.; Maskell, K. & Johnson, C.A. (eds.) 2001 *Climate Change 2001: The Scientific Basis (Contribution of Working Group 1 to the Third Assessment Report of the IPCC)* Cambridge University Press.

Hutchinson, J.N. 1969 A reconsideration of the coastal landslides at Folkestone Warren, Kent. *Géotechnique* 19, 6–38.

Hutchinson, J.N. 1992 Keynote paper: landslide hazard assessment. *In Landslides: Proceedings 6th International Symposium on Landslides (Bell, D.H. ed.)* Balkema, Rotterdam. 1805–1841.

Hutchinson, J.N.; Bromhead, E.N. 2002 Keynote paper: Isle of Wight landslides. *in Instability: Planning and Management (McInnes, R.G. & Jakeways, J. eds.).* Thomas Telford, London. 3–70.

Hutchinson, J.N. & Chandler, M.P. 1991 A preliminary landslide hazard zonation of the Undercliff of the Isle of Wight. *In Slope Stability Engineering (Chandler, R.J. ed.)* Thomas Telford, London. 197–205.

Hutchinson, J.N.; Bromhead, E.N. & Chandler, M.P. 1991 Investigations of landslides at St. Catherines Point, Isle of Wight. *In Slope Stability Engineering (Chandler, R.J. ed.)* Thomas Telford, London. 169–179.

Hutchinson, J.N.; Bromhead, E.N. & Chandler, M.P. 2002 Landslide movements affecting the lighthouse at Saint Catherine's Point, Isle of Wight. *In Instability: Planning and Management (McInnes, R.G. & Jakeways, J. eds.)* Thomas Telford, London 291–298.

Lee, E.M. 2005 Coastal cliff recession risk: a simple judgement-based model. *Quarterly Journal of Engineering Geology and Hydrology* 38, 89–104.

Lee, E.M. & Clark, A.R. 2002 *Investigation and Management of Soft Rock Cliffs.* Thomas Telford, London.

Lee, E.M.; Moore, R. & McInnes, R.G. 1998 Assessment of the probability of landslide reactivation: Isle of Wight Undercliff, UK. *Proceedings 8th International Congress IAEG.* Balkema, Rotterdam. 1315–1321.

Leroueil, S. 2004 Geotechnics of slopes before failure. *Landslides: Evaluation and Stabilisation (Lacerda, E.; Fontoura & Sayäo, eds.)* Taylor & Francis Group, London. 2, 863–884.

McCann, D.M. & Forster, A. 1990 Reconnaissance geophysical methods in landslide investigations. *Engineering Geology* 29, 59–78.

McCauley, M.L. 1976 Microsonic detection of landslides. *Transportation Research Record, Transportation Research Board* 581, 25–30.

Novosad, S.; Blaha, P. & Kneijzlik, J. 1977 Geoacoustic methods in slope stability investigation. *Bulletin International Association of Engineering Geology* 16, 229–231.

Polemio, M. & Petrucci, O. 2000 Rainfall as a landslide triggering factor: an overview of recent international research. *In Landslides in Research, theory and Practice (Bromhead, E.N., Dixon, N. & Ibsen, M-L eds.)* Thomas Telford, London. 3, 1219–1226.

Potts, D.M.; Kovacevic, N. & Vaughan, P.R. 1997 Delayed collapse of cut slopes in stiff clay. *Géotechnique* 40, 953–982.

Rouse, W.C.; Styles, P. & Wilson, S.A. 1991 Microseismic emissions from flow-slide type movements in South Wales. *Engineering Geology* 31, 91–110.

Selley, R.C. 2006 The Holsworthy 'ghost': does the answer lie in the rocks ? *Magazine of the Geologists' Association* 5, No. 2, 22.

Sheldrake, R. 2000 *Dogs That Know When Their Owners Are Coming Home and Other Unexplained Powers of Animals.* Crown Publishing, New York.

Van Genuchten, P.M.B. 1988 Intermittent sliding of a landslide in varved clays. *Proceedings 5th International Symposium on Landslides, Lausanne.* 1, 471–476.

Vaughan, P.R. 1994 Assumption, prediction & reality in geotechnical engineering. *Géotechnique* 23, 531–539.

Vaughan, P.R. & Walbanke, H.J. 1973 Pore pressure changes and the delayed failure of cutting slopes in overconsolidated clay. *Géotechnique* 23, 531–539.

Zvelebil, J. 1991 Some problems in the interpretation of monitoring records of rock slope movements. *in Field Measurements in Geomechanics (G. Sφrum, ed.)* Balkema, Rotterdam 1, 387–393.

Landslides and Climate Change – McInnes, Jakeways, Fairbank & Mathie (eds)
© 2007 Taylor & Francis Group, London, ISBN 978-0-415-44318-0

Application of landslide studies for risk reduction in the Andean community of Reinaldo Espinoza, southern Ecuador

C.E. Ibadango & S. Escarate
Dirección Nacional de Geología DINAGE, Quito, Ecuador

L. Jackson & W. Sladen
Geological Survey of Canada, MAP:GAC, Canada,

ABSTRACT: The 45 ha Reinaldo Espinoza community (population: 329) is situated in the highlands along the south-western limits of the Andean city of Loja, Loja province, Ecuador. The community is underlain partly by a landslide that predates urbanization. The area is underlain by monoclinally dipping conglomerate and mudstone beds of Miocene/Pliocene age. The mudstone is largely derived from volcanic ash and it has a high plasticity index. The landslide is largely composed of failed mudstone and has been sporadically active. Movement in the years 1994, 1999, and 2001 damaged or destroyed infrastructure and houses but fortunately people were not injured. Considering the hazard to and vulnerability of the Reinaldo Espinoza community, the Multinational Andean Project: Geosciences for the Andean Communities (MAP:GAC) chose this site as a pilot project for the development of investigative techniques by using a community approach. The project is lead by the Direccion Nacional de Geologia (DINAGE) in cooperation with the Universidad Tecnica Particular de Loja (UTPL), the Programa Regional de Desarrollo del Sur (PREDESUR), the Municipality of Loja, the local Civil Defence Agency, and the Reinaldo Espinoza community. The project consists of two components: geotechnical and sociological. The geotechnical investigation includes topographic and geological mapping, monitoring of ground movement using differential GPS, and geotechnical test drilling and borehole instrumentation: inclinometers and piezometers monitor subsurface movement and pore water pressure, respectively. In addition, a rain gauge gathers rainfall data for correlation of precipitation with pore pressure and rate of movement. The sociological component brings awareness of the landslide hazard to the community. It also includes interviews with the residents to determine the history and nature of past landslide movements. This information is integrated into the geotechnical investigation. The overall objective is to determine and implement the best mitigation measures to reduce risk to the people living in the Reinaldo Espinoza area. To date, a preliminary map of the geology of the landslide and surrounding area has been completed. Test drilling has indicated that the landslide follows a pre-existing valley. The results of the GPS monitoring indicate that the landslide continues to creep since monitoring began in 2004. Historical accounts indicate a relationship between periods of destructive movements and high precipitation. The sociological work has resulted in the strengthening of the community organization.

1 INTRODUCTION

The suburb of Reinaldo Espinoza is partly underlain by a landslide that predates urbanisation. Movements in 1994, 1999, and 2001 damaged or destroyed homes and infrastructure (Ibadango, et al., 2005). Considering the high probability of future movement and vulnerability of the community, the Multinational Andean Project:Geoscience for the Andean Community (MAP:GAC) chose this site as a pilot project for the development of investigative techniques by using a community approach.

The Direccion Nacional de Geologia (DINAGE) in cooperation with the Universidad Tecnica Particular de Loja (UTPL), the Programa Regional de Desarrollo del Sur (PREDESUR), the Municipality of Loja, and the local Civil Defence Office are carrying out a detailed investigation. Geological, topographic and geotechnical investigations were completed at the end of March 2006, including drilling of boreholes to obtain subsurface data and install inclinometer and piezometer instrumentation. In addition, a rain gage was installed for monitoring the rainfall at the site.

All the implemented instrumentation is being used to collect data to determine the depth of the failure plane or planes of the landslide and establish the relationship between the rainy season (precipitation), groundwater pressure, and movement.

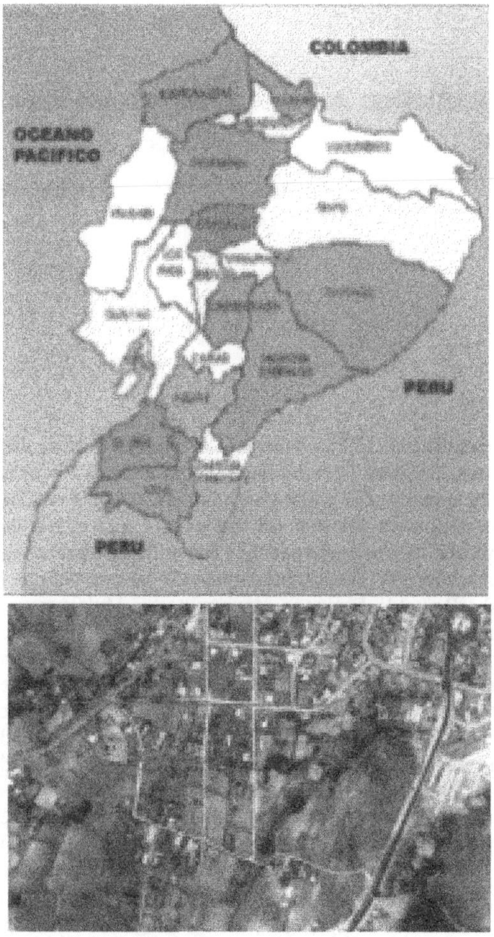

Figure 1. Location of Reinaldo Espinoza suburb.

The socio-economical investigation identified the main public services and the activities of the people living in the area in order to develop a contingency plan for future movements. Community participation is regarded to be very important in diminishing risk to people and their property.

2 LOCATION, POPULATION, INFRASTRUCTURE AND LANDUSE

Reinaldo Espinoza suburb is located in the southwest of the greater Loja area (population 405,000), in Loja Province, southern Ecuador, near the Peruvian border (Fig. 1). Reinaldo Espinoza covers 45 ha. Its population is 329; 150 women and 179 men (Fig. 2). Public services include a primary school, health centre and police station. Most of the houses have electricity, potable water, sewage service, and telephones. Roads are unpaved. Street names were assigned to them as a part of this project in 2005 with community cooperation. Land use in and around Reinaldo Espinoza is largely agricultural. There is significant economic pressure for development of this area for housing.

Figure 2. Panorama of Reinaldo Espinoza from E to W.

Figure 3. Conglomerate outcrop in Reinaldo Espinoza area.

3 GEOLOGIC AND GEOMORPHIC SETTING

Reinaldo Espinoza is underlain by sedimentary rocks of the Loja Basin, corresponding to the Oligocene-Miocene Quillollaco Formation and Quaternary colluvial surficial deposits (BGS & CODIGEM, 1994). The Quillollaco Formation is composed of poorly sorted conglomerate and mudstone. Conglomerate is composed of clasts of schist, quartzite, and vein quartz within a clayey sand matrix. The thickness of the conglomerates may exceed 45 meters (Ibadango, et al., 2005; Figure 3). There are some intercalations of shale and siltstone within the conglomerates, varying in thickness between 1 and 2 meters.

The overburden is derived from conglomerates and sandstones; banded textures corresponding to fine-grained soils composed of clays and silts, which have been corroborated by field observations (Ibadango, et al. 2005). Many areas of the suburb have been built over pre-existing un-engineered fill deposits and in some cases, houses span hillside cuts and fills because of the steep settings of building sites.

Clay-rich colluvium covers a large portion of the central and southern part of Reinaldo Espinoza. It is stiff and contains organic material and shows grey colour with brown and black mottles, with moderate plasticity. Colluvium is compact and exhibits deformation and cracks in some places.

The topography of the area consists of parallel ridges and valleys underlain by dip and scarp slopes created by the erosion of tilted or folded mudstone and conglomerate. Slopes between 5 and 30 degrees characterise the regional topography (Sauer, 1965). Slopes around the Reinaldo Espinoza landslide range between 2 and 15 degrees.

4 LANDSLIDE MORPHOLOGY

The Reinaldo Espinoza landslide has a length of 213 m. The landslide is 104 m wide at the crown, 82 m at mid-slope and 35 m at the toe. The total area is 23,850 m^2; its perimeter is 768 m. The average slope from the middle to the top is 13% and from the middle part to the toe is 7% (Fig 2). According the observations during the drilling campaign, the failure plane could be approximately at 5 m depth, resulting in a landslide volume of 119,250 m^3. It is an active, composite slide-earth flow (Cruden and Varnes, 1996) involving sand, clay and silt.

The main scarp at the crown of the landslide is spans a gently dipping contact between overlying conglomerate and underlying mudstone. The scarp indicates a vertical displacement of 3.5 m. A 200 m^3 steel water reservoir tank is inappropriately located above the scarp on conglomerate (Fig. 2).

The lower half of the landslide is presently inactive. It is composed of mixed masses of conglomerate and mudstone. A road cut reveals a disturbed layer of mudstone, which is probably one of the failure planes (Fig. 4).

Transverse cracks, scarps and thrust ridges are common in the parts of the landslide that affected houses and the land in 2001 (Fig. 5).

5 RECENT HISTORY OF REINALDO ESPINOZA LANDSLIDE

The landslide predates the earliest air photos taken in 1984. Interviews with residents revealed that three main episodes of landslides activity occurred in the last 15 years. All events caused damage to or destruction of buildings and infrastructure and were coincident with the rainy season (March to July) following two to three days of heavy precipitation.

Figure 4. Lithology at the toe of the old landslide.

Figure 5. Transverse cracks in middle section of slide showing a vertical displacement of 0.8 m.

Although events prior to 1980 are known of, the earliest event that resulted in damage occurred in 1994, shortly after construction in area started. Cracks in one home showed displacement of up to 90 cm. Movement described as an earth flow in 1999 displaced threes, caused cracking in the order of 30–40 cm, and destroyed a house after moving it about 5 m. Similar movements are described for the 2001 event, during which a section of a house is described as being displaced about 2 m over 15 to 20 days. Rates of movement during these events ranged from less than 1 cm/day to 0.5 m/day, corresponding to a slow to moderate moving landslide (Cruden and Varnes, 1996). Evidence of tilted fence posts and cracks in newly reconstructed buildings indicates that very slow movement has been ongoing since 2001 (Fig. 6).

Figure 6. Cracks affecting the floor of one reconstructed house.

Reinaldo Espinoza residents were concerned that the landslide movement was the result of leakage from a 200 m^3 water storage tank that was constructed along the ridge crest above the crown of the landslide (Fig. 2). Investigation of the possible role of this water tank in landslide activity was a significant aspect or MAP:GAC study.

6 METHODS

GPS surveying, geologic, and socio-economic investigations were carried out between 2004 and completed 2006. GPS surveying, geologic and socio-economic investigations were carried out between 2004 and 2006. A net of surveyed monuments was established across the Reinaldo Espinoza suburb. Their positions are being monitored using periodic differential GPS with the objective to more precisely determine the limits of movement. The first survey was carried out in May 2004. The geotechnical drilling and instrumentation were completed in March 2006. This work, included drilling of four holes using a mud-rotary drill rig, two at Site 1 (upper portion of slide), one at Site 2 (middle portion of slide), and one at Site 3 (Lower portion of slide) (Fig. 1). In addition, one Standard Penetration Test (SPT) hole was made at each Site 2 and 3. This work involved staff from PREDESUR, DINAGE, and the GSC as well as students from the UTPL.

Stratigraphic information obtained from the mudrotary borehole cuttings and changes in the rate of drill advance are shown for Site 1 in Fig. 7. Split spoon samplers obtained from the SPT holes provided slightly disturbed samples that were used to verify the interpretation of the lithology in the upper 6 m of the adjacent mud-rotary hole. Moisture content, Atterberg limits, and grain seize analysis determined from the split spoon cores are included in the respective SPT log for the two sites, Fig. 8a and 8b.

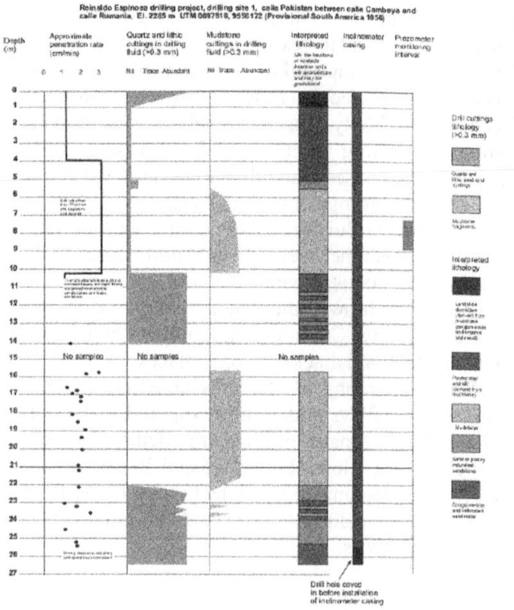

Figure 7. Stratigraphic information at site 1 (Pakistan street).

Flexible casings (17–27 m depth) were installed at each site. These are being monitored for deformation with inclinometer surveys. Casagrande-type piezometers were installed at Sites 1 (9.6 m) and 3 (6.4 m). In addition, a rain gage was installed for monitoring the precipitation at the site.

The objective of the instrumentation is to determine the depth of the failure plane or planes of the landslide through the inclinometer survey (Fig. 9) and establish the relationships between movement, the rainfall and the variations of the groundwater levels. Data collected to date only reflects static groundwater levels during the dry season.

The socio-economic investigation identified the main public services that are present in the community and the activities of the people living there in order to develop a contingency plan for future landslides that involves the community in its planning and execution in order to minimise the risk of people and their properties.

6.1 *Analysis of the preliminary data*

Inclinometer- Preliminary data obtained between June and September 2006 do not reveal any significant variation. The first surveys have been carried out monthly up to the end of September. From the end of September through the end of the rainy season, data are being collected every fifteen days.

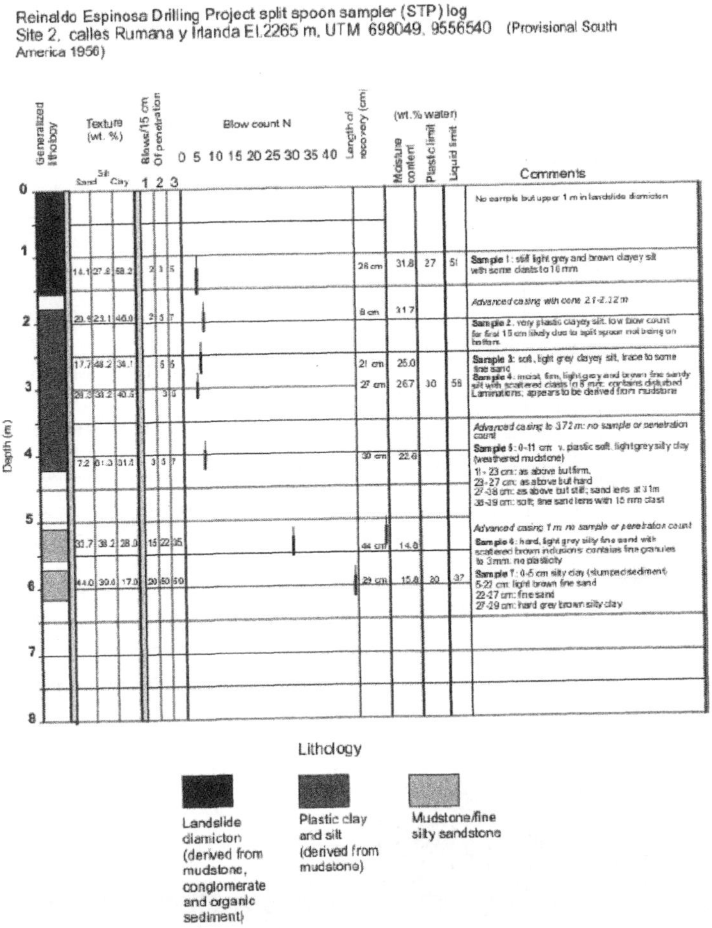

Figure 8a. SPT log at site 2.

Piezometer – Data during the largely dry period between April and October 2006 are presented in Figures 10 and 11, where pressure head is plotted against time. Although the records are broadly similar, there have been interesting individual differences. At Site 1 the pressure was constant during April and May and increased slightly at the end of May, and began to decrease at the end of August.

At Site 3 the pressure head was constant during April and May decreased dramatically by 0.5 m during the last week of May, maintaining constant until the final week of June to start decreasing moderately until October 2006.

Differential GPS – May 2004 and August 2006 data indicate only small variations in some points across the area. The largest movement detected was about 5 cm at station VILA (coordinates 697825, 9556190 WGS 84), which is located on the roof of a house affected by the landslide in 2001, (Fig. 12).

Socio-economic – Interviews with residents provided insight on the timing of the movements and link with increased precipitations. This information is covered an earlier section on recent history of Reinaldo Espinoza landslide. It also enabled the identification of key structures and services that are vulnerable.

7 FACTORS CONTROLLING LANDSLIDE MOVEMENT

Based upon data collected to date, the main factors contributing to the instability of the area are:

1. Pre-existing geomorphology – Much of the landslide fills a buried narrow, steep sided ravine. The erosion of this ravine may initially have destabilised the hillside and caused initiation of the landslide.
2. Climate – According the field observations across the Loja Basin and the testimonies of people living

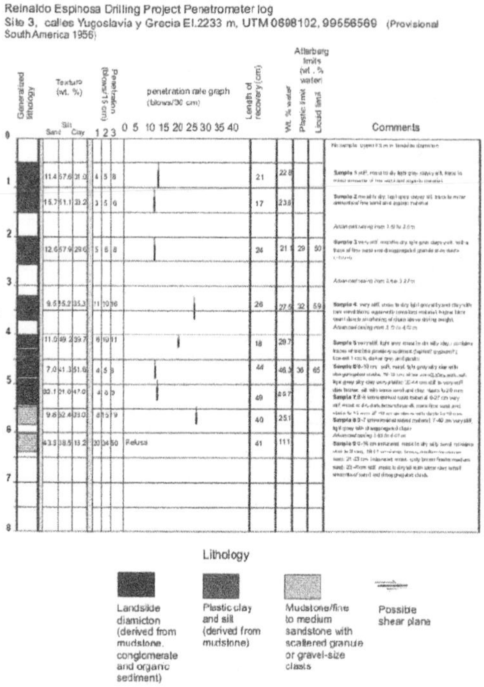

Figure 8b. SPT log at site 3.

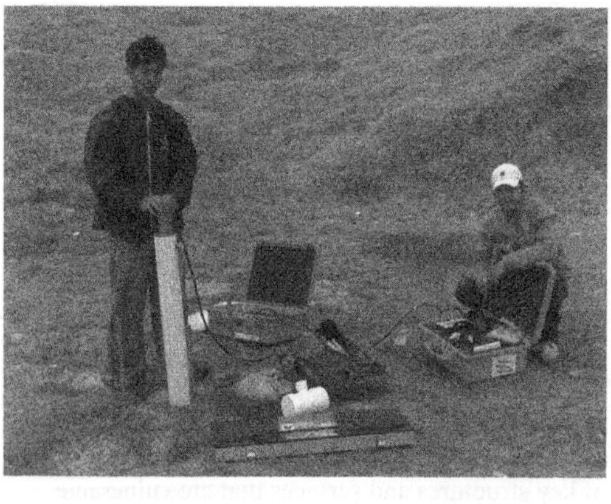

Figure 9. Periodical monitoring using inclinometer system Site 1 (Pakistan Street).

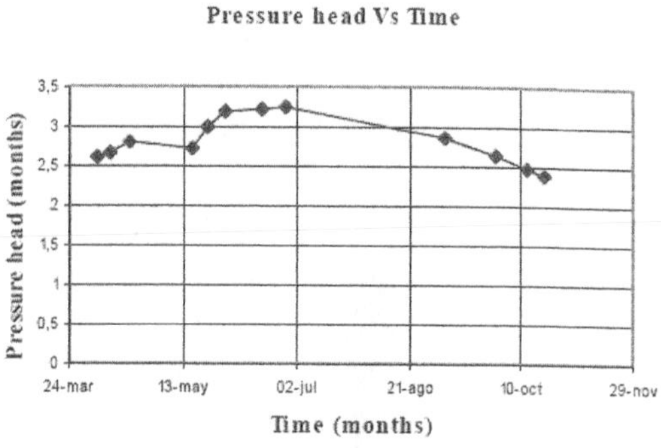

Figure 10. Pressure head Vs Time (Upper site).

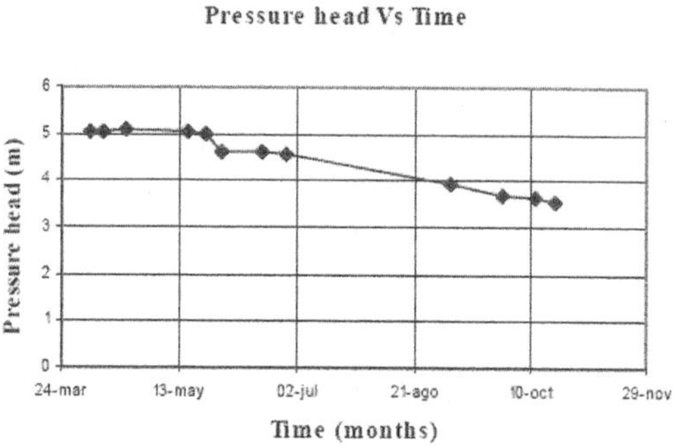

Figure 11. Pressure head Vs Time (Lower site).

Figure 12. Differential GPS monitoring at VILA.

in the area, mass movement processes peak during or at the end of the rainy season. At Reinaldo Espinosa, mass movement process occurred at the end of the rainy season in 1994, 1999 and 2001 apparently of the increase of the pore water pressure and decrease in soil strength.

Figure 13 shows the total yearly precipitations registered in the Argelia pluviometer station (approximately 3 Km to the SE of Reinaldo Espinoza) during the last 40 years. It is possible to see the total yearly precipitations, over 1000 mm correspond to the years 1967, 1972, 1983, 1984, 1985, 1990, 1993, 1994, 1999, 2000, and 2004.

– Stratigraphy and lithology – The presence of mudstone intercalations create weak layers within the conglomerate. Exposure of a deformed mudstone

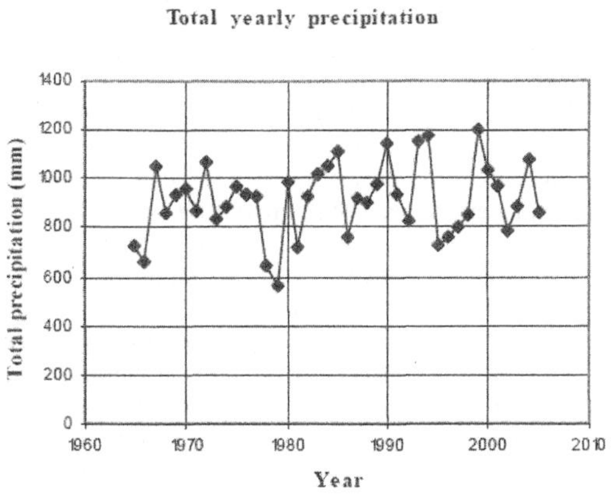

Figure 13. Total yearly precipitation 1965–2005.

layer near the toe indicates that at least a portion of the landslide has moved in the bedrock where conglomerate is sliding on a mudstone layer. Further weakening of this layer probably contributes to continued movement along this interface at a slow rate. The earth flow type movement occurring closer to the surface is in unconsolidated, clay-rich material of high plasticity.

- Hydrogeology – High moisture contents seem to characterise parts of the unstable area as signs of seepage, denoted by lush vegetation in summer, indicate that the water table or at least a perched water is close to the surface in places. Furthermore, moisture above the plastic limit was encountered in a zone of very soft silt and clay.
- Human activity – Excavations and fills have been developed across the area to facilitate road and house construction, modifying the original topography and increasing the susceptibility of landslide occurrence. Water lines, sewers and septic tanks may also contribute to increased moisture in the ground. Although there is not evidence linking the water tank at the crown of the slide to slope movement, it is poorly located as it will eventually be undermined as the landslide scarp retrogresses.

8 LANDSLIDE VULNERABILITY

The vulnerability of the Reinaldo Espinoza community to economic losses and loss of life caused by landslides has been taken into account because most of the existing housing is placed around the border of the most active area of landslide and movement rates have been slow. However a few houses are built completely within or straddle the edge of the unstable area and some landowners are considering constructing houses in this area.

Figure 14. An interpretive display set up at one of the drill sites for education and promotion of the investigative and community participation work in Reinaldo Espinoza.

The urbanisation is vulnerable for the mass movement process described above. If presently inactive parts of the landslide begin to move or new areas begin to fail, additional homes and public works including a primary school, health centre, roads, water supply pipelines, the 200 m^3 steel water tank, sewers, electrical and telecommunication cables and their pylons would be at risk thus increasing the community's vulnerability.

9 CONTINGENCY AND EMERGENCY RESPONSE

Institutions such as: DINAGE, PREDESUR, the Municipality of Loja, the local Civil Defence Office, and the community, are carrying out the contingency planning. Community-based activities developed during the last two years for Reinaldo Espinoza include:

- Geohazards workshops (Mass Movement Processes),
- Community communication workshops,
- Community organisation,
- Design of material for education and promotion of the work developed in Reinaldo Espinoa area to prevent and avoid the landslides risks (Fig. 14),
- Creation and implementation of an evacuation plan,
- Identification of safe sites for evacuation,
- Labelling of the streets names, and
- Community works (Mingas) to maintain a clean community and to prevent the surface water accumulation across the area (Fig. 15)

As a result of these activities it is important to consider the community participation during all the process. These initiatives have underscored the

Figure 15. Community works (Mingas) to maintain a clean suburb.

importance of including community participation at every stage of this study.

10 PLANNED ACTIVITIES

- Three dimensional study of the landslide using geophysical methods:
- Evaluation of landslide stabilisation options.
- Design of the best mitigation option for stabilisation.
- Simulation of the evacuation plan.

11 CONCLUSIONS

- The Reinaldo Espinoza suburb is partly underlain by a complex landslide.
- The onset of movement corresponds with the end of the rainy season, at which time the soil is likely saturated and the factor of safety within the slope falls to 1.
- Although a potential failure plane was intersected by the lowest of the three deep boreholes, inclinometer surveys of flexible casings installed in these boreholes will eventually determine the depths and number of failure planes intersected by the boreholes.
- The preliminary data from the DGPS suggest that the rate of landslide movement has been very slow since monitoring was initiated in 2004. Maximum displacement has been 5 cm during a period of two years.
- Inclinometer data shows no variation during the first summer (dry) season.
- Piezometer levels reveal drop during the summer (dry) season consistent with reduced rainfall.
- The investigation of the Reinaldo Espinoza suburb was developed with the collaboration of many institutions and most importantly the community. Community communication is very important to the execution this geohazard investigation and the implementation of its findings.
- Inclusion of the sociological component to the investigation resulted in the strengthening of the community organisation.

REFERENCES

BGS & CODIGEM, 1994, Geological and Metal Occurrences Maps of the Southern Cordillera Real and El Oro Methamorphic Belts, Ecuador.

CRUDEN, D.M. and VARNES, D.J. 1996. Landslide types and processes. In "Landslides – Investigation and Mitigation", Transportation Research Board Special Report No. 247, Turner and Schuster (eds.), National Academia Press, Washington, DC, p. 36–70.

IBADANGO, C.E., SOTO, J., TAMAY, J.,

ESCUDERO, P., and PORTER, M., 2005, Mass Movements in the Loja Basin – Ecuador, South America, in O. Hungr, R. Fell, R. Couture, E. Eberhardt (eds.) Landslide Risk Management, Proceedings of the International Conference on Landslide Risk Management, Vancouver, Canada, 31 May-3 June 2005, paper I010 (accompanying CD-ROM),A.A. Balkema, Leiden. Vancouver – Canada.

SAUER, W., 1965, Geología del Ecuador, Quito – Ecuador.

VARNES, D.J., 1978, Landslides Analysis and Control. Edited by R.L. Schuster and R.J. Krizek. National Academy of Sciences.

Landslides and Climate Change – McInnes, Jakeways, Fairbank & Mathie (eds)
© 2007 Taylor & Francis Group, London, ISBN 978-0-415-44318-0

The effect of waste water disposal on coastal slope stability and the problem of climate change

P.D.J. Watson
Department of Civil Engineering, University of Portsmouth, Portsmouth, UK

E.N. Bromhead
School of Engineering, Kingston University, Kingston-upon-Thames, UK

ABSTRACT: The effects of water on the stability of slopes have long been recognised by Geotechnical Engineers, such that an essential design requirement in slope stabilisation is to reduce the volume of water present in the slope and hence raise the overall effective stress. Despite this long term appreciation of the effects of groundwater on slope stability, there appear to be numerous instances where the stability of slopes is affected adversely through human interference with the groundwater balance. Examples of this anthropogenic cause of slope instability include: leakage of water from supply pipes, irrigation, leakage from ponds or reservoirs, leakage from waste water disposal pipes, or disposal of waste water in soakaways. In rural or semi-rural areas, the provision of mains drainage (in contrast to mains water supply) is a low priority. Dwellings in remote coastal locations are particularly susceptible to the combination of mains water supply and local disposal of waste water. These properties therefore input a volume of water into the slope, and adversely affect the groundwater balance. In contrast, those properties on mains water and mains drainage have little or no effect on the total amount of water in the slope, and are hydraulically neutral. Even in urban locations, drainage of paved or surface areas is disposed of via soakaways. With climate change now a reality, the paper also examines how increased winter rainfall can impose an additional load on overall slope stability. The paper then reviews case records where waste water disposal has been identified as a major agent in promoting instability. Many of these case records are from coastal locations. The effectiveness of the installation of mains drainage as a means of reducing the incidence of instability is considered. A field reconnaissance has been carried out, and a number of locations where the preliminary conclusion that waste water disposal is a factor in nearby instability have been identified. Some of these locations are described.

1 INTRODUCTION

A dwelling constructed with a thatched roof and unpaved access, occupied by people who draw water for washing, drinking and cooking from a local well, and who dispose of their wastes via middens or earth closets, is effectively neutral with respect to the water balance of the land occupied. The thatched roof holds water, which is released by evaporation in much the same way as grass or shrub covered terrain, and contrasts strongly with roofs constructed of slates, tiles, or the more modern equivalents: metal ("corrugated iron") or plastic sheeting. Such impermeable roofing materials cause a virtually complete run-off from the covered area, see Table 1 below. Where mains water is not available, this runoff is stored in water butts, and when applied to domestic purposes or garden watering ("irrigation") is disposed of at the ground surface. Here, it is available for evapotranspiration removal in due course and is still largely neutral in terms of the water balance of the site.

Impermeable coverings to land also inhibit evapotranspiration, and further complicate water balance computation. A typical cross-section through a landslide complex detailing the likely water inputs and outputs are shown in Figures 1 and 2.

Table 1. Typical impermeability factors (After Tebbutt 1998).

	Impermeability Factor
Developed land	
Watertight roof	0.70–0.95
Asphalt pavement	0.85–0.90
Concrete flagstones	0.50–0.85
Macadam road	0.25–0.60
Gravel drive	0.15–0.30
Undeveloped land	
Flat ground	0.10–0.20
Sloping ground	0.20–0.40
Steep rocky slope	0.60–0.80

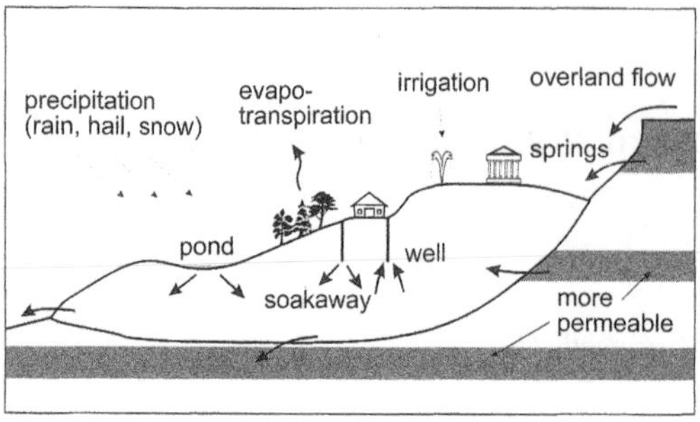

Figure 1. Cross-section of landslide complex with typical water inputs and outputs.

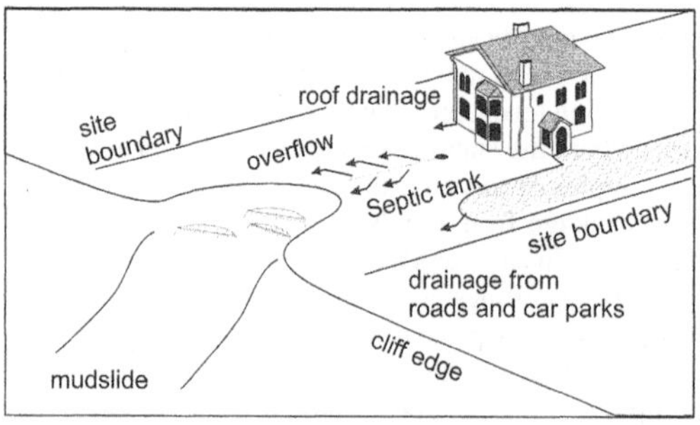

Figure 2. How soakaway drainage affects adjacent slopes.

In Britain, the system of charging for mains water supply has traditionally been based on some notional taxable value ("rateable value") of the household, and has not been for consumption per se. This charging mechanism has encouraged the use of mains water for non-potable purposes. After the installation of a mains water supply, householders may continue to use runoff water for gardens (especially where the water supply is metered – increasingly so in some areas of Britain), but there has been a tendency to put rainwater runoff from roofs into soakaways for cosmetic reasons. It is then not so neutral in terms of the site water balance. A similar detrimental effect on the water balance is obtained when paved areas such as car parks, drives and access roads collect surface water, which is disposed of via soakaways. This effect can be shown by reference to typical impermeability factors, as shown in Table 1 below.

These factors show that the net effect of impermeability is usually small for domestic dwellings, where the total impermeable area of drives and hard standings is less that the roof area. Impermeable areas cause increased runoff, and concentrate this, leading to erosion (surface discharge) or landsliding (if put into soakaways). This problem is not dealt with here.

Table 2. Typical UK domestic water consumption per person (Tebbutt 1998).

	Annual consumption in litres per person per day		
Use	1978	1992	1998
Toilet flushing	32	35	50
Drinking, cooking and dishwashing	33	45	38
Baths and showers	17	25	35
Washing machines	12	15	20
Garden watering	1	5	7
Car washing	1	2	2
Total	96	127	142

2 THE SCALE OF THE PROBLEM

2.1 *Lifestyle effects*

Lifestyle changes from the middle nineteenth century have seen the need for potable water supplies to every dwelling translated into a requirement for the provision of essentially unlimited quantities of water via a centralised treatment and distribution system. People wash themselves and their clothes more frequently than in the past. A major element of change was the perceived requirement to move sanitary facilities indoors: the water closet being the only way to achieve this in a socially acceptable way. Once the plumbing infrastructure was in place to provide a water supply for this and other domestic water uses, the march of technology took hold. Modern domestic white goods such as washing machines are huge consumers of water. Early washing machines used a single charge of water: modern automatic machines run through several cycles and are more wasteful. Cooling water (e.g. for air conditioning) is sometimes discharged to waste in a similar way, and with climate change forecasting hotter summers for the UK, this could become an additional source of waste water.

Overall, the scale of the problem may be judged by consideration of the water usage per person in a typical domestic dwelling. This is summarised in Table 2, and relates to British conditions.

This table shows that water consumption in the U.K. rose by nearly 50% in the 20 years to 1998, from 96 litres per person day to 142 litres per person day, of which only 3 litres is used for drinking.

In 1994, the Institution of Water and Environmental Management forecast that domestic demand for water in the UK would increase by a further 20 per cent by 2020, to about 175 litres per person day (Latham 1994); although recent studies by OFWAT have shown water demand to have peaked at 140 litres per person per day (OFWAT, 2005).

Once used, all of this water must be disposed of. In urban areas, a centralised collection and treatment system can usually be provided comparatively cheaply, as the main piping system and treatment works have numerous users. Flow under gravity is usually possible even if deep collection points need to be constructed. However, this is not so for isolated or rural locations. Piped potable water supplies are provided, even though the costs of provision are larger than in an urban setting, because of the perceived requirement for supply on public health grounds. The pumping costs are, of course, concealed in the overall costs of supply, and since the pumping is usually from a central location, additional marginal costs are rarely identifiable. This is not the case with waste disposal. The water-borne wastes are not easily transportable through narrow bore (cheap) pipes, the system is not pressurised, and pumping costs would be more likely to fall on individual consumers (producers).

As a consequence, biological treatment on-site is widely used in many comparatively isolated rural locations. Waste water is stored in an underground reservoir ("septic tank" or "cess pit") where an anaerobic digestion process acts on it. Surplus water from the system is permitted to soak into the surrounding ground, where soil bacteria complete the transformation of the sewage to harmless compounds. Accumulations of solids are periodically removed from the digester. Cess drainage is not intrinsically a great producer of water, but where domestic drainage is added, the water burden may be considerable. It is however, highly effective for low occupancy dwellings in isolated locations. Systems may become overloaded where the occupancy of habitations is larger (e.g. hotels, barracks, camp sites, and where washing water is disposed of by the same pipework. Impacts are severe adjacent to slopes (Figure 2). The water consumptions ofTable 2, for an average domestic dwelling site occupied by an average family, correspond to a local increase in effective rainfall to tropical levels.

Extensive systems of pipework are very vulnerable to damage from any ground movement, and may therefore leak copiously for very long periods of time. This is particularly the case if the leaks are not serious enough to cause interruptions in supply, or for the water not to emerge at the ground surface. There has been much criticism in the UK of the problems of leaks from water supply even in areas where landslide ground movements are not the source of the problem. Cases where adverse alterations to the water balance (by septic tank drainage or water leakage) have contributed to local slope instability are discussed below.

2.2 *Climate effects*

A study of ground movement in the Ventnor area by Ibsen (1995) showed that the most dramatic period of ground movement in recent times in Ventnor occurred during the winter of 1960–1961. This coincided with extremely high autumn rainfalls in 1960. Ibsen's study concluded that land failures occur in the year after excess rainfall is recorded, i.e. an increase in rainfall is followed by an increase in the number of landslide events. Brunsden (1973) explained this relationship in study of the pore pressures within a slope. In essence, the increase in the water regime causes a rise in the water table and pore pressures, and a decrease in the shearing resistance of the soil with the elimination of surface tension. An increase in the groundwater level also increases the weight of the soil unit. These factors promote the landslide process.

Thus climate is related to landsliding in providing effective precipitation, but also by temperature influences and the evapotranspiration regime.

3 CASE HISTORIES

3.1 *Areas where pipe leakage is a problem*

During the period 1950–1990, Luccombe village, Isle of Wight, experienced three major phases of landsliding, in 1950–1951, 1961–1962, and 1987–1988. Following the loss of 4 residential properties in the January 1988 landslip, Luccombe Residents Association appealed for help to safeguard their properties. The Department of the Environment commissioned an extensive investigation of the landslide complex at Luccombe and identified four main factors as responsible for reactivation of the landslide movements. These were the recession of the coastal cliffs; the construction and development activities in Luccombe village since 1927; water supply and sewerage leakage; and prolonged periods of intense rainfall (Moore & Longman, 1991). The conclusion of the DoE report suggested however that interference and poor maintenance of the water supply and drainage systems were by far the largest contributory factors in landslide reactivation. Leakage of supply water was monitored at 4,350 litres per day or 29 per cent of the total supplied. Conversely, the sewerage system only carried a minor amount of the water used away from the area, resulting in up to 15,000 litres per day of water entering the landslide complex via sewerage leakage or septic tank discharge.

During construction of the village in 1927, level plots for houses were formed by means of cut and fill, with similar works being carried out in laying the roads. Water, drainage and service pipes were laid, ponds built and gardens terraced. All of these operations interfered with the natural drainage of the site. Since then, there had been little maintenance of the surface water drainage, with rubble drains infilled with debris, and drainage outlets discharging directly onto the landslide. In conclusion, the report stated there had been large human interference with the natural drainage of the area, and recommended that:

- The village needed a sewerage scheme to prevent further discharge via septic tanks.
- The water supply network needed to be improved and maintained.
- Drainage ditches should be dug used and regularly maintained.
- Land needed clearing and grading to prevent ponding.

These recommendations were subsequently carried out and completed in the early 1990's. Little or no movement has occurred since then, despite heavy winter rainfall subsequently. Similar improvements to the sewerage/drainage schemes have recently – 2000/2001 – been carried out in the Ventnor area of the Isle of Wight to reduce the amount of water leaking into the slope.

3.2 *Areas where septic tank drainage is a problem*

The exposed rear scarp of the 1971 landslide at Warden Point, Isle of Sheppey intercepted the septic tank drainage of the Post Office. This location rapidly evolved into a secondary mudslide in the rear scarp of the main slide. Activity ceased when the Post Office was abandoned. Retreat of the rear scarp of the main landslide has continued, although at a slow rate, and in 1997/98 the zone of influence of the cess drainage of the adjacent house was encountered. This has re-stimulated the mudslide.

West of the 1971 landslide, there was a row of substandard construction bungalows. A rotational slide in 1976 began to uncover the rear gardens of these bungalows, where septic tank drainage was employed. A small mudslide at the margin of the 1976 landslide enlarged progressively, and ate back into the cliff edge and rear scarp of the main slide, following the line of cess pits. The mudslide was contaminated with foul water. Progressive abandonment of the houses was followed by a diminution of activity of the mudslide, which is now dormant, and being consumed by rear scarp retreat generally.

3.3 *Areas where occupied properties are located on, or adjacent to landslides*

At Blackgang Gore Cliff, the 1978 landslide (Bromhead *et al.*, 1991) destroyed three houses: Sandrock Spring, Sandrock Spring Cottage and Cliff Cottage. All three employed cess drainage, but had a mains water supply. The nearby South View House which does not have mains water, was off the area of reactivation of movement and was undamaged. Subsequent surveys (Chandler, 1984) showed that this area was stationary. This slide was a reactivation of an area that

Figure 3. Mudslide at Warden Point, Isle of Sheppey. Septic tank drainage from the exposed pipe at bottom left of photograph is the source of this mudslide.

had been stable for hundreds of years. A later landslide in the Gore Cliff area occurred in January 1993, and destroyed two further properties with local water disposal. Also at Blackgang, Isle of Wight, a large mudslide retrogressed more rapidly than the adjacent cliff into a garden of The Coach House, (a multi-occupancy small hotel). At the height of activity, pools of discoloured water (Figure 4) were seen in the slide debris, and a sewage odour was obvious. On abandonment of the building, the activity of the slide appears to have ceased.

4 ADDITIONAL ANTHROPEGENIC EFFECTS ON LOCAL SLOPE STABILITY

Figure 2 shows, in addition to the concentrated effect of septic tank discharges, inputs to the ground water system from runoff both from roofs and paved areas. These have a high impermeability factor, and are laid to falls so that the discharge hydrograph has a high, early, peak. Surface erosion and gullying may also result from disposal of this water. Ponds (e.g. as shown in Figure 1) may be the manifestation of the local groundwater table, but if they are artificially constructed and regularly refilled, they are a major source of water inputs. Balancing ponds, empty most of the time, but used under heavy rainfall to control the discharge from surface water drainage, may contribute to an adverse

Figure 4. Views of slide retrogressing towards "The Coach House", Blackgang, Isle of Wight.

water balance. Drainage networks of shallow land drains, which usually have a beneficial effect on stability, may have an adverse effect if their outfalls are blocked. This happens due to neglect, wheel rutting, construction work etc. (Bromhead & Ibsen, 1997).

All these supplementary conditions can have an additional effect on the local slope stability, and should be taken into account in any slope stability analysis of the area.

5 CONCLUSIONS

In conclusion it can be stated that:

1. Significant inputs of water have a pronounced effect on local slope instability.
2. Septic tank drainage causes a large net input of water into the groundwater system.
3. Mains water pipe leakage causes a net input of water into the ground water system.
4. Mudslides are the dominant landslide type, with retreat of the slide head.
5. Mudslide activity ceases rapidly on abandonment of occupied dwellings.
6. Climate change will affect slope stability.

REFERENCES

Bromhead, E.N. and Ibsen, M-L. 1997. Land-use and climate-change impacts on landslide hazards in SE Britain. In: Cruden, D. M., and Fell, R. (Editors): *Landslide risk assessment*: 165–175 Rotterdam: Balkema.

Bromhead, E.N., Chandler, M.P. and Hutchinson, J.N., 1991. The recent history and geotechnics of landslides at Gore Cliff, Isle of Wight. *Int. Conf. on Slope Stability Engineering – developments and applications, Thomas Telford, London.*

Brunsden, D. 1973. The application of systems theory to the study of mass movement. *Geol. Appl. Idrogeol.* 8(1): 185–207.

Chandler, M.P. 1984. *The coastal landslides of the Isle of Wight.* Ph.D. Thesis, Imperial College, University of London.

Ibsen, M.L. and Brunsden D. 1995. The nature, use and problems of historical archives for the temporal occurrence of landslides, with specific reference to the south coast of Britain, Ventnor, Isle of Wight. *Geomorphology* 15: 241–258.

Latham, B. 1994. An Introduction to Water Supply in the United Kingdom. *Booklet 4 – The Institution of Water and Environmental Management*, 78. London.

Moore, R. and Longman, F. 1991. The impact, causes and management of landsliding at Luccombe village, Isle of Wight. In Chandler, R. J. (ed.), *Slope Stability Engineering – developments and applications*: 225–230. Thomas Telford: London.

OFWAT, 2005. *Sustainable Development – The Government's approach – delivering UK sustainable development together.* http://www.sustainable-development.gov.uk/.

Tebbutt, T.H.Y. 1998. *Principles of Water Quality Control.* 5th Edition: 107–116. Oxford: Butterworth Heinemann.

Session 8
Risk governance – making better planning policies and decisions

The most effective way of combating losses is to incorporate natural hazards into the land-use planning framework, placing constraints on development of land where there is deemed a high degree of risk, and by guiding development to the most suitable locations. Robust and transparent evidence is required to support planning policy development. In view of the impacts of climate change, planners and developers should be working together with engineers to ensure that planning guidance takes full account of natural hazards now and in the future.

The Needles, Isle of Wight, UK
Courtesy Wight Light Gallery, Ventnor, Isle of Wight, UK.

Landslides and Climate Change – McInnes, Jakeways, Fairbank & Mathie (eds)
© 2007 Taylor & Francis Group, London, ISBN 978-0-415-44318-0

The planning response to climate change

David Brook OBE
Independent consultant, London, United Kingdom

ABSTRACT: With its introduction of large uncertainties, climate change may appear to be the ultimate nightmare to a planning system that aims for certainty through lines on maps. Recent changes to the planning system in England include the adoption of a spatial approach that is wider than conventional land-use planning and the placing of sustainable development as a core principle of planning. As a result regional and local development plans can more readily take account of climate change. Policy guidance is available and we know what is likely to happen as a result of climate change, though not the scale on which things will happen. This personal evaluation of the planning response to climate change concludes that measures to mitigate climate change and to adapt to the urban heat island have multiple benefits in terms of amenity and the environment and basically represent what is already good planning. Flooding, coastal erosion and landsliding are now firmly within the planner's horizon. Implementation of current policy, to consider these weather-related issues as they are now and to test the sensitivity of development design against likely changes should lead to an appropriate planning response to the need to adapt to climate change.

1 INTRODUCTION

For 60 years, the English planning system has aimed to achieve a measure of certainty in its control of development and the use of land in the public interest through its forward planning and development control procedures.

Planning shapes the places where people live and work and good planning ensures that the right development occurs in the right place at the right time. The system is designed to reconcile the benefits of development, in the homes, investments and jobs delivered, with the economic and environmental costs that development can impose, while recognizing social equity. Sustainable development is now the core principle underpinning planning and the overall aim is to enhance the quality of life now and for future generations.

The System aims to achieve certainty for developers in knowing what development will be allowed where and operates over a limited time-scale of 10–15 years. However, the longer-term influence in shaping where people live and work is recognized. The recognition over the last decade or so that human-induced climate change is a reality has meant that the system will have to learn to deal with large uncertainties on a longer time-scale than hitherto.

2 THE PLANNING SYSTEM

The English planning system was introduced under the Town and Country Planning Act 1947. While this has been amended over the years and amendments consolidated in the Town and Country Planning Act 1971 and the Town and Country Planning Act 1990 (the 1990 Act), the system remains essentially plan-led, with national policies and regional and local development plans acting as the framework for sustainable development. Within that framework, decisions can be complex and often inherently political and the framework needs to foster consistent, predictable and prompt decision-making.

The planning system comprises two main areas of activity:

- Developing strategic policies and plans (development planning);
- Determining individual planning applications (development control).

Under the 1990 Act, as amended by the Planning and Compensation Act 1991, decisions on planning applications should conform to the development plan, where it contains relevant policies, unless other material considerations indicate otherwise. The development plan should conform to government plans and policies at national level. Under the Planning and Compulsory Purchase Act 2004 (the 2004 Act), it now comprises the regional spatial strategy, prepared by regional planning bodies and issued by the Secretary of State and development plan documents issued by local planning authorities as part of their local development framework.

2.1 *National planning policies*

There is no national development plan but advice on policies and procedures is issued by government in planning policy guidance notes (PPGs) and circulars. Existing PPGs are being reviewed and replaced by planning policy statements (PPSs) to provide clear concise and well focused statements of national planning policy, accompanied where necessary by technical annexes and good practice guidance to advise on the implementation of those policies.

Policies in PPGs/PPSs and circulars are strategic guidance to guide regional planning bodies and local planning authorities in preparing their development plans and dealing with planning applications and to guide developers. They are not mandatory but carry considerable weight in the determination by the Planning Inspectorate and the Secretary of State of appeals against refusal of planning permission.

2.2 *Regional planning*

England is divided into 8 planning regions plus London, within which the regional planning body was responsible for preparing draft regional planning guidance applying the principles set out in national policies to a framework for the preparation by local planning authorities of their development plans. The regional planning bodies comprise representatives from local planning authorities within the region and other bodies.

Under the 2004 Act, the latest version of the regional planning guidance constitutes the regional spatial strategy, which now constitutes part of the development plan, and the regional planning body is responsible for preparing, consulting on and carrying out a sustainability appraisal of revisions in whole or in part. This regional spatial strategy adopts a spatial planning approach that goes wider than traditional land-use planning, bringing together and integrating policies for the development and use of land with other policies and programmes that influence the nature of places and the way they function.

The regional spatial strategy provides a broad spatial development strategy for the region with policies and proposals for the whole of the region or for specific areas, where necessary. It should conform to national policies and any departures must be fully justified. It should also be consistent with and supportive of other regional frameworks and strategies. The regional spatial strategy covers a 10–15-year time-scale but the need is recognized to look beyond the end of this period in certain instances because relevant forecasting horizons are longer-term. Regional planning bodies are required to report annually on the achievement of targets in the regional spatial strategy, identify remedial action and trigger further revisions where necessary.

2.3 *Local development planning*

The strategic planning function was transferred to the regional spatial strategy with the abolition under the 2004 Act of county structure plans. Detailed site-specific local development planning remains the responsibility of local planning authorities. However, local plans are being replaced by local development documents, which will comprise the local development framework – an informal non-statutory term for the portfolio of documents that collectively delivers the spatial planning strategy for individual authorities. Like the regional spatial strategy, these take a spatial planning approach that is wider than traditional land-use planning and are subject to sustainability appraisal.

The local development framework comprises development plan documents that are, along with the regional spatial strategy, part of the development plan and supplementary planning documents, which expand policies contained in a development plan document or provide additional detail. It also includes the statement of community involvement, the local development scheme, the annual monitoring report on achievement of targets and any need for change and any local development orders or simplified planning zones.

Development plan documents include a core strategy, site-specific allocations of land, area action plans for areas where significant change or conservation is needed and the adopted proposals map. The core strategy sets out the spatial vision and strategic objectives for an authority's area, the spatial strategy, core policies and a monitoring and implementation framework. All other development plan documents must conform with the core strategy. The adopted proposals map expresses geographically the adopted development plan policies of the authority, identifying areas of protection, locations and sites for particular land-use and development proposals included in any development plan document and areas in which specific policies apply.

2.4 *Planning applications and development control*

All development, as defined in the 1990 Act, is subject to the grant of planning permission on application to the local planning authority. However, agricultural land-use changes are not classified as development and certain developments have deemed planning permission under the Town and Country

Planning (General Permitted Development) Order 1995. Permitted development rights may be removed and an application for planning permission be required by a direction from the local planning authority where it considers their imminent implementation is likely to cause harm.

Apart from certain types of major application, including minerals and waste management that are classed as county matters, district councils are the local planning authority that determines applications. Local planning authorities must consider all material considerations when determining applications and decisions should generally be in accordance with the development plan. Permission may be granted subject to such conditions as the local planning authority think fit, provided they are relevant, precise, enforceable, related to planning and reasonable in all other ways. Applicants may appeal to the Secretary of State against refusal of permission, against non-determination within the statutory time-scale or against any conditions attached to a planning permission.

3 PLANNING AND CLIMATE CHANGE – NOW

Some statements of national policy now include reference to the need for planners to be aware of climate change and how they can deal with it through the planning system. The Office of the Deputy Prime Minister (2004a) has also published guidance on 'Planning and climate change – a guide to better practice'.

Much of the advice on dealing with climate change is relatively indirect through sustainable development being at the core of the planning system. PPS1 (Office of the Deputy Prime Minister 2005), for example, outlines the key principles for achieving sustainable development through the planning system. These include the need for development plans to 'contribute to global sustainability by addressing the causes and impacts of climate change through policies which reduce energy use, reduce emissions...promote the development of renewable energy resources and take climate change impacts into account in the location and design of development.'

The introduction of a spatial planning approach, which goes wider than conventional land-use planning and the requirement for both regional and local development plans to be subject to sustainability appraisal also increases the chance of climate change being taken more into account. Under the 2004 Act, policies in regional spatial strategies have to be related to the development and use of land but should not be restricted to policies that can be implemented through the grant or refusal of planning permission. PPS11 (Office of the Deputy Prime Minister 2004b) highlights the need for regional spatial strategies to be consistent with and supportive of other regional strategies and programmes, including any regional sustainability or climate change strategy.

Local development documents must be in general conformity with the regional spatial strategy and Annex B of PPS 12 (Office of the Deputy Prime Minister 2004c) gives advice on the treatment of climate change impacts and adaptation. Policies should cover:

- the need for new development to avoid areas at risk, now or over the lifetime of the development;
- the physical and environmental constraints on the use of land, given that these may increase with climate change;
- the likely increase in pressure on resources with climate change;
- the way that the distribution of species and habitats may alter with climate change and the effects on biodiversity and areas designated for protection;
- the need to consider possible adaptation options for vulnerable areas.

It is understood that the government is currently in process of preparing a specific planning policy statement on climate change, which it is assumed will cover both mitigation and adaptation. The planning measures dealing with the former are largely concerned with reducing emissions of greenhouse gases. Such policies include encouraging patterns of development which reduce energy and water use, reduce the need to travel by private car or reduce the impact of moving freight and promoting the development of renewable energy resources. Along with higher densities of development to improve the efficiency of land-use, these are standard good planning and design issues. As a result a number of authorities have introduced policies requiring development that is, as far as possible, energy-neutral.

Policies to deal with adaptation to the impacts of climate change are much more significant in terms of changes that we already know are likely to occur. To date, the only formal guidance on adaptation policies was that contained in PPG 25 (Department for Transport, Local Government & the Regions 2001), which has now been replaced by PPS25 Development and flood risk (Department for Communities and Local Government 2006a), with a companion practice guide (Department for Communities and Local Government in prep). This outlines the likely impacts of climate change on flood risk in both fluvial and coastal situations and advises on the allowances that should be made for climate change in assessing flood risk during the lifetime of a development.

PPS25 also strongly encourages the use of sustainable drainage systems (SUDS). While this is not specifically directed at climate change impacts it will help to deal with one of the major impacts identified by the Department of Trade and Industry's 'Foresight future flooding' project (Evans et al. 2004a, b), that of increased urban flooding due to more intense rainfall overwhelming drainage systems.

4 IMPACTS OF CLIMATE CHANGE

Scenarios, known as the UKCIP02 scenarios, based on climate model projections from a range of 4 different emission scenarios have been published by the United Kingdom Climate Impact Programme (Hulme et al. 2002). The key results with implications for planning are:

- UK climate will be warmer, by between 2°C and 5°C by the 2080s;
- high summer temperatures will be more frequent and very cold winters will be increasingly rare;
- winters will become wetter and summers drier;
- snowfall will decrease throughout the UK;
- heavy winter precipitation will become more frequent;
- relative sea level will continue to rise around most of the UK shoreline; and
- sea-levels currently regarded as extremes will be experienced more frequently.

Changes in wind speed are predicted to be relatively small but the level of uncertainty is very high. Similarly there may be an increase in storminess, leading to a greater frequency of storm surges but, again, the level of uncertainty is high.

4.1 *Implications for planning and development*

The main implications for planning of these projected changes in climate relate to:

- increases in the frequency and severity of flooding due to higher sea levels, greater intensity of winter rainfall and possible increases in storm surges;
- increased coastal erosion due to higher sea levels and possible increases in storm surges;
- possible increase in landslides due to increased winter rainfall and summer droughts;
- water resource problems, particularly in the south-east, due to decrease in summer rainfall;
- increase in the urban heat-island effect due to higher temperatures.

The 'Foresight future flooding' project (Evans et al. 2004a, b) examined the implications of climate change, among other driving factors on flooding and coastal erosion over the next century under 4 different economic scenarios. Assuming no change in current flood management policies and expenditure, annual average damages due to fluvial and coastal flooding were projected to increase from £1B to £20B by the 2080s in the worst-case scenario. Adopting a portfolio of responses, including considerable emphasis on the use of the planning system to avoid risk of flooding, this increase would be reduced to £2B.

Damages due to intra-urban flooding are projected to rise from £270,000 to £7.8B by the 2080s in the worst-case scenario. With an integrated portfolio of responses, damages would be reduced to £4.2B. Damages due to coastal erosion are projected to rise by 3–9 times to £126 M in the worst-case scenario. Both flooding and coastal erosion will need to be taken fully into account in future planning decisions on the location of development.

Higher sea levels and increased flooding will lead to greater erosion of the toes of slopes at the coast and in river valleys. Wetter winters with more intense rainfall may lead to trigger levels for slope movements being exceeded more frequently. There is thus likely to be an increase in landsliding, particularly at the coast but not restricted to it.

Changes in land cover due to the effect of climate change on habitats, agriculture and forestry, together with changes in land-management practices may significantly affect rates of infiltration and run-off. These could affect both the frequency of local flooding due to increased run-off as well as having implications for local slope stability due to increased erosion potential and to decreased evapo-transpiration leading to higher water levels in slopes.

The south-east of England is an area already under water stress in that resource availability is barely sufficient for the population that needs to be served. Hotter drier summers will increase water demand for recreational and cooling purposes and will increase evaporation from soils and water bodies leading to decreased water availability. More frequent droughts of more extended duration will extend the period of soil moisture deficits in excess of zero, reducing the opportunity for groundwater recharge and reducing water availability. Future development and the provision of infrastructure will need to take into account the availability of water resources.

Hotter summers will also lead to an increase in the urban heat-island effect with consequent effects on human comfort in urban areas. As a result more time may be spent out of doors and there is likely to be increased energy demand for air conditioning.

In addition to the direct effects of climate change in the United Kingdom, changes elsewhere may have significant effects on planning issues in the UK. For example, current holiday destinations in the Mediterranean may become less favoured than holidays at home. This will inevitably lead to pressure for development in coastal holiday areas, precisely where risks from coastal erosion and flooding are projected to increase.

5 PLANNING AND CLIMATE CHANGE – THE FUTURE

Most of the adaptation measures necessary to respond to the projected impacts of climate change are deliverable through the planning system. Exceptions are measures on pre-existing buildings and those which affect only the interior of buildings. These will need to be addressed through advice and encouragement or by other mechanisms, such as the Building Regulations.

Guidance is available in government policy statements and research publications. Checklists for development are included in Office of the Deputy Prime Minister (2004a) and South East Climate Change Partnership (2005).

The planning response needed will vary according to the individual impacts outlined above.

5.1 *Increase in risk of flooding*

The guidance in PPS25 gives advice on how climate change can be taken into account in considering development in the light of increases in risks of flooding. This is beginning to be taken on board in regional and local development plans through the carrying out of broad-scale appraisals and assessments that look at flood risk on a fairly wide catchment scale and feed this into the sustainability appraisal of these plans.

At the local scale, development plans take account of these assessments, including an assessment of the impacts of climate change, in allocating land for particular uses. Allocations are based on the application of a risk-based sequential test, which gives priority to development in areas of lowest risk and identifies the types of development that might be appropriate in higher-risk areas where such development cannot be avoided in the interests of overall sustainability, taking account of existing settlements.

While there has been only gradual take-up of the concept of flood-risk assessment for individual developments, those submitting applications for planning permission are beginning to accompany them with appropriate assessments that look at existing risk and how it is likely to change in the future.

Some developments have been designed to reduce the risk, both now and in the future, to an acceptable level. For example, proposals for development alongside the River Kennet to the west of Reading involve raising land to the level of the 0.1% annual probability of flooding (i.e. into the lowest-risk zone of the sequential test) with floor levels of houses raised above that to take account of the predicted impacts of climate change. A development on the south bank of the River Thames, downstream of the Thames Barrier, comprises apartment blocks on columns to raise residential accommodation above the likely flooding level, taking account of climate change. Ground-floor levels beneath these blocks are used for car parking and secure access is provided to higher ground for when the flood occurs.

The introduction of uncertainty into future changes in risk adds to what is already a fairly large uncertainty in flood-risk estimation. It is therefore necessary that flood zones, on the basis of which the sequential test in PPS25 is applied, are based on existing flood risks, with climate change being considered insofar as it is appropriate for the life of the development concerned. Regional flood-risk appraisals, sub-regional and local strategic and site-specific flood-risk assessments will also need to be consistent with and supportive of flood-risk management plans, such as catchment flood-management and shoreline-management plans, which now take full account of the potential impact of climate change.

It will also be necessary to consider more fully the prospects for flood-resistant and flood-resilient construction, where development may have to take place for wider sustainability reasons in areas of flood risk. Those techniques relating to the construction and external features of buildings can be applied by planning condition but the design and installation of internal fixtures and fittings can only be encouraged through the planning system. The development of the 'Code for sustainable homes' and its application through the Building Regulations (Department for Communities and Local Government 2006b,c) will be necessary to achieve full implementation of the necessary techniques for flood-resilience.

A more general take-up of sustainable drainage systems will be needed to counter the likely increases in intensity of rainfall. Current developments under the Government's flood and coastal management strategy Making Space for Water (Department for Environment, Food & Rural Affairs 2004) include the development of integrated drainage-management plans in urban areas. The National SUDS Working Group (2004), a group of central and local government, water and construction industry stakeholders, has developed an 'Interim code of practice for sustainable drainage'. This advises on the means of securing long-term management arrangements for SUDS through the use of planning agreements. Technical and good practice guidance (e.g. Martin et al. 2000, 2001; Wilson et al 2004) highlights among other factors the need to consider the potential impacts of increased infiltration on local slope stability in choosing the drainage techniques appropriate for new development.

5.2 *Increases in coastal erosion and landsliding*

Current guidance on coastal planning is contained in PPG20 (Department of the Environment 1992). This contains little if any mention of climate change but it does advise on the avoidance of areas subject to active coastal erosion. Both the Foresight future flooding project Evans et al. 2004a, b) and the Future Coast research carried out for the Department for Environment, Food and Rural Affairs (Halcrow 2002) give indications of the likely extent of coastal erosion over the next century. These are now being taken on board in the second round of shoreline-management plans prepared by the coastal groups of local authorities and the Environment Agency.

Local planning authorities should thus use these shoreline-management plans and the research results as a basis for defining active areas of instability now and in the future, to enable the advice in PPG20 and PPG14 Development on unstable land (Department of the Environment 1990) to be followed. Only by doing so can the pressures arising from increased British tourism activities as a result of climate changes elsewhere be adequately resisted where it is necessary to do so.

Inland landslides are also likely to increase, particularly in terms of reactivation of old landslides. While the current advice in PPG14 and its Annex 1 Landslides and planning (Department of the Environment 1996) does not specifically mention climate change, it does emphasize the need to be aware of these problems and to take them into account for the lifetime of the development.

Local planning authorities are beginning to recognize the need to take account of landsliding in their decisions, in contrast to the situation 15–20 years ago, when this was largely considered to be a matter for the developer and not the planning system (Brook 1991). It is considered that the existing advice is sufficient to enable planners to deal with any impacts of climate change on landsliding. The critical feature is the need to consider the lifetime of development, both in identifying possible hazard areas in development plans and in deciding applications for planning permission.

5.3 *Water-resource issues*

The problems of water-availability, particularly in south-east and eastern England will increase as a result of climate change. This will require greater co-operation than has been evident to date between local planning authorities, building control officers, the Environment Agency and the water industry. In particular, there is now the opportunity for regional and local plans to take more account of water plans and strategies, which extend over longer time-scales than is usual in land-use planning. Samuels et al (2006) summarizes both the problems and gives examples of good practice on the integration of sustainable water management in land-use planning.

The European inter-regional Espace (European Spatial Planning Adaptation to Climate Events) project has produced a useful toolkit for delivering water management through the planning system in the light of adaptations needed to cope with climate change (Land Use Consultants 2005). This deals with both the flooding issues already mentioned and the problems arising from water resource shortages in south-east England. It outlines a number of principles needed to integrate climate change adaptation into water management through the planning system and highlights appropriate implementation measures through development plan documents and supplementary planning documents, the use of planning conditions, planning obligations and informatives and the need for local planning authorities to be involved in the design process to enable the earliest consideration of climate change adaptation. It recognizes also the importance of developing to the full a sustainable buildings code, applicable through the Building Regulations, to further the efficiency of water-use in new development. A consultation document on 'Water efficiency in new buildings' has recently been issued by the Department for Communities and Local Government (2006d).

5.4 *The urban heat island*

The main aim of adapting to the likely increase in the urban heat-island effect is to achieve cooling of both the internal and external environment. The internal environment is primarily affected by the thermal mass of the building and by ventilation. These are matters of design, which are basically for the Code for sustainable homes and the Building Regulations rather than the planning system. The use of green roofs to achieve both evaporative cooling and reduction in urban run-off can be applied through the planning system. The London Climate Change Partnership (2006) cites some useful overseas examples.

The external environment is more readily addressed by the planning system through careful design of open space, both green areas and water features, within the urban area. However the success of any measures depends on the management of these open spaces and particularly of urban trees. There will also be benefits in terms of amenity and biodiversity, which are common features of conventional planning even without the climate-change element.

6 DISCUSSION AND CONCLUSIONS

English planners have tried for 60 years to give a measure of certainty in the balance between the demands of development and protection of amenity and the environment by putting lines on maps. They are now faced with a future of climate change which introduces large measures of uncertainty over a longer time-scale than is conventionally considered in forward planning.

However, planners have always recognized that decisions taken now will affect the longer-term pattern of development. They have also worked closely with other regulatory authorities to ensure that climatic factors affecting the physical integrity of development are taken into account.

Climate change as a consideration has only developed over the last 15 years or so. However, it has developed quickly and we now know much more about what is likely to happen in the future. The large uncertainties relate not to what is likely to happen but the scale on which it will happen.

Government guidance has developed since about 2000 to include consideration of climate change in both forward planning and in development control. It is gradually being taken on board by regional planning boards and local planning authorities.

A new draft supplement to PPS1 on Planning and climate change (Department for Communities and Local Government 2006e) has been issued for consultation. This sets out how spatial planning should contribute to reducing emissions and stabilizing climate change and take account of the unavoidable consequences of a changing climate. Together with the changes to the planning system under the 2004 Act, particularly the adoption of a spatial approach and the introduction of mandatory sustainability appraisal of regional and local development plans, this will assist planners in taking account of climate change.

The 2004 Act also introduced mandatory annual reporting on how regional spatial strategies and local development frameworks are meeting their objectives and whether revision is required in whole or in part. Together with the monitoring of sustainability indicators – such as the number of houses built in flood-risk areas – this gives the system the flexibility to respond to the advance of scientific knowledge on the scale of climate impacts on development and the consequent reduction in the levels of uncertainty.

Regional and local development plans can now include policies which affect where people live but may not be deliverable through the planning system. They should also be consistent with and supportive of other regional strategies and programmes. There is thus increased opportunity for collaboration between those strategies such as the shoreline management plans, catchment flood management plans and the river basin management plans that will be produced under the European Union Water Framework Directive.

What response is needed depends on the issues under consideration. Mitigation has probably been addressed to a greater degree than adaptation because the planning measures considered appropriate have dual benefits in terms of local amenity and the primary objective of creating pleasant places to live and work. In addition, the recent consultations by the Department for Communities and Local Government can only increase the profile of this issue in both the planning and building control systems.

As far as adaptation is concerned, dealing with flooding issues is now well advanced. Both regional and local development plans contain policies to take account of climate change in considering development and flood risk. It is also being included in the strategic flood-risk assessments which feed into those plans and in the assessments prepared by developers to accompany applications.

The greater use of sustainable drainage systems is being widely advocated in development plans. It offers multiple benefits in terms of reducing flooding, reducing the heat island effect, improving water quality, amenity and biodiversity and increasing groundwater recharge to reduce the water supply problems likely to be faced in future.

Coastal erosion and landsliding have been recognized as material planning considerations to a greater degree than previously. The policies now being adopted will enable these issues to be properly addressed in the light of likely future changes.

Most of the measures necessary for adaptation to climate change are deliverable through the planning system but there are some elements that can only be addressed through the Building Regulations, particularly through the adoption within those regulations of a sustainable building code.

What is required as a planning response to climate change is to ensure that planning and design take due account of weather-related features that affect developments now and to test the sensitivity of the design against the climate change scenarios and relevant advice on impacts as they are developed.

ACKNOWLEDGEMENTS

While this paper presents a personal view of the response of the planning system to climate change, it has benefited from discussions over a time with colleagues in the then Office of the Deputy Prime Minister and the Department for Environment, Food and Rural Affairs and with other stakeholders. Thanks are particularly due to Dr Brian Marker OBE, Robert Shaw of the Town and Country Planning Association and Professor John Handley of the University of Manchester, who reviewed this paper in its initial version.

REFERENCES

Brook, D. 1991. Planning aspects of slopes in Britain. In Chandler, R.J. (ed.), *Slope stability engineering, developments and applications; Proc Int. Conf. on Slope stability organised by Instn Civil Engrs & held at Shanklin, Isle of Wight, 15–19 April 1991*. London: Thomas Telford, 85–93.

Department for Communities & Local Government 2006a. *Planning policy statement 25: Development and flood risk.* Norwich: The Stationery Office.

Department for Communities & Local Government 2006b. *The code for sustainable homes.* London: Department for Communities & Local Government.

Department for Communities & Local Government 2006c. *Building a greener future: towards zero-carbon development. Consultation.* London: Department for Communities & Local Government.

Department for Communities & Local Government 2006d. *Water efficiency in new buildings: a consultation document.* London: Department for Communities & Local Government 2006.

Department for Communities & Local Government 2006e. *Consultation. Planning policy statement: Planning and climate change; supplement to Planning policy statement 1.* London: Department for Communities & Local Government.

Department for Communities & Local Government In preparation. *Development and flood risk – a practice guide companion to PPS25.*

Department for Environment, Food & Rural Affairs 2004. *Making space for water – developing a new Government strategy for flood and coastal erosion risk management in England: a consultation exercise.* London: Department for Environment, Food & Rural Affairs.

Department for Transport, Local Government & the Regions 2001. *Planning policy guidance note 25: Development and flood risk.* Norwich: The Stationery Office.

Department of the Environment 1990. *Planning policy guidance note 14: Development on unstable land.* London: HMSO.

Department of the Environment 1992. *Planning policy guidance note 20: Coastal planning.* London: HMSO.

Department of the Environment 1996. *Planning policy guidance note 14: Development on unstable land – Annex 1: Landslides and planning.* London: HMSO.

Evans, E., Ashley, R., Hall, J., Penning-Rowsell, E., Saul, A., Sayers, P., Thorne, C. & Watkinson, A. 2004a. *Foresight. Future Flooding. Scientific Summary: Volume I Future risks and their drivers.* London: Office of Science and Technology.

Evans, E., Ashley, R., Hall, J., Penning-Rowsell, E., Sayers, P., Thorne, C. & Watkinson, A. 2004b. *Foresight. Future Flooding. Scientific Summary: Volume II Managing future risks.* London: Office of Science and Technology.

Halcrow 2002. *Prediction of future coastal evolution for SMP review: final report.* London: Department for Environment, Food & Rural Affairs.

Hulme, M., Turnpenny, J. & Jenkins, G. 2002 *Climate change scenarios for the United Kingdom: the UKCIP02 briefing report.* Norwich: Tyndall Centre for Climate Change Research.

Land Use Consultants 2005. *A toolkit for delivering water management climate change adaptation through the planning system:.* London: Land Use Consultants for the Environment Agency and the South–East England Regional Assembly.

London Climate Change Partnership 2006. *Adapting to climate change: lessons for London.* London: Greater London Authority.

Martin, P., Turner, B., Waddington, K., Dell, J., Pratt, C., Campbell, N., Payne, J. & Reed, B. 2000. *Sustainable urban drainage systems – design manual for England and Wales.* CIRIA Report C522, London: CIRIA.

Martin, P., Turner, B., Dell, J., Payne, J., Elliott, C. & Reed, B. 2001. *Sustainable urban drainage systems – best practice manual for England, Scotland, Wales and Northern Ireland. CIRIA Report C523.* London: CIRIA.

National SUDS Working Group 2004. *Interim code of practice for sustainable drainage systems.* London: National SUDS Working Group.

Office of the Deputy Prime Minister 2004a. *Planning and climate change: a guide to better practice.* London: Office of the Deputy Prime Minister.

Office of the Deputy Prime Minister 2004b. *Planning policy statement 11: Regional spatial strategies.* Norwich: TSO.

Office of the Deputy Prime Minister 2004c. *Planning policy statement 12: Local development frameworks.* Norwich: TSO.

Office of the Deputy Prime Minister 2005. *Planning policy statement 1: Delivering sustainable development.* Norwich: TSO.

Samuels, P., Woods-Ballard, B., Hutchings, C., Felgate, J., Mobbs, P., Elliott, C. & Brook, D. 2006. *Sustainable water management in land use planning. CIRIA Report C630.* London: CIRIA.

South East Climate Change Partnership 2005. *Adapting to climate change: a checklist for development – guidance on designing developments in a changing climate.* London: Greater London Authority.

Wilson, S., Bray, R. & Cooper, P. 2004. *Sustainable drainage systems – hydraulic, structural and water quality advice. CIRIA Report C609.* London: CIRIA.

Landslides and Climate Change – McInnes, Jakeways, Fairbank & Mathie (eds)
© 2007 Taylor & Francis Group, London, ISBN 978-0-415-44318-0

Landslide management in a changing climate – coordinating the community response

R.G. McInnes
Centre for the Coastal Environment, Isle of Wight Council, Ventnor, Isle of Wight, UK

ABSTRACT: Since 1993 the Isle of Wight Council has endeavoured to encourage a coordinated approach to the management of ground instability problems through its 'Landslip Management Committee'. By bringing together professional interests, including local authority coastal management, planning, building control and highways staff alongside estate agents, construction industry representatives, service companies and insurers, a united professional approach to instability problems can be achieved. Local residents can also play a significant role by implementing good practice relating to property maintenance and land management. The challenge is to raise interest and awareness with homeowners who may have little knowledge of the history or extent of ground instability in their area. Climate change also presents significant challenges for those responsible for managing landslides and it is necessary to provide information to residents which explain potentially worsening scenarios. The paper describes how the Council is transmitting climate change and landsliding issues to non-technical audiences.

1 INTRODUCTION

Landslide management has been practiced for the last fifteen years on the Isle of Wight and forms an effective means of helping to address the impacts of ground movements around the Island's 110 km coastline. This process has involved the interpretation of field and desktop studies and the results of ground investigations. The results have helped to address the health, safety, economic and social issues in locations such as the Ventnor Undercliff and other vulnerable frontages along the Island's northern coast. The landslide management approach avoids the shortcomings of the 'emergency response' and instead concentrates on preplanning and preparation, allowing longer-term, more sustainable planning decisions to be made which are in line with the planning policy framework. With the support of a 'landslide management strategy' it is possible to address the different mitigation opportunities and to put 'policy into practice'.

In recent years scientific research has provided a wealth of additional technology and techniques that have assisted the understanding of ground instability problems. At a local level lessons learnt from a number of major landslide events on the Isle of Wight have also highlighted the role that human activity can play in instigating ground movements. As development pressures have increased the occupation of more stable and commercially attractive locations, there has been a demand also to extend new development into adjacent areas where there may be greater physical constraints because of problems such as instability. Such developments, particularly in the absence of effective planning policies, may lead to costly mistakes posing risks to life and property. These can usually be avoided if appropriate planning and landslide management strategies are in place, which have taken full account of ground conditions in their formulation.

The main focus for the Isle of Wight Council's landslide risk management strategies have been those frontages where development and natural hazards interact. These locations include the Isle of Wight Undercliff, which extends for 12 km along the south coast of the Isle of Wight, from Luccombe in the east to Blackgang in the west, including the town of Ventnor and the villages of Blackgang, Niton, St Lawrence, Steephill, Bonchurch and Luccombe.

It has been concluded (Jones & Lee 1994) that landslide problems are not 'acts of god' unpredictable, entirely natural events that can at best only be resolved by avoidance or large-scale civil engineering works. The role of human activity in initiating or reactivating many coastal slope problems should not be underestimated. In areas such as the Isle of Wight where urban development has taken place on coastal landslides, the problems tend to be related to slow ground

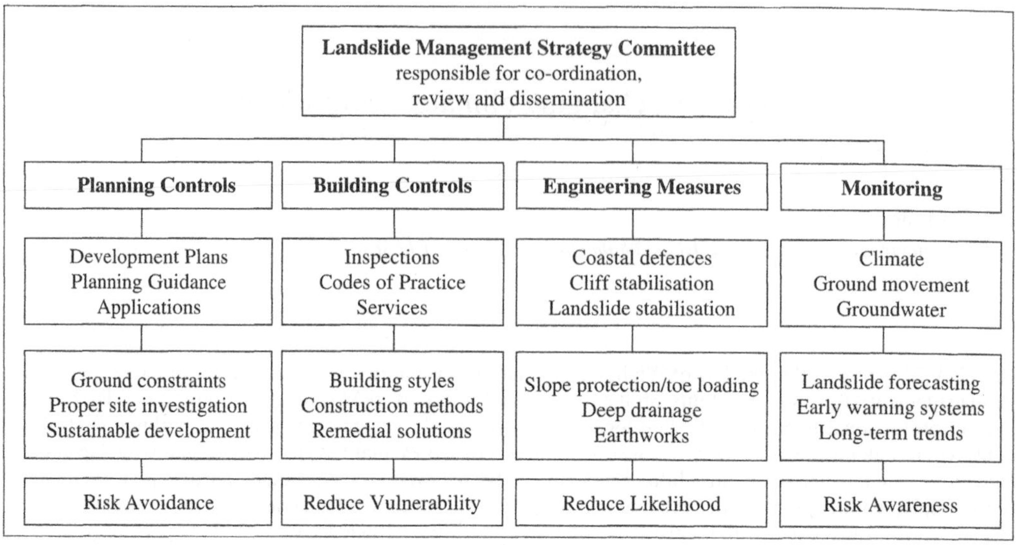

Figure 1. Isle of Wight landslide management strategy.

movement and progressive damage to property, services and infrastructure (Doornkamp *et al.* 1991). In such circumstances many problems can be reduced if there is a programme of active landslide management in place, where the local community is able to come to terms with the situation and learn to 'live with landslides' (Noton 1991, Lee 1997, Lee 2000, McInnes 2000).

A study of 'Coastal landslip potential' commissioned by the former Department for the Environment (DOE 1991), used the coastal town of Ventnor as a case study for developing national planning policy guidance for development on unstable land. As a result of this important research project, a fundamental change in approach was adopted. Following detailed geomorphological mapping on a 1:2,500 scale, the Undercliff landslide systems were identified (Lee & Moore 1991). This information, together with a survey of the results of past site investigations assisted in identifying the nature and extent of landslide systems, the types of contemporary movement taking place, and the magnitude and frequency of events and their impact on development. Furthermore, the information gathered assisted in assessing the nature of land use at risk and the vulnerability of structures to ground movements of different intensities (Lee & Moore 1991).

All this information was incorporated within a geographical information system which allowed the factors influencing the distribution and frequency of contemporary movements to be summarized on a ground behaviour map. A simplified version of this map was prepared to offer guidance to the planning process and to assist in the development of the landslide management strategy.

In addition an analysis was made of historical records of past landslide events and rainfall records, which contributed to a technical report. This report encouraged the development of a 'Landslide management strategy' for the Isle of Wight Undercliff.

The Strategy (Figure 1) aims to reduce the likelihood of future ground movements by seeking to control the factors that cause ground movement and by limiting the impact of future movement through the adoption of appropriate planning and building controls (Lee & Moore 1991). Ground instability in the Undercliff has, in fact, been addressed in a number of ways but the key tasks have been:

- preventing unsuitable development through sound planning controls and building control measures;
- monitoring ground movements and weather conditions using a range of automatic and manual recording instruments and stations;
- seeking to improve ground conditions through a range of measures aimed at controlling water in the ground as well as coast protection schemes which reduce marine erosion at the toe of the landslide;
- a major awareness-raising programme for the benefit of both professionals and the general public living and working in the area.

1.1 *Involving stakeholders in landslide management*

To implement the Landslide Management Strategy a Management Committee comprising key professionals from the Council (coastal management, highways, planning and building control), the water authority and other service industries, surveyors, estate agents and insurers meet twice a year. The Committee assesses progress on implementation of the Strategy and exchanges information on new initiatives being led by the Isle of Wight Council and others.

1.2 *Consultation and information exchange with Undercliff residents*

There has been a long history of consultations over ground conditions within the Ventnor Undercliff and, therefore, many residents are aware of the geological situation. The town is extremely attractively located with development on the various landslide benches offering panoramic sea views over the Victorian town, the adjacent spectacular coastline and the English Channel and the property market is healthy.

On completion of the DOE study in 1991 the first of a range of publications were produced to disseminate the findings of the study. In addition, a shop was opened in the town centre where the Council's geotechnical consultants were able to deal with questions from interested or concerned residents and businesses on a one to one basis.

A range of display boards were assembled covering the following themes:

- What is the history of ground movement in the area?
- What is the scale of the problem?
- Why is there a ground movement problem?
- What causes ground movement?
- How can we define landslide hazard?
- How can the landslide problems be managed most effectively?
- What is the local authority doing to help?
- What can developers do?
- What can property owners do?
- What can estate agents, solicitors and insurers do?
- What does the future hold if the community works together with the local authority?

The display was accompanied by a four page explanatory leaflet entitled 'Land Stability in Ventnor and You'. The temporary information centre provided the opportunity for interested residents not only to read the display boards, but also to ask questions and discuss any problems or concerns that they had with the local authority technical staff or its consultants, confidentially if required.

The information being disseminated by experts at the Geological Information Centre was not entirely straightforward. However, the final paragraph from the DOE summary report provided a basis for explanation:

'There is no reason why there should not be confidence in Ventnor from a building, insurance or financial development point of view. This is true so long as sensible use is made of the technical information presented in the report and obtained from future monitoring exercises, and that the proposed landslide management strategies are implemented.' (Doornkamp *et al.* 1991).

Table 1. (McInnes 2006).

Advice to homeowners in areas of coastal instability
Contents of four page information leaflet
– Background to the problem; aims of the leaflet
– The management of slopes and walls
– The control of water
– Maintenance and development of property
– Property insurance
Contents of display panels in Coastal Centre
– Why is there a ground movement problem?
– What is the impact of ground movement?
– How can the landslide problem be managed?
– What does the future hold? (in the context of climate change)
Key elements of advice provided on the display panels
– Current arrangements for managing landslide risk
– The role of the Council as Planning Authority
– How coast protection can reduce landslide risk
– Landslide monitoring and warning systems
– Improving drainage and reducing leakage
– How homeowners can help reduce risks
– How new technology can assist landslide management
– The role of the Council as an information provider

As a result of the success of the temporary information centre as a dissemination point, a permanent display opened in 1998 within the Isle of Wight Coastal Visitors' Centre based in Ventnor (Table 1). The town has been able to turn the geological situation to its advantage and capitalize on the interest in geological, coastal and environmental tourism and education.

Political support for improving knowledge and understanding of ground conditions in the Undercliff has been particularly strong. The Centre for the Coastal Environment has aided the process by commissioning or producing a series of information leaflets that have been distributed to every homeowner in the area together with more comprehensive reports which provide a wealth of information on the range of landslide management measures that have been promoted by the Council.

Over the last ten years four different information leaflets have been circulated to all 2,600 property owners and a range of reports and technical information have been provided with financial support from the Council.

In September 2000 a slope stability study in Cowes and Gurnard on the northern coast of the Isle of Wight was completed (Moore *et al.* 2000). This study successfully followed the model developed through the

Ventnor studies and produced geomorphology, ground behaviour and planning guidance maps in addition to the technical report. The Cowes study was important because it illustrated the transferability of the approach developed in the Undercliff.

The European Union LIFE Environment Programme (L'Instrument Financier de L'Environnement – European Commission 1997 and 2001) has also provided financial support for landslide risk management initiatives.

In particular a LIFE Environment study led by the Centre entitled 'Coastal change, climate and instability' (McInnes *et al.* 2000) allowed the landslide management work on the Isle of Wight Undercliff to be taken forward.

As part of the Landslide Management Strategy the feedback from local residents is regarded as very important. During the course of the LIFE project a survey was conducted which showed that a high percentage of residents (over 60%) had lived in the Undercliff over ten years and the majority were aware of ground instability issues at the time of moving into the area (82%). It is interesting to note that approximately 50% of those who intended to move to the Undercliff had obtained ground instability information from surveyors, consulting engineers or estate agents.

The Council was encouraged to note that of those who sought its advice on ground instability, some 90% found the advice very helpful or helpful and 66% of those who responded to the Council survey had read the key report on ground movement in the Undercliff. It was very pleasing to note that all those who responded had found this report to be either very informative (55%) or informative (45%). It should be noted, however, that over the previous four years some 25% of those responding had moved into the area, which indicated a significant turn-over in occupation of residential properties. As a result this demonstrated the need to continue to provide up to date information for residents on a regular basis.

One particular concern as far as property owners are concerned has been difficulty in obtaining property insurance in certain parts of the Undercliff, often due to a lack of knowledge over the true extent and nature of ground instability conditions. Certainly the DOE study assisted the process by indicating those areas where the risk is greatest. It was encouraging to note, therefore, from the resident's survey that 76% of those questioned had been able to obtain full insurance including subsidence cover in 2003, rising to 94% by 2006. The Centre for the Coastal Environment believes that a significant contributory factor to this statistic has been the availability of better information and guidance for local residents as well as for insurers over the intervening period. The residents survey was updated in 2006 as part of the 'Response' LIFE project (McInnes *et al.* 2006) and the results provide further support for the Council in terms of the value of this kind of information for non-technical users (see Table 2).

1.3 *Addressing the impacts of climate change*

Whilst the results of the residents' surveys have proved encouraging it is recognised by the Council that there are significant problems still to address. A ground investigation commenced in 2002 and further investigations were completed in 2005, which allowed the completion of a Landslide Quantitative Risk Assessment for central Ventnor in order that the Council can plan how to address increasing levels of risk arising from the predicted impacts of climate change.

A vital component of the Landslide Management Strategy is the monitoring programme, which includes recording meteorological conditions as well as analysis of data from a wide range of sub-surface instrumentation. In total nearly 200 instruments record ground water levels, settlement, cracking and tilt along the Undercliff's 12 km frontage. The programme benefits particularly from climate records which commenced in 1839 and it has been possible to compare the relationship between winter rainfall, in particular, with ground instability events. Valuable historical data of this kind alongside good local knowledge forms an important tool in terms of assisting prediction, allowing the Council to be better placed to address changing conditions in terms of rainfall patterns in the future.

The effects of climate change in unstable coastal zones are set out in Table 3 below; whilst a graphical presentation of the management problems is provided in Figure 2. These problems necessitate particular responses by coastal managers which will be increasingly focused on adapting to changing conditions alongside improved educational programmes.

A range of external influences as well as personal perceptions exist with respect to climate change and these are tabulated in Table 4.

As climate change rises rapidly up national political agendas the resulting media coverage must be tempered by the publication and dissemination of non-technical information such as the publication 'Responding to the risks from climate change' (McInnes *et al.* 2006).

1.4 *Transferability of the 'landslide management' approach*

There is already significant awareness amongst some of the relatively well informed and comparatively well-off residents of some of the Isle of Wight coastal zones, elsewhere in the United Kingdom and other parts of Europe. The challenge will be to transmit the message of landslide risk management globally.

Table 2. Residents' survey results.

ADVICE TO HOMEOWNERS ON GROUND INSTABILITY IN THE VENTNOR UNDERCLIFF, ISLE OF WIGHT

Comparative Residents' Survey Results 2000 and 2006

			2000	2006
1.	Number of residents participating in the survey.		475	490
2.	Location of properties within the Undercliff:	Bonchurch	28%	28%
		Ventnor	65%	66%
		Steephill	2%	1%
		St Lawrence	3%	3%
		Niton	2%	2%
			100%	100%
3.	Length of residency in the Undercliff (years):	0-5	25%	27%
		5-10	15%	20%
		10-20	30%	35%
		>20	30%	18%
4.	Were you aware of ground instability issues when considering buying a property in the Undercliff?		41%	59%
5.	Did you seek any professional advice in relation to the stability of the property you were purchasing?		25%	41%
6.	Did you seek any information relating to ground instability from the Council's Centre for the Coastal Environment?		13%	29%
7.	How helpful did you find the advice and displays at the Coastal Visitors' Centre?:			
		Very helpful	90%	95%
		Helpful	10%	5%
		Not particularly helpful	-	-
8.	Have you read the Council's eighty page report 'The Undercliff – A review of ground behaviour' (1995)?		48%	61%
9.	How useful did you find the report?	Very helpful	80%	86%
		Helpful	20%	14%
		Not particularly helpful	-	-
10.	The Council published a four page leaflet 'Advice to homeowners' in 2003. Did you receive the leaflet?			
		Yes	-	72%
		No	-	28%
11.	The Council has been improving advice and information for stakeholders with financial support from the EU LIFE Environment Programme; 'Coastal change, climate and instability' (2000-2003) and 'Response' (Responding to the risks from climate change) (2003-2006). Do you believe it is important for the Council to be involved in EU/International networks to improve information exchange?			
		Very important	86%	94%
		Important	10%	5%
		Not particularly important	4%	1%
12.	Have you visited the displays about ground instability at the Coastal Visitors' Centre, Ventnor?		45%	52%
13.	Have you been able to obtain property insurance cover for landslip?		76%	94%
14.	Do you regard ground instability as an issue of concern to you?			
		Very significant	82%	94%
		Significant	12%	5%
		Not particularly significant	6%	1%

Table 3. The impacts of climate change on coastal landslide systems and the management response (McInnes 2006).

Physical impacts
– Increased coastal erosion at the toe of coastal landslides;
– Increased coastal cliff and slope instability;
– Steepening of beaches in front of hard defences;
– Increased clay shrinkage as a result of drier summers;
– Easy ingress from increased winter rainfall.
Management responses
– 'Adaptation' policies to facilitate relocation of development away from coast;
– Strengthening defences to protect key coastal assets;
– Education programmes on risks from climate change to suit needs of stakeholders;
– Improve landslide management involving all sections of the community – led by local government;
– Strengthen planning policies to prevent new development in areas of landslide risk.

The key elements of a landslide management communication strategy should involve:

- explaining the nature and extent of the instability hazard;
- modifying the hazard to the community by means of engineering works (including coastal defence, drainage measures or slope stabilization) and improved building practice where appropriate. The purpose of these measures must be explained to local residents in advance of works commencing;
- implementing effective planning controls to guide development to more sustainable locations and to control the way new development takes place;
- co-ordinating the community response to the problems by following the 'Landslide management' approach described above.

Climate change and sea level rise present significant challenges to future instability management. On the coast an understanding of long-term evolution will allow the identification of areas where management problems are likely to arise in the future (Defra 2006). The arrangements in place for strategic examination

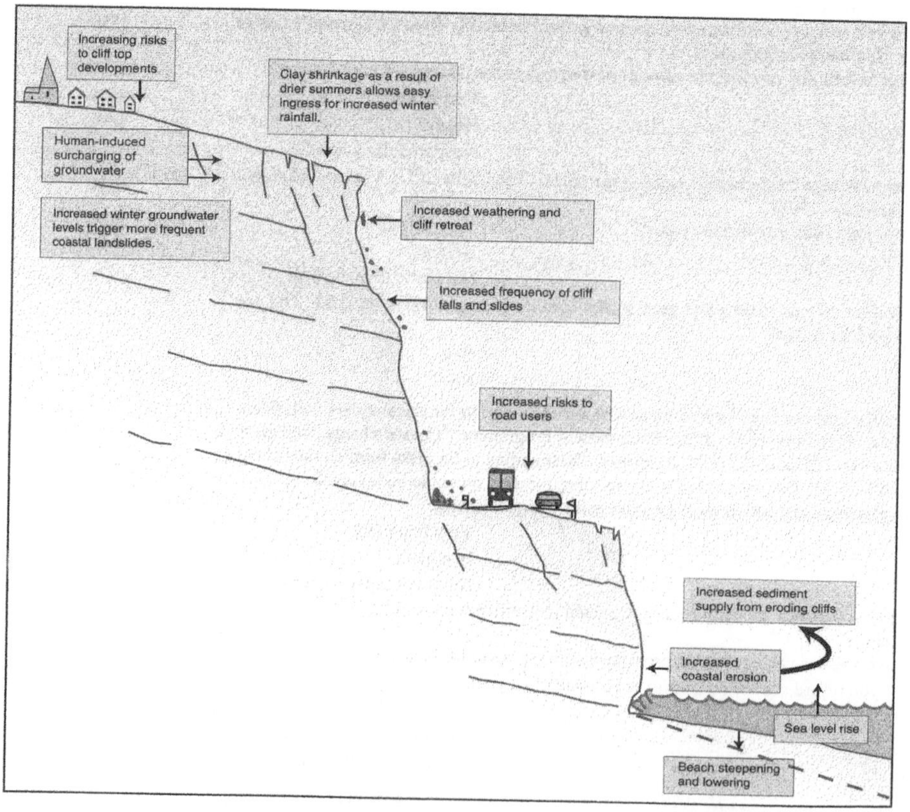

Figure 2. Some of the physical impacts of climate change on a developed coastal zone (McInnes, 2006).

Table 4. Factors influencing public perception of climate change in the Isle of Wight Undercliff (McInnes 2006).

Media influence:	• Television coverage of natural disasters – landslides, hurricanes, El Nino effects. • Newspaper coverage of events, newsletters and debates
Influence of public bodies:	• Publications, guidance, advice to homeowner's leaflets. • Updated planning and building control policies. • Introduction of 'adaptation to coastal change' strategies.
Influence of external organizations:	• Concerns of the insurance industry – increasing insurance premiums or unavailability of insurance. • Results of independent research into changes in geographical range of species, crops – possible extinctions; changes in weather patterns.
Perception of local residents:	• Observations of increasing frequency and severity of landslide events. • Changing weather patterns e.g. drier summer, more disturbed weather, wetter winters. • Changes in insect and plant life, including new species from France. • Shared concerns over climate change impacts amongst local community.

of coastal issues within the framework of guidance provided in the UK form a good model which could be replicated elsewhere. This would assist in avoiding poor planning and sitting of developments that have, in the past, made people and property more vulnerable to natural hazards.

The relevance of natural and man-made risks to the planning system is already recognized in guidance published by a number of countries. It is in the interest of insurance companies and the other stakeholders to ensure that guidance is realistic and appropriate in the way that it considers risk.

Local authorities can play a pivotal role by co-ordinating landslide management strategies 'on the ground', and they are best placed to maintain the momentum following the development of a strategy.

It is hoped that the advice and information provided in this paper and through the LIFE Response study (McInnes *et al.* 2006) will prove of practical assistance in reducing the impact of instability on local communities and economies in areas affected by instability problems.

REFERENCES

Department for Environment, Food and Rural Affairs (Defra), 2006. *Shoreline management plans guidance.* Crown copyright, London.

Department of the Environment (DOE), 1991. *PPG14 – Development on unstable land.* London: HMSO.

Doornkamp, J., Lee, E.M. & Moore, R. 1991. *Coastal landslip potential assessment: Isle of Wight Undercliff.* Technical Report by Geomorphological Services Ltd for the Department of the Environment, UK. Research contract PECD/7/1/272. London. Department of the Environment, HMSO.

European Commission, 1997. *LIFE Environment – Information Package 1997–1999.* Brussels. European Commission DGXI (XI.B.2).

European Commission, 2001. *LIFE Environment in action – 56 new success stories for Europe's environment.* Luxembourg : European Commission.

Fell, R., Ho, K.K.S., Lacasse, S. & Leroi, E. 2005. *A framework for landslide risk assessment and management.* In *Landslide Risk Management,* Proceedings of International Conference Vancouver, 2005. Balkema.

Halcrow Group Limited. 2006. *Coastal instability risk, Ventnor, Isle of Wight: Interpretive report and Quantitative Risk Assessment.* Report for Isle of Wight Council.

Jones, D.K.C. & Lee, E.M., 1994. *Landsliding in Great Britain.* London: HMSO.

Lee, E.M. 1995. *Coastal Planning and Management: A review of Earth Science information needs.* London: HMSO.

Lee, E.M. & McInnes, R.G. 2000. Landslide hazard mapping for planning and management: the Isle of Wight Undercliff, UK, in *Living with Natural Hazards, Proceedings of the CALAR Conference, Vienna.*

Lee, E.M. & Moore, R. 1991. *Getting the message across: Ground movement and public perception.* Birmingham. Report for South Wight Borough Council.

Leroi, E., Bonnard, Ch., Fell, R. & McInnes, R.G. 2005. *Risk assessment and management.* In *Landslide Risk Management.* Proc. Int. Conf. Vancouver, 2005. Balkema.

Marker, B. & McInnes, R.G. 2006. *Guidelines on landslides – communicating results.* Workshop on Landslide Susceptibility, hazard and zoning. Barcelona.

McInnes, R.G. 2000. *Managing ground instability in urban areas – A guide to best practice.* Newport, Isle of Wight: Isle of Wight Centre for the Coastal Environment.

McInnes, R.G. & Jakeways, J. 2000. *Coastal change, climate and instability*. EU LIFE Environment project for the European Commission, Ventnor.

McInnes, R.G., Jakeways, J. & Fairbank, H. 2006. EU LIFE *'Response'* project Final Report for European Commission, Ventnor, Isle of Wight.

Moore, R. & Lee, E.M. 1991. *Ventnor Information Centre – Getting the message across, ground movement and public perception*. Report for South Wight Borough Council, Ventnor, UK.

Moore, R., Lee, E.M. & Brunsden, D. 2000. *Cowes ground stability study*. Report by Halcrow for Isle of Wight Council.

Moore, R., Lee, E.M. & Clark, A.R. 1995. *The Undercliff of the Isle of Wight – A review of ground behaviour*. Report by Rendel Geotechnics for South Wight Borough Council, Ventnor, Isle of Wight.

Landslides and Climate Change – McInnes, Jakeways, Fairbank & Mathie (eds)
© 2007 Taylor & Francis Group, London, ISBN 978-0-415-44318-0

Author index